T0191513

Advances in Intelligent Systems and Computing

Volume 328

Series editor

Janusz Kacprzyk, Polish Academy of Sciences, Warsaw, Poland
e-mail: kacprzyk@ibspan.waw.pl

About this Series

The series "Advances in Intelligent Systems and Computing" contains publications on theory, applications, and design methods of Intelligent Systems and Intelligent Computing. Virtually all disciplines such as engineering, natural sciences, computer and information science, ICT, economics, business, e-commerce, environment, healthcare, life science are covered. The list of topics spans all the areas of modern intelligent systems and computing.

The publications within "Advances in Intelligent Systems and Computing" are primarily textbooks and proceedings of important conferences, symposia and congresses. They cover significant recent developments in the field, both of a foundational and applicable character. An important characteristic feature of the series is the short publication time and world-wide distribution. This permits a rapid and broad dissemination of research results.

Advisory Board

Chairman

Nikhil R. Pal, Indian Statistical Institute, Kolkata, India
e-mail: nikhil@isical.ac.in

Members

Rafael Bello, Universidad Central "Marta Abreu" de Las Villas, Santa Clara, Cuba
e-mail: rbellop@uclv.edu.cu

Emilio S. Corchado, University of Salamanca, Salamanca, Spain
e-mail: escorchado@usal.es

Hani Hagras, University of Essex, Colchester, UK
e-mail: hani@essex.ac.uk

László T. Kóczy, Széchenyi István University, Győr, Hungary
e-mail: koczy@sze.hu

Vladik Kreinovich, University of Texas at El Paso, El Paso, USA
e-mail: vladik@utep.edu

Chin-Teng Lin, National Chiao Tung University, Hsinchu, Taiwan
e-mail: ctlin@mail.nctu.edu.tw

Jie Lu, University of Technology, Sydney, Australia
e-mail: Jie.Lu@uts.edu.au

Patricia Melin, Tijuana Institute of Technology, Tijuana, Mexico
e-mail: epmelin@hafsamx.org

Nadia Nedjah, State University of Rio de Janeiro, Rio de Janeiro, Brazil
e-mail: nadia@eng.uerj.br

Ngoc Thanh Nguyen, Wroclaw University of Technology, Wroclaw, Poland
e-mail: Ngoc-Thanh.Nguyen@pwr.edu.pl

Jun Wang, The Chinese University of Hong Kong, Shatin, Hong Kong
e-mail: jwang@mae.cuhk.edu.hk

More information about this series at http://www.springer.com/series/11156

Suresh Chandra Satapathy · Bhabendra Narayan Biswal
Siba K. Udgata · J.K. Mandal
Editors

Proceedings of the 3rd International Conference on Frontiers of Intelligent Computing: Theory and Applications (FICTA) 2014

Volume 2

 Springer

Editors
Suresh Chandra Satapathy
Department of Computer Science
  and Engineering
Anil Neerukonda Institute of Technology
  and Sciences
Vishakapatnam
India

Bhabendra Narayan Biswal
Bhubaneswar Engineering College
Bhubaneswar, Odisha
India

Siba K. Udgata
University of Hyderabad
Hyderabad, Andhra Pradesh
India

J.K. Mandal
Department of Computer Science
  and Engineering
Faculty of Engg., Tech. & Management
University of Kalyanai
Kalyanai, West Bengal
India

ISSN 2194-5357   ISSN 2194-5365   (electronic)
ISBN 978-3-319-12011-9   ISBN 978-3-319-12012-6   (eBook)
DOI 10.1007/978-3-319-12012-6

Library of Congress Control Number: 2014951353

Springer Cham Heidelberg New York Dordrecht London

Preface

This AISC volume-II contains 88 papers presented at the Third International Conference on Frontiers in Intelligent Computing: Theory and Applications (FICTA-2014) held during 14–15 November 2014 jointly organized by Bhubaneswar Engineering College (BEC), Bhubaneswar, Odisa, India and CSI Student Branch, Anil Neerukonda Institute of Technology and Sciences, Vishakhapatnam, Andhra Pradesh, India. It once again proved to be a great platform for researchers from across the world to report, deliberate and review the latest progresses in the cutting-edge research pertaining to intelligent computing and its applications to various engineering fields. The response to FICTA 2014 has been overwhelming. It received a good number of submissions from the different areas relating to intelligent computing and its applications in main track and five special sessions and after a rigorous peer-review process with the help of our program committee members and external reviewers finally we accepted 182 submissions with an acceptance ratio of 0.43. We received submissions from eight overseas countries including India.

The conference featured many distinguished keynote addresses by eminent speakers like Dr. A. Govardhan, Director, SIT, JNTUH, Hyderabad, Dr. Bulusu Lakshmana Deekshatulu, Distinguished fellow, IDRBT, Hyderabad and Dr. Sanjay Sen Gupta, Principal scientist, NISC & IR, CSIR, New Delhi. The five special sessions were conducted during the two days of the conference.

Dr. Vipin Tyagi, Jaypee University of Engg and Tech, Guna, MP conducted a special session on "Cyber Security and Digital Forensics", Dr. A. Srinivasan, MNAJEC, Anna University, Chennai and Prof. Vikrant Bhateja, Sri Ramswaroop Memorial Group of Professional colleges, Lucknow conducted a special session on "Advanced research in 'Computer Vision, Image and Video Processing". Session on "Application of Software Engineering in Multidisciplinary Domains" was organized by Dr Suma V., Dayananda Sagar Institutions, Bangalore and Dr Subir Sarkar, Former Head, dept of ETE, Jadavpur University organized a special session on " Ad-hoc and Wireless Sensor Networks".

We take this opportunity to thank all Keynote Speakers and Special Session Chairs for their excellent support to make FICTA 2014 a grand success.

The quality of a referred volume depends mainly on the expertise and dedication of the reviewers. We are indebted to the program committee members and external reviewers who not only produced excellent reviews but also did in short time frames. We would also like to thank Bhubaneswar Engineering College (BEC), Bhubaneswar having coming forward to support us to organize the third edition of this conference in the series. Our heartfelt thanks are due to Er. Pravat Ranjan Mallick, Chairman, KGI, Bhubaneswar for the unstinted support to make the conference a grand success. Er. Alok Ranjan Mallick, Vice-Chairman, KGI, Bhubaneswar and Chairman of BEC deserve our heartfelt thanks for continuing to support us from FICTA 2012 to FICTA 2014. A big thank to Sri V Thapovardhan, the Secretary and Correspondent of ANITS and Principal and Directors of ANITS for supporting us in co-hosting the event. CSI Students Branch of ANITS and its team members have contributed a lot to FICTA 2014. All members of CSI ANITS team deserve great applause.

We extend our heartfelt thanks to Prof. P.N. Suganthan, NTU Singapore and Dr. Swagatam Das, ISI Kolkota for guiding us. Dr. B.K. Panigrahi, IIT Delhi deserves special thanks for being with us from the beginning to the end of this conference. We would also like to thank the authors and participants of this conference, who have considered the conference above all hardships. Finally, we would like to thank all the volunteers who spent tireless efforts in meeting the deadlines and arranging every detail to make sure that the conference can run smoothly. All the efforts are worth and would please us all, if the readers of this proceedings and participants of this conference found the papers and conference inspiring and enjoyable.

Our sincere thanks to all press print & electronic media for their excellent coverage of this conference.

November 2014

Volume Editors
Suresh Chandra Satapathy
Siba K. Udgata
Bhabendra Narayan Biswal
J.K. Mandal

Organization

Organizing Committee

Chief Patron

Er. Pravat Ranjan Mallick,
    Chairman   KGI, Bhubaneswar

Patron

Er. Alok Ranjan Mallick,   KGI, Bhubaneswar
    Vice-Chairman   Chairman, BEC, Bhubaneswar

Organizing Secretary

Prof. B.N. Biswal,   BEC, Bhubaneswar
    Director (A &A)

Honorary Chairs

Dr. P.N. Suganthan   NTU, Singapore
Dr. Swagatam Das   ISI, Kolkota

Steering Committee Chair

Dr. B.K. Panigrahi   IIT, Delhi, India

Program Chairs

Dr. Suresh Chandra Satapathy   ANITS, Vishakapatnam, India
Dr. S.K. Udgata   University of Hyderabad, India
Dr. B.N. Biswal, Director (A &A)   BEC, Bhubaneswar, India
Dr. J.K. Mandal   KalayanI University, West Bengal, India

International Advisory Committee/Technical Committee

P.K. Patra, India   Igor Belykh, Russia
Sateesh Pradhan, India   Nilanjan Dey, India
J.V.R Murthy, India   Srinivas Kota, Nebraska
T.R. Dash, Kambodia   Jitendra Virmani, India
Sangram Samal, India   Shabana Urooj, India
K.K. Mohapatra, India   Chirag Arora, India
L. Perkin, USA   Mukul Misra, India
Sumanth Yenduri, USA   Kamlesh Mishra, India
Carlos A. Coello Coello, Mexico   Muneswaran, India
S.S. Pattanaik, India   J. Suresh, India
S.G. Ponnambalam, Malaysia   Am,lan Chakraborthy, India
Chilukuri K. Mohan, USA   Arindam Sarkar, India
M.K. Tiwari, India   Arp Sarkar, India
A. Damodaram, India   Devadatta Sinha, India
Sachidananda Dehuri, India   Dipendra Nath, India
P.S. Avadhani, India   Indranil Sengupta, India
G. Pradhan, India   Madhumita Sengupta, India
Anupam Shukla, India   Mihir N. Mohantry, India
Dilip Pratihari, India   B.B. Mishra, India
Amit Kumar, India   B.B. Pal, India
Srinivas Sethi, India   Tandra Pal, India
Lalitha Bhaskari, India   Utpal Nandi, India
V. Suma, India   S. Rup, India
Pritee Parwekar, India   B.N. Pattnaik, India
Pradipta Kumar Das, India   A. Kar, India
Deviprasad Das, India   V.K. Gupta, India
J.R. Nayak, India   Shyam lal, India
A.K. Daniel, India   Koushik Majumder, India
Walid Barhoumi, Tunisia   Abhishek Basu, India
Brojo Kishore Mishra, India   P.K. Dutta, India
Meftah Boudjelal, Algeria   Md. Abdur Rahaman Sardar, India
Sudipta Roy, India   Sarika Sharma, India
Ravi Subban, India   V.K. Agarwal, India
Indrajit Pan, India   Madhavi Pradhan, India
Prabhakar C.J., India   Rajani K. Mudi, India
Prateek Agrawal, India   Sabitha Ramakrishnan, India

Sireesha Rodda, India
Srinivas Sethi, India
Jitendra Agrawal, India

Suresh Limkar, India
Bapi Raju Surampudi, India
S. Mini, India and many more

Organizing Committee

Committee Members from CSI Students Branch ANITS

M. Kranthi Kiran, Asst. Professor, Dept. of CSE, ANITS
G.V. Gayatri, Asst. Professor, Dept. of CSE, ANITS
S.Y. Manikanta, ANITS
Aparna Patro, ANITS
T. Tarun Kumar Reddy, ANITS
A. Bhargav, ANITS
N. Shriram, ANITS
B.S.M. Dutt, ANITS
I. Priscilla Likitha Roy, ANITS
Ch. Geetha Manasa, ANITS
S.V.L. Rohith, ANITS
N. Gowtham, ANITS
B. Bhavya, ANITS and many more

Committee Members from BEC

Sangram Keshari Samal, HOD, Aeronautical Engineering
Manas Kumar Swain, Prof. Computer Science
A.K. Sutar, HOD, Electronics and Telecommunication Engineering
Pabitra Mohan Dash, HOD, Electrical & Electronics Engineering
Rashmita Behera, Asst. Prof. Electrical & Electronics Engineering
Utpala Sahu, Asst. Prof. Civil Engineering
Sonali Pattnaik, Asst. Prof. Civil Engineering
V.M. Behera, Asst. Prof. Mechanical Engineering
Gouri Sankar Behera, Asst. Prof. Electrical & Electronics Engineering
Debashis Panda, Prof. Department of Sc &H and many more

Contents

Section I: Network and Information Security, Grid Computing and Clod Computing

Security and Privacy in Cloud Computing: A Survey   1
Mahesh U. Shankarwar, Ambika V. Pawar

Analysis of Secret Key Revealing Trojan Using Path Delay Analysis for Some Cryptocores 13
Krishnendu Guha, Romio Rosan Sahani, Moumita Chakraborty, Amlan Chakrabarti, Debasri Saha

Generating Digital Signature Using DNA Coding 21
Gadang Madhulika, Chinta Seshadri Rao

DNA Encryption Based Dual Server Password Authentication 29
P.V.S.N. Raju, Pritee Parwekar

Dynamic Cost-Aware Re-replication and Rebalancing Strategy in Cloud System 39
Navneet Kaur Gill, Sarbjeet Singh

Signalling Cost Analysis of Community Model 49
Boudhayan Bhattacharya, Banani Saha

Steganography with Cryptography in Android 57
Akshay Kandul, Ashwin More, Omkar Davalbhakta, Rushikesh Artamwar, Dinesh Kulkarni

A Cloud Based Architecture of Existing e-District Project in India towards Cost Effective e-Governance Implementation 65
Manas Kumar Sanyal, Sudhangsu Das, Sajal Bhadra

A Chaotic Substitution Based Image Encryption Using APA-transformation   75
Musheer Ahmad, Akshay Chopra, Prakhar Jain, Shahzad Alam

Security, Trust and Implementation Limitations of Prominent IoT Platforms   85
Shiju Sathyadevan, Boney S. Kalarickal, M.K. Jinesh

An Improved Image Steganography Method with SPIHT and Arithmetic Coding   97
Lekha S. Nair, Lakshmi M. Joshy

Image Steganography – Least Significant Bit with Multiple Progressions . . .   105
Savita Goel, Shilpi Gupta, Nisha Kaushik

On the Implementation of a Digital Watermarking Based on Phase Congruency   113
Abhishek Basu, Arindam Saha, Jeet Das, Sandipta Roy, Sushavan Mitra, Indranil Mal, Subir Kumar Sarkar

Influence of Various Random Mobility Models on the Performance of AOMDV and DYMO   121
Suryaday Sarkar, Meghdut Roychowdhury, Biswa Mohan Sahoo, Souvik Sarkar

Handling Data Integrity Issue in SaaS Cloud   127
Anandita Singh Thakur, P.K. Gupta, Punit Gupta

Load and Fault Aware Honey Bee Scheduling Algorithm for Cloud Infrastructure   135
Punit Gupta, Satya Prakash Ghrera

Secure Cloud Data Computing with Third Party Auditor Control   145
Apoorva Rathi, Nilesh Parmar

Effective Disaster Management to Enhance the Cloud Performance   153
Chintureena, V. Suma

Implementation of Technology in Indian Agricultural Scenario   165
Phuritshabam Robert, B. Naveen Kumar, U.S. Poornima, V. Suma

Section II: Cyber Security and Digital Forensics

JPEG Steganography and Steganalysis – A Review   175
Siddhartha Banerjee, Bibek Ranjan Ghosh, Pratik Roy

Enhanced Privacy and Surveillance for Online Social Networks   189
Teja Yaramasa, G. Krishna Kishore

Neuro-key Generation Based on HEBB Network for Wireless
Communication 197
Arindam Sarkar, J.K. Mandal, Pritha Mondal

KSOFM Network Based Neural Key Generation for Wireless
Communication 207
Madhumita Sengupta, J.K. Mandal, Arindam Sarkar, Tamal Bhattacharjee

Hopfield Network Based Neural Key Generation for Wireless
Communication (HNBNKG) 217
J.K. Mandal, Debdyuti Datta, Arindam Sarkar

Section III: Advanced research in ŚComputer Vision, Signal, Image and Video Processing

Automatic Video Scene Segmentation to Separate Script
and Recognition 225
Bharatratna P. Gaikwad, Ramesh R. Manza, Ganesh R. Manza

Gesture: A New Communicator 237
Saikat Basak, Arundhuti Chowdhury

A Novel Fragile Medical Image Watermarking Technique for Tamper
Detection and Recovery Using Variance 245
R. Eswaraiah, E. Sreenivasa Reddy

MRI Skull Bone Lesion Segmentation Using Distance Based Watershed
Segmentation 255
Ankita Mitra, Arunava De, Anup Kumar Bhattacharjee

Extraction of Texture Based Features of Underwater Images Using RLBP
Descriptor 263
S. Nagaraja, C.J. Prabhakar, P.U. Praveen Kumar

Summary-Based Efficient Content Based Image Retrieval in P2P
Network 273
Mona, B.G. Prasad

Dynamic Texture Segmentation Using Texture Descriptors and Optical
Flow Techniques 281
Pratik Soygaonkar, Shilpa Paygude, Vibha Vyas

Design and Implementation of Brain Computer Interface Based Robot
Motion Control 289
Devashree Tripathy, Jagdish Lal Raheja

Advanced Adaptive Algorithms for Double Talk Detection in Echo
Cancellers: A Technical Review 297
Vineeta Das, Asutosh Kar, Mahesh Chandra

A Comparative Study of Iterative Solvers for Image De-noising 307
Subit K. Jain, Rajendra K. Ray, Arnav Bhavsar

Assessment of Urbanization of an Area with Hyperspectral Image Data 315
Somdatta Chakravortty, Devadatta Sinha, Anil Bhondekar

Range Face Image Registration Using ERFI from 3D Images 323
Suranjan Ganguly, Debotosh Bhattacharjee, Mita Nasipuri

Emotion Recognition for Instantaneous Marathi Spoken Words 335
Vaibhav V. Kamble, Ratnadeep R. Deshmukh, Anil R. Karwankar,
Varsha R. Ratnaparkhe, Suresh A. Annadate

Performance Evaluation of Bimodal Hindi Speech Recognition under
Adverse Environment 347
Prashant Upadhyaya, Omar Farooq, M.R. Abidi, Priyanka Varshney

Extraction of Shape Features Using Multifractal Dimension for
Recognition of Stem-Calyx of an Apple 357
S.H. Mohana, C.J. Prabhakar

An Approach to Design an Intelligent Parametric Synthesizer for
Emotional Speech 367
Soumya Smruti, Jagyanseni Sahoo, Monalisa Dash, Mihir N. Mohanty

Removal of Defective Products Using Robots 375
Birender Singh, Mahesh Chandra, Nikitha Kandru

Contour Extraction and Segmentation of Cerebral Hemorrhage from
MRI of Brain by Gamma Transformation Approach 383
Sudipta Roy, Piue Ghosh, Samir Kumar Bandyopadhyay

An Innovative Approach to Show the Hidden Surface by Using Image
Inpainting Technique 395
Rajat Sharma, Amit Agarwal

Fast Mode Decision Algorithm for H.264/SVC 405
L. Balaji, K.K. Thyagharajan

Recognizing Handwritten Devanagari Words Using Recurrent Neural
Network 413
Sonali G. Oval, Sankirti Shirawale

Homomorphic Filtering for Radiographic Image Contrast Enhancement
and Artifacts Elimination 423
Igor Belykh

A Framework for Human Recognition Based on Locomotive Object
Extraction 431
C. Sivasankar, A. Srinivasan

Abnormal Event Detection in Crowded Video Scenes................... 441
V.K. Gnanavel, A. Srinivasan

**Comparative Analysis and Bandwidth Enhancement with Direct Coupled
C Slotted Microstrip Antenna for Dual Wide Band Applications** 449
Rajat Srivastava, Vinod Kumar Singh, Shahanaz Ayub

**Quality Assessment of Images Using SSIM Metric and CIEDE2000
Distance Methods in Lab Color Space** 457
T. Chandrakanth, B. Sandhya

Review Paper on Linear and Nonlinear Acoustic Echo Cancellation 465
D.K. Gupta, V.K. Gupta, Mahesh Chandra

PCA Based Medical Image Fusion in Ridgelet Domain 475
Abhinav Krishn, Vikrant Bhateja, Himanshi, Akanksha Sahu

**Swarm Optimization Based Dual Transform Algorithm for Secure
Transaction of Medical Images** 483
Anusudha Krishnamurthi, N. Venkateswaran, J. Valarmathi

**Convolutional Neural Networks for the Recognition of Malayalam
Characters** 493
R. Anil, K. Manjusha, S. Sachin Kumar, K.P. Soman

**Modeling of Thorax for Volumetric Computation Using Rotachora
Shapes** 501
*Shabana Urooj, Vikrant Bhateja, Pratiksha Saxena, Aime lay Ekuakille,
Patrizia Vergalo*

A Review of ROI Image Retrieval Techniques 509
Nishant Shrivastava, Vipin Tyagi

**A Novel Algorithm for Suppression of Salt and Pepper Impulse Noise in
Fingerprint Images Using B-Spline Interpolation** 521
P. Syamala Jaya Sree, Prasanth Kumar Pattnaik, S.P. Ghrera

Spectral-Subtraction Based Features for Speaker Identification 529
Mahesh Chandra, Pratibha Nandi, Aparajita kumari, Shipra Mishra

**Automated System for Detection of Cerebral Aneurysms in Medical CTA
Images** 537
M. Vaseemahamed, M. Ravishankar

**Contrast Enhancement of Mammograms Images Based on Hybrid
Processing** 545
Inam Ul Islam Wani, M.C. Hanumantharaju, M.T. Gopalkrishna

An Improved Handwritten Word Recognition Rate of South Indian Kannada Words Using Better Feature Extraction Approach 553
M.S. Patel, Sanjay Linga Reddy, Krupashankari S. Sandyal

An Efficient Way of Handwritten English Word Recognition 563
M.S. Patel, Sanjay Linga Reddy, Anuja Jana Naik

Text Detection and Recognition Using Camera Based Images 573
H.Y. Darshan, M.T. Gopalkrishna, M.C. Hanumantharaju

Retinal Based Image Enhancement Using Contourlet Transform 581
P. Sharath Chandra, M.C. Hanumantharaju, M.T. Gopalakrishna

Detection and Classification of Microaneurysms Using DTCWT and Log Gabor Features in Retinal Images 589
Sujay Angadi, M. Ravishankar

Classifying Juxta-Pleural Pulmonary Nodules 597
K. Sariya, M. Ravishankar

The Statistical Measurement of an Object-Oriented Programme Using an Object Oriented Metrics 605
Rasmita Panigrahi, Sarada Baboo, Neelamadhab Padhy

Section IV: Application of Software Engineering in Multidisciplinary Domains

Applicability of Software Defined Networking in Campus Network 619
Singh Sandeep, R.A. Khan, Agrawal Alka

Author-Profile System Development Based on Software Reuse of Open Source Components 629
Derrick Nazareth, Kavita Asnani, Okstynn Rodrigues

Software and Graphical Approach for Understanding Friction on a Body 637
Molla Ramizur Rahman

An Investigation on Coupling and Cohesion as Contributory Factors for Stable System Design and Hence the Influence on System Maintainability and Reusability 645
U.S. Poornima, V. Suma

Application of Component-Based Software Engineering in Building a Surveillance Robot 651
Chaitali More, Louella Colaco, Razia Sardinha

Comprehension of Defect Pattern at Code Construction Phase during
Software Development Process 659
Bhagavant Deshpande, Jawahar J. Rao, V. Suma

Pattern Analysis of Post Production Defects in Software Industry 667
Divakar Harekal, Jawahar J. Rao, V. Suma

Secure Efficient Routing against Packet Dropping Attacks in Wireless
Mesh Networks 673
T.M. Navamani, P. Yogesh

Section V: Ad-hoc and Wireless Sensor Networks

Key Management with Improved Location Aided Cluster Based Routing
Protocol in MANETs 687
Yogita Wankhade, Vidya Dhamdhere, Pankaj Vidhate

Co-operative Shortest Path Relay Selection for Multihop MANETs 697
Rama Devi Boddu, K. Kishan Rao, M. Asha Rani

Intelligent Intrusion Detection System in Wireless Sensor Network 707
*Abdur Rahaman Sardar, Rashmi Ranjan Sahoo, Moutushi Singh,
Souvik Sarkar, Jamuna Kanta Singh, Koushik Majumder*

Secure Routing in MANET through Crypt-Biometric Technique 713
Zafar Sherin, M.K. Soni

Remote Login Password Authentication Scheme Using Tangent Theorem
on Circle 721
Shipra Kumari, Hari Om

A Survey of Security Protocols in WSN and Overhead Evaluation 729
Shiju Sathyadevan, Subi Prabhakaranl, K. Bipin

DFDA: A Distributed Fault Detection Algorithm in Two Tier Wireless
Sensor Networks 739
Kumar Nitesh, Prasanta K. Jana

Secured Categorization and Group Encounter Based Dissemination
of Post Disaster Situational Data Using Peer-to-Peer Delay Tolerant
Network 747
Souvik Basu, Siuli Roy

Lifetime Maximization in Heterogeneous Wireless Sensor Network Based
on Metaheuristic Approach 757
Manisha Bhende, Suvarna Patil, Sanjeev Wagh

Lightweight Trust Model for Clustered WSN 765
Moutushi Singh, Abdur Rahaman Sardar, Rashmi Ranjan Sahoo,
Koushik Majumder, Sudhabindu Ray, Subir Kumar Sarkar

An Efficient and Secured Routing Protocol for VANET 775
Indrajit Bhattacharya, Subhash Ghosh, Debashis Show

**Authentication of the Message through Hop-by-Hop and Secure the
Source Nodes in Wireless Sensor Networks** 785
B. Anil Kumar, N. Bhaskara Rao, M.S. Sunitha

Rank and Weight Based Protocol for Cluster Head Selection for WSN 793
S.R. Biradar, Gunjan Jain

Author Index 803

Security and Privacy in Cloud Computing: A Survey

Mahesh U. Shankarwar and Ambika V. Pawar

CSE Department, SIT, Symbiosis International University, Pune, India
{mahesh.shankarwar,ambikap}@sitpune.edu.in

Abstract. Cloud Computing is continuously evolving and showing consistent growth in the field of computing. It is getting popularity by providing different computing services as cloud storage, cloud hosting, and cloud servers etc. for different types of industries as well as in academics. On the other side there are lots of issues related to the cloud security and privacy. Security is still critical challenge in the cloud computing paradigm. These challenges include user's secret data loss, data leakage and disclosing of the personal data privacy. Considering the security and privacy within the cloud there are various threats to the user's sensitive data on cloud storage. This paper is survey on the security and privacy issues and available solutions. Also present different opportunities in security and privacy in cloud environment.

Keywords: Cloud Computing, Security, Privacy, Cloud storage, Sensitive data, Data loss, Data leakage.

1 Introduction

Cloud Computing has recently emerged as new paradigm for hosting and delivering services over the Internet. Cloud Computing is the use of computing resources such as hardware and software that are delivered as service over the internet.

The US National institute of standard and technology has given the complete definition of cloud computing that is "Cloud computing is a model for enabling ubiquitous, convenient, on-demand network access to a shared pool of configurable computing resources (e.g., networks, servers, storage, applications, and services) that can be rapidly provisioned and released with minimal management effort or service provider interaction. This cloud model is composed of five essential characteristics, three service models, and four deployment models" [1].

Cloud computing is recently receiving great deal of attention among users. Another way of defining the cloud computing is to examine its five essential characteristics as mention in table I, on demand self-service, broad network access, resource pooling, rapid elasticity and measuredservices [1].In *On-demand self-service,* user gets services provided by cloud as per his requirement without any human interaction. *Broad network access*makescloud services available over the internet so user can get access to any cloud service using network through any client. *Resource pooling*characteristic makes resources available over the cloud and it get access from anywhere by multiple consumer. There is no need of knowing where the resources are stored. *Rapid elasticity* tells thatCapabilities of cloud services as per consumer

© Springer International Publishing Switzerland 2015

S.C. Satapathy et al. (eds.), *Proc. of the 3rd Int. Conf. on Front. of Intell. Comput. (FICTA) 2014*
– *Vol. 2,* Advances in Intelligent Systems and Computing 328, DOI: 10.1007/978-3-319-12012-6_1

demand can be rapidly and elastically provisionedand which are available in unlimited manner to consumer at any time. *Measured service* monitor, control, and report, providing transparency for both the provider and consumer of the utilized service.

The cloud computing provides different services; these services put forwarded three models as given in table I, software as service, platform as service, infrastructure as service [1]. In *SaaS,* The consumer of cloud services can use application which is already running on a cloud infrastructure. These applications can be accessible from any location. The example of saas is salesforce.com a CRM application. In *PaaS,* The cloud provider provides the platform as service to the consumer where he can manage its application and use it without managing cloud infrastructure. Example is GoogleApps.*IaaS*type of service, cloud provider provide infrastructure where the consumer can manage its platform along with application for its purpose. The best example of Iaas is amazon web services. There are different types of deployment models listed by the NIST (2009) mention in table I: public cloud, private cloud and hybrid cloud. The cloud infrastructure owned by single personal, single organization or single business unit for their own purpose is called*Private cloud.* The cloud infrastructure is provisioned for open use by the general public. It may be owned, managed, and operated by a business, academic, government organization, or some combination of them is called. *Public clouds while* thecloud infrastructure which is combination of two or more distinct cloud infrastructures (private, public) called*Hybrid cloud.*

In the general architecture of Cloud the front end is nothing but different types of clients who uses the services provided by cloud while back end is cloud platform. The architecture having different layer in which layer is made of one particular service model of cloud. The bottom layer of architecture is Iaas which manages infrastructure services physically or virtually and delivers in the form of storage and network. The middle layer is platform as service, which provides environment and platform from various purposes. The top most layer is software a service where all application is provided to client.

Table 1. Introduction to Cloud Computing

Five Essential Characteristics	1. On demand self-service 2. Broad network access 3. Resource pooling 4. Rapid elasticity 5. Measured Services
Three Service Models	1. Software as Service(Saas) 2. Platform as service(Paas) 3. Infrastructure as service(Iaas)
Deployment Models	1. Public Cloud 2. Private Cloud 3. Hybrid Cloud

According to shibashini and kavitha[2]many organizations Small as well as medium started using the cloud computing rapidly because of fast accessing to applications and reduced cost of infrastructure. But due to constantly increase in the popularity of cloud computing there is an ever growing risk of security and privacy becoming main and top issues. According to IDC survey in august 2008[33], conducted by 244 executives and their colleagues' security is the main challenge to cloud computing. There are many security issues to cloud computing such as data storage secuirity, confidentiality, data access from third party, data loss or data theft, data transmission and so many.

Cloud computing raises the range of important issues of privacy. Risk of comprise to confidential information through third party access to sensitive information. This can pose a significant threat to ensuring the protection of intellectual property (IP), and personal information. Privacy issues exist in cloud domain and it will remain for long time, many laws have been published yet to protect privacy of individual's data or business secret information. Nevertheless these acts are expired and not applicable to today's domain. Since privacy problems still become more hazardous. The issues ranges from malicious insiders, misuse of cloud computing and many more.

The rest of the paper has been arranged as follows. The different security and privacy issue arises in cloud computing, which is described in section 2. The available methods or existing solutions are described in section 3. The open research problem to be work given in 4. Conclusion from survey is given in section 5.

2   Challenges to Security and Privacy

2.1   Privacy Issues

Privacy: - It is protection oftransmitted data from passive attacks. The objective is to ensure that sensitive data of customer is not being accessed by or disclosed by any unauthorized person.

A. Misuse of Cloud computing
Provider gives the unlimited access of network and storage in limited cost or sometimes they just provide the free trial so type of things causes different harms to cloud computing paradigm [3][4][5].
B. Malicious Insiders
Generally provider may not reveal about their employees access to assets or resources this helps the attacker to gain access over asset or data [6][7][8].
C. Trans border data flow and data proliferation
Several companies involves data proliferation but in uncontrolled and unmanaged by owner of data. Vendors are free to worry about the use of copy data in several datacenters. This is very difficult to ensure that duplicate of the data or its backups are not stored or processed in a certain authority, all these copies of data are deleted if such a request is made[3][4].

D. Dynamic provision
In the cloud the consumer store its private or secret data but there is no one who can take the responsibility of security of the consumer data. There is needof dynamic provision of data in cloud[3][4][9].

2.2   Security Issues

Security:-It is protection of sensitive data from vulnerable attacks. There is various risk factors are involved with cloud computing which are explained as bellows.

A. Multitenancy
Multitenancy is nothing but one program can run on multiple machines at same time but these causes vulnerabilities in case of cloud infrastructure[6][10][11].
B. Access
The sensitive information in the cloud has many threats. Attacker may hack the data which is present in the cloud and which can be access and used later [6][11][12].
C. Availability
In cloud whatever the data user store so it should be available to that user at anytime and anywhere but in case of cloud there is problem of backup recovery in case of failure. This resulted as loss of confidence among the consumers [12][13][14].
D. Trust
In case of cloud there is still lots of issues are related to the cloud. There is still no trust relationship between the cloud provider and the consumer of the cloud one who uses the cloud. Consumer cannot trust completely on the provider about their secret data to be store on cloud [6][11][15].
E. Audit
In order to implement internal monitoring control CSP there is need of external audit mechanism but still cloud fails to provide auditing of the transaction without effecting integrity [6][16][17].

3   Existing Solutions and Open Issues to Security and Privacy

3.1   Client Based Privacy Manager

There are different issues related to the privacy and security of data are there so Miranda M. and S. Pearson proposed client based Manager which helps to reduce the risk related to data leakage and loss of sensitive data using obfuscation and de-obfuscation techniques. They provided the architecture with different features such as obfuscation, data access; feedback etc. The basic idea is to store encrypted form of client's private data in cloud. The advantage of these methods is that it preserves privacy of data by customizing the end user service problems. The limitation of this work is that there is requirement of honest co-operation of service provider and also vendors not add extra services for privacy protection [18].

Table 2. Challenges to Cloud Computing

Category	Issues
Privacy Issues	1. Misuse of Cloud Computing 2. Malicious Insiders 3. Trans border data flow and data proliferation 4. Dynamic provision
Security Issues	1. Audit 2. Trust 3. Availability 4. Multitenancy 5. Access

Anonymity Based Methods

The Wang J. and et al. provided anonymity based method for privacy of data in cloud. They provided the algorithm anonymises data before storing to cloud. Whenever provider requires data it uses domain knowledge it has to analyze anonymous data to get the needed knowledge. Advantages of this method are it is simple and flexible and differ from the traditional cryptography technology. It is limited for number of services [19].

Fully Homomorphic Encryption

The C.Gentry presented encryption method for preserving the privacy of data. This method of encryption enables the computation on the encrypted data which is stored in cloud. Cloud provider is not aware of data processing function as well as result of computation. It is a powerful tool for privacy maintenance but it fails to use practically [20].

Preventing Data Leakage from Indexing

The A. Squicciarini and et al. explored the issues arises because of indexing in cloud. Indexing causes the data leakageso to reduce the data leakage and the loss they present three tier data protection architecture as well as portable data binding technique to ensure strong users' privacy requirements. This method gives different levels of privacy of customer data. The architecture helps to prevent leakage by data indexing. The limitation of this method is only solving problem from indexing [21].

Public Auditability and Data Dynamics for Storage Security

The concept explained in this paper guarantees correctness of data in cloud and explore problem of public auditability and data dynamicsfor data integrity checking. The Qian Wang and et al. improved the existing proof of storage models by merklehash tree and also explore the technique of bilinear aggregate signature. It helps to preserve the privacy and security for batch auditing. Highly efficient and secure. Require third party auditor for operation. [22].

Table 3. Existing solutions advantages and limitations to Cloud Computing

Existing Schemes	Advantages	Limitations
Client based Privacy manager	Preserve Privacy	Require honest co-operation of service provider
Anonymity based methods	Simple and Flexible	Limited for number of services
Fully Homomorphic encryption	Powerful tool	Fails to use practically
Preventing data leakage from Indexing	Prevent leakage by data indexing	Indexing based
Public Auditability and data Dynamics for Storage Security	Highly efficient and secure	Require third party auditor for operation.
Privacy Persevering Repository	Achieves the confidentiality and availability	Not secure enough to achieve all issues of security
Privacy Preserving System	Preserve the privacy	Providing machine readable access rights
Privacy-Preserved Access Control	Maintain the privacy of data without disclosing	Challenging because of data outsourcing and untrusted cloud server
Securing the storage Data Using RC5	Very easy to use and secure	Challenges of cracking the algorithm
Fog Computing	Protects against misuse of data	Not resolve all issues
Keeping data Private while Computing	Efficient for privacy preservation as well robust to network delay.	only concentrates on NP problems
Security using Elliptic Curve Cryptography	Preserve the confidentiality and authentication	Fails to preserve all the security issues
Privacy Preserving Public Auditing for Storage Security	No leakage of data with better performance	Require some amount of improvement
Service oriented identity Authentication	Preserve privacy of the data.	Not support for multiservice and not practical concept

Privacy Persevering Repository
In this paper theR.Mishra and et al.constructed the repository where the data sharing services can perform updation of data as well as control the access by limiting the usage allowance to the data. The Repository helps to restrict data owner for his work on data to be done on server without disclosing data. Advantages of this are it achieves the confidentiality of data along with high availability. Not secure enough to achieve all issues of security [23].

Privacy Preserving System
The Greveler U, Justus b. and et al.provided the architecture for database storage in cloud which preserves the privacy of users' data. By encrypting and assigning secure identities to the each and every request and response and they send as the XML request to database with the maintenance of machine readable access rights. An advantage of it is that easier to handle the encryption schemes and helps to preserve the privacy but limitation is providing machine readable access rights [24].

Privacy-Preserved Access Control
The Zhou M, Mu Y and et al. proposed a method of access control by considering privacy of data. In case of this method every user is provided with some attributes, which defines their access rights. The method having a two-tier encryption so at the first tier, the data owner performs local attribute-based encryption on the data that is requested. In the second tier server implements the Server re-encryption mechanism (SRM). The SRM dynamically re-encrypts the encrypted data in the cloud. So because of re-encryption the privacy of users data is maintain and providing full accesscontrol to the owner of the data without disclosing data to cloud providerdue to data outsourcing and untrusted cloud server becomes challenging issue in cloud storage system [25].

Securing the Storage Data Using RC5
The J.Singh and et al. used the RC5 encryption algorithm to secure storage data. A system uses the encryption and decryption keys of user's data and stores it on remote server. Each storage server has an encrypted file storage system which encrypts the data and store. It is very easy to use and secure. Ensures the data is stored on trusted server but there are challenges of cracking the algorithm [26].

Fog Computing
The Salvatore J. and et al. used the offensive Decoy technology called as fog computing. Methods tells about monitoring the access rights by detecting data access patterns and when unauthorized access is found then it gets verified using challenge questions and launches disinformation attack by returning the large amount of information to the attacker. Protects against misuse of data [27].

Keeping Data Private While Computing
The Y. Brun and N. Medvidovicaddress the problem of distributing computation by preserving the privacy of data. Uses the sTyle approach which separates the computation into the small computation and distribute it in such way that it hard to reconstruct the data. This approach is efficient for privacy preservation as well robust to network delay.But it only concentrate on NP problems not all types of problems[28].

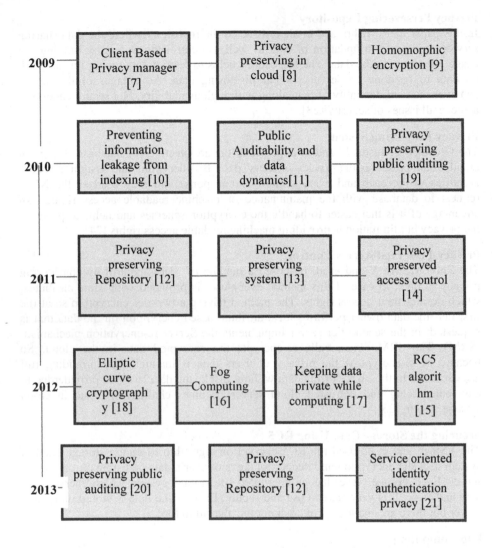

Fig. 1. Taxonomy of existing methods and solutions

Security Using Elliptic Curve Cryptography
The V.Gampala and et al. proposed the data security using the elliptic curve cryptography. This type of encryption algorithm helps to preserve the confidentiality along withauthentication between cloud and the user but fails to preserve all the security issues [29].

Privacy Preserving Public Auditing for Storage Security
According to the Wang C and et al. the cryptography fails to secure efficiently storage data so they proposed the technique of auditing with concept of third party auditing (TPA).uses the Homomorphic authenticator which gives the guarantee about the TPA.The technique consist of 4 algorithms along with two phases of it. In which first

phase consist of sigGen and the keyGen algorithm which verify the metadata while the second phase consist of genproof and verify proof algorithm for auditing. Audit report gives the harm present in the data [30].The next paper is an extension to this paper with little bit of improvement in this paper. New protocol is proposed. This gives no leakage of data with better performance [31].

Service Oriented Identity Authentication
The Xiaohui Li and et al. proposed method which is an improvement over the coarse grained authentication method. The method defines access control as process and extending the user information into fuzzy set as an authentication condition. It guarantees the global minimal sensitive information discloser. Preserve privacy of the data. It does not support for multiservice and not practical concept [32].

4 Open Research Problem

In this paper different existing methods and available solutions surveyed. Paper presented existing open research problem in cloud security and privacy.

Opportunities
Thepaper [18] require honest co-operation of service provider but we never 100% believe onservice provider because there are more challenges of misusing of data.so security system must be independent from service provider. The technique [20] fails to use practically and technique [32] is just theoretical concept so method should be practical enough to work. The system should solve all security and privacy issues because many methods [23, 27, 29, and 32] only concentrate on particular issue at a time. The system [19][32] should be strong enough to support for multiservice provider environment and provide more number of services. This is some open issues to be work on.

Possible Approach
The paper [34] is providing more security and privacy to sensitive data in cloud using multicloud environment in which enhanced security is provided by dividing file into multiple chunks and storing those chunks into multiple clouds to protect security and privacy.

The basic idea is to protect file from threat firstly encrypt that file using encryption technique and after encryption process, split that file into multiple chunks so that no one get complete file. Then the different chunks of file should be store on different cloud. This process implements the concept of multiple clouds storage along with enhanced security using encryption technique.

5 Conclusion

This paper gives an insight of different threats in cloud computing environment with respect to security and privacy of user's sensitive data in the cloud environment. Researchers have proposed different methods to tackle issues using different approaches which somewhat helps to minimize the problem over the data security and privacy in the cloud. We have discussed about the advantages and limitations of existing methods to completely solve security and privacy issue. These are the open issues to work on.

References

1. Mell, P., Grance, T.: The NIST Definition of Cloud Computing. National Institute of Standards and Technology, Information Technology Laboratory (2011), http://csrc.nist.gov/groups/SNS/cloud-computing/
2. Subashini, S., Kavitha, V.: A survey on security issues in service delivery models of cloud computing. Journal of Network and Computer Applications 34(1), 1–11 (2011)
3. Top Threats to the Cloud Computing V1.0, Cloud Security Alliance, http://www.cloudsecurityalliance.org/topthreats/2010
4. Babu, J., Kishore, K., Kumar, K.E.: Migration from Single to Multi-Cloud Computing. International Journal of Engg. Research and Tech. 2(4) (April 2013)
5. Chandran, S., Angepat, M.: CloudComputing: Analyzing the Risk involved in Cloud Computing Environment (2011)
6. Munir, K., Palaniappan, S.: Security threats/attacks present in cloud environment. IJCSNS 12(12) (2012)
7. Munir, K., Palaniappan, S.: Secure Cloud Architecture. ACIJ 4(1) (2013)
8. Sravani, K., Nivedita, K.L.A.: Effective service security schemes in cloud computing. IJCER 3(3)
9. Chen, D., Zhao, H.: Data Security and Privacy Protection Issues in Cloud Computing. In: IEEE International Conference on Computer Science and Electronics Engineering (2012)
10. Behl, A.: Emerging Security Challenges in Cloud Computing. In: IEEE International Conference Information and Communication Technologies, WICT (2011)
11. Mahmood, Z.: Data Location and Security Issues in Cloud Computing. In: IEEE International Conference on Emerging intelligent Data and Web Technologies (2011)
12. Attas, D., Batrafi, O.: Efficient integrity checking technique fro securing client data in cloud computing. IJECS-IJENS 11(5) (2011)
13. Arockiam, L., Parthasarathy, et al.: Privacy in Cloud Computing: Survey. CS&IT (2012)
14. Hashizume, K., et al.: An analysis of security issues for cloud computing. Journal of Internet Services and Applications, A Springer Open Journal, 1–13 (2013)
15. Sharma, P., Sood, S.K., Kaur, S.: Security Issues in Cloud Computing. In: Mantri, A., Nandi, S., Kumar, G., Kumar, S. (eds.) HPAGC 2011. CCIS, vol. 169, pp. 36–45. Springer, Heidelberg (2011)
16. Popovic, K., Hocenski, Z.: Cloud Computing security issues and challenges. In: MIPRO, Proceedings of the 33rd International Convention (2010)
17. Siani, Miranda: Security Threats in cloud computing. In: 6th International Conference on Internet Technologies and Secure Transactions (2011)
18. Miranda, M., Pearson, S.: A Client-based Privacy Manager for cloud Computing. In: Proceeding of the Fourth International ICST Conference on Comm. and Middleware, COMSWARE 2009 (2009)
19. Wang, J., Zhao, Y., et al.: Providing Privacy preserving in cloud computing. In: International Conference on Test and Measurement, vol. 2, pp. 213–216 (2009)
20. Gentry, C.: Fully Homomorphic encryption using ideal lattices. In: STOC, pp. 169–178 (2009)

21. Squicciarini, A., Sundareswaran, S., Lin, D.: Preventing Information Leakage from Indexing in the Cloud. In: 2010 IEEE 3rd International Conference on Cloud Computing, pp. 188–195 (2010)
22. Wang, Q., Wang, C., et al.: Enabling public Auditability and Data Dynamics for Storage Security in Cloud Computing. IEEE (2010)
23. Mishra, R., Dash, S., Mishra, D.: Privacy preserving Repository for securing data across the Cloud. IEEE (2011)
24. Greveler, U., Justus, B., et al.: A Privacy Preserving System for Cloud Computing. In: 11th IEEE International Conference on Computer and Information Technology, pp. 648–653 (2011)
25. Zhou, M., Mu, Y., et al.: Privacy-Preserved Access Control for Cloud Computing. In: International Joint Conference of IEEE TrustCom 2011/IEEE ICESS 2011/FCST 2011, pp. 83–90 (2011)
26. Singh, J., Kumar, B., Khatri, A.: Securing the Storage Data using RC5 Algorithm. Internatinal Journal of Advanced Computer Research 2(4(6)) (2012)
27. Stolfo, S.J., Salem, M.B., Keromytis, A.D.: Fog Computing: Mitigating Insider Data Theft Attacks in Cloud. In: IEEE CS Security and Privacy Workshop (2012)
28. Brun, Y., Medvidovic, N.: Keeping Data Private While Computing in Cloud. In: IEEE Fifth International Conference on Cloud Computing (2012)
29. Gampala, V., Inuganti, S., Muppidi, S.: Data Security in Cloud Computing using Elliptic Curve Cryptography. International Journal of Soft Computing and Engg 2(3) (July 2012)
30. Wang, C., Wang, Q., et al.: Privacy-Preserving Public Auditing for Storage Security in Cloud Computing. In: Proceedings of the IEEE INFOCOM 2010 (2010)
31. Wang, C., Chow, S., et al.: Privacy-Preserving Public Auditing for Secure Cloud Storage. IEEE Transactions on Computers 62(2), 362–375 (2013)
32. Li, X., He, J., Zhang, T.: A Service Oriented Identity Authentication Privacy Protection Method in Cloud computing. International Journal of Grid and Distributed Computing 6(1) (February 2013)
33. IDC, It cloud services user survey, pt.2: Top benefits & challenges (2008), http://blogs.idc.com/ie/?p=210
34. Bohli, J.M., Gruschka, N., Jensen, M., Iacono, L.L., Marnau, N.: Security and Privacy-Enhancing Multi-cloud Architectures. IEEE Transactions on Dependable and Secure Computing 10(4) (July/August 2013)

Analysis of Secret Key Revealing Trojan Using Path Delay Analysis for Some Cryptocores

Krishnendu Guha[1], Romio Rosan Sahani[2], Moumita Chakraborty[1], Amlan Chakrabarti[1], and Debasri Saha[1]

[1] A.K. Choudhury School of Information Technology, Kolkata, India
[2] Institute of Radio Physics and Electronics University of Calcutta Kolkata, India
{mail2krishnendu,romios6}@gmail.com,
moumitachakraborty_it@yahoo.co.in, acakcs@caluniv.ac.in,
debasri_cu@yahoo.in

Abstract. The design outsourcing of the IC supply chain across the globe has been witnessed as a major trend of the semiconductor design industry in the recent era. The increasing profit margin has been a major boost for this trend. However, the vulnerability of the introduction of malicious circuitry (Hardware Trojan Horses) in the untrusted phases of chip development has been a major deterrent in this cost effective design methodology. Analysis, detection and correction of such Trojan Horses have been the point of focus among researchers over the recent years. In this work, analysis of a secret key revealing Hardware Trojan Horse is performed. This Trojan Horse creates a conditional path delay to the resultant output of the cryptocore according to the stolen bit of secret key per iteration. The work has been extended from the RTL design stage to the pre fabrication stage of ASIC platform where area and power analysis have been made to distinguish the affected core from a normal core in 180nm technology node.

Keywords: Hardware Trojan Horses (HTH), ASIC, DES, AES.

1 Introduction

The recent era has witnessed a breach of security in the untrusted phases of chip development. Several phases are involved in the design of an Integrated Circuit (IC) from the RTL design stage to the fabrication stage which may be actuated in different geographical locations. This trend of outsourcing of design and fabrication in the semiconductor industry is prevalent as it sufficiently increases the profit margin. While on the flip side, such a strategy of design outsourcing of IC supply chain across the globe provides an efficient platform for the introduction of malicious circuitry which causes the chip to malperform at a later stage of design. Such additional circuitry may also produce additional functionality which the designer is unaware of and pose a danger to the security and reliability of the chip. Such circuitry is called Hardware Trojan Horses (HTH). Hardware Trojans can be implemented as hardware modifications to ASIC's (Application Specific Integrated Circuits), commercial-off-the-shelf (COTS)

© Springer International Publishing Switzerland 2015
S.C. Satapathy et al. (eds.), *Proc. of the 3rd Int. Conf. on Front. of Intell. Comput. (FICTA) 2014*
– *Vol. 2*, Advances in Intelligent Systems and Computing 328, DOI: 10.1007/978-3-319-12012-6_2

parts, microcontrollers, network processors, digital signal processors (DSP) or a modification in firmware like FPGA bitstreams [1].

In ASIC design process, the chip is designed by tools developed by companies like Cadence, Synopsys, Mentor Graphics, etc which are considered trusted. However, the IP blocks, models and standard cells used by the designer during the design process and by the foundry during the post design process are considered untrusted [1]. Moreover the fabrication stage is also not secure enough and poses a threat of Trojan insertion. Such untrusted stages of chip development provide the gateway for the implantation of malicious circuitry by adversaries. The effects of hardware Trojans can range from unsuccessful functional operation of IC to the detoriation of reliability and expected lifetime of ICs [1, 2]. It may also cause the leakage of secret information through secure communication channels [3]. The research in this domain of Hardware Trojans is in its infancy but can be broadly classified into three groups, i.e., analysis of the effects of HTH in normal cores, detection of HTH in these cores and finally correction of the behavior of the cores from the effects of HTH.

Though several works exist in the analysis and detection of HTH in cryptocores, but to the best of the knowledge gathered, no Trojan has been classified which reveals the secret key of the cryptocore through the analysis of the path delay in the resultant output. In this work, we have analyzed how a Trojan Horse can reveal the secret key of certain cryptocores through conditional path delay of the resultant output. The design has been imported from the RTL level to the ASIC level where area and power analysis is made in 180nm technology node. Through the results obtained, we can confer that as the circuit complexity increases while the Trojan size is negligible, side channel analysis remains ineffective to detect small Trojans revealing secret key.

This paper is organized as follows. A brief description and a review of the taxonomy and detection methodologies of Hardware Trojan Horses is given in Section 2, followed by a brief discussion of AES and DES in Section 3, the cryptographic cores used in the present analysis. In Section 4, we give the detail of how the Trojan works on the cryptocores to reveal the secret key while the implementation and results of our work is presented in Section 5. In Section 6, we discuss the future scope of our work and finally we conclude this paper in Section 7.

2   Hardware Trojan Horses

Hardware Trojan Horses (HTH) can be described as simple malicious circuitry or modification to original circuitry which are made in the untrusted phases of chip development and has the capability to exploit the hardware. Hardware Trojans can be inserted at any stage of design and manufacturing flow, i.e. specifications, design, fabrication, testing, assembly and packaging. When such circuitry is activated, i.e. triggered, which may be external like signal sensing through an antenna or internal like a certain number of predefined attempts, it causes the chip to malperform or deviate from its normal operation (payload). It may also reveal the secret key in cryptographic applications which can cause a severe threat to security. Detection methodology of Trojans encompasses logic testing techniques and side channel analysis techniques. Logic testing based approaches tries to create rare events at internal nodes to trigger Trojan circuitry while in side channel analysis, a physical parameter like power or critical path is selected based on which a Trojan infected circuitry is tried to be detected and distinguished from a perfect circuitry. However, the limitations of side channel analysis are prominent when the circuit size is large but Trojan size is small.

The authors in [1, 2] present a detailed study of classification of Hardware Trojans and a survey of published techniques of Trojan detection. The work in [3] deals with the effect of hardware trojans in wireless cryptographic ICs where the Trojan objective is to reveal the secret key with an analog counterpart in addition to the normal digital circuitry. The authors also propose statistical analysis techniques to detect such Trojans. The direct modification of bitstream by Hardware Trojan Horses in FPGA platform which in recent times can serve as an alternative to ASIC in some application is dealt in [4]. The authors in [5] investigate the sensitivity of a power supply transient signal analysis method for detecting Trojans in the presence of measurement noise and background switching activity. The authors also focus on determining the smallest detectable Trojan, i.e., the least number of gates a Trojan may have and still be detected. The authors in [6] propose a non-destructive side channel approach that characterizes and compares transient power signature using principle component analysis to achieve the hardware Trojan detection. Detection of small Trojans under large noise and variations is effective from the work. Detecting trojans through leakage current analysis is dealt in [7] which also demonstrate the effectiveness of a trojan detection methodology based on the analysis of a chip's steady state current measured simultaneously from multiple places on the chip. A technique for improving hardware trojan detection and reducing trojan activation time is proposed in [8] as time to activate a hardware trojan circuit is a major concern from the authentication viewpoint. In this work, the authors analyze time to generate a transition in functional trojans and this transition is modeled by geometric distribution and the number of clock cycles required for generating a transition. The procedure increases transition probabilities of nets beyond a specific threshold.

Normal side channel analysis techniques suffer from decreased sensitivity towards small trojans especially due to the large process variations present in modern nanometer technologies. A novel noninvasive, multiple-parameter side-channel analysis based trojan detection approach is discussed in [9] where the authors use an intrinsic relationship between dynamic current and maximum operating frequency of a circuit to isolate the effect of a trojan from process noise. Combination of side channel analysis and logic testing approach to provide high overall detection coverage for hardware trojan circuits of varying types and sizes is also discussed in [9]. Segmentation and gate level characterization technique of detection of hardware trojan is proposed in [10]. Segmentation method is used to divide the large circuit into small sub circuits using input vector selection. Based on selected segments, gate level leakage power is traced to detect Hardware Trojans.

Detection of Trojans in the layout level of design is dealt in [11] where the comparison is made between optical microscopic pictures of the silicon product and the original view from a GDSII layout database reader. Moreover, analysis is also made to introduce a hardware Trojan horse without changing placement or routing of the cryptographic IP logic. The authors in [12] propose a architecture that reorders scan cells based on their placement during physical design to reduce circuit switching activity by limiting it into a specific region.

3   Cryptocores

3.1   Advanced Encryption Standard (AES)

Advanced Encryption Standard (AES) [13, 14] is a 128 bit block non- Feistel cipher which takes as input a 128 bit plain text and a 128 bit key and generates a 128 bit cipher text as output. AES implemented in this work comprises of a pre round transformation along with 10 identical rounds (Round 10 is slightly different) and a Key Expansion module which produces 10 round keys for each round. AES provides four types of transformations namely, Substitute Bytes, Shift Rows, Mix Columns and Add Round Key. The pre round module uses only Add Round Key transformation while round 10 does not utilize the Mix Columns transformation. All other rounds use all the four transformations. To create a round key for each round, AES uses a Key Expansion module which creates ten 128 bits round keys for the ten rounds from the 128 bit input cipher key.

3.2   Data Encryption Standard (DES)

Data Encryption Standard (DES) [13, 14] is a 64 bit block cipher which takes as input a 64 bit plain text and a 56 bit key and generates a 64 bit cipher text as output. It consists of an initial and final permutation rounds which are keyless and predetermined. A full DES cipher consists of 16 Feistel rounds. The round key generator module of the DES cipher takes as input a 64 bit key which is reduced to 56 bits after parity drop. This round key generator generates 16 round keys for the sixteen rounds of the DES cipher after a Left Shift operation and a Compression D box at each round. Each Feistel round consists of a mixer and a swapper, both of which are invertible. The DES function is enclosed inside the mixer which is the heart of the DES cipher. The DES function comprises of an Expansion D box, a XOR operation with the key for the respective round obtained from the round key generator, eight S boxes which provide the real confusion of the cipher and finally a Straight D box. Round 16 is slightly different from the other rounds as it comprises only the mixer but not the swapper.

4   Design and Analysis of Trojan on Cryptocores

The Trojan module steals one bit of the secret key of the cryptocore in each iteration. The key bit is either '0' or '1'. If the key bit is '0', then no path delay is made to the resultant output of that iteration but if the key bit for the corresponding iteration is '1', then an additional path delay of 2ns is made to the resultant output. The small path delay can be caused due to circuit deformities and may go unnoticed to the normal user while an adversary can analyze the path delay in the respective iterations to obtain the secret key. The number of iterations required to reveal the secret key depends on the size of the secret key. The cryptocore can be represented by an 'm' bit block cipher utilizing an 'n' bit key. For AES, 'm=128 bits' while the secret key 'n' is of

128 bits and hence, after 128 iterations the secret key will be revealed. Similarly for DES, 'm=64 bits' while the secret key 'n' is of 56 bits and hence after 56 iterations, the key is revealed. The diagrammatic representation of the effect of Trojan on Cryptocores has been shown in Fig. 1.

For Trojan detection, the conventional methods of logic analysis are ineffective as the affected output is not different from the normal output. Hence, refuge must be sought to side channel analysis. In this work, after the import of the RTL design into the ASIC platform, power and the gates utilized of the affected cryptocores is analyzed from the normal cryptocores.

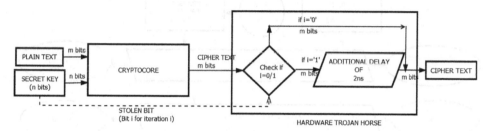

Fig. 1. Effect of Trojan on Cryptocores

5   Implementation and Results

The RTL design or the verilog code of the cryptocores have been written in Xilinx ISE 14.4 platform and simulated in ISim Simulator to obtain the perfect output. Then the RTL design of the Trojan Horse have been added as a separate module to the original cryptocore module which steals a key bit in each iteration of the synthesis and provides a conditional delay to the resultant output. Then the synthesized netlist of the golden and the affected cores are generated via Leonardo Spectrum of Mentor Graphics where we also obtain the total area of each circuitry. Finally, we import the netlist of the perfect and the affected cores in the Pyxis Schematic Platform of Mentor Graphics. In this ASIC platform, we compare the golden and the affected cores through power analysis. The whole methodology has been diagrammatically described in Fig. 2. Fig. 3 gives the simulation result of a golden cryptocore where no delay is present in the ideal case. Fig. 4 shows the condition when the secret key bit is '0' while Fig. 5 shows the condition when the secret key bit is '1' when an additional path delay of 2 ns is added to the resultant output. Table 1 gives the analysis of area (number of gates utilized) and power in ASIC platform of golden and affected cryptocores simulated in 180 nm technology library.

Analyzing Table 1, we can find that the Hardware Trojan circuitry for AES cryptocore enhances the gate count by only 725 gates which accounts to the increase in power consumption by 0.015% only. While for DES cryptocore, the increase in gate count is 362 due to the addition of Trojan circuitry which increments the power consumption by 0.009%. Though the logic used for the Hardware Trojan Horse is the same for the two cryptocores, yet the difference in hardware resources is caused due to the difference in cipher key size of the two cryptocores, i.e. 128 bits for AES and 56 bits for DES. From the results, we can infer that as the circuit size increases while the Trojan size is negligible, side channel analysis based detection technology which uses power as a significant detection parameter seems to be unworthy as the increase in power is negligible.

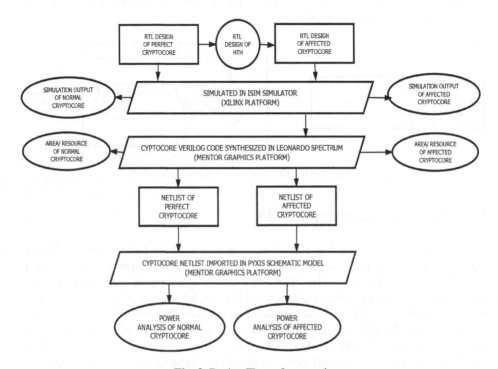

Fig. 2. Design Flow of our work

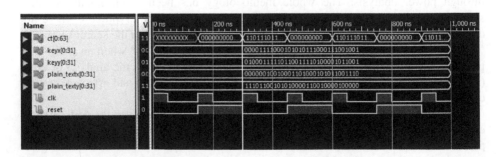

Fig. 3. Simulation Output for Normal Cryptocore

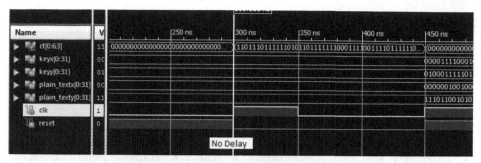

Fig. 4. Affected output when Stolen Key Bit= '0'

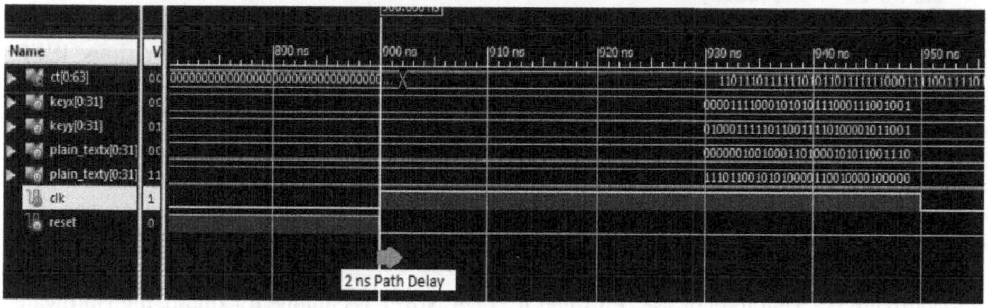

Fig. 5. Affected output when Stolen Key Bit= '1'

Table 1. Analysis of Area and Power of Normal and Affected Cryptocores in ASIC platform

Cryptocores	Area (Number of Gates Utilized)			Power Consumption (UW)		
	Normal	Affected	Increase in Gates	Normal	Affected	% increase
AES	153852	154577	725	520.9377	521.0164	0.015
DES	18153	18515	362	229.0403	229.0604	0.009

6   Future Work

The analysis of the effects of Hardware Trojans for several other systems will be analyzed in near future. Detection of the affected portion and a correction mechanism is a prosperous scope for future work.

7   Conclusion

In this work, we have analyzed the effects of a Hardware Trojan Horse, which can reveal the secret key of a cryptocore by conditionally varying the path delay of the resultant output. Thus the reliability and security of the core is at stake. However, as we can find from Table 1, the added hardware resources due to the Trojan circuitry increments area and power to a very nominal amount, we can infer that side channel analysis based detection methodology is not full proof. Thus, work needs to be carried out for a better detection technique using circuit partitioning.

References

1. Tehranipoor, M., Koushanfar, F.: A survey of hardware Trojan taxonomy and detection. IEEE Design and Test of Computers 27(1), 10–25 (2010)
2. Bhunia, S., Abramovici, M., Agrawal, D., Hsiao, M.S., Plusquellic, J., Tehranipoor, M.: Protection Against Hardware Trojan Attacks: Towards a Comprehensive Solution. IEEE Design and Test of Computers 30(3), 6–17 (2013)
3. Jin, Y., Makris, Y.: Hardware Trojans in Wireless Cryptographic ICs. IEEE Design and Test of Computers 27(1), 26–35 (2010)
4. Chakraborty, R.S., Saha, I., Palchaudhuri, A., Naik, G.K.: Hardware Trojan Insertion by Direct Modification of FPGA Configuration Bitstream. IEEE Design and Test of Computers 30(2), 45–54 (2013)
5. Rad, R., Plusquellic, J., Tehranipoor, M.: A Sensitivity Analysis of Power Signal Methods for Detecting Hardware Trojans Under Real Process and Environmental Conditions. IEEE Transactions on Very Large Scale Integration (VLSI) Systems 18(12), 1735–1744 (2010)
6. Wang, L., Xie, H., Luo, H.: Malicious Circuitry Detection Using Transient Power Analysis for IC Security. In: International Conference on Quality, Reliability, Risk, Maintenance, and Safety Engineering (QR2MSE), pp. 1164–1167 (2013)
7. Aarestad, J., Acharyya, D., Rad, R., Plusquellic, J.: Detecting Trojans through leakage current analysis using multiple supply pad. IEEE Transactions on Information Forensics and Security 5(4), 893–904 (2010)
8. Salmani, H., Tehranipoor, M., Plusquellic, J.: A Novel Technique for Improving Hardware Trojan Detection and Reducing Trojan Activation Time. IEEE Transactions on Very Large Scale Integration (VLSI) Systems 20(1), 112–125 (2012)
9. Narasimhan, S., Du, D., Chakraborty, R.S., Paul, S., Wolff, F.G., Papachristou, C.A., Roy, K., Bhunia, S.: Hardware Trojan Detection by Multiple-Parameter Side-Channel Analysis. IEEE Transactions on Computers 62(11), 2183–2195 (2013)
10. Wei, S., Potkonjak, M.: Scalable Hardware Trojan Diagnosis. IEEE Transactions on Very Large Scale Integration (VLSI) Systems 20(6), 1049–1057 (2012)
11. Bhasin, S., Danger, J., Guilley, S., Thuy, X., Sauvage, L.: Hardware Trojan Horses in Cryptographic IP Cores. In: Workshop on Fault Diagnosis and Tolerance in Cryptography, pp. 15–29 (2013)
12. Salmani, H., Tehranipoor, M.: Layout-Aware Switching Activity Localization to Enhance Hardware Trojan Detection. IEEE Transactions on Information Forensics and Security 7(1), 76–87 (2012)
13. Forouzan, B., Mukhopadhyay, D.: Cryptography and Network Security, 2nd edn. McGraw Hill, ISBN: 978-0-07-070208-0
14. Stallings, W.: Cryptography and Network Security, 3rd edn. Prentice Hall (2003) ISBN: 0-13-11 1502-2

Generating Digital Signature Using DNA Coding

Gadang Madhulika and Chinta Seshadri Rao

Department of Computer Science and Engineering,
Anil Neerukonda Institute of Technology and Sciences Sangivalasa,
Bheemunipatnam[M],Visakhapatnam, India
gadangmadhulika@gmail.com, seshadri.rao.cse@anits.edu.in

Abstract. This work focuses on signing the data with a signature using DNA coding for limited bandwidth systems or low computation systems. The proposed process has two modules. In first module, the sender generates a digital signature by signing the message using the DNA Coding Sequence. The second module is a hybrid of public key cryptography. A DNA symmetric key is generated, encrypted with DNA public key and shared with the intending recipient. Then the messages are exchanged using shared symmetric key. In both the modules, the work uses a simple non-linear function XOR to encrypt and decrypt the DNA coding sequence. The computation time required to perform the XOR operation matches the capabilities of limited bandwidth systems and suits our work. In addition the work also achieves high security in two levels, one is the secret matching of plain text letters to DNA Codon Sequence and the second is increase in the complexity of computation for breaking the algorithm using brute-force attack to square of the complexity achieved with 128-bit binary key, for the same length of DNA key.

Keywords: DNA Coding Sequence, Message Digest (MD), Digital Signature, Public Key Cryptography, XOR, Nucleotides.

1 Introduction

Security is by far a means and seriously considered in communication of data, images and video. Public Key Cryptography provides secure communications over devices connected with wired or wireless networks. The practical problem in implementation of Public Key Cryptography using RSA is the setup-time of key. It takes many minutes to generate 1024-bit RSA key [1]. The limited bandwidth of few systems, improvisation of factorization algorithms and computing power to generate keys [2] of 4096-bits or 8192-bits may limit RSA use on low computational systems. ECC (Elliptic Curve Cryptography) can be used effectively for limited bandwidth devices but for the cost of generating signatures 320-bits [3] long, and signature verification is 40 times slower than using RSA [4]. DNA Cryptography is an upcoming and promising area that helps provide security. The inherent complex structure of DNA makes it difficult to break the algorithm built on it. A. Gehani T.H. LaBean and J.H. Reif, presented DNA-based cryptography technique in "DNA-based cryptography"

© Springer International Publishing Switzerland 2015

S.C. Satapathy et al. (eds.), *Proc. of the 3rd Int. Conf. on Front. of Intell. Comput. (FICTA) 2014*
– *Vol. 2*, Advances in Intelligent Systems and Computing 328, DOI: 10.1007/978-3-319-12012-6_3

[5]. Xiang Wang, Qiang Zhang [6], demonstrated RSA algorithm using public-key cryptography DNA sequence as keys. This paper focuses on generating signatures using Public Key Cryptography with the help of DNA Coding Sequence. The private key and public key are carefully chosen that the process is reversible on both sides. The work uses a simple non-linear function XOR for encryption instead of complex algorithms like RSA that benefits systems with limited or little bandwidth in terms of computation and speed. Alice computes the Message Digest of the message (plain text) and converts to the DNA Sequence. Alice signs the message (i.e., DNA Sequence and plain text) using his Private Key (128 nucleotides) and sends to Bob. Bob decrypts the cipher message using the Public Key (128 nucleotides) of Alice and converts the DNA Coding Sequence to the plain text. Bob re-computes the Message Digest on the plain text and verifies with the Message Digest sent by Alice. Bob acknowledges the signature of Alice after the Message Digest matches. The paper is organized in to four chapters; the second chapter describes the related work on using DNA over cryptography, the work of using DNA on Public Key Cryptography is presented in the proposed method, the results are placed in the fourth chapter, the conclusion and future enhancements is discussed in the last.

2   Related Work

DNA computing was pioneered by Adleman [7].The DNA computing is used to solve Hamiltonian path problem and NP complete problem [8].D. Boneh, C. Dunworth ,R. Lipton used DNA computing methods to crack the DES [9].C.T. Celland V. Risca and Bancroft C ,demonstrated a method that encodes a message by secretly hiding them in DNA strands[10].An encryption scheme that uses DNA synthesis, PCR amplification, DNA digital coding and PCR traditional cryptography was proposed by G. Cui,L. Qin,Y. Wang and X. Zhang [11]. Naveen Jarold K, P. Karthigai kumar, N. M. Sivamangai, Sandhya R, Sruthi B Ashok [12] ,suggested using a DNA message format which used a DNA-codon a 3-alphabet set to represent a character. Public-key Cryptography also called as the asymmetric Cryptography, is a better choice to use in public networks or unsecured channels (as symmetric key cryptography is a better choice to use in secure channel).The keys are used for two different purposes one for encryption and other for decryption [13]. Qiang Zhang, Ling Guo, Xianglin Xue, Xiaopeng Wei [14], demonstrated DNA sequence addition operation to encrypt an image. Olega Torana, Monica Borda [15], demonstrated XOR operation on DNA tiles with matching sticky ends using One-Time-Pad.

3   Proposed Work

3.1   Digital Signature

The proposed work is designed and implemented using the DNA Coding Sequence of length 9. The message format is generally alphanumeric in nature sometimes including few special symbols. The characters in the message (alphabets, numbers or symbols) are converted to a DNA Coding Sequence using a special function block and stored in a database. Alice needs to send a message to bob. The private and public keys of Alice and Bob are in DNA Sequence. The private and public keys are carefully chosen to be complementing each other. Alice takes the message, computes the Message Digest (MD). The message along with the Message Digest is converted to DNA Coding Sequence using Table 1.We take 00,01,10,11 for C,A,T,G. Alice encrypts the DNA sequence using XOR operation with private key on message and the Message Digest using Table 2. The resulting text is called cipher text. The cipher text is sent to Bob. Bob reads the cipher text and performs the reverse process of encryption on the cipher text to retrieve the DNA Coding Sequence using Alice public key. Bob complements the DNA Coding Sequence to obtain the original DNA Coding Sequence. This sequence is converted to the plain text using Table 1. The plain text contains the message and the Message Digest sent by Alice. Bob re-computes the Message Digest and verifies with the one sent by Alice. The schematic view as shown in figure 1. If the Message Digest matches, Alice is authenticated and the message integrity is not lost. Authentication, integrity and non-repudiation are security signs of digital signatures. This module generates digital signature using DNA coding sequence.

3.2   Sharing Symmetric Key Using Public Key Cryptography

Alice constructs a message which is key (128 nucleotides), converts it to DNA coding sequence. Alice encrypts the DNA coding sequence using the public key of Bob and sends the message. Bob on receiving the encoded message decrypts the message using his private key to obtain the original message which is key. Once the key is exchanged, a session is established between Alice and Bob. The exchanged Key is used to encrypt and decrypt messages for further communication. The Public and Private keys of Alice and Bob are generated randomly and distributed using a secure channel .The key is 128 nucleotides long.

Table 1. Code for DNA Coding Sequence

character	codon	character	codon	character	codon
	TGAGATCCT	<	GGCTCAAGA		
-	GTGGCGAAG	=	ATCGCCATT		
!	AGATATAAT	>	ACTACGTGC		
"	TAAAGGGAG	0	CATAGTCAA		
#	TCCCCCATT	1	GCAATTAGT		
$	CAAGTATAC	2	CTTCCTGAC		
%	GTGCTAACG	3	TCAGACGAA		
&	CGTCGAGCT	4	AGCGCTGAT	character	codon
(	CAACTGGTG	5	TTTAATACG	L	CCACCCGAA
)	GCTCCGTCG	6	GGCTCCTCC	M	AATTTGTTC
*	GTTGAGAGA	7	TATGTGCTC	N	CAGCGTAAC
,	CAATCAGAA	8	GCGTGAGTG	O	CAACAATAG
.	ATTCAAGTC	9	CAGTTACCA	P	ATTAATTCT
/	GAAGTAGCC	A	ATATGGCGG	Q	AGAGAATTT
:	TGCGTATAA	B	GCCCCGTGA	R	ACTAGGTTG
;	CAGCTATCG	C	GCAAACGCA	S	GTTGGGTCT
?	ACATATGAT	D	CTTCAATCG	T	GATTCACTC
@	GCACGCAAC	E	GGCTCGCCC	U	TAGTGTAGG
^	AGCGAGTAG	F	CCCTATATC	V	TTCTAATTT
_	GACAGTCGA	G	TAGATATCT	W	AACTGGATC
]	AGACCTGTC	H	TGAGTTAGT	X	CCGGTTGAC
~	GATGAGCCT	I	CAAGGATAT	Y	GCGTCCTCT
+	GTCTCCACG	J	CTATCCCCC	Z	CTACATTTA

Table 2. XOR operation for DNA sequence

$\bigoplus$	C	A	T	G
C	C	A	T	G
A	A	C	G	T
T	T	G	C	A
G	G	T	A	C

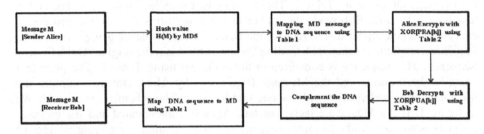

Fig. 1. Block diagram for generating DNA Digital Signature

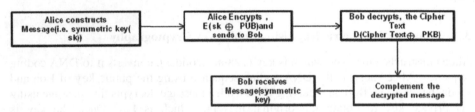

Fig. 2. Block diagram Sharing symmetric key

3.1.1   Algorithm 1: Digital Signature

The aim is to prove authentication, integrity of the message and non-repudiation of sender sending the message .The algorithm below describe the process.

> INPUT: Message "M"
> Step 1:   Message Digest (MD) ⟵——— MD5 (M)
> Step 2:   for i = 0 to n   // n is length of Message Digest
>   ch ← MD [i]
>   DNA Code ← ch   //using Table 1
> Step 3:   Cipher Text ← E (PK$_A\bigoplus$ (DNA (MD)))
> Step 4:   DNA Code ←   D (PU$_A \bigoplus$ (Cipher Text))
> Step 5:   DNA Code ← ~ (DNA Code (M))
> Step 6:   Message Digest (MD)←Mapping DNA Code using Table 1
> Step 7:   Message Digest ← Bob Message (MD)
> Step 8:   Bob verifies the Signature of Alice.
> OUTPUT: DNA Digital Signature

3.1.2   Algorithm2: Sharing Symmetric Key Using Public Key Cryptography

The Symmetric key is exchanged between parties of communication using public key cryptography.

INPUT: Symmetric key 'k' (128-nuclotides)
Step 1: for k=0 to n   /*n is symmetric key length*/
      ch ← SK[k]   /*SK (symmetric key)*/
      Cipher Key ← E (PU$_B$ ⊕(DNA (ch)))
Step3: Decrypted Key ← D (PK$_B$ ⊕(Cipher Key))
Step4: Symmetric key 'k' ← C (Decrypted Key)
OUTPUT: Symmetric key (shared between parties of communication)

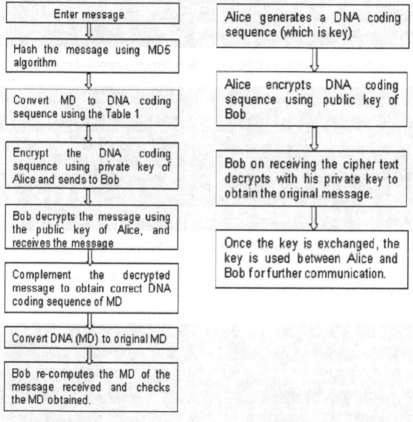

Fig. 3. Digital Signature   **Fig. 4.** Sharing Symmetric Key

4   Results

From figure 5 it shows a given message is initially converted into a hash value and then each digit of given data is matched with DNA sequence (9 nucleotide) as shown in the table1.The DNA codon message digest is encrypted with private key of Alice using XOR operation. The cipher text is send to bob, bob on receiving the cipher text, he decrypts the cipher text with public key of Alice and complement it, and by mapping with the table1 ,he will receive the MD. The MD send by the sender is same as the receiver received then authentication is proved.

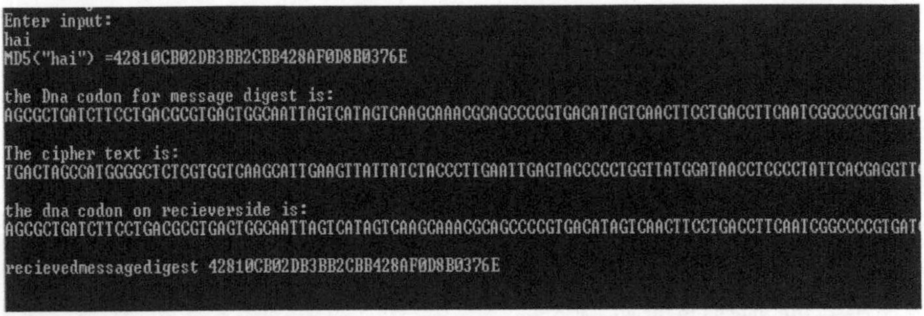

Fig. 5. Generating and verifying Digital Signatures by sender and receiver

Fig. 6. Generating and exchange of Symmetric key

The above figure 6 shows, the key exchange by sender and receiver after Signature is verified.

Fig. 7. Encryption and decryption of message using the symmetric key (128 nucleotide)

The above figure 7 depicts the sender encrypting the famous message JUNE6_INVASION: NORMANDY with the exchanged symmetric key and receiver decrypts the cipher text using the same key.

5   Conclusion

The present work is towards the development of generating secure signatures (in simple manner and making it hard to break) for limited bandwidth or low computational systems. The complexity of the algorithm is measured in withstanding the brute-force attack and chosen-cipher text attack. The high randomness of DNA and the DNA codon generation makes it difficult to obtain the plain text from the cipher text. Generally, the complexity of 128-bit binary key in brute-force attack is measured to be 2^{128}. The DNA key uses 128-nucleotides length each having four possibilities of A, G, C, T measures to 4^{128} i.e., $2^{2 \times 128}$ increasing to square of the complexity achieved by binary key. With the analysis in hand, it takes 2^{128} years on Cray 1 Super computer to break the encryption and finding the key. This work may be extended to generate digital certificates or generating keys and distributing them securely.

References

1. Boneh, D., Modadugu, N., Kim, M.: Generating RSA keys on a handheld using an untrusted server. In: Roy, B., Okamoto, E. (eds.) INDOCRYPT 2000. LNCS, vol. 1977, pp. 271–282. Springer, Heidelberg (2000)
2. Schneier, B.: Applied Cryptography: protocols, Algorithms, and Source Code in C, 2nd edn. Wiley, New York (1996)
3. The Elliptic Curve Digital Signature Algorithm (ECDSA), ANSI X9.62, American National Standards Institute (1998)
4. De Win, E., Mister, S., Preneel, B., Wiener, M.: On the performance of signature schemes based on elliptic curves. In: Buhler, J.P. (ed.) ANTS 1998. LNCS, vol. 1423, pp. 252–266. Springer, Heidelberg (1998)
5. Gehani, A., LaBean, T.H., Reif, J.H.: DNA-based cryptography. In: DNA Based Computers V, vol. 54, pp. 233–249. American Mathematical Society, Providence (2000)
6. Wang, X., Zhang, Q.: DNA computing-based cryptography. In: 4th International Conference on "Bio-Inspired Computing", BIC-TA 2009, pp. 1–3. IEEE (2009)
7. Adleman, L.: Molecular Computation of Solutions to Combinatorial Problems. Science 266, 1021–1024 (1994)
8. Lipton, R.J.: Using DNA to solve NP-complete problems. Science 268, 542–545 (1995)
9. Boneh, D., Dunworth, C., Lipton, R.: Breaking DES using a molecular Computer, pp. 37–65. American Mathematical Society (1995)
10. Celland, C.T., Risca, V., Bancroft, C.: Hiding messages in DNA microdots. Nature 399, 533–534 (1999)
11. Cui, G., Qin, L., Wang, Y., Zhang, X.: An encryption scheme using DNA technology. In: IEEE 3rd International Conference on Bio-Inspired Computing: Theories and Applications (BICTA 2008), Adelaid, SA, Australia, pp. 37–42 (2008)

12. Naveen, J.K., Karthigaikumar, P., Sivamangai, N.M., Sandhya, R., Asok, S.B.: Hardware Implementation of DNA Based Cryptography. In: IEEE Conference on Information and Communication Technologies, ICT, pp. 696–700 (2013)
13. Katayangi, K., Murakami, Y.: A new product-sum public-key cryptosystem using message extension. IEICE Transactions on Fundamentals E84-A(I0), 2482–2487 (2001)
14. Zhang, Q., Guo, L., Xue, X., Wei, X.: An Image Encryption Algorithm Based on DNA Sequenc Addition Operation. In: 4th International Conference on Bio-Inspired Computing, BICTA, pp. 75–79 (2009)
15. Torana, O., Borda, M.: 8th International Conference on Communications (COMM), pp. 451–456. IEEE (2010)

DNA Encryption Based Dual Server Password Authentication

P.V.S.N. Raju and Pritee Parwekar

Department of Computer Science and Engineering,
Anil Neerukonda Institute of Technology and Sciences, Sangivalasa,
Bheemunipatnam[M], Visakhapatnam, India
ieg.rajup@gmail.com, pritee.cse@anits.edu.in

Abstract. Security-authentication is a crucial issue in networking for establishing communication between clients and servers, or between servers. Authentication is required whenever a secure exchange of information is sought between two computers. In a normal password authenticated key exchange all clients passwords are stored in a single server. If the server is compromised because of hacking or even insider attack, passwords in the server are all disclosed. This paper proposes two server password authenticated key exchange between two servers which is used to authenticate single client and thereby making loss of passwords to hackers much more difficult. The paper proposes the DNA for Encryption and Decryption along with ElGamal Encryption technique. This would prevent the intruder from using information obtained from one server towards accessing vital login in information.

Keywords: Password authenticated key exchange, Elgamal Encryption, DNA Encryption.

1 Introduction

In today's world of computers and IT instruments passwords is the main method of authenticating genuine users for personal computers, mobile phones, ATM's etc. A computer user may require passwords for many purposes like checking the E-mails, databases, networks etc. The user information will be stored in servers. To login into the particular server the user has to be authenticate by means of password. If any intruder attacks or hijack the particular server all the passwords will be known. This is a big threat to the privacy of the user.

Earlier passwords are transmitted over a public channel by encrypting it using a hash value which makes that hash value will be accessed by the attacker. If that is done he can predict the password by using different password combinations against the original password hash value.

DNA refers to Deoxyribo Nucleic Acid. Every cell in a human body contains set of DNA. It is unique for each and every individual. DNA can be applied in cryptography[1]. Adleman [2] proposed DNA for computation. No one can read the message after encrypting it using DNA. A DNA strand [3] consists of four different

© Springer International Publishing Switzerland 2015
S.C. Satapathy et al. (eds.), *Proc. of the 3rd Int. Conf. on Front. of Intell. Comput. (FICTA) 2014*
– *Vol. 2*, Advances in Intelligent Systems and Computing 328, DOI: 10.1007/978-3-319-12012-6_4

nucleotides named as adenine(A), cytosine(C), guanine(G), thymine(T). Every single string will be paired up with its complementary string to become a double helix. A only pairs with its complement T and G only pairs with its complement C. In binary form A will be considered as 01, T as 10, C as 00, G as11.These values will be used while encryption [4] and decryption [4] of the message. Using these four nucleotides A,T,G,C can encrypt the message and decrypt the message. In this paper we are proposing a new system that is two server password only authenticated key exchange using DNA. In this we are using Elgamal encryption and DNA encryption techniques.

2   Related Work

There are two models for password based authentication key exchange.

(a) The first model is PKI -based model [5], assumes that the client keeps the server's public key in addition to share a password with the server. In this model, the client send the password to the server by public key encryption.

(b) The other model is password-only model. In this the password is used as a secret key to encrypt random numbers for key exchange.

The password based authentication protocols assumes that all the passwords will be stored in a single server. This server will be used to authenticate the client. If that particular server is compromised then the intruder will know all the details of the client. To overcome this issue two server password based authentication protocols[6][7][8][9] were introduced. In this two servers are cooperate each other to authenticate the client based on the password. Because of this method even the intruder compromised one server he cannot predict the client information from the compromised server.

The present solutions for two server password authenticated key exchange (PAKE) [10] are Symmetric and Asymmetric. In symmetric two servers will equally contribute to authenticate the client. In asymmetric one server authenticates the client with the help of another server.

There are few authentication methods based on two server password based authentication key exchange. They are as follows.

Mukesh et al.'s[11] proposed finger print based two server authentication key exchange. In this the user's password is replaced with the random string generated by finger print template. Brainard et al.'s[12] proposed two server password system in which one server exposes to the user and the other one is invisible to the user. Katz et al.'s proposed two server password only authenticated key exchange, this is the provably secure two server protocol for the important password-only setting. Yang et al.'s[13] proposed an efficient password only two server authenticated key exchange system. This scheme is a password-only variant of the one introduced by Brainard et al.'s.[12]

3   Proposed System

The system consists of one client and two servers. The client will send a password(pw) to one of the server during registration. The server in turn will divide the password into 2 parts pwd1 and pwd2 (pw=pwd1+pwd2). The pwd1 will be encrypted using Elgamal encryption. Then after encryption pwd1 will become E(pwd1). Then E(pwd1) will be again encrypted using DNA. After encryption E(pwd1) will become E1(E(pwd1)). This will be stored in server1 database. The same process will be repeated in the server2 i.e pwd2 will be sent to server2. pwd2 will be encrypted as E(pwd2) again this will be encrypted using DNA as E1(E(pwd2)).

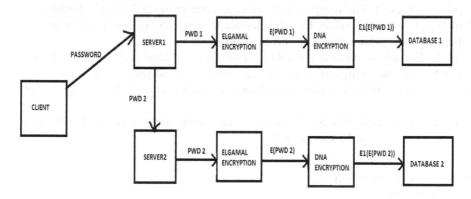

Fig. 1. Encryption process

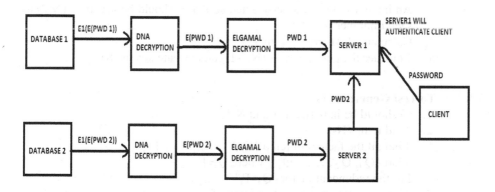

Fig. 2. Decryption process

With the client logging into the system, the password will be authenticated at one of the server. First at each server database we will have E1(E(pwd1)) and E1(E(pwd2)). These two will be decrypted using DNA. Then, these will become E(pwd1) and E(pwd2). Again these will be decrypted using Elgamal decryption. It will be decrypted as pwd1 and pwd2. Using pwd1 and pwd2(pw=pwd1+pwd2) it will authenticate the client/user.

For DNA encryption [14] we are using A(Adenine), T(Thymine), G(Guanine), C(Cytosine) codons. For Encryption we are using XNOR operation and for Decryption we are using XOR operation

3.1 ElGamal Encryption/Decryption Algorithm [10]

The receiver should select a largest prime number(N),select a generator number(G).The G should be tested under some conditions to check whether G is acceptable as generator number or not. If accepted then continue else choose another number and test it. The receiver then select a secret number(S) which should be less than N-2. Using these values he will calculate D. Then (N,G,D) will be considered as public key and (S) should be considered as private key

In the Encryption process a number(K) should be selected such that it lies between 1 and N-2 and the plain text should be represented with M such that it lies between 0 and N-1 and calculate Y,Z using the above parameters. That (Y,Z) will be the cipher text.

In the Decryption process R should be calculated using Y,N and S parameters. Recover the plain text M using R,Z,N.

3.1.1 Key Generation
At Receiver side

1. A large prime number should be generated randomly (say N)
2. Select a number(say generator G)
3. An integer (S) should be selected so that it should be less than (N-2) as secret number
4. Calculate D where $D=G^S \bmod N$
5. Consider the public key as (N,G,D) and private key as (S)

3.1.2 To Test Generator G
1. G should be in between 1 and N-1
2. Find temp=N-1
3. Find all the factors of temp $\{F_1,F_2,\ldots\ldots F_n\}-\{1\}$
4. Find $\{Q_1, Q_2,\ldots\ldots Q_n\}$ where $Q_i=F_i$
   For the redundant factors $Q_i=F_i^{freq}$
5. Generator number is G if and only if
   $W_i= G^{temp/Q_i} \bmod N$, $W_i \mathrel{!}=1$, for all Q_i

3.1.3   Encryption

Sender should perform the following steps

1. Receive public key (N,G,D) from the receiver
2. An integer (K) should be selected so that it should be lies between 1 and (N-2)
3. Plain text should be represented as an integer M where M lies between 0 and (N-1)
4. Calculate Y where $Y=G^K modN$
5. Calculate Z where $Z=(D^K*M)modN$
6. Find the cipher text C where C=(Y, Z)
7. The sender should send C to the receiver

3.1.4   Decryption

Sender should perform the following steps

1. Obtain the cipher text C from receiver
2. Calculate R where $R=Y^{N-1-S}modN$
3. Recover the plain text as follows
$$M=(R*Z)modN$$

3.2   DNA Encryption/Decryption Algorithm

Generate the random DNA sequence using A,T,G,C for all possible characters. Convert the text into A,T,G,C sequence which was generated earlier and consider it as (X).Generate a public key using DNA codons and consider it as Y. Perform Exclusive NOR operation between X and Y and consider the result as (Z). It is encrypted text using DNA.

In the Decryption process consider Y^1(complement of Y) as private key. Perform the Exclusive OR between Z and Y^1, that would be the Decrypted text using DNA.

3.2.1   Random DNA Generation

1. Generate Random DNA sequence using A,T,G,C characters for all possible characters.
2. Store those Random DNA Sequences in a Database

3.2.2   Encryption

1. Receive the cipher text after ElGamal encryption.
2. Replace the characters in cipher text with the DNA sequence and consider it as (X).
3. Generate a random DNA sequence (Y) and consider it as a public key.
4. Perform XNOR operation between X and Y and consider it as (Z).
5. Z is the Encrypted text using DNA

Table 1. XNOR Table for A,T,C,G

	A	T	C	G
A	G	C	T	A
T	C	G	A	T
C	T	A	G	C
G	A	T	C	G

3.2.3   Decryption

1. Generate a private key Y^1(Complement of Y).
2. Consider Z(Encrypted text using DNA).
3. Perfom XOR[15s] operation between Z and Y^1 and will obtain the decrypted text.

Table 2. XOR Table for A,T,C,G

	A	T	C	G
A	C	G	A	T
T	G	C	T	A
C	A	T	C	G
G	T	A	G	C

Table 3. Complemented Table for A,T,C,G

Value	Complemented Value
A	T
T	A
C	G
G	C

Here A is considered as 01
T is considered as 10
C is considered as 00
G is considered as 11

4   Result Analysis

From Figure 3 it shows that the client will send the username and password. The server has to authenticate the user for the respective username and password.

Fig. 3. Client Authentication

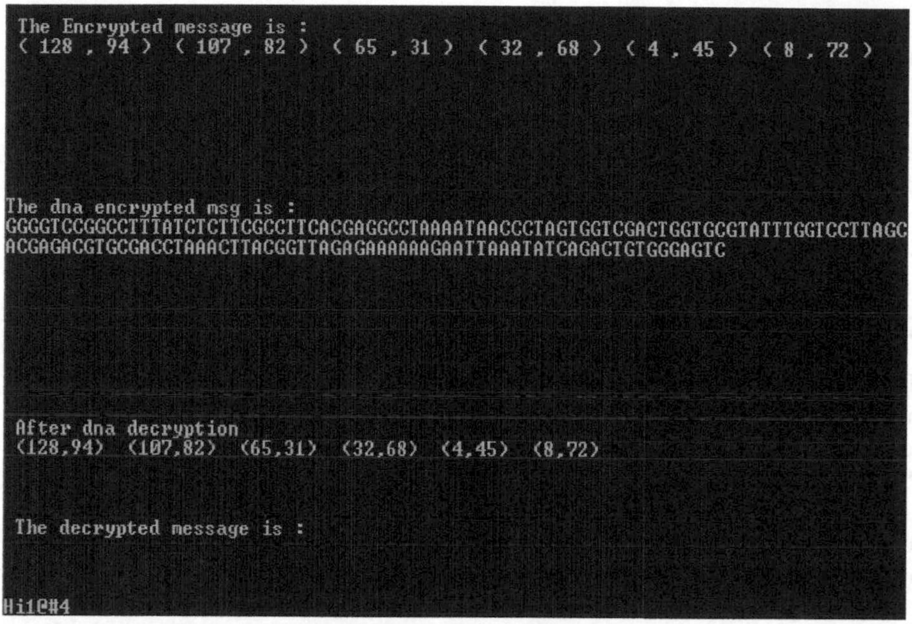

```
The Encrypted message is :
< 128 , 94 >  < 107 , 82 >  < 65 , 31 >  < 32 , 68 >  < 4 , 45 >  < 8 , 72 >




The dna encrypted msg is :
GGGGTCCGGCCTTTATCTCTTCGCCTTCACGAGGCCTAAAATAACCCTAGTGGTCGACTGGTGCGTATTTGGTCCTTAGC
ACGAGACGTGCGACCTAAACTTACGGTTAGAGAAAAAAGAATTAAATATCAGACTGTGGGAGTC




After dna decryption
(128,94)  (107,82)  (65,31)  (32,68)  (4,45)  (8,72)


The decrypted message is :

Hi1@#4
```

Fig. 4. Server1 Authentication

The above figure shows the encryption and decryption of the password for the particular username in server1

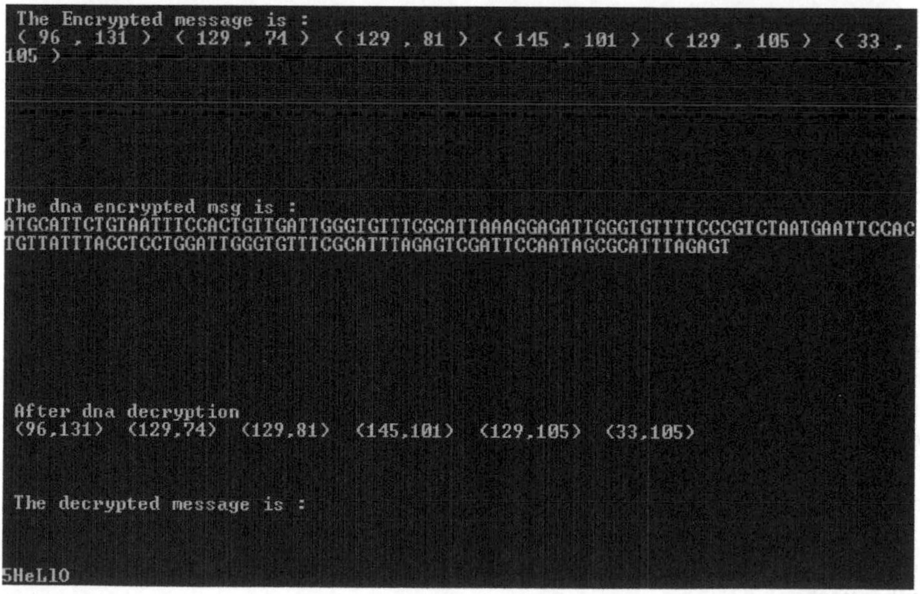

```
The Encrypted message is :
< 96 , 131 >  < 129 , 74 >  < 129 , 81 >  < 145 , 101 >  < 129 , 105 >  < 33 ,
105 >




The dna encrypted msg is :
ATGCATTCTGTAATTTCCACTGTTGATTGGGTGTTTCGCATTAAAGGAGATTGGGTGTTTTCCCGTCTAATGAATTCCAC
TGTTATTTACCTCCTGGATTGGGTGTTTCGCATTTAGAGTCGATTCCAATAGCGCATTTAGAGT




After dna decryption
(96,131)  (129,74)  (129,81)  (145,101)  (129,105)  (33,105)


The decrypted message is :

5HeL10
```

Fig. 5. Server2 Authentication

The above figure shows the encryption and decryption of the password for the particular username in server2

The existing works are based on encrypting the given text for one time using Elgamal Encryption which may not be completely secure. So in order to increase the security the encryption here has been carried out in two phases using two different encryption models, where in first phase the text is encrypted using Elgamal and followed by DNA encryption in the next phase. There by strengthening the Authentication and Confidentiality.

4.1   Security against Brute-Force Attack

Hackers use brute-force attacks to crack the passwords using a software which tries different character combinations in quick succession. The attacker uses a high-performance computer, which performs a large number of calculations and thus checks a large number of combinations in a short period.

This method is successfully used because many users use short passwords, which consists of characters of the alphabet, which reduces the number of possible combinations and makes it easier to guess the password.

In our application we are using numbers(10 different ones:0-9), letters(52 different ones: A-Z and a-z) and special characters(29 different ones) to create a password.

The number of different combinations can be calculated with the following formula

$$\text{Different combinations} = \text{number of possible characters}^{\text{password length}} \text{[16]}$$

In our application we consider the minimum length of the password is 10. So the possible combinations are $91^{10} = 38,941,611,811,810,745,401$. If a single high performance computer manages 2 billion keys per second then the time taken to manage our password is 38,941,611,811,810,745,401/2,000,000,000 =19470805905 seconds which is approximately equal to 6174 years.

By this our application is secured against Brute-Force attack.

5   Conclusion

Computer servers are prone to hacker attacks which would compromise all the passwords stored in the system. In this paper we have proposed a two server authenticated key thereby reducing considerably the risk of losing information on passwords. Here the login information is received by one server and authenticated through an DNA based encrypted exchange of information between the two servers to authenticate the password. The encryption and decryption process has been explained and we consider it to be quite robust. As a future work we intend to compare it with other systems and also explore option of multiple server authentication.

References

1. Jacob, G., Murugan, A.: DNA based Cryptography: An Overview and Analysis. Int. J. Emerg. Sci. 3(1), 36–42 (2013) ISSN: 2222-4254
2. Adleman, L.M.: Molecular computation of solutions to combinational problems. Science 266, 1021–1024 (1994)
3. Leier, A., Richter, C., Banzhaf, W.: Cryptography with DNA binary strands. Biosystems 57, 13–22 (2000)
4. Naveen, J.K., Karthigaikumar, P., Sivamangai, N.M.: Hardware implementation of DNA based cryptography. In: Proceedings of 2013 IEEE Conference on Information and Communication Technologies, ICT 2013 (2013)
5. Gong, L., Lomas, T.M.A., Needham, R.M., Saltzer, J.H.: Protecting Poorly-Chosen Secret from Guessing Attacks. IEEE J. Selected Areas in Comm. 11(5), 648–656 (1993)
6. Brainard, J., Jueles, A., Kaliski, B.S., Szydlo, M.: A New Two-Server Approach for Authentication with Short Secret. In: Proc. 12th Conf. USENIX Security Symp., pp. 201–214 (2003)
7. Katz, J., MacKenzie, P., Taban, G., Gligor, V.: Two-Server Password-Only Authenticated Key Exchange. In: Ioannidis, J., Keromytis, A.D., Yung, M. (eds.) ACNS 2005. LNCS, vol. 3531, pp. 1–16. Springer, Heidelberg (2005)
8. Yang, Y., Bao, F., Deng, R.H.: A New Architecture for Authentication and Key Exchange Using Password for Federated Enterprise. In: Proc. 20th IFIP Int'l Information Security Conf. (SEC 2005), pp. 95–111 (2005)
9. Yang, Y., Deng, R.H., Bao, F.: A Practical Password-Based Two-Server Authentication and key Exchange System. IEEE Trans. Dependable and Secure Computing 3(2), 105–114 (2006)
10. Yi, X., Ling, S., Wang, H.: Efficient Two-Server Password-Only Authenticated Key Exchange. IEEE Transactions on Parallel and Distributed Systems 24(9) (2013)
11. Mukesh, R., Damodaram, A., Subbiah Bharathi, V.: A robust fingerprint based twoserver authentication and key exchange system. In: 3rd International Conference on Communication Systems Software and Middleware and Workshops, Bangalore, pp. 167–174 (2008)
12. Kaliski, B., Szydlo, M., Brainard, J., Juels, A.: Nightingale: A new two-server approach for authentication with short secrets. In: Proceedings of the 12th USENIX Workshop on Security, pp. 1–2. IEEE Computer Society (2003)
13. Yang, D., Yang, B.: A Novel Two-Server Password Authentication Scheme with Provable Security. In: IEEE Transaction 2010 10th IEEE International Conference on Computer and Information Technology, CIT 2010 (2010)
14. Wang, X., Zhang, Q.: DNA computing-based cryptography. Key Laboratory of Advanced Design and Intelligent computing (Dalian university), Ministry of Education, Dalian, 116622, China
15. Madhulika, G., Rao, C.S.: Generating digital signature using DNA coding. In: Satapathy, S.C., Biswal, B.N., Udgata, S.K., Mandal, J.K. (eds.) Proc. of the 3rd Int. Conf. on Front. of Intell. Comput. (FICTA) 2014. AISC, vol. 328, pp. 21–28. Springer, Heidelberg (2015)
16. BruteForceAttacks,
   http://www.password-depot.com/know-how/bruteforceattacks.htm

Dynamic Cost-Aware Re-replication and Rebalancing Strategy in Cloud System

Navneet Kaur Gill and Sarbjeet Singh

Computer Science and Engineering,
UIET, Panjab University,Chandigarh, India
navneetgill31@gmail.com, sarb_j@yahoo.com

Abstract. Cloud computing is a "pay per use" model, where the user or clients pay for the computational resources they use. Furthermore, in cloud failures are normal. Therefore cost is an important factor to be considered along with availability, performance and reliability. Also it is not necessary that the benefits accrued from the replication will be greater than the cost incurred. Thus, this paper proposes an algorithm named Dynamic Cost-aware Re-replication and Re-balancing Strategy (DCR2S). This algorithm optimizes the cost of replication using the concept of knapsack problem. The proposed algorithm is evaluated using CloudSim. Experimental results demonstrate the effectiveness of proposed algorithm.

Keywords: Re-replication, re-balancing, replication cost, knapsack algorithm, cloud computing.

1   Introduction

Cloud computing is a model based upon "pay per use" concept. In cloud computing all the computational resources (like storage, data) are shared among the users [1].SLA is signed between the user and the service provider. This agreement defines QoS parameters (like availability, bandwidth, performance etc.). In cloud failures are normal rather than exceptional, therefore high availability, high performance and high fault tolerance are important factors to be considered.

Concept of replication is used in order to have high availability, high performance and high fault tolerance. Replication is the process to store multiple copies of a data files at data centers for performance and availability reasons. As cloud is pay per use model, therefore the user will pay for using the cloud storage. User will choose that service provider who will assure him maximum availability of his data file. As a result, replication is used to achieve maximum availability. It is not necessary that the benefits accrued from the replication will be greater than the cost incurred. Hence cost of replication is important concept to be considered.

We proposed an algorithm for heterogeneous cloud system. In this algorithm, we address 3 main issues of replication along with optimizing the cost of replication. Whenever the cost of replication becomes more than the user budget than the replicas

© Springer International Publishing Switzerland 2015

S.C. Satapathy et al. (eds.), *Proc. of the 3rd Int. Conf. on Front. of Intell. Comput. (FICTA) 2014*
– *Vol. 2*, Advances in Intelligent Systems and Computing 328, DOI: 10.1007/978-3-319-12012-6_5

placed at higher cost data center are re-replicated at lower cost data centers. The re-replication is done until the availability is greater than or equal to the desired availability defined in SLA.

The rest of paper is organized as follows. The related work is presented in section 2. System model is described in section 3. The proposed algorithm i.e. Dynamic Re-replication and Rebalancing Strategy is described in section 4. Simulation details are described in section 5. Finally, section 6 concludes the work with future scope.

2   Related Work

Replication is a concept of storing multiple copies at multiple sites. In [2], 3 important issues of replication (which, when and where to replicate) are discussed. Till now, two type of algorithms are proposed: static algorithms [3,4,6] and dynamic algorithms [5,7-10]. In static replication algorithms the number of replica is predetermined. It does not adapt according to the environment. However, in dynamic replication strategy the number of replicas needed and their placement all changes dynamically according to the environment. In [3,6], re-replication is done whenever the number of replicas decreases then the required number. Block is added to the priority queue if its number of replicas decreases. The block with minimum number of replicas will have highest priority and will be replicated first. In [4], placement algorithm is proposed that uses p-median model to minimize the average response time. Replica maintenance algorithm is used in order to maintain performance by relocating replicas on different candidate sites. In [5], popularity of a data file is determined by assigning different weights to different historical data. Recently accessed records will have higher weights as compared to other. In [7], the blocking probability and the capacity of a data node is considered, while placing the replicas. However, in [8], hierarchal algorithm is used to place the replicas. In [9], aging algorithm is used to find popularity of data file and also probabilistic sampling is used to place replicas in order to improve data locality. In [12], a strategy is proposed, which uses replication in order to optimize the energy consumption along with maintaining high availability and bandwidth consumption. Research has been done on replication of files but very little has been done considering the effect of replication on the cost paid by the user. This motivates us to prepare an algorithm that can optimize cost along with high availability of system.

3   System Model

Multi-tier hierarchical architecture supports efficient data and resource sharing [10,11]. Similar architecture is used, but in our case it is heterogeneous in nature rather than homogeneous. As shown in Fig. 1, data centers in one tier will have different configuration, then the data centers in other tiers. Super data centers stores the original copy of each data file. In order to meet availability, replicas are spread from super to main data centers and from main to ordinary data centers.

As shown in Fig. 1, User sends a file access request to the broker, which furthermore interacts internally with the Replica catalog. Replica catalog contains information where the replicas of requested data file are placed. Replica catalog will send the list of data centers having that file to the broker. Broker will analyze the received list of data centers, and will schedule the request to the nearest data center. Basic unit of storage is block.

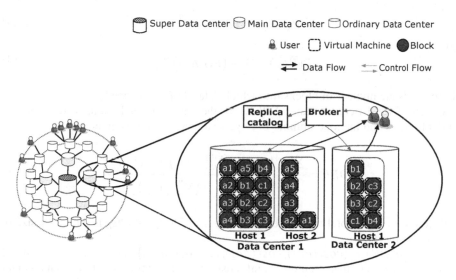

Fig. 1. Magnified replication scenario in multi-tier heterogeneous hierarchical cloud system

Let $DC = \{dc_1, dc_2 .... dc_x\}$ be a set of x data centers in a cloud environment. Let $F = \{f_1, f_2, ..., f_s\}$ be the set of s different data files. Let $B = \{B_1, B_2, ..., B_s\}$ be the set of s different blocks of s different files stored at a data center .Here $B_k = \{b_{k_1}, b_{k_2}, ..., b_{k_{n_k}}\}$ be the set of blocks that belong to data file f_k . Data file f_k is stripped into n_k blocks of fixed size.

Probability of block availability $P(ba_k)$ is the probability that the block is available [11]. If a block b_k has br_k number of replicas, then available and unavailable probability of block in heterogeneous system is given by Eq. (1) and Eq. (2) respectively.

$$P(BA_k) = 1 - \prod_{i=1}^{br_k}(1 - p(ba_k)_i) ,$$ (1)

and

$$P(\overline{BA_k}) = \prod_{i=1}^{br_k}(1 - p(ba_k)_i) .$$ (2)

where $p(ba_k)_i$ is the available probability of block b_k of file f_k at data center dc_i .

Since a file can be replicated at more than one data center, therefore the blocks of a file will have different block available probabilities at different types of data centers. Super data centers have higher cost, reliable hardware; hence it will have highest block available probability. Whereas ordinary data centers have lower cost, less reliable hardware, hence it will have lowest block available probability.

$$p(ba_k)_{Super\,DC} > p(ba_k)_{Main\,DC} > p(ba_k)_{Ordinary\,DC}.$$

Similarly using Eq. (1), Eq. (2) available and unavailable probability of file f_k is obtained by Eq. (3) and Eq. (4)

$$P(FA_k) = \left(P(BA_k)\right)^{n_k},$$ (3)

and

$$P(\overline{FA_k}) = 1 - \left(P(BA_k)\right)^{n_k}.$$ (4)

where n_k is the number of blocks into which data file f_k is stripped.

System Byte Effective Rate (SBER) is the ratio of total number of probably available bytes and the total number of requested bytes [11]. It is given by Eq. (5)

$$SBER = \frac{\sum_{k=1}^{s}\left(an_k \times \left(\sum_{h=1}^{n_i} bs_h\right) \times P(FA_k)\right)}{\sum_{k=1}^{s}\left(an_k \times \left(\sum_{h=1}^{n_i} bs_h\right)\right)}.$$ (5)

Where an_k is the total number of user access to file f_k and bs_h is the block size of block b_h of file f_k.

Let $Cost(DC) = \{cost(dc_1), cost(dc_2), \ldots, cost(dc_x)\}$ be a set of cost of replication at each data center. Total cost of replicating file at all data center $\left(Cost_k(DC)\right)$ is the sum of cost of replicating file on each data center. As more than one replica can be created at a data center, the number of replicas of a data file at data center is considered. It is given by (6)

$$Cost_k(DC) = \sum_{i=1}^{x}\left(cost(dc_i) \times br_k(dc_i)\right).$$ (6)

where $br_k(dc_i)$ is the number of replicas of data file f_k at data center dc_i.

4   Dynamic Cost-Aware Re-replication and Rebalancing Strategy

DCR2S (Dynamic Cost-Aware Re-Replication and Rebalancing Strategy) has 3 phases, which are discussed below.

Phase 1: Determine which and when to replicate a data file

This phase discovers the appropriate data file for replication and also decides when it should be replicated using the concept of temporal locality. According to this concept, recently accessed data file has more probability of being accessed again in the future. Access history of each data file is examined to determine its respective popularity. For a data file, replication operation is provoked as soon as its popularity crosses a dynamic threshold. Replica factor(RF_k) is used to examine the access history of a data file f_k [11]. Replica factor(RF_k) is given by Eq. (7)

$$RF_k = \frac{\sum_{t_i=t_s}^{t_c} \left(an_k(t_i, t_{i+1}) \times a^{(t_s-t_c)^d} \right)}{br_k \times \sum_{i=1}^{n_k} bs_i}. \tag{7}$$

Where $a > 1$, $d \in \{1,2, ...\}$ is the rate of decay, $an_k(t_i, t_{i+1})$ is the number of user accesses to a file k from time t_i to t_{i+1}, t_s is the start time, t_c is the current time, br_k is the number of replicas of a file f_k having n_k number of blocks of size bs_i.

System replica factor (RF_{sys}) is used to define a dynamic threshold value .Using Eq. (7), RF_{sys} is given by Eq. (8)

$$RF_{sys} = \frac{\sum_{k=1}^{s} \left(\sum_{t_i=t_s}^{t_c} \left(an_k(t_i, t_{i+1}) \times a^{(t_s-t_c)^d} \right) \right)}{\sum_{k=1}^{s} \left(br_k \times \sum_{i=1}^{n_k} bs_i \right)}. \tag{8}$$

The replication operation will be provoked for data file f_k on the basis of Eq. (9)

$$RF_k \geq \begin{cases} (1 + \alpha) \, RF_{sys} & \textit{only single file is accessed} \\ (1 + \alpha) RF_{kold} & \textit{2 or more than 2 files are accessed} \end{cases}. \tag{9}$$

where RF_{kold} is the old replica factor of file f_k

Phase 2: Determining the Suitable Number of New Replicas

This phase determines the suitable number of additional replicas required in order to meet the availability requirements. The new probability of file availability$(P_{new}(FA_k))$ is given by Eq. (10) and can be calculated using old probability of file availability$(P_{old}(FA_k))$ [11].

$$P_{new}(FA_k) = P_{old}(FA_k) + \frac{RF_k}{\sum_{i=0}^{num_f} RF_i} \times (1 - P_{old}(FA_k)). \tag{10}$$

where num_f is the total number of data files selected to be replicated.

In heterogeneous system, data file replicas stored at different datacenters will have different available probability. In order to determine the suitable number of new replicas for simplicity average available probability$(bavg_k)$ is considered which is calculated by (11)

$$p(bavg_k) = \frac{\sum_{i=1}^{br_k} p(ba_i)}{br_k}. \tag{11}$$

The suitable number of new replicas $br_k(new)$ is calculated using (10), (11) and is given by Eq. (12)

$$br_k(new) = \left| \frac{ln\left(1 - \left(P_{new}(FA_k)\right)^{\frac{1}{n_k}}\right)}{\ln(1 - p(bavg_k))} - br_k(old) \right| . \tag{12}$$

where $br_k(old)$ is the old number of replicas of data file f_k .

Phase 3: Placement of New Replicas

This phase is related to the placement of new replicas. First of all, the number of new replicas $br_k(dc_i)$ is determined at each directly attached datacenter (dc_i) [11] using Eq. (13)

$$br_k(dc_i) = \left| \frac{RF_k(dc_i)}{RF_k} \times br_k(new) \right| . \tag{13}$$

where $RF_k(dc_i)$ is the file replica factor specifically at datacenter dc_i whereas RF_k is the overall replica factor of file f_k

Estimate the total cost of replication, when these new replicas will be placed along with old replicas using Eq. (6). If the estimated cost is more than the maximum budget, then use Knapsack Problem.

In knapsack problem, each data center (dc_i) represents an item and the corresponding cost of replication at that data center $cost(dc_i)$ represents the weight of an item. Thus total number of items is equal to x i.e. the total number of data centers. Let budget B is the maximum weight or capacity of knapsack. The value of the item (v_i) is decided by a service provider. Usually, the ordinary data centers will have highest value because they are closer to the user and are also cheaper as compared to other two types. On the other hand, main data centers will have least value because they are costly as compared to ordinary data centers. Although super data center have highest cost but the value of super data center will be more than the ordinary data centers because they have higher availability. It means that we want to optimize the cost keeping in mind the availability.

Goal is to determine selected and unselected set of items or datacenters in order to maximize the total value while keeping the total weight less than or equal to maximum budget.

Let $M[i, b]$ be the subset of selected items, where $i = 1$ to x and $b = 0$ to B. For $i = 1$ to x initialize $M[i, 0]$ to 0. Similarly for $b = 0$ to B initialize $M[0, b]$ to 0.

Eq. (14) represents the required recursive equation for knapsack algorithm.

$$M[i, b] = \begin{cases} 0 & if\ i = 0 \\ M[i-1, b] & cost(dc_i) > b \\ max\{M[i-1, b], M[i-1, b - cost(dc_i)] + v_i\} & otherwise \end{cases} \tag{14}$$

Knapsack algorithm will reject certain datacenters. The corresponding replicas of a data file that were placed on the rejected datacenters will be re-replicated at lower cost data center i.e. ordinary data centers. Remember that the availability should be greater than or equal to the least availability guaranteed by SLA.

5   Simulation

We used CloudSim as a simulator to evaluate the performance for our DCR2S algorithm. 21 data centers are created in order to have multi-tier hierarchical model as shown in Fig. 1. Configuration details of data centers are given in Table 1.

Table 1. Configuration Details of Data Centers

Data Center Type	No. of Data Center	No. of Hosts	No. of PE per Host	MIPS	Available Probability	Cost of Replication
Super Data Center	1	20	16	800	0.6 ~ 0.9	500
Main Data Center	5	10	4 ~ 8	300 ~400	0.3 ~ 0.6	300
Ordinary Data Center	15	5	1 ~ 2	100 ~200	0.1 ~ 0.3	100

3 different data files of size 200Mb, 150Mb, and 100Mb respectively are used. Each file has fixed size block of size = 64Mb.Initialy, super data center has one replica of each data file. 1000 tasks or cloudlets are submitted to 274 virtual machines using poisson distribution. Each cloudlet will request for one or two data files.

The proposed algorithm is evaluated on the basis of cost of replication, availability of block, number of replicas, availability of file and SBER.

In Fig. 2, a comparison of different cost of replication is shown. When DCR2S algorithm is not used, the cost of replication increases continuously as the number of replicas increases. However, when DCR2S algorithm is used, cost of replication becomes constant for a certain more number of replicas. Thus, DCR2S algorithm can create more number of replicas from a given budget. As a result, our proposed algorithm optimizes the cost of replication.

In Fig. 3, a data file having same block availability is used to make a comparison between different budget values and the available probability of file. It shows that with an increase in number and the budget, the available probability of file will be higher.

In Fig. 4, comparison of cost of replication and the available probability of file with an increase number of replicas is shown. If the cost factor is considered without using DCR2S and the cost of replication becomes equal to the budget then new replicas can't be created anymore. Hence, the available probability of file can't be increased anymore. However, when DCR2S is used, new replicas can still be created, even if the cost of replication becomes equal to the budget. Thus the available probability of file will also increase. Hence, DCR2S algorithm can optimize the cost of replication resulting in increased number of replicas that can be created, which further leads to increase in the available probability of a data file.

In Fig. 5, comparison of system byte effective rate is shown. For lower values of block available probabilities, there is slight increase in SBER. However, for higher values of block available probabilities, there is tremendous increase in the system byte effective rate with increase in number of replicas. Hence, DCR2S improves the system byte rate effectively.

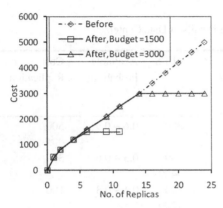

Fig. 2. Cost of replication comparison

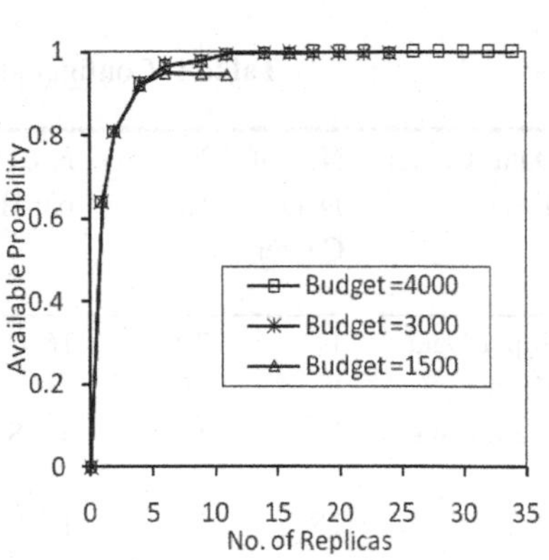

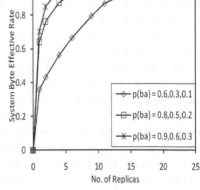

Fig. 3. Available probability of file with different value of budgets

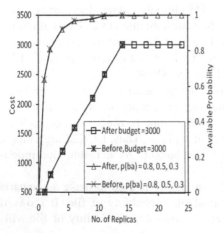

Fig. 4. Cost of replication and available probability of file comparison

Fig. 5. System Byte Effective Rate Comparison

6   Conclusions and Future Scope

In this paper, we proposed a dynamic cost-aware re-replication and re-balancing strategy in cloud system. First we designed the system, to understand the relation between number of replicas, cost of replication, availability. Then, we proposed an algorithm. It determines which file needs to be replicated and when to replicate it. It also determines the suitable number of replicas and then places theses replicas in such a way that the cost should not increase more than the budget along with high system byte effective rate and bandwidth consumption. Knapsack algorithm is used for optimizing the cost of replication. We used CloudSim for simulation purpose. Experimental results conclude that DCR2S can optimize the cost of replication and also achieve high system byte effective rate and is effective in heterogeneous cloud system architecture. In future, we are planning to introduce consistency protocols in order to have high consistency rates.

References

1. Mell, P., Grance, T.: The NIST definition of cloud computing. Communications of the ACM 53(6), 50 (2010)
2. Goel, S., Buyya, R.: Data Replication Strategies in Wide Area Distributed Systems. In: Proceedings of ICS 2002, pp. 211–241. Idea Group Inc., Hershey (2006)
3. Tanenbaum, A.S., Steen, H.V.: Distributed Systems: Principles and Paradigms. Prentice-Hall (2006)
4. Ghemawat, S., Gobioff, H., Leung, S.T.: The Google File System. ACM SIGOPS Operating Sytems Review 37(5), 29–43 (2003)
5. Rahman, R.M., Barker, K., Alhajj, R.: Replica Placement Design with Static Optimality and Dynamic Maintainability. In: 6th IEEE International Symposium on Cluster Computing and the Grid, pp. 434–437. IEEE Press, Singapore (2006)
6. Chang, R.S., Chang, H.P.A.: Dynamic Data Replication Strategy using Access-Weights in Data Grids. In: IEEE Conference on Computer Systems and Application, pp. 414–421. IEEE Press, Doha (2008)
7. Shvachko, K., Hairong, K., Radia, S., Chansler, R.: The Hadoop Distributed File System. In: Proc. the 26th Symposium on Mass Storage Systems and Technologies, pp. 1–10. IEEE Press, Incline Village (2010)
8. Wei, Q., Veeravalli, B., Gong, B., Zeng, L., Feng, D.: CDRM: A Cost-effective Dynamic Replication Management Scheme for Cloud Storage Cluster. In: IEEE International Conference on Cluster Computing, pp. 188–196. IEEE Press, Heraklion (2010)
9. Mansouri, N., Dastghaibyfard, G.: A new Dynamic Replication algorithm for Hierarchy networks in Data Grids. In: International Conference on P2P, Parallel, Grid, Cloud, Internet Computing, pp. 187–192. IEEE Press, Barcelona (2011)
10. Abad, C., Lu, Y., Campbell, R.: DARE: Adaptive Data Replication for Efficient Cluster Scheduling. In: IEEE International Conference on Cluster Computing, pp. 159–168. IEEE Press, Auxtin (2011)
11. Sun, D.W., Chang, G.R., Gao, S., Jin, L.Z., Wang, X.W.: Modeling a Dynamic Data Replication Strategy to Increase System Availability in Cloud Computing Environments. Journal of Computer Science and Technology 27(2), 256–272 (2012)
12. Boru, D., Kliazovich, D., Granelli, F., Bouvry, P., Zomaya, A.Y.: Energy-efficient data replication in cloud computing datacenters. In: IEEE Globecom Workshops (GC Wkshps), pp. 446–451. IEEE Press, Atlanta (2013)

Signalling Cost Analysis of Community Model

Boudhayan Bhattacharya[1] and Banani Saha[2]

[1] Sabita Devi Education Trust – Brainware Group of Institutions, 398 Ramakrishnapur Road, Barasat, Kolkata -700124, West Bengal, India
[2] University of Calcutta, Rashbehari Shiksha Prangan, 92 Acharya Prafulla Chandra Road, Kolkata - 700009, West Bengal, India
mailforboudhayan@gmail.com, bsaha_29@yahoo.com

Abstract. Data fusion is generally defined as the application of methods that combines data from multiple sources and collect that information in order to get conclusions. This paper analyzes the signalling cost of different data fusion filter models available in the literature with the new community model. The signalling cost of the Community Model   has been mathematically formulated by incorporating the normalized signalling cost for each transmission. This process reduces the signalling burden on master fusion filter and improves throughput. A comparison of signalling cost of the existing data fusion models along with the new community model has also been presented in this paper. The results show that our community model incurs improvement with respect to the existing models in terms of signalling cost.

Keywords: Master Fusion Filter, Reference Sensor, Local Filter, Community Model, Data Transmission Time, Normalized Signalling Cost, Signalling Cost.

1   Introduction

Data fusion is a state-estimation process based on data from multiple systems or data sources. The research work published to date does not provide any systematic approach towards calculating the signalling cost between different data fusion filter levels, in terms of data fusion methods. In this paper the signalling cost of community model[1][2] has been compared with the existing models like Waterfall Model, Multisensor Integration Fusion Model, Behavioural Knowledge-based Data Fusion Model, Omnibus Model, Dasarathy Model and JDL Model[10]-[14] and the results show that the community model requires less signalling cost due to the verification with reference sensor in each stage. This paper is organized as follows - Section 2 deals with the classical data fusion models and their contributions. Section 3 introduces the Community model and deduces a formula to calculate the signalling cost based on the model and the corresponding algorithm for calculation of the signalling cost. Section 4 deals with the performance analysis of the new model along with the other existing models in terms of signalling cost.

© Springer International Publishing Switzerland 2015   49
S.C. Satapathy et al. (eds.), *Proc. of the 3rd Int. Conf. on Front. of Intell. Comput. (FICTA) 2014*
– Vol. 2, Advances in Intelligent Systems and Computing 328, DOI: 10.1007/978-3-319-12012-6_6

2   Related Work

The discussions within the GDR-PRCISIS[3] work group on information fusion, formulated the basic architectural specifications for data fusion. European work group FUSION, worked on fusion from various perspectives in 1996 to1999[4]. The Multi-Sensor Data Fusion (MSDF) was introduced for military applications[5][6] and then are applied to civil industries[7]-[9]. The popular filter models available in literature are: Waterfall Model, Multisensor Integration Fusion Model, Behavioural Knowledge-based Data Fusion Model, Omnibus Model, Dasarathy Model and JDL Model[10] -[14].

The Waterfall Model and Multisensor Integration Fusion Model take the data from all navigation sensors[10]-[12]. Behavioural Knowledge-based Data Fusion Model uses outputs of one filter as inputs to a subsequent filter[10]. The Omnibus Model and Dasarathy Model are two-stage filtering models where all parallel local filters combine their own sensor systems with a common reference system[10]-[14]. The JDL Model is a generic model which combines two approaches of data fusion - Measurement Fusion and State Fusion[10][11]. The signalling cost is a parameter to measure the transmission of a signal. It is used in all the areas where data transmission is required to analyze or fuse data[15][16]. This paper provides a scheme for the calculation of the signalling cost of the community model.

3   Signalling Cost Estimation of Community Model

Data fusion models are based on either parity vector/space techniques[17]-[19] or comparison method[20]-[22]. Using any of these techniques, the minimum and maximum residual magnitude can be measured[23]. These methods compute and compare each parity equation and deploy a least-squares estimator for the estimation of the measured states. Naturally it is a time-expensive procedure for a large number of redundant sensors.

In this model, the signalling cost is a combination of data transmission time and normalized signalling cost. For each level of filters, the signalling cost is incurred in the transition between the filter level and the reference sensor and the cumulated result of these calculations are forwarded to the master fusion filter for the typical calculation of the signalling cost. If a particular level is visited more than once, then the required data are not considered as the predicament of the filter levels which are already predesignated. More over the effective cost decreases drastically as only the desired data are being forwarded. The normalized signalling cost actually deals with the average of the signalling cost that is incurred due to various transmission.

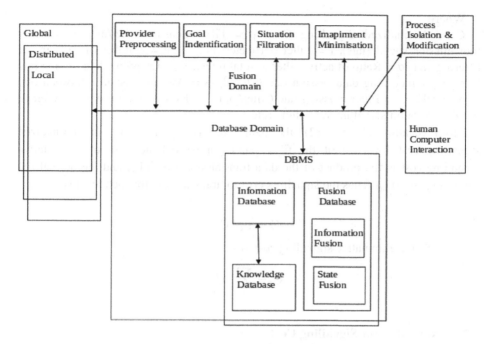

Fig. 1. Community Model

3.1   Mathematical Formulation for Signalling Cost

Here the mathematical formulation of the community model is deduced to calculate the signalling cost for the data fusion using the Dynamic Transformation Model.

The general n-level LF transmission time for the single master fusion filter domain[1] is :–

$$T_R = \sum_{1}^{N} L_T * 1/A[S(H*\rho v P_n/\pi)/F + N*N_R] \tag{1}$$

Where

A = Alignment, S = Scale factor stability, H = Bias uncertainty, ρ = Data fusion density per LF, v = Average velocity of the light, P_n = Perimeter of each LF, F = Flattening of the LF surface, N = Total number of LF, N_R = Random noise

The normalized signalling cost is another standard parameter which is used to calculate the signalling cost from a point to another. Here the transmission and the reception points are assumed to be filters which are used to fuse different types of data for a particular objective. The formula to deduce normalized signalling[24] cost is :–

$$C_n = \rho v l \sqrt{n} /\pi + \rho v l/\pi(n - \sqrt{n})\alpha + \rho(l/4)^2 n(1 - \alpha)(\lambda_d - \lambda_a) +$$
$$d_{L1L2}(n - 1)\rho(l/4)^2 n(1 - \alpha)\lambda_a/d_{L1Ln}(R_c R_w + R_l) \tag{2}$$

Where

C_n = Normalized Signalling Cost, ρ = Filter density (filter/m^2), v = Filter transmission speed (m/s), l = Filter perimeter (m),   n = Number of cells in the filtering area, α = Ratio of active filter to total filters, λ_d = Outgoing data session rate (1/s), λ_a = Incoming data session rate (1/s), d_{L1L2} = Average distance between two filter levels, R_c = Filter crossing rate (filter/s), R_w = Hop weight ratio, R_l = Average filter registration signalling refreshing rate

From equation (1) and (2), the data transmission time and the normalized signalling cost for each of the filter pairs can be obtained. So the cumulative signalling cost is the product of the data transmission time (T_R) and the normalized signalling cost (C_n). Thus the signalling cost for transmission for each level is –

$$C_s = T_R * C_n \tag{3}$$

So the cumulative signalling cost is -

$$C_T = \sum_1^N C_s \tag{4}$$

3.2   Algorithm for Signalling Cost

The signalling cost is calculated on the basis of the application area. Initially, the Data Transmission Time (DTT)[1] and the normalized signalling cost[21] is determined and based on these two, the signalling cost is calculated. The output of the process may vary based on the input parameters and objectives. The following algorithm describes the procedure to calculate the signalling cost for the community model. The reference sensor takes the standard parameter values[24] as input for verification at a subsequent stage. Then the data transmission time and the normalized signalling cost are initialized to zero. After that, the procedure Impairment Minimisation (IM)[1] is called. If the result matches with the reference sensor, then the signalling cost is calculated and if any of the parameters does not match with the reference sensor values then those parameters are updated and the total transmission time and normalized signalling cost are calculated till the Nth level of filter is reached. The following symbols are used for calculating the signalling cost -

T_R =Data transmission time
D = Data for master fusion filter
δ = Impairment for Data

RS_i = Referece Sensor considering different parameters $RS_{(i)}$ like A= Alignment consideration, S= Scale factor stability, H=Bias uncertainity, ρ = Data fusion density per local filter, V= Average velocity of light, P = Perimeter of master fusion filter, N= Number of filter levels

The rest of the parameters are described in sub-section 3.1.

Procedure: Signalling Cost

1. Begin
2. input RS_A, RS_S, RS_H, RS_ρ, RS_V, RS_p, RS_N,
3. set $T_R \leftarrow 0$
4. set $C_n \leftarrow 0$
5. input L_T, A, S, H, ρ, V, P_n, P, N, N_R, C_n, ρ, v, l, n, α, λ_d, λ_a, d_{L1L2}, R_c, R_w, R_l
6. call IM
7. if (δ != $\emptyset$) || D !=RS
8.   update A$\leftarrow$ RS_A, S$\leftarrow$ RS_S, H$\leftarrow$ RS_H, $\rho\leftarrow$ RS_ρ, V$\leftarrow$ RS_V,
         P$\leftarrow$ RS_p, N$\leftarrow$ RS_N
9. endif
10. i $\leftarrow$ 1
11.   while (i <= N)
12. $T_R = T_R + L_T * 1/A[S(H*\rho v P_n/\pi)/F + N*N_R]$
13. $C_n = \rho v l\sqrt{n} /\pi + \rho v l/\pi(n - \sqrt{n})\alpha + \rho(1/4)^2 n(1 - \alpha)(\lambda_d - \lambda_a) +$
             $d_{L1L2}(n - 1)\rho(1/4)^2 n(1 - \alpha)\lambda_a / d_{L1L,n}(R_c R_w + R_l)$
14. $C_s = T_R * C_n$
15.   i $\leftarrow$ i + 1
16. $C_T = \displaystyle\sum_1^N C_s$
17.   end while
18. end procedure

4   Performance Analysis

The data association for signalling cost looks for data that belongs to same cluster. It consists of detecting and associating noisy measurements, the origins of which may be unknown. Signalling cost determination is quite difficult in data fusion filtering. In this situations either deterministic (based on Classical Hypothesis), or probabilistic (based on Bayesian Hypothesis) models are used. In this paper, the deterministic model has been used for the calculation. The values taken here are standard parameters[24]. It is found that the formulation of signalling cost based on community model[1] yields better result than that of Waterfall Model, Multisensor Integration Fusion Model, Behavioural Knowledge-based Data Fusion Model, Omnibus Model, Dasarathy Model and JDL Model[10]-[14].

Considering the number of LF levels as 1, the formulation of equation (3) yields 97 msg/s for community model and in case of other existing models, the values are 122, 113, 109, 116, 118 and 107 msg/s for Waterfall Model, Multisensor Integration Fusion Model, Behavioural Knowledge-based Data Fusion Model, Omnibus Model, Dasarathy Model and JDL Model respectively and the comparison is shown in the form of bar chart in figure 2.

Table 1. The Standard Parameters

Parameters	Value
Bias uncertainty (o/h)	10-40
Scale factor stability (ppm)	100-500
Alignment (arcs)	200
Random noise (o/h/$\sqrt{Hz}$)	1-5
Flattening (f)	1/298.257223563
R_w	8
R_c	0.5
R_l	0.5
d_{L1L2}	16
d_{L1Ln}	$\sqrt{n}$
V	28.9 m/s
P	0.0002 users/m2
A	5
λ_a	0.0008/s
λ_d	0.0008/s

1. Waterfall Model
2. Multi-sensor Integration Fusion Model
3. Behavioural Knowledge-based Data Fusion Model
4. Omnibus Model
5. Dasarathy Model
6. JDL Model
7. Community Model

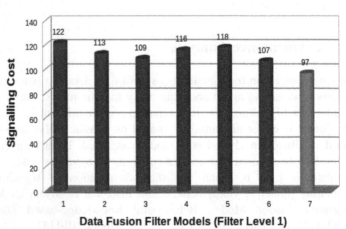

Fig. 2. Comparison of LF Signalling Cost (Filter Level 1)

Considering the number of LF levels as 1 to 8, the signalling cost of each of these models are calculated for each of these 8 levels and the results are shown pictorially in the following figure 3:

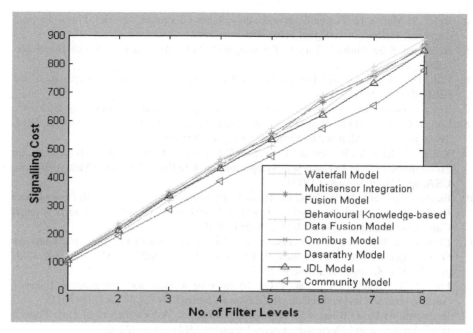

Fig. 3. Comparison of LF Signalling Cost

The above two graphs reveal that, the community model when compared to other existing models shows the least signalling cost irrespective of the filter levels.

5   Conclusion

This paper tries to put the new Community Model scheme to upgrade the data fusion methodology. It also attempts to give an overview of the signalling cost of the Community Model scheme for the benefit of the data fusion in decision support system. The performance evaluation based on the formulation of Signalling Cost shows that the new Community Model is more efficient compared to other existing models. More over the cost does not include the verification cost as each filter stage individually verifies it's transmitted data with that of reference sensor.

References

1. Bhattacharya, B., Saha, B.: Community Model - A New Data Fusion Filter Paradigm sent in EAIT 2014 (2014)
2. Bhattacharya, B., Saha, B.: Community Model Architecture – A New Data Fusion Paradigm for Implementation. International Journal of Innovative Research in Computer and Commmucation Engineering 2(6), 4774–4783 (2014)
3. Bar-Shalom, Y., Fortmann, T.E.: Tracking and Data Association. Academic Press, San Diego (1988)

4. Bloch, I., Maître, H.: Fusion de données en traitement d'images: modèles d'information et décisions. Traitement du Signal 11(6), 435–446 (1994)
5. Blackman, S.S.: Multiple Targets Tracking with Radar Applications. Artech House Inc. (1986)
6. Hall, D.L., Llinas, J.: An Introduction to Multisensor Data Fusion. Proceedings of the IEEE 85(1), 6–23 (1997)
7. http://www.data-fusion.org/article.php (last accessed on May 29, 2014)
8. Luo, R.C., Kay, M.G.: Multisensor Integration and Fusion in Intelligent System. IEEE Trans. on System, Man and Cybernetics 19(5), 901–931 (1989)
9. Kokar, M., Kim, K.: Review of Multisensor Data Fusion: Architecture and Techniques. In: Proceedings of The International Symposium on Intelligent Control, Chicago, Illinois, USA, pp. 261–266 (August 1993)
10. Esteban, J., Starr, A., Willetts, R., Hannah, P., Bryanston-Cross, P.: A Review of Data Fusion Models and Paradigms: Towards Engineering Guidelines. Journal Neural Computing and Applications 14(4), 273–281 (2005)
11. Elmenreich, W.: A Review on System Architectures for Sensor Fusion Applications. In: Obermaisser, R., Nah, Y., Puschner, P., Rammig, F.J. (eds.) SEUS 2007. LNCS, vol. 4761, pp. 547–559. Springer, Heidelberg (2007)
12. Luo, R.C., Chang, C.C., Lai, C.C.: Multisensor Fusion and Integration: Theories, Applications, and its Perspectives. IEEE Sensors Journal 11(12), 3122–3138 (2011)
13. Bedworth, M.D., O'Brien, J.C.: The Omnibus Model: A New Model of Data Fusion? IEEE Aerospace and Electronic Systems Magazine 15(4), 30–36 (2000)
14. Nakamura, E.F., Loureiro, A.A.F., Frery, A.C.: Information Fusion for Wireless Sensor Networks: Methods, Models, and Classifications. ACM Computing Surveys 39(3), 9/1–9/55 (2007)
15. Zhang, X., Castellanos, J.G., Campbell, A.T.: P-MIP: Paging Extension for Mobile IP, Columbia University, pp. 127–141 (2002)
16. Wang, M., Georgiades, M., Tafazolli, R.: Signalling Cost Evaluation of Mobility Management Schemes for Different Core Network Architectural Arrangements in 3GPP LTE/SAE. In: Vehicular Technology Conference 2008, pp. 2253–2258 (May 2008)
17. Karatsinides, S.E.: Enhancing Filter Robustness in Cascaded GPS-INS Integrations. IEEE Trans. on Aerospace and Electronic Systems 30(4), 1001–1008 (1994)
18. Bell, W.B., Gorre, R.G., Cockrell, L.D.: Cascading Filtered DTS Data into a Loosely Coupled GPS/INS System. In: Proceedings of IEEE PLANS 1998, pp. 586–593 (1998)
19. Carlson, N.A.: Federated Filter for Fault-Tolerant Integrated Systems. In: Proceedings of 1988 IEEE PLANS, pp. 110–119 (1988)
20. Felter, S.C.: An overview of decentralized Kalman filter techniques. In: Proceedings of IEEE Southern Tier Technical Conference, pp. 79–87 (1990)
21. Liggins, M.E., Chong, C.-Y., Kadar, I., Alford, M.G., Vannicola, V., Thomopoulos, S.: Distributed Fusion Architectures and Algorithms for Target Tracking. Proceedings of the IEEE 85(1), 95–107 (1997)
22. Evans, F.A., Wilcox, J.C.: Experimental Strapdown Redundant Sensor Inertial Navigation System. Journal of Spacecraft and Rockets 7(9), 1070–1074 (1970)
23. Potter, J.E., Deckert, J.C.: Minimax Failure Detection and Identification in Redundant Gyro and Accelerometer System. Journal of Spacecraft 10(4), 236–243 (1973)
24. Spitzer, C.R.: The Avionics Handbook. CRC Press LLC (2001)

Steganography with Cryptography in Android

Akshay Kandul, Ashwin More, Omkar Davalbhakta, Rushikesh Artamwar,
and Dinesh Kulkarni

Computer Department, PVG's COET Affiliated to University of Pune, India
{Akshaykandul,davalbhakta.o.s,ashwinmore21092,
hrushi.na,dineshakulkarni}@gmail.com

Abstract. The paper introduces work on developing secure data communication system. It includes the usage of two algorithms RSA and AES used for achieving cryptography along with LSB for achieving steganography both on Android platform. The joining of these three algorithms helps in building a secured communication system on 'Android' platform which is capable of withstanding multiple threats.

The input data is encrypted using AES with a user defined key prior to being embedded in image using LSB algorithm. The key used for encryption is further wrapped by means of receivers public key and that wrapped product key is further kept hidden so as to pass it to receiver that assure all the security purposes (RSA), making this a reliable communication channel for sensitive data. All stages assure that the secret data becomes obsolete and cannot be break easily; the steganographic algorithm thus introduces an additional level of security.

Keywords: stego-image, steganalysis, public key, private key.

1   Introduction

A basic idea behind implementing project in ANDROID is to achieve secure data transfer on mobile networks in finer manner. However, the two basic terms associated with this concept are 'Cryptography' and 'Steganography'. Cryptography is the art of protecting information by transforming it (encrypting it) into an un-readable format, called cipher text. Only those who possess a secret key can decipher (or decrypt) the message into plain text. Encrypted messages can sometimes be broken by cryptanalysis, also called code breaking, although modern cryptography techniques are virtually unbreakable.

1.1   Advanced Encryption Standards (AES)

However, one of the widely used and secure standards of cryptography is 'Advanced Encryption Standard' (AES). The implementation of AES along with UTF − 8 standard encoding is most securing platform for achieving privacy of confidential data.

S.C. Satapathy et al. (eds.), *Proc. of the 3rd Int. Conf. on Front. of Intell. Comput. (FICTA) 2014*
– Vol. 2, Advances in Intelligent Systems and Computing 328, DOI: 10.1007/978-3-319-12012-6_7

The Advanced Encryption Standard comprises three block ciphers, AES-128, AES-192 and AES-256. AES has a fixed block size of 128 bits and a key size of 128, 192, or 256 bits. The block-size has a maximum of 256 bits, but the key-size has no theoretical maximum. The cipher uses number of encryption rounds which converts plain text to cipher text. The output of each round is the input to the next round. The output of the final round is the encrypted plain text known as cipher text [1].

1.2   Ron Rivest, AdiShamir, Leonard Adleman (RSA):

Public-key/two key/asymmetric/RSA cryptography involves the use of two keys:

- **a Public-key**, which may be known by anybody, and can be used to encrypt messages, and verify signatures
- **a Private-key**, known only to the recipient, used to decrypt messages, and sign (create) signatures

RSA is asymmetric because those who encrypt messages or verify signatures cannot decrypt messages or create signatures [3].

1.2.1   How RSA Works
- Begin by select two prime numbers P and Q.
- Then we multiply P and Q to get N.
- Also obtain another number M totient function of N.

$$M = (P-1) (Q-1)$$

- Calculate E such that E and M are relatively prime number

$$GCD (E, M) = 1$$

1.2.2   Public Key and Private Key Pair Generation
- Public key is pair of two number N&E
- This public key is used by others to encrypt message
- A private key is needed to decrypt the encrypted data using our Public key.
- Private key is pair of N & D where D=(KM+1) /E [3]

1.2.3   Encryption and Decryption Using RSA
- Encryption is done by using Public Key

    - Let S is our message
    - Encryption => (S^E)mod N = S'

- To Decrypt use private key

    - Let S' is our encrypted message
    - Decryption => (S'^d)mod N=S [3]

However, the second part of proposed project is to achieve data hiding of encrypted text behind the digital media, digital images in our case.

The proposed project put forward a work related to implementation of image steganography in smart phones supporting ANDROID operating system.

1.3   Steganography (LSB)

Least significant bit (LSB) insertion is a common and simple approach to embed information in an image file. In this method the LSB of a byte is replaced with an M"s bit. This technique works good for image steganography. To the human eye the stego image will look identical to the carrier image. For hiding information inside the images, the LSB (Least Significant Byte) method is usually used.[4]

The algorithm provides the good means for achieving the secure transfer of data from one system to other. However, the data can be encrypted by making the use of AES and can be hidden inside the images using LSB algorithm. The decryption of same data can be done by using the same algorithms just by working it in reverse order [5].

The important part in proposed project is supporting security essentials of data throughout its transfer and at the level of base station also. This can be achieved by applying keys and passwords for each operation in the application [6][7].

2   Problem Definition

To develop an application which will avail the services of steganography along with cryptography and transferring image from one end to the other to achieve secure data transfer over 'Android' smart phones by means of public private key concepts.

3   Related Work

The basic aim of our project is to design an application which will implement 'steganography with cryptography' in ANDROID based smart phones.

In today's world, development of such application on mobile platform is definitely beneficial in terms of carrying out secured transfer of confidential data.

At the present times, there exists an application which works somewhat similar to our aim, but; its existence is related to operating system called as 'WINDOWS' only. But, it is not available on ANDROID operating system. On Android based smart phones, there exists some applications which supports only cryptography. Thus, users are allowed to send and receive confidential data achieving good security results in WINDOWS based systems only. Thus, this application is available on desktop computers or laptops [8].

The implementation of same application in case of another operating system called ANDROID and making it available on each system which supports the same OS is challenging task [8].

4   Proposed System Architecture

The basic aim behind developing this application is to achieve data security at cellular level by means of performing steganography along with cryptography on ANDROID platform.

The process includes three main phases:

- Public key and private key generation and transfer
- Encryption phase
- Decryption phase

4.1   Public Key and Private Key Generation and Transfer

This phase basically includes the generation of RSA public and private keys which are generated at both ends i.e. on the user device as well as on the device of the receiver.

However, in order to carryout encryption we have used two algorithms, AES and RSA in which the encryption of AES key (private key) is carried out by means of public key of the receiver. So, in order to carry out this process, there exists a need such that the public key of receiver should be known at the user side.

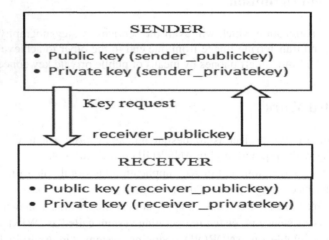

Fig. 1. Key Generation Phase

In order to achieve successful transfer of public key from receiver device to user device the 'SMS (Short Message Service)' is used as a carrier.

The above figure shows the exact process of key transfer.

The message 'GPK' is sent from user device to receiver device where GPK stands for 'GET PUBLIC KEY'. On the reception of this message at receiver device, one message is automatically generated and transferred to user device and this message contains the public key of receiver.

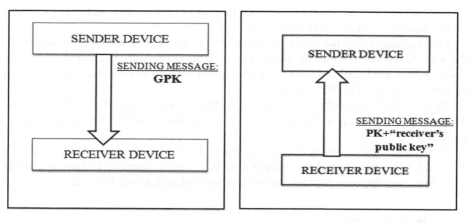

Fig. 2. Key Transfer Phase

At sender device the broadcast receiver is placed which recognizes the actual message of interest as the received message will have fixed format which is 'PK+ 'receiver's public key'' where PK stands for 'PUBLIC KEY'.

In this way, the keys are generated and transferred. The concept of broadcast receiver in android plays important role in this process.

4.2   Encryption Phase

This phase basically runs as follows:

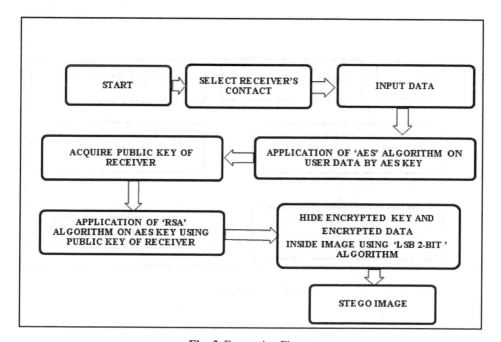

Fig. 3. Encryption Phase

The user is allowed to input the desired text which is to be transmitted and to select image behind which data is to be hidden.

The data entered by the user is encrypted by means of AES key which is private to the user device meanwhile; the public key of desired receiver is acquired. The same public key is used further by RSA for encryption of AES key and the finalized form of that key is passed to the receiver.

Then, the LSB algorithm is applied for hiding the data inside the pixels of image. The output of this step is the stego image i.e. the image on which steganography is carried out.

Thus, the whole content that is supposed to be hided inside image includes finalized form of the key that is supposed to be passed to receiver end and encrypted data. Some mechanisms are included to separation of both at receiver side.

4.3   Decryption Phase

This phase basically involves regeneration of original text from stego image.

This phase basically runs as follows:

The user is supposed to browse the image. The application is designed in such a way that it will make correct use of LSB again in reverse manner to extract the hidden bits from input image given to it.

Thus output of this step is encrypted key and encrypted data. The key obtained is in encrypted form and is decrypted using receivers RSA private key. The original AES key is thus obtained and is used to decrypt the data already acquired in above step.

Thus, the finally generated output data resembles the original data.

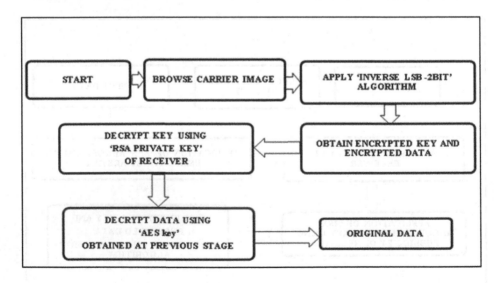

Fig. 4. Decryption Phase

5   Future Scope

The proposed project has implemented steganography along with cryptography in ANDROID platform. Same can be designed on other operating systems if not available. After all, the main aim is to achieve secure data transfer.

Also, in proposed project we have worked on all these concepts for images supporting .png format at the beginning. The same project can be worked in the future for making it run on images having other formats such as .jpg/.jpeg.

6   Applications

The project is being implemented in ANDROID platform. Thus, it can provide a unique communication medium for the transfer of secret messages form source system to destination system.

The benefits of such type of steganography are:
Secure communication

    - Terrorism Organization- Cyber Terrorism
  - Anti- Terrorism Organization- ATS
    - Investigation Agencies: CBI, FBI
Copyright Protection

7   Conclusion

In the present world, the data transfers using internet is rapidly growing because it is so easier as well as faster to transfer the data to destination. So, many individuals and business people use to transfer business documents, important information using internet. Security is an important issue while transferring the data using internet because any unauthorized individual can hack the data and make it useless or obtain information un- intended to him.

The proposed approach in this project uses a new steganographic approach called image steganography. The application creates a stego image in which the personal data is embedded and is protected with a password which is highly secured. The main intention of the project is to develop a steganographic application that provides good security over ANDROID operating system. The proposed approach provides higher security and can protect the message from stego attacks.

Thus, the project being implemented provides a medium for secure transfer of data between two mobile systems supporting ANDROID platform.

References

1. Agrawal, B., Agrawal, H.: Implementation of AES and RSA Using Chaos System. International Journal of Scientific & Engineering Research 4(5) (May 2013) 1413ISSN 2229-5518 IJSER © (2013)

2. Rayarikar, R., Upadhyay, S., Pimpale, P.: SMS Encryption Using AES Algorithm on Android. International Journal of Computer Applications (0975-8887) 50(19) (July 2012) (© 2012 by IJCA Journal)
3. Kota, C.M., Aissi, C.: Implementation of the RSA algorithm and its cryptanalysis. In: ASEE Gulf-Southwest Annual Conference, Session- IVB4, The University of Louisiana at Lafayette, March 20-22. American Society for Engineering Education (2002)
4. Shirli-Shahreza, M., Shirali-Shahreza, M.H.: Text Steganography in SMS. In: International Conference on Convergence Information Technology. IEEE (November 2007), doi:10.1109/ICCIT.2007.100, ISBN: 0-7695-3038-9, INSPEC Accession Number: 9893347
5. Rafat, K.F.: Enhanced Text steganography, Computer, Control and Communication. In: 2nd International Conference on IC4 2009, pp. 1–6. IEEE (February 2009), doi:10.1109/IC4.2009.4909228, E-ISBN: 978-1-4244-3314-8, Print ISBN: 978-1-4244-3313-1, INSPEC Accession Number: 10626971
6. Tyagi, V., Kumar, A., Patel, R., Tyagi, S., Gangwar, S.S.: Image steganography with cryptography using LSB. Journal of Global Research in Computer Science 3(3) (March 2012) ISSN-2229-371X
7. Gupta, S., Gujraland, G., Aggarwal, N.: Enhanced LSB for image steganography. IJCEM International Journal of Computational Engineering & Management 15(4) (July 2012) ISSN (Online): 2230-7893
8. Rughani, P.H., Pandya, H.N.: Steganography on ANDROID based smart phones. IJMAN International Journal of Mobile & Adhoc Network 2(2), 150–152 (2012)

A Cloud Based Architecture of Existing e-District Project in India towards Cost Effective e-Governance Implementation

Manas Kumar Sanyal, Sudhangsu Das, and Sajal Bhadra

Department of Business Administration, Kalyani University
Manas_sanyal@rediffmail.com,
{Iamsud,Sajal.bhadra}@gmail.com

Abstract. The e-District project is a comprehensive and web enabled service portal have been designed by Government of India. E-District project is acting as an electronic gateway into the Government's portfolio of services for common citizens. The e-District portal has redefined the process of public services by catering the services at common citizen's door step at any time. The Government of India is implementing e-District project for each state in India by hosting the e-District project in each State Data Center, which is leading huge amount of cost due to building individual host environment for each state e-District project. In this study, authors have proposed a cost effective private cloud based architecture in e-District project with the help of multiple virtual machine, which would be created from high end physical server machine. The configuration and number of physical machine is subject to current capacity plan. Apart from cost, this proposed architecture would help to share common services among different state e-District projects seamlessly. Also, it will provide better security management, better control in maintenance, flexibility for making disaster recovery plan.

Keywords: e-District, Cloud Computing, ICT, Pay-Per-Use, e-Governance.

1   Introduction

According to announcement of NEGP plan of Government of India (GI) on 18th May 2006, all public services will be accessible to common people locally through common service delivery outlets and ensure efficiency, transparency and reliability of such services at affordable costs. GI has declared 27 mission mode projects and 8 components to reach out these services to common people. The e-District project is one of the very important e-Governance initiatives of GI.

The e-District project is getting very popularity among the common citizens due to the below perceptions of Citizens about the Government Services.

© Springer International Publishing Switzerland 2015   65
S.C. Satapathy et al. (eds.), *Proc. of the 3rd Int. Conf. on Front. of Intell. Comput. (FICTA) 2014*
– *Vol. 2*, Advances in Intelligent Systems and Computing 328, DOI: 10.1007/978-3-319-12012-6_8

- Long queues for submission of applications
- Improper or insufficient information on how to fill the application and what all documents are required to be attached with the application
- Very long lead time for processing of application
- Submission of same information again and again at different levels of processing
- Difficult to track the status of application
- Improper and untimely reply/response to queries raised by them
- No obvious person to help those most in need to find their way around the system

Benefits of Cloud Based Architecture

It has been realized that there is enormous reason to choose cloud for addressing the e-Governance implementation challenges created by this complexity across the country like-

- **Doing More with Less:** Reduce capital and operational expenses.
- **Reducing Risk:** Ensure the right levels of security across all important data and processes.
- **Higher Quality Services:** Improve quality of services and deliver new services that help the Government to introduce new IT initiative and reduce costs.
- **Breakthrough Agility:** Increase ability to quickly deliver new services to capitalize on opportunities while containing costs and managing risk
- **Flexibility & Extensibility:** Every process is automated, could be integrated with any system easily without any manual intervention and changing.
- **Efficiency:** Cloud Computing is an evolution in Information technology which is changing economics of IT, Automating service delivery, Radically developing standardization and rapidly deploying new capabilities.

  More ever, Cloud is a transformation for consumption and delivery of IT with the goal of simplifying to manage complexity with more effectively. Clouds include Public, Private and Hybrid as delivery model and it categorized as Infrastructure as a Service (**IaaS**), Platform as a Service (**PaaS**), Software as a Service (**SaaS**). In this context, Authors have done extensive study and ended with the suggestion for implementing IaaS in e-District implementation across the country towards cloud implementation.

  The key advantages to delivering e-District through cloud are:

  Reduced Cost: It will reduce infrastructure and operation cost.

  Workload Optimized: Infrastructure optimized for specific workloads, services in ways that deliver orders of magnitude better performance, scale and efficiency

  Service Management: It will provide visibility, control and automation across the different e-District project for different state and IT infrastructure to accelerate the delivery of high quality services.

2   Literature Review

According to National e-Governance Plan (NeGP, 2006), Government of India (GI) aimed to transform all the Government activities from manual to computerized system in order to meet the commitment of providing fully ICT based Government services. To make it happen, GI has initiated 27 Mission Mode Projects (MMPs) including 10 Central MMPs, 10 State MMPs and 7 Integrated MMPs.

The Department of Electronics and Information Technology, GI (2012) has detailed the e-District projects which are aimed to accelerate the citizen centric Government services in Rural India. They have published document entitled "Integrated Framework for Delivery of Services", has described that the application including Service Scope and Coverage, Business Re-Engineering Process and Technical Details like underline network, software, Database, Software delivery model etc.

In order to make successful and rapid implementation of e-Governance projects, it is very important to reduce IT roll out and implementation cost for all government IT initiatives. In this context, Prasad A., Chaurasia S., Singh A. and Gour D (2010) has strongly suggested to introduce service oriented grids/cloud approach for implementing generic government functioning. They also concluded in their study that SOA and CMMS models of service oriented grids/clouds can provide government service in more reliable, faster and transparent way.

With the similar thought Sharma R. and Kanungo P. (2011) also has recommended for opting cloud computing in case of e-Governance projects with the parallel of others business solution. They have proposed an intelligent & energy efficient Cloud computing architecture based on distributed data-centers to support application and data access from local data-center with minimum latencies. It has been found that the proposed architecture is efficient for business entrepreneurs, suitable to apply for e-Governance and provides a green eco-friendly environment for Cloud computing.

In other study, Zissis D. and Lekkas D. (2011) has claimed that applying cloud computing in e-Governance can increase participation and sophistication of government services, especially in e-Voting system. Taking a step further, they have proposed a high level electronic governance and electronic voting solution, supported by cloud computing architecture and cryptographic technologies.

In a study Mukherjee K. and Sahoo G. (2010) has highlighted that how to do better utilization of IT resources by using cloud computing in e-Governance projects. Authors have pointed out that the existing non-cloud based e-Governance framework cannot address all kind of users but proposed cloud computing would be intelligent.

While it has been accepted that cloud computing is the best cost effective approach to introduce IT solutions for e-Governance project, Rastogi A. (2010) has proposed a model base approach to adopt cloud computing in e-Governance projects. In this prototype model based approach, author have suggested six different steps to launch cloud computing in e-Governance projects like-1) Learning, 2) Organizational Assessment, 3) Cloud prototype, 4) Cloud assessment, 5) Cloud Rollout Strategy, 6) Continuous Impartment.

3   Methodology

In favor of covering analysis and interpretation of this study, authors took help of Microsoft Visio for designing the new proposed cloud based framework of e-District project. The main brainstorms of this designing came from analysis of the past literature and modern industrial IT solution approaches for similar kind of business requirements. Literatures have been used for changing the mind set of authors in order to introducing cloud computing in e-Governance projects. For getting more confident and to explore way forward cloud design approach, authors conducted survey among IT industries. The literature has strongly suggested that moving towards cloud computing would help a lot to reduce e-District IT infrastructure cost. Existing industrial project knowledge helped to get idea for designing the new cloud architecture in e-District project.

4   Reason to OPT IaaS in e-District Project

Clouds provide actual advantage of technology, helping to think to roll out IT solutions which were never possible due to huge cost involvement. IaaS cloud solution is about to delivering competitive and sustainable high performance e-District application.   It's not about just cutting cost – it's about doing things previously not possible.The most cost-efficient way to traverse and mine Internet. In fig1, it has been depicted that how IT evolution is going towards cloud and encouraging people to go for cloud based IT solution.

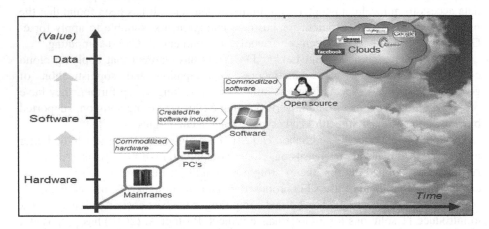

Fig. 1. IT evolution based on Time

Along with this, world economic is also forcing industry to opt cloud computing. In Fig 2 & Fig 3, it has been highlighted data center cost and Amazon web service band width cost , which is highly indicating to minimize the IT cost to save industry. It only could be possible by cloud solutions.

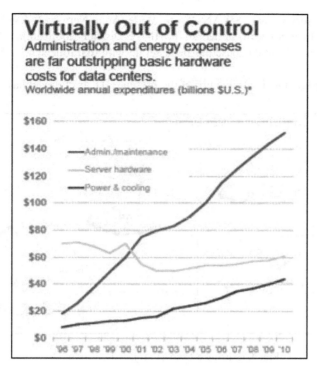

Fig. 2. Data Center Cost, Source: IDC

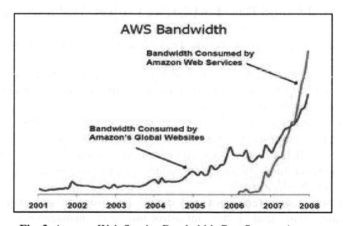

Fig. 3. Amazon Web Service Bandwidth Cost Source: Amazon

5   Existing e-District Project Delivery Model

The present e-District solution is designed with a central data center. All district level users like common citizens, Government officers, CSC/Kiosk operator is connecting to the central server over SWAN (State Wide Area Network), SWAN has been utilized as a backbone of network for the e-District application. The Disaster Recovery for server and database has been set up with real time sync in Head Quarter Data center.

In Fig 4, the technical architecture of the present e-District model has been shown:

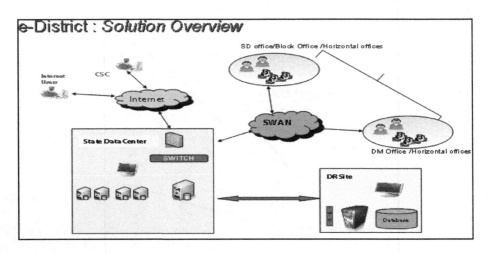

Fig. 4. Existing e-District project delivery model

6   Proposed Cloud Based e-District Project Architecture

The basic needs in order to create Private IaaS Cloud for e-District solution are:

- Sharing physical infrastructure among all state e-District application for reducing infrastructure cost. Also, using existing infrastructure can minimize the infrastructure procurement
- Sharing common application services among all the state e-Districts solution
- Centralized infrastructure procurement and management
- Provide collaboration in between multiple virtual machine
- On demand load balancing to reduce process overheads to a particular underline hardware.
- Centralized data repository and disaster recovery for all state e-District solution will lead better infrastructure cost management, scalability, reliability and manageability.

These needs are given a thought to authors in favor to bring the private cloud strategy in place.

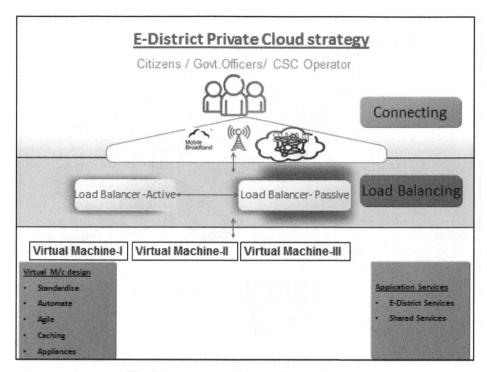

Fig. 5. Proposed cloud based e-District Strategy

According to the Fig 5, e-District portal users request would be redirected to Load Balancer (LB) to get connected with the application. The LB will pass the same HTTP request to a specific Virtual Machine (VM) web server based on the load among the different VM for that particular moment in order to serve the user request. The passive LB must be provisioned as a backup of active LB for avoiding any application downtime. The preferable design of VM will be with the capability of standardization, Automation, Agile, Caching and Appliances.   The numbers of dedicated VM's for a particular state e-District portal must be aligned with the capacity plan of existing stand along deployment scenario.

According to the above private cloud strategy of e-District application, authors are proposing the following system architecture towards building Infrastructure as a Service (IaaS) base cloud solution.

In the fig 6, it has been shown that every e-District portal is being isolated with network device and each portal has been provisioned with three virtual machines though actual no. of virtual machine will depend on the actual load for that particular state e-District portal. Each virtual machine must have some virtual storage as secondary data storage device. In this designed, 192.168.3.0/24 node has been used as physical server cluster. The no. of physical server also will grow based on the capacity requirement of the environment. Here, physical storage will be SAN (Storage Area Network) which also can grow with the requirement of the environment. All the virtual storage is getting space from this physical storage. According to the design, there will be proper Firewall in place with router configuration for ensuring security.

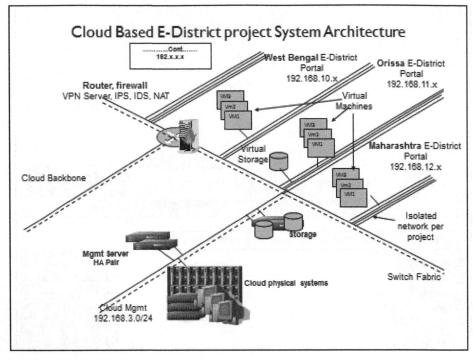

Fig. 6. Proposed cloud based e-District Model

7   Conclusion

Authors have concluded that with regard to the increasing trend of IT infrastructure expenses and expedite the new enhancement roll out process; Cloud solution in e-District project would help in India to improve new functionality delivery in a timely, efficient and cost-effective manner. It would be beneficial for new state for sure for rolling out e-District project. Apart from this, new application also could be hosted in same environment subject to capacity of the environment.

References

1. Prasad, A., Chaurasia, S., Singh, S., Gour, D.: Mapping Cloud Computing onto Useful e-Governance. International Journal of Computer Science and Information Security (IJCSIS) 8(5) (2010)
2. Tsai, W., Sun, S., Balasooriya, J.: Service-Oriented Cloud Computing Architecture. In: Seventh International Conference on Information (2010)
3. Guo, Z., Song, M., Song, J.: Governance Model for Cloud Computing. In: Management and Service Science (MASS). IEEE (2010)
4. Sharma, R., Kanungo, P.: An Intelligent Cloud Computing Architecture Supporting e-Governance. In: Proceedings of the 17th International Conference on Automation & Computing, University of Huddersfield, Huddersfield, UK, September 10 (2011)

5. Masud, A.H., Huang, X.: Cloud Computing for Higher Education: A Roadmap. In: Proceedings of the 2012 IEEE 16th International Conference on Computer Supported Cooperative Work in Design (2012)
6. Rastogi, A.: A Model based Approach to Implement Cloud Computing in E-Governance. International Journal of Computer Applications (0975 – 8887) 9(7) (November 2010)
7. Zissis, D., Lekkas, D.: Securing e-Government and e-Voting with an open cloud computing architecture. Government Information Quarterly 28, 239–251 (2011)
8. Mukherjee, K., Sahoo, G.: Cloud Computing: Future Framework for e-Governance. International Journal of Computer Applications (0975 – 8887) 7(7) (October 2010)
9. Cellary, W., Strykowski, S.: E-Government Based on Cloud Computing and Service-Oriented Architecture. In: Proceedings of the 3rd International Conference on Theory and Practice of Electronic Governance, ICEGOV 2009, Bogota, Columbia, November 10-13 (2009)
10. Dwivedi, S.K., Bharti, A.K.: E-Governance in India – Problems and Acceptability. Journal of Theoretical and Applied Information Technology (2010)
11. Shirin, M.: e-Governance for Development. Plangrave Macmillan, London (2009)
12. Official Website of Bankura, http://www.bankura.nic.in/census.htm
13. Department of Electronics And Communication Technology, http://deity.gov.in/
14. Department of Information Technology & Electronics, Government of West Bengal, http://www.itwb.org
15. Official Website of Jalpaiguri, http://www.jalpaiguri.gov.in
16. IDC, http://www.in.idc.asia/
17. Amazon, http://aws.typepad.com/aws/2008/05/

A Chaotic Substitution Based Image Encryption Using APA-transformation

Musheer Ahmad, Akshay Chopra, Prakhar Jain, and Shahzad Alam

Department of Computer Engineering, Faculty of Engineering and Technology,
Jamia Millia Islamia, New Delhi 110025, India

Abstract. In this paper, we propose a new chaotic substitution based image encryption algorithm. Our approach combines the merits of chaos, substitution boxes, APA-transformation and random Latin square to design a cryptographically effective and strong encryption algorithm. The chaotic Logistic map is incorporated to choose one of thousand S-boxes as well as row and column of selected S-box. A keyed Latin square is generated using a 256-bit external key. The selected S-box value is transformed through APA-transformation which is utilized along with Latin square to substitute the pixels of image in cipher block chaining mode. Round operations are applied to fetch high security in final encrypted content. The performance investigations through statistical results demonstrate the consistency and effectiveness of proposed algorithm.

Keywords: Image encryption, substitution box, chaotic map, Latin square, APA-transformation.

1 Introduction

In modern cryptography, a substitution-box is a fundamental element of symmetric-key algorithms. According to "*The Design of Rijndael*" the aim of S-boxes in encryption techniques is to induce essential nonlinearity to prevent linear and differential cryptanalysis [1]. Mathematically, an 8×8 S-box is a function of the form: $S: GF(2^8) \rightarrow GF(2^8)$ which takes 8-bits as inputs and generates 8-bit as output. In other words, an S-box is combination of Boolean function f_i i.e. $S(x) = (f_1(x), f_2(x), ..., f_8(x))$. The effectiveness of encryption depends on the aptness of S-box in confusing the data. Hence, designing new and efficient S-boxes is of great importance. Because of the critical role of Rijndael S-box in AES, many researchers focused on S-box improvements. A noteworthy development in the past few years for S-boxes has been done mainly to increase the nonlinearity of S-boxes [2]. In [3], Cui and Cao addressed the problem of simple algebraic structure of AES S-box and a new S-box structure named Affine-Power-Affine (APA) is designed to enhance the algebraic complexity of AES S-box that also inherits other cryptographic strength. The aim of encryption is to create confusion in plain-text, and the S-box plays a significant role in achieving this task. Recent research study shows that, instead of

© Springer International Publishing Switzerland 2015
S.C. Satapathy et al. (eds.), *Proc. of the 3rd Int. Conf. on Front. of Intell. Comput. (FICTA) 2014*
– *Vol. 2*, Advances in Intelligent Systems and Computing 328, DOI: 10.1007/978-3-319-12012-6_9

number-theoretic based approaches, the pseudo-random and nonlinear properties of chaotic systems are tapped for the design of efficient cryptographic S-boxes [a-b]. Since, the images are most commonly and inevitably used medium of our communication. The modern technological advances in communications facilitate their sharing, distribution and transmission over the open networks at a very fast rate. To ensure a secure image-based communication over insecure networks to resist them against malicious access and usage, the cryptographically secure image encryption algorithms are applied. But, due to the high spatial auto-correlation and redundancy in multimedia image data, the S-box exhibits poor substitution performance despite of having high nonlinearity [4]. The new trend in the direction of image encryption is to explore the features of S-boxes in a way to cope with the intrinsic auto-correlated featured multimedia images so as to provide sufficient security, from cryptographic viewpoint, to them. The notable contributions in this direction are the proposals suggested in [4-8].

In this paper, a new image encryption algorithm is synthesized by combining the merits of chaotic S-boxes, APA-transformation, keyed Latin square and chaos. The proposed algorithm exhibits the characteristics of confusion and diffusion which are indispensable for any cryptographically strong encryption method. The algorithm has the advantages of having high visual indistinguishability, high entropy, high plaintext sensitivity, low auto-correlation and large key space measures.

The rest of this paper is structured as follows: Section 2 gives a brief review on preliminary concepts employed in the design. In Section 3, the details of the proposed chaotic substitution based image encryption algorithm is provided and discussed. The performance and strength of the new encryption algorithm is quantified and analyzed in Section 4. Moreover, the work is also compared with recent algorithm proposed by Anees *et al.* in the same section to demonstrate the effectiveness of our algorithm. We finally conclude the work in Section 5.

2   Preliminaries

2.1   Chaotic Logistic Map

The 1D Logistic map proposed by May [9] in 1967 is one of the simplest nonlinear chaotic discrete systems that exhibit chaotic behavior; its state equation is governed by the following iterative formula.

$$x(n+1) = \lambda \times x(n) \times (1 - x(n)) \tag{1}$$

Where x is map's variable and $x(0)$ acts as its initial condition, λ is system parameter and n is number of iterations needs to be applied. The research shows that the map exhibits chaotic dynamics for $3.57 < \lambda < 4$ and $x(n) \in (0, 1)$ for all n. This chaotic map is used to select various values in random fashion and to make the whole algorithmic process key dependent.

2.2   APA-transformation

The Rijndael AES S-box has been considered as secure against linear and differential cryptanalyses. However, it has simple algebraic structure with only 9 terms in its algebraic expression, which makes AES S-box susceptible to algebraic attacks [10]. In 2007, Cui and Cao considered the problem and introduced a new structure called Affine-Power-Affine (APA) that amplified the algebraic complexity [3]. Due to the APA-transformation, the algebraic complexity of improved AES S-box increases from 9 to 253. AES S-box is the combination of a power function $P(x)$ (the multiplicative inverse modulo the polynomial $x^8 + x^4 + x^3 + x + 1$) and an affine transformation $A(x)$ [1, 3]:

$$P(x) = \begin{cases} x^{-1} & x \neq 0 \\ 0 & x = 0 \end{cases}$$

$$A(x) = \begin{bmatrix} 1 & 0 & 0 & 0 & 1 & 1 & 1 & 1 \\ 1 & 1 & 0 & 0 & 0 & 1 & 1 & 1 \\ 1 & 1 & 1 & 0 & 0 & 0 & 1 & 1 \\ 1 & 1 & 1 & 1 & 0 & 0 & 0 & 1 \\ 1 & 1 & 1 & 1 & 1 & 0 & 0 & 0 \\ 0 & 1 & 1 & 1 & 1 & 1 & 0 & 0 \\ 0 & 0 & 1 & 1 & 1 & 1 & 1 & 0 \\ 0 & 0 & 0 & 1 & 1 & 1 & 1 & 1 \end{bmatrix} \times \begin{bmatrix} x_0 \\ x_1 \\ x_2 \\ x_3 \\ x_4 \\ x_5 \\ x_6 \\ x_7 \end{bmatrix} \oplus \begin{bmatrix} 1 \\ 1 \\ 0 \\ 0 \\ 0 \\ 1 \\ 1 \\ 0 \end{bmatrix}$$

Where x_i's are the coefficients of x (8-bit elements of S-box in GF(2^8)). The APA transformation of an element x of AES S-box is defined as [3]:

$$S(x) = A \circ P \circ A$$

Cui and Cao has shown that compared to AES S-box, the APA S-box has stronger algebraic complexity and inherits other cryptographic characteristics of AES S-box.

2.3   Keyed Latin Square

Latin square refers to a 2D matrix of size $n \times n$ which contains n distinct symbols (we use $0, \ldots, n-1$ to represent the symbols) that are appear once per each row and once per each column. Latin square was first developed by *Leonhard Euler* and termed as "Latin square" because the symbols that were used by him were in Latin. *Sudoku* puzzle is a special case of Latin square with additional block constraint. In 2013, Wu *et al.* [11] highlighted the weaknesses of chaos-based image encryption techniques and proposed a Latin square based image cipher. An example of Latin square of size 5×5 is shown in Fig. 1. For a detail description of Latin square, the readers are referred to [11, 12]. The algorithm presented by Wu *et al.* for generation of a keyed Latin square of size 256×256 is adopted. Thus, the advantages of keyed Latin square are explored along with the chaos and S-boxes in this work.

2	0	4	1	3
3	2	1	4	0
1	4	0	3	2
4	3	2	0	1
0	1	3	2	4

Fig. 1. A Latin square of size 5×5

2.4   Substitution Boxes

To provide efficient and better pixel substitution for auto-correlated pixels of image, we utilize 1000 dynamic chaotic S-Boxes designed by Ahmad *et al.* [13]. The dynamic S-boxes are generated using classical Fisher-Yates shuffle technique and chaos. The method of their generation is simple and key dependent. It has been demonstrated and investigated by Ahmad *et al.* that all generated S-boxes have better cryptographic strength as compared to most of the existing chaos-based S-boxes. Each pixel of the plain-image is treated with a randomly selected element of a randomly chosen S-box. Moreover, the random selection is not only key dependent but also pending image information dependent through round operations and CBC mode of encryption. Therefore, even if adjacent pixels are same, they are substituted with different elements of different S-boxes. The way these S-boxes are used for substitution of pixels ensures the abolition of any type of, either low or high, auto-correlation among the pixels.

3   Proposed Algorithm

In this algorithm, we encrypt the image by combining and exploring the concepts of chaos, substitution-boxes, affine-power-affine transformation and keyed Latin square. The S-Box and the element of S-Box are randomly selected using the digits of chaotic map variable. The pixels of the image are then confused by performing XOR operations on the pixel value, the output of the APA S-Box, the previous cipher pixel value and the value of Latin square. After all the pixel values have been modified, the obtained image is rotated and flipped about the left diagonal. This procedure is repeated for a pre-defined number of rounds and at the end of all rounds, encrypted image is obtained. The steps of proposed image encryption algorithm are as follows:

A.1.   Input plain-image P (having size $M{\times}N$) the secret key components β (number of rounds), $x(0)$, $C(0)$ and Latin square key K.

A.2.   Let r be the count variable for round operation and initially $r = 1$.

A.3.   Update the key components $x(0)$, $C(0)$ and K using current count r.

A.4.   Reshape the image into a 1D array P and $i = 1$.

A.5.   Generate 256×256 Latin square using key K and reshape it to 1D array L.

A.6.   Iterate chaotic map of Eqn.(1) once and capture the map variable x of the form $x = 0.d_1d_2d_3 \ldots d_{15}$ ($0 \leq d_i \leq 9$). Calculate the following:

$$k = (a_1)mod(1000)+1$$
$$l = (a_2)mod(16)+1$$
$$m = (a_3)mod(16)+1$$

Where

$$a_1 = d_1d_2d_3d_4d_5$$
$$a_2 = d_6d_7d_8d_9d_{10}$$
$$a_3 = d_{11}d_{12}d_{13}d_{14}d_{15}$$

A.7.   Take APA-transformation of value obtained from k^{th} S-Box at the index (l, m). Let this value be Φ.

A.8.   Calculate the cipher image pixel as

$$C(i) = C(i-1) \oplus P(i) \oplus \Phi \oplus L(q)$$

Where $q = (i)mod(256 \times 256) + 1$

A.9.   Evaluate $t = (C(i))mod(4) + 1$ and iterate the chaotic map for t times and discard the values.

A.10. If $i < M \times N$ then set $i = i + 1$ and go to step **A.6.**

A.11. Reshape cipher image C into 2D matrix and perform two 90° anti-clockwise rotations and then flip about its left diagonal.

A.12. If $r < \beta$ then set $r = r + 1$, $P = C$ and go to step **A.3** else terminate.

The decryption process is similar to the encryption process but all the operations are applied in reverse order. We start with the last round first and move in backward direction. Before the start of each round, we flip the image matrix about its left diagonal and then rotate the matrix twice by 90° clockwise. Then we apply the operations, as in the encryption process, to all the pixels of the image and this is continued till all rounds are complete. Thus, at the end original image is obtained.

4   Simulation Results

The proposed scheme is simulated on Matlab 2011 with three plain-images each of size 256×256 which are shown in Figure 2 and their histograms are shown in Figure 3. The initial value used for encrypting the images are: $x(0) = 0.23456$, $\lambda = 3.99$, $\beta = 4$, $K = $ '12A34F56E78D90C31B72AF4835DC0981237654CD185A3FEB01CAE7259018FD14'. The corresponding encrypted images by the proposed algorithm are shown in Figure 4 and histograms are shown in Figure 5. It is evident that images in Figure 4 have high visual indistinguishability and distortion. The encrypted images, by the Anees *et al.* encryption method, with their corresponding histograms are depicted in Figure 6 and 7 respectively. The histogram of an image depicts how the pixels in the image are distributed. A perfectly encrypted or noise-image has a flat and uniform histogram.

As can be seen that proposed algorithm provides better encryption effect than the Anees at al algorithm. Reason being the encrypted images shown in Figure 4 are more distorted visually than the ones shown in Figure 6. Moreover, the histograms of our encrypted images shown in Figure 5 are more uniform and flat akin to noise image than the histograms of images encrypted with Anees *et al.* algorithm.

In proposed algorithm, the components of secret key includes $x(0)$, λ, β, $C(0)$, K and key used for dynamic S-boxes generation. With a 15 digit precision, the size of key space comes out as $6 \times 256 \times 1000 \times 10 \times 2^{256} \times (10^{14})^4 \approx 2^{465}$. Thus, the key space is large enough to resist the brute-force attack.

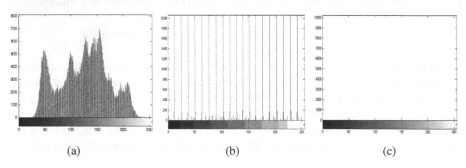

(a)   (b)   (c)

Fig. 2. Plain-images: (a) image#1 *Lena* (b) image#2 *Gray-strips* (c) image#3 *Black*

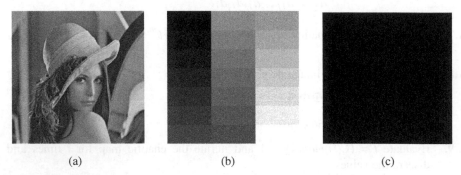

(a)   (b)   (c)

Fig. 3. Histograms of plain-images: (a) *Lena* (b) *Gray-strips* (c) *Black*

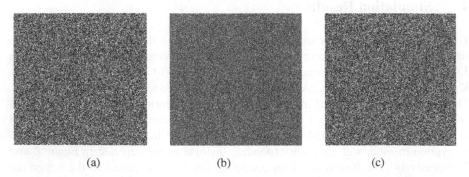

(a)   (b)   (c)

Fig. 4. Encrypted images with proposed algorithm: (a) *Lena* (b) *Gray-strips* (c) *Black*

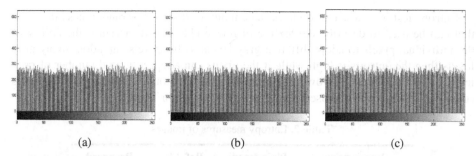

(a)   (b)   (c)

Fig. 5. Histograms of encrypted images in Figure 4 : (a) *Lena* (b) *Gray-strips* (c) *Black*

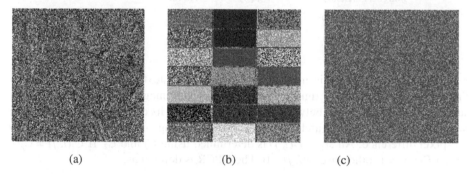

(a)   (b)   (c)

Fig. 6. Encrypted images with Anees *et al.* algorithm: (a) *Lena* (b) *Gray-strips* (c) *Black*

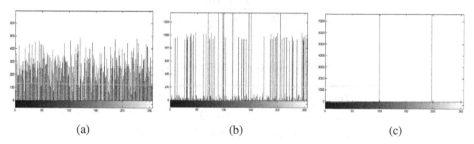

(a)   (b)   (c)

Fig. 7. Histograms of encrypted images in Figure 6 : (a) *Lena* (b) *Gray-strips* (c) *Black*

Table 1. Correlation coefficients in different images

Image	Plain	Ref. [4]	Proposed
Image #1 (Lena)	0.95679	0.00846	0.00673
Image #2 (Strips)	0.99979	0.3944	-0.00128
Image #3 (Black)	1	0.01712	0.00623

The correlation test is conducted to assess the amount of auto-correlation among adjacent pixels in an image. An efficient encryption algorithm should be able to eliminate completely any type of correlation among adjacent pixels in images [11, 14]. The correlation measures for images under consideration are listed in Table 1.

Entropy test is conducted to check the amount of disorder or randomness in image that can be used to describe the texture of image [11, 15]. It measures the ability of the individual pixels to adopt different gray levels. If the pixels can adopt many gray levels then the entropy is high while if the pixels can adopt a limited number of gray levels then the entropy is low. The entropy measures for images encoded in 8-bits (ideal value for a perfect noise-image is 8) are provided in Table 2.

Table 2. Entropy measures of images

Image name	Plain-image	Ref. [4]	Proposed
Image #1 (Lena)	7.4439	7.8403	7.9965
Image #2 (Strips)	4.3923	6.1582	7.9958
Image #3 (Black)	0.0000	0.9185	7.9972

Net pixels change rate is used to quantify the plaintext sensitivity i.e. effect of changing a single pixel in original image on encrypted image [11]. It also depicts the randomness and difference between the plain-image and its encrypted image. We take two encrypted images, C_1 and C_2, whose corresponding original images have only one-pixel difference. An array $D(i, j)$ is determined using C_1 and C_2. If $C_1(i, j) \neq C_2(i, j)$ then $D(i, j) = 1$, otherwise $D(i, j) = 0$. The NPCR is defined as:

$$NPCR = \frac{\sum_{i=1}^{M} \sum_{j=1}^{N} D(i, j)}{M \times N} \times 100$$

Table 3. Entropy measures of images

Image name	Ref. [4]	Proposed
Image #1 (Lena)	0.00153	34.37
Image #2 (Strips)	0.00153	36.72
Image #3 (Black)	0.00153	99.60

5   Conclusion

In this work, a new image encryption algorithm is suggested which combines and explores the features of chaos, S-boxes, APA-structure and Latin square. The role of chaotic Logistic map is to select various elements in random fashion and to make the whole process under the control of key. The key holds the over all security of algorithm which must be hidden from an unintended user. The key space of algorithm is extremely high to perplex the work of attacker. The simulation and statistical results confirm the excellent performance of proposed algorithm.

References

1. Hussain, I., Shah, T., Gondal, M.A., Khan, W.A., Mahmood, H.: A group theoretic approach to construct cryptographically strong substitution boxes. Neural Computing and Applications 23(1), 97–104 (2012)
2. Daemen, J., Rijmen, V.: The Design of RIJNDAEL: AES—The Advanced Encryption Standard. Springer, Berlin (2002)
3. Cui, L., Cao, Y.: A new S-box structure named Affine-Power-Affine. International Journal of Innovative Computing, Information and Control 3, 751–759 (2007)
4. Anees, A., Siddiqui, A.M., Ahmed, F.: Chaotic substitution for highly auto-correlated data in encryption algorithm. Communication in Nonlinear Science and Numerical Simulation 19(9), 3106–3118 (2014)
5. Huang, C., Nien, H., Chiang, T., Shu, Y., Changchien, S., Teng, C.: Chaotic S-Box based pixel substituting for image encryption. Advanced Science Letters 19(5), 1525–1529 (2013)
6. Hussain, I., Shah, T., Gondal, M.A.: An efficient image encryption algorithm based on S8 S-box transformation and NCA map. Optics Communications 285(14), 4887–4890 (2012)
7. Hussain, I., Shah, T., Gondal, M.A.: Image encryption algorithm based on total shuffling scheme and chaotic S-box transformation. Journal of Vibration and Control (2013), doi:10.1177/1077546313482960
8. Wang, X., Wang, Q.: A novel image encryption algorithm based on dynamic S-boxes constructed by chaos. Nonlinear Dynamics 75(3), 567–576 (2014)
9. May, R.M.: Simple mathematical model with very complicated dynamics. Nature 261, 459–467 (1967)
10. Murphy, S., Robshaw, M.J.B.: Essential algebraic structure within the AES. In: Yung, M. (ed.) CRYPTO 2002. LNCS, vol. 2442, pp. 1–16. Springer, Heidelberg (2002)
11. Wu, Y., Zhou, Y., Noonan, J.P., Agaian, S.: Design of image cipher using latin squares. Information Sciences 264, 317–339 (2014)
12. Latin Square, http://en.wikipedia.org/wiki/Latin_square (last access on June 12, 2014)
13. Ahmad, M., Khan, P.M., Ansari, M.Z.: A simple and efficient key-dependent S-box design using fisher-yates shuffle technique. In: Martínez Pérez, G., Thampi, S.M., Ko, R., Shu, L. (eds.) SNDS 2014. CCIS, vol. 420, pp. 540–550. Springer, Heidelberg (2014)
14. Ahmad, M., Farooq, O.: A Multi-Level Blocks Scrambling Based Chaotic Image Cipher. In: Ranka, S., Banerjee, A., Biswas, K.K., Dua, S., Mishra, P., Moona, R., Poon, S.-H., Wang, C.-L. (eds.) IC3 2010. CCIS, vol. 94, pp. 171–182. Springer, Heidelberg (2010)
15. Ahmad, M., Farooq, O.: Secure Satellite Images Transmission Scheme Based on Chaos and Discrete Wavelet Transform. In: Mantri, A., Nandi, S., Kumar, G., Kumar, S. (eds.) HPAGC 2011. CCIS, vol. 169, pp. 257–264. Springer, Heidelberg (2011)

Security, Trust and Implementation Limitations of Prominent IoT Platforms

Shiju Sathyadevan, Boney S. Kalarickal, and M.K. Jinesh

Amrita Center for Cyber Security Systems and Networks,
Amrita Vishwa Vidyapeetham, Kollam, India
{shiju.s,jinesh}@am.amrita.edu,
p2csn12025@student.am.amrita.edu

Abstract. Internet of Things (IoT) is indeed a novel technology wave that is bound to make its mark, where anything and everything (Physical objects) is able to communicate over an extended network using both wired and wireless protocols. The term "physical objects" means that any hardware device that can sense a real world parameter and can push the output based on that reading. Considering the number of such devices, volume of data they generate and the security concerns, not only from a communication perspective but also from its mere physical presence outside a secure/monitored vault demands innovative architectural approaches, applications and end user systems. A middleware platform/framework for IoT should be able to handle communication between these heterogeneous devices, their discoveries and services it offers in real time. A move from internet of computers to internet of anything and everything is increasing the span of security threats and risks. A comparative study of existing prominent IoT platforms will help in identifying the limitations and gaps thereby acting as the benchmark in building an efficient solution.

Keywords: Internet of Things, Heterogeneous devices, Middleware.

1   Introduction

Internet of things is considered as the future technology, where every day physical object is able to communicate with each other through an extended network using wired and wireless protocols. By allowing various devices to communicate helps in creating applications spanning across diverse domains like medical, transportation, home automation, business, agriculture, animal husbandry etc... IERC (European Research Cluster on the Internet of Things) defines IoT as "a dynamic global network infrastructure with self-configuring capabilities based on the standard and interoperable communication protocols where physical and virtual 'things' have identities, physical attributes, virtual personalities and use intelligent interfaces, and are seamlessly integrated into the information network" [1].

IoT is composed of a collection of heterogeneous and independent physical objects sharing the same spectrum of existing Internet infrastructure. According to the

© Springer International Publishing Switzerland 2015
S.C. Satapathy et al. (eds.), *Proc. of the 3rd Int. Conf. on Front. of Intell. Comput. (FICTA) 2014*
– *Vol. 2*, Advances in Intelligent Systems and Computing 328, DOI: 10.1007/978-3-319-12012-6_10

statistics given by IMS Research, by the end of 2020, the number of web-connected devices will be more than 20 billion and they will generate a huge amount of data as much as that generated by a single human brain. Today, most of the business sectors automate their business process and as a result of that huge amount of sensor data are transferred every second. All these sensor devices use heterogeneous protocols to communicate among themselves increasing the complexity of the IoT environment. Security issues are going to be yet another concern as the participating devices are not confined within a secure area. Also, most of the devices are equipped with less processing resources that enforcing reliable and secure communication mechanisms might not be practical. Such limitations, demands the need to have an efficient middleware that can provide secure, scalable and interoperable services to meet the challenges.

This paper provides a detailed study of different IoT platforms and framework, identifying what their challenges and limitations are. A comparative study of four cloud based IoT platforms and one IoT middleware is being done in this paper. The remainder of the paper is organized as follows. Second section gives a comparison of IoT middleware platforms and frameworks. Related work in the field of comparative study is presented in the third section. Fourth section discusses about different platforms and framework employed in the current paper. Implementation limitations and problems of above mentioned platforms and framework are discussed in the fifth section. Analysis of different platforms are explained in sixth section.

2   IoT Middleware Platform / Framework Comparison

For devices to communicate with each other breaking the barriers of heterogeneity, a platform or a middleware is a much needed component. But relying on such a distributed cloud based middleware solution will mean that it is built to be highly scalable, reliable and resilient. Every device that communicates either with the middleware or with other devices will have to have the traffic flowing through the middleware at all times which increases the need for reliable, consistent bandwidth for the internet pipeline. In order to cut back on this there should be intelligent gateways that can manage a cluster of devices to a greater degree locally thereby reducing the need to push the entire traffic at all times to the middleware.

Current IoT deployments can be basically classified into two categories;

- Middleware that is deployed as a web based solution i.e. it can be hosted in the cloud (Fig. 1).
- Middleware using a backbone network (Fig. 1).

In the former case the middleware is provided as a platform hosted in the cloud where as in the latter the middleware firstly need to set up a backbone network and each participating device should run an embedded software module as part of its firmware, to establish device level communication.

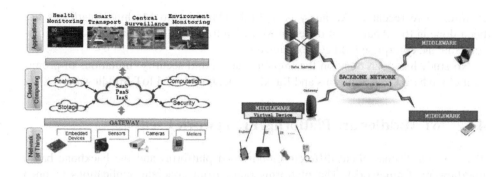

Fig. 1. Cloud Based IoT Platform and Backbone Network Based IoT Middleware Architecture

There are so many advantages and disadvantages for these two architectures. If it is a cloud based architecture, then the user can achieve device to device communications by registering and configuring devices in few mouse clicks, but in backbone based network the user need to handcraft code for every device and all such devices should be intelligent enough to set up a network and must have the capability of running middleware software modules. The scalability of cloud based middleware is huge as it relays purely on the public internet. Once it is deployed it can be accessed from any-where by referring to IP assigned to it. Since cloud based IoT platform uses public internet the security issues associated with it is relatively at the high end compared with a backbone based network. All type of prominent network based attack likes Man in the Middle attack, DoS (Denial of Service) attack, Spoofing attacks, Masque-rading devices etc. are very much applicable to those devices trying to communicate using cloud based middleware. Moreover, achieving device to device communication in Backbone network based IoT middleware is difficult for a novice user as it in-volves complex programming.

3   Related Work

A detailed study of IoT middleware has been proposed in [3]. This paper gives an insight view of two areas. It classifies the middleware systems into different catego-ries so that it satisfies IoT requirements based on the domain in which it is being used. It then discusses about the technical challenges of designing middleware systems for IoT. The basic functional components of various IoT middlewares has been proposed in [4]. This helps developers to understand how current IoT middleware system works and also finding out existing issues and gaps. The paper also compares and also cate-gorize different IoT middleware solutions based on their features and provides the application domain for each of these middleware solutions. Security challenges and privacy issues of IoT middleware has been discussed in [5]. The paper categorizes IoT into different topics like communication, sensors, actuators, storage, devices, processing, localization and identification. Technologies used and security issues associated with each topic are explained. Also the paper list out some 'topics' that requires more research. An analysis of M2M (Machine to Machine) platforms has been done in [6]. Analysis of different M2M/Platforms based on object, people, environment and enterprise has been explained.

No study has been done on the implementation and security limitations of prominent cloud based IoT Platforms and Backbone network based IoT Middleware.

4   IoT Middleware Platforms/Framework Used

This section discuss about different cloud based platforms and the backbone based middleware (framework). The platforms range from concrete applications to open source.

4.1   CARRIOTS

Carriots [7] is a simple M2M platform with focus on three main objectives: Simple, fast and cheap M2M application development, Scalability and Easy to plug in customer products.

4.2   XIVELY

Xively [8] is a public cloud specifically build for the Internet of Things. Xively's IoT public cloud, with web based tools and rich development resources allow clients to focus on critical resources rather than on enabling.

4.3   THINGSPEAK

ThingSpeak [9] is an open platform designed to enable connection between things and people. With open source ThingSpeak code the user can contribute, change and implement it in his/her own local network.

4.4   RUNMYPROCESS

RunMyProcess [10] platform is meant especially for the enterprise market. Dragging and dropping of functions from platform's Graphical User Interface allows user to develop complex applications.

4.5   HYDRA

Hydra [11] is a network embedded middleware platform for network embedded systems that allows developers to develop cost-effective, high-performance applications for heterogeneous physical network.

5   Problem Matrix Identified

In order to test out the various platforms that is being analyzed we used a device that can sense temperature, ambient Light and humidity at various locations within our research campus facility.

5.1   Handling Heterogeneous Devices

One of the critical challenges that need to be addressed by IoT middleware is how to make heterogeneous devices talk to each other, i.e. to support interoperability. Interoperability is the ability of the middleware to support prominent wired and wireless protocols currently existing in the technology world [12]. The middleware should also provide a mechanism to integrate new protocols that will be available in future.

In analyzing the four cloud based platform, interoperability is achieved by converting the data into REST API format i.e. at every device end the developer needs to convert the device data into a server known format. The problem with this approach is that if there is thousands of devices, it is impractical to write code for every device. Fig. 2, shows the conversion diagram.

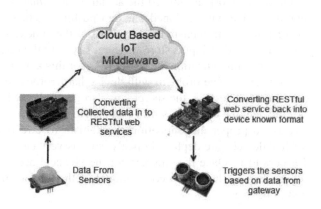

Fig. 2. Interoperability issue with cloud Based IoT platform

In case of backbone based Hydra, if the developer needs to plug in his/her device into an existing network he/she should encode data in the Hydra compatible SOAP (Simple Object Access Protocol) format.   For this the device need to be Hydra enabled by embedding the enabler code snippet onto participating devices.

5.2   Abstraction

Abstraction deals with hiding the implementation details to the developer. An ideal middleware for an intelligent environment such as the IoT should provide abstractions at various levels, such as heterogeneous input and output hardware devices, hardware and software interfaces, data streams, physicality and the development process. In cloud based platform the device data need to be converted in to server known format. It incurs extra level of coding from the developer. In case of Hydra, if the user wants his or her device to communicate, he/she needs need to write the necessary functionality to the Linksmart module. Comparing cloud based IoT platform and IoT middleware framework abstraction is very less in the latter one. In order to achieve true abstraction then any participating device should not have to have any firmware dependency to convert the data to a standard format as expected by the middleware. Instead it should be able to transmit the data in its native format and it is up to the middleware or intermediate intelligent gateway devices to convert them into a middleware compatible format. This in a way will hide all the complexities from the middleware users/developers and will provide easy means to interface devices to the middleware.

5.3   Device Authentication

Any device that need to join an IoT network should first get Authenticated and establish trust. Trust of the device is considered as a critical factor for any kind of Cyber Physical Systems [13]. Trust is a combination of many characteristics, mainly reliability, safety, security, privacy and usability. Such an authentication step will ensure that the devices joining the network are indeed the actual device what it claims to be. Replacing the genuine device with a cloned one and pushing malicious and erroneous data to server is one of the mechanism used by most of the attackers to create confusion. Usually the firmware developers stores a secret key onto the non-volatile memory of the device. During authentication the server asks for this key and the client sends its Secret key to the server. The problem with device hard corded secrete key is that, when an attacker creates a replica, the secret key also get replicated and hence could create a clone of the genuine device. As far as the above mentioned IoT platforms are concerned, there is no proper authentication mechanism to effectively identify the device. Most of IoT devices are deployed openly, so it is very easy for an attacker to get a physical access to it. Once the attacker got the physical access he/she will use two methodologies. Fig. 3, shows two conventional methodologies used by attackers to break the trust of devices.

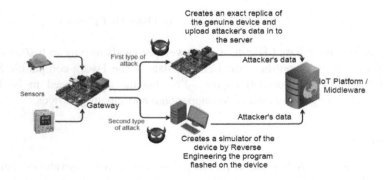

Fig. 3. Methodologies used by an attacker to break device trust

The first methodology is that the attacker will find a similar kind of device. With the help of hardware debugging tools like "AVR Dragon", the attacker will extract the firmware program of genuine device and flash it into new devices. This extraction process causes the secret key also to be copied into new device and the cloned device will be successfully authenticated by IoT Platform/Middleware.

In second methodology, the attacker will extract the firmware program as is done in the first case and reverse engineering the obtained hexadecimal code to get the secret key. Once the secret key is obtained the attacker can write his/her own http request based on the platform's encoding style and successfully send the malformed data to the platform. This methodology requires good expertise from the attacker's side. Fig. 4, shows malformed data successfully uploaded by an attacker using a simulator.

Fig. 4. Attacker's data successfully uploaded at server

5.4 Security and Privacy

Security of IoT devices and the data generated is very critical [14] as most of the IoT application have direct impact on human lives especially medical and vehicular applications. In cloud based IoT platform the device is communicating directly with the server, hence all the security issues applicable to any network device is applicable to IoT device also. Some of the frequent attacks are Man In The Middle attack, DOS (Denial Of Service), DDOS (Distributed Denial Of Service), eavesdropping etc... Privacy of device can be classified into communication privacy and physical privacy. Communication privacy can be achieved by cryptographic encryption, but physical privacy is difficult to achieve as anyone can bring his, her devices near the genuine device and can grab the information. It is also possible to get physical access to most of the devices and hence can be cloned by an attacker to inject malicious values to create confusion. It is even trickier when the device that need to be authenticated is in motion at all times. How is it possible for intelligent devices to form a secure network to share authentication information among themselves without flooding the network so that traffic to the middleware can be minimized? None of the platforms mentioned in this paper will address these security concerns.

5.5   Storage Protection

There is a direct relationship between data created from IoT enabled devices and data Protection. Most, if not all, of this data can never be recreated; an image and soil sample from last year, will never be the same as it was on the day it was collected. Therefore, data protection is potentially even more critical than it is for more conventional data. In case of cloud based platforms the data of the user are submitted to the server and it is out of the control of the user. Every platform provides third party application integration exposing the data to even larger spectrum of risk factors. The data collected by the servers are stored in plain text, so anyone who compromise the server can easily read the data stored there.

5.6   Prototyping Device Limitation

Device trust, security measures, interoperability are all very much dependent on the prototyping device used. If the device used is intelligent then complex security algorithms can be executed on them to achieve cryptography. With the help of PKI (Public Key Infrastructure) software level trust can be achieved. Most of the devices used in IoT are dump devices, with constrained resources such as low onboard RAM, low CPU cycles, hence less processing power, and to a large extend are battery powered and hence the processing intensity should be minimized to save power. In case of Hydra middleware, it is not possible for the user to interface a proprietary device as the devices need to have the Linksmart component built into it so as to Hydra enable the device. Other alternative is to use Hydra's own proprietary devices, if those suits the user needs.

5.7   Lack of Good Programming Abstractions for Concurrent Processing of Data Streams

Internet of Things are continually generating event data from embedded sensors, including producing real-time data streams. In order to take advantage of this scenario, these events must be concurrently processed by applications running in computing systems ranging from embedded to server systems. There is a lack of fundamental research and development in proper programming abstractions for such systems. Good programming abstractions would allow us to easily take advantage of true concurrency offered by multi-cores for concurrent data processing. None of the platforms support or provide such functionality.

6   Analysis

Fig. 5, shows the analysis chart of four cloud based IoT platforms and one IoT framework. The analysis shows that cloud based IoT platforms consistently fail across all problem matrix parameters. From a security perspective such platforms are highly vulnerable and hence is rated highly unreliable in terms of monitoring critical parameters. In case of IoT framework like Hydra, since using a separate network the security issues are less. But security is almost only confined within Hydra enabled Linksmart devices. Device trust is paramount problem in both cloud based and backbone network based, but in Hydra, software level trust is achieved through PKI (Public Key Infrastructure). Every cloud based platform stores data in plain text, and there is no encryption of data stored. Only physical protection of data is ensured. In Hydra it is the responsibility of the developer to maintain a secure storage, which makes it more complex. In cloud based IoT platforms, devices can be registered with few mouse clicks, but in Hydra the user need to use Linksmart (Hydra) compatible devices as he or she needs to write entire protocols and frameworks to achieve interoperability and device communication. Generally the complexity of programming is very high in backbone network based IoT middleware.

	Carriots	Xively	ThinkSpeak	RunMyProcess	Hydra
ABSTRACTION	✗	✗	✗	✗	✗
INTEROPERABILITY	✗	✗	✗	✗	✗
SECURITY & PRIVACY	✓	✓	✓	✓	✓
TRUST	✗	✗	✗	✗	✓
STORAGE SECURITY	✗	✗	✗	✗	✗
HARDWARE LIMITATION	✓	✓	✓	✓	✓
ABSTRACTION FOR CONCURRENT PROCESSING OF DATA STREAMS	✗	✗	✗	✗	✗

Fig. 5. Analysis of Different IoT platforms and IoT Framework

It is not a practical solution to define a new unified standard, protocol or communication mechanism in order to streamline the overall roll out of IoT solution across all types of devices. Such ambiguities should be something that need to be tacked at the level of middleware and other supporting intelligent devices. But security aspect is something that poses the highest threat to its acceptance, stability and sustenance because of the application criticality of the devices in their respective domains. Security is one critical area where there need to be a collaborative effort where in which a standardized process need to be defined across all application domains that can scale as per the processing capability and energy efficiency of the device. Based on the capability of a device security solutions ranging from simple to complex can be embedded within the device; this can be from light to heavy weight cryptographic algorithms, authentication techniques, non-replicable device identify etc. Such a process can only be achieved using a combination of hardware and software solutions where in which a device will have an unique way of identifying itself, which will also be fool proof against any cloning attempts. If able to incorporate such a mechanism will then ensure that any device that is communication enabled should be able to identify/authenticate itself to an intelligent Gateway device or middleware at any point of time so as to establish a trustworthy, secure communication channel.

7   Conclusion

Today IoT is gaining so much importance in the world of Information and Communication Technology. It is being used in every sectors like business, agriculture, vehicle, medicine, robotics and the list is endless. Wherever there is data, there is also dire necessity in collecting and processing the same in order to carve out meaning information from the same. Because of the operational, behavioral and the positioning (location where a device is being placed) aspect of all participating devices, it is even more challenging in defining an optimal solution for the integrating all devices in an IoT environment. Yet another worries to add on is the enormous amount of data that is being generated erratically by these devices. Data so generated need to be securely transmitted, processed and should be kept out of the reach of attackers. The challenge of harnessing these huge amounts of data is the main problem to be tackled by all the technologies related to IoT.

This paper will serve as a roadmap for IoT solution architects in designing and developing a middleware that will address the core challenges in this arena, thereby coming up with a Secure, Scalable and Interoperable IoT middleware.

References

1. IERC, About IoT,
   http://www.internet-of-things-research.eu/about_iot.htm
2. Atzori, L., Iera, A., Morabito, G.: The Internet of Things: A survey. Computer Networks 54, 2787–2805 (2010)
3. Chaqfeh, M.A., Mohamed, N.: Challenges in middleware solutions for the internet of things. In: 2012 International Conference on Collaboration Technologies and Systems (CTS), May 21-25, pp. 21–26 (2012)
4. Bandyopadhyay, S., Sengupta, M., Maiti, S., Dutta, S.: Role of Middleware for Internet of things: A Study. International Journal of Computer Science & Engineering Survey (IJCSES) 2(3) (August 2011), doi:10.5121/ijcses.2011.230794
5. Mayer, C.P.: Security and Privacy Challenges in the Internet of Things. In: Workshops der Wissenschaftlichen Konferenz Kommunikation in Verteilten Systemen, WowKiVS 2009 (2009)
6. Castro, M., Jara, A.J., Skarmeta, A.F.: An Analysis of M2M Platforms: Challenges and Opportunities for the Internet of Things. In: 2012 Sixth International Conference on Innovative Mobile and Internet Services in Ubiquitous Computing (IMIS), July 4-6, pp. 757–762 (2012)
7. Carriots: M2M Cloud Based IoT platform,
   https://www.carriots.com/documentation/
   carriots_ecosystem#introduction
8. Xively: Public cloud, https://xively.com/dev/docs/api/
9. ThingSpeak: Open source platform,
   http://community.thingspeak.com/documentation/api/
10. RunMyProcess: Enterprise platform, https://www.runmyprocess.com
11. Hydra: A network embedded middleware,
    http://www.hydramiddleware.eu/articles.php?article_id=68

12. Kominers, P.: Interoperability Case Study: Internet of Things (IoT) (April 1, 2012), Berkman Center Research Publication No. 2012-10. Available at SSRN:
    http://ssrn.com/abstract=2046984
    or http://dx.doi.org/10.2139/ssrn.2046984
13. Sha, L., Gopalakrishnan, S., Liu, X., Wang, Q.: Cyber-Physical Systems: A New Frontier. In: IEEE International Conference on Sensor Networks, Ubiquitous and Trustworthy Computing, SUTC 2008, June 11-13, pp. 1–9 (2008)
14. Weber, R.H.: Internet of Things – New security and privacy challenges. Computer Law & Security Review 26(1), 23–30 (2010),
    http://dx.doi.org/10.1016/j.clsr.2009.11.008 ISSN 0267-3649

An Improved Image Steganography Method with SPIHT and Arithmetic Coding

Lekha S. Nair and Lakshmi M. Joshy

Computer science and Engineering, Amrita School of Engineering,
Amrita Vishwa Vidyapeetham, Clappana P.O., Kollam, India
lekhaas@gmail.com,
lakshmijoshy@gmail.com

Abstract. The paper proposes a Steganography scheme which focuses on enhancing the embedding efficiency. There are only limited ways on which one can alter the cover image contents. So, for reaching a high embedding capacity, in the proposed method , the data is compressed using SPIHT algorithm and Arithmetic Coding. After which the information is embedded into the cover medium . The proposed method suggests an efficient strategy for hiding an image into a cover image of same size without much distortion and could be retrieved back successfully. The advantage of the system is that the cover medium size is reduced to the same size of the input image where in normal cases it is twice or even more. Also the cover image could be recovered from the original stego-image.

Keywords: DWT, SPIHT ALGORITHM, ARITHMETIC Coding, Rhombus Prediction Scheme.

1 Introduction

Digital steganography is the art of discretely hiding data within data. In general Steganography aims to hide data well enough that the unintended recipients do not suspect the presence of hidden data in the steganographic medium . To implement this, steganography algorithm embeds secret information into cover mediums, that includes text files, audio files ,image files and video files.

The information could be concealed in different ways such as encoding every single bit of the cover region or selecting random embedding regions to hide data or randomly scattering it through out the cover medium. General tendency is that, noisy region are preferred to embed secret message since it draws less attention.

A steganographic technique must possess two important properties: good imperceptibility and sufficient embedding capacity. The first property ensures that the embedded messages are not easily detectable , and the second explains its efficiency in communication, despite that it only alters the most insignificant components.

© Springer International Publishing Switzerland 2015   97
S.C. Satapathy et al. (eds.), *Proc. of the 3rd Int. Conf. on Front. of Intell. Comput. (FICTA) 2014*
– *Vol. 2*, Advances in Intelligent Systems and Computing 328, DOI: 10.1007/978-3-319-12012-6_11

2   Related Work

The design of a Steganographic system can be categorized into spatial domain techniques[6][7]where the data embedding processes are directly employed to the image pixel values. The major advantages of these techniques is that it is much simpler and computationally fast. But, the main drawback is that it can add visual distortion to the system,the signal to noise ratio is reduced. In frequency domain the input image is decomposed into multi scale coefficients initially[8]. Then data is embedded to these coefficients depending on their different characteristics. The main advantage of this method is that the signal to noise ratio is higher , it has a high ability to withstand noises. But this method is very complex and hence very slow.

C.K. Chan, L.M. Cheng [6] used LSB of their cover image to hide their data.It is the most conventional method used which is much simpler but has larger impact. Here the secret message is encoded at the least significant bit level which are selected randomly or through the aid of embedding algorithm. The LSB insertion method is used for first LSB to fourth LSBs.But for the four LSB technique it adds more visual distortion for the cover image. So we could conclude that it exhibits a trade off between payload and visual distortion. The drawback is that it requires a much larger size cover image so as to conceal the entire secret message.

H. C. Wu, Wu, Tsai, and Hwang [7]proposed a Steganographic scheme that combines both pixel-value differencing and LSB substitution to embed secret data into still images . Their scheme embeds more secret data into edged areas in the cover image and has a better image quality by using pixel-value differencing (PVD) method alone. To increase the capacity, the secret data is hidden in the smooth areas by using an LSB method with the edged areas still using the PVD method. The experimental results reveal that the proposed method results in stego-images with an better quality and provide a large embedded secret data capacity . The drawback is that both LSB and PVD approaches can be easily detected.

LSB method has a very high impact in steganography but it didn't succeed much in deceiving HVS. The drawback of this is that every bit is used for embedding information so if a lossy compression is used then, the data could not be regenerated back. LSB method with high payload often resulted in visual distortion which could invite different attack to the system. So, Steganography was extended to Frequency level.

The main advantage of frequency domain technique is that it hides information into areas that are less exposed to compression ,cropping and image processing. F5 algorithm[8] was introduced by Westfeld. It is based on subtraction and matrix encoding (also known as syndrome coding). F5 embeds only into non-zero AC DCT coefficients by decreasing the absolute value of the coefficient by 1. A shrinkage occurs, when the same bit has to be re-embedded in case the original coefficient is either 1 or -1 as at the decoding phase all zero coefficients will be skipped whether they were modified or not.

3   Solution Approach

The main objective of the proposed system is to hide information image into a cover image of same size as that of the secret image. The paper mainly focuses on gaining an efficient embedding capacity. Mostly, in different steganographic approaches it is to be noted that the cover images appear in larger sizes. Most often the cover images seems to have twice or four times the size of input secret images. Also it is to be highlighted that there exists a trade off between payload and HVS distortion. For a high payload the cover medium should be effectively larger to accommodate the whole data. The distortion caused by this could be a honey pot for attackers, which results in tampering the data and there by destroying the whole system. So, proposed method aids to seal the secret data into a cover image of same size however big the secret data is. For attaining this, priorly the data need to be compressed using SPIHT algorithm[2] and arithmetic coding[1][3]. Then the compressed bit stream could be hidden to the cover image using the Rhombus Prediction scheme[4].

3.1   Data Compression

The paper focuses on optimizing the embedding efficiency. Normally cover size required to embed the information is much larger when compared with that of input secret image. The proposed methods aims to reduce the cover image size without settling for the embedding capacity. The input image is compressed to reduce the redundancy of image data and is embedded efficiently.

SPIHT Algorithm. SPIHT image compression [2]is one of the efficient lossless image compression algorithm because it guarantees the highest image quality ,it gives high PSNR values when compared with other traditional methods. Also considering the computational level it gives out the best of it by fast coding and decoding.

The input image is decomposed using DWT transformation and corresponding DWT coefficients obtained are compressed using SPIHT algorithm. To obtain the significant information the practical implementation requires three list LIS(List of Insignificant Set),LIP(List of Insignificant Pixel) and LSP(List Of Significant Pixel). During the sorting pass , the coefficients in the LIP are checked for its significance using significant function. The significant coefficients are maintained in the LSP and corresponding bit stream is generated considering its sign and its descendants. This is performed for all the coefficients. The threshold value is decreased for the next pass and the coefficients which are insignificant in the previous pass are tested, and those that become significant are moved to the LSP. Similarly, sets are sequentially evaluated following the LIS order, and when a set is found to be significant it is removed from the list and partitioned. The new subsets with more than one element are added back to the LIS, while the single-coordinate sets are added to the end of the LIP or the LSP, depending whether they are insignificant or significant, respectively. The LSP contains the coordinates of the coefficients that are visited in the refinement pass.

$$S_n(\tau) = \begin{cases} 1, max_{(i,j)\epsilon\tau}\{| \; C_{i,j} \; |\} \geq 2^n \\ 0, otherwise \end{cases}$$

The main advantage of SPIHT is that it is fully progressive. The image's PSNR will be directly related to the amount of the file received from the transmitter. This means that the image quality will only increase with the percentage of the file received. After the SPIHT transformation some regularities will exist in the file. These regularities may allow us to further compress the file. With this in focus we investigated the addition of arithmetic compression to a SPIHT encoded image.

Arithmetic Coding. Arithmetic coding[3] [4]is an entropy coding method which provides a better compression for the bit stream generated by the SPIHT algorithm . The bit stream obtained from SPIHT is converted into symbols by taking every two bits and encoding it into a symbol to efficiently code with the arithmetic coder. The frequency of each symbol is used for encoding using Arithmetic coder ,so higher the frequency of a symbol higher will be the compression level. Using SPIHT, the image is compressed and is extended to a desired level by combining with arithmetic coding. The bit stream generated in the SPIHT is converted into symbols an then Arithmetic coding is performed on to this and corresponding binary sequence is generated[9]

3.2   DATA HIDING

The main goal in data hiding is that it should not impair Human visual System and the distortion should be less.The data is embedded using Rhombus prediction Scheme.

Rhombus Prediction Scheme[5]. This method uses cell to embed data.It employs prediction errors to encode data into the cover image. The main idea behind this technique is efficient utilization of the correlation of neighbouring pixels.Difference values of the neighbouring pixels in a cell highly correlate with average values of neighbouring pixels. The pixels of the image are divided into

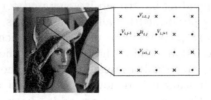

Fig. 1. Prediction pattern [5]. The pixel value u of the Cross set can be predicted by using the four neighbouring pixel values of the Dot set and expanded to hide one bit of data.

two sets: the Cross set and Dot set (see Fig. 1). The Cross set is used for embedding the data and Dot set for computing the predictors. So,this scheme will be called the Cross embedding scheme.

Here, a rhombus pattern of prediction scheme is used with four neighbouring pixels (i.e., $V_{i,j1}$, $V_{i+1,j}$, $V_{i,j+1}$, and $V_{i1,j}$).The pixel in the centre ,u_{ij},of the rhombus is used to hide the data.The centre pixel can be predicted using the neighbouring pixels.
Predicted value ,

$$u'_{i,j} = Average(V_{i,j1}, V_{i+1,j}, V_{i,j+1}, V_{i1,j})). \tag{1}$$

Then, a prediction error,d_{ij}, is calculated.The prediction error,dij, increases the embedding efficiency since the retrieval of the data is difficult without knowing the error value.

$$d_{i,j} = u_{i,j} - u'_{i,j} \tag{2}$$

This prediction error can be expanded to hide the information. The prediction error after expansion is called Modified prediction error,D_{ij}.

$$D_{ij} = 2 * d_{i,j} \tag{3}$$

By expansion, the error value is modified to an even value so that the LSB is zero and it eases the inclusion of the data bit. The data hiding is done in the Modified prediction error.

$$D_{ij} = 2d_{i,j} + b \tag{4}$$

After embedding the data bit into D_{ij},the original pixel value $u_{i,j}$is changed to $U_{i,j}$ as

$$U_{i,j} = u'_{i,j} + D_{ij} \tag{5}$$

The decoding processes of the scheme is reverse of the encoding process. During data hiding,the neighbouring pixels set are not modified, so the predicted values $u'_{i,j}$ are also not changed. Using the predicted value $u'_{i,j}$ and the modified pixel value $U_{i,j}$the decoder can exactly recover the embedded bit and original pixel value.

Modified Prediction Error is computed as

$$D_{ij} = U_{i,j} - u'_{i,j} \tag{6}$$

The embedded bit value is computed as

$$b = D_{ij} mod2. \tag{7}$$

The original prediction error is computed as

$$d_{ij} = D_{ij}/2 \tag{8}$$

The original pixel's value is computed as

$$u_{i,j} = u'_{i,j} + d_{ij} \tag{9}$$

Since the error term depends on the neighbouring pixel values ,the data could be accurately retrieved. The main advantage of this method is ,there is no quality trade off between payload and high embedding efficiency. The data could be retrieved accurately without distorting the cover medium. So, the secret message and the cover medium is reconstructed.

The first row and the first column of the image is not used in the Rhombus prediction scheme. So the information required for the decoding the image is embedded in the first row of the image. The data is hidden using the classic LSB method which includes maximum bits,frequencies of the symbols used for arithmetic coding and the level of the decomposition for DWT transform.

4   Decoding

In decoding the data is retrieved back successfully. The decoding process involves the reverse process of encoding. The decoding of the Rhombus prediction is done first. The so obtained bit stream is then decoded using Arithmetic Decoding and SPIHT. The result of this will be coefficients of DWT then inverse of this Inverse DWT is performed and the image is efficiently retrieved.

5   Performance Evaluation and Experimental Results

Gray scale images with variable sizes were used to validate the proposal. The system was implemented on Matlab in windows platform. Different comparison metrics are used to analyse the performance of the system. The most important metric used is PSNR to obtain signal to noise ratio.It is found that PSNR values are much higher.

The method is carried out on 200X200, 256X256,512x512 and 1024x1024 pixel images and similar size cover size images are taken for data embedding.

Same input image is used and made into different sizes to match the cover image size.Steg images where seen with less distortion.

Table 1. PSNR and MSE for different size images

Input Image in Pixels	PSNR	MSE
Image200x200	32.12	0.05
Image256x256	34.54	0.032
Image512x512	33.76	0.041
Image1024x1024	38.24	0.029

Fig. 2.   a)Secret Image b)Coverimage of 256x256 c)Steg Image of 256x256
d)Coverimage of 512x512 e) Steg Image of 512x512 f) Coverimage of 1024x1024 g)Steg
Image of 1024x1024

The PSNR values and MSE values of cover and Steg image were compared.
Higher the PSNR, closer the distorted image is to the original. In general, a
higher PSNR value should correlate to a higher quality image.

Table 2. Input Image and its compressed bit rates in different stage

Input Image in kilo bits	After SPIHT Compression	After SPIHT And Arithmetic Coding
Image200x200 (320kb)	40kb	32kb
Image256x256(524kb)	65kb	56kb
Image512x512(2097kb)	262kb	168kb
Image1024x1024(8388kb)	1048kb	1038kb

The Histogram of Cover image and Steg image is shown for 512x512 pixel
image. Both the histogram images appears similar , and it is evident that the
Steg image is more closer to the cover image. Similarly the images (256x256
,1024x1024) shows similar histograms for cover and Steg images.

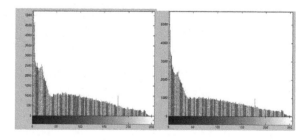

Fig. 3. Histogram of Cover image and steg image(512x512 pixels)

6   Conclusion

The paper proposed a new Steganographic scheme to hide an image into a same sized cover image. The image is compressed into desired level using SPIHT and Arithmetic Coding and is then hidden to the cover image using Rhombus Prediction scheme. Different experiments were conducted on the different size images to validate the proposal. Image quality was retained with high PSNR values. Similarity in the Histograms of the Cover and Steg images confirms the closeness of these images. This abides by the objective of our study on Image Steganography.

References

1. Liu, K., Belyaev, E., Guo, J.: VLSI Architecture of Arithmetic Coder Used in SPIHT. IEEE Transactions on Very Large Scale Integration (VLSI) Systems 20(4), 697–710 (2012)
2. Said, A., Pearlman, W.A.: A New, Fast and Efficient Image Codec Based on Set Partitioning in Hierarchial trees. IEEE Transactions on Circuits and Systems for Video Technology 6(3), 243–249 (1996)
3. Langdon Jr., G.G.: An introduction to arithmetic coding. IBM Journal of Research and Development 28(2), 135–149 (1984)
4. Witten, I.H., Neal, R.M., Cleary, J.G.: Arithmetic coding for data compression. Communications of the ACM 30(6), 520–540 (1987)
5. Sachnev, V., Kim, H.J., Nam, J., Suresh, S., Shi, Y.Q.: Reversible Watermarking Algorithm Using Sorting and Prediction. IEEE Transactions on Circuits and Systems for Video Technology 19(7), 989–999 (2009)
6. Chan, C.K., Cheng, L.M.: Hiding data in images by simple LSB substitution. Pattern Recognition 37(3), 469–474 (2004)
7. Wu, H.C., Wu, N.I., Tsai, C.S., Hwang, M.S.: Image steganographic scheme based on pixel-value differencing and LSB replacement methods. IEE Proceedings Vision, Image and Signal Processing 152(2), 611–615 (2005)
8. Westfeld, A.: F5-A steganographic algorithm. In: Moskowitz, I.S. (ed.) IH 2001. LNCS, vol. 2137, pp. 289–302. Springer, Heidelberg (2001)
9. Introduction to Data Compression Khalid Sayood

Image Steganography – Least Significant Bit with Multiple Progressions

Savita Goel[1], Shilpi Gupta[2], and Nisha Kaushik[3]

[1] Computer Service Centre, IIT Delhi, India
[2] Department of Computer Science & Engineering,
ASET, Amity University, Noida, India
[3] Department of Computer Science & Engineering
Satyam College of Engineering, Ghaziabad, India
savita@cc.iitd.ac.in, sgupta5@amity.edu,
neishakaushik24@gmail.com

Abstract. In this paper we have proposed a new technique of hiding a message on least significant bit of a cover image using different progressions. We have also compared the cover image and stego images with histograms and computed the CPU Time, Mean Square Error (MSE), Peak Signal Noise Ratio (PSNR), Structural Similarity (SSIM) index and Feature similarity index measure (FSIM) of all images with our method and earlier approaches and make an empirical study. Experimental results concluded that our method is more efficient and fast as compared to classical LSB and LSB using prime.

Keywords: Data Hiding, Steganography, Least Significant Bit (LSB), Cover Image, Stego Image.

1   Introduction

We are living in an internet era, where in every field massive amount of digital information is generated and shared in the form of text, images, audio, video etc., it becomes essential that confidential data must be shared in such a way that information can only be retrieved by authentic recipient. Data hiding become a hot topic of research due to increase in demand of security. There are four popular techniques under the umbrella of Data hiding- Cryptography, Steganography, Digital Watermarking.

Cryptography only scrambles the message from plain text to cipher text which can't be easily understood. A receiver can retrieve the original text back only by using valid key [17]. There is a limitation of cryptography that it can only work on one kind of digital media i.e., text. If intruder knows about the key then message can be easily decrypted. Steganography is a technique of hiding any secret information within another digital media. The word Steganography is derived from Greek word which means" Concealed Writing" [2] .The importance of steganography lies in the fact that it hides the existence of the secret data that are intended to be protected. Digital watermarking [1, 24] is a method to embed information to any digital data using

© Springer International Publishing Switzerland 2015
S.C. Satapathy et al. (eds.), *Proc. of the 3rd Int. Conf. on Front. of Intell. Comput. (FICTA) 2014*
– *Vol. 2*, Advances in Intelligent Systems and Computing 328, DOI: 10.1007/978-3-319-12012-6_12

steganography principles in such a way that visible or invisible data remain integral. This is used to reinforce copyright laws of digital data.

Over the last decade researchers have proposed many theories and applications of steganography in text, IP Datagram, image, audio and video [3, 4]. In most steganography [10] techniques data is first broken down into smaller bits and they are embedded into suitable locations of digital artifact in such a way that there is no major difference between the original image and the modified image. Only the legitimate person knows about the presence of data. Data embedding methods has been broadly categorized in spatial domain and frequency/transform domain. Main components of image steganography are:

- Cover image (c) in which message has to be embedded.
- Secret message (m) which the sender intends to hide.
- Steganography techniques
- Stego image (s) image after embedding the secret message.

In this paper, we have focused on embedding a message in an image in a spatial domain. Least Significant Bit (LSB) algorithm is the most popular approach in this domain. We have proposed a modified LSB by using the concept of Progressions. Results of proposed approach have been compared with various variations of LSB.

Rest of the paper is organized as follows. In Section 2 we have discussed related work in spatial domain. Section 3 provides brief overview of proposed scheme. This is followed by empirical study of proposed approach and Classical LSB methods. In Section 5 concluding remarks and future scope has been discussed.

2   Related Work

Abbas et al [20] have given a detailed survey of image steganography. In this section a brief overview of some popular methods of image steganography in spatial domain has been presented.

The most popular spatial domain based steganography technique use either the LSB or Bit Plane Complexity Segmentation (BPCS) algorithm [5]. LSB can be further subdivided in two fields (1) Bit Substitution [7, 11, 13] (2) Bit Distortion. In LSB based Techniques LSBs of the cover file are directly changed with message bits. Manipulation in the Least Significant Bits takes advantage of fact that human eye does not perceive the difference between small changes in the color of pixel. If a pixel was represented by grey scale value 233 and it is changed to 232 or 234 then human visual system will not be able to recognize this change. There are a few drawbacks of LSB techniques because this is extremely sensitive to scaling, rotation, cropping, addition of noise, or lossy compression. Due to these manipulations on stego image, attackers can easily destroy the message [14]. Still LSB techniques are widely used because of its simplicity. BPCS [21, 23] Steganography uses multiple bit-planes, and so can embed a much higher amount of data, though this is dependent on the individual image.

Fridrich et al [6] claimed that changes as small as flipping the LSB of one pixel in a JPEG image can be reliably detected. Avcibas et al [8] used binary similarity measures for steganography. Fridrich et al [9] also had done steganalysis on JPEG images. Tsai et al. [12] divide the image into blocks of 5×5 where the residual image is calculated using linear prediction then the secret data is embedded into the residual values, followed by block reconstruction. Chin chen et al [15] used run length encoding (RLE) and proposed 2 approaches named as bitmap file by run length (BRL) and general file by run length (GRL). They used 2 consecutive pixels of cover image for hiding data. Xin Liao et al [19] used 3 bit common LSB substitution method and attempted to hide message on edge pixels of cover image. Hassan et al selected corner, center and quadrant wise pixel both in clock and anti-clock wise direction to hide message in colored images.

Battisti et al. [16] proposed a method of embedding data into digital media by decomposition of Fibonacci number sequence which allowed different bit plane decomposition when compared to the classical LSB scheme. LSB data hiding using prime numbers is a data hiding technique proposed by Dey et.al [18]. The main idea of the work was to use the prime number decomposition and generate new set of bit planes and embed information in these newly generated bit planes with minimal distortion. Dey et.al also proposed data hiding by decomposition of a pixel value in sum of natural numbers. By this decomposition technique more number of planes can be generated. Ghazanfari et al [22] proposed LSB+ and an improved approach named as LSB++.

3   Proposed Approach

An **Arithmetic progression** (AP) is a sequence of numbers such that the common difference between the consecutive terms is constant. If the initial term of an arithmetic progression is a_1 and the common difference of successive members is d, then the nth term of the sequence (a_n) is given by:

$$a_n = a_1 + (n-1)d,$$

and in general

$$a_n = a_m + (n-m)d.$$

The behavior of the arithmetic progression depends on the common difference d. If the common difference is:

Positive, the members (terms) will grow towards positive infinity. Negative, the members (terms) will grow towards negative infinity.

Geometric progression is a sequence of numbers in which each number is obtained from the previous number by multiplying by a constant

The nth term of a geometric sequence with initial value a and common ratio r is given by $a_n = ar^{n-1}$.

A LSB Using Progression-Based Embedding Algorithm
Input: Choose the cover image (c), Secret message (m), Key (d/r).

Find least significant bits of each gray pixel from cover image
This goes to each byte, if the least significant bit is not the bit of the message position, flip it, else do nothing
Apply a progression scheme on height and width to get the position.
Output: Stego image(s)

A LSB-Based Extracting Algorithm
Input -: key (d/r), Secret image(s)

The extraction algorithm is exactly the reverse of the embedding algorithm. From the Stego-image(s), each pixel with embedded data bit is converted to its corresponding progression decomposition .the secret message bit is extracted from the bit-plane. Later, all the bits are combined to get the secret message.

4   Results and Analysis

As input the 8bit gray scales "Lena "image in different image format is used. We have embedded a message of length 1400 bits and it can vary according to the size of the image. Later we ask for a numeric key which is common difference. The secret bit is hidden in common bit planes by using our progression approach. In this we have use CPU time and also applied standard measures to calculate the quality of images as listed below-

CPU Time: as mentioned in Table 1 we have compared our progression approach with the classical LSB and then calculate the CPU time and results shows that our approach takes less time as compared to classical LSB.
Perceptual Quality: progression approach provides high perceptual quality as the difference between the cover image & stego image is unnoticed by human eye.
Peak Signal Noise Ratio (PSNR) & Mean Square Error(MSE) :- Higher the value of PSNR, lesser the value of MSE [25] and better the quality of the stego image.

$$PSNR = 10 \log_{10} \left(\frac{R^2}{MSE} \right)$$

Histogram: It is a measure of the number of amount of pixels with respect to particular pixel value in Figure 2 we have compared the histograms of cover image with stego image using different approaches and analyze that there is slight difference between the histograms.
Ratio of Squared Norms (L2Rat): It is again quality metrics which determines the ratio of the squared norm of the signal or image approximation   to the input signal or images.
Structural Similarity Index Measure (SSIM): SSIM [26] considers image degradation as perceived change in structural information.it is designed to improve on methods like PSNR & MSE.
Feature Similarity Index Measure (FSIM): It *is* proposed based on the fact that human visual system understands an image mainly according to its low-level features.

From the Table 1 we can conclude that PSNR is high for LSB using progressions. Figure 1 shows that our method has less perceptual distortion in terms of structural and feature based similarity. Cover image, stego image, their histograms and structural similarity image before and after embedding message has been shown in Figure 2.

Table 1. Empirical Analysis of Quality Measures

Image Type	Steganography Methods	PSNR	MSE	CPU Time	SSIM	FSIM	L2RAT
BMP	Classical LSB	52.72	0.3477	1.482	0.9996	0.9996	1.0016
	LSB using Prime	53.44	0.2942	1.388	0.9996	0.9996	0.9998
	LSB using A.P	68.69	0.0088	0.8112	1.0000	1.0000	1.0001
	LSB using G.P	61.96	0.0413	1.0296	1.0000	0.9999	1.0002
GIF	Classical LSB	52.82	0.3398	0.7956	0.9998	0.9998	1.0009
	LSB using Prime	62.26	0.0386	1.0920	1.0000	1.0000	0.9999
	LSB using A.P	68.69	0.0088	0.7956	1.0000	1.0000	1.0000
	LSB using G.P	61.98	0.0413	0.78	1.0000	1.0000	1.0002
JPEG	Classical LSB	70.87	0.0053	2.121	1.0000	1.0000	1.0000
	LSB using Prime	70.90	0.0053	6.942	1.0000	1.0000	1.0000
	LSB using A.P	70.85	0.0053	1.5288	1.0000	1.0000	1.0000
	LSB using G.P	70.86	0.0053	1.3572	1.0000	1.0000	1.0000

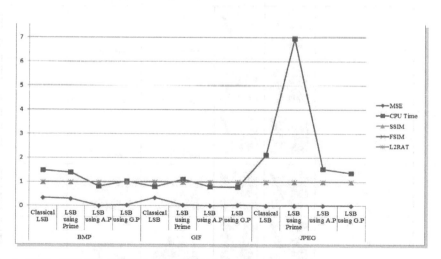

Fig. 1. MSE, CPU Time, SSIM, FSIM, L2RAT vs Steganography methods for different file formats

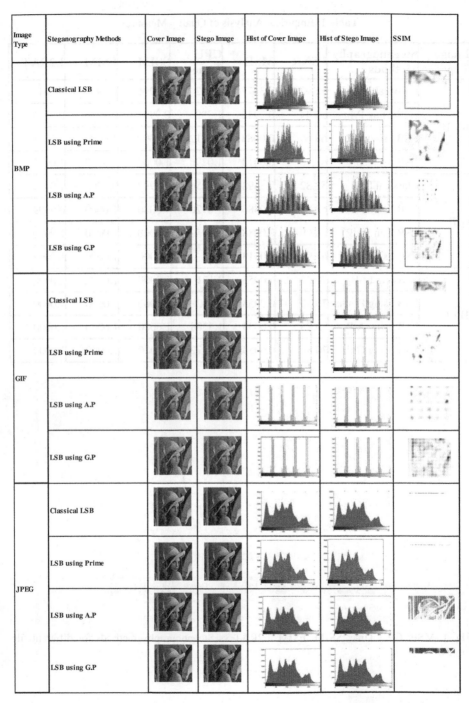

Fig. 2. Comparison of Results obtained before and after embedding a message

5   Conclusion

In this paper we have presented two improved approaches over classical LSB by using Progressions. Results obtained by proposed approach are better in comparison of classical LSB. Although LSB based techniques are not robust against various attacks but still provide an easy way to embed large amount of data, high PSNR and perceptual quality. Proposed approach is also independent of different file formats. Proposed approach is limited to grey scale images, in future we would like to extend the same approach for RGB images.

References

1. Wolfgang, R., Delp, E.: A watermark for digital images. In: Proceedings of the IEEE International Conference on Image Processing, pp. 219–222 (1996)
2. Johnson, N.F., Jajodia, S.: Exploring steganography: seeing the unseen. IEEE Computer 31(2), 26–34 (1998)
3. Marvel, L.M., Retter, C.T.: A Methodology for Data Hiding Using Images. In: Proceedings of IEEE Military Communications Conference (MILCOM), pp. 1044–1047 (1998)
4. Anderson, R.J., Petitcolas, F.A.P.: On the Limits of Steganography. IEEE Journal of Selected Areas in Communications 16(4), 474–481 (1998)
5. Eason, R.O., Kawaguchi, E.: Principle and applications of bpcs-steganography. In: Proceedings of SPIE, vol. 3528, pp. 464–473 (1998)
6. Fridrich, J., Goljan, M., Du, R.: Reliable Detection of LSB Steganography in Grayscale and Color Images. In: Proceedings of ACM, Special Session on Multimedia Security and Watermarking, Ottawa, Canada, pp. 27–30 (2001)
7. Lin, C.F., Wang, R.Z., Lin, J.C.: Image hiding by optimal lsb substitution and genetic algorithm. Pattern Recognition 34, 671–683 (2001)
8. Avcibas, I., Memon, N., Sankur, B.: Image Steganalysis with Binary Similarity Measures. In: Proceedings of the International Conference on Image Processing, pp. 24–28 (2002)
9. Fridrich, J., Goljan, M., Hogeg, D.: Steganalysis of JPEG Images: Breaking the F5 Algorithm. In: Petitcolas, F.A.P. (ed.) IH 2002. LNCS, vol. 2578, pp. 310–323. Springer, Heidelberg (2003)
10. Provos, N., Honeyman, P.: Hide and seek: an introduction to steganography. IEEE Security and Privacy 1(3), 32–44 (2003)
11. Hsiao, J.Y., Chang, C.C., Chan, C.-S.: Finding optimal least-significant-bit substitution in image hiding by dynamic programming strategy. Pattern Recognition 36, 1583–1595 (2003)
12. Wu, D.C., Tsai, W.H.: A steganographic method for images by pixel-value differencing. Pattern Recognition Letters 24, 1613–1626 (2003)
13. Chan, C.K., Cheng, L.M.: Hiding data in images by simple lsb substitution. Pattern Recognition 37, 469–474 (2004)
14. Kong, X., Wang, Z., You, X.: Steganalysis of Palette Images: Attack Optimal Parity Assignment Algorithm. In: Proceedings of 5th IEEE International Conference on Information, Communications and Signal Processing, pp. 860–864 (2005)
15. Chang, C.C., Lin, C.Y., Wang, Y.Z.: New image steganographic methods using run-length approach. Information Sciences 176(22), 3393–3408 (2006)

16. Battisti, F., Carli, M., Neri, A., Egiaziarian, K.: A Generalized Fibonacci LSB Data Hiding Technique. In: 3rd International Conference on Computers and Devices for Communication (CODEC 2006) TEA, Institute of Radio Physics and Electronics, University of Calcutta, December 18-20 (2006)
17. Forouzan, A.: Cryptography and Network Security, 1st edn. McGraw-Hill, USA (2007)
18. Sandipan, D., Ajith, A., Sugata, S.: An LSB Data Hiding Technique Using Prime Numbers. In: The Third International Symposium on Information Assurance and Security, Manchester, UK. IEEE CS Press (2007)
19. Li, Z., Chen, X., Pan, X., Zeng, X.: Lossless data hiding scheme based on adjacent pixel difference. In: Proceedings of the International Conference on Computer Engineering and Technology, pp. 588–592 (2009)
20. Cheddad, A., et al.: Digital image steganography: Survey and analysis of current methods. Signal Processing 90(3), 727–752 (2010)
21. Bhattacharyya, S., Khan, A., Nandi, A., Dasmalakar, A., Roy, S., Sanyal, G.: Pixel mapping method (PMM) based bit plane complexity segmentation (BPCS) steganography. In: 2011 World Congress on Information and Communication Technologies (WICT), pp. 36–41 (2011)
22. Ghazanfari, K., Ghaemmaghami, S., Khosravi, S.R.: LSB++: An improvement to LSB+ steganography. In: IEEE Region Conference TENCON, pp. 364–368 (2011)
23. Bansod, S.P., Mane, V.M., Ragha, L.R.: Modified BPCS steganography using Hybrid cryptography for improving data embedding capacity. In: 2012 International Conference on Communication, Information & Computing Technology (ICCICT), pp. 1–6 (2012)
24. Kamble, M.V.: Information Hiding Technology- A Watermarking: Golden Research Thoughts (April 2012)
25. Wang, Z., Bovik, A.C.: Mean squared error: love it or leave it? - A new look at signal fidelity measures. IEEE Signal Processing Magazine 26(1), 98–117 (2009)
26. Wang, Z., Bovik, A.C., Sheikh, H.R., Simoncelli, E.P.: Image quality assessment: From error visibility to structural similarity. IEEE Transactions on Image Processing 13(4), 600–612 (2004)

On the Implementation of a Digital Watermarking Based on Phase Congruency

Abhishek Basu[1], Arindam Saha[1], Jeet Das[1], Sandipta Roy[1], Sushavan Mitra[1], Indranil Mal[2], and Subir Kumar Sarkar[3]

[1] RCC Institute of Information Technology, West Bengal, India-700015
[2] Acharya Prafulla Chandra College, West Bengal India-700131
[3] Jadavpur University, West Bengal, India-700032
idabhishek23@yahoo.com, indranil.mal@gmail.com,
su_sircir@yahoo.co.in

Abstract. In this paper a human visual system (HVS) model guided least significant bit (LSB) watermarking approach for copyright protection is proposed. The projected algorithm can embed more information into less featured surrounded areas within the host image determined by phase congruency. Phase congruency offers a dimensionless quantity which is an excellent measure feature points with high information and low in redundancy within an image. The region with fewer features indicates the most trivial visible aspects of an image, so alteration within these areas will be less noticeable to any viewer. Furthermore the algorithm will be tested by means of imperceptibility and robustness. Thus a new spatial domain image watermarking scheme will be projected with higher bit capacity.

Keywords: HVS, LSB, Copyright protection, Phase congruency, image watermarking.

1 Introduction

Requirement of effective tools for preventing duplication are supreme necessity in a world where there is an increasing amount of duplication and piracy going on. Various algorithms have been devised and implemented to address the concerned issues [1-2]. Digital Watermarking takes an important place in these regard and can serve as an effective tool for copyright protection of digital documents. Digital Watermarking is the process by which information is hided inside an object, such that the information can be extracted out later, in some unique ways, to make an assertion by the owner of the object [3]. Digital watermarking is an attractive area for researchers for last few decades [4-7]. However the principle requirements of watermarking are imperceptibility, robustness and hiding capacity and there are tradeoffs among them [8]. As result watermarking is still an open area for research [9-12].

The proposed scheme in this paper is a phase congruency based digital watermarking technique which employs the less prominent image feature regions for watermark hiding. Phase congruency presents a dimensionless quantity which helps in

© Springer International Publishing Switzerland 2015
S.C. Satapathy et al. (eds.), *Proc. of the 3rd Int. Conf. on Front. of Intell. Comput. (FICTA) 2014*
– *Vol. 2*, Advances in Intelligent Systems and Computing 328, DOI: 10.1007/978-3-319-12012-6_13

measuring the important feature points with high information (or entropy) and low in redundancy within an image [13]. The idea of Phase congruency provides a simple but realistic model of how HVS identify and distinguish features in an image.

In proposed technique, the bits of watermark are concealed in the areas which are least significant to the human visual system. These areas are marked out with the help of phase congruency. Thus the imperceptibility is better and since the information is hided more than once the robustness and data capacity are also high.

The rest of the paper is planned as follows: Section 2 explains concise description on watermarking technique. Section 3 presents experimental results and discussions. Finally, Section 4 concludes the paper.

2   Watermark Embedding and Extraction

The proposed digital image watermarking approach aims to attain the tradeoffs between robustness and the imperceptibility of the information during embedding the watermark [14-15]. Moreover increase the data hiding capacity by hiding multiple copies inside the original image.

Figures 1 and 2 are presenting the encoding as well as decoding process, correspondingly. In the encoder segment, a phase congruency map or feature map is produced from the original image, and adaptive bit depths are estimated for every pixel based on that, with bit depth value for pixels and binary watermark, the watermarked Images are formed by adaptive LSB replacement technique. The identical practice is applied in the decoder part to get the pixel wise bit depth value and that facilitate to extract the binary watermark from LSB region of watermarked Image following multiplicity estimation. The feature map, adaptive bit depth estimation as well as encoder and decoder are explained here.

2.1   Feature Map

Grayscale images are used to implement the system and resulting feature maps is used to calculate the human fixation. A set of feature maps is thus obtained by means of following technique.

Let I as the gray scale image with size of $C \times D$ and represented as

$$I = \{X(a, b)|\ 0 \le a \le C, 0 \le b \le D, X(a, b) \in \{0, 1 \dots 255\}\} \tag{1}$$

Phase congruency (PC) is used to find out the significant local image features, such as edges, junctions or other textures and it is defined as [16].

$$PC(x) = \frac{max}{\phi(x) \in [0, 2\Pi]} \left(\frac{\sum_n A_n \cos[\phi_n(x) - \phi(x)]}{\sum_n A_n} \right) \tag{2}$$

A_n and $\phi_n(x)$ are the amplitude and local phase at location x of the nth Fourier component, $\phi(x)$ is the amplitude weighted mean of local phases at location of all Fourier components.

In proposed approach, PC computed through the technique developed by Kovesi [17] as it is complicated to calculate the value of PC from equation no 2. In image I, PC value of a pixel at location (x,y) is computed as.

$$PC(x,y) = \frac{\sum_{i=1}^{m}\sum_{j=1}^{n} W_i\,(x,y) f_P\{A_{ji}(x,y)[\Delta\phi_{ji}(x,y)]-T_i\}}{\sum_{i=1}^{m}\sum_{j=1}^{n} A_{ji}\,(x,y)} \tag{3}$$

where m and are the numbers of orientations and scales, $A_{ji}(x)$ and $\phi_{ji}(x)$ the amplitude and local phase deviation for the jth scale logarithmic Gabor function at the ith orientation, T_i the estimated noise at the ith orientation, $W_i(x)$ the weighting function at the ith orientation, and $f_p(u)=(u>0)u$. The local phase deviation $\Delta\phi(x,y)$ is computed as.

$$\Delta\phi_{ji}(x,y) = \cos[\phi_n(x,y) - \phi(x,y)] - |\sin[\phi_n(x,y) - \phi(x,y)]| \tag{4}$$

Where $\phi(x,y)$ the amplitude weighted mean of local phases at location (x,y).
Now this PC value for each pixel will create the feature map.

2.2   Adaptive Bit Depth Estimation

From the produced feature map, an adaptive watermark strength map is made by using high embedding strength in less information feature areas and a low embedding strength in areas of major information feature areas. To get better watermark invisibility and to improve the strength of the embedded watermark, the adaptive bit depth (ABD) capacity is obtained from feature regions (fr).

$$ABD(a,b) = round\left((7 - rMSB) \times \frac{(max(fr)-fr(a,b))}{(max(fr)-min(fr))}\right) \tag{7}$$

Where $1 \leq rMSB \leq 7$ and rMSB represents the MSBs of each pixel used to compute the visual perception. Function round (z) proceeds the nearest integer to the argument z and max (z) and min (z) returns the maximum and minimum of the array z respectively. The factor $ABD(a, b)$ is the final visual perception factor of the pixel I (a, b) and $1 \leq ABD(a, b) \leq (7-rMSB)$ where rMSB planes are discarded for better imperceptibility.

2.3   Encoder and Decoder

Let I of equation 1 is the original grayscale image and W_m be the binary watermark with size of $P \times Q$ and illustrated as:

$$W_m = \{Y(i,j)|0 \leq i < P, 0 \leq j < Q, Y(i,j) \in \{0,1\}\} \tag{8}$$

The adaptive LSB watermarking with index n is defined as a pair of functions (f_{enc}^n, f_{dec}^n):

$$f_{enc}^n: I \times W_m \rightarrow I_w \text{ and } f_{dec}^n: I_w \rightarrow W_m \tag{9}$$

Which satisfies, for all $x \varepsilon I$, $w \varepsilon W_m$ and $y \varepsilon f_{enc}^n (x, w)$

Figure 1 and 2 illustrates the encoding scheme and decoding method correspondingly.

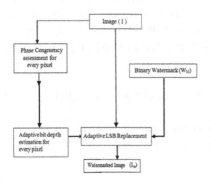

Fig. 1. Watermark Encoder   **Fig. 2.** Watermark Decoder

3   Results and Discussion

This fragment reports the experimental outcomes which have been estimated and applied to evaluate the result through comparative study with the earlier schemes and proposed technique. To reveal the performance of the projected technique some regularly available gray scale (8 bit) test images of size 256x256 and binary watermark image of size 32x32 are used form open database. The test gray scale images are given in Figure 3 and the binary watermark images are given in Figure 4.

Fig. 3. Gray Scale Images

A

Fig. 4. Binary Watermark

Figure 5 presents the phase congruency estimation or feature map for cover images from assessed values of the pixel and Figure 6 stands for adaptive bit depth images based on feature map; different shades represent different bit depth value from 0 to 3. Figure 7 illustrates the watermarked Images.

Fig. 5. Phase Congruency for the images / Feature Map

Fig. 6. Bit depth images based on Feature Map

Fig. 7. Watermarked images

Every intentional or unintentional processing related to image will cause a considerable loss of information or quality. Table 1 tabularize the average result for a few well-recognized quality measures which are essential to verify the invisibility of the implanted watermarks in grayscale test images in terms of imperceptibility.

Table 1. The performance results of Imperceptibility

Sl No.	Parameters	Result
1.	Peak Signal to Noise Ratio	55.64
2.	Mean Square Error	14.52
3.	Average Difference	0.448
4.	Structural Content	0.998
5.	Maximum Difference	7
6.	Universal Image Quality Index	0.872
7.	Normalized Absolute Difference	0.045
8.	Structural Content	0.998
9.	Structural SIMilarity (SSIM) index	0.993
10.	Total Perceptual Error	0.056

Robustness of the proposed scheme is evaluated; under significant intentional and unintentional signal processing impairments. The performance results against different category of attacks are presented using Table 2. Figure 8 present recovered watermarks after different image impairments.

Table 2. Results for Robustness

Name of the Attack	MI	PCC	PSNR	NCC	BER	JC
Crop	0.2005	0.3942	51.731	0.6591	0.0881	0.3412
Scaling	0.1950	0.1127	50.970	0.1909	0.0821	0.1455
Noise(Salt & Pepper)	0.5801	0.9593	60.442	0.9733	0.0117	0.9348
Contrast	0.1725	0.4550	50.341	0.9545	0.0301	0.4529
Erode	0.407	0.4011	50.182	0.5676	0.0712	0.4667
Noise (Gaussian)	0.2326	0.5877	51.872	0.8864	0.0806	0.4756

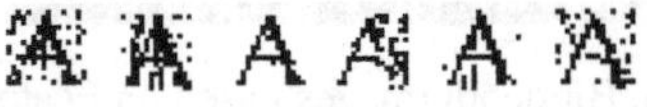

[i]   [ii]   [iii]   [iv]   [v]   [vi]

[i] Crop, [ii] Scaling, [iii] Noise (salt & pepper), [iv] Contrast, [v] Erode, [vi] Noise (Gaussian)

Fig. 8. Recovered watermarks after different image impairments

To substantiate the performance of the scheme, an assessment among different algorithms is presented in Table 3. The outcome authenticates that the proposed approach shows better imperceptibility, larger capacity and lower execution time.

Table 3. The Performance Comparison

Sl.No	Method	PSNR (dB)	Capacity (bpp)	Time (sec)
1.	Proposed Method	55.64	3	0.9591
2.	Optimal LSB substitution by dynamic prog. [18]	38.34	3	1.25030
3.	Genetic Algorithm [19]	38.32	3	19.2937
4.	Mielikainen's method [20]	33.05	2.2504	-

4   Conclusion

The algorithm proposed in this paper has the appropriate trade-off between imperceptibility and robustness. The experimental results suggest that the algorithm stands against various attacks. Moreover data hiding capacity of the algorithm is also better. Comparison with some state of the art technique suggests that the proposed idea can be an effective tool for copyright protection and authentication. Adaptive

LSB replacement along with phase congruency is used for incorporating the bits of the logo, but in future some other techniques can also be used which does not damage the integrity of the main algorithm.

References

1. Kim, H.J., Macq, B.: Introduction to the special issue on authentication, copyright protection, and information hiding. IEEE Transactions on Circuits and Systems for Video Technology 13(8) (August 2003)
2. Garzia, F.: Handbook of Communications Security. WIT Press (2013)
3. Mohanty, S.P.: Digital Watermarking: A Tutorial Review, Report, Dept. of Electrical Engineering, Indian Institute of Science, Bangalore, India (1999),
   http://www.cs.unt.edu/~smohanty/research/
   Reports/MohantyWatermarkingSurvey1999.pdf
4. Macq, B.R., Pitas, I.: Special Issue on Watermarking. Signal Process. 66(3), 281–282 (1998)
5. Swanson, M.D., Kobayashi, M., Tewfik, A.H.: Multimedia Data Embedding and Watermarking Technologies. Proc. IEEE 86, 1064–1087 (1998)
6. Acken, J.M.: How Watermarking Adds Value to Digital Content. Commun. ACM 41(7), 74–77 (1998)
7. Low, S.H., Maxemchuk, N.F., Lapone, A.M.: Document Identification for Copyright Protection using Centroid Detection. IEEE Trans. on Commun. 46, 372–383 (1998)
8. Basu, A., Das, T.S., Sarkar, S.K.: Robust Visual Information Hiding Framework Based on HVS Pixel Adaptive LSB Replacement (HPALR) Technique. International Journal of Imaging and Robotics 6(A11), 71–98 (2011) ISSN: 2231-525X
9. Basu, A., Sarkar, S.K.: On the Implementation of Robust Copyright Protection Scheme Using Visual Attention Model. Information Security Journal: A Global Perspective (Taylor & Francis Journal) 22(1), 10–20 (2013)
10. Keskinarkaus, A., Pramila, A., Seppänen, T.: Image watermarking with feature point based synchronization robust to print–scan attack. Journal of Visual Communication and Image Representation 23(3), 507–515 (2012)
11. Jin, X., Kim, J.: Imperceptibility Improvement of Image Watermarking Using Variance Selection. In: Kim, T.-H., Mohammed, S., Ramos, C., Abawajy, J., Kang, B.-H., Ślęzak, D. (eds.) SIP/WSE/ICHCI 2012. CCIS, vol. 342, pp. 31–38. Springer, Heidelberg (2012)
12. Jaseena, K.U., John, A.: Text Watermarking using Combined Image and Text for Authentication and Protection. International Journal of Computer Applications 20(4) (2011)
13. Morrone, M.C., Burr, D.C.: Feature detection in human vision: a phase-dependent energy model. Proc. R. Soc. Lond. B Biol. Sci. 235(1280), 221–245 (1988)
14. Liu, R., Tan, T.: A SVD-Based Watermarking Scheme for Protecting Rightful Ownership. IEEE Transactions on Multimedia 4, 121–128 (2002)
15. Aliwa, M.B., El-Tobely, T.E.-A., Fahmy, M.M., Nasr, M.E.S., El-Aziz, M.H.A.: Fidelity and Robust Digital Watermarking Adaptively Pixel based on Medial Pyramid of Embedding Error Gray Scale Images. IJCSNS (International Journal of Computer Science and Network Security) 10(6), 284–314 (2005)
16. Morrone, M.C., Burr, D.C.: Feature detection in human vision: A phase dependent energy model. Proc. R. Soc. Lond. B, Biol. Sci. 235(1280), 221–245 (1988)

17. Kovesi, P.D.: Image features from phase congruency. Videre: Journal of Computer Vision Research 1, 1–26 (1999)
18. Lai, C.-C.: A digital watermarking scheme based on singular value decomposition and tiny genetic algorithm. Digital Signal Processing 21, 522–527 (2011)
19. Mielikainen, J.: LSB Matching Revisited. IEEE Signal Processing Letters 13(5), 285–287 (2006)
20. Xu, H., Wang, J., Kim, H.J.: Near-optimal solution to pair-wise LSB matching via an immune programming strategy. Information Sciences 180, 1201–1217 (2010)

Influence of Various Random Mobility Models on the Performance of AOMDV and DYMO

Suryaday Sarkar[1], Meghdut Roychowdhury[2], Biswa Mohan Sahoo[3], and Souvik Sarkar[4]

[1] Netaji Subhas Engineering College, Kolkata, West Bengal, India
[2] Techno India, Salt Lake, Kolkata, West Bengal, India
[3] Department of CSE, Amity University, Greater Noida Campus, India
[4] IBM, Hyderabad, India
suryadaysarkar@ymail.com, meghdut.tig@gmail.com,
bmsahoo@gn.amity.edu, souviksarkar@in.ibm.com

Abstract. A Mobile Adhoc Network (MANET) is a collection of autonomous self organizing mobile devices that communicate with each other by creating a network in a given area. The moving behavior of each mobile device is the MANET is determined by the mobility model which is a crucial component in its performance evaluation. Here is the present work; we have investigated the influences of various random mobility models on the performance of Adhoc on-demand Multipath Distance Vector (AOMDV) Routing Protocol and Dynamic MANET On-demand (DYMO) Protocol. In order to validate our work, three different mobility scenarios are considered: Random waypoint (RWP), Random Walk with Wrapping (RWP-WRP) and Random Walk with reflections (RWP-REF). Experimental results establish the fact that the performance of the routing protocols is significantly influenced by the different parameters like number of nodes, and to end delay and packet delivering ratio.

Keywords: MANET, Mobility Models, Routing Protocols, AOMDV, DYMO.

1   Introduction

Ad HOC networks require no centralized administration or fixed network infrastructure such as base stations and can be quickly and inexpensively set up as needed. A MANET can be envisioned as a collection of mobile routers, each equipped with a wireless trans-receiver, which are free to move about arbitrarily. The router mobility and the variability of other connecting events result in a network with a potentially rapid and unpredictable changing topology. In a MANET, the device assumption is that two devices willing to communicate may be outside the wireless transmission range of each other but still be able to communicate if other devices in the network are willing to forward packets from them [shown in the Fig. 1]

In the Fig. 1, devices F and A are outside the transmission range of each other but still be able to communicate via the intermediate device D. In the simulation of

© Springer International Publishing Switzerland 2015   121
S.C. Satapathy et al. (eds.), *Proc. of the 3rd Int. Conf. on Front. of Intell. Comput. (FICTA) 2014*
– *Vol. 2*, Advances in Intelligent Systems and Computing 328, DOI: 10.1007/978-3-319-12012-6_14

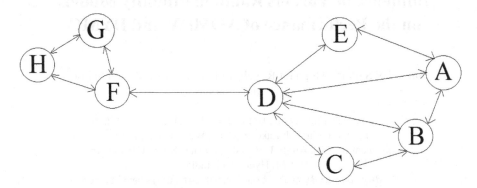

Fig. 1. A Mobile Ad-hoc Network

routing protocols, mobility models play a very important role. Hence, to get accurate results, these mobility models should reverberate with real life scenarios.

AOMDV is a reactive routing protocol. AOMDV employs the multiple loop-free and link disjoint path "technique [1]. In this routing protocol, only disjoint nodes are considered in all the paths, thereby achieving path disjoint. For route discovery Route Request packets are propagated throughout the network thereby establishing multiple paths at destination node and at the intermediate nodes. Since AOMDV is a multipath algorithm we have many paths running between the nodes of different groups resulting in higher energy consumption [2].

DYMO routing protocol is an improved version of AODV routing protocol. In this protocol routes are established on demand. Whenever a source node wants to set up a route to the destination the Route Request (RREQ) messages are flooded throughout the network. Only those nodes that have not broadcasted previously will forward the messages during broadcasting. The Route Request message contains its own address, sequential number, a hop count and the destination node address.

Mobility Models

Mobility models usually describe the movement pattern of nodes and how their speeds and directions are altered over the time [3]. We will use these different mobility scenarios: random Waypoint, Random Walk with Wrapping and Random walk with reflections.

2   Simulation Environment

2.1   Simulation Setup

The MANET network simulations are implemented using NS-2 simulator [5] and the scenario Generation tool. Using scenario generating programs, the mobility models are computed. The results are then transferred into the NS-2 simulation models.

Each node is then assigned a particular trajectory. The simulation period for each scenario is 900 seconds and the simulated mobility network area is 800m ×500m rectangle with number of nodes varying from 5 to 25. The nodes are initially located at the center of the simulator area in each simulation scenario. Every node starts moving after the first 10 seconds of simulation time. In all the simulations, the MAC layer protocol IEEE 802.11 has been used with the data rate of 11 Mbps.

2.2   Mobility Metrics

In order to evaluate the performance of the different protocols, we have considered the packet delivery protocols, protocol control overhead and average end-to-end delay as matrices.

  (a) **Packet Delivery Ratio (PDR):** It is defined as the ratio of the number of packets sent from the source to the number of packets received at the destination.
  (b) **Protocol Control Overhead (PCO):** It is defined as the ratio of the number of protocol control packets transmitted to the number of data packets received.
  (c) **Average end-to end delay (AEED):** It is the average time delay for data packets from the source node to the destination node.

3   Results and Discussion

In the present work we have investigated the influence of various random mobility models, namely RWP, RWP-WRP and RWP-REF (mobility models) [5] on the performance of AOMDV and DYMO. The nodes are configured with a constant pause interval of 100 seconds. The Fig. 2 represents the PDR with RWP, RWP-WRP and RWP-REF mobility models. The Fig.2 depicts that the mobility models drastically influence the protocol performance. It is evident from the Fig. 2 that RWP produces the highest throughput whereas the throughput of other two models drastically falls. It is also very clear in the Fig. 2 that Random-Walk-Reflection model provides packet delivery ratio. We find that RWP-WRP [5] gives the highest delay with node density as well as mobility. The average end-to-end delay on Random waypoint is less as compared with two other models.

The influence on the routing overhead is very less with Random walk model with wrapping while the other two models suffer a lot from routing overhead packets. We also see that Random Walk with wrapping achieve the lowest routing overhead with node density as well as mobility. DYMO [3] protocol has better performance over the AOMDV [3] protocols in all the cases we have considered here.

Table 1. Packet delivery ratio of different routing protocols with mobility models

Number of Nodes			5	10	15	20	25
PDR%	RWP	AOMDV	99.6	99.62	99.6	99.55	99.58
		DYMO	99.7	99.63	99.7	99.6	99.72
	RWP	AOMDV	99.48	99.5	99.48	99.42	99.45
	WRP	DYMO	99.55	99.45	99.6	99.45	99.55
	RWP	AOMDV	99.53	99.57	99.53	99.54	99.51
	REF	DYMO	99.60	99.5	99.65	99.5	99.6

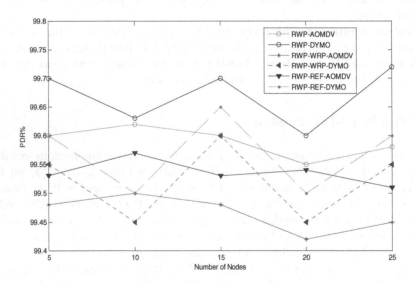

Fig. 2. Packet delivery ratio Vs Number of nodes with different mobility

4   Conclusion

In summary, we can say that we have studied here the influences of various random mobility models on the performance of AOMDV and DYMO. We have calculated the packet delivery ratio, end-to-end delay and routing overhead to evaluate the performance of AOMDV and DYMO. The numerical values indicate that the performance of a routing protocol varies widely across different mobility models.

References

1. Gowrishankar, S., Basavaraju, T.G., Sarkar, S.K.: Effects of Random Mobility Models Pattern in Mobile Ad Hoc Networks. International Journal of Computer Science and Network Security 7(6), 160–164 (2007)
2. Rhee, I., et al.: On the Levy-Walk nature of Human mobility. In: INFOCOM, Arizona, USA (2008)
3. Gowrishankar, S., Basavaraju, T.G., Sarkar, S.K.: Analysis of AOMDV and OLSR Routing protocols under Levy-Walk Mobility model and Gauss-Morkov model for Ad Hoc Networks. International Journal of Computer Science and Engineering 2, 979–986 (2010)
4. http://www.isi.edu/nsnam/ns
5. http://www.cise.ufl.edu/~helmy/papers/ Survey-Mobility-Chapter-1.pdf

Handling Data Integrity Issue in SaaS Cloud

Anandita Singh Thakur, P.K. Gupta, and Punit Gupta

Department of Computer Science and Engineering
Jaypee University of Information Technology, Waknaghat, Solan, 173234, India
{anandita275,punitg07}@gmail.com, pradeep1976@yahoo.com

Abstract. Cloud computing is a technology that being widely adopted by many organizations like Google, Microsoft etc in order to make the resources available to multiple users at a time over the internet. Many issues are identified due to which cloud computing is not adopted by all users till now. The aim of this paper is to analyze the performance of the encryption algorithms in order to improve the data integrity in SaaS cloud. The proposed modified algorithm encrypts the data from different users using cryptographic algorithms namely: RSA, Bcrypt and AES. The algorithm is selected by user based on the level of security needed to be applied to the user's data. Performance analysis of the given framework and algorithm is done using CloudSim. From our obtained results this could be easily found that the time taken for encryption of data using the discussed framework and proposed algorithms is much less in comparison to various other techniques.

Keywords: AES, Bcrypt, Cloud computing, Data integrity, RSA.

1 Introduction

Cloud computing is an emerging technology where the resources, information and software are shared and are provided to the users over the internet as per their demands. In cloud, multiple systems can interact with each other at a time and move computing tasks from their system to the cloud [1]. The computing services are delivered over the internet. These services allow individuals and businesses to use software and hardware that are managed by third parties at remote locations [2]. Cloud computing has three service models namely [3]: Infrastructure as a Service (IaaS), Platform as a Service (PaaS), and Software as a Service (SaaS).

1.1 Issues in Cloud Computing

Though cloud computing has been adopted by the industries, it still has certain drawbacks. The foremost issue in cloud computing is security and privacy related to the data of the users. Since multiple users access the information in cloud at a time, the integrity and privacy of the information is at high risk. It is important to ensure that the information being processed on cloud is secure and no tampering of information is done when previously unknown parties may be present [4].

© Springer International Publishing Switzerland 2015
S.C. Satapathy et al. (eds.), *Proc. of the 3rd Int. Conf. on Front. of Intell. Comput. (FICTA) 2014*
– Vol. 2, Advances in Intelligent Systems and Computing 328, DOI: 10.1007/978-3-319-12012-6_15

Security issues: Security is termed as the prevention of any unauthorized access, unauthorized deletion or amendment of the information. The main dimensions of security that should be kept in mind for providing user satisfaction are availability, confidentiality and integrity [5].

Privacy issues: The ability of individual or group to seclude themselves or information about themselves and selectively reveal them is termed as privacy. Privacy issues vary according to different cloud scenario. The privacy issues are defined as [5]: legal requirements, control and processing of data, and data replications in a consistent state.

Trust issues: Trust is a measureable belief that is used to make trustworthy decisions based on experience. It has attributes like reliability, confidence, dependability, honest etc.

This paper discusses about the various issues in SaaS. The focus is not on application's portability but on migration of data and enhancement of security functionalities. This paper is categorized into various sections where Section 1 is an Introduction and provides the details about various kind of issues in cloud computing, Section 2 presents the detailed view of literature survey related to previously discussed framework and algorithms, Section 3 discusses about the algorithm implemented to evaluate the performance of the proposed framework, Section 4 presents the results of performance analysis of framework using the proposed modified algorithm, and at last, in Section 5 we have concluded the work.

2   Related Works

Cloud has many benefits like cost reduction, scalability, flexibility, multi-tenancy, availability. Even after having so many benefits users are not fully comfortable in adapting to this new technology due to many issues. In [6] Chalse et.al provides a detailed analysis of the cloud security problem. In [7] R and Saxena provided a scheme that gave a proof of data integrity in which the correctness of the user data can be done by the user. This proof can be incorporated in SLA and is agreed upon by both the cloud and the customer. The proposed scheme does not involve the encryption of the whole data. Few bits of data per data block are encrypted; hence the computational overhead on the clients is reduced. Raju et al. [8] introduced a protocol for integrity checking of cloud storage that would provide integrity protection of user information. The prediction of information consistency was missing in the existing systems and with the help of the proposed protocol this problem is resolved. In [9] Kumar et al. assess that how cloud providers can gain trust of their customers and provide them with security, privacy and reliability on the data when processing of sensitive data is done by the third party in remote machines located in various countries.. In [10] Eswaran et al. proposed an approach to secure the data and ensure the integrity of data in cloud using cryptographic keys. Proof of retrieve (POR) ability is used in order to verify the integrity of data. The owner of file must encrypt the data in file before storing it in cloud in order to prevent unauthorized access to the file. In [11] Rana et. al have discussed the combined framework for IaaS and PaaS and also simulated this framework using CloudSim.

3   Proposed Algorithm

This algorithm analyzes the performance of different encryption algorithms over the framework as discussed in [11]. This framework is categorized into three different layers: platinum, gold and silver. Depending upon the type of information in file, the level of security is applied. The sensitive information is stored in platinum layer and file is encrypted using RSA algorithm. The silver layer provides less security than Platinum layer and the file that is stored here is encrypted by using Bcrypt algorithm. The gold layer provides least level of security and the file stored here is encrypted using AES algorithm.

```
/* Variables used are:
            Consumer => u; Platinum => p,
            Gold => g; Silver => s     */
    Begin
        If u chooses p
            Then call module m1
        If u chooses g
            Then call module m2
          If u chooses s
            Then call module m3
    End
    Module m1:
    Begin
1. User's data is encrypted using RSA.
2. User sends the encrypted data to CSP where it is stored
3. TPA verifies the data stored at CSP using digital signature
4. If signature matched
        Verification passed, data is valid
    Else
        Verification failed
    End
    Module m2:
    Begin
1. User's data is encrypted using Bcrypt algorithm
2. User sends the encrypted data to CSP where it is stored
3. TPA verifies the data stored at CSP using digital signature
4. If signature matched
        Verification passed, data is valid
    Else
        Verification failed
    End
    Module m3:
    Begin
1. User's data is encrypted using AES algorithm
2. The encrypted data is sent to the CSP where it is stored
3. TPA verifies the data stored at CSP using digital signature
4. If signature is matched
        Verification passed, data is valid
    Else
        Verification failed
    End
```

4   Performance Analyses

To analyze the performance of proposed algorithm, we have used the framework proposed by Gupta & Thakur [12]. In previous scheme [13] all files were provided with same level of security irrespective of the type of data in it whereas in [12] the framework provides different levels of security on files of different sizes. Here, Table 1, Table 2 and Table 3 present the comparison between the encryption time of previous scheme and proposed scheme when the number of requests keeps on increasing. Fig. 2(a), Fig. 2(b) and Fig. 2(c) represents this process.

Table 1. Performance analysis of previous scheme v/s proposed framework when number of requests are 8

No. of REQUESTS	PREVIOUS SCHEME [13] (TIME in ms)	PROPOSED SCHEME (TIME in ms)
1	3074	3074
2	3074	2917
3	3075	796
4	3009	1108
5	2995	1108
6	2905	796
7	3079	921
8	3420	827

Table 2. Performance analysis of previous scheme v/s proposed scheme when number of requests increased to 16

No. of REQUESTS	PREVIOUS SCHEME [13] (TIME in ms)	PROPOSED SCHEME (TIME in ms)
1	3248	3006
2	3291	3110
3	3106	900
4	3341	1249
5	3570	1289
6	3320	1160
7	3334	1276
8	3456	900
9	3418	1000
10	3534	958
11	4018	2987
12	3987	3540
13	3765	3491
14	4211	4010
15	4300	2667
16	4696	3800

Table 3. Performance analysis of previous scheme v/s proposed scheme when number of requests increased to 32

No. of REQUESTS	PREVIOUS SCHEME [13] (TIME in ms)	PROPOSED SCHEME (TIME in ms)
1	3896	3694
2	3424	3341
3	3308	3208
4	3410	3400
5	3540	3459
6	3691	3576
7	3330	3200
8	3120	3110
9	3600	3567
10	3498	1000
11	3900	3741
12	4009	4003
13	3774	3669
14	4219	4198
15	4166	4047
16	4333	4216
17	3681	3538
18	3908	3497
19	3724	3623
20	3805	3551
21	3500	3348
22	4460	4234
23	4000	3811
24	4790	4626
25	4546	4377
26	3980	3694
27	2196	987
28	2897	1248
29	3071	2381
30	4298	3106
31	4571	3963
32	5168	4685

By using Table1, Table2, and Table3 we have recorded the average time taken for all three cases for our framework and performance is evaluated as shown in Table 4.

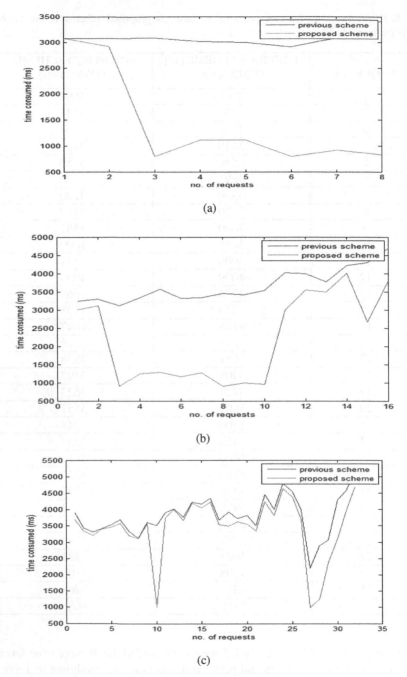

Fig. 1. Encryption time for previous scheme and proposed scheme: (a) no. of requests = 8 (b) no. of requests =16 (c) no.of requests = 32

Table 4. Average time taken by previous schemes and proposed scheme when requests varies from 8 to 32

No. of REQUESTS	PREVIOUS SCHEME [13] AVG TIME (ms)	PROPOSED SCHEME AVG TIME (ms)
8	3105.91	1086.3
16	3870.19	2519.88
32	3899.72	3149.19

The average time taken by both the schemes is shown in Fig. 2 below. It is clearly seen that the time taken by proposed scheme is less than the time taken by previous scheme.

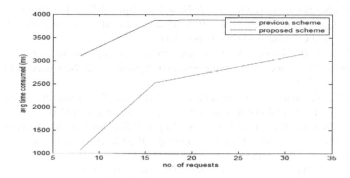

Fig. 2. Average time taken

If the file is modified then the data in file will get encrypted using RSA algorithm but it will not be decrypted as shown in Table 5 below

Table 5. Encryption and decryption time when file is modified

FILE SIZE	ALGORITHM	ENCRYPTION	DECRYPTION
1	AES	1037	3184
2.90	Bcrypt	698	1786
3.98	RSA	1123	-
4.65	Bcrypt	695	1781

5   Conclusion

Security is major issue in cloud computing. To maintain the integrity of data a framework is proposed in which data is stored to different cloud based on the type of information. If the file is not modified, the data is decrypted and hence the integrity of file is maintained. Performance analysis of the framework and algorithm shows that the time taken for encryption of files is less incomparison to old scheme [13] where all files are encrypted using RSA algorithm only and irrespective of the level of security needed to store different types of data.

References

1. Shaikh, F.B., Haider, S.: Security Threats in Cloud Computing. In: 6th International Conference on Internet Technology and Secured Transactions, UAE, pp. 11–14 (2011)
2. Kumar, S., Goudar, R.H.: Cloud Computing – Research Issues, Challenges, Architecture, Platforms and Applications: A Survey. International Journal of Future Computer and Communication 1(4) (December 2012)
3. Zhang, Q., Cheng, L., Boutaba, R.: Cloud computing: state-of-the-art and research challenges. Springer (2010)
4. Naruchitparames, J., Giines, M.H.: Enhancing Data Privacy and Integrity in the Cloud. IEEE (2011)
5. Sun, D., Chang, G., Sun, L., Wang, X.: Surveying and Analyzing Security, Privacy and Trust Issues in Cloud Computing Environments. Published by Elsevier Ltd. (2011)
6. Chalse, R., Selokara, A., Katara, A.: A New Technique of Data Integrity for Analysis of the Cloud Computing Security. In: 2013 5th International Conference on Computational Intelligence and Communication Networks (2013)
7. Sravan Kumar, R., Saxena, A.: Data integrity proofs in cloud storage. In: 2011 Third International Conference on Communication Systems and Networks (COMSNETS), pp. 1–4. IEEE (2011)
8. Raju, R.V., Vasanth, M., Udaykumar, P.: Data integrity using encryption in cloud computing. Journal of Global Research in Computer Science 4(5) (2013)
9. Kumar, P., Sehgal, V.K., Chauhan, D.S., Gupta, P.K., Diwakar, M.: Effective Ways of Secure, Private and Trusted Cloud Computing. IJCSI International Journal of Computer Science Issues 8(3(2)) (2011)
10. Eswaran, S., Abburu, S.: Identifying Data Integrity in the Cloud Storage. IJCSI International Journal of Computer Science Issues 9(1) (March 2012)
11. Rana, P., Gupta, P.K., Siddavatam, R.: Combined and Improved Framework of Infrastructure as a Service and Platform as a Service in Cloud Computing. In: Babu, B.V., Nagar, A., Deep, K., Pant, M., Bansal, J.C., Ray, K., Gupta, U. (eds.) 2012 Second International Conference on Soft Computing for Problem Solving (SocProS 2012). AISC, vol. 236, pp. 1–8. Springer, Heidelberg (2014)
12. Thakur, A.S., Gupta, P.K.: Framework to Improve Data Integrity in Multi Cloud Environment. International Journal of Computer Applications 87(10) (February 2014)
13. Mehta, A., Sahu, R.K., Awasthi, L.K.: Robust Data Security for Cloud while using Third Party Auditor. International Journal of Advanced Research in Computer Science and Software Engineering 2(2) (February 2012)

Load and Fault Aware Honey Bee Scheduling Algorithm for Cloud Infrastructure

Punit Gupta[1] and Satya Prakash Ghrera[2]

[1,2] Department of Computer Science Engineering,
Jaypee University of Information Technology Himachal Pradesh, India
punitg07@gmail.com, spghrera@rediffmail.com

Abstract. Cloud computing a new paradigm in the field of distributed computing after Grid computing. Cloud computing seems to me more promising in term of request failure, security, flexibility and resource availability. Its main feature is to maintain the Quality of service (QoS) provided to the end user in term of processing power, failure rate and many more .So Resource management and request scheduling are important and complex problems in cloud computing, Since maintaining resources and at the same time scheduling the request becomes a complex problem due to distributed nature of cloud. Many algorithms are been proposed to solve this problem like Ant colony based, cost based, priority based algorithms but all these algorithm consider cloud environment as non fault, which leads to degrade in performance of existing algorithms. So a load and fault aware Honey Bee scheduling algorithm is proposed for cloud infrastructure as a service(IaaS). This algorithm takes into consideration fault rate and load on a datacenter to improve the performance and QoS in cloud IaaS environment.

Keywords: Cloud, QoS, Cloud IaaS, Fault, System load, Network load, Datacenters.

1   Introduction

With the evolution of cloud and grid computing distributed system and distributed computing has shown a great benefit to the industry and society by providing the services which are cheaper and more reliable. With grid computing we only have the benefit of computing the bigger task in smaller span of time, but with cloud computing any task can be computed easily, much faster by using the computation power of server and in more secure way. Cloud computing is best example of Distributed computing. It have all the properties of distributed systems namely atomicity, consistency, isolation and durability. It provides all type of services that may be software or platform based or simply hosting services. Services provided by cloud can be classified into three categories SaaS (Software as a service), PaaS (Platform as a service), IaaS[1]. SaaS provides services over cloud in terms of applications which are used to process data.

S.C. Satapathy et al. (eds.), *Proc. of the 3rd Int. Conf. on Front. of Intell. Comput. (FICTA) 2014*
– *Vol. 2*, Advances in Intelligent Systems and Computing 328, DOI: 10.1007/978-3-319-12012-6_16

Cloud IaaS has the four basic features which are as follows [1]:

1. **Pay per Use:** Traditionally organizations procure their own computing resources to build IT infrastructure to archive their business goals. Cloud computing will change this trend in building IT infrastructure through providing infrastructure as a service. Cloud computing allows quick implementation of computing resources without huge amount of upfront capital investment [1].

2. **Elastic Computing:** In the model of cloud computing, computing resources form a shared pool, in which a resource can join and depart from the pool dynamically. Users request the resources on demand, and return the resources to the pool after competition of use. The resources can be re-allocated to other users. Elastic computing feature supports high scalability of computing resources in cloud computing environment.

3. **Virtualization:** All types of computing resources can be virtualized with the development of cloud computing technology. VLAN is an example of network virtualization. Server virtualization is the most common virtualization type in computing resources and supported by most all cloud product vendors. Distributed file system often use storage virtualization to tackle with the heterogeneous storage devices.

4. **Cost:** It plays an important role in cloud computing, because as the definition of cloud says that it will provide the service at cheaper rates.

But the main problem over cloud IaaS is resource management and request scheduling because in this the VM once allocate and running runs for long span of time. So we need to allocate resources in such a way that the request is fulfilled at least cost and highest quality of service which can be provided to a user. There are many cloud IaaS frameworks that provide cloud computing services and virtualization services to the user like OpenNode [18], CloudStack [20], Eucalyptus [17], CloudSigma [21], EMOTIVE(Elastic Management of Tasks in Virtualized Environments) and Archipel. Many solution have been proposed over the time based on priority, cost, rank based which is used in OpenNebula and round robin and power aware scheduling algorithm used in Eucalyptus and many more. But they do not take in to consideration the QoS parameters of the datacenters like fault rate, initialization time, MIPS and many more. So here a QoS aware honey bee algorithm is proposed to provide higher QoS to the user and at the same time taking into consideration least cost to be provided to the user.

2   Related Work

2.1   Scheduling Algorithm

Scheduling can be defined as the process for allocation of resources provided is such a way to have best resource utilization and provide quality of services (QoS) in term of resources provided to the user. It can also be defined as the process of providing best QoS by allocating the best resources provided in minimum span of time. Problem lies in process of finding the best resources.

Taking into consideration scheduling algorithm proposed for distributed system like grid computing and cloud computing. Many scheduling algorithms are been proposed based on power efficiency, cost and priority. Some of the similar algorithm based on cost was proposed by Zhiyang [1], a cost based scheduling algorithm for cloud IaaS, which takes into consideration the cost of resources of the datacenter which are cost of the CPU, memory and no. of cores. Based on these, they calculated the cost according to the user request. Cost is been calculated at different datacenters and the datacenter with optimal cost is allocated to user. Similar improved cost based algorithm proposed by Sadhasivam[2], which take into consideration more dependent parameters like MIPS and CPU of datacenter for cost. Murata [3] proposed an algorithm based on the history which uses vector cloud computing i.e it uses the previous history of resources allocated to schedule the new request. Other then these Danilo Ardagna[7] proposed an energy aware resource allocation algorithm which take into consideration the energy consumption of a datacenter to priorities them based on their energy efficiency. This paper also discourses about various energy aware load balancing and resource allocation algorithm.All these above proposal are based on cost and profit in many term but the proposal given by Elghoneimy,E [6] discourses about the cloud Software as a Service (SaaS) platform provided by Hadoop and proposed a clustered scheduling algorithm for SaaS. Proposed algorithm uses map reduction to reduce the task and find the best suited resource based on the profit based on graph.

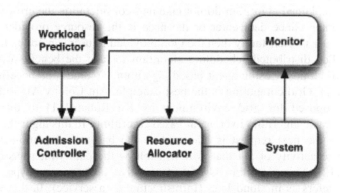

Fig. 1. Autonomic workload and energy efficiency control [7].

Other then this genetic algorithm for scheduling of resources proposed by Chenhong Zhao [5] describes how a genetic algorithm can be used for resource allocation in cloud. Priority based scheduling algorithm are proposed in grid computing environment which is similar to cloud environment. Weifeng Sun [8] proposed a priority based scheduling algorithm in grid computing using few parameters to evaluate the grid parameters. In this a directed acyclic graph is drown

based on the priority and then scheduling is done based on the task having the highest priority. They have compared this algorithm with Min-max and Min-min algorithm. Similar to this algorithm   Zhongyuan Lee [9] proposed apriority based scheduling algorithm for cloud environment .In this priority of the request or task dynamically changes based on time spend in queue. Huang Qiyi [10] proposed a multiple QoS based priorities scheduling algorithm for cloud. In this they have taken into consideration multiple parameters related to a request some of them are cost, time delay, deadline time, number of processors requested and MIPS.
C=pm/MIPS

$$C = \frac{pm}{MIPS} \tag{1}$$

Where C is the cost of resource.pm refers to the budget or the dead line for a task. Expected time to complete task is referred as:

$$Expected\ time = \frac{\sum_k length(t) + \sum_q length(t)}{resource.rate} \tag{2}$$

Taking into consideration the scheduling algorithm used in real implemented clouds. In Eucalyptus, Greedy and Round-robin algorithms are implemented and also scheduling algorithm on power saving for datacenters [13] [14]. In OpenNebula, ranking scheduling policy is being implemented and rank is based on the free CPU's [15]. But Amazon uses trust and reliability based scheduling algorithm.

All above explained algorithm do not take into consideration the error of the fault rate at datacenter. Since datacenter or the node is the resource provider and if the resource provider is itself faulty then the QoS provided to user can not be fulfilled. So we have studied distributed scheduling algorithm to take the benefit of distributed architecture of cloud by using agent based algorithm. To study the benefits of agent based algorithm, Grid computing is the best example. Ant Colony Algorithm for job scheduling proposed for grid environment by Ku Ruhana[11] to overcome the drawbacks of basic algorithm over agent based algorithm . In this algorithm there is a drawback that the next request is allocated on the basis of allocation of previous request. Due to activity of ant that is it will follow the path which has the highest pheromone [12]. This algorithm is best if we have same request type, but if the request type defers as in cloud IaaS (Infrastructure as a service). In this we need to evaluate the profit function for each request [12].

3   Proposed Work

As from above scheduling algorithms they only take into consideration either the power or simple resource allocation algorithm, to overcome that and map them to real paradigm the cost based algorithm was proposed and rank based algorithm were proposed[1][2].

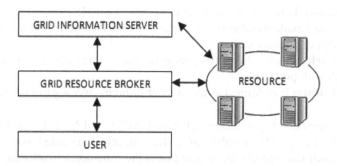

Fig. 2. Ant Colony System Architecture

There are many load balancing algorithms previously used in grid computing and cloud computing environment for load balancing [16]. But they do not take into consideration fault and QoS of datacenter in cloud environment. So to overcome all the problems a honey bee based fault and load aware honey bee scheduling algorithm (FLHB) is proposed. Proposed algorithm is an optimization algorithm inspired by the natural foraging behavior of honey bees to find the optimal solution. The algorithm requires a number of parameters to be set, namely: number of scout bees (n), number of sites selected out of n visited sites (m), number of best sites out of m selected sites (e), number of bees recruited for best e sites, number of bees recruited for the other (m-e) selected sites, initial size of patches which includes site and its neighborhood and stopping criterion.

Steps of proposed algorithm are as follows:

Step1. Initialize scout bees equal to number of datacenters.
Step2. Recruit scout bees for selected sites (more bees for best e sites) and evaluate fit nesses value for datacenter.
Step3. Assign bees to search randomly and evaluate their fit nesses for a request.
Step4. Stop when all bees have arrived, else wait.
Step5. Select the fittest bee from each datacenter.
Step6. Assign remaining bees to search randomly and   evaluate their fit nesses for each request.
Step7. End While no request in queue.

In first step, the bee algorithm starts with the scout bees (n) being placed randomly in the search space. In step 2, the algorithm conducts searches in the neighborhood of the selected sites, assigning more bees to search near to the best 'e' sites i.e. search for new datacenters. In step 3 the fit nesses of the datacenters visited by the scout bees are evaluated. In step 4 waiting until all bees are arrived. In step 5, 6 bees that have the highest fit nesses are chosen as "selected bees" and sites visited by them are chosen for allocation of resources. In step 7 repeat all above steps until there is request in queue. Most complicated part of this algorithm is fitness value calculation. Proposed algorithm take into consideration parameters of datacenter which are used for calculating fitness value for a datacenter are as follows.

   a. **Initiation Time:** How long it takes to deploy a VM.

   b. **System Load:** Number of busy or allocated Machine Instruction per Second (MIPS) of a datacenter.

   c. **Network Load:** Allocated network bandwidth out of total available bandwidth provided.

   d. **Fault Rate:** It is defined as the number of faults over a period of time.

In above mentioned parameters allocated MIPS (MP) and Bandwidth of a datacenter changes as the number of virtual machine allocated on a datacenter changes, but fault rate, initialization time that is the time taken to allocate resource at datacenter also increases as the load increases. So to calculate fitness of a datacenter can be calculated as: Fitness (FT), allocated MIPS (MP), Fault rate (FR), Initialization time (IT), Network load (N_L).

$$FT = \alpha 1 \frac{1}{N\_L} + \alpha 2 \frac{1}{FR} + \alpha 3 \frac{1}{MP} \tag{3}$$

$$\alpha 1 < 1, \alpha 2 < 1 \ \& \ \alpha 3 < 1 \tag{4}$$

$$\alpha 1 + \alpha 2 + \alpha 3 = 1 \tag{5}$$

Where Fault rate (FR) is given as:

$$FR(t) = f(MP , N\_L) \tag{6}$$

Where FR(t) refers to as the fault over the time t, which is function of system load and network load over the time t. $\alpha 1$, $\alpha 2$ and $\alpha 3$ represent the ratio in which these parameters contribute. As discoursed above a last step is to compare the fitness function of all the datacenters and allocating the request at the datacenter with highest fitness function values above all other datacenters.

4   Experimental Results

Proposed algorithm is simulated using cloudsim simulator [18] is used. Cloudsim basically support cost estimation, and FIFO algorithm for scheduling the resource sequentially. Firstly the Cloudsim API does not support fault rate at datacenter. So firstly fault is added as a parameter of datacenter which responds to fault occurring at the datacenter. This cloudsim API is used for simulation of cloud IaaS. So it includes all the cloud IaaS request parameters. Honey bee scheduling algorithm is implemented in cloudsim replacing basic FCFS (first come first serve) scheduling algorithm. Comparative study is done between basic load aware honey bee (BLHB) which allocates request on datacenter with least system load and proposed algorithm.

Table 1.

Server Name	Fault rate FR(t)
Server1	0.143
Server2	0.125
Server3	0.5

Table 2.

	Request count				
	60	100	200	300	400
FLBH	8	15	28	48	68
BLHB	13	23	43	71	89

Simulation is done with 3 datacenters with corresponding fault rate for each datacenter as shown in table 1. Table 2 shows the faults occurring using BLHB and proposed FLBH algorithm. Simulation is done over 60, 100, 200, 300, 400 requests to study the performance of proposed algorithm with increasing request rate.

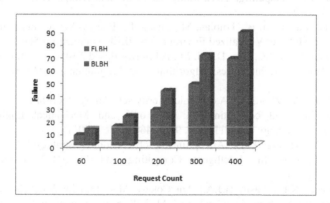

Fig. 3. Comparison of failure counts

Fig 3 shows the number of request failed using proposed and basic honey bee algorithm. This graph shows the algorithm when tested with 60, 100, 200, 300, 400 requests. So the result shows the improvement of proposed algorithm over BLHB in fault aware environment

5   Conclusion

In this paper different scheduling algorithm from the field of grid computing and cloud computing are been discussed with their drawbacks. To overcome the drawbacks a honey bee cost efficient algorithm is proposed which perform better them scheduling algorithm implemented in Cloudsim. For future work this trust model may be compared with other models and see the improvement in the QoS.

References

[1] Yang, Z., Yin, C., Liu, Y.: A Cost-Based Resource Scheduling Paradigm in CloudComputing. In: PDCAT 2011, pp. 417–422 (October 2011)
[2] Selvarani, S., Sadhasivam, G.S.: Improved cost-based algorithm for task scheduling incloud computing. In: Computational Intelligence and Computing Research (ICCIC), pp. 1–5 (December 2010)
[3] Murata, Y., Egawa, R., Higashida, M., Kobayashi, H.: History-Based Job Scheduling Mechanism for the Vector Computing Cloud. In: Applications and the Internet (SAINT), pp. 125–128 (July 2010)
[4] Huang, Q.-Y., Huang, T.-L.: An optimistic job scheduling strategy based on QoS for Cloud Computing. In: Intelligent Computing and Integrated Systems (ICISS), pp. 673–675 (October 2010)
[5] Zhao, C., Zhang, S., Liu, Q., Xie, J., Hu, J.: Independent Tasks Scheduling Based on Genetic Algorithm in Cloud Computing. In: Wireless Communications, Networking and Mobile Computing, pp. 1–4 (September 2009)
[6] Elghoneimy, E., Bouhali, O., Alnuweiri, H.: Resource allocation and scheduling in cloud computing. In: Computing, Networking and Communications (ICNC), pp. 309–314 (September 2012)
[7] Ardagna, D., Panicucci, B., Trubian, M., Zhang, L.: Energy-Aware Autonomic Resource Allocation in Multitier Virtualized Environments. IEEE Transactions 5(1), 2–19 (2012)
[8] Sun, W., Zhu, Y., Su, Z., Jiao, D., Li, M.: A Priority-Based Task Scheduling Algorithm in Grid. In: Parallel Architectures, Algorithms and Programming (PAAP), pp. 311–315 (2010)
[9] Lee, Z., Wang, Y., Zhou, W.: A dynamic priority scheduling algorithm on service request scheduling in cloud computing. In: Electronic and Mechanical Engineering and Information Technology (EMEIT), vol. 9, pp. 4665–4669 (2011)
[10] Huang, Q.-Y., Huang, T.-L.: An optimistic job scheduling strategy based on QoS for Cloud Computing. In: Intelligent Computing and Integrated Systems (ICISS), pp. 673–675 (2010)
[11] Ku-Mahamud, K.R., Nasir, H.J.A.: Ant Colony Algorithm for Job Scheduling in Grid Computing. In: Mathematical/Analytical Modelling and Computer Simulation (AMS), pp. 40–45 (2010)
[12] Sagayam, R., Akilandeswari, K.: Comparison of Ant Colony and Bee Colony Optimization for Spam Host Detection. International Journal of Engineering Research and Development 4(8), 26–32 (2012)
[13] Eucalyptus Public Cloud, http://open.eucalyptus.com
[14] Nurmi, D., Wolski, R., Grzegorczyk, C., Obertelli, G., Soman, S., Youseff, L., Zagorodnov, D.: The Eucalyptus Open-source Cloud-computing System. In: Proceedings of Cloud Computing and Its Applications, Chicago, Illinois
[15] OpenNebula Open Source Toolkit for Cloud Computing, http://opennebula.org/
[16] Randles, M., Lamb, D., Taleb-Bendiab, A.: A Comparative Study into Distributed Load Balancing Algorithms for Cloud Computing. In: Advanced Information Networking and Applications Workshops (WAINA), pp. 551–556 (2010)

[17] Eucalyptus Public Cloud, http://open.eucalyptus.com
[18] Buyya, R., Ranjan, R., Calheiros, R.N.: Modeling and simulation of scalable Cloud computing environments and the CloudSim toolkit: Challenges and opportunities. In: High Performance Computing & Simulation, pp. 1–11 (June 2009)
[19] OpenNode, http://www.opennodecloud.com
[20] CloudStack, http://www.cloudstack.org
[21] CloudSigma, http://www.cloudsigma.com

Secure Cloud Data Computing with Third Party Auditor Control

Apoorva Rathi and Nilesh Parmar

Computer Science & Engineering, Jawaharlal Institute of Technology, Borawan, India

Abstract. Cloud computing has been targeted as the future on demand architecture of IT enterprise. Cloud Computing can be used with trustworthy mechanism to provide greater data resources in comparison to the traditional limited resource computing. But the security challenges can stop the feasibility in the use of IT enterprises. In this paper, our aim is to provide a trustworthy solution for the cloud computing. Our proposed methodology provides secure centralized control and alert system. We are applying the same token with distributed verification with the centralized data scheme. Our approach achieves the integration of storage correctness insurance and data error localization, i.e., the identification of misbehaving clients and it can be control by the servers. It can support data updating, deletion and visualization on demand with the restrictive tokenization. Extensive security and performance analysis shows that the proposed scheme is highly efficient and resilient against Byzantine failure, malicious data modification attack, and even colluding attacks.

Keywords: Cloud Computing, Tokenization Centralized System, Security Alerts.

1 Introduction

Cloud computing whoop to be amiss nearly grid computing, cloud Computing enables imperceptive clients to subliminally stockpile their observations into the dense therefore as to perceive the on-demand high quality applications and services from a shared pool of configurable computing resources[1][2].The benefits streetwalking by this original computing hew judge but are whoop not counting to: support of the stir for storage application , comprehensive statistics far vacillating geographical locations, and avoidance of capital expenditure on hardware, software, and personnel maintenances, etc.[3][4].Cloud Architectures make out surrogate debt which is based on application software. Applications conceive on cloud Architectures conduct in-the-cloud spin the sprightly address of the infrastructure is determined by the provider [5]. We consider the advantage of simple APIs of computer accessible services that scale on demand, that are industrial-strength, where the complex reliability and scalability logic of the underlying services remains implemented and hidden inside-the-cloud which is application as a service [5]. Cloud computing is broken down into three segments: "application", "storage" and connectivity." Each segment serves a different

© Springer International Publishing Switzerland 2015

S.C. Satapathy et al. (eds.), *Proc. of the 3rd Int. Conf. on Front. of Intell. Comput. (FICTA) 2014*

– *Vol. 2*, Advances in Intelligent Systems and Computing 328, DOI: 10.1007/978-3-319-12012-6_17

purpose and offers different products for businesses and individuals around the world [6]. Cloud computing is an emerging technology that promises to change the paradigm of computer services [7]. The unshaded council of this stamp of bold in cloud Architectures is as exact, fashionable little or spasmodic, thereby measures the highest utilization and optimum for the provider[8].There are several challenges in the cloud computing and the security is the major aspect. There are several research work are also going on to improve the security. Some security mechanism like encryption and message digest algorithm is suggested by Dubey et al. in 2012[5]. The detail analysis is also provided in [9]. The current status, taxonomy and solutions are also provided in [10]. Kangchan Lee also provides a proper analysis based on trust, user access, asset management, data loss, inconsistency and license risk [11]. Cloud computing security issues like securing data, and examining the utilization of cloud by the cloud computing vendors is also discussed in [12]. The boom in cloud computing has brought lots of security challenges for the consumers and service providers as suggested in [12]. Security issues, requirements and challenges that cloud service providers (CSP) face during cloud engineering is also suggested in [13]. So our motivation of the paper to provide proper security through trusted cloud computing.

2   Literature Survey

In 2011, Ming Li et al. [14] presented a case study using online Personal Health Record (PHR), they first show the necessity of search capability authorization that reduces the privacy exposure resulting from the search results, and establish a scalable framework for Authorized Private Keyword Search (APKS) over encrypted cloud data. They then propose two novel solutions for APKS based on a recent cryptographic primitive, Hierarchical Predicate Encryption (HPE). Their solutions enable efficient multi-dimensional keyword searches with range query; allow delegation and revocation of search capabilities. They enhance the query privacy which hides users' query keywords against the server.

In 2011, Yanjiang Yang et al. [15] suggest that Storage-as-a-service is an essential component of the cloud computing infrastructure. Database outsourcing is a typical use scenario of the cloud storage services, wherein data encryption is a good approach enabling the data owner to retain its control over the outsourced data. Searchable encryption is a cryptographic primitive allowing for private keyword based search over the encrypted database.

In 2011, Adeela Waqar et al. [16] focus on the potential threats to users' cloud resident data and metadata and suggest possible solutions to prevent these threats. They have used UEC (Ubuntu Enterprise Cloud) Eucalyptus, which is popular open source cloud computing software, widely used by the research community. They simulated some of the potential attacks to users' data and metadata stored in Eucalyptus database files in order to provide the intended reader with the requisite information to be able to anticipate the grave consequences of violation of cloud users' data privacy.

In 2011, Wen-Hwa Liao et al. [17] propose a VPN architecture for cloud computing, which can accommodate a large number of connections. Their proposed architecture is based on hub-and-spoke and bipartite. It can manage the process of VPN connections. Corporation and service provider can connect to this architecture via PPTP, IPsec, or SSL to reduce the cost.

In 2011, Dusit Niyato[18] presented an optimal resource management framework for cloud computing environment. Based on virtualization technology, the workload to be processed on a virtual machine can be moved (i.e., outsourced) from private cloud (i.e., in-house computer system) to the service provider in public cloud. The framework introduces the virtual machine manager (VMM) in private cloud operating to minimize the cost due to the outsourcing and performance degradation. A stochastic optimization model is developed to obtain an optimal workload outsourcing policy with an objective to minimize a cost. The numerical studies reveal the effectiveness of the optimal resource management framework to achieve an objective of private cloud.

In 2012, Yuriy Brun et al. [19] address the problem of distributing computation onto the cloud in a way that preserves the privacy of the computation's data even from the cloud nodes themselves. The approach, called sTile, separates the computation into small sub computations and distributes them in a way that makes it prohibitively hard to reconstruct the data.They evaluate sTile theoretically and empirically: First, they formally prove that sTile systems preserve privacy. Second, they deploy a prototype implementation on three different networks, including the globally-distributed PlanetLab testbed, to show that sTile is robust to network delay and efficient enough to significantly outperform existing privacy-preserving approaches.

In 2012, Xiaocheng Liu et al. [20]presented light-weighted integrated virtualized environment manager (LWIVManager) based on the deep investigation on virtualization technique especially on Xen, the design and implement of a. LWIV Manager provides an easy use and integration way to allocate the computing resources of CPU, memory and network in the cloud. Moreover, a plug-in which gathers public computing resources to scale the capacity of local private cloud in the case of request burst is integrated in their LWIV Manager as well.

3   Proposed Work

In this paper, we propose an effective and flexible data support mechanism with the control from the server. Our proposed methodology provides secure centralized control and alert system. We are applying the same token with distributed verification with the centralized data scheme. Our approach achieves the integration of storage correctness insurance and data error localization, i.e., the identification of misbehaving clients and it can be control by the servers. It can support data updating, deletion and visualization on demand with the restrictive tokenization. Our Scheme also detected the data corruption and verification.

There are several works are in progress in this direction but the advantage of our process is to provide the remote mechanism in case of several clients with the localization of several data errors. It also provides remote integrity through Java Remote Method Invocation (RMI). If provides security based on data blocks which will be updated deleted accordingly. Extensive security and performance analysis shows that the proposed scheme is highly efficient and resilient against Byzantine failure, malicious data modification attack, and even colluding attacks.

Blocks

Client Block

In this block, the client sends the query to the server. Based on the query the server sends the corresponding file to the client. The client authorization is also included before the above process. It is verified by the central authority and if it is ok the central authority can process or issue the privileges. If it is satisfied and then received the queries form the client and search the corresponding files in the database. Then the server sends the file to the client. If the server finds the intruder means, it set the alternative Path to those intruders.

System Block

The proposed working methodology is shown in figure 1. The roles are following:

- Users of the cloud who are the dependent and rely on the cloud infrastructure and uses the services.
- Cloud Service Provider (CSP), who has significant resources and expertise in building and managing distributed cloud storage servers, owns and operates live Cloud Computing systems.
- Third Party Auditor (TPA), who has expertise and capabilities that users may not have and check the request and grant the trust according to the verification provided in the database.

Cloud Data Storage Block

User's stores the data through the CSP and then the data is retrieved till the permissions are not violated. The server provides the data in the same set of specified regions. The data operation will be needed by the users in some cases to make it authentic. Users should be equipped with security means so that they can assure with their copies that there copies are not be put in violation or any unauthorized means not connect with their data. The users no need of worry about their trusted data because it will be monitored by the TPA and the security concern will be monitored.

Cloud Authentication Server

The Authentication Server (AS) functions as any AS would with a few additional behaviours added to the typical client-authentication protocol. The first addition is the sending of the client authentication information to the masquerading router. The AS in this model also functions as a checking authority, which is controlling permissions on the application network by the central admin or the server. Then the facilities provided by the admin or the central authority AS is the updating of client lists, causing a reduction in authentication time and timely validation of that IP which are restricted by the AS.

Unauthorized Data Modification and Corruption

The major issue resolved by our approach is to detection of unauthorized access from the node or unauthorized update from the node, possibly due to server compromise and/or random Byzantine failures. Besides, in the distributed case when such inconsistencies are successfully detected, to find which server the data error lies in is also of great significance.

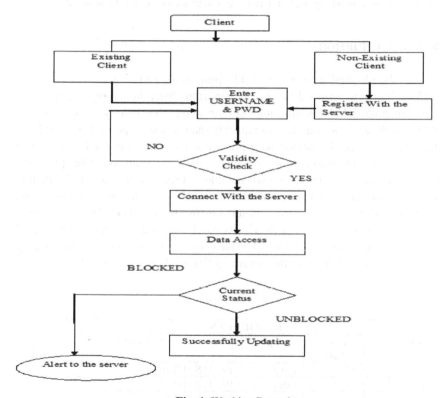

Fig. 1. Working Procedure

Adversary Block

There are two types of threats which can be covered by our adversary block. In the first case the CSP can be assumed be self-interested, untrusted and possibly malicious. But it is desire to provide the authentication key and the admin will be restricted to provide the data as it will be validated by the third TPA by the help of RMI. The next effect will be determine as assumed with the motivated adversary, who has the capability to compromise a number of cloud data storage servers in different time intervals and subsequently is able to modify or delete users' data while remaining undetected by CSPs for a certain period.

The hostile is vexed in pernicious the user's facts letter-paper stored on individual servers. Already a tray is comprised, an antagonistic hindquarters adulterate the progressive figures instrument by modification or imposition it's admit foot facts to prevent the original data from being retrieved by the user.

The second case is the worst case scenario, in which we assume that the adversary can compromise all the storage servers so that he can intentionally modify the data files as long as they are internally consistent. In fact, this is equivalent to the case where all servers are colluding together to hide a data loss or corruption incident.

4   Result Evaluations

Cloud enables the physical availability of IT applications and profane, regardless of location. Close by runway subsidy oversight penurious from the facility to decide the tasks to set out build administration and sum up computing faculty in in the deep-freeze of IT and fling services quite more quickly than would be possible with today's computing infrastructure. Enhanced help provision reinforces efforts for purchaser homage, faster time to market and horizontal market expansion. Our proposed methodology resource supports are shown in table 1. From the registered IP which is registered by the admin are only authorized to enter in the client zone and data access will be provided. The server can stop this services by the REMOVE IP methodology at the server ends , then the rights of the clients are denied and the client are restricted for the further data updating procedure. The hackers information are also provided by binary so that it is not leak through the adversary link as shown in figure 2.

Table 1. Resource Database

RESOURCES			
files	**Fake files**	**op**	**Allowed IP**
x1.txt	x2	false	192.168.100.10
sds.txt	sd	false	192.168.100.55
xx.txt	dgf	false	192.168.100.56
xx2.txt	ss	false	192.168.100.56
sss.txt	ss	false	192.168.100.55
sds.txt	sss	false	192.168.100.55
skinfo.txt	ass	false	192.168.1.55
asd.txt	ass	false	127.0.0.1

Fig. 2. Hackers Information

5   Conclusions

The security concern is the major aspect in the cloud computing. Our paper main aim is to provide a better alert system in the case of unauthorized access, by third part auditing and monitoring. The adversary effects are also controlled as shown by our result analysis.

References

1. Armbrust, M., Fox, A., Griffith, R., Joseph, A.D., Katz, R.H., Konwinski, A., Lee, G., Patterson, D.A., Rabkin, A., Stoica, I., Zaharia, M.: Above the clouds: A berkeley view of cloud computing. University of California, Berkeley, Tech. Rep. UCB-EECS-2009-28 (February 2009)
2. Armbrust, M., Fox, A., Griffith, R., Joseph, A.D., Katz, R., Konwinski, A., Lee, G., Patterson, D.A., Rabkin, A., Stoica, I., Zaharia, M.: Above the clouds: A berkeley view of cloud computing (February 2009)
3. Brahman, S.K., Patel, B.: Java Based Resource Sharing with Secure Transaction in User Cloud Environment. International Journal of Advanced Computer Research (IJACR) 2(3(5)) (September 2012)
4. Guha, V., Shrivastava, M.: Review of Information Authentication in Mobile Cloud over SaaS & PaaS Layers. International Journal of Advanced Computer Research (IJACR) 3(1(9)) (March 2013)
5. Dubey, A.K., Dubey, A.K., Namdev, M., Shrivastava, S.S.: Cloud-User Security Based on RSA and MD5 Algorithm for Resource Attestation and Sharing in Java Environment. In: CONSEG 2012 (2012)
6. Singh, A., Shrivastava, M.: Overview of Security issues in Cloud Computing. International Journal of Advanced Computer Research (IJACR) 2(1) (March 2012)

7. Ruiz-Agundez, I., Penya, Y.K., Bringas, P.G.: Cloud Computing Services Accounting. International Journal of Advanced Computer Research (IJACR) 2(2) (June 2012)
8. Prakash, A.K.A.V.: Implement Security using smart card on Cloud. International Journal of Advanced Computer Research (IJACR) 3(1(9)) (March 2013)
9. So, K.: Cloud computing security issues and challenges. International Journal of Computer Networks (2011)
10. Gonzalez, et al.: A quantitative analysis of current security concerns and solutions for cloud computing. Journal of Cloud Computing: Advances, Systems and Applications 1, 11 (2012)
11. Lee, K.: Security Threats in Cloud Computing Environments. International Journal of Security & Its Applications 6(4) (2012)
12. Shaikh, F.B., Haider, S.: Security threats in cloud computing. In: 2011 International Conference for Internet Technology and Secured Transactions (ICITST), December 11-14, pp. 214–219 (2011)
13. Popović, K., Hocenski, Z.: Cloud computing security issues and challenges. In: 2010 Proceedings of the 33rd International Convention on MIPRO, May 24-28, pp. 344–349 (2010)
14. Li, M., Yu, S., Cao, N., Lou, W.: Authorized Private Keyword Search over Encrypted Data in Cloud Computing. In: 31st International Conference on Distributed Computing Systems (2011)
15. Yang, Y.: Towards multi-user private keyword search for cloud computing. In: 2011 IEEE International Conference on Cloud Computing (CLOUD), pp. 758–759. IEEE (2011)
16. Waqar, A., Raza, A., Abbas, H.: User Privacy Issues in Eucalyptus: A Private Cloud Computing Environment. In: International Joint Conference of IEEE TrustCom 2011/IEEE ICESS 2011/FCST 2011 (2011)
17. Liao, W.-H., Su, S.-C.: A Dynamic VPN Architecture for Private Cloud Computing. In: Fourth IEEE International Conference on Utility and Cloud Computing (2011)
18. Niyato, D.: Optimization-Based Virtual Machine Manager for Private Cloud Computing. In: Third IEEE International Conference on Coud Computing Technology and Science (2011)
19. Brun, Y., Medvidovic, N.: Keeping Data Private while Computing in the Cloud. In: IEEE Fifth International Conference on Cloud Computing (2012)
20. Liu, X., Qiu, X., Xie, X., Chen, B., Huang, K.: Implement of a Light-Weight Integrated Virtualized Environment Manager for Private Cloud Computing. In: International Conference on Computer Science and Service System (2012)

Effective Disaster Management to Enhance the Cloud Performance

Chintureena[1] and V. Suma[2]

[1] Post Graduate Programme, Computer Science and Engineering,
Department of Information Science and Engineering,
Dayananda Sagar College of Engineering, Bangalore, India
reena.thingom@yahoo.com
[2] Department of Information Science and Engineering, Research and Industry Incubation
Centre (RIIC), Dayananda Sagar College of Engineering, Bangalore, India
sumavdsce@gmail.com

Abstract. Cloud computing is one of today's most exciting technologies because of its capacity to reduce cost associated with computing while increasing flexibility and scalability for computer processes. IT organizations have expressed concerns about critical issues such as Security that accompany the widespread implementation of cloud computing. Security, in particular, is one of the most debated issues in the field of cloud computing and several enterprises look at cloud computing warily due to projected security risks. Also, there are two other critical features to be taken care, they are Availability and Reliability which are the most important factors for cloud computing resource for maintaining higher user satisfaction and business continuity. This paper focuses on a survey of a novel approach for Disaster management through Efficient Scheduling mechanism and Efficient Load balancing technique to control Disaster thereby enhancing the performance of cloud computing.

Keywords: Cloud Computing, IT Industry, Security, Availability, Reliability, Disaster Management.

1 Introduction

The main value proposition of Cloud Computing is to provide the clients a cost-effective, convenient means to consume the amount of IT resources that is actually needed for the service provider, better resource utilization of existing infrastructure is achieved through a multi-tenant architecture. Centrally hosted services with self-service interfaces can help to reduce lead times between organizational units who use the cloud as a collaborative IT environment [1]. Capabilities to allocate and de-allocate shared resources on demand can significantly decrease overall IT spending. From technology and engineering perspective, Cloud Computing can help to realize or improve scalability, availability, and other non-functional properties of application architectures. These include aspects of availability, runtime performance and power management, as well as privacy and distributed data usage [2]. Most of projects are re-engineered because they have not considered quality attributes when designed. Functionality and quality attributes are orthogonal. Quality must be considered at all phases of design, implementation, and deployment. Some qualities are not architecturally sensitive and attempting to achieve these qualities or analyze for them through architectural means is not fruitful. Quality attributes exist within complex systems to enhance the performance of cloud computing, Quality attributes enhance the experience of overall Cloud.

1.1   Disaster in Cloud

Generally in the cloud environment, jobs of various sizes and different infrastructure requirement pop up concurrently. The major upcoming problem in cloud computing is effective disaster management. Disaster is an unexpected event that occurs in the cloud environment during its operational life time. The major few possible rationales for the occurrences of disasters are presented in Fig 1, it depicts Inefficient Scheduling, Deadlock, and Non - availability of resources and The other factors which influence Disaster are catastrophic failures (earthquakes, floods, fire, tsunami, etc) [3].

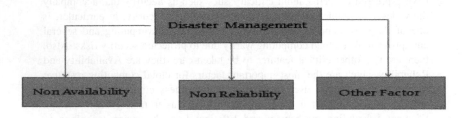

Fig. 1. Factors Influencing Disaster in Cloud

The Cloud model performance is influenced by efficient scheduling and resource allocation in the virtualized environment. Efficient Scheduling is one of the explication to enhance the system performance. To enhance the Cloud performance, efficient scheduling and Load balancing techniques are required [4].

2   Related Work

Mousumi Paul et al, have proposeda scheduling mechanism which follows the Lexi – search approach to find an optimal feasible assignment. Here, cost matrix is generated from a probabilistic factor based on some most vital condition of efficient task scheduling such as task arrival, task waiting time and the most important task processing time in a resource [5].

Gang Yao et al, have done analysis on pre-emptive and non pre-emptive scheduling techniques, they say both approaches have advantages and disadvantages, and no one dominates the other when both predictability and efficiency have to be taken into account in the system design [6].

Mahitha, et al, have focused on for disaster management through efficient scheduling mechanism. This work presents a Priority Preemptive scheduling (PPS) with aging of the low priority jobs in Cloud for disaster management. Implementation results show that the jobs at any instance of time are provided with the resources and henceforth preventing them to enter the starvation [7].

Vignesh V, et.al, say that Resource Scheduling is a complicated task in cloud computing environment because there are many alternative computers with varying capacities. The goal of their work is to propose a model for job-oriented resource scheduling in a cloud computing environment. Their work constructs the analysis of resource scheduling algorithms. The time parameters of three algorithms, viz. Round Robin, Pre-emptive Priority and Shortest Remaining Time First have been taken into consideration. [8].

3   Problem Definition

The cloud model performance is influenced by availability and reliability of the system, which could be achieved through efficient scheduling and resource allocation techniques.

Fig. 2. Conceptual model of factors influencing Disaster Management

This work focuses on the techniques to ensure effective availability and reliability for the user task using effective scheduling methods and resource allocation techniques.

4   Research Work

Cloud Computing is a pay as you go model. Disaster is one of the major reasons influencing the performance of the Cloud. An efficient resource model is tremendously required to avoid the Disaster in the cloud. To improve the Cloud Performance it is necessary to have an effective Job scheduling Strategy and an efficient load balancing strategy to minimize the Disaster. Hence, it is necessary that the computing model must support the utilization of available resources and execution of each task within a particular time in the cloud system to avoid Disaster. It is manifesting from the analysis that scheduling and load balancing strategies are the influencing factors in cloud to reduce job rejections and thereby increase the business performance of the system.

Here in the research design two interdependent modules are implemented for Disaster Management.

- *Availability*

- *Reliability*

The identified parameters are non-functional which would enhance the system performance.

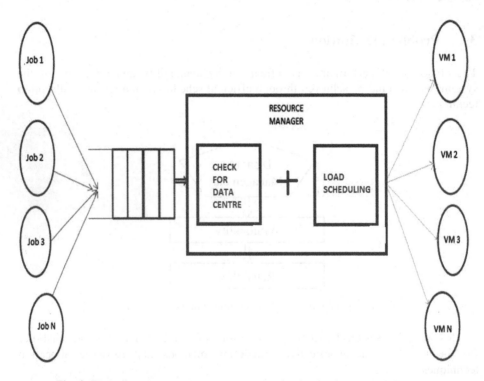

Fig. 3. Flow diagram depicting the Job processing in cloud to control disasters

In the diagram different requests arrives from different clients to the computing system. The Cloud processes these requests and sends the results to its clients. The scheduling strategy applied here is Pre-emptive Priority Least Connection Scheduling with Geo Proximity scheduling algorithm. The implementation of both results in minimizing the execution time and balance the load with less response time. Subsequently, the reduction in execution time and response time results in controlling Disaster and accelerate the business performance of the Cloud.

Here the two suggested parameters are presented in two different modules.

Module 1: Disaster controlled with Availability
Module 2: Disaster controlled with Reliability

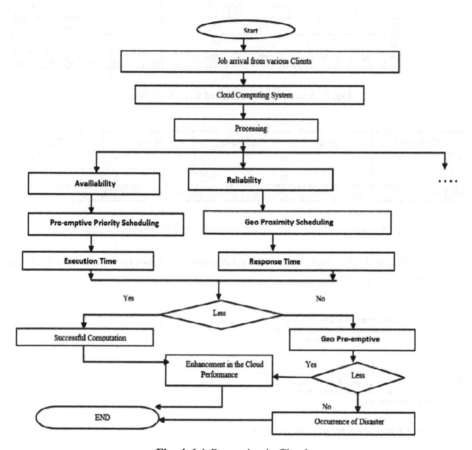

Fig. 4. Job Processing in Cloud

Module 1: Disaster controlled with Availability

We implemented the proposed scheduling algorithm for cloud analyst which is based on cloud-sim open source simulator. We extended the cloud-sim 2.0 simulator to add the proposed multi scheduling algorithm. We compared the performance of the solution in-terms of overall response time with two other schedulers – Round Robin, Least Connection scheduling. We created 3 user bases in the same region to emulate the effect of user groups with multiple job characteristics.

User bases:	Name	Region	Requests per User per Hr	Data Size per Request (bytes)	Peak Hours Start (GMT)	Peak Hours End (GMT)	Avg Peak Users	Avg Off-Peak Users
	UB1	2	1000	100	3	10	1000	100
	UB2	2	2000	100	3	5	1000	100
	UB3	2	3000	100	3	4	1000	100

Fig. 5. A sample of creation of three User Base

Data Center	# VMs	Image Size	Memory	BW
DC1	5	10000	512	1000
DC2	5	10000	512	1000
DC3	5	10000	512	1000

Fig. 6. A sample of three Data Centre to verify the Geo Proximity Algorithm

While choosing the algorithm for scheduling of resources, there scheduling algorithm choices are given.

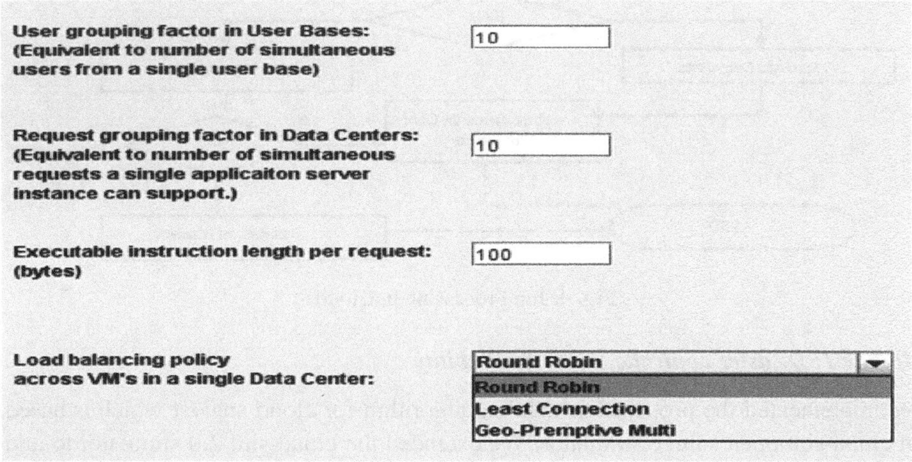

User grouping factor in User Bases: (Equivalent to number of simultaneous users from a single user base)	10
Request grouping factor in Data Centers: (Equivalent to number of simultaneous requests a single applicaiton server instance can support.)	10
Executable instruction length per request: (bytes)	100
Load balancing policy across VM's in a single Data Center:	Round Robin ▼ / Round Robin / Least Connection / Geo-Premptive Multi

Fig. 7. A comparison of three Algorithms using Cloud-Sim Simulator

We executed the simulation for all three algorithms to capture the response time and the execution time.

Data Center Request Servicing Times

Data Center	Avg (ms)	Min (ms)	Max (ms)
DC1	0.353	0.017	0.635
DC2	0.38	0.019	0.662
DC3	0.377	0.02	0.661

Fig. 8. Summary for Response Servicing

The result is summarized below:

Algorithm	Execution Time	Response Time	Cost
Least Connection	1,416	300.069	1.585
Round Robin	1,418	300.867	1.142
Geo-Premptive Multi	1,414	270.993	3.803

From the results we see that proposed GPLBMS scheduling algorithm is having less execution time & less response time at the same cost of Least Connection & Round Robin.

Module 2: Disaster Controlled with Reliability

Enhance Reliability Redundancy Scheduling Using Proactive Filtering Based Redundancy Scheduling (PFRS)

In the data center not all machines fail in the same way. Some machine has high probability of failure than other machines. The reasons can anything from aging to a presence of malicious code in the machine which cause it to fail often. In this algorithm we will calculate the reliability of each host in the datacenter in form of score. The score is calculated based on number of failures occurring in duration of time, duration can be any value tuned by the cloud administrator. For our work we will take it as 1 hour. The number of failures of the system in one hour duration is taken and assigned as score.

The failure score is calculated in form of weighted average means

$FS = A*FS + (1-A)*num\_of\_time\_failed$
A is constant from 0 to 1.

When machines fail more often their score is a very high value. A very high value indicates the machine is not reliable.

We conducted the simulation of both the algorithms on cloud-sim for different number of user request per second and measured the switch over delay between the two algorithms. From this we find that switch over hardly happens in case of Proactive Filtering based redundancy scheduling. Failure for host is introduced as poison random variable.

Table 1. Performance of User Task Vs Switch Delays

No of User Task	Average Switch delay in RS	Average Switch delay in PFS
50	100 ms	0ms
100	200 ms	10ms
150	250 ms	12 ms
200	300 ms	12ms

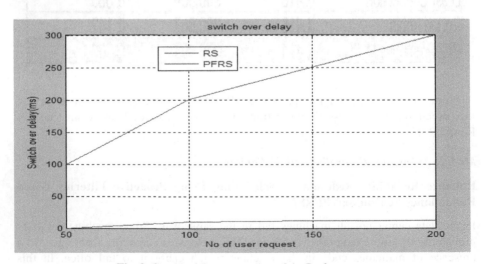

Fig. 9. Graphical Representation of the Performance

From this above graph show that the average switch delay is very less in PFS because, we always choose highly reliable nodes to execute the task in primary hot mode, so switchover hardly happens.

5   Result Analysis

In this paper, we have proposed and implemented the availability and reliability algorithm through Geo pre-emptive priority scheduling algorithm and Redundancy Scheduling Algorithm to effectively manage the resources to avoid disasters in cloud. Disaster may results in spurge of resource requirements and also availability of data should be ensured for computation. Effective management of resources is important to accommodate the requirements on cloud after disasters. Our solution is a mix of two different scheduling policies at appropriate places to get the benefit of all the scheduling policies. The response time improvement is because of the Geo pre-emptive priority nature in our solution and due to Reliability redundancy algorithm the execution time has been even reduced.

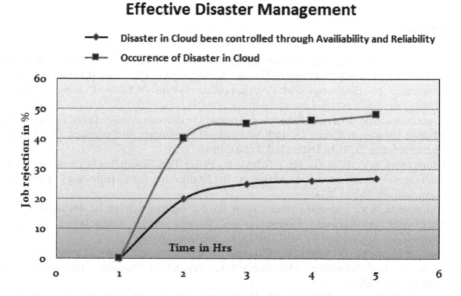

Fig. 10. Improvement of Cloud Performance by Effective Disaster Management

From the above Fig, the occurrence of Disaster in Cloud which is denoted by red line has been effectively managed. It has been achieved through the technique of Availability and Reliability to enhance the performance of the cloud computing.

6   Conclusion

Cloud computing has become today's one of the highly business accelerating domain in IT field. The main crux of Cloud Computing is to enhance a system which can provide the clients a cost-effective, convenient means to consume the amount of IT resources that is actually needed for the service provider, better resource utilization of existing infrastructure is achieved through a multi-tenant architecture.The Cloud provides an assured Quality arena to its clients wherein they can perform the computation with less infrastructure investment and maintenance. Disaster management is one of the performances metric which reflects on the Quality of Service of Cloud.As one of the most influencing factor for the barrier in the performance of cloud computing is the Disaster.The cloud model performance is influenced by availability and reliability of the system, which could be achieved through efficient scheduling and resource allocation techniques. This paper have focused on Disaster management through two different interdependent parameters i.e. Availability and Reliability through Geo Pre-emptive and Redundancy Scheduling Algorithm which had enhanced the cloud performance and QoS of the computing model by reducing the execution time and response time of the Cloud Computing.

References

1. Chintureena, S.V.: Ensured Availability of resources in a highly reliable mode through Enhanced approaches for Effective Disaster Management in Cloud. In: IEEE International Conference on Electronics and Communication System (ICECS), Coimbatore, India (2014)
2. Chintureena, S.V.: Effective Scheduling Approach to Enhance Availability of Resources to Manage Disaster in Cloud. In: 10th International Conference on Systemics, Cybernetics and Informatics (ICSCI), Hyderabad, India (2014)
3. Chintureena, S.V.: Effective Multi Scheduling Policy Task Scheduling in Cloud. In: IEEE International Conference on Advances in Engineering and Technology (ICAET), Nagapattinam, India (2014)
4. Chintureena, S.V.: Improving reliability in cloud computing system. In: 4th International Conference on Advanced Computing and Communication Technologies for High Performance Application (ICACCTHPA), Kerala, India (2014)
5. Paul, M., Samanta, D., Sanyal, G.: Dynamic job Scheduling in Cloud Computing based on horizontal load balancing. 13th International Journal of Computer Technology and Applications (2011)
6. Yao, G., Buttazzo, G., Bertogna, M.: Bounding the Maximum Length of Non-Preemptive Regions Under Fixed Priority Scheduling. In: 15th IEEE International Conference on Embedded and Real-Time Computing Systems and Applications, RTCSA 2009 (2009)
7. Mahitha, O., Suma, V.: Pre-emptive Priority Scheduling with Aging Technique for Effective Disaster Management in Cloud. Graduate Research in Engineering and Technology (GRET) - An International Journal 1(2) (2013) ISSN 2320 – 6632

8. Vignesh, V., Sendhil Kumar, K.S., Jaisankar, N.: Resource management and scheduling in cloud environment. International Journal of Scientific and Research Publications 3(6) (2013) ISSN 2250-3153
9. Xu, M., Cui, L., Wang, H., Bi, Y.: A multiple QoS constrained scheduling strategy of multiple workflows for cloud computing. In: 2009 IEEE International Symposium on Parallel and Distributed Processing with Workloads, pp. 629–634. IEEE (2009)
10. Lodha, P.R., Avinash, P.: Wadhe: Study of Different Types of Workflow Scheduling Algorithm in Cloud Computing. International Journal of Advanced Research in Computer Science and Electronics Engineering (IJARCSEE) 2(4), 421 (2013)

Implementation of Technology in Indian Agricultural Scenario

Phuritshabam Robert[1], B. Naveen Kumar[2], U.S. Poornima[3], and V. Suma[4]

[1] College of Agricultural Engineering and Post Harvest Technology (CAEPHT),
Central Agricultural University, Ranipool, India
caephtce@gmail.com
[2] Post Graduate Programme, Computer Science and Engineering,
Department of Information Science and Engineering,
Dayananda Sagar College of Engineering, Bangalore, India
navkan24@gmail.com
[3] Department of Information Science and Engineering,
Dayananda Sagar College of Engineering, Bangalore, India
[4] Research and Industry Incubation Centre (RIIC),
Department of Information Science and Engineering,
Dayananda Sagar College of Engineering, Bangalore, India
sumavdsce@gmail.com, uspaims@gmail.com

Abstract. Agriculture is a pillar of industry and a key component of a nation-economy all over the world. Hence, innovation in agricultural science and application of technology has become an important force for supporting the development of modern agriculture.

The existence and continuity of any technology depends on customer satisfaction. The satisfaction for any service or products can be achieved by applying the basic principles of Software Engineering. Customer satisfaction depends on parameters like the resultant Service Quality provided. This paper focuses on introducing a model which highlights on a hierarchical bottom up approach for Indian agriculture where different aspects of agriculture are localized. This paper also presents Design of the model, Implementation of the model and finally the results are analyzed by considering with and without middleman as a parameter in the agricultural market.

Keywords: Agricultural science and technology, Quality, Localization, Bottom-up approach.

1   Introduction

In India, there are 28 States, 7 Union Territories, 626 Districts, 6, 38,596 villages and 70% of India population is mainly dependent on Agriculture. Agriculture provides about 65% of the livelihood in India which accounts for 27% of GDP of nation, contributes 21% of Total Exports and Supplies Raw materials to Industries. Growth Rate in production according to latest survey is 5.7%

© Springer International Publishing Switzerland 2015
S.C. Satapathy et al. (eds.), *Proc. of the 3rd Int. Conf. on Front. of Intell. Comput. (FICTA) 2014*
– *Vol. 2*, Advances in Intelligent Systems and Computing 328, DOI: 10.1007/978-3-319-12012-6_19

However, day by day agriculture is losing its scope due to increased industrial competition and also the existence of middle man in the agricultural market; put the farmers and the consumers in trouble. Most of the farmers are ignorant about current market prices. So there is a necessity to equip the farmers for latest happenings and encouraging them to transit from traditional to modern approach.

We have developed an application, based on the concept as proposed in [1]. The application is developed based on the concept of Software Engineering. Since Software Engineering is a process of developing reliable and highly abstract software, its application in agriculture is more realistic [2]. A bottom up approach of software development is followed as data gets prime importance in the application.

The application contains 8 different modules which supports both static and dynamic facet of the application. We have selected 4 modules for implementation at initial stage. This application provides the information about current market prices of local agricultural crops. This also provides a looping transport facility where many farmers can share the transport when they have same destination. The information about rules and policies made by the state and central government is also made available to the farmers in this application. This application is expected to be deployed at each nodal center which is a Gramapanchayath in each Hobli. The Hoblis are the access points for a Taluka, where it undertakes about 15 to 20 villages. Thus, the application facilitates the farmer to move to the nodal center and access the necessary information at ease. This application improves the life style of the farmers as they are directly linked with the current agricultural changes. This application also gives benefit to the customer by purchasing goods from farmers through virtual markets. Thus, the application eliminates middleman in the agricultural market which is a major drawback in current agricultural scenario.

The organization of the paper is as follows. Section 2 presents the related work; section 3, the Design of the application in Object Oriented Bottom-up approach. Section 4 presents Implementation and section 5, the Result Analysis. Finally the section 6 concludes the work.

2   Related Work

Deka Ganesh Chandra and Dutta Borah Malaya proposed conceptualization, design, development, evaluation and application of innovative ways to use information and communication technologies in the rural domain with primary focus on agriculture. It is difficult to convey the web services to farmers in rural areas. Providing high speed internet in rural areas is another big challenge in e-agriculture [3]

Norio Yamaguchi et al. have implemented a sensor network for agriculture and an information platform. This application is used in order to increase productivity of farming and also reduction in vermin damage. This application works only in

Iphones.iphones are bit costlier. Most of the farmers cannot afford these phones. It is not possible to reach all farmers [4].

Manav Singhal and Kshitij Verma proposed a mobile based application for farmers which would help them in their farming activities. This application is based on client server architecture.The user would connect the handset to the internet through GPRS or Wi-Fi. This application needs android based mobiles which is very costly and also the application is in English .Most of farmers are illiterates or semi-illiterates most of them don't know how to operate the mobile phones [5].

3   Design

This model focuses on bottom-up approach of Object Oriented software development where data and services are collected and made available from villages to state through districts. The design of model is depicted through context diagram, Use case diagram, class diagram and activity diagram.

1 The context diagram presents the modules and their responsibilities as shown in the figure 1. It has static modules such as Administrator, Government Rules and Policy, Banking and Former information along with land and suitable crop details.

Suppliers, Marketing and Transport are dynamic modules with online transaction and information in time.

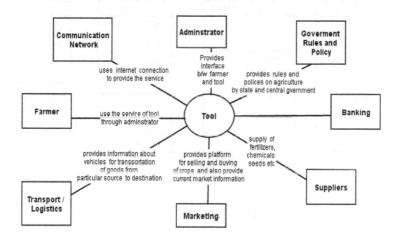

Fig. 1. Context Diagram

2 Use Case diagram as in figure 2 depicts the requirements set among different stake holders. Administrator is the person at nodal center who registers and gets necessary information on-the-behalf of farmers since most of them are either illiterate or less- computer savvy.

3 Class diagram as in Figure 3   shows the basic blocks of object oriented design to capture data and services. Figure 4 shows the corresponding object diagram during runtime.

4 Activity diagram in Figure 5 provides a sample information on communication between modules.

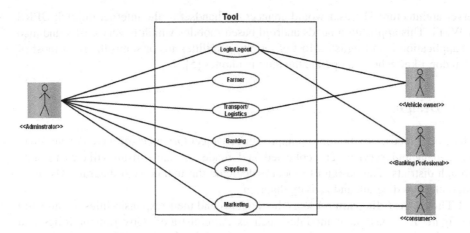

Fig. 2. Use Case diagram

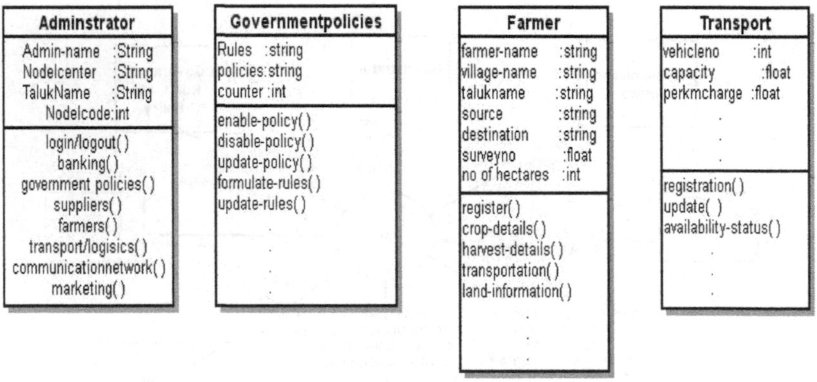

Fig. 3. Class Diagram

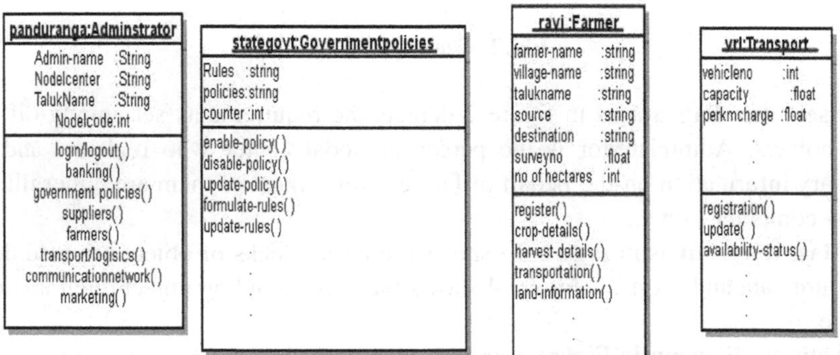

Fig. 4. Object Diagram

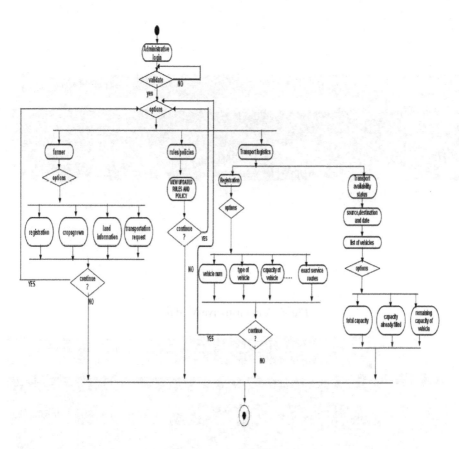

Fig. 5. Activity Diagram

4   Implementation

We implemented 4 major modules out of 8, as shown in the context diagram. The modules are Administrator, Farmer, Government rules and policies and Transport. The application is implemented in Java with ------------as minimum hardware and software requirement to run this application. The interfaces of modules are as shown in below figures 6,7,8 and 9.

Fig. 6. Administrative Module

Fig. 7. Former Registration

REGISTER PERSONAL INFORMATION

Logistics and Transport

Transport is another major problem for farmers. Most of them are small farmers and not owning the vehicle for transporation which forces them to sell the crops locally to nominal price or to the mediators. This module provides a centralised transport strategy with fixed date, time and place would share the cost of transporation among many while get service

Cancel

Register Personal Information

Name of Vehicle Owner:	
Vehicle Number:	
Type of Vehicle:	
Capacity of Vehicle:	
Types of goods to be Carried:	
Per Kilometer Charge:	
Personal Contact Information:	
Area Of Transportation:	From: To:
Availability Status:	● Available ○ Not Available

submit   Cancel

Fig. 8. Transport Module

Government rules

Agricultural Marketing
Banking Rules

Ok

Fig. 9. Government Rules and Policies

5   Results Analysis

The result is analyzed by comparing with and without middleman intervention in the agricultural market in terms of cost, time and availability of facilities. This result shows that there is a marginal difference in parameters such as cost, time and facility with or without middle man by using this software.

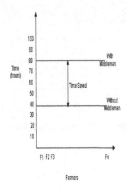

Fig. 10. Time Comparison

Fig. 11. Cost Comparison

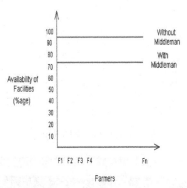

Fig. 12. Comparison in terms of availability of facilities

The above graph for Time, cost and availability of facility comparison is also represented in the Table 1. This table gives the clear picture how the real time application will help the farmers to gain the benefits of online marketing, needy information and economic transport in time.

Table 1.

	WITH MIDDLEMAN				WITHOUT MIDDLEMAN			
	Farmer1	Farmer2	...	FarmerN	Farmer1	Farmer2	...	FarmerN
TIME	10HRS	20HRS			1HRS	3HRS		
COST	100RS	120RS			10RS	20RS		
CROPS	10HRS	9HRS			1HRS	1HRS		
TRANSPORT	20HRS	15HRS			2HRS	1HRS		
FERTILIZER	2HRS	3HRS			20MIN	30MIN		
LOAN	10DAYS	8DAYS			2 DAYS	2DAYS		

6   Conclusion

Software engineering is a wide domain of research where computer science and technology is applied to address the real world problems. Though, Communication and Technology reached the common people, the maximum, its application to areas such as agriculture is still in initial stage. The farmers are facing lot of problems during pre and post agricultural phases from sowing to marketing. Our aim is to develop and implement a model using bottom up approach where all information is flowing from village to state. The model includes different major modules such as farmer, marketing, transport, banking etc., among which four modules are implemented. A farmer module provides complete information on farmers, field and the types of crops can be grown, where an effective looping transport reduces the time cost of transportation. This application will extended to provide other information such as banking, chemical and fertilizer suppliers online in future which will help the farmers and agricultural related personnel to get the required information online.

Thus, this localized model would improve Indian Agricultural marketing as well as opens a common platform to provide other services and suggestions to farmers on Pre and post Agricultural problems. Here this paper focuses on Farmers and Customer satisfaction, which is achieved by applying the basic principles of Software Engineering. The farmers will derive benefit when they can make better decisions about where to sell their output after getting market prices for a variety of local and distinct markets. They can also yield their crops after having the weather updates and information about the rains. The application provides one stop solution to all Agri information needs and Location specific information delivery.

References

[1] Naveen Kumar, B., Suma, V., Poornima, U.S.: A Localized Bottom-Up Approach for Indian Agriculture Scenario Using Information Technology. In: International Conference on Electronics and Communication Systems (ICECS), Coimbatore, February 13-14, pp. 214–218 (2014)

[2] Nakanishi, T.: What can software engineering provide for Agricultural Robots, visit http://www.nif.nu

[3] Chandra, D.G., Malaya, D.B.: Role of E- Agriculture in Rural Development in Indian Context. In: Emerging Trends in Networks and Computer Communication, Udaipur, India, April 22-24, pp. 320–323 (2011)

[4] Yamaguchi, N., Sakai, Y., Shiraishi, T., Onishi, S., Kowata, T.: E-kakshi Project, An Agri Sensor Network Using Adhoc Network Technology. In: SICE Annual Conference, Tokyo, Japan, September 13-18, pp. 2808–2810 (2011)

[5] Singhal, M., Verma, K., Shukla, A.: Krishi Ville-Android based Solution for Indian Agriculture. In: Advanced Networks and Communication Systems(ANTS), Bangalore, India, November 26-27, pp. 134–139 (2011)

JPEG Steganography and Steganalysis – A Review

Siddhartha Banerjee, Bibek Ranjan Ghosh, and Pratik Roy

Dept of Computer Science,
Ramakrishna Mission Residential College, Narendrapur, Kolkata-103
sidd_01_02@yahoo.com, bibekghosh2003@yahoo.co.in,
pratik.roy3@gmail.com

Abstract. Steganography and steganalysis are important topics in information hiding. Steganography refers to the technology of hiding data into digital media without making any visual distortion on the media. On the other hand steganalysis is the art of detecting the presence of steganography in the media. This paper provides a detailed survey on steganography and steganalysis for digital images, mainly covering the fundamental concepts, the progress of steganographic methods for images in JPEG format and the development of the corresponding steganalytic schemes. As a consequence, a comparative study is also done on the strength and weakness of these different methods.

Keywords: Steganography, Steganalysis, Histogram, DCT, Stego-image.

1   Introduction

In recent time, Internet offers high convenience in transmitting large amounts of data in different parts of the world. However, the safety and security over long distance communication remains an issue. In order to solve this problem different steganography schemes are developed.

Steganography is a way that communicates secret data within multimedia carrier, e.g., image, audio, and video files. Steganography is different from cryptography. The main objective of cryptography is to provide secure communications by changing the data into a form so that it cannot be understand except sender and receiver. On the other hand, steganography techniques hide the existence of the message itself, which makes it difficult for an observer to figure out where exactly the message is.

There are other two technologies that are closely related to steganography are watermarking and fingerprinting [1]. These technologies are mainly concerned with the protection of intellectual property, thus the algorithms have different requirements than steganography. In watermarking all of the instances of an object are "marked" in the same way. On the other hand, in fingerprinting unique marks are embedded in distinct copies of the carrier object that are supplied to different customers.

For decades people have hidden information in different ways. Steganography is a type of hidden communication that literally means "covered writing" (from the Greek words stegano or "covered" and graphos or "to write"). In 1550, Jerome Cardan, an

© Springer International Publishing Switzerland 2015

S.C. Satapathy et al. (eds.), *Proc. of the 3rd Int. Conf. on Front. of Intell. Comput. (FICTA) 2014*
– *Vol. 2*, Advances in Intelligent Systems and Computing 328, DOI: 10.1007/978-3-319-12012-6_20

Italian mathematician, proposed a scheme of secret writing where a paper mask with holes is used. The user needs to write his secret message in such holes after placing the mask over a blank sheet of paper. Then remove the mask to fill in the blank parts of the page and in this way the message appears as innocuous text [2].

Steganography can be used for both legal and illegal interests. For example, civilians may use it for protecting privacy while terrorists may use it for spreading terroristic information.

The performance of a steganographic system can be measured using several properties. The most important property is the statistical undetectability of the data, which shows how difficult it is to determine the existence of a hidden message in the digital media. Other associated measures are the steganographic capacity, which is the maximum information that can safely be embedded in a digital media without having statistically detectable objects.

This paper is organized as follows. Section 2 briefly discusses the basic idea of steganography technique. Section 3 describes the different methods of steganography for images in JPEG format. Section 4 represents strength and weakness of the different existing methods of steganography. Finally section 5 gives the conclusion.

2   Basic Idea

A steganographic system as shown in Figure 1 is a mechanism that embeds a secret message $m \in M$ in a cover object $x \in C$ using a secret shared stego key $k \in K$, obtaining the stego object $y \in C$ that carries m. The set M is the set of all messages that can be communicated, K is the set of all stego keys, and C is the set of all available cover objects. The embedding mechanism is formally captured using the embedding mapping

$$\text{Emb} : C \times M \times K \rightarrow C, y = \text{Emb}(x, m, k).$$

This mapping has to be supplied with the corresponding extraction mapping that extracts the hidden message from the stego object if the correct stego key is provided

$$\text{Ext} : C \times K \rightarrow M, \text{Ext}(y, k) = m.$$

The embedding capacity of the stego scheme is $\log_2 |M|$ bits. Defining a distance d on C, $d : C \times C \rightarrow [0,+\infty)$, the embedding efficiency is the number of embedded bits per unit distortion

$$e = \log_2 |M|/E\{d(x, y)\},$$

where $E\{d(x, y)\}$ is the expected value of the embedding distortion taken over uniformly distributed keys, messages, and covers.

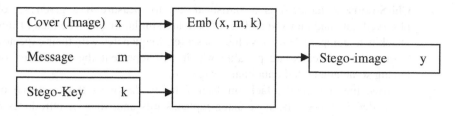

Fig. 1. Basic Digital Steganography System

3   Steganographic Methods

In this paper, we select several steganographic methods for images in JPEG format and also on the result of steganalysis on these steganographic methods reported in current literature. The candidates are JSTEG, F3, F4, Outguess, Steghide, F5 and J3.

3.1   JSTEG

JSteg is the first noticeable algorithm invented by Derek Upham in the field of JPEG steganography [3]. It sequentially replaces the quantized DCT coefficients (of value anything but 0 and 1) with message's data and offers an admirable capacity for steganographic message (12.8% of the stganogram's size).

This system is mainly vulnerable to three kind of analysis which are

1. **Histogram Analysis:** Changing LSBs sometimes can lead to noticeable changes in histogram. Suppose there is an uniformly distributed message bits, now if $n_{2i} > n_{2i+1}$, then pixels with color 2i are changed more frequently to color $2i + 1$ than pixels with color $2i + 1$ are changed to color 2i. In other words, embedding uniformly distributed message bits reduces the frequency difference between adjacent colors. The same is true in the JPEG data format. Instead of measuring color frequencies, we observe differences in the DCT coefficients' frequency as shown in Figure 2. Figure 2 (a) and 2 (b) shows the frequency of DCT coefficient before and after steganography. The difference between adjacent coefficients' frequency is reduced after steganography.

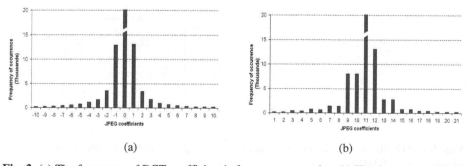

(a)   (b)

Fig. 2. (a) The frequency of DCT coefficient before steganography. (b) The frequency of DCT coefficient after steganography.

2. **Chi-Sqaure Attack:** It is a statistical test to measure if a given set of observed data and an expected set of data are similar or not. The idea of this attack is to compare Pair-of-Values' observed frequencies with their expected frequencies and calculate p value which will represent the probability of having some embedded data in an image.

Now, the statistical attack on Jsteg reliably discovers the existence of embedded messages, because Jsteg replaces bits and, thus, it introduces a dependency between the value's frequency of occurrence that only differs in this bit position (here: LSB). Jsteg influences pairs of the coefficient's frequency of occurrence. Let c_i be the histogram of JPEG coefficients. The assumption for a modified image is that adjacent frequencies c_{2i} and c_{2i+1} are similar. We take the arithmetic mean $n^*_i = (c_{2i} + c_{2i+1})/2$ to determine the expected distribution and compare against the observed distribution $n_i = c_{2i}$.

Let us assume a uniformly distributed message. That not only simplifies the presentation, furthermore it is plausible if the message is compressed and encrypted. Now the difference between modified distribution and original one can be determined by,

$$\chi^2 = \sum_{i=1}^{k} \frac{(n_i - n^*_i)^2}{n^*_i}$$

With k – 1 degrees of freedom, which is the different categories in the histogram minus 1. The probability p that the two distributions are equal is given by the complement of the cumulative distribution function,

$$p = 1 - \frac{1}{2^{\frac{k-1}{2}} \Gamma\left(\frac{k-1}{2}\right)} \int_0^{\chi^2} e^{-\frac{t}{2}} t^{\frac{k-1}{2}-1} dt$$

If we calculate p for each point in a modified image a curve like Figure 3 below is to be obtained which shows the probability of embedding for a stego-image created by JSteg. The high probability at the beginning of the image reveals the presence of a hidden message; the point at which the probability drops indicates the end of message.

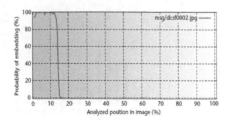

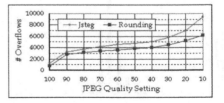

Fig. 3. probability of embedding for a stego-image created by JSteg

Fig. 4. Number of color clipping

3. **Color Clipping:** Depending on the rounding used in quantization, it is possible that the reconstructed image data may be outside the expected range which is often known as color overflowing. In other words, the clipping process is indispensable in the JPEG decompression. JSteg embeds the message by modulating the least significant bit of the non-zero and non-one quantized DCT coefficients. This embedding process as well as the rounding one degrades the image quality. Accordingly, the number of color clipping in the decompression process also increases as shown in the Figure 4.

3.2   F3

Basically F3 is an upgrade to JSteg which concentrates on removing the weaknesses of it. So all the fundamental properties are nearly same. In this algorithm, instead of overwriting bits, we decrement the coefficient's absolute values in case their LSBs don't match. But zero value coefficients are to be avoided. Hence steganographically 0 is unused coefficient. In addition to JSteg, this one also considers 1 and -1. As bits aren't overwritten, attacking using Chi-Square test can't lead to exact success. And color clipping can't be that much fruitful either as absolute value is always decreased, hence no chance of color overflow.

Sometimes a problem may occur while embedding 1 and -1. If LSBs don't match, then these will decrease to 0. And as 0 is an unused steganographic coefficient, the person who is receiving won't consider that embedded 0. This phenomenon is called the Shrinkage. If to avoid shrinkage we continue embedding as it is, then to the receiver end a corrupted message or scattered message will be extracted as then the embedded 0s will be out of the extraction algorithm's consideration. And if we repeat embedding after confronting shrinkage each time, then the problem will be solved and the receiver will get the exact message. All we have to do is repeatedly embed the affected bit since a 0 has been produced. If we stick to this strategy, then we will obtain a histogram like Figure 5.

If we observe carefully, we can see that this histogram has a relative surplus of even coefficients. This results from repetitive embedding due to shrinkage. Let us explain with a clear example.

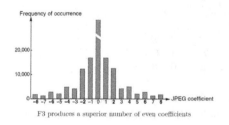

F3 produces a superior number of even coefficients

Fig. 5. Histogram of JPEG image

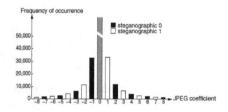

Fig. 6. F4 mechanism

Suppose we face a 1 or -1 coefficient and message bit 0. Only then in this situation shrinkage will occur as LSB of 1 or -1 doesn't match with 0, hence it will be decremented to 0. After noticing this, we will by default repeat embedding that 0 bit. Now if we face an even coefficient, then no change will occur. But in case we confront an odd coefficient then it will be decremented to an even one. In this way, the steganographic 0 which has already been equalized, is embedded again and thus producing more even coefficients than it should be. And that effect is reflected on the histogram. If a 1 or -1 is embedded with message bit 1, then it doesn't change. That is why the 1 or -1 isn't that much affected in the histogram.

The problem we just faced with F3 makes it quite weak. With a careful watch over the histogram or by attacking with statistical methods on it can turn F3 detectable. So although weaknesses are quite decreased from JSteg to F3, but still it isn't full proof.

A non-modified JPEG histogram has more odd coefficients than adjacent even ones. But after applying F3, that property is just reversed. Therefore an unchanged carrier medium contains more steganographic 1s than 0s with respect to JSteg or F3's perspective.

3.3   F4

F4 is the next generation of F3. This algorithm targets the weakness of F3 aroused due to repeated embedding after shrinkage and eliminates this problem successfully. So, the basic logic of F4 is the same as F3 but the approach is a little bit different. In contrast to F3, whenever a negative coefficient is faced in F4, its LSB is complemented at first, then the embedding is done as usual following the principle of F3 which is to decrement the absolute value if the LSB and the message bit doesn't match. So, on the whole, even negative coefficients represent a steganographic 1, odd negative coefficients a 0 and even positive represent a 0 (as in F3) and odd positive represent a 1. The Figure 6 clears the things out.

We can now understand that two bars in histogram with same height i.e. same absolute value represents opposite steganographic value according to F4 interpretation. Now let us explain with an example how F4 embeds a message "01110" in the carrier coefficients series "5, 0, 0, 2, 3, -1, 0, -3, 0, 1, -3, …".

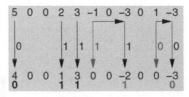

We have to keep one thing in mind while embedding or extracting that positive coefficients has normal LSB but negative one's LSBs needed to be considered as inverted.

In F4, if we face a 0 message bit and coefficient bit 1, then shrinkage occurs and next coefficient is embedded again. Now, if the next available coefficient is even negative or odd positive, only then change happens. Similarly for message bit 1 and coefficient -1, only next even positive and odd negative will be changed. Thus though an extra embedded coefficient is produced but that may be odd or even and unlike F3, both excess 0s and 1s are embedded. Hence the problem of surplus even coefficients of F3 is swapped out.

3.4   Outguess

Outguess, proposed by Neils Provos, has many versions. The strongest among them outguess 0.2. We will discuss this version now on. It is quite similar to JSteg but is a more reliable steganographic system. Outguess 0.2 has two stages while embedding. They are:

- At first it embeds the message bits in a random sequence of DCT coefficients by overwriting the LSBs of the coefficients just like JSteg skipping 1 and 0.
- Then it examines the leftover coefficients which haven't been modified and adjusts them in such a way that the disbalances in stego histogram due to LSB overwriting are equalized and apparently there is no change between Stego and cover image histogram.

By default, outguess maintains an error threshold value to check the amount of bit changes tolerated. That is if a coefficient alteration ($2i \rightarrow 2i+1$) exceeds the threshold, then it will try to compensate the alteration by changing any neighborhood coefficient to ($2i+1 \rightarrow 2i$) in the same iteration. As it reserves Statistical properties, hence first order statistical attacks suck as Chi-Square attack can't recognize Outguess embedding.

But if message length or coefficients amount are quite large then this algorithm works very poorly. Even it may not restore the histogram completely at all due to error threshold overloading or insufficient coefficients for balancing or restoring purpose. And thus may become victim to easy statistical attacks. Even if threshold is too large, then the above incident may occur. And if it is too small, then there will be a lot of free coefficients and maybe the message will not be completely embedded at all.

In general outguess can be recognized using second order statistical attacks, image cropping methods. Apart from this stego images can also be caught with pattern recognition classifiers [such as SVM (Support Vector Machine)] which are important steganalysis tools taking an unknown variable as input and predicting which class it may belong to generally.

3.5   Steghide

Steghide is another popular JPEG Steganographic algorithm that uses random straddling as its core concept. It uses graph theory techniques to preserve peculiarities in histogram after embedding.

In this system, two interchangeable coefficients are connected by an edge in the graph with coefficients as vertices of the graph. The message is then embedded by swapping the two coefficients connected in the graph. As coefficients are swapped out instead of replacing LSBs, it is difficult to detect any distortion using first order statistical analysis.

Though the algorithm is quite strong, but it only provides 6% capacity of the stegeanographic cover picture which is quite low with respect to other existing algorithms.

Steghide is immune against first order statistical attacks and of course visual attacks. But 2nd order attacks, image cropping techniques can reveal presence of Steghide steganography.

3.6   F5

F5 is a modified version of F4 and it removes all the possible weaknesses of its ancestor. The core embedding process is the same but in F5, security and efficiency is uplifted in a greater amount than that of F4 and all other algorithms described before. F5 uses two additional methods to improve all needs of a good steganographic system.

Permutative Straddling: One thing is quite clear that to prevent attacks, continuous straddling (described in the previous section) should be avoided at all costs. The preferred way is to scatter the message within the carrier as regularly as possible and thus keeping the embedding density almost same everywhere. Some well-known systems perform this in their own manner, but most of them are quite time-consuming and can delay a lot in case of a large message and carrier trying to exhaust the capacity completely.

Permutative Straddling is an easy concept. What F5 actually do is estimate the overall capacity of the carrier (it can only be estimated, not assured because we don't know where shrinkage may occur) and then randomly shuffles the available coefficients using permutation. This permutation depends on a key retrieved from a password given to the user. Then the message is embedded in the permuted sequence of coefficients as it is done in F4. So, only the correct password can unlock the permutation sequence by getting which, exact extraction can be performed. The permutation has linear time complexity O (n).

Matrix Encoding: Ron Crandall introduced matrix encoding as a new technique of increasing efficiency of embedding. And F5 is most probably the first implementation of this technique. Matrix Encoding can increase the efficiency when there is a lot of capacity unused within a carrier.   Suppose there is a uniformly distributed message and uniformly distributed values at the positions to be changed. One half of the message changes the coefficients, the other half one doesn't. Then we have an embedding frequency of 2 bits per change. Even 2 isn't exact because we can't be sure where F4 causes shrinkage and thus efficiency decreases to 1.5.

The following example shows what matrix encoding does in details. Suppose two bits x_1, x_2 are to be embedded in three modifiable bit places a_1, a_2, a_3. Then we will encounter any of these four cases:

$$x_1 = a_1 \oplus a_3, \ x_2 = a_2 \oplus a_3 \Rightarrow \text{change nothing}$$
$$x_1 \neq a_1 \oplus a_3, \ x_2 = a_2 \oplus a_3 \Rightarrow \text{change } a_1$$
$$x_1 = a_1 \oplus a_3, \ x_2 \neq a_2 \oplus a_3 \Rightarrow \text{change } a_2$$
$$x_1 \neq a_1 \oplus a_3, \ x_2 \neq a_2 \oplus a_3 \Rightarrow \text{change } a_3.$$

In all four cases, we at most change one bit. Now in general, suppose we have a code word 'a' with n modifiable bit places for k secret message bits 'x'. From now on, we will denote this code by an ordered triplet (d_{max}, n, k) which means to embed k bits in n modifiable bit places changing up to d_{max} bits. The above example is a (1, 3, 2) code as there are 2 bits to be embedded x_1, x_2 and 3 available places a_1, a_2, a_3 changing only 1 time at most. F5 always considers d_{max} to be 1. For (1, n, k) the code words have length $n=2^k-1$. If we neglect shrinkage, we get the change density,

$$D(k) = \frac{1}{n+1} = \frac{1}{2^k}$$

and the embedding ratio as

$$R(k) = \frac{k}{n} = \frac{k}{2^k - 1}$$

Using these two, we can define the embedding efficiency W(k) which denotes the average no. of bits embed per change,

$$W(k) = \frac{R(k)}{D(k)} = \frac{2^k}{2^k - 1} \cdot k$$

The embedding efficiency of (1, n, k) code is always greater than k. Table 1 below shows the relationship between rate, density and efficiency:

Table 1. Relationship between rate, density and efficiency

k	n	change density	embedding rate	embedding efficiency
1	1	50.00 %	100.00 %	2
2	3	25.00 %	66.67 %	2.67
3	7	12.50 %	42.86 %	3.43
4	15	6.25 %	26.67 %	4.27
5	31	3.12 %	16.13 %	5.16
6	63	1.56 %	9.52 %	6.09
7	127	0.78 %	5.51 %	7.06
8	255	0.39 %	3.14 %	8.03
9	511	0.20 %	1.76 %	9.02

This implies that rate decreases with increasing efficiency and vice versa. Hence larger messages can't have higher efficiency. We can find an optimal k for every message to embed and every carrier medium having sufficient capacity such that it just fits into the carrier. For example, if a message with 1000 bits is to be embedded into a carrier medium with 50000 bits capacity. Then rate, R=1000:50000=2%. This value is between rate of k=8 and k=9 from the table above. We choose k=8, and are able to embed 50000:255=196 code words with a length n=255. The (1,255,8) could embed 196.8= 1568 bits [4]. But if we choose k=9, then similarly we get the triplet as (1, 511, 9) and could embed (50000:511).9=97.9=873 bits which will be insufficient to embed 1000 bits.

Implementing the above two principles, F5 shifts himself to the next level of steganography. It has many pros like:

1. High Steganographic Capacity (>13%) and can be increased more.
2. High Efficiency due to Matrix Encoding.
3. Prevents visual attacks.
4. Resistant to statistical attacks like chi-square attack.
5. Its source code is publicly available.

But F5 isn't invincible completely. Second order stronger statistical attacks can reveal presence of F5 steganography within a well embedded message. Or some cropping methods can too (like one proposed by Fridrich et al).

3.7   J3

J3 is another famous approach which is quite compact and strong among the algorithms which use random straddling as its main working principle. In this method, the message is generally at first encrypted via AES, and then the embedding follows. A pseudo-random no. generator is used to embed. This algorithm always makes changes to the coefficients in a pair wise fashion. That is, (2n+1) will always be changed to 2n, and 2n will change to (2n+1) if LSBs don't match. For negative values, same work is done taking its absolute value. But the strategy followed for 1 and -1 is a little bit different in this case. A -1 coefficient is considered to be a steganographic 0, and 1 coefficient is considered to be strganographic 1 and changing them doesn't produce 0 coefficient. For example, message bit 0 will change 1 to -1, not 0. Similarly, a message bit 1 will change -1 to 1. Thus shrinkage is avoided in this algorithm. All 0s are skipped as usual.

Before embedding a data bit in a coefficient, the algorithm determines whether there are sufficient no. of other member coefficients of the pair left to balance the histogram strangeness. If not, it stores the coefficient index in the JPEG header array which is also known as Stop Point for that specific pair. Once stop point is reached, then those coefficients aren't embedded any more as the unused ones must balance the histogram. The header bits are embedded at last as all stop points will be known only at the end of complete operation, not before that.

The extraction algorithm is quite simple. User will generate the random no. using the password given to him/her. And then he decodes the header part for knowing the stop point for coefficient pairs. Then he will just need to decode the message visiting the coefficients keeping the embedding strategy in mind and he will avoid coefficients if their stop point has reached. Thus message is extracted.

The Figure 7 shows the results of J3 embedding in histogram:

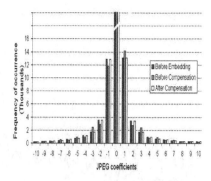

Fig. 7. The results of J3 embedding in histogram

The before compensation bars show that odd coefficients have increased quite significantly and even ones are reduced (apart from 1 & -1). This is , in histogram, there are almost double 2x coefficients than 2x+1 coefficients. For example, there are more 2 coefficients than 3. That is hist(2x) almost equal to 2*hist(2x+1). Hence, even ones are embedded to odd coefficients more than odds are converted to even. That is why, the histogram shoes such peculiarity. But after compensation, that is removed and zero deviation from original is found.

4   Comparative Study

All the algorithms discussed above like Jsteg, F3, F4 provide quite a high capacity and are immune to visual attacks. JSteg, F3 are statistically detectable due to histogram peculiarities. These algorithms somehow lack security. As no secrecy is maintained in default algorithms, anyone who is aware of the embedding can extract the message by linearly capturing the coefficient LSB except 0 coefficients.

As message is embedded linearly within the JPEG picture, if the message is quite short, then the whole message will concentrate start of the file. Hence the left capacity remains unused. This is one main demerit of Continuous embedding. As a result of this, hidden text can be easily extracted without sender's consent.

J3 has very high capacity of embedding. Almost 40% - 70% of non-zero coefficients embed data. And this percentage is greater than that of any other algorithms faced before. And shrinkage is also avoided which is an important plus point. J3 is also immune against 1st order statistical analysis and visual attacks as F5 is. But stronger 2nd order attacks, image cropping techniques and SVM can reveal it. Grossly it has 2% less detection chance than other algorithms described before. If equal data is embedded, then J3 has 3% - 4% less detection ratio. And thus it is the strongest algorithms among all approaches discussed.

These methods are implemented using C language in UNIX platform and comparison between these various methods is shown in the Figure 8.

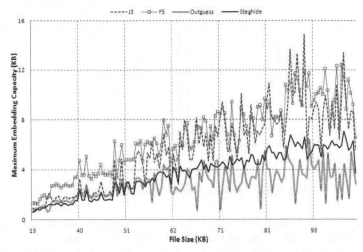

Fig. 8. Comparison between various steganography methods

From the graph, we can conclude that J3 algorithm performs better when the image size is large. Peaks and valleys in the graph are due to the varying texture of images. Valleys occur when images don't contain much variation in them and are usually plain textured. This leads to good compression ratio and hence a large number of zero coefficients, which doesn't leave many coefficients in which to embed data. J3 has a better data capacity than Outguess and Steghide when the image size is small, and it performs better than F5 in some cases with larger image size. J3 uses stop points to minimize the wastage of any unused coefficients and leaves just the right amount to balance the histogram. OutGuess performs the worst in embedding capacity since it stops embedding data when a certain threshold is reached. Information obtained from the graph is shown in Table 2.

Table 2. Performance of different methods

File Size (KB) →	40	51	62	71	81	93
Algorithm ↓						
J3	3-3.4	5-5.1	5.1-5.3	7.9-8.0	9.3-9.6	8.5-8.7
F5	5-5.2	4.9-5.2	5.4-5.7	8.4-8.6	9.6-9.9	8.9-9
Outguess	2-2.1	2.9-3	2.0-2.1	2.1-2.2	5.0-5.3	4.2-4.3
Steghide	2-2.5	3-3.2	3.1-3.3	4.2-4.5	5.0-5.3	5.9-6
	↑ Maximum Embedding size(KB) ↑					

5   Conclusion

In this paper we review the fundamental concepts and notions as some typical techniques in steganography and steganalysis for digital images in JPEG format. This paper reviewed the main steganographic techniques. Each of these techniques tries to satisfy the three most important factors of steganographic design (imperceptibility or undetectability, capacity, and robustness).

References

1. Anderson, R.J., Petitcolas, F.A.P.: On the limits of steganography. IEEE Journal of selected Areas in Communications 16(4), 474–481 (1998)
2. Sadkhan, S.B.: Cryptography: Current status and future trends. In: Proc. IEEE Conference on Information & Communication Technologies, pp. 417–418 (2004)
3. Kumar, M.: High payload histogram neutral JPEG steganography. In: Eigth Annual Conference on Privacy Security and Trust, pp. 46–53 (2010)
4. Li, B., He, J., Huang, J., Shi, Y.Q.: A Survey on Image Steganography and Steganalysis. Journal of Information Hiding and Multimedia Signal Processing 2(2), 142–172 (2011)
5. Fridrich, J., Pevný, T., Kodovský, J.: Statistically Undetectable JPEG Steganography: Dead Ends, Challenges, and Opportunities. In: MM&Sec 2007, Dallas, Texas, USA, September 20-21, pp. 3–14 (2007)
6. Andrews, C.E., Joseph, I.T.: An Analysis of Various Stegonographic Algorithms. International Journal of Advanced Research in Electronics and Communication Engineering (IJARECE) 2(2), 116–123 (2013)

Enhanced Privacy and Surveillance for Online Social Networks[*]

Teja Yaramasa and G. Krishna Kishore

Dept. of Computer Science and Engineering,
Velagapudi Ramakrishna Siddhartha Engineering College Vijayawada, India
tejayaramasa@gmail.com, gkk@vrsiddhartha.ac.in

Abstract. An Online Social Network (OSN) is a platform to build social networks or social relations among people. The OSN's allow users to share interests, activities, social details and professional details. Some of the OSN's that are currently being used are Facebook, Twitter, Orkut etc. The major problem of social networks is providing privacy to the users. Social privacy, institutional privacy and surveillance are the key problems that are being faced by the OSN users. We developed a novel method to provide institutional privacy and surveillance to the OSN users. We introduced a new algorithm HSurveillance, which effectively implements the surveillance in OSN. The institutional privacy is provided to the users using locking mechanism. We believe that the proposed method will resolve the key security and privacy problems experienced by the OSN users.

Keywords: OSN, Privacy, Surveillance, Institutional privacy, Social privacy, Hashtable.

1   Introduction

An online social network (OSN) is a platform to build social networks or social relations among people is a system that protects informational privacy by removing or minimizing personal data thereby preventing unwanted processing of personal data. This is performed without losing the functionality of the information system. The OSN's require users to agree to Terms of Use policy before they may use their services. But these Terms of Use declarations that users must agree to often contain conditions for permitting the website hosts to store and sell data of users or even share it with third parties.

Generally users have reasonable expectations of privacy in OSN's. Even in the transparent world created by OSN's, users have legitimate privacy that may be violated. There are three main privacy problems in OSN's. The first problem is 'surveillance' that arises when the personal information and social interactions of users are used by other users, governments and service providers without the

---

[*] 3rd International Conference on Frontiers in Intelligent Computing, Theory and Application (FICTA-2014), 14th-15th November, Bhubaneswar, India.

© Springer International Publishing Switzerland 2015   189
S.C. Satapathy et al. (eds.), *Proc. of the 3rd Int. Conf. on Front. of Intell. Comput. (FICTA) 2014*
– *Vol. 2*, Advances in Intelligent Systems and Computing 328, DOI: 10.1007/978-3-319-12012-6_21

knowledge of actual users. The second problem is 'social privacy' that is related to the privacy issues regarding the social details of the users. The third problem is 'institutional privacy' that arises when the user's data is used by the host or service providers.

Surveillance is a key feature that is not present in most of the existing OSN's. If the users data is subject to theft it is difficult to know who stole the data. So, by providing surveillance to the OSN's it is easy to identify the users who stole it. In some social networking websites like Facebook etc., the users don't know who is visiting their profiles and what other users are accessing from their profiles. LinkedIn is an OSN in which surveillance is present but it just specifies only who visited the users profile but it doesn't mention what data is being accessed from the profile. So the surveillance is provided to only a certain limit in LinkedIn. So, surveillance is an important feature that enhances the security and privacy in OSN's.

The users losing control and oversight over the collection and processing of their information in OSN's is known as institutional privacy. Institutional privacy also refers to the usage of user's data by the service providers or host. The existing service providers don't give an option to the users whether to use their data or not. They just include it in the terms and conditions, which are not seen by most of the users. The service providers store, use and sell the users data to third parties and advertisers which makes the user's data vulnerable. So there is lack of privacy to the users. Majority of the existing OSN's doesn't provide institutional privacy. So there is a much need for providing the users with institutional privacy.

Social privacy refers to the control for a user to what data he can share and with whom to share his data. Most of the existing OSN's like Facebook, LinkedIn etc., provide good social privacy to the users. In Facebook, the users have permissions to share their data with public, friends, and groups or only with particular users. This article focuses less on the social privacy issue. Most of the current OSN's provide privacy based on different technical privacy solutions called 'Privacy Enhancing Technologies (PET's) [1].

2   Related Work

Mohsen Jamali and Hassan Abolhassani [2] define social network analysis as the mapping of relations and flows between people, groups and organizations. Social network analysis provides a visual analysis of human relations. It also provides a mathematical analysis. Surveillance and institutional privacy are the important aspects that are lacking in the existing OSN's. Graphs and matrices are used to represent social relationships [2]. There are different statistical Models for social network analysis like Markov Random Fields (MRFs) [3] and Exponential Random Graphical Models (ERGMs) [4] [5].

Leucio Antonio Cutillo et al. [6] discuss about the inherent handling of user personal data in OSN's with centralized architecture. The centralized architecture of existing OSN's is a major privacy issue. It can be avoided by using peer to peer architectures. A new approach is proposed in this paper in which the system mainly contains three components i.e. matryoskshas, a peer to peer substrate, (e.g. Distributed Hash Table (DHT)) and a trusted identification service [6]. Each user of an OSN can store his information in a basic distributed structure which is provided by the matryoshka. The global access to a user's data is provided by peer-to-peer substrate. Authentication is performed by the trusted identification service.

Leucio Antonio Cutillo et al. [9] discusses about the analysis of privacy in OSN's from the Graph Theory Perspective. An OSN application whether it is centralized or de-centralized, its privacy degree strongly depends on the topological properties of the Social Graph which represents relationships. Sensitive data can easily be gathered, stored, replicated and correlated [7] [8]. Graphs are used to represent communication networks or social networks. A social graph G (V, E) has a set V of users and a set E of edges representing social relationships. An adjacency matrix A represents a graph. In graph theory deg (v) represents the degree of a vertex. The node degree represents a straight forward relation with privacy since when v establishes a relationship with a new friend. The probability of linking to a misbehaving user increases when there is an increase of degree. Let 'p_{mal}' denote the probability a new friend 'n' of a malicious user 'v'. The number of malicious friends $F_{mal}(v)$ of v follows the binomial distribution

$$F_{ma}(v) \sim B(p_{mal}, \deg(v))$$

The probability of having at least one misbehaving friend in [9] is given by (1)

$$p_v = 1 - p_{mal}^{\deg(v)} \tag{1}$$

The clustering coefficient of node v in [9] is given by (2)

$$c(v) = \frac{2e_{\deg(v)}}{\deg(v)(\deg(v) - 1)} \tag{2}$$

The clustering coefficient of overall graph G in [9] is given by (3)

$$C(G) = \frac{\sum_{v \in V} c(v)}{\| V \|} \tag{3}$$

Random walks [10] in a graph have an important property: when the random walk approximates its steady state distribution after a sufficient number of hops, the start point and end point of the walk are uncorrelated. This number of hops is called mixing time. The mixing time in [9] is given by (4)

$$T(\varepsilon) = \max_{x \in V} T_x(\varepsilon) \tag{4}$$

Ahmad Kamran Malik and Schahram Dustdar [11] explained about the Sharing and Privacy-Aware Role Based Access Control (SP-RBAC) model in OSN's. Different collaborative groups and relationships facilitate users in controlling the privacy and sharing levels of their information in OSN's. The users share their personal data without setting any privacy options. This raises security and privacy issues. A user's privacy concerns are only a weak predictor of their membership to the social network [12]. Sharing of restricted information can cause serious repercussions [13].

ASP-RBAC model is modeled by extending the well-known Role Based Access Control (RBAC) model [14]. There are different types of collaborative groups as per the user need. SP-RBAC model has three components namely, Core SP-RBAC, Hierarchical SP-RBAC and Constrained SP-RBAC [11].

Weimin Luo et al. [15] discusses about the analysis of security in OSN's. The security issues that are primarily considered in OSN are password protection, protection for private information of the user. Personal information can be updated in two modes: private mode and public mode. Due to limiting preferences in private information, most users expose their personal data on web that forces it to be easily attacked by the hackers and attackers. Personal information on social networks may expose more information about a particular user from different websites [16]. The general attacking techniques are phishing, spam, worm, viral marketing, XSS, plug-in, third party applications. In phishing attacks, the attacker pretends to be a legitimate user and sends requests the other users by using his own URLs, which gains access to personal information of other users on their acceptance to the request [17, 18]. Gilbert Wondracek et al. [19] performed a practical attack to de-anonymize the social network users. The building block of this attack is the browsing history of a user.

Ed Novak and Qun Li [20] discusses about the privacy provided in the existing OSN's. In most of the OSN's, the users generate a massive amount of data. The social network provider must protect the user's data from other users. There are three types of users. They are directly connected users, unconnected or indirectly connected users and general public. The unconnected users are those who are a hop away from each other. For example unconnected users include friends of friends or friends of friends of friends. The general public is the entire users of the OSN. The OSN's frequently utilize some sort of application system. So the third parties use applications to access the user content typically only available to the OSN provider. The FAITH Architecture [21] is a system that protects users from the third party applications.

Lerone Banks and Shyhtsun Felix Wu [22] shows that all friends in a social network are not created equal and discusses about the interaction intensity based approach to privacy in OSN's. The basic problem that exists within the OSN's is that the inability of the network to automatically distinguish the relationship quality between a user and its friends. Interaction intensity is a feasible approach for setting privacy preferences. Seda Gurses et al. [23] explain about social privacy and surveillance in OSN's. All the existing OSN's have two major privacy issues, namely, surveillance and institutional privacy. The OSN users do not have any surveillance on their profiles. So the users don't know who are visiting their profiles and what data they are accessing. The institutional privacy issue arises when the host or service providers sell the user's data to advertisers and third parties.

3   Proposed Method

To overcome the drawbacks of the existing OSN's, the proposed system is included with enhanced surveillance and institutional privacy features. So the surveillance in an OSN allows the users to know who are visiting their profiles and also what data they are accessing. If the user's data is subject to malicious access, with surveillance it is easy to identify the suspected users who accessed the data. In all the existing OSN's the service providers sell the user's data to advertisers, third party applications etc. This makes the user's data vulnerable to external threats. So in the proposed system an option is given to the users to choose whether the service providers can use their data or not. So, the proposed system provides an enhanced privacy to the OSN's.

3.1   Methodology

The proposed system is implemented in three modules: Social network implementation phase, Institutional privacy initialization during registration phase and Surveillance initialization phase on self profile and content. Figure 1 gives an overview of all the phases. The modules are explained in detail below

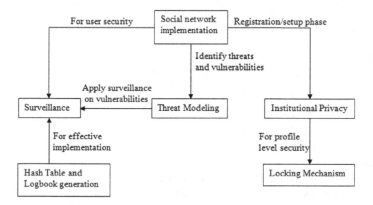

Fig. 1. Work flow of the proposed method

1. **Social network implementation phase:** In this phase, a basic OSN is implemented with features like news feed, friends, wall, photos, videos, sharing, messages, chat, notifications, groups, and pages. In general, the design principles which are followed in this social network are scope and objectives of the social network, user groups, standards, free expression, dynamic content, easy to use multiple communication channels.

   a) Threat Modeling: Threat modeling is an important process for developing a secure web application. Threat modeling is applied to the current OSN to identify threats and vulnerabilities. In general, threat modeling is applied to web applications for the following purposes [24].

   - Identifying Security Objectives
   - Decomposing the application
   - Threats identification
   - Vulnerabilities identification

2. **Institutional Privacy Initialization during Registration Phase:** During the
   setup phase, the user is provided with an option whether the host or service
   provider could use your data or not. If the user gives permission to the hosts to
   use their data then the data is accessed and sold to the third parties. If the user
   doesn't give permission to the host to use their data, then a master lock will be
   applied on the respective user's profile in the administrator's database, so that
   they cannot access the user's data.
3. **Surveillance Initialization Phase on Self-profile and Content:** In this phase,
   surveillance is provided to the users profile and content in the current OSN. The
   OSN has many features but surveillance is applied to the vulnerable features that
   are identified by the threat modeling in the first phase. An algorithm
   HSurveillance is designed to provide surveillance to different features of the
   OSN.

```
Algorithm HSurveillance
{
Input:
          Ui : User i
          Uj : User j
          Pi : Profile of Ui
          Pj : Profile of Uj
          T   : Timestamp
Output:
          Hashtable H for each user
          Logbook LB for each user
Method:
     for i=1 to n do
               Assign hashcode Hito user Ui;
               Add a tuple <Ui,Hi> to the hash table
          H; Create a logbook LBi for Ui;
     endfor
        for i=1 to n do
               forall users j not equal to i do
               if  Ui  visits  profile  Pj  of  user  Uj  at
                  time T then add a tuple <T,Ui,Pj> to the
                  LBi;
               endif
               endfor
          endfor
}
```

The HSurveillance algorithm takes users and their profiles as input with current
times-tamp. Initially the algorithm assigns hash code Hi to each user Ui during
registration. Then a tuple <Ui,Hi> is added to the hash table H. A Logbook LBi is
created for each user i of the social network. If a user Ui visits the profile Pj of user Uj

at time T, then a tuple <T,Ui,Pj> is added to the logbook LBi of user Ui. The users are provided with surveillance link on their profiles. Whenever the users click on the surveillance link, and then all the records of the other users containing the hash code of current user are displayed. So, the user can identify other users who are accessing their profiles and data.

4   Conclusion

In this paper, we were able to identify different security aspects of the OSN's and problems faced by the users of OSN's. We developed a novel method to provide institutional privacy and surveillance to the OSN users. We introduced a new algorithm HSurveillance, which effectively implements the surveillance in OSN. The institutional privacy is provided to the users using locking mechanism. With this surveillance scheme, the users not only know who visited their profiles but also the data accessed from their profile. The users acquire more control over their data with institutional privacy. We believe that the proposed method will resolve the key security and privacy problems experienced by the OSN users.

Acknowledgement. The authors wish to thank G N Sunand Kumar and Y Murali for their support in algorithm development and manuscript proof reading.

References

1. Pelkola, D.: A Framework for Managing Privacy-Enhancing Technology. IEEE Software 29(3) (June 2012)
2. Jamali, M., Abolhassani, H.: Different Aspects of Social Network Analysis. In: IEEE/WIC/ACM International Conference on Web Intelligence (2006)
3. Frank, O., Strauss, D.: Markov graphs. Journal of the American Statistical Association 81, 832–842 (1986)
4. Wasserman, S., Pattison, P.: Logit models and logistic regression for social networks: I. An introduction to Markov graphs and p! Psychometric Society 61, 401–425 (1994)
5. Anderson, C., Wasserman, S., Crouch, B.: A p! primer: logit models for social networks. Social Networks 21, 37–66 (1999)
6. Cutillo, L.A., Molva, R., Strufe, T.: Privacy Preserving Social Networking Through Decentralization. In: IEEE WONS 2009, Sixth International Conference on Wireless On-Demand Network Systems and Services, February 2-4 (2009)
7. Bilge, L., Strufe, T., Balzarotti, D., Kirda, E.: All Your Contacts Are Belong to Us: Automated Identity Theft Attacks on Social Networks. In: WWW, April 20-24 (2009)
8. Jagatic, T., Johnson, N., Jakobsson, M., Menczer, F.: Social phishing. Communications of the ACM 50(10), 94–100 (2007)
9. Cutillo, L.A., Molva, R., Onen, M.: Analysis of Privacy in Online Social Networks from the Graph Theory Perspective. In: IEEE, Global Telecommunications Conference (GLOBECOM 2011), December 5-9 (2011)
10. Mitzenmacher, M., Upfal, E.: Probability and Computing: Randomized Algorithms and Probabilistic Analysis. Cambridge University Press (2005)

11. Malik, A.K., Dustdar, S.: Sharing and Privacy-Aware RBAC in Online Social Networks. In: IEEE International Conference on Privacy, Security, Risk, and Trust, and IEEE International Conference on Social Computing (2011)
12. Acquisti, A., Gross, R.: Imagined Communities: Awareness, Information Sharing, and Privacy on the Facebook. In: Danezis, G., Golle, P. (eds.) PET 2006. LNCS, vol. 4258, pp. 36–58. Springer, Heidelberg (2006)
13. Wang, Y., Komanduri, S., Leon, P.G., Acquisti, G.N.A., Cranor, L.F.: I regretted the minute I pressed share: A Qualitative Study of Regrets on Facebook. In: Symposium on Usable Privacy and Security (SOUPS), July 20-22 (2011)
14. Ferraiolo, D., Sandhu, R., Gavrila, S., Kuhn, D.R., Chandramouli, R.: Proposed nist standard for role-based access control. ACM Transactions on Information and System Security 4(3), 224–274 (2001)
15. Luo, W., Liu, J., Liu, J., Fan, C.: An Analysis of Security in Social Networks. In: Eighth IEEE International Conference on Dependable, Autonomic and Secure Computing, December 12-14 (2009)
16. Gross, R., Acquisti, A.: Information Revelation and Privacy in Online Social Networks. In: Workshop on Privacy in the Electronic Society (WPES), November 7 (2005)
17. Rajalingam, M., Alomari, S.A., Sumar, P.: Prevention of Phishing Attacks Based on Discriminative Key Point Features of WebPages. International Journal of Computer Science and Security (IJCSS) 6(1) (2012)
18. Jakobsson, M., Stamm, S.: Invasive Browser Sniffing and Countermeasures. In: 15th International World Wide Web Conference, May 23-26 (2006)
19. Wondracek, G., Holz, T., Kirda, E., Kruegel, C.: A Practical Attack to De-anonymize Social Network Users. In: IEEE Symposium on Security and Privacy (SP), May 16-19 (2010)
20. Novak, E., Li, Q.: A Survey of Security and Privacy in Online Social Networks. College of William and Mary Computer Science Technical Report (2012)
21. Lee, R., Nia, R., Hsu, J., Levitt, K.N., Rowe, J., Wu, S.F., Ye, S.: Design and implementation of faith, an experimental system to intercept and manipulate online social informatics. In: IEEE International Conference on Advances in Social Networks Analysis and Mining (ASONAM), pp. 195–202 (July 2011)
22. Banks, L., Wu, S.F.: All Friends Are Not Created Equal: An Interaction Intensity Based Approach to Privacy in Online Social Networks. In: IEEE International Conference on Computational Science and Engineering, August 29-31 (2009)
23. Gurses, S., Diaz, C.: Two tales of privacy in online social networks. IEEE Security & Privacy 11(3), 29–37 (2013)
24. Shostack, A.: Experiences Threat Modeling at Microsoft. Microsoft (2008)

Neuro-key Generation Based on HEBB Network for Wireless Communication

Arindam Sarkar, J.K. Mandal, and Pritha Mondal

Department of Computer Science & Engineering, University of Kalyani,
Kalyani-741235, Nadia, West Bengal, India
{arindam.vb,jkm.cse}@gmail.com

Abstract. In this paper a key generation technique for encryption/decryption, based on a single-layer perceptron network (Hebb Network), for wireless communication of information or data has been proposed. Two HEBB Neural networks have been used at both the sender and receiver ends. Both the networks have a Random Number Generator (RNG) that generates identical inputs at both ends. As both the networks are synchronized they generate same output pair for same input pair which is used as the secured secret key to encrypt the plain text through some reversible computation to form the cipher text. The receiver generate plain text by performing identical operation. The key is never transmitted during encoding across the network. This process ensures the integrity and confidentiality of a message transmitted via any medium as the secret key is unknown to any intruder thus imparts a potential solution to Man-in-the-middle attack.

Keywords: Cryptography, Intruder, Single-layer Perceptron, Secret-key, Encryption, Decryption, Wireless communication, Neural Network, HEBB Network.

1 Introduction

Cryptography [1] is the technique of securing a message in transit by encoding a plain text message into an encoded form that is non readable. Cryptography concerns with the confidentiality of the message to be transmitted, i.e. it converts the message from a comprehensible form to an intermediate form and vice versa at the rceiver, so that it is non-readable by the hackers. It is done using a secret key. There are two methods of cryptography: Symmetric key cryptography[1] and Asymmetric key cryptography[1]. Symmetric key cryptography uses the same key for both encryption and decryption and the key is shared between the two parties. Here the main problem lies in key distribution. Asymmetric key cryptography uses two different keys, one for encryption (public key)and other for decryption(private key). The public key is distributed publicly, and is used to encrypt the message by anyone who wishes to send a message to the owner of that key. The private key is kept secret and is used by the receiver to decrypt the message. But it is slower than symmetric key cryptography so it is not feasible in case of bulk messages or energy dependent encryption.

© Springer International Publishing Switzerland 2015   197
S.C. Satapathy et al. (eds.), *Proc. of the 3rd Int. Conf. on Front. of Intell. Comput. (FICTA) 2014*
– *Vol. 2*, Advances in Intelligent Systems and Computing 328, DOI: 10.1007/978-3-319-12012-6_22

In this paper HEBB Neural Network [3] is used to implement the process. Two synchronized HEBB networks are used at both sender and receiver ends which generate same set of updated weights for each same pair of inputs. This updated weight is used as the secret key. Both the sender and receiver can encrypt or decrypt messages using this key. As the key is self generated, so no need to share the key, hence key distribution problem is resolved. As this is symmetric key cryptography so it can also be used for bulk messages and the complexity is also much less as compared to asymmetric key cryptography as it uses only one key.

2   Literature Survey

Cryptographic techniques have two basic components: an algorithm (or methodologies) and a key. Modern cryptographic techniques include many different types of algorithms. All the algorithms have some relative advantages and disadvantages. The survey [4, 5, 6] of such literature identify loopholes of existing techniques so that a better system can be developed. Cryptographic system includes symmetric key algorithms such as: Diffie-Hellman key exchange algorithm [1], DES [1], AES [1] etc. and asymmetric key algorithms such as: RSA [2].

Diffie-Hellman algorithm is used to transmit or share the cryptographic key. The advantage of this algorithm is that the key is never transmitted over the medium. But the main drawback of this algorithm is that it suffers the "MAN-IN-THE-MIDDLE" attack[1]. So the authentication of the message is compromised and can only be used to share the secret keys but cannot be used to encrypt the message.

RSA is the first practicing asymmetric key cryptosystem. It is advantageous because it uses asymmetric key cryptography i.e. two different keys are used for encryption and decryption. It also deals with authentication of the message i.e. the receiver can verify that the message is sent from a genuine sender and not from any intruder.

DES (Data Encryption Standard) is a block cipher. It is the oldest symmetric key algorithm and no real weakness has been found the only threat is still Brute force attack. It is fast and simple to use. But its main disadvantage is the 56-bit key size. AES (Advanced Encryption Standard) has suppressed the use of DES. In spite of being a block cipher it is quite different than DES. It uses 128 bit key. In this proposal a HEBB network based cryptosystem for wireless communication has been proposed to eliminate the problem of Man-in-the-Middle attack.

3   Proposed Technique

The problem domain is the area where the key distribution problem is discussed. The process of cryptography is combination of two process firstly generation of secret key and then encrypting the data or message by using the secret key. The problem in case of symmetric key cryptography is that both the parties have to share the same secret key that results into the key distribution problem. There is another problem that is the Man-In-The-Middle attack that hampers the integrity, authentication and confidentiality of the message. Asymmetric key cryptography solves the problems of symmetric key cryptography, but it is slow and as it uses two different keys for encryption and decryption so the complexity is quite high. It is costly too as compared to symmetric key cryptography.

The Key distribution problem and the Man-In-The-Middle attack can be resolved by combining the modern Soft computing tools (i.e. Artificial Neural Network, Genetic Algorithm etc.) and Cryptographic techniques. By using HEBB Neural Network the key distribution problem can be solved. Both the sender and receiver will generate their own secret key so there is no need to share to distribute the key over the network So the key becomes immune to hacking. So the intruder can never get the key, and without the key the intruder can neither read nor tamper the content of the message.

In this paper the an algorithm has been used for the key generation which is discussed in section 3.1.

3.1   Key Generation

Here HEBB Neural Network has been used at both sender and receiver sides to generate the secret key. The following algorithm is used by the two HEBB Networks:

Training Algorithm:
Training algorithm for Hebb Network is given below:

> *Input:* - *Random input vectors for both HEBB Network.*
> *Output:* - *Secret key through synchronization of input and output neurons as vectors.*
> *Method: Through synchronization of HEBB Network secret key is generated*
> *Step0: First weights are initialized. Basically in this network they are set to 0, i.e. $Wi=0$ for $i=1$ to n, where "W" is weight and "n" is the total number of input neurons.*
> *Step1: Steps 2-4 have to be performed for each input and target output pair.*
> *Step2: Input units activations are set. Generally, the activation function of input layer is identity functions: $Xi=Si$ for $i=1$ to n.*
> *Step3: Output units activations are set: $Y=t$.*
> *Step4: Weight adjustments are performed:*
> $$Wi \ (new) = Wi \ (old) + XiY$$

In this paper the following algorithm has been used for the encryption. The algorithm is discussed in section 3.2.

3.2   Encryption Technique

Input: - *Random File of different sizes and may be of different formats and the secret Key.*
Output: - *Encrypted File.*
Method: - *Exclusive-OR (XOR) operation*
Step1: *Read a file.*
Step2: *Convert the contents of the file to decimal values and store in an integer array.*
Step3: *Read the KEY.*
Step4: *Perform the Exclusive-OR operation between the Key and the message.*
Step5: *Convert the decimal number to its corresponding ASCII character, which is the encrypted output of the entered message.*

In this paper the following algorithm has been used for the Decryption technique. The algorithm is discussed in section 3.3.

3.3   Decryption Technique

Input: - *Encrypted File and the secret Key.*
Output: - *Original File.*
Method: - *Original file gets generated by the Exclusive-OR (XOR) operation.*
Step1: *Read the encrypted file.*
Step2: *Convert the contents of the files to its equivalent decimal number.*
Step3: *Read the key.*
Step4: *Perform the Exclusive-OR operation between the key and the received encrypted message that generates a decimal number.*
Step5: *Convert the decimal number to equivalent ASCII character, which is the decrypted output or the original message.*

4   Example

4.1   Encryption

Consider a message Hello; Now each Character is converted to decimal values i.e. H=72, e=101, l=108, l=108, o=111; Now the key generated through tuning(say) is [1, 0, 1, 0]; Perform the Exclusive-OR operation between the key and the message; the encrypted text is =>**Iemlo.**

4.2   Decryption

The received encrypted text is **Iemlo**; Now each character is converted into equivalent decimal values i.e. I=73, e=101, m=109, l=108, o=111; Perform the Exclusive-OR operation between the key(which is also generated at destination during tuning) and the encrypted message; the decrypted text is Hello, i.e. the original message.

5   Results and Analysis

Results of the implementation of the technique discussed has been presented in terms of Chi-square testing, source file size vs. encryption time along with source file size vs. decryption time, various formats of files vs. encryption time and various formats of files vs. decryption time.

5.1   Encryption Time vs. File Size

Files of different sizes are taken and respective encryption time is calculated. The table 1 shows the encryption time against file size and figure1 shows the graphical representation of the same. Table1 shows the encryption time of files ranges from 80KB to 400KB.

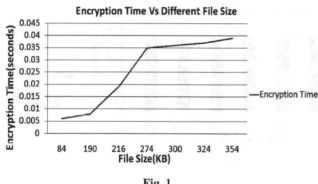

Table 1.

File size(KB)	Encryption time(Sec)
84	0.006
190	0.008
216	0.019
274	0.035
300	0.036
324	0.037
354	0.039

Fig. 1.

5.2   Decryption Time vs. File Size

Files of different sizes are taken and respective decryption times are calculated. The table 2 shows the decryption time against file size that of figure 2 shows the pictorial representation. Table2 shows the decryption time of files ranges from 80KB to 400KB.

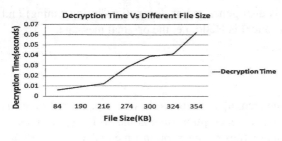

Fig. 2.

Table 2.

File size(KB)	Encryption time(Sec)
84	0.006
190	0.009
216	0.012
274	0.028
300	0.039
324	0.041
354	0.062

5.3   Various Types of Files vs. Encryption Time

Files of different format(.cpp, .sys, .txt, .dll, .exe, .odt, .odp, .odg) of same size are taken and their encryption time have been calculated and shown in the following graph in figure 3.

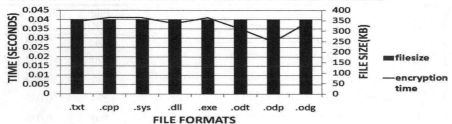

Fig. 3.

5.4   Various Types of Files vs. Decryption Time

Files of different format such as .cpp, .sys, .txt, .dll, .exe, .odt, .odp, .odg of same sizes are taken and their decryption times are calculated and shown in the below graph in figure 4 which shows a consistent results irrespective of types of files.

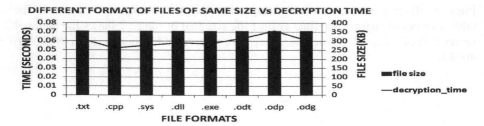

Fig. 4.

5.5   Chi-Square Test vs. Different File Format with Various File Size

Table 3 and figure 5 shows the chi-square value of .com files of different sizes

Table 3.

File Size(Bytes)	Chi-Square value
11776	3973
13312	5095
15872	5750
18944	7451
25600	10920
69886	20356

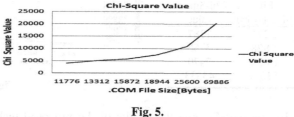

Fig. 5.

Table 4 and figure 6 shows the chi-square value of .dll files of different sizes

Table 4.

File Size(Bytes)	Chi-Square value
11776	5006
28517	12295
35720	29930

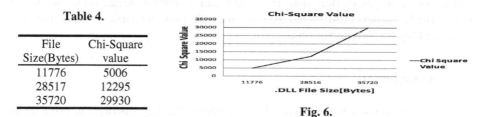

Fig. 6.

Table 5 and figure 7 shows the chi-square value of .exe files of different sizes

Table 5.

File Size(Bytes)	Chi-Square value
16408	2923
20240	5246
79144	25718

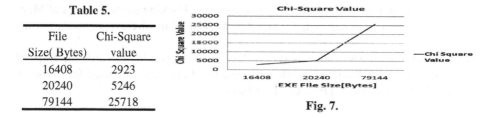

Fig. 7.

Table 6 and figure 8 shows the chi-square value of .pdf files of different sizes

Table 6.

File Size(Bytes)	Chi-Square value
18020	1772
33169	3824
57218	10960

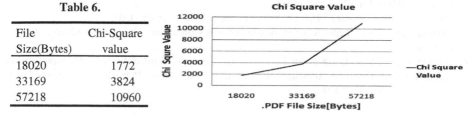

Fig. 8.

Table 7 and figure 9 shows the chi-square value of .txt files of different sizes

Table 7.

File Size(Bytes)	Chi-Square value
5271	1287
9828	3356
13862	7470
53611	43333

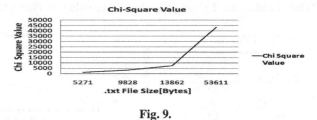

Fig. 9.

In this specific area the results of the proposed technique has been analyzed. In cryptographic system a sender must send the secret key to the receiver over the insecure channel. So there is a chance of attack at the time of key distribution that leads to the key distribution problem. To solve this issue Hebb Network based key generation is proposed. Here both the sender and receiver generate their own secret key so the key is never distributed over the network. Both the parties can use their self generated key for the encryption-decryption process.

6   Conclusion

This paper presents a novel approach to overcome the key distribution problem in symmetric key cryptography by using HEBB Neural Network based key generation technique. Here the secret key is self generated at both the sender and receiver end, i.e. there is no need to distribute the key over the insecure channel. The final updated weights are used as the key. As the key is immune to Hacking so the chances of Man-In-Middle attack is also reduced. This model can be used in wireless communication.

Acknowledgement. The authors express deep sense of gratuity towards the Dept of CSE University of Kalyani where the computational resources are used for the work and the PURSE scheme of DST, Govt. of India.

References

1. Kahate, A.: Cryptography and Network Security. Tata McGraw-Hill Publishing Company Limited (2003), Eighth reprint (2006)
2. http://people.csail.mit.edu/rivest/Rsapaper.pdf
3. Sivanandam, S.N., Deepa, S.N.: Principles of Soft Computing. Wiley-India (2008), Reprint (2012)
4. Arindam, S., Mandal, J.K.: Artificial Neural Network Guided Secured Communication Techniques: A Practical Approach. LAP Lambert Academic Publishing (June 04, 2012) ISBN: 978-3-659-11991-0

5. Arindam, S., Karforma, S., Mandal, J.K.: Object Oriented Modeling of IDEA using GA based Efficient Key Generation for E-Governance Security (OOMIG). International Journal of Distributed and Parallel Systems (IJDPS) 3(2) (March 2012), doi:10.5121/ijdps.2012.3215, ISSN : 0976- 9757 [Online] ; 2229 - 3957 [Print]. Indexed by: EBSCO, DOAJ, NASA, Google Scholar, INSPEC and WorldCat, 2011
6. Mandal, J.K., Arindam, S.: Neural Session Key based Triangularized Encryption for Online Wireless Communication (NSKTE). In: 2nd National Conference on Computing and Systems (NaCCS 2012), Department of Computer Science, The University of Burdwan, Golapbag North, Burdwan −713104, West Bengal, India, March 15-16 (2012) ISBN 978-93-808131-8-9

KSOFM Network Based Neural Key Generation for Wireless Communication

Madhumita Sengupta, J.K. Mandal, Arindam Sarkar, and Tamal Bhattacharjee

Department of Computer Science & Engineering,
University of Kalyani, Kalyani-741235, Nadia, West Bengal, India
{madhumita.sngpt,jkm.cse,arindam.vb}@gmail.com

Abstract. In this paper a single layer perceptron (KSOFM Network) based neural key generation for encryption/decryption for wireless communication has been proposed. Identical KSOFM network has been used in both sender and receiver side and the final output weight has been used as a secret key for encryption and decryption. The final output matrix of the KSOFM network is taken and the minimum value neuron in the matrix is considered as the secret key. Depending upon the input and output neuron, different keys has been generated for each session which helps to form secret session key. In sender side the plain text is encrypted with the secret key to form the cipher text done by the EX-OR operation between the secret key and the plain text. Receiver will use the same secret key to decrypt the cipher text to plane text. The secret key is not sent via any medium so it minimizes the man-in-the-middle attack. Moreover various tests are performed in terms of chi-square test, which shows comparable results with the said proposed system.

Keywords: Kohonen Self-Organizing Feature Map (KSOFM Network), Encryption, Decryption, Secret Key, Cryptography, Wireless Communication.

1   Introduction

Cryptography [1] is the art and science of attaining security by transforming a plain text to cipher text which is non readable. Cryptography is concerned solely with message confidentiality i.e. transformation of messages from a readable form into a non readable one and back again at the other end, present it non readable by eavesdroppers or interceptors without secret knowledge.

The two types of cryptography [1] are Symmetric key cryptography and Asymmetric key cryptography. In case of Symmetric key cryptography both the parties use the same key to encrypt and decrypt the message. The key is to be shared between them, this arises the key distribution problem. Whereas in case of Asymmetric key cryptography two keys are used, one is public key and the other one private key. Public key is used for encryption and private key is used for decryption. The main limitation of this type is much slower then symmetric key cryptography. Not viable for large size messages.

© Springer International Publishing Switzerland 2015   207
S.C. Satapathy et al. (eds.), *Proc. of the 3rd Int. Conf. on Front. of Intell. Comput. (FICTA) 2014*
– *Vol. 2*, Advances in Intelligent Systems and Computing 328, DOI: 10.1007/978-3-319-12012-6_23

In this paper a neural secret session key generation technique is proposed using KSOFM Network [3]. Using identical KSOFM Network in both sender and receiver side help us to get a same updated weight in both side from which a secret key is taken and used for encryption and decryption in wireless communication. No sharing of the secret key help us to overcome the man-in-the-middle attack. As one key is used for both encryption/decryption purposes so it is viable for large size messages. Different format of files has been used in this proposed technique.

2   Literature Survey

In Cryptography, the main problem is key distribution problem, while wireless communication the intruder can hack the key and decrypt the cipher text at the time when the key is shared over the public channel by staying in between the sender and the receiver. The survey [4, 5, and 6] is aimed at finding the loopholes of the existing systems to develop a better one.

Diffie-Hellman key exchange [1] states that every single data/information transmitting from both sender and receiver is captured by an intruder who is residing in between them. Diffie-Hellman key swap over technique suffers from this problem. Intruders can perform as sender and receiver concurrently and try to whip secret session key at the time of exchanging key via public channel.

RSA [2] is one of the first practicable public-key cryptosystems and is widely used for secure data [2] transmission. RSA technique use public key for encryption and private key for decryption. The main drawback is the speed it takes for encryption/decryption is much high. In case of wireless communication speed is the main factor.

3   Proposed Technique

In order to solve key distribution problem and speed factor one needs to examine the problem domain which is essentially the area of expertise. In cryptography main problems lie on key distribution and speed factor. In symmetric key cryptography main problem is key distribution and in case of Asymmetric key cryptography the main problem is speed it takes for encryption/decryption is too high. Man-in-the-middle attack [1] in Diffie-Hellman key exchange algorithm is another problem in cryptography.

By using this proposed technique key distribution problem and man-in-the-middle attack is minimized, as identical KSOFM Network is used in both sender and receiver side to generate identical updated weight which is used for generating secret session key for encryption/decryption for wireless communication. Encryption and decryption time is very low in this technique to prove this various tests have been conducted.

In this paper thea novel algorithm has been used for the key generation. The algorithm is discussed in section 3.1.

3.1   Key Generation

In the proposed methodology the KSOFM Network is used in both sender and receiver side. Both KSOFM network generates secret key through synchronization of both networks. The following algorithm is used in both sides to generate same secret key.

Input: - *Random input vectors for both KSOFM Network.*
Output: - *Secret key through synchronization of input and output neurons as vectors.*
Step 1.   *Initialize the weights W_{ij} randomly. This can be chosen as the components of the input vector. If information related to distribution of clusters is known, the initial weights can be taken to reflect that prior knowledge. Set topological neighbourhood parameters: As clustering progresses, the radius of the neighbourhood decreases. Initialize the learning rate alpha: It should be a slowly decreasing function of time.*
Step 2.   *Perform steps 2-8 when stopping condition is false.*
Step 3.   *Perform step 3-5 for each input vector x.*
Step 4.   *Compare the square of the Euclidean distance i.e., for each j=1 to m,*
$$D\ (j) = \sum\sum (X_i\text{-}W_{ij})^2$$
Step 5.   *Find the winning unit index j, so that D (j) is minimum. (In Steps 3 and 4, dot product method can also be used to find the winner, which is basically the calculation of net input, and the winner will be one with the largest dot product.)*
Step 6.   *For all unit of j within a specific neighborhood of j and for all i, calculate the new weights:*
$$W_{ij}\ (new) = W_{ij}\ (old) + \alpha\ [X_i\text{-}W_{ij}\ (old)]$$
$$Or\ W_{ij}(new) = (1\text{-}\alpha)\ W_{ij}\ (old) + \alpha X_i$$

Step 7.   *Update the learning rate α using the formula $\alpha\ (t+1) = 0.5\alpha t$.*
Step 8.   *Reduce radius of topological neighborhood at specified time intervals.*
Step 9.   *Test for stopping condition of the network.*

By using this algorithm the secret key is generated at both the sender and receiver and encryption through this secret key has been done using encryption algorithm as discussed in section 3.2.

3.2   Encryption Technique

The encryption technique that takes random file of any size as input, and encrypt that with the secret key to produce the encrypted file as output.

Input: -   *Random file of different sizes and may be of different formats and the secret Key.*
Output: - *Encrypted file*

Method: - *Exclusive-OR (XOR) operation*
Step 1: *Read the entered file and the format of the file.*
Step 2: *Convert the file into equivalent decimal values and store in an integer array say A[].*
Step 3: *Read the key.*
Step 4: *Next perform the Exclusive-OR (XOR) operation between the key and the file, which generates the encrypted values of the decimal numbers of the file.*
Step 5: *Convert this decimal number to equivalent ASCII character, which is the encrypted output of the entered file of the same format.*

Example
Consider a message hello; Now the decimal values for h= 104, e=101, l=108, l=108, o=111 and stored into an array say A[]; Now the key say is K[]=[1, 5, 4]. Now A[] has 5 integer numbers and the key K[] has 3 integer numbers, so bit padding is needed to perform the action; Next perform the Exclusive-OR (XOR) operation between the key and the message; the encrypted text is=> `deog

In this paper the following algorithm has been used for the Decryption purpose. The algorithm is discussed in section 3.3.

3.3   Decryption Technique

Input: - *Encrypted File and the secret Key.*
Output: - *Original File.*
Method: - *Exclusive-OR (XOR) operation*
Step 1: *After getting the encrypted file.*
Step 2: *Convert the file to equivalent decimal numbers.*
Step 3: *Read the key.*
Step 4: *Next perform the Exclusive-OR (XOR) operation between the key and the encrypted file, which generates a decimal number.*
Step 5: *Convert this decimal number to equivalent ASCII character, which is the decrypted output i.e. the original file sent by the sender.*

Example
The encrypted text is `deog is received by the receiver; Now the text are converted into decimal values for `= 96, d=100, e=101, o=111, g=103 and stored in an array say B[]; Now the key say is K[]=[1, 5, 4]; A[]has 5 integer numbers and the key K[] has 3 integer number so bit padding is needed to perform the action; Next perform the Exclusive-OR (XOR) operation between the key and the encrypted message; The decrypted text is hello. This is the original message.

4   Results and Analysis

In these section results of the proposed technique is presented and analyzed in terms of encryption/decryption time, chi-square test with different formats of file with different sizes, encryption time vs. file size, decryption time vs. file size, and various file types vs. encryption time and various file types vs. decryption time.

4.1   Encryption Time vs. File Size

Files of different sizes have been taken and respective encryption time has been calculated. The table and the graph shown in table 1 and figure 1 depict the time of encryption against increasing file size. Table1 shows the encryption time of files ranges from 80Kb to 400Kb. Figure 1 shows that with increasing file size encryption time increases.

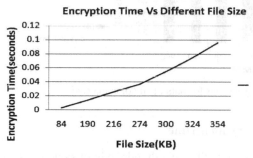

Fig. 1. Encryption time vs. file size

Table 1. Encryption time of 80Kb to 400Kb file size

File Size(KB)	Encryption Time(Sec)
84	0.003
190	0.014
216	0.026
274	0.037
300	0.055
324	0.075
354	0.096

4.2   Decryption Time vs. File Size

Files of different sizes have been taken and respective decryption times have been calculated. Table 2 shows the values of decryption times against file sizes and Figure 2 shows the corresponding pictorial representation.Table2 shows the decryption time of files ranges from 80KB to 400KB. Figure 2 shows that with increasing file size decryption time increases.

Fig. 2. Decryption time vs. file size

Table 2. Decryption time of 80 Kb to 400Kb file size

File Size(KB)	Encryption Time(Sec)
84	0.003
190	0.015
216	0.028
274	0.035
300	0.067
324	0.078
354	0.101

4.3   Various Types of Files vs. Encryption Time

Files of different formats such as .cpp, .sys, .txt, .dll, .exe, .odt, .odp and .odg of same sizes have been taken and their respected encryption time have been calculated and shown in figure 3. Figure 3 shows that .odg file format takes minimum time to encrypt and that of .cpp takes maximum time to encrypt.

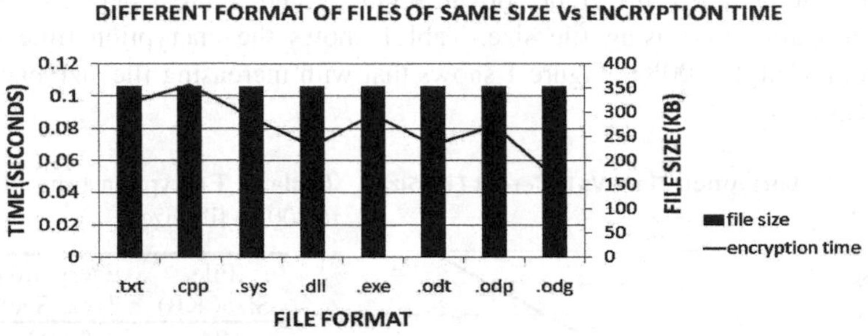

DIFFERENT FORMAT OF FILES OF SAME SIZE Vs ENCRYPTION TIME

Fig. 3. Encryption time of various formats of file of same size

4.4   Various Types of Files vs. Decryption Time

Files different formats such as .cpp,.sys,.txt,.dll,.exe,.odt,.odp,.odg of same sizes have been taken and their respected decryption time have been calculated and shown as graphical representation in figure 4. Figure 4 shows that .odt file format takes minimum time to decrypt and .dll takes maximum time to decrypt which shows non coherences between encryption and decryption process.

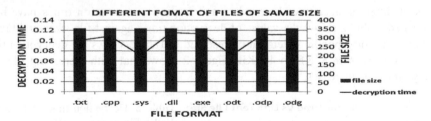

Fig. 4. : Decryption time of various formats of file of same size

4.5   Chi-Square Test vs. Different File Format with Various File Size

Table 3 and figure 5 shows the chi-square value of .com files of different sizes. Figure 5 shows that in increasing file size Chi-Square values also increases.

Table 3. Chi-square value vs. different file format

FILE SIZE(Bytes)	Chi-Square value
11776	5312
14848	6314
16896	7203
22016	8266
50648	12860
69886	17657

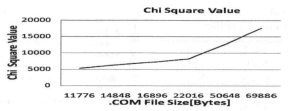

Fig. 5. Chi square value of different file format

Table 4 and figure 6 shows the chi-square value of .dll files of different sizes. Figure 6 shows that with increasing file size Chi-Square Value increases.

Table 4. Chi-square value of .dll files

FILE SIZE(Bytes)	Chi-Square Value
23040	8404
35624	10713
42496	15035
78836	25487

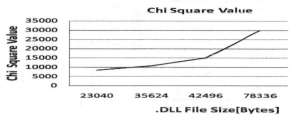

Fig. 6. Chi-square value of .dll files

Table 5 and figure 7 shows the chi-square value of .exe files of different sizes. Figure 7 shows that with increasing file size Chi-Square Value increases.

Table 5. Chi-square value of .exe files

FILE SIZE(Bytes)	Chi-Square Value
14480	3834
28272	6737
64512	20420

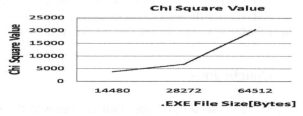

Fig. 7. Chi-square value of .exe files

Table 6 and figure 8 shows the chi-square value of .pdf files of different sizes. Figure 8 shows that with increasing file size Chi-Square Value increases.

Table   6.   Chi-square value of .pdf files

FILE SIZE(Bytes)	Chi-Square Value
42176	2062
42701	2126
80651	2294
98264	2580

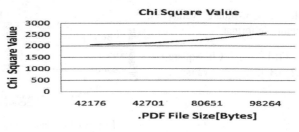

Fig. 8. Chi-square value of .pdf files

Table 7 and figure 9 shows the chi-square value of .txt files of different sizes. Figure 9 shows that with increasing file size Chi-Square Value increases.

Table   7.   Chi-square value of .txt files

FILE SIZE(Bytes)	Chi-Square Value
10236	7241
36672	22377
45170	51129

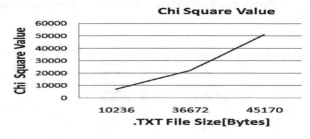

Fig. 9. Chi-square value of .txt files

In the result and analysis section the results of the proposed technique has been analyzed and discussed. In Cryptography to perform the decryption procedure a user key has to be transmitted over the insecure channel to the receiver. So there is a problem in key distribution. To overcome this problem a neural network based key generation technique is proposed. The security issue can be boost up by using this technique over the existing key exchange algorithm. In this technique two parties say sender and receiver do not have to share their keys over a public channel but use their updated final weight as a Secret Key for Encryption/Decryption process.

5   Conclusions

This paper presents a novel approach to overcome the key distribution problem by generating neural secret key by using the KSOFM Network algorithm. Proposed approach boost up the security issue over the existing key exchange algorithms. The secret key is not shared over the insecure public channel the updated final weight in both sides is taken as the secret key for encryption/decryption process. Man-in-the –middle attack is minimized in this propose technique.

Acknowledgement. The authors express deep sense of gratuity towards the Dept of CSE University of Kalyani where the computational resources are used for the work and the PURSE scheme of DST, Govt. of India.

References

[1] Kahate, A.: Cryptography and Network Security. Tata McGraw-Hill Publishing Company Limited (2003), Eighth reprint (2006)

[2] http://people.csail.mit.edu/rivest/Rsapaper.pdf

[3] Sivanandam, S.N., Deepa, S.N.: Principles of Soft Computing. Wiley-India (2008), Reprint (2012)

[4] Arindam, S., Mandal, J.K.: Artificial Neural Network Guided Secured Communication Techniques: A Practical Approach. LAP Lambert Academic Publishing (June 04, 2012) (2012) ISBN: 978-3-659-11991-0

[5] Arindam, S., Karforma, S., Mandal, J.K.: Object Oriented Modeling of IDEA using GA based Efficient Key Generation for E-Governance Security (OOMIG). International Journal of Distributed and Parallel Systems (IJDPS) 3(2) (March 2012), doi:10.5121/ijdps.2012.3215, ISSN: 0976- 9757 [Online]; 2229 - 3957 [Print]. Indexed by: EBSCO, DOAJ, NASA, Google Scholar, INSPEC and WorldCat (2011)

[6] Mandal, J.K., Arindam, S.: Neural Session Key based Triangularized Encryption for Online Wireless Communication (NSKTE). In: 2nd National Conference on Computing and Systems (NaCCS 2012), Department of Computer Science, The University of Burdwan, Golapbag North, Burdwan −713104, West Bengal, India, March 15-16 (2012) ISBN 978-93-808131-8-9

Hopfield Network Based Neural Key Generation for Wireless Communication (HNBNKG)

J.K. Mandal, Debdyuti Datta, and Arindam Sarkar

Department of Computer Science & Engineering, University of Kalyani, W.B, India
{jkm.cse,arindam.vb}@gmail.com, debdyutidatta@yahoo.com

Abstract. In this paper, a key generation and encryption/decryption technique based on Hopfield Neural network has been proposed for wireless communication. Hopfield Neural networks at both ends forms identical input vector, weight vector which in turn produces identical output vector which is used for forming secret-key for encryption/decryption. Using this secret-key, plain text is encrypted to form the cipher text. Encryption is performed by *Exclusive-OR* operation between plaintext and secret-key. Decryption is performed at the receiver through *Exclusive-OR* operation between cipher text and identical secret-key generated. Receiver regenerate the original message sent by the sender as encrypted stream. In HNBNKG technique sender and receiver never exchange secret-key. This technique ensured that, when message is transmitting between sender-receiver nobody can regenerate the message as no key is exchanged.

Keywords: Hopfield Neural network, Secret-key, Encryption, Decryption, Wireless communication.

1 Introduction

There are many techniques developed for secured data transfer. Algorithms have benefits and shortcoming also. Types of cryptographic techniques are Symmetric key cryptography and Asymmetric key cryptography. Symmetric key cryptography uses same key for encryption and decryption. The main problem of this technique is that the key have to be shared via some channel between sender and receiver at the beginning. Now, if the channel is not secure then eavesdroppers can hack the key and get the encrypted message. Asymmetric key cryptography uses two different keys for encryption and decryption, for this reason this is much more slow and complex technique. This technique is not useful for decrypting bulk messages as speed is a very important factor in case of wireless communication.

In this HNBNKG technique Hopfield network [2] is used. Using one Hopfield network(sender) output vector generates a secret-key. Another network (receiver) have been synchronised with the sender side network to generate the same output vector. This output can be used to form the secret-key between the sender and receiver. In this technique private key cryptography have been used but with some exception, no sharing of key occurred here i.e. secret key is never transmitted. So this

© Springer International Publishing Switzerland 2015   217
S.C. Satapathy et al. (eds.), *Proc. of the 3rd Int. Conf. on Front. of Intell. Comput. (FICTA) 2014*
– *Vol. 2*, Advances in Intelligent Systems and Computing 328, DOI: 10.1007/978-3-319-12012-6_24

HNBNKG technique will be safe and fast also. As Symmetric key cryptography i.e. only one key is used and no key is exchanged, so complexity will be much less in comparison with Asymmetric key cryptography.

2   Literature Survey

The term Cryptography consists of two different components: one is cryptographic algorithm and other is cryptographic key generation. Modern cryptographic algorithms are of two types: Symmetric-key algorithm (AES, DES) and Asymmetric-key algorithm (RSA). All these methods have advantages as well as some shortcomings. The purpose of this survey [1,3,4,5] is to understand these benefits and loopholes, in order to develop a better system.

RSA [1] cryptosystem uses public key cryptosystem i.e. two different keys for encryption and decryption. The main drawback of using RSA technique is the encryption speed. There are many secret key encryption techniques available which are faster than any available public key encryption techniques .DES[1] is a block cipher algorithm. DES technique was not implemented for software, so it runs comparatively slowly. As in DES only one key is used for encryption and decryption, if we lost the key for decryption we cannot get the message at the receiver. AES[1] is also a symmetric block cipher algorithm like DES, which uses a common key for encryption/decryption. Deffie-Hellman key exchange algorithm was developed to generate same private cryptographic key at both ends, so that no transmission of key is needed. The main loophole of this algorithm is that it involves expensive exponential operations and it is a bit slow and suffers from Man-in-the-Middle attack. This technique is used for forming secret key only. It has authentication problem also. In this paper a Hopfield Neural network based encryption/decryption has been proposed.

3   Proposed Technique

In a secure communication environment the main problem engages distribution of secret key which will be used for encryption and decryption. When key is being exchanged between two parties over public channel, eavesdroppers can listen to the channel to get the secret key. A technique has been evolved to solve this particular problem.

In Man-In-The-Middle attack [1] sender and receiver thinks that they are exchanging information between each other, but actually they are linked with the eavesdropper. Diffie-Hellman key exchange algorithm [1] faces this Man-In-The-Middle attack. Eavesdropper pretends to be sender and receiver and keeps contacting both the parties to get the secret key for encryption/decryption.

The Man-In-The-Middle attack problem has been addressed in HNBNKG by not exchanging the secret key for encryption/decryption over insecure medium. The neural weight synchronization strategy generates same output at both sides. This output can then be used to form secret key between sender and receiver for encryption and decryption.

The main problems of Symmetric key cryptography are: key sharing and Man-In-The-Middle attack. These problems have been addressed in HNBNKG by creating a fusion of cryptographic tools and soft-computing techniques. Here Hopfield neural network has been used at both ends for forming same secret key used in encryption/decryption. Sharing of secret key is not needed any more with the help of HNBNKG. If sender sends any message encrypted with the secret key receiver can easily decrypt it with the identical self-generated secret key. Man-in-the-middle can only listen to these messages but can never grasp what the key is.

In this paper the algorithm that has been used for the key generation is discussed in section 3.1. The encryption technique and decryption technique is discussed in section 3.2 and section 3.3.

3.1   Key Generation

Input: - Random input vectors for both Hopfield Network.

Output: - Secret key.

Method: Secret key through synchronization of input and output neurons as vectors.

Step1: Input two vectors randomly. They are vect_zero and vect_one. Then convert these two vectors into bipolar format using the formulae

$$fn\text{-}sgn(a)=(a\text{-}0.5)\times 2 . \qquad (1)$$

Step 2: Multiply vect_zero[i] with vect_zero[j] and vect_one[i] with vect_one[j]. Then add these two matrices.

Step 3: Set the diagonal elements of the matrix to zero.

Step 4: Take inputs as input vector and the convert them to bipolar form using the above formula (1).

Step 5: Formulate the result vector using two recursive loops. The result vector is formulated by

$$result\_vec[i]=result\_vec[i]+input\_vect[j]\times weight[i][j] . \qquad (2)$$

Step 6: Apply conditions to limit the result vector.

    a. If result vector is greater than zero then set result vector as 1.

    b. If result vector is equal to zero then the result vector is equal to the input vector.

    c. Else the result vector is -1.

Step 7: The current result vector will be set as the new input to the network.

3.2   Encryption Technique

Input: - *File of any type and any size, secret key.*

Output: - *Encrypted result*

Method:- *Exclusive-OR(XOR) operation between input file and the key*

Step 1: *Enter a sting i.e. message.*

Step 2: *Convert the string to binary values.*

Step 3: *Convert the binary values to equivalent decimal numbers.*

Step 4: *Output vector is a 128 bit number, set this as key.*

Step 5: *Convert this 128 bit key to equivalent decimal number.*

Step 6: *Next perform the Exclusive-OR (XOR) operation between the key and the message, which generates decimal number.*

Step 7: *Convert this decimal number to equivalent ASCII character, which is the encrypted output of the entered message 256 bit binary number, set this as the key.*

3.3   Decryption Technique

Input: - *Encrypted result, secret key.*

Output: - *Original file.*

Method: *Exclusive-OR (XOR) operation between encrypted file and the key.*

Step 1: *After getting the encrypted message, convert them to equivalent binary values.*

Step 2: *Convert the binary values to equivalent decimal numbers.*

Step 3: *Get the 128 bit key, same as that, which was used for encryption*

Step 4: *Convert this 128 bit key to equivalent decimal number.*

Step 5: *Next perform the Exclusive-OR(XOR) operation between the key and the encrypted message , which generates a decimal number.*

Step 6: *Convert this decimal number to equivalent ASCII character, which is the decrypted output i.e. the original message sent by the sender.*

4   Example

For simplicity a 4 bit input vector is used here. Let ,vect_zero[i]={1, 0, 0, 0} , vect_one[i]={0, 1, 1, 0}. Convert vect_zero[i] and vect_one[i] to bipolar using fn_sgn(x)=2(-0.5);x =2(1-0.5)=1;x=2(0-0.5)= -1. So, vect_zero[i]={1, -1, -1, -1}, vect_one[i]={-1, 1, 1, -1}. Initialize weight[i][j]=0. Calculate weight[i][j] ,where $i \neq j$, weight[i][j]=((vect_zero[j]×vect_zero[i])+(vect_one[j]×vect_one[i])) .

Now, vect_zero[j] × vect_zero[i] =

$$
\begin{bmatrix} 1 \\ -1 \\ -1 \\ -1 \end{bmatrix}
\begin{bmatrix} 1\ -1\ -1\ -1 \end{bmatrix}
=
\begin{bmatrix}
1 & -1 & -1 & -1 \\
-1 & 1 & 1 & 1 \\
-1 & 1 & 1 & 1 \\
-1 & 1 & 1 & 1
\end{bmatrix}
$$

vect_one[j]×vect_one[i]=

$$
\begin{bmatrix} -1 \\ 1 \\ 1 \\ -1 \end{bmatrix}
\begin{bmatrix} -1\ 1\ 1\ -1 \end{bmatrix}
=
\begin{bmatrix}
1 & -1 & -1 & 1 \\
-1 & 1 & 1 & -1 \\
-1 & 1 & 1 & -1 \\
1 & -1 & -1 & 1
\end{bmatrix}
$$

So,Weigh [i][j] =

$$
\begin{bmatrix}
2 & -2 & -2 & 0 \\
-2 & 2 & 2 & 0 \\
-2 & 2 & 2 & 0 \\
0 & 0 & 0 & 2
\end{bmatrix}
$$

Next enter the input vector [1 1 1 1]. Convert them to bipolar usingfn_sgn(x)=2(x-0.5) . So input vector is : [1 1 1 1]. Initialize result_vec=0. Calculate the result vector :

result_vec[i]=result_vec[i]+input_vec[j]×weight[i][j].

input_vec[j]×weight[i][j]=

$$
\begin{bmatrix} 1\ 1\ 1\ 1 \end{bmatrix}
\times
\begin{bmatrix}
2 & -2 & -2 & 0 \\
-2 & 2 & 2 & 0 \\
-2 & 2 & 2 & 0 \\
0 & 0 & 0 & 2
\end{bmatrix}
$$

$$
=
\begin{bmatrix} -2\ 2\ 2\ 2 \end{bmatrix}
$$

Next, apply condition : if(result_vec[i]>0)set, result_vec[i] = 1; if(result_vec[i]<0) set, result_vec[i]=-1; if(result_vec[i]==0) set, result_vec[i]=input_vec[i]; So, the final result vector is : result_vec=   [-1 1 1 1]. Next Set input_vec = result_vec . In 2nd iteration input is=[-1 1 1 1] and the result_vec will be=[-1 1 -1 1]. The iterations will be continued this way. After 6 iterations the first input [1 1 1 1] has been generated as output . So this forms a cycle. The 6 steps are :

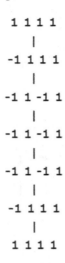

```
1 1 1 1
   |
-1 1 1 1
   |
-1 1 -1 1
   |
-1 1 -1 1
   |
-1 1 -1 1
   |
-1 1 1 1
   |
1 1 1 1
```

Iterations continue until the output vector will be the same as the first input vector to the network. So Hopfield network forms a cycle for any input vector. In this project a 16 bit random input vector is used. This seed to the rand function is unique, will be known only to the sender and receiver. So this will produce unique but identical input vectors at both the ends. This input vectors produces a 16 bit output vector after first iteration. The after certain number of iterations(depends on the input vector), the output from the network became same as the input vector of the network. This is the stopping condition of this network. So in this way both the sender and receiver generate identical 16 bit output. In this project this is used as secret key. This 16 bit bipolar output vector has been then converted to binary values."-1" = "11111111","1" = "00000001".As an example, suppose the 1st bit of output vector is "-1" ,then it is converted to 8 bit binary value "11111111" and if this bit is "1", then it is converted to 8 bit binary value "00000001".This will also act as a secret thing, as only the sender and receiver knows what is set for 1 and what is set for -1. This process finally generates a 16*8=128 bit secret key.

5   Result and Analysis

The result of the implementation of the HNBNKG technique is discussed here. The implementation result of the proposed technique has been presented in terms of encryption decryption time, source file size vs. encryption time along with source file size vs. decrypted time, Chi-Square testing of different format files with different sizes and various file types vs encryption time.

Table 1. Encryption / Decryption time Vs File size

File Size [bytes]	Encryption time [sec]	Decryption time[sec]
128	0.005	0.005
256	0.009	0.010
512	0.021	0.020
1024	0.033	0.032
2048	0.089	0.087
4096	0.336	0.340

Table 1 shows the encryption/decryption time with respect to different file sizes ranges from 128- 4096 bytes.

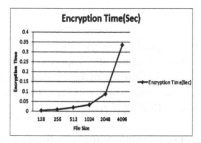

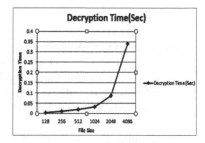

Fig. 1. source size vs encryption time   **Fig. 2.** source size vs decryption time

Fig 1 and fig 2 shows the graphical representation of file size vs encryption/decryption time. It takes different time to find the secret key, so these graphs are non-linear.

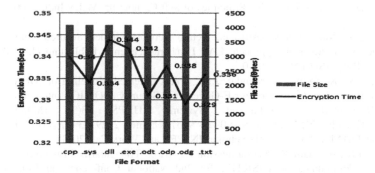

Fig. 3. Encryption time vs files of different category but of same sizes

Table 2. File size vs Chi-Square value

.EXE Source Size[bytes]	Chi-Square value	.COM Source Size[bytes]	Chi-Square value	.DLL Source Size[bytes]	Chi-Square value	.TXT Source Size[bytes]	Chi-Square value
2809	327	4217	2145	1027	396	858	996
3208	362	7023	4603	2048	3336	2816	2676
4203	3930	11072	8741	8864	4222	5478	4911
7168	9056	11776	12366	12888	7115	8580	7633
11776	11843	14848	15146	13400	7166	13862	22648
14336	17919	18944	20262	15960	9502	16443	14068

Table 2 shows various types of file with different sizes and their corresponding Chi-Square values.

6   Conclusion

In this paper a secret key has been generated using Hopfield network which is a single layer perceptron network. As Hopfield network has a feature of forming a cycle of input and outputs so, output from any intermediate iteration can be chosen as the secret key and this logic will be only known to the sender and receiver. For these reason chances of attack is very low in this proposed HNBNKG technique.

Acknowledgement. The authors express deep sense of gratuity towards the Dept of CSE University of Kalyani where the computational resources are used for the work and the PURSE scheme of DST, Govt. of India.

References

1. Kahate, A.: Cryptography and Network Security. Tata McGraw-Hill Publishing Company Limited (2003), Eighth reprint (2006)
2. Sivanandam, S.N., Deepa, S.N.: Principles of Soft Computing. Wiley-India (2008), Reprint (2012)
3. Arindam, S., Mandal, J.K.: Artificial Neural Network Guided Secured Communication Techniques: A Practical Approach. LAP Lambert Academic Publishing (June 04, 2012) (2012) ISBN: 978-3-659-11991-0
4. Arindam, S., Karforma, S., Mandal, J.K.: Object Oriented Modeling of IDEA using GA based Efficient Key Generation for E-Governance Security (OOMIG). International Journal of Distributed and Parallel Systems (IJDPS) 3(2) (March 2012), doi:10.5121/ijdps.2012.3215, ISSN: 0976- 9757 [Online]; 2229 - 3957 [Print]. Indexed by: EBSCO, DOAJ, NASA, Google Scholar, INSPEC and WorldCat (2011)
5. Mandal, J.K., Arindam, S.: Neural Session Key based Triangularized Encryption for Online Wireless Communication (NSKTE). In: 2nd National Conference on Computing and Systems (NaCCS 2012), Department of Computer Science, The University of Burdwan, Golapbag North, Burdwan –713104, West Bengal, India, March 15-16 (2012) ISBN 978-93-808131-8-9

Automatic Video Scene Segmentation to Separate Script and Recognition

Bharatratna P. Gaikwad, Ramesh R. Manza, and Ganesh R. Manza

Department of CS and IT, Dr. B.A.M. University, Aurangabad (MS), India
{bharat.gaikwad08,manzaramesh,ganesh.manza}gmail.com

Abstract. Text or character detection in images or videos is a challenging problem to achieve video contents retrieval. In this paper work we propose to improved VTDAR (Video Text Detection and Recognition) Template Matching algorithm that applied for the automatic extraction of text from image and video frames. Video Optical Character Recognition using template matching is a system model that is useful to recognize the character, upper, lower alphabet, digits& special character by comparing two images of the alphabet. The objectives of this system model are to develop a model for the Video Text Detection and Recognition system and to implement the template matching algorithm in developing the system model. The template matching techniques are more sensitive to font and size variations of the characters than the feature classification methods. This system tested the 50 videos with 1250 video key-frames and text line 1530. In this system 92.15% of the Character gets recognized successfully using Texture-based approaches to automatic detection, segmentation and recognition of visual text occurrences in images and video frames.

Keywords: Video Processing, text detection, localization, tracking, segmentation, Template Matching, OCR.

1   Introduction

The rapid growth of video data leads to an urgent demand for efficient and true content-based browsing and retrieving systems. In response to such needs, various video content analysis schemes are using with one or a combination of image, audio, and textual information in the video [1]. All are types of video file formats are available on internet, cell phones and easily downloaded. A variety of approaches to text information extraction from images and video have been proposed for specific applications including page segmentation , address block location , license plate location , and content-based image/video indexing. In the extraction of this information involves the detection, localization, tracking, extraction, enhancement and recognition of text from the images and video frames are provided. Text in images and video frames carries important information for visual content understanding and retrieval [2]. Optical character recognition (OCR) is one of the most popular areas of research in pattern recognition because of its immense

© Springer International Publishing Switzerland 2015   225
S.C. Satapathy et al. (eds.), *Proc. of the 3rd Int. Conf. on Front. of Intell. Comput. (FICTA) 2014*
– *Vol. 2*, Advances in Intelligent Systems and Computing 328, DOI: 10.1007/978-3-319-12012-6_25

application potential. The two fundamental approaches to OCR are template matching and feature classification. In the template matching approach, recognition is based on the correlation of a test character with a set of stored templates. In the feature classification method, features are extracted from a standard character image to generate a feature vector. A decision tree is formed based on the presence or absence of some of the elements in the feature vector. When an unknown character pattern is encountered, this tree is traversed from node to node till a unique decision is reached. The template matching techniques are more sensitive to font and size variations of the characters than the feature classification methods. However, selection and extraction of useful features is not always straight forward [5]. Several software is available for editing and shows the videos types as .AVI,.FLV,.DAT,.3GP,.MPEG,.MP4 etc. Extracting text information from videos generally involves three major steps:

- Text detection: Find the regions that contain text.
- Text segmentation: Segment text in the detected text regions. The result is usually a binary image for text recognition.
- Text recognition: Convert the text in the video frames into ASCII characters.

1.1   Video Text Detection and Analysis

Video Processing:-
Shot: Frames recorded in one camera operation form a shot.
Scene: One or several related shots are combined in a scene.
Sequence: A series of related scenes forms a sequence.
Video: A video is composed of different story units such as shots, scenes, and sequences arranged according to some logical structure (defined by the screen play). These concepts can be used to organize video data. The video consists of sequence of images (video frames). In the first step, we convert video into all frames and saved as JPEG images.

A.   Pre Processing
A scaled image was the input which was then converted into a gray scaled image. This image formed the first stage of the pre-processing part. This was carried out by considering the RGB color contents of each pixel of the image and converting them to grayscale. The conversion of a colored image to a gray scaled image was done for easier recognition of the text appearing in the images as after grayscale conversion, the image was converted to a black and white image containing black text with a higher contrast on white background [12].

B.   Detection and Localization
In the text detection stage, since there was no prior information on whether or not the input image contains any text, the existence or nonexistence of text in the image must be determined. However, in the case of video, the number of frames containing text is much smaller than the number of frames without text. The text detection stage seeks to detect the presence of text in a given image. Text localization methods can be categorized into two types: region-based and texture-based. Select a frame containing

text from shots elected by video framing, this stage used region Based Methods for text tracking. Region based methods use the properties of the color or gray scale in a text region [1], [19], [24].

C.   Tracking and Segmentation
When text was tack, the text segmentation step deals with the separation of the text pixels from the background pixels indirectly separate single character from whole text. The output of this step is a binary image where black text characters appear on a white background. This stage included extraction of actual text regions by dividing pixels with similar properties into contours or segments [2], [9], [22].

D.   Recognition
This stage included actual recognition of extracted characters , The result of recognition was a ratio between the number of correctly extracted characters and that of total characters and evaluates what percentage of a character were extracted correctly from its background. For each extraction result of correct character [4], [21], [25].

1.2   Survey of Literature

1) Jie Xi and et.al. has work on Text detection, tracking and recognition to extract the text information in news and commercial videos. He has used Techniques morphological opening procedure on the smoothed edge map. They got the text detection rate is 94.7% and the recognition rate is 67.5% [7].
2) Palaiahna kote Shivakumara and et.al. has work on elimination of non-significant edges from the segmented text portion of a video frame to detect accurate boundary of the text lines in video images. They got percentage 93% [8].
3) Rainer Lienhart and et.al. has worked on the text localizing and segmenting text in complex images and videos, It is able to track each text line with sub-pixel accuracy over the entire occurrence in a video. They got percentage text recognition 69.9% [9][10].

2   Methodology

2.1   Figures Canny Edge Detector

Among the several textual properties in an image, edge-based methods focus on the 'high contrast between the text and the background'. The edges of the text boundary are identified and merged, and then several heuristics are used to filter out the non-text regions. Usually, an edge filters (e.g. canny operator) is used for the edge detection, and a smoothing operation. The Canny method finds edges by looking for local maxima of the gradient of I. The gradient is calculated using the derivative of a Gaussian filter. The method uses two thresholds, to detect strong and weak edges, and includes the weak edges in the output only if they are connected to strong edges [13],[15]. This method is therefore less likely than the others to be fooled by noise, and more likely to detect true weak edges [3],[16].

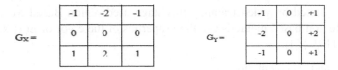

$$G_x = \begin{array}{|c|c|c|} \hline -1 & -2 & -1 \\ \hline 0 & 0 & 0 \\ \hline 1 & 2 & 1 \\ \hline \end{array} \qquad G_Y = \begin{array}{|c|c|c|} \hline -1 & 0 & +1 \\ \hline -2 & 0 & +2 \\ \hline -1 & 0 & +1 \\ \hline \end{array}$$

Fig. 1. Canny Edge detection operator (a) x direction (b) y direction

1. Compute f_x and f_y

$$f_x = \frac{\partial}{\partial x}(f * G) = f * \frac{\partial}{\partial x}G = f * G_x \tag{1}$$

$$f_y = \frac{\partial}{\partial y}(f * G) = f * \frac{\partial}{\partial y}G = f * G_y \tag{2}$$

$G(x, y)$ is the Gaussian function $G_x(x, y)$ is the derivate of

$G(x, y)$ with respect to x:

$$G_x(x, y) = \frac{-x}{\sigma^2}G(x, y) \tag{3}$$

$G_y(x, y)$ is the derivate of $G(x, y)$ with respect to y:

$$G_y(x, y) \frac{-y}{\sigma^2}G(x, y) \tag{4}$$

2. Compute the gradient magnitude $magn(i, j)$

$$= \sqrt{f_x^2 + f_y^2} \tag{5}$$

3. Apply non − maxima suppression.

4. Apply hysteresis thresholding / edge linking .

The canny edge detection algorithm is easy to implement, and more efficient than other algorithms. From this edge detected images, text region is identified [3],[23].Text in images and video frames can exhibit many variations with respect to the following properties are character font size, width ,hight,alignment,edge,color etc.

2.2 Design a System and Implementing VTDAR Algorithm

The Video Text Detection and Recognition template matching worked on this following Algorithm (VTDAR)

a) Load the video (E.g. Avi, Mpeg etc.).
b) Then video is converted into frames with frames name from"img-1 to img-N "till the video will be come to an end.
c) Template is made of Upper case, Lower case, Special character & digit with size 24x42 size.
d) Applying OCR techniques, select the frame among one of them (E.g.img-50).
e) Image is Converted to gray scale and then converted to binary by using CC algorithm.

f) Applying edge detection& Binarization algorithm for focuses on text region.
g) Then top-down: extracting texture features of the image and then locating text regions.
h) Bottom-up: separating the image into small regions and then grouping character regions into text regions.
i) Applying simultaneously by space vector for maintain space between two lines as per Image
j) The character image from the detected string is selected.
k) Segmentation: Each character was automatically selected and thresholding using methods.
l) After that, the image to the size of the first template is rescaled.
m) After rescale the image to the size of the first original image then comprising letters with template matching techniques are used and the matching metric is computed.
n) Then the highest match found is stored. If the template image is not match, it might be getting recognized as some other character.
o) The index of the best match is stored as the recognized character.
p) All recognized character showing on Word file.

Architecture of Video Scene Segmentation and Recognition system

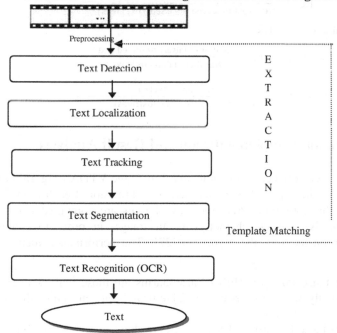

Fig. 2. System for character detection and recognition from video/image

Character Recognition:- Among the 256 ASCII characters, only 94 are used in document images or frame and among these 94 characters, only 80 are frequently used. In the present scope of experiment, we have considered 80 classes recognition problem. These 80 characters are listed in Table 1. These include 26 capital letters, 26 small letters, 10 numeric digits and 18 special characters table 1. These include 26 capital letters, 26 small letters, 10 numeric digits and 18 special characters [26].

Table 1. Videos frame template classes

1	2	3	4	5	6	7	8	9	0
A	B	C	D	E	F	G	H	I	J
K	L	M	N	O	P	Q	R	S	T
U	V	W	X	Y	Z	a	b	c	d
e	f	g	h	i	j	k	l	m	n
o	p	q	r	s	t	u	v	w	x
y	z	"	;	,	.	#	&	@	(
)	-	%	!	:	'	$	?	+	/

$$\text{Recall} = \frac{\text{Correct Detected}}{(\text{Correct Detected} + \text{Missed Text Lines})} \tag{6}$$

Whereas precision is defined as:

$$\text{False alarm rate} = \frac{\text{Number of falsely detected text}}{\text{Number of detected text}} \tag{7}$$

$$\text{Precision} = \frac{\text{Correct Detected}}{(\text{Correct Detected} + \text{False Positives})} \tag{8}$$

3   Experimental Implementation and Result Analysis

There are several performance evaluations to estimate the VTDAR algorithm for text extraction. Most of the approaches quoted here used Precision, Recall to evaluate the performance of the algorithm. Precision, Recall rates are computed based on the number of correctly detected characters in an image, in order to evaluate the efficiency and robustness of the algorithm [27]. The performance metrics are as follows:

False Positives: False Positives (FP) / False alarms are those regions in the image which are   actually not characters of a text, but have been detected by the algorithm as text.

False Negatives: False Negatives (FN)/ Misses are those regions in the image which are actually text characters, but have not been detected by the algorithm.

Precision Rate: Precision rate (P) is defined as the ratio of correctly detected characters to the sum of correctly detected characters plus false positives.

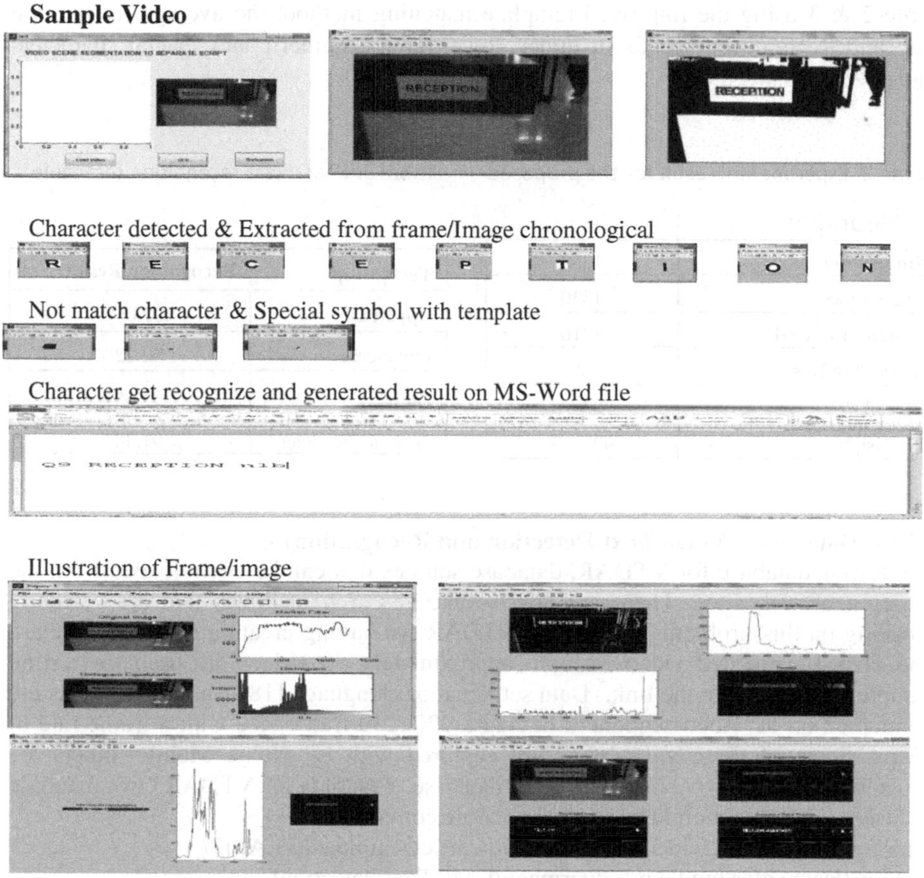

Sample Video

Character detected & Extracted from frame/Image chronological

Not match character & Special symbol with template

Character get recognize and generated result on MS-Word file

Illustration of Frame/image

Fig. 3. a) Median filter & Histogram b) Sobel vertical Edge filter & Histogram, Threshold c) Sobel Horizontal Edge d) Computes global threshold

Recall Rate: Recall rate (R) is defined as the ratio of the correctly detected characters to sum of correctly detected characters plus false negatives.

Test Result and Analysis: The following table compares recognition result of improved template matching method and traditional template matching method, the test result is shown in below table 2&3.

4   Performance Evaluation VTDAR Algorithms

Every character set the text box as per the character ,digit, special character size is detected correctly, all character is completely surrounded by a box, some character is not match with template data set then showing other character ,so a detected text box is considered as a false alarm, if no text appears in that box. The text localization algorithm achieved a recall of 90.96% and a precision of 92.15%.As seen from the table 2 & 3 using the improved template matching method, the average recognition rate and Recognition speeds of upper, lower letters, numeric and special characters have been enhanced.

Table 2. Experimental result for the proposed Algorithm

Images/Frames	1250
Text Lines	1530
Correct Detected	1410
False Positives	120
Recall (%)	90.96%
Precision (%)	92.15%

Table 3. Characters recognition test table

Test group	Recognition Result
Uppercase	92.05%
Lowercase	92.22%
Digits	93.05%
Special Character	91.30%

Video Databases (Video Text Detection and Recognition):-

We testing database for VTDAR, database sources that can be down loaded from the location of web sites. There are also several research institutes that are currently working on this problem. Testing for VTDAR we having creating our own database and also downloaded videos sample approximately 50 videos are used for testing sample .The below in the link. Data set, location, language [18]. Our experiments are conducted on 50 video sequences having a 320x240 frame size with a frame rate of 25fps. Some of these video clips are captured with E7 Nokia Mobile; others are downloaded from above listed standard database. Test data for VTDAR Own database by using E7 Nokia mobile https://sites.google.com/site/bharatgaikawad2012/videos, http://documents.cfar.umd.edu/LAMP/, http://www.cfar.umd.edu/~doermann/UMDTextDetectionData.tar.gz

Comparison Study of Video text detection and recognition (VTDAR) with Tesseract and Transym OCR .On the basis of below table 4, we have compared the recognition rate of VTDAR with Tesseract and Transym OCR .Tesseract and Transym both tools are proprietary Optical Character Recognition, we tried to tested all set 1250 of videos frames or images with all 3 different techniques or tools. Empirical result of VTDAR precision rate is 92.15 %. Experiments show that the recognition rates of VTDAR are compared to branded tools as Tesseract 3.02, Transym OCR 3.3 which is approximately similarly.

Table 4. Percentage of recognition with different Techniques/Tools

Sr.No.	Technique s/Tools	No. of Test Frames/images	Text Lines	No. of Correct Detected	False Positives	% of Recognition
1	VTDAR	1250	1530	1410	120	92.15%
2	Transym OCR 3.3	1250	1530	1425	105	93.13%
3	Tesseract 3.02	1250	1530	1450	80	94%

Graphically illustration of Character recognition result as following in figure 4.

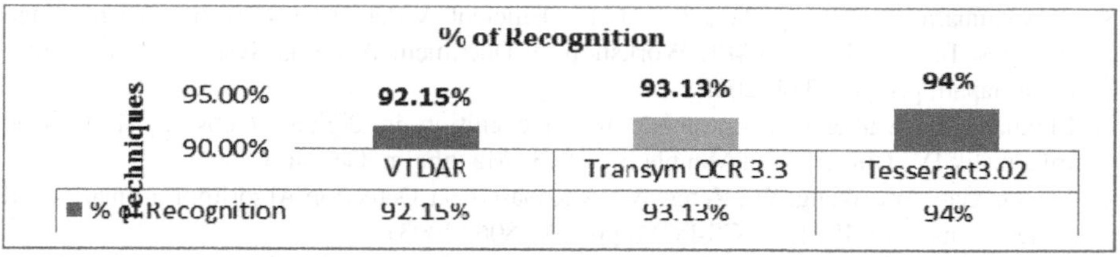

% of Recognition		
95.00% 92.15%	93.13%	94%
90.00% VTDAR	Transym OCR 3.3	Tesseract3.02
■ % of Recognition 92.15%	93.13%	94%

Fig. 4. Rate of percentage for character recognition from video frames or image

5   Conclusions

There are many cases this system are useful for video text information extraction system, vehicle license plate extraction, text based video indexing, video content analysis and video event identification .In this work, we have new approach for character recognition system based on template matching. This system tested the 50 videos with 1250 video frames and 1530 text lines .The system is texture-based approaches to automatic detection, segmentation and recognition of visual text occurrences in images and video frames. The characters are recognized automatically on run-time basis, In a few cases in which 7.85% characters could not get detected but some other character get recognized . The overall empirical performance of this system recognizing rate is 92.15%successfully. Empirically show that the recognition rates of VTDAR are compared to branded tools as Tesseract3.02, Transym OCR 3.3 which is approximately similarly.

References

1. Hua, X.-S., Wenyin, L., Zhang, H.-J.: Automatic Performance Evaluation for Video Text Detection. In: Sixth International Conference on Document Analysis and Recognition (ICDAR 2001), Seattle, Washington, U.S.A, September 10-13, pp. 545–550 (2001)
2. Junga, K., Kimb, K., Jain, A.K.: Text information extraction in images and video: a survey. Published by Elsevier Ltd. (2003)
3. Canny, J.: A Computational Approach to Edge Detection. IEEE Trans. Pattern Analysis and Machine Intelligence 8, 679–714 (1986)
4. Kim, H.K.: Efficcient automatic text location methodand content-based indexing and structuring of video database. J. Visual Commun. Image Representation 7(4), 336–344 (1996)
5. Zhong, Y., Jain, A.K.: Object localization using color, texture, and shape. Pattern Recognition 33, 671–684 (2000)
6. Antani, S., Kasturi, R., Jain, R.: A survey on the use of pattern recognition methods for abstraction, indexing, and retrieval of Images and video. Pattern Recognition, 945–965 (2002)

7. Jie, X., Hua, X.-S., Chen, X.-R., Wenyin, L., Zhang, H.: A Video Text Detection and Recognition System. In: IEEE International (2009)
8. Shivakumara, P., Huang, W., Tan, C.L.: Efficient Video Text Detection Using Edge Features. In: The Eighth IAPR Workshop on Document Analysis Systems (DAS 2008), Nara, Japan, pp. 307–314 (2008)
9. Lienhart, R., Stuber, F.: Automatic text recognition in digital videos. In: Praktische Informatik IV, University of Mannheim, 68131 Mannheim, Germany
10. Ye, Q., Gao, W., Wang, W., Zeng, W.: A Robust Text Detection Algorithm in Images and Video Frames. In: IEEE ICICS-PCM, pp. 802–806 (2003)
11. Aghajari, G., Shanbehzadeh, J., Sarrafzadeh, A.: A Text Localization Algorithm in Color Image via New Projection Profile. In: IMECS, Hong Kong (2010)
12. Ghorpade, J., Palvankar, R.: Extracting Text from Video. Signal & Image Processing, An International Journal (SIPIJ) 2(2) (2011)
13. Gaikwad, B., Manza, R.R.: Critical review on video scene segmentation and Recognition. International Journal of Computer Information Systems (IJCIS) 3(3) (2011)
14. Manza, R.R., Gaikwad, B.P.: A Video Edge Detection Using Adaptive Edge Detection Operator. CiiT International Journal of Digital Image Processing (2012), doi: DIP012012006, ISSN: 0974–9691 & Online: ISSN: 0974-9586
15. Manza, R.R., Gaikwad, B.P., Manza, G.R.: Use of Edge Detection Operators for Agriculture Video Scene Feature Extraction from Mango Fruits. Advances in Computational Research 4(1), 50–53 (2012)
16. Manza, R.R., Gaikwad, B.P., Manza, G.R.: Used of Various Edge Detection Operators for Feature Extraction in Video Scene. In: Proc. of the Intl. Conf. on Advances in Computer, Electronics and Electrical Engineering, ICACEEE 2012 (2012) ISBN: 978-981-07-1847-3
17. Sumathi, C.P., Santhanam, T., Priya, N.: Techniques and challenges of automatic text extraction in complex images: a survey. Journal of Theoretical and Applied Information Technology 35(2) (2012)
18. Spitz, A.L.: Determination of the Script and Language content of Document Images. IEEE Transactions on Pattern Analysis and Machine Intelligence 19(3) (1997)
19. Sharma, S.: Extraction of Text Regions in Natural Images. Masters Project Report (Spring 2007)
20. Mollah, A.F., Majumder, N.: Design of an Optical Character Recognition System for Camera based Handheld Devices. IJCSI 8(4(1)) (2011)
21. Su, Y.-M., Hsieh, C.-H.: A Novel Model-Based Segmentation Approach To Extract Caption Contents On Sports Videos. In: IEEE International Conference on Multimedia & Expo, pp. 1829–1832 (2006)
22. Leon, M., Vilaplana, V., Gasull, A., Marques, F.: Caption Text Extraction for Indexing Purposes Using a Hierarchical Region-Based Image Model. In: Proceedings of the 16th IEEE International Conference on Image Processing, pp. 1869–1872 (2009)
23. Zhong, Y., Zhang, H., Jain, A.K.: Automatic Caption Localization in Compressed Video. In: International Conference on Image Processing, vol. 2, pp. 96–100 (1999)
24. Liu, X., Wang, W.: Extracting Captions From Videos Using Temporal Feature. In: Proceedings of the International Conference on ACM Multimedia, pp. 843–846 (2010)

25. Lilo, B., Tang, X., Liu, J., Zhang, H.: Video Caption Detection and Extraction Using Temporal Information. In: International Conference on Image Processing, vol. 1, pp. I297–I300 (2003)
26. Gaikwad, B.P., Manza, R.R., Manza, G.R.: Video scene segmentation to separate script. In: Advance Computing Conference (IACC). IEEE xplore IEEE (2013) 978-1-4673-4527-9
27. Gaikwad, B.P., Manza, R.R., Manza, G.R.: Automatic Video Scene Segmentation to Separate Script for OCR. International Journal in Computer Application (IJCA) (2014) ISBN: 973-93-80880-06-7

Gesture: A New Communicator

Saikat Basak and Arundhuti Chowdhury

Dept. of MCA, Techno India College of Technology,
Kolkata, India
{saikatbsk,arundhatichowdhury}@gmail.com

Abstract. This paper is an illustrative approach for developing a visual interface for a gesture recognition system using color based blob detection. The software developed using the prescribed framework serves as a gesture interpretation system and is used to emulate the computer mouse with finger gestures. The objective is to develop an intuitive way to interact with computers and other digital devices and yet make it easy to use and cost effective at the same time. Although vision interfaces for gesture recognition have been researched and developed for some time, this approach has its own uniqueness and is more effective in many circumstances. The prescribed framework minimizes hardware requirements as it only requires a webcam other than the computer itself. The minimization of hardware requirements make it cost effective and easier to obtain. The predefined gestures are simple yet intuitive as these are inspired by certain every day gestures or movements that are used to interact with several tools and equipment in our daily life.

Keywords: vision interface, vision framework, gesture, finger gesture, mouse control.

1   Introduction

The point and click method is arguably the most common way to interact with today's digital devices, such as the personal computer. And nowadays the emergence of computer vision technologies has enabled us to send control instructions to the computers by means of gestures. This paper aims to use computer vision to analyze different sets of gestures or actions done using the human fingers and interpret them as meaningful instruction to be feed to the computer.

Computer vision is nowadays being used in gaming industry or in the fields of perceptual computing and the aim is to use gestures to interact with the computer or the gaming device. Color based hand tracking systems [1], integrated person tracking with color and pattern detection [2], 3D hand tracking [3] are some of the examples where computer vision is used. In this paper a gesture interpretation system capable of emulation computer mouse is proposed.

The proposed system or framework is capable of interpreting predefined sets of gestures into mouse control instructions i.e. click and scroll. In other words, the aim is to emulate mouse driven point and click events by using finger gestures.

S.C. Satapathy et al. (eds.), *Proc. of the 3rd Int. Conf. on Front. of Intell. Comput. (FICTA) 2014*
– *Vol. 2*, Advances in Intelligent Systems and Computing 328, DOI: 10.1007/978-3-319-12012-6_26

2   Previous Work

2.1   SixthSense

SixthSense, a wearable gesture interface that links the physical world around us with digital information using natural hand gestures is introduced in [4]. This technology introduces the use of color markers to detect and track hand gestures. Also a mini-projector is used to augment the world around the user.

2.2   FingerMouse

FingerMouse [5] is a freehand pointing alternative to the mouse. A vision system constantly monitors the hand and tracks the fingertip of the pointing hand. The screen cursor is moved using hand gesture and the mouse clicks are registered using the keyboard.

3   Proposed Method

The primary goal of gesture recognition research is to create a system which can identify specific human gestures and use them to convey information or for device control. A gesture, in general, might consist of certain meaningful movement using the hand, face or any other body part. It might carry a body language that can be interpreted as an instruction or can be used as a medium to convey information. Gestures have always been used as a method or tool for nonverbal communication. Following is the framework that has been used to capture and identify specific finger gestures to control the computer mouse.

3.1   Framework

Methods and applications of vision based user interfaces are well studied in [7]. The proposed framework captures real time images using a webcam to get the required information. The captured images are in BGR format. The images are flipped because mirror images captured by webcams are not intuitive. Then the color space of the images are converted into HSV color space. The regions of the color markers were masked using predefined color range to generate a binary image. An opening operation is then applied to the image to reduce noise. Later the blobs of the color markers are detected and the gestures are recognized using the detected blobs. Once the gestures are detected and analyzed system calls are sent to control the mouse. A framework diagram is shown in Fig. 1.

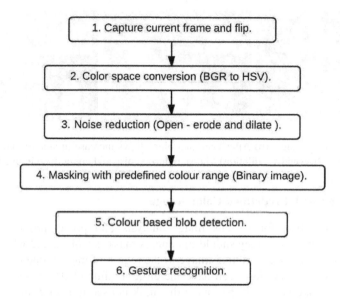

Fig. 1. Framework for gesture recognition using color markers

3.2   Color Space Conversion

The RGB color space is any additive color space based on the RGB color model. A particular RGB color space, is defined by the individual chromaticity of the red, green, and blue additive primaries, and can produce any chromaticity that is the triangle defined by those primary colors.

On the other hand, HSV uses cylindrical-coordinate representations of points in an RGB color model. The representations rearrange the geometry of RGB in an attempt to be more intuitive and perceptually relevant than the Cartesian (cube) representation; thus making the HSV color space relatively robust to light variations, so it is effective to detect the color markers. Comparison of several color models have been studied in [8].

3.3   Noise Reduction

For noise reduction an opening operation is used. Opening consists of an erode operation followed by a dilate operation. A general idea of noise reduction is using a blur operation but opening is used because it does not change the size of the area of the object, blur does, Fig. 2. Erosion with small square structuring elements shrinks an image by removing a layer of pixels from both the inner and outer boundaries of regions. Thus the small details are eliminated. Larger structuring elements have a more pronounced effect, the result of erosion with a large structuring element being similar to the result obtained by iterated erosion using a smaller structuring element of the same shape. Erosion removes small-scale details from a binary image but simultaneously reduces the size of regions of interest, too. Dilation adds a layer of pixels to both the inner and outer boundaries of regions. Resulting is an opening.

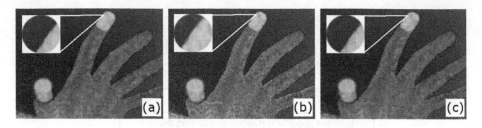

Fig. 2. (a) The HSV image, (b) After Gaussian blur, shows increase in area of pixels (c) After Opening (erode followed by a dilation), shows reduced noise and no increase in pixels

3.4   Masking with Predefined Color Range

Masking is an option for various image analysis and processing functions that specifies on which regions they should operate. A mask is a black and white image of the same dimensions as the original image or the region of interest one is working on. Each of the pixels in the mask can have therefore a value of 0 (black) or 1 (white). When executing operations on the image the mask is used to restrict the result to the pixels that are 1 in the mask. In this way the operation restricts to some parts of the image. Region with any specific color can be extracted from the image using this method. The color markers are masked using predefined color range threshold. The original image taken using the webcam is shown in Fig. 3. At first this image is masked with the following HSV range to detect the red color marker,

$$61<H<179, 133<S<255, 67<V<255$$

The blue marker is detected using a mask on the same image with the following HSV range,

$$61<H<179, 133<S<255, 67<V<255$$

Here, Hue is calculated within a range of 0 to 180. Both Saturation and Value are calculated within a range of 0 to 255. The result mask is shown in Fig. 4.

Fig. 3. Original image

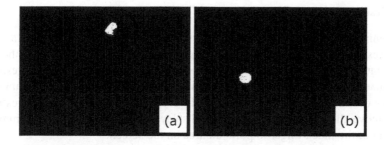

Fig. 4. Detected blobs are shown in masked images (a) and (b)

3.5   Blob Detection

The color blobs (concentration of the white pixels) are detected using image moments. An image moment is a particular weighted average of the pixel intensities. Image moments are useful to describe objects after segmentation. Simple properties of the image which are found via image moments include area or the total intensity, its centroid, and information about its orientation. Object detection using image moments are well surveyed in [9] and [10].

X coordinate of the center of the detected blob is calculated using eq. (1) and eq. (2) is used to calculate the Y coordinate.

$$x = moment10 / moment00 \tag{1}$$

$$y = moment01 / moment00 \tag{2}$$

Here, (x, y) be the position co-ordinates of the center of the detected blob, moment10 is 1st order spatial moment around x-axis, moment01 is 1st order spatial moment around y-axis, moment00 is 0th order central moment. Here similar methods are applied to detect two different color blobs. The result of blob detection is shown is Fig. 5. The blobs are sorted by color. The blob of the marker in the index finger is labeled b1 and the other is labeled b2.

Fig. 5. Detected blobs

3.6   Gesture Recognition

To recognize gestures, the gestures are categorized in two – "one finger" and "two finger" gestures. "One-finger gesture" is gesture by one finger, usually the index finger, controls the movement of the mouse cursor on screen. "two-finger gesture" is a combination of gestures by two fingers, usually the index and the thumb, and comprises of 'click', 'right click' and 'scroll' commands. The "one-finger gesture" is used to move the mouse pointer on the display screen. The gesture is simply the movement of the index finger to point an area on the screen.

The "one-finger gesture" is executed when the number of detected blob is 1. The coordinates of the center of the detected blob that represents the index finger is used to calculate where in the screen the mouse cursor should be. The aforementioned coordinates are simply mapped to the usually larger screen size by calculating the ratio of the size of the image taken from the camera and the size of the display screen.

The "two-finger gesture" comprises of 'click gesture', 'right click gesture' and 'scroll gesture'. We assume x-value and y-value of center of b1 is x_{b1} and y_{b1}. Also, x-value and y-value of center of b2 is x_{b2} and y_{b2}, respectively. Let r be a constant, d be the distance between the position of the center co-ordinates of b1 and b2. Let us take a circle around the center of b1 with radius r. If the center of b2, represented by (x_{b2}, y_{b2}), is within this circle it is the 'click gesture', Fig. 6. The 'left mouse button pressed' and 'left mouse button released' events are controlled using the 'click gesture'.

For the 'right click gesture' a counter c and a timer t which is a constant, has been introduces. Let us assume that the position of the center of the blob b1 in current frame (n) be $(x_{b1}, y_{b1})_n$. The position of b1 in frame (n+1) is $(x_{b1}, y_{b1})_{n+1}$. We also assume that the displacement of the center of b1 in two consecutive frames be D. Let δ be a constant. Then the 'right click gesture' is determined using the following algorithm,

```
// Right click gesture
Let c = 0
If ('left mouse button pressed' is FALSE) Then
   If (D < ) Then
      c = c+1
      If (c == t) Then
        Perform 'right click button event'
        c = 0
      End If
   Else
      c = 0
   End If
End If
```

In simple words, if the center of the blob b1 stays within a boundary of radius δ in consecutive frames for a certain time identified by t, then it is a 'right click gesture'.

For the 'scroll gesture', the image is separated in three regions, region A, B and C, Fig. 7(a). If the position of the center of the second blob b2 is inside region A then the 'downward scroll event' is triggered, Fig. 7(b). When center of b2 is inside region B, the 'upward scroll event' is triggered. Fig. 7(c) shows gesture for downward scroll.

```
// Scroll gesture
If ('left mouse button pressed' is FALSE && 'right click
button event' is FALSE) Then
   If (xb2> xb1&& yb2< yb1) Then
     Perform 'upward scroll event'
   Else If (xb2> xb1&& yb2> yb1) Then
     Perform 'downward scroll event'
   End If
End If
```

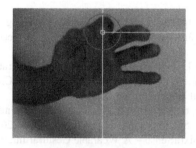

Fig. 6. Click gesture

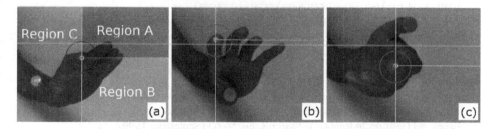

Fig. 7. (a) Regions in the image, (b) Scroll down gesture, (c) Scroll up gesture

4   Software Implementation

The framework has been implemented using OpenCV (Open Computer Vision Library) and tested in a Gnu/Linux system.

5   Conclusion

Using this framework a gesture interpretation system capable of controlling the computer mouse is developed. The framework is presented for gesture recognition using color markers. The framework works independent of any hardware other than a built in or an USB webcam and the computer itself, making the system cost effective. The system makes our interaction with computers more intuitive. Simple predefined gestures make the software easier to adopt and use. This kind of framework can be used to build gesture based input systems for other digital appliances like televisions, media centers. Gestures can be introduced to control web browsers, media players, and interactive games.

References

1. Imagawa, K., Lu, S., Igi, S.: Color-Based Hands Tracking System for Sign Language Recognition. In: Proc. 3rd Int. Conf. on Face and Gesture Recognition, Nara, Japan (April 1998)
2. Darrell, T., Gordon, G., Harville, M., Woodfill, J.: Integrated Person Tracking Using Stereo, Color, and Pattern Detection. In: Proc. Conf. on Computer Vision and Pattern Recognition, Santa Barbara, California (June 1998)
3. Ahmad, S.: A usable real-time 3d hand tracker. Conference Record of the Asilomar Conference on Signals, Systems and Computers, pp. 1257–1261 (1994)
4. Mistry, P., Maes, P.: SixthSense – A Wearable Gestural Interface. In: The Proceedings of SIGGRAPH Asia 2009, Sketch, Yokohama, Japan (2009)
5. Mysliwiec, T.A.: Fingermouse: A Freehand computer pointing interface. Technical report VISLab-94-01, University of Illinois at Chicago (1994)
6. Quek, F.K.H., Mysliwiec, T., Zhao, M.: Fingermouse: A freehand pointing interface. In: Proc. Int. Workshop on Automatic Face-and Gesture-Recognition, Zurich, Switzerland, pp. 372–377 (June 1995)
7. Porta, M.: Vision-based User Interfaces: Methods and Applications. Int. J. Human-Computer Studies 57(1), 27–73 (2002)
8. Zarit, B.D., Super, B.J., Quek, F.K.H.: Comparison of Five Color Models in Skin Pixel Classification. In: International Workshop on Recognition, Analysis, and Tracking of Faces and Gestures in Real-time Systems, pp. 58–63 (September 1999)
9. Flusser, J.: On the independence of rotation moment invariants. Pattern Recognition - The Journal of the Pattern Recognition Society (May 19, 1999)
10. Flusser, J., Suk, T.: Rotation Moment Invariants for Recognition of Symmetric Objects. IEEE Transactions on Image Processing 15(12) (December 2006)

A Novel Fragile Medical Image Watermarking Technique for Tamper Detection and Recovery Using Variance

R. Eswaraiah[1,*] and E. Sreenivasa Reddy[2]

[1] Vasireddy Venkatadri Institute of Technology, Guntur, Andhra Pradesh, India
eswar_507@yahoo.co.in
[2] Acharya Nagarjuna University, Guntur, Andhra Pradesh, India
edara_67@yahoo.com

Abstract. In this paper, we propose a novel fragile block based medical image watermarking technique to produce high quality watermarked medical images, verify the integrity of ROI, accurately detect the tampered blocks inside ROI using both average and variance and recover the original ROI without loss. In the proposed technique, the medical image is segmented into three sets of pixels: ROI, Region of Non Interest (RONI) and border pixels. Later, authentication data along with ROI information is embedded inside border. ROI recovery data is embedded inside RONI. Results of experiments disclose that proposed method produced high quality watermarked medical images, identified the presence of tampers inside ROI with 100% accuracy and recovered the original ROI without any loss.

Keywords: watermarking, ROI, RONI, tamper detection, recovery.

1 Introduction

Telemedicine eliminates distance hurdle and provides access to medical services. It allows transmission of medical data like medical images through network, ensures handy and faithful interactions between patients and medical staff. This exchange must maintain integrity of medical images besides results in more cost and transmission time while transferring patient data and his medical image independently [1]. Watermarking is used to deal with the above two conflicts.

Watermarking techniques are classified into two categories based on medium for hiding data: spatial domain and frequency domain. In spatial domain watermarking techniques [2-5], data is masked directly into host image. In frequency domain techniques [6-8], data is masked into transformed host image.

The other taxonomy of watermarking techniques is: reversible and irreversible. In reversible watermarking techniques image reconstruction is lossless [7-9, 17]. While in irreversible watermarking techniques lossless recovery of original image is not possible [6]. The apt technique for medical images is reversible watermarking [11].

---

* Corresponding author.

© Springer International Publishing Switzerland 2015   245
S.C. Satapathy et al. (eds.), *Proc. of the 3rd Int. Conf. on Front. of Intell. Comput. (FICTA) 2014*
– *Vol. 2*, Advances in Intelligent Systems and Computing 328, DOI: 10.1007/978-3-319-12012-6_27

Watermarking techniques are classified as robust, fragile and hybrid depending on context. Robust watermarking techniques [6, 7, 8, 12] are used in applications that requires protection of copyright information of images. Fragile watermarking techniques [2, 3, 4, 5, 13] are used in applications which require detection of tampers. Hybrid watermarking techniques [14-16] are used in applications where both privacy and integrity control of images is desired.

Most of the medical images contain two parts: ROI and RONI. ROI part of medical image is more important for diagnosis. The visual quality of ROI of medical image must not degrade while hiding data into ROI part besides any tampering to ROI must be identified and the original ROI has to be recovered to avoid misdiagnosis and retransmission of medical image. The recovery data of ROI is usually embedded into RONI [3, 4, 5, 15, 8, 10, 16, 18]. The tampered area of ROI of received watermarked medical image is replaced with the recovery data embedded inside RONI when a tamper is detected inside ROI.

Till dated many block-based watermarking techniques have been developed for identifying tampers inside ROI of medical images and recovering original ROI. The shortcomings of reviewed watermarking techniques for tamper detection and recovery are summarized as:

1. To check directly whether the ROI or the entire image is tampered; some methods [2, 3, 4, 5, 7] are not using any authentication data of either ROI or entire medical image. So, all blocks in the ROI or entire image are checked one after the other to detect the presence of tampers. This checking process leads to wastage of time when the image is not tampered.
2. To recover the original ROI; some methods [2, 3, 4, 5, 19] are replacing the pixels in a tampered ROI block with the average of pixels in its original block and some methods [6, 8, 16] are using compressed form of block for recovering the tampered block. So, original ROI cannot be recovered exactly.
3. The ROI of watermarked medical image is highly distorted as some methods [8, 16] are embedding large data inside ROI.
4. Accurate identification of tampered blocks is not possible as some methods [2, 3, 4, 5, 7, 8, 10, 16] are identifying tampered blocks based on average intensity of blocks.
5. Finally, some methods [9, 15, 20] are not specifying how to recover the original ROI when it is tampered.

Most of the reviewed schemes are identifying the tampered blocks in watermarked medical images based on average intensity of the blocks. If the values of pixels in a block are modified without changing its average value then these schemes fail in identifying the changes or tampers in the block. For example; in a watermarked medical image, if the values of pixels of a block are as shown in Fig. 1 then the average intensity of the block will be 72. By changing the values of pixels as shown in Fig. 2 there is a possibility to get the same average intensity for the block. It is not possible to detect modifications done in the tampered block precisely by comparing only the average values of original and modified blocks. Hence there is a necessity to develop a system that can detect the tampers accurately besides keeping pixels values changed and average intensity value unchanged.

In this paper, we propose a novel block based fragile watermarking technique to achieve the following objectives.

1. Accurately detecting tampered blocks inside ROI using both average and variance values of blocks.
2. Recovering original ROI with no loss when it is tampered.
3. Identifying tampered blocks inside ROI and recovering original ROI with simple mathematical calculations.
4. Avoiding the process of checking ROI blocks of watermarked medical image for the presence of tampers when the ROI is not tampered.
5. Avoiding distortion in the ROI of watermarked medical image by not embedding any data in ROI.

The rest of the paper is organized as follows. Proposed method is explained in section 2. Section 3 presents the experimental results. The conclusion is in section 4.

65	74	61	79
75	82	70	66
80	73	67	78
72	68	78	66

Fig. 1. A 4×4 block of pixels in a medical image

62	72	65	83
65	76	75	70
72	77	81	74
77	70	82	63

Fig. 2. The modified 4×4 block of the medical image

2   Proposed Method

We propose a novel medical image watermarking method to meet the above mentioned objectives. In this method, the medical image is segmented into three sets of pixels: ROI, RONI and border pixels as shown in Fig. 3. Next, hash value of ROI is calculated using SHA-256 and is used as authentication data of ROI. ROI and RONI are divided into non overlapping blocks of size 4×4 and 9×8 respectively. With the assumption that the number of blocks in ROI is less than the number of blocks in RONI, each ROI block is mapped to a corresponding RONI block. For each block in ROI, authentication data is generated by calculating both average and variance of pixels inside the block. Variance value of the block is used to detect changes in the block when the pixels inside the block are modified such that the average intensity remains same. For example, the average and variance values of the block shown in Fig. 1 are 72 and 53 respectively. The average and variance values become 72 and 62 if the pixels of the block are modified as shown in Fig. 2. For each block in ROI, recovery data is generated by collecting the bits of pixels inside the block. Authentication and recovery data of each ROI block are both embedded into LSBs of pixels in the mapped block of RONI.

For each 4×4 block in ROI of 8-bit medical images, the sizes of authentication and recovery data are 16 bits (8 bits for average and 8 bits for variance) and 128 bits (collection of bits of pixels inside the ROI block) respectively. As shown in Fig. 4, 2 LSBs in each pixel of mapped RONI block are used to embed this authentication and recovery data. In 12-bit and 16-bit medical images; the size of authentication data is 24 and 32 bits respectively, the size of recovery data is 192 and 256 bits respectively. The authentication as well as recovery data of each ROI block is embedded into 3 LSBs (in case of 12-bit images), 4 LSBs (in case of 16-bit images) of pixels in the mapped RONI block. Finally, hash value and information of ROI are embedded into LSBs of border pixels.

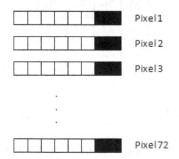

Fig. 3. Division of medical image into three regions

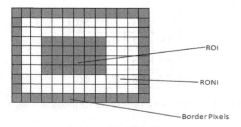

Fig. 4. Embedding authentication and recovery data

2.1 Embedding Algorithm

1. Segment the medical image into three sets of pixels: ROI, RONI and border pixels.
2. Using SHA-256 technique calculate hash value of ROI (h1).
3. Divide the ROI part into non overlapping blocks of size 4×4.
4. Divide the RONI part into non overlapping blocks of size 9×8.

5. Assuming that the number of blocks in ROI is less than the number of blocks in RONI, map each ROI block to a block in RONI.
6. For each ROI block, generate authentication data by calculating both average and variance of pixels in the block.
7. For each ROI block, collect bits of 16 pixels as recovery data.
8. Depending on bit depth, embed authentication and recovery data of each ROI block into 2 or 3 or 4 LSBs of pixels in mapped RONI block.
9. Embed both hash value (h1) and information of ROI into LSBs of border pixels.

At receiver's end, information and hash value of ROI are both extracted from LSBs of border pixels in the received watermarked medical image. ROI and RONI parts are identified in the received medical image using the extracted ROI information. Hash value of ROI is calculated and is compared with the extracted hash value to detect the presence of tampers inside ROI of received medical image. If there is a match between the extracted and re calculated hash value then the received medical image is authentic and is directly used for making diagnosis decisions. Otherwise, ROI and RONI of received image are divided into non overlapping blocks of size 4×4 and 9×8 respectively. The authentication and recovery data of each ROI block are extracted from LSBs of related mapped RONI block. The tampered ROI blocks are identified by comparing authentication data of each ROI block with its extracted authentication data. The extracted recovery data is used to recover the original ROI block when a block in ROI is identified as tampered.

2.2   Extraction Algorithm

1. Extract both information and hash value of ROI (h1) from LSBs of border pixels.
2. By using information of ROI, identify ROI and RONI pixels.
3. Using SHA-256 technique calculate hash value of ROI (h2).
4. Compare h1 with h2.
5. If h1==h2 then stop the extraction procedure, otherwise go to step 6.
6. Segment ROI and RONI into blocks of size 4×4 and 9×8 respectively.
7. To identify tampered ROI blocks, repeat steps 8 to 10 for each ROI block (B).
8. Calculate average and variance values of block B representing as a2 and v2.
9. Depending on bit depth, extract average (a1), variance (v1) and recovery data (r) of ROI block B from the 2 or 3 or 4 LSBs of corresponding mapped RONI block.
10. Mark the ROI block B as tampered if either a1≠ a2 or v1≠v2.
11. To get the original ROI, replace each tampered ROI block B with the recovery data (r) extracted from the corresponding mapped RONI block.

3   Discussion of Experimental Results

About hundred medical images of different bit depths (8, 12 and 16) and of different modalities like CT scan, MRI scan, Ultrasound and so on are considered for conducting experimental results. For measuring the quality of generated watermarked medical images, Peak Signal to Noise Ratio (PSNR) and Weighted Peak Signal to Noise Ratio (WPSNR) were used as metrics. To measure the similarity between the original medical image and the watermarked medical image Mean Structural SIMilarity index (MSSIM) metric is used. Total Perception Error (TPE) metric is used to measure the visual degradation in watermarked medical image. Fig. 5 shows three medical images used in experiments. It also shows watermarked and reconstructed medical images.

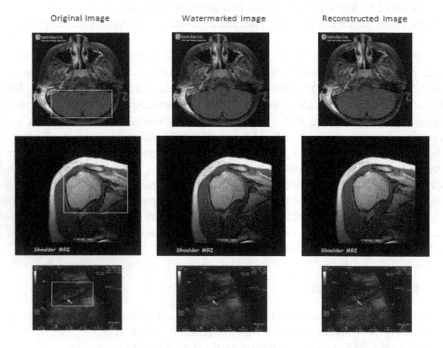

Fig. 5. Original, watermarked and reconstructed medical images. From top to bottom: CT scan, MRI scan and Ultrasound images

Table 1. Results of Embedding in Three Medical Images

Modality	Size of Image	Bit depth	Size of ROI	Number of blocks in ROI	PSNR	WPSNR	MSSIM	TPE
CT	336×406	12 bits	136×164	1394	53.36	54.15	0.9575	0.0549
MRI	480×512	16 bits	192×204	2448	51.52	53.44	0.9327	0.0828
US	309×255	8 bits	124×104	806	56.82	58.13	0.9854	0.0346

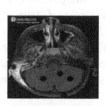
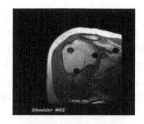

Fig. 6. Watermarked medical images (from left to right: CT scan, MRI scan and Ultrasound) with tampers inside ROI

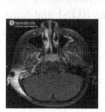
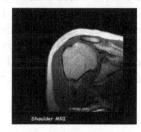

Fig. 7. Recovered medical images (from left to right: CT scan, MRI scan and Ultrasound)

Table 2. Comparison between reviewed methods and proposed method

Scheme	ROI-Based	Embedding distortion inside ROI	Accurate identification of tampered blocks	Recovery of tamped blocks inside ROI or image
Zain	No	-	No	With average intensity of blocks
Wu	Yes	No	Yes	With JPEG compressed form of ROI
Chiang	Yes	No	No	With average intensity of blocks
Liew[3, 4]	Yes	Yes	No	With average intensity of blocks
Agung	Yes	Yes	No	With average intensity of blocks
Qershi[8]	Yes	Yes	No	With compressed form of ROI
Qershi[10]	Yes	No	No	With original pixels of blocks
Qershi[16]	Yes	Yes	No	With compressed form of ROI
Proposed method	Yes	No	Yes	With original pixels of blocks

The results of experiments conducted on the three medical images are depicted in Table 1. The performance of proposed scheme is tested in terms of detecting tampered areas inside ROI and recovering original ROI by modifying some of the pixels inside ROI of the watermarked medical images as shown in Fig. 6. The proposed scheme accurately identified all tampered locations inside ROI and recovered the original ROI with no loss as shown in Fig. 7. The comparison between the proposed scheme and the reviewed schemes is shown in table 2.

Proposed method can be used with medical images of different bit depth and of modalities. It is suitable for medical images whose ROI is up to 20% of entire image. The RONI and border parts are not recovered exactly as LSBs of pixels in RONI as well as border are set to bit 0 after extracting embedded data from them. This limitation does not affect the efficiency of the proposed method as RONI and border parts of medical images are insignificant to make diagnosis decisions. This method can recover original ROI only when ROI is modified and the RONI and border of the watermarked medical image are not modified by intruders.

4   Conclusion

The proposed method is apparent from the values of PSNR, WPSNR, MSSIM and TPE which produces high quality watermarked medical images. No embedding distortion inside ROI of watermarked medical images because no data is embedded into ROI. This method precisely identifies and localizes tampered blocks inside ROI using average and variance values of blocks. Original ROI with no loss is recovered as the pixels in tampered blocks are replaced with original pixel values. The proposed method uses simple mathematical calculations for generating authentication and recovery data, identifying tampered blocks inside ROI and recovering original ROI. This method does not check for the presence of tampers inside ROI when the extracted hash of ROI matches with recalculated hash of ROI.

References

1. Coatrieux, G., Montagner, J., Huang, H.: Ch. Roux.: Mixed reversible and RONI watermarking for medical image reliability protection. In: 29th IEEE International Conference of EMBS, pp. 5653–5656. IEEE Press, Lyon (2007)
2. Zain, J.M., Fauzi, A.R.M.: Medical image watermarking with tamper detection and recovery. In: Proceedings of the 28th IEEE EMBS Annual International Conference, pp. 3270–3273. IEEE Press, New York (2006)
3. Liew, S.C., Zain, J.M.: Reversible medical image watermarking for tamper detection and recovery. In: 3rd IEEE International Conference on Computer Science and Information Technology (ICCSIT), pp. 417–420. IEEE Press, Chengdu (2010)
4. Liew, S.C., Zain, J.M.: Reversible medical image watermarking for tamper detection and recovery with Run Length Encoding compression. World Academy of Science, Engineering and Technology, 799–803 (2010)

5. Agung, T., AdiWijaya, B.W., Fermana, F.B.: Medical image watermarking with tamper detection and recovery using reversible watermarking with LSB modification and Run Length Encoding compression. In: IEEE International Conference on Communication, Networks and Satellite, pp. 167–171. IEEE Press, Bali (2012)
6. Wu, J.H.K., Chang, R.-F., Chen, C.-J., Wang, C.-L., Kuo, T.-H., Moon, W.K., Chen, D.-R.: Tamper detection and recovery for medical images using near-lossless information hiding technique. Journal of Digital Imaging 21, 59–76 (2008)
7. Chiang, K.-H., Chang-Chien, K.-C., Chang, R.-F., Yen, H.-Y.: Tamper detection and restoring system for medical images using wavelet-based reversible data embedding. Journal of Digital Imaging 21, 77–90 (2008)
8. Al-Qershi, O.M., Khoo, B.E.: Authentication and data hiding using a reversible ROI-based watermarking scheme for DICOM images. In: Proceedings of International Conference on Medical Systems Engineering (ICMSE), pp. 829–834 (2009)
9. Deng, X., Chen, Z., Zeng, F., Zhang, Y., Mao, Y.: Authentication and recovery of medical diagnostic image using dual reversible digital watermarking. Journal of Nanoscience and Nanotechnology 13(3), 2099–2107 (2013)
10. Al-Qershi, O.M., Khoo, B.E.: ROI-based tamper detection and recovery for medical images using reversible watermarking technique. In: International Conference on Information Theory and Information Security (ICITIS), pp. 151–155 (2010)
11. Luo, X., Cheng, Q., Tan, J.: A lossless data embedding scheme for medical images in application of e-Diagnosis. In: Proceedings of the 25th Annual International Conference of the IEEE EMBS, pp. 852–855 (2003)
12. Eggers, J.J., Bauml, R., Tzschoppe, R., Girod, B.: Scalar SOSTA scheme for information hiding. IEEE Transactions on Signal Processing 151(4), 1003–1019 (2003)
13. Memon, N.A., Gilani, S.A.M.: NROI watermarking of medical images for content authentication. In: Poceedings of 12th IEEE International Mutitopic Conference (INMIC 2008), Karachi, Pakistan, pp. 106–110 (2008)
14. Giakoumaki, Pavlopulos, S., Koutouris, D.: Multiple image watermaking applied to health information management. IEEE Transactions on Information Technology and Biomedicine 10(4), 722–732 (2006)
15. Memon, N.A., Chaudhry, A., Ahmad, M.: Hybrid watermarking of medical images for ROI authentication and recovery. International Journal of Computer Mathematics 88(10), 2057–2071 (2011)
16. Al-Qershi, O.M., Khoo, B.E.: Authentication and data hiding using a hybrid ROI-based watermarking scheme for DICOM images. Journal of Digital Imaging 24(1), 114–125 (2011)
17. Lei, B., Tan, E.-L., Chen, S., Ni, D., Wang, T., Lei, H.: Reversible watermarking scheme for medical image based on differential evolution. Expert Systems with Applications 41, 3178–3188 (2014)
18. Nyeem, H., Boles, W., Boyd, C.: Ultilizing least significant bit-planes of RONI pixels for medical image watermarking. In: International Conference on Digital Image Computing: Techniques and Applications (DICTA), pp. 1–8. IEEE Press, Hobart (2013)
19. Kim, K.-S., Lee, M.-J., Lee, J.-W., Oh, T.-W., Lee, H.-Y.: Region-based tampering detection and recovery using homogeneity analysis in quality-sensitive imaging. Computer Vision and Image Understanding 115, 1308–1323 (2011)
20. Tan, C.K., Ng, J.C., Xu, X., Poh, C.L., Guan, Y.L., Sheah, K.: Security Protection of DICOM Medical Images Using Dual-Layer Reversible Watermarking with Tamper Detection Capability. Journal of Digital Imaging 24(3), 528–540 (2011)

MRI Skull Bone Lesion Segmentation Using Distance Based Watershed Segmentation

Ankita Mitra[1], Arunava De[2], and Anup Kumar Bhattacharjee[1]

[1] Department of Electronics and Communication,
National Institute of Technology, Durgapur, India
ankimitra.2009@gmail.com, akbece12@yahoo.com
[2] Department of Information Technology,
Dr. B. C. Roy Engineering College, Durgapur, India
arunavade@yahoo.com

Abstract. The objective of separating touching objects in an image is a very difficult task. The task is all the more difficult when the touching objects are healthy tissues and unhealthy tissues of lesions in human brain.

A gray level MR image may be considered as a topographic relief and thus Watershed segmentation is used. Watershed refers to a ridge that divides areas drained by different river systems. A catchment basin is interpreted as a geographical area draining into a river or reservoir. The concept of watershed and catchment basins are used for analyzing biological tissues.

An MR image segmentation method is developed using Distance and Watershed Transforms.

Keywords: Watershed, Watershed lines, Catchment basins, Distance Transform.

1 Introduction

This paper proposes a scheme for segmentation of Magnetic Resonance Images (MRI). The objective is to identify the healthy and the non-healthy tissues of the human brain. In this article we deal with multimodal MRI images. MRI images contain multiple objects and background and hence it contains multiple thresholds.

We propose to segment the MR image of the brain with lesions. Before the segmentation can be done the MR image has to be de-noised. A low pass filter mask was used for de-noising purpose.

Image segmentation using Watershed Transform is an efficient method provided we solve its main drawback namely of over-segmentation. Pre-processing of the image has to be done to prevent over-segmentation. The solution for preventing over-segmentation is discussed by [1]. The Watershed transform together with Distance transforms are used in this article for the segmentation purposes.

2   Related Work

Reference[2] used texture gradient for segmenting images. It uses a non decimated form of complex wavelet transform to extract texture information. Reference [3] segments overlapping objects using locally constrained watershed transform introduced by Beare[4]. This type of watershed transform offers an alternative to other methods, such as distance function flooding, in segmenting complex backgrounds.

Reference [5] provides a robust method for the removal of non-cerebral tissue in T1- weighted magnetic resonance (MR) brain image. This procedure called Skull stripping is an important step in neuroimaging. Reference [6] utilizes MLP neural networks and watershed algorithm to segment MRI of liver. Reference[7] proposes a new watershed framework which allows for segmenting spatio-temporal images. It is based on discrete mathematical morphology. This is an automated approach to segment the left ventricular myocardium in 4D cine-MRI sequence. Reference [8] proposed an automatic spleen segmentation algorithm.

3   MRI DATA Acquisition

We have taken a set of slices of human brain of patients for testing purposes. We have tested our method in a number of patients. The patients include those who are currently undergoing chemotherapy and also those who are cured of the brain lesions. The lesions can be anything starting from tuberculosis to brain tumor.

The images are viewed using centricity dicom viewer provided by GE Medical. For our experiments we consider the images in Axial T2 view which mainly gives the pathology of the disease. All the examinations were done on the same 1.5-T MRI imaging device. The slice thickness is 5.0 mm and the gap between two slices was 1.5 mm. Each slice is having a resolution of 256×256.

4   Proposed Algorithm

Segmentation using this method works better if we can identify the foreground objects and background locations. Our segmentation follows the below basic procedure:

1. **Input Image Is De-noised Using Filter Mask:** We read the image using DiCOM viewer provided by GE. The input image is in Di-Com format. A 3×3 mask is used for filtering purpose. The mask is moved over the ROI, if the neighborhood pixels display a similar value we increase the size of the mask. The mask is grown in all the four directions depending on the similarities of the pixel values. If the neighborhood pixels have different value then we again start afresh in the new region of the image. After execution of this procedure we get a filtered image. This filtered image is then used for further processing.
2. **MR Image of Skull into Binary:** The input image is de-noised using a filter mask. The de-noised image is then converted into binary image.

3. **Compute the Segmentation Function:** Convert the grayscale image into a binary image. Find the Negative of the MR image. The MR images have more dark pixels and less light pixels as a result the watershed transform cannot be applied since it is only applied to images where light pixels are high and dark pixels are low. Thus we convert the binary MR image into its negative so that the Watershed transform can be applied. Compute the City Block distance transform.

   In 2-D, the city block distance between (x_1, y_1) and (x_2, y_2) is

   $$|x_1 - x_2| + |y_1 - y_2|.$$

4. **Suppress the Shallow Minima:** An image may have many regional minima and maxima but they will have a single global minima or maxima. To create marker images that are used in Morphological reconstruction of the image we need to determine the peaks or valleys. We can identify the region of the image where the change in intensity is extreme, i.e. the difference between the intensities of the given pixel is more than (or lesser than) its neighbor by a certain threshold. Small fluctuations in the intensity results in regional maxima and minima but we are interested in only significant changes and not these small changes (smaller minima and maxima) which are caused by background texture and hence the need for suppression of these. An illustration of Maxima and Minima is depicted in Fig.1.

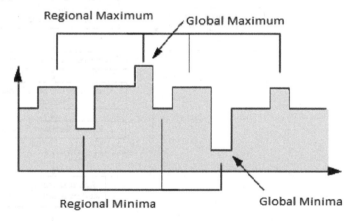

Fig. 1. Illustration of Maxima and Minima of an image

5. **Watershed Transformation** is computed for modified segmentation function. We assume the image to be segmented as a surface. Let us consider the bright areas in the image are having high pixel values and dark areas are having low pixel values. Then the bright areas result in watershed lines whereas dark areas form the catchment basins. Thus the images are analyzed in terms of watershed lines and catchment basins. The catchment basins of the MR image of the brain have to be identified.

Topography may be defined as the configuration of a surface including its relief and the position of its natural and man-made features, whereas relief is a part of topography referring to the elevations and depression characteristics of a surface. A image can be considered as a topographic relief where the altitude of the relief is the intensity level of the pixel. A drop of liquid falling on a topographic relief flows along a path to reach finally to a local minimum. The limits of the catchment basins of the liquid correspond to the watershed of the relief.

A function Q is a watershed of a function S if and only if Q<=S and Q preserves the contrast between the regional minima of S. The minimal altitude one has to climb to order to go from one regional minima N_1 to another regional minima N_2 is defined as the contrast between N_1 and N_2.

Watershed transform is applied to brain MR images to separate the healthy and non-healthy tissues. Since the tissues of the healthy and non-healthy tissues touch each other, it is truly a very difficult task. It finds catchment basins and watershed ridge lines in the MR image treating it as a surface where light pixels are high and dark pixels are low.

6. **Display:** Skull lesion of the Human brain is displayed.

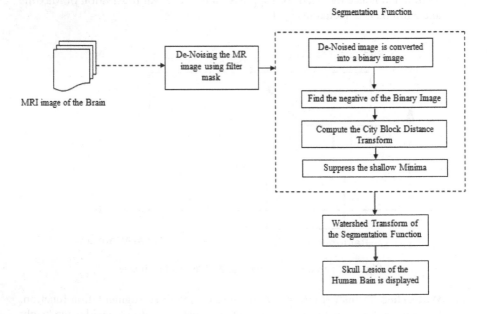

Fig. 2. Flow Diagram of the proposed method

5   Results

The flow diagram of the proposed technique is discussed in Fig.2. We have used MR image of brain of a patient with a Bone tumor of Skull. Various slices of brain were taken. We analyze the different MR slices of brain of a single patient . The MR image datasets of fig.3(a),4(a) and 5(a) refer to the same patient taken at different times and in different view-points. The proposed method gives satisfactory results as depicted in Fig.3(b), Fig.4(b) and Fig.5(b).

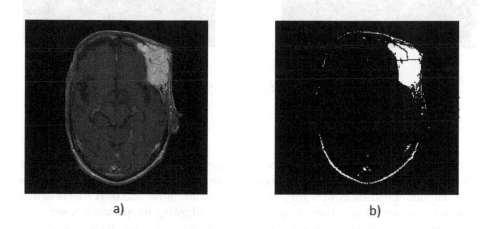

a)   b)

Fig. 3. Data set-1 a) MRI of Brain b) Segmented Image

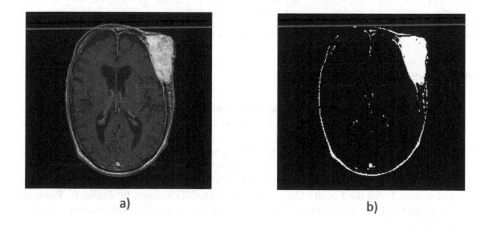

a)   b)

Fig. 4. Data set-2 a) MRI of Brain b) Segmented Image

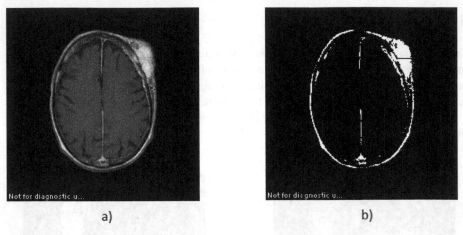

a)   b)

Fig. 5. Data set-3 a) MRI of Brain b) Segmented Image

6   Conclusion

In this work we have described an automatic approach to segment skull bone lesions. This method uses a combination of distance transform and watershed transform. The regional minima which is primarily due to the texture of the image is suppressed so that only the significant changes are visible. By adjusting the parameters we could overcome the problem of over-segmentation in the watershed transform thereby obtaining a automatic algorithm for Skull bone lesion segmentation. The lesion of the skull is correctly segmented using this method.

We applied this algorithm to set of MR images and we got satisfactory results. Therefore, this automated schema described in this article can be used in clinical routine. Future work will aim at calculating the accuracy of segmentation.

References

1. Beucher, S., Meyer, F.: The morphological approach to segmentation: the watershed transformation. In: Dougherty, E. (ed.) Mathematical Morphology in Image Processing, pp. 433–481. Marcel Dekker Inc. (1992)
2. Hill, et al.: Image segmentation using a texture gradient based watershed transform. IEEE Transactions on Image Processing 12(12), 1618–1633 (2003)
3. Béliz-Osorio, N., Crespo, J., García-Rojo, M., Muñoz, A., Azpiazu, J.: Cytology Imaging Segmentation Using the Locally Constrained Watershed Transform. In: Soille, P., Pesaresi, M., Ouzounis, G.K. (eds.) ISMM 2011. LNCS, vol. 6671, pp. 429–438. Springer, Heidelberg (2011)
4. Beare, R.: A locally constrained watershed transform. IEEE Trans. Pattern Anal. Mach. Intell. 28(7), 1063–1074 (2006)
5. Hahn, H.K., Peitgen, H.-O.: The Skull Stripping Problem in MRI Solved by a Single 3D Watershed Transform. In: Delp, S.L., DiGoia, A.M., Jaramaz, B. (eds.) MICCAI 2000. LNCS, vol. 1935, pp. 134–143. Springer, Heidelberg (2000)

6. Masoumi, H., et al.: Automatic liver segmentation in MRI images using an iterative watershed algorithm and artificial neural network. Biomedical Signal Processing and Control 7, 429–437 (2012)
7. Cousty, J., et al.: Segmentation of 4D cardiac MRI: Automated method based on spatio-temporal watershed cuts 28(8), 1229–1243 (2010)
8. Behrad, A., et al.: Automatic spleen segmentation in MRI images using a combined neural network and recursive watershed transform. In: Proc. of 10th Symposium on Neural Network Application in Electrical Engineering, Belgrade, Serbia, pp. 63–67 (2010)

Extraction of Texture Based Features of Underwater Images Using RLBP Descriptor

S. Nagaraja, C.J. Prabhakar, and P.U. Praveen Kumar

Department of P.G. Studies and Research in Computer Science,
Kuvempu University, Karnataka, India
{nagarajas27,psajjan}@yahoo.com, praveen577302@gmail.com

Abstract. In this paper, we present an approach for extraction of texture features of underwater images using Robust Local Binary Pattern (RLBP) descriptor. The literature survey reveals that the texture parameters that remain constant for the scene patch for the whole underwater image sequence. Therefore, we proposed technique to extract the texture features and these features can be used for object recognition and tracking. The underwater images suffer from image blurring and low contrast and performance of feature extractors is very less if we employ directly. Thus, we propose a novel image enhancement technique which is combination of different individual filters such as homomorphic filtering, curvelet denoising and LBP based Diffusion. We employ DoG based feature detector, for each detected interest point, the texture description is extracted using RLBP feature descriptor. The proposed feature extraction technique is compared and evaluated extensively with well known feature extractors using datasets acquired in underwater environment.

Keywords: Feature descriptor, Feature matching, DoG, RLBP, Underwater image pre-processing.

1 Introduction

Feature detection and matching of stable and discriminative features are fundamental problems of research in computer vision and machine learning. Feature detector and descriptor are playing an important role in applications like image matching, image stitching, and image registration. In underwater images, extracting the invariant features is a big challenging task because of optical properties of the water. The absorption and scattering processes of the light in the water influence the overall performance of underwater imaging systems. Even the same object in the pair of images which are captured from same viewpoint can be drastically different due to varying water conditions. Color and geometrical variations are very high between pair of underwater images due to the optical properties of the water. As a result, the description of the same point of an object may be completely different between underwater images taken under the same conditions. The current state-of-the art feature descriptors such as SIFT, SURF, KLT and DAISY, which extracts geometric features and cannot be employed directly because of geometric variations of underwater images. Therefore, it is required to extract the features which remain constant in the whole sequence.

© Springer International Publishing Switzerland 2015
263
S.C. Satapathy et al. (eds.), *Proc. of the 3rd Int. Conf. on Front. of Intell. Comput. (FICTA) 2014*
– *Vol. 2*, Advances in Intelligent Systems and Computing 328, DOI: 10.1007/978-3-319-12012-6_29

In general, the key points finding in a given image is very challenging problem due to variation in illumination, scale and orientation. The key points are selected based on the distinguishable property of the locations, such as corners, junctions and blobs. In past two decades, many researchers proposed techniques for detection of key points. The most widely used detectors can be classified into two categories such as Hessian-based and Harris-based detectors. Harris and Stephens in 1988 [1] presented a Harris feature detector; it is based on the auto-correlation matrix. This technique is not neither to scale nor to affine transformations [2] due to the fact that the Harris corner detector does not provide a good basis for matching images of different sizes because it is very sensitive to changes in image scale. Schmid et al. [3] showed, the Harris corner detector performs excellent in the evaluation of single-scale interest point detectors. Lindeberg [4] proposed a technique for automatic scale selection based on Hessian matrix as well as Laplacian. Mikolajczyk et al. [2] have further extended this technique to create robust and scale invariant feature detector based on the combination of Harris-Laplacian and Hessian-Laplacian. In this method Harris measure to select the location and Laplacian to select the scale. R. Garcia et al. [5] detected key points in underwater turbid images, using a combination of Hessian and Harris point detectors.

The literature survey reveals that the texture parameters that remain constant for the scene patch for the whole underwater image sequence. One of the most popular texture descriptors is Local Binary Pattern (LBP) [6], which is computationally simple, efficient, good texture discriminative property and has been highly successful for various computer vision problems like face image analysis, dynamic texture recognition, motion analysis, texture analysis [7] etc. LBP has a property that favors its usage in interest region description such as computationally simplicity and tolerance against illumination changes. LBP operator is structural and statistical texture descriptor in terms of the characteristics of the local structure, so that it is most powerful for texture analysis. LBP have some drawbacks, namely it is sensitive to noise and sometimes it produces same binary code for different structural patterns, which will reduce its discriminability. In order to overcome these demerits Y. Zhao et al. [8] proposed another variant of the LBP, called Completed Robust Local Binary Pattern, which is robust on flat image areas usually found in underwater environment. The Robust Local Binary Pattern (RLBP) descriptor outperforms the existing local descriptor for the most part of the test cases, particularly for images with severe illumination variations and noise.

In this paper, we propose feature extraction technique for underwater images based on detection and extraction of texture based features using RLBP descriptor. We detect the keypoints using DoG based feature detector and the texture information of the feature points are stored using RLBP descriptor. In order to extract reliable and robust features from underwater images, it is necessary to enhance the quality of underwater image prior to feature extraction. We propose a novel image enhancement technique which is combination of different individual filters such as homomorphic filtering, curvelet denoising and LBP-Diffusion which are employed sequentially on degraded underwater images. The rest of the paper is organized as follows: Section 2 describes proposed image enhancement technique. Section 3 provides theoretical background of proposed feature extraction technique. The experimental results are presented in Section 4. Finally, section 5 draws the conclusion.

2   Enhancement of Underwater Images

In this section, we present an underwater image enhancement technique, which is a combination of different individual filters and employed sequentially on degraded underwater images.

2.1   Homomorphic Filtering

Homomorphic filtering [9] is used to correct non-uniform illumination and to enhance contrasts in the image. It is a frequency filtering technique, preferred to others techniques because it corrects non-uniform lighting and sharpens the image features at the same time. The mathematical representation of filter function is given below:

$$H(w_x, w_y) = (r_H - r_L) \cdot \left(1 - \exp\left(-\left(\frac{w_x^2 - w_y^2}{2\delta_w^2}\right)\right)\right) + r_L, \tag{1}$$

where $r_H = 2.5$ and $r_L = 0.5$ are the maximum and minimum coefficients values and δ_w a factor which controls the cut-off frequency. Computations of the inverse Fourier transform to come back in the spatial domain and then taking the exponent to obtain the filtered image.

2.2   Curvelet Denoising

The Curvelet Transform introduced by Candes and Donoho in 1999 a multi-scale representation suited for objects which are smooth away from discontinuities across curves [10]. Curvelets via wrapping [11] is used for removing noise [12] from underwater image. Let signal be $\{f_{i,j}; i, j = 1 \dots N\}$, where N is some integer power of 2. It is corrupted by additive noise and signal $g_{i,j}$ can be mathematically expressed as

$$g_{i,j} = f_{i,j} + \varepsilon_{i,j}, \qquad i, j = 1 \dots N, \tag{2}$$

where $\{\varepsilon_{i,j}\}$ are independent and identically distributed as normal $N(0, \sigma^2)$ and independent of $\{f_{i,j}\}$. Underwater image denoising is done through soft thresholding techniques, operator with threshold T is defined as

$$D(g_{i,j}, T) = sgn(g_{i,j}) * \max(g_{i,j} - T, 0). \tag{3}$$

Soft thresholding is more efficient as it shrinks towards zero. BayesShrink approach is used for calculating threshold for soft thresholding. It achieves near minmax rate over a large number and yield visually more pleasing images.

2.3   LBP Based Diffusion

Mandava et al. [13] combined LBP and non linear diffusion [14], LBP is used to classify the textons of image around a pixel into flat, spot, corner and edge regions. Depending on different types of regions, a variable weight is assigned into the diffusion equation; it is adaptively encouraging the strong diffusion in flat or spot regions and less on the edge or corner regions. When diffusing more on flat regions, the diffusion preserves edges and local details. The diffusion process is mathematically described as

$$\frac{\partial}{\partial t} I(x, y, t) = \nabla \cdot (c(x, y, t) \nabla I), \tag{4}$$

where $I(x, y, t)$ is the image, t is the iteration steps and $c(x, y, t)$ is the so called diffusion function and is monotonically decreasing function of the image gradient magnitude. The discrete diffusion structure is

$$I_{i,j}^{n+1} = I_{i,j}^n + (\nabla t) \cdot \begin{bmatrix} c_N(\nabla_N I_{i,j}^n) \cdot \nabla_N I_{i,j}^n + c_S(\nabla_S I_{i,j}^n) \cdot \nabla_S I_{i,j}^n + \\ c_E(\nabla_E I_{i,j}^n) \cdot \nabla_E I_{i,j}^n + c_W(\nabla_W I_{i,j}^n) \cdot \nabla_W I_{i,j}^n \end{bmatrix}. \tag{5}$$

The N, S, E and W are referred to as north, south, east and west respectively, express the direction of the local gradient, and the local gradient is designed using nearest-neighbor differences

$$\nabla_N I_{i,j} = I_{i-1,j} - I_{i,j}, \nabla_S I_{i,j} = I_{i+1,j} - I_{i,j}, \nabla_E I_{i,j} = I_{i,j+1} - I_{i,j}, \nabla_W I_{i,j} = I_{i,j-1} - I_{i,j}. \tag{6}$$

3   Extraction of Texture Features

We detect interest points using Difference of Gaussian (DoG) based feature detection technique which is part of SIFT. For each interest point, a texture descriptor is built using RLBP technique to distinctively describe the local region around the interest point. Finally, the feature descriptors are matched using Nearest Neighbour Distance Ratio (NNDR) to measure the similarity.

3.1   Detection of Interest Points Using DoG

The scale invariant features are detected from a reference image using Difference of Gaussian technique [16].

Scale-space extrema detection: In this step Interest or key points are detected. Difference-of-Gaussian pyramid is computed from the differences between the adjacent levels in the Gaussian pyramid. Then, key points are obtained from the points at which the Difference-of-Gaussian values assume extrema with respect to both the spatial coordinates in the image domain and the scale level in the pyramid.
Keypoint Localization: Scale-space extrema detection produces too many keypoint candidates, some of which are unstable. A detailed fit to the nearby data for accurate scale, location and ratio of principal curvatures is done. This information allows points to be rejected that have low contrast or poorly localized along an edge.

3.2   RLBP Descriptor

In RLBP, the value of centre pixel in 3x3 local area is replaced by its average local gray value of the neighborhood pixel values, in which the RLBP is calculated. Compared to the centre gray value of LBP, the average local gray value of RLBP is more robust to noise and illumination variants. The Average Local Gray Level (ALG) is defined as

$$ALG = \frac{\sum_{i=1}^{8} g_i + g}{9},$$ (7)

where g represents the gray value of the centre pixel and $g_i(i=0,...,8)$ denotes the gray value of the neighbour pixel. Then the LBP process is applied by using ALG as the threshold instead of the gray value, named Robust Local Binary Pattern (RLBP). This can be defining as follows:

$$RLBP_{P,R} = \sum_{P=0}^{P-1} s(g_p - ALG_c)2^P = \sum_{P=0}^{P-1} s\left(g_p - \frac{\sum_{i=1}^{8} g_{ci} + g_c}{9}\right)2^P,$$ (8)

where g_c represents the gray value of the centre pixel and $g_p (p = 0,...,P-1)$ denotes the gray value of the neighbour pixel on a circle of radius R, P is the total number of the neighbours and $g_{ci}(i = 0,...,8)$ denotes the gray value of the neighbour pixel of g_c.

ALG ignores the specific value of an individual pixel. While sometimes the specific information of the central pixel is needed. To make a balance between anti-noise and information of individual pixel, we define a Weighted Local Gray Level (WLG) as follows:

$$WLG = \frac{\sum_{i=1}^{8} g_i + \alpha g}{8 + \alpha},$$ (9)

where g and g_i are defined in Eq. (7), α is parameter set by user. It should be notice that WLG is equivalent to the conventional ALG if α is set as 1. Now the RLBP can be calculated as follows:

$$RLBP_{P,R} = \sum_{P=0}^{P-1} s(g_p - WLG_c)2^P = \sum_{P=0}^{P-1} s\left(g_p - \frac{\sum_{i=1}^{8} g_{ci} + \alpha g_c}{8 + \alpha}\right)2^P,$$ (10)

where g_p, g_c, g_{ci} are defined as in Eq. (8), α is the parameter of WLG.

In order to incorporate spatial information into the descriptor, the input region is divided into cells with Cartesian grid 4 x 4 (16 cells) is used. For each cell a RLBP histogram is built. The resulting descriptor is a 3D histogram of RLBP feature locations and values. The number of different feature values ($2^{N/2}$) depends on the neighborhood size N of the chosen RLBP. The final descriptor is built by concatenating the feature histograms computed for the cells to form M x N x $2^{N/2}$ dimensional vector, where M is the gird size and N is the RLBP neighborhood size.

We compute the similarity between the descriptors using NNDR based technique. The threshold T is set empirically. The ambiguity distance compares the distance between the closest and the second closest descriptor. The distance ratio is threshold to avoid matching when the nearest neighbors are very similar. Considering $d_{L,1}$ first closest descriptor and $d_{L,2}$ second closest descriptor, ambiguity distance can be defined as

$$NNDR = \frac{\sqrt{\sum_{i=1}^{128}(d_R(i) - d_{L,1}(i))^2}}{\sqrt{\sum_{i=1}^{128}(d_R(i) - d_{L,2}(i))^2}} < T$$ (11)

4   Experimental Results

We conducted several experiments in order to validate our method for underwater video sequence captured in small water body in which the object was kept at a depth of 5 feet from the surface level of the water. The underwater image capturing setup consists of a waterproof camera which is Canon-D10. The camera is moved around an object at a distance between from 1m to 2m, near the corner of the water body to capture the video sequence. We have captured three set of video sequences (dataset1, dataset2 and dataset3) in three different water conditions with various turbidity levels. From each video sequence, we select two test frames which are not consecutive (frame 1 and frame 20) for experimentation.

4.1   Evaluation of Image Enhancement Technique

To evaluate the performance of our proposed method, we compared both quantitatively and qualitatively with other techniques [15] and [9] by considering first frame of each dataset. The experimental results show that our approach based on curvelet transform yields improved PSNR (Peak Signal Noise Ratio) values compared to wavelet based image denoising techniques shown in Table 1.

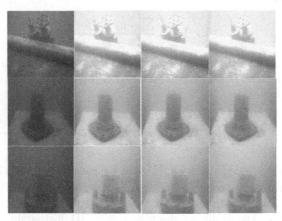

Fig. 1. Result of Image enhancement method: First column: original images, Second column: after homomorphic filtering, Third column: after curvelet denoising, Last column: after LBP- Diffusion

We also employed image contrast measure to evaluate quantitatively and evaluation results are shown in Table 2. Generally, the higher the contrast value means that the clearer the image. We compute the contrast measurement for both original and enhanced images using the equation given below:

$$C(I) = \frac{\sqrt{\frac{1}{N}\sum_{v=1}^{N}\sum_{\chi=r,g,b}(I_v^\chi - \bar{I}^\chi)^2}}{\sum_{\chi=r,g,b}\bar{I}^\chi}, \tag{12}$$

where I is the image with N pixels, $C(I)$ is the contrast, χ is the index of the chromatic band (red, green and blue channels) and $\bar{I}^\chi$ is mean value, which is calculated for each channel independently using Eq. 13.

$$\bar{I}^{\chi} = \frac{1}{N}\sum_{v=1}^{N} I_v^{\chi} \tag{13}$$

We compared the contrast measure of our approach with Bazielle approach [15] for original image and enhanced underwater image (Fig. 1). Our approach yields better contrast value for enhanced image.

Table 1. Comparison of curvelet with wavelet for denoising based on PSNR and MSE (*Mean Square Error*)

Dataset	Frames	Methods	PSNR(dB)	MSE
1		Haar Wavelet	45.780	1.7170
		Sym4 Wavelet	47.609	1.1277
		Db4 Wavelet	47.579	1.1355
		Curvelet	**52.782**	**0.3426**
2		Haar Wavelet	48.233	0.9765
		Sym4 Wavelet	50.238	0.6154
		Db4 Wavelet	50.212	0.6192
		Curvelet	**52.307**	**0.3822**
3		Haar Wavelet	49.982	0.6529
		Sym4 Wavelet	51.370	0.4743
		Db4 Wavelet	51.347	0.4767
		Curvelet	**52.218**	**0.3901**

Table 2. Comparison of our approach with Bazielle approach [15] based on contrast of the image

Data sample	Original Image	Bazielle et al. Approach	Our Approach
dataset1	2.58	2.34	**5.53**
dataset2	2.50	5.12	**5.15**
dataset3	2.46	1.79	**3.33**

4.2 Evaluation of Feature Extraction Technique

The proposed method is compared with standard feature extractors such as SIFT [16], SURF [17], KLT [18], [19] based on the repeatability measure. The repeatability measurement is computed as a ratio between the number of point-to-point correspondences that can be established for detected points and the mean number of points is detected between two images.

We used three set of frames namely dataset1, dataset2 and dataset3, which are enhanced using proposed image enhancement method. The descriptors are matched using NNDR approach and results of feature matching for three datasets are shown in Fig. 2. The descriptors having a value below a threshold is selected as the matched point, threshold value is 0.8 for all the datasets. Experimental results clearly show that, our proposed method yields good repeatability rate compared to KLT and Hessian method shown in Table 3. Hence, our proposed method is efficient technique for detecting feature points for underwater images.

Table 3. Comparison of repeatability measure of our approach for feature detection with other feature detectors

Approaches	Frame 1 keypoints	Frame 2 keypoints	Number of Matches	Repeatability	Processing Time (Sec)
KLT	86	71	39	0.4968	14.8
Hessian	92	84	52	0.5909	11.4
Our approach	99	88	67	**0.7165**	**10.2**

The comparison of results of proposed RLBP feature descriptor with other standard descriptors is presented using recall and 1-precision in the Table 4. The comparative results clearly show that the value of recall and 1-precision of proposed method is very high compared to SIFT and SURF method.

Table 4. Recall and Precision values for feature matching of dataset1, dataset 2 and dataset 3

Datasets	Methods	Recall	1-Precision
Dataset1	SIFT	0.62	0.58
	SURF	0.68	0.61
	Our Approach	**0.71**	**0.64**
Dataset2	SIFT	0.59	0.56
	SURF	0.67	0.60
	Our Approach	**0.73**	**0.65**
Dataset3	SIFT	0.62	0.58
	SURF	0.64	0.57
	Our Approach	**0.69**	**0.59**

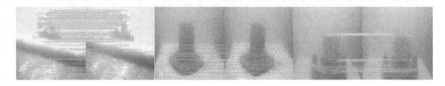

Fig. 2. The result of feature points matching using proposed method for enhanced images of dataset1, dataset2 and dataset3

5   Conclusion

In this paper, we proposed feature extraction technique for underwater images based RLBP descriptor. Since there is no benchmark database is available for underwater image sequence, we conducted experiments using our own captured image sequences. The proposed image enhancement technique comprises three filters such as homomorphic filter, curvelet denoising and LBP-diffusion, which are employed sequentially for enhancing the quality of degraded underwater images. Experimentally we showed that curvelet transform yields higher PSNR value compared to wavelet and it also observed that our approach yields very high contrast value for enhanced underwater images. For the enhanced underwater images, interest points are detected using DoG based feature detector and texture description is extracted using RLBP descriptor. The experimental results shows that, the combination DoG with RLBP descriptor yields high repeatability, recall and precision rate compared to other techniques. The processing time of our technique is low compared to other state-of-the-art techniques. The proposed technique can be employed to extract the texture features in order to match the objects or track the objects in video sequence of underwater environment.

Acknowledgments. The authors would like to thank the anonymous reviewers for their valuable comments and suggestions which helped lot to improve previous version of the paper.

References

1. Harris, C., Stephens, M.: A Combined Corner and Edge Detector. In: Proceedings of Alvey Conference, pp. 147–151 (1988)
2. Mikolajczyk, K., Schmid, C.: Scale & Affine Invariant Interest Point Detectors. International Journal of Computer Vision 60(1), 63–86 (2004)
3. Schimd, C., Mohr, R., Bauckhage, C.: Evaluation of Interest Point Detectors. International Journal of Computer Vision 37(2), 151–172 (2000)
4. Lindeberg, T.: Scale-space theory: A basic tool for analysing structures at different scales. Journal of Applied Statistics 21(2), 224–270 (1994)
5. Garcia, R., Gracias, N.: Detection of Interest Points in Turbid Underwater Images. In: IEEE Oceans, pp. 1–9 (2011)
6. Ojala, T., Pietikainen, M., Harwood, D.: A comparative study of texture measures with classification based on feature distributions. Pattern Recognition 29(1), 51–59 (1996)
7. Garcia, R., Xevi, C., Battle, J.: Detection of matchings in a sequence of underwater images through texture analysis. In: International Conference on Image processing, vol. 1, pp. 361–364 (2001)
8. Zhao, Y., Jia, W., Hu, R.X., Min, H.: Completed Robust Local Binary Pattern for Texture Classification. Neurocomputing 106, 68–76 (2013)
9. Prabhakar, C.J., Praveen Kumar, P.U.: An Image Based Technique for Enhancement of Underwater images. International Journal of Machine Intelligence 3(4), 217–224 (2011)
10. Candes, E.J., Donoho, D.L.: Curvelets-A Surprisingly Effective Nonadaptive Representaion for Objects with Edges. Vanderbilt University Press, Nashville (2000)
11. Candes, E.J., Demanet, L., Donoho, D.L., Ying, L.: Fast Discrete Curvelet Transform. SIAM Multiscale Model. Simul. (2006)
12. Starck, J.L., Candes, E.J., Donoho, D.L.: The Curvelet Transform for Image Denoising. IEEE Transactions on Image Processing 11(6), 670–684 (2002)
13. Mandava, A.K., Regentova, E.E.: Speckle Noise Reduction Using Local Binary Pattern. Procedia Technology 6, 574–581 (2012)
14. Perona, P., Malik, J.: Scale-space and Edge Detection using Anisotropic Diffusion. IEEE Transactions on Pattern Analysis and Machine Intelligence 12(7), 629–639 (1990)
15. Bazeille, S., Quidu, I., Jaulin, L., Malkasse, J.P.: Automatic Underwater Image Pre-Processing. In: Proceedings of the European Conference on Propagation and Systems, Brest, France (2006)

16. Lowe, D.G.: Distinctive Image Features from Scale-Invariant Keypoints. International Journal of Computer Vision 60(2), 91–110 (2004)
17. Bay, H., Ess, A., Tuytelaars, T., Gool, L.V.: Speeded-Up Robust Features (SURF). Computer Vision and Image Understanding 110(3), 346–359 (2008)
18. Shi, J., Tomasi, C.: Good features to track. In: Proceedings of the IEEE Conference on Computer Vision and Pattern Recognition, pp. 593–600 (1994)
19. Tomasi, C., Kanade, T.: Detection and tracking of point features. Technical Report CMU-CS-91-132, Carnegie Mellon University, Pittsburg, PA (April 1991)

Summary-Based Efficient Content Based Image Retrieval in P2P Network

Mona* and B.G. Prasad

BNM Institute of Technology,
Department of Computer Science and Engineering, Bangalore, India
{mona.bnmit,drbgprasad}@gmail.com

Abstract. The World Wide Web provides an enormous amount of images which is generally searched using text based methods. Searching for images using image content is necessary to overcome the limitations of text based search. Generally, in Unstructured P2P systems like Gnutella a complete blind search is used that floods the network with high query traffic. In this paper, we present a P2P system that uses "'informed search"' in which peers try to learn about the information maintained at their neighbours in order to minimise the query traffic. Here the images are first clustered using K-means clustering technique and then each peer is made to exchange its cluster information with its neighbouring peers using PROBE and ECHO. Typically one summary table per peer is maintained in which neighbouring peers data information is stored. When processing queries, these summaries are used to choose the most probable peer that is likely to contain information relevant to the query. If none of its neighbours has a match then standard random-walk algorithm is used for query propagation.

Keywords: Peer-to-Peer(P2P) Network, K-means Clustering, Image Retrieval, Searching.

1 Introduction

Earlier techniques for searching and retrieval for images in the network were text-based that has lot of limitations like manual text annotations, which is cumbersome and not suitable for large set of images. Recent techniques rely on searching, based on the content of the image that led to Content based Image Retrieval (CBIR), in which the features (like Color, Shape, and texture) are extracted automatically from the images. The extracted features are compared with the query image's features for a similarity check that results in a set of images that closely match with that of the query image.

Here the problem of CBIR in P2P networks has been considered. In a P2P system files like text documents, images, and other multimedia files are stored

---

* Corresponding author.

© Springer International Publishing Switzerland 2015   273
S.C. Satapathy et al. (eds.), *Proc. of the 3rd Int. Conf. on Front. of Intell. Comput. (FICTA) 2014*
– *Vol. 2,* Advances in Intelligent Systems and Computing 328, DOI: 10.1007/978-3-319-12012-6_30

at different machines as against centralized server system in which all the files are stored at a central server. The centralized storage and retrieval has lot of drawbacks like single point of failure, lack of load distribution and non-scalability. These are overcome in a P2P system which is decentralized and self-administered. The load can be distributed and it also permits nodes to join and leave the network resulting in a dynamic network.

In this paper, we present a P2P system in which

1 Images are clustered using K-means algorithm
2 The clusters are distributed across the network.
3 Peers are made to exchange the cluster information with each other which results in each peer maintaining summaries of data information at their neighbours.
4 A peer submits a query image and the system tries to search for a promising peer that has images similar to query image, by using the local cluster information as well as the information maaintained in the summary table.

P2P systems are classified into two types, viz:Structured and Unstructured networks. Structured P2P network uses Distributed Hash Table (DHT) in which files are assigned to a specific peer, which is a variant to traditional hash table's assignment of each key to a particular array slot. In structured P2P network data cannot be placed at random peers which is the case in unstructured networks also, dynamism with respect to peers joining the network and leaving is limited as it affects the proper functioning of DHTs. But a node can efficiently route a search query to some peer that has the desired file, even if the file is extremely rare. Examples include Content Addressable Networks (CAN)[2], Chord[3].

In Unstructured P2P networks, peers act as equals, whereby the roles of client and the server can be merged. The links between peers are formed arbitrarily and the overlay network is formed as a random graph. In Unstructured networks the desired information can be at any of the peers, so usually flooding mechanism is used to forward the query messages in the network like Gnutella[16]. There are several methods that have been used to improve the performance of the basic flooding based search such as Time-To-Live (TTL)in order to reduce network load by limiting the scope of the search. Other mechanisms for reducing network load includes random walk and degree-based random walk, in which the query is forwarded to a randomly chosen neighbour or degree number of neighbours respectively and this process is repeated until the required object is found. But all these methods blindly search for the data item. Another method is informed search which forwards the queries to relevant peers based on information collected from neighbouring peers.

2   Related Work

Much work has been done on search in P2P systems. The first P2P file systems such as Napster[7] and Gnutella[16]used simple name-based query term matching for resource selection and document retrieval. Content Based Image Retrieval in a P2P network, incorporating different techniques and measures have been addressed in [13][12][9][5]. But all these methods induce lot of query traffic as all the nodes get involved in query processing before actually retrieving the relevant results. There are a few that try to minimise the query traffic in which only a subset of nodes get involved in query processing [6] [14] [1].

2.1 Probabilistic Retrieval

Image retrieval system can be considered as a mapping from images to image classes[4],[8].Which means Given the feature space X, a retrieval system is simply a map $g : X \rightarrow 1 \ldots M$, $x \rightarrow y$ from X to the index set of the M classes in the database where $x = T(z)$ is a feature vector $x \in X$ and $y \in Y$ be $Y = 1 \ldots M$, is the label of an image class. For a given query x, the system should retrieve all images in class y with $g(x) = y$. The retrieval system should minimize the retrieval error (i.e. probability of retrieving images from a class not equal to $g(x)$) which can be best described using bayes classifier.

$$g * (x) = arg \max_i P(Y = i | X = x) \tag{1}$$

$$= arg \max_i P(X = x | Y = i) P(Y = i) \tag{2}$$

arg max P($\ldots$) is the value of x that leads to the maximum value of P(x). Also, the peer selection problem can be viewed as a classification problem:

$$g * (x) = arg \max_a P(A = a | X = x) \tag{3}$$

$$= arg \max_a P(X = x | A = a) P(A = a) \tag{4}$$

Here, $P(X = x | A = a)$ is the probability of finding x in a given peer a, and $P(A = a)$ is the prior probability, which is proportional to the number of documents contained in peer a in our case.

3 Experimental Setup

In our work, we have used the peersim simulator for evaluating the P2P system. It is highly scalable, is written in Java and has two simulation engines, one is cycle-based and the other event driven. Though the Event based engine is less efficient, is more realistic than cycle driven engine[10]. For our evaluation purpose we have implemented the system using event based engine.

3.1 Feature Extraction

For evaluating the system we have combined both color moments of an RGB image and HSV that finally results in 41-dimensional vector for each image. Color moments have significant applications in image retrieval and are used to check for similarity between two images. The HSV color space is quantized into 8 intervals in the hue dimension, 2 intervals in the saturation dimension, and 2 intervals in the value dimension which results in a feature vector of 8H*2S*2V. In addition, 9 more bins are used for Mean(M), Variance(V) and Standard-Deviation(SD) for the RGB color image is extracted that gives us 9 features 3M+3SD+3V. Each image therefore is represented as a 41-dimensional feature vector.

3.2   K-means Clustering

The set of images in the data set are clustered using K-means algorithm which is an unsupervised learning algorithm. The 41-dimensional feature vectors act as data points to K-means algorithm. From the set of data points d_i, we choose K random points to represent the initial group centroids. Each data point is assigned to the group that has the closest centroid. Then, when all datapoints have been assigned, k cluster centers are re-computed again as the mean vector of each cluster. This process is repeated until the cluster centroids do not change.This produces a separation of the datapoints into groups where inter-cluster similarity is less and intracluster similarity is very high[15].

3.3   Mapping of Clusters to Peers

Once the images are clustered, we map these clusters to the peers. Each peer has the global cluster information. Then a mapping is carried out, in which each peer is assigned with some of the image clusters. Usually $NoOfPeers > NoOfClusters$, which is why we go on reinitializing the cluster centroid set S during mapping. So, one cluster can get mapped to more than one Peer. The mapping is carried out in the manner as shown in algorithm 1:

After the clusters are mapped to the peers, they exchange summaries of data they contribute to the network with their neighbours using Probe and Echo which was described by E.J.H.Chang in[11]. This algorithm finally results in a minimum spanning tree of a given graph. The network of peers is considered as a Graph and this simple algorithm minimises the number of messages sent. Each peer receives information only once because of the simplicity of the algorithm that minimises the number of messages sent. An initiator for Probe and Echo is chosen randomly that sends its local cluster information to its 1-hop neighbours.The PROBE and ECHO steps carried out at the initiator node is shown as algorithm2:

Algorithm 1. Mapping Function

Data: A set S of cluster centroids S := C1,C2,C3 ... Ck
Data: Set of peers N := P1, P2 ....
Data: $S^* := S$
Result: $M: N \leftarrow S$

foreach $P_i \in N$ **do**
$\quad M[P_i] \leftarrow \phi$;
$\quad CPeer \leftarrow Rand()\%4$;
$\quad$ **for** $i \leftarrow CPeer$ **to** 0 **do**
$\quad\quad$ **if** $S^* < 0$ **then**
$\quad\quad\quad$ $S^* = S$;
$\quad\quad$ Choose $C_{rand} \in S^*$ randomly.
$\quad\quad$ $M[P_i] \leftarrow C_{rand}$;
$\quad\quad$ Remove C_{rand} from the set S^*;

Algorithm 2. Initiator for PROBE

Data: C_{rp} Remote peer's set of clusters
Data: C_l Local set of clusters
Data: N set of neighbours
Result: Initiator sends a PROBE message

begin
$\quad$ Randomly choose an initiator I ;
$\quad$ $engaged \longleftarrow True$;
$\quad$ $ReceivedCounter \longleftarrow 1$;
$\quad$ $send(PROBE, C_l)$ to N ;

When a Peer i receives a PROBE message then it udates its summary table with the cluster information that it has received from its neighbour and marks that neighbour as predecessor. Peer i then issues PROBE message to all its neighbours except the predecessor. When a Peer i receives ECHO messages from its neighbouring peers it just updates the $ReceivedCounter$. If the $ReceivedCounter$ matches the $NoOfNeighbours$ count and if Peer i happens to be the Initiator then the algorithm stops. Otherwise, the ECHO message is sent to the predecessor node.

3.4   Search Protocol

Whenever a query Q is generated at a peer, the 41-dimensional features are extracted from the Q image.The extracted features are compared with the local cluster information of the peer. If the peer has the matching information locally, then the images are ranked using minimum Euclidian distance to the Query Image Q.If there is no match found with local cluster centroids, then a search is made in the local summary table in which the peer would have maintained centroid information of its 1-hop neighbours. If none of its neighbours have the matching data, then a random neighbour which is not visited is chosen and the query is forwarded to it. This process is repeated for all neighbors till a promising peer is found or else the search stops.The steps that take place when a Query Q is received at a Peer is shown in algorithm 3.

Algorithm 3. Searching

> **Data**: Query Image Q
> **Data**: G Global cluster centroids
> **Data**: C Local cluster centroids
> **Data**: Set of peers $N = P_1, P_2 \ldots$
> **begin**
> > Received(Q)
> > $P_i[visited] \leftarrow 1$;
> > Extract features from Q.
> > Compare Q with G.
> > Find a closest match $g \in G$ to Q.
> > Search for g in C.
> > **if** C *contains* g **then**
> > > found=true;
> > > rank the images with minimum Euclidean distance to Q.
> >
> > **if** *!found* **then**
> > > Search in the Summary table for g.
> > > **if** $n \in N$ *contains* g **then**
> > > > Forward Q to n.
> > >
> > > **else**
> > > > randomly choose a neighbour $n \in N$ where $n[visited] = 0$.
> > > > Forward Q to n.
>
> **end**

---

4   Evaluation

For experimental purpose, 5000 images of size 256*384*3 have been used out of which 1000 are from Wang dataset and 4000 are from Benchathlon CBIR benchmarking dataset (www.benchathlon.net). 41-dimensional features are extracted from each image and the image set is clustered into 100 global clusters. The system is tested by varying the number of peers and the TTL value. TTL here is used to limit the number of hops. The results for different TTL values and peer count is shown below.

From Table 1 and Table 2 it is clear that even for a very low TTL of 3 the hit rate with summary maintenance is quite very high compared to that of normal search results(without summaries) and the graph in Figure1 and 2 also shows the performance difference for varying TTL.

Table 1. Network size of 1000 nodes and 100 Queries

TTL	Hit Rate with Data Summary	Hit Rate without Data Summary
3	38	16
5	65	18
10	86	21

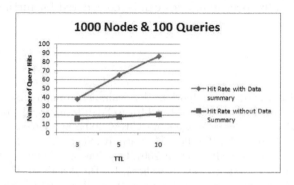

Fig. 1. Query Hits for Network size of 1000 nodes and 100 Queries

Table 2. Network size of 1500 nodes and 150 Queries

TTL	Hit Rate with Data Summary	Hit Rate without Data Summary
3	74	9
5	89	17
10	134	28

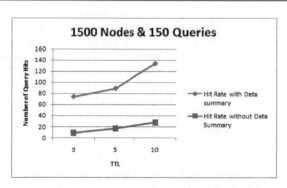

Fig. 2. Query Hits for Network size of 1500 nodes and 150 Queries

5   Conclusion

Our evaluation shows that for efficient CBIR in P2P networks, having knowledge about neighbouring peers is advantageous. The results clearly show that an informed search gives better results compared to blind search used in Unstructured P2P networks.Here in this paper, we have clustered the images and then performed exchange of data among the peers to achive better efficiency. Since the queries are not flooded in the network the query traffic also is less. To further improve the efficiency, peers can be clustered based on their content and a representative chosen for each cluster that results in a semantic overlay network. Additionally, peer dynamics can be considered for further work.

References

1. Müller, W., Eisenhardt, M., Henrich, A.: Scalable Summary Based Retrieval in P2P Networks. In: Proceedings of the 14th ACM International Conference on Information and Knowledge Management, CIKM 2005, pp. 586–593 (2005)
2. Ratnasamy, S., Francis, P., Handley, M., Karp, R., Schenker, S.: A scalable content-addressable network. In: SIGCOMM 2001, San Diego, CA, USA (2001)
3. Stoica, I., Morris, R., Karger, D., Kaashoek, F., Balakrishnan, H.: Chord: A scalable Peer-To-Peer lookup service for internet applications. In: SIGCOMM 2001, San Diego, CA, USA (2001)
4. Vasconcelos, N.: Bayesian Models for Visual Information Retrieval. PhD thesis, MIT (June 2000)
5. Müller, W., Eisenhardt, M., Henrich, A.: Effcient content-based P2P image retrieval using peer content descriptions. In: Internet Imaging V. Proceedings of the SPIE, vol. 5304 (2003)
6. Ng, C.H., Sia, K.C.: Peer clustering and firework query model. In: Proceedings of 11th World Wide Web Conference (May 2002)
7. The napster homepage, http://www.napster.com
8. Cox, I.J., Miller, M.L., Minka, T.P., Papathomas, T.V., Yianilos, P.N.: The bayesian image retrieval system, pichunter, theory, implementation, and psychophysical experiments. IEEE Transactions on Image Processing 9, 20–37 (2000)
9. King, I., et al.: Distributed content-based visual information retrieval system on peer-to-peer networks. ACM Trans. Info. Sys. 22, 477–501 (2004)
10. http://www.peersim.sourceforge.net
11. Chang, E.J.H.: Echo Algorithms: Depth Parallel Operations on General Graphs. IEEETransactions on Software Engineering 8(4), 391–401 (1982)
12. Su, C.-R., Chen, J.-J., Chang, K.-L.: Content-Based Image Retrieval on reconfigurable Peer-to-Peer networks. In: International Symposium on Biometrics and Security Technologies (ISBAST), pp. 203–211 (July 2013)
13. Zhang, L., Wang, Z., Feng, D.: Content-Based Image Retrieval in P2P Networks with Bag-of-Features. In: 2012 IEEE International Conference on Multimedia and Expo Workshops (ICMEW), July 9-13, pp. 133–138 (2012)
14. Ghanem, S.M., Ismail, M.A.: Omar, System Design of a Super-Peer Network for Content-Based Image Retrieval. In: 2010 IEEE 10th International Conference on Computer and Information Technology (CIT), pp. 2486–2493 (2010)
15. Duda, R.O., Hart, P.E.: Pattern Classification. David G. Stork, Wiley (1973)
16. The Gnutella Protocol Specification v0.4,
   http://www.limewire.com/developer/gnute-lla_protocol_0.4.pdf

Dynamic Texture Segmentation Using Texture Descriptors and Optical Flow Techniques

Pratik Soygaonkar[1], Shilpa Paygude[1], and Vibha Vyas[2]

[1] Computer Engineering Department, Maharashtra Institute of Technology, Pune, India
pratiksoygaonkar@gmail.com, shilpa.paygude@mitpune.edu.in
[2] Electronics and Telecommunication Engineering Department,
College of Engineering, Pune, India
vsv.extc@coep.ac.in

Abstract. The texture which is in motion is known as Dynamic texture. As the texture can change in shape and direction over time, Segmentation of Dynamic Texture is a challenging task. Furthermore, features of Dynamic texture like spatial (i.e., appearance) and temporal (i.e., motion) may differ from each other. However, studies are mostly limited to characterization of single dynamic textures in the current literature. In this paper, the segmentation problem of image sequences consisting of cluttered dynamic textures is addressed. For the segmentation of dynamic texture, two local texture descriptor based techniques and Lucas-Kanade optical flow technique are combined together to achieve accurate segmentation. Two texture descriptor based techniques are Local binary pattern and Weber local descriptor. These descriptors are used in spatial as well as in temporal domain and it helps to segment a frame of video into distinct regions based on the histogram of the region. Lucas-Kanade based optical flow technique is used in temporal domain, which determines direction of motion of dynamic texture in a sequence. These three features are computed for every section of individual frame and equivalent histograms are obtained. These histograms are concatenated and compared with suitable threshold to obtain segmentation of dynamic texture.

Keywords: Dynamic texture, Segmentation, Optical flow, Spatial, Temporal.

1 Introduction

A large class of scenarios commonly experienced in real world exhibit characteristic motion with certain form of regularities. Dynamic texture refers to image sequences of these motion patterns. A flock of flying birds, water streams, and fluttering leaves etc. serve to illustrate examples of such motion patterns. Characterization of visual processes of dynamic textures has vital importance in research areas of computer vision, electronic entertainment, and content-based video coding with a number of potential applications in recognition (automated surveillance and industrial monitoring), synthesis (animation and computer games), and segmentation (robot navigation and MPEG-4).

© Springer International Publishing Switzerland 2015

S.C. Satapathy et al. (eds.), *Proc. of the 3rd Int. Conf. on Front. of Intell. Comput. (FICTA) 2014*
– *Vol. 2*, Advances in Intelligent Systems and Computing 328, DOI: 10.1007/978-3-319-12012-6_31

Segmentation is considered as one of the basic problem in computer vision [2]. Meanwhile, as compared with static texture, Segmentation of dynamic texture is very challenging because of their unknown spatiotemporal extension, stochastic nature of the motion fields and the different moving particles. Segmentation of dynamic texture is to separate the different groups of particles showing different random motion. One major limitation of the existing dynamic texture segmentation techniques is their inability to characterize visual processes consisting of multiple dynamic textures. For example, a flock of flying birds in front of a water fountain, highway traffic moving in opposite directions, image sequences containing both windblown trees and fire, and so forth. In such cases, existing dynamic texture segmentation techniques are inherently limited.

In this paper, a new method based on both appearance and motion information for the segmentation of dynamic textures is introduced. For the appearance of Dynamic texture, we use local spatial texture descriptors to describe the spatial mode of Dynamic texture; for the motion of Dynamic texture, we use the optical flow and local temporal texture descriptors to represent the movement of objects. For computation of optical flow, we used Lucas-Kanade which is simple and efficient. We employ both the appearance and motion modes for the Dynamic texture segmentation as dynamic textures might be different from their spatial feature (i.e., appearance) and/or temporal feature (i.e., motion). Combining the spatial and temporal modes of dynamic texture, we exploit the use of discriminate features of both the appearance and motion for the robust segmentation of various Dynamic textures.

2   Literature Survey

We review the methods of dynamic texture segmentation in this section. There are two types of statistical models, which can be applied to Dynamic texture segmentation i.e. (1) generative models and (2) discriminative models. A generative model can be defined as a model for randomly generating observable data, typically given some hidden parameters. For example, Doretto et al. [2] modeled the spatiotemporal dynamics by Gauss–Markov models, and inferred the model parameters. Vidal and Ravichandran [8] proposed to model each moving dynamic texture with a time varying linear dynamical system (LDS) plus a 2-D translational motion model. The generalized principle component analysis (GPCA) is used by Cooper et al. [9] which is represents the single temporal texture by a low dimensional linear model as well as segmented dynamic texture. Chan and Vasconcelos [10] presented a statistical model for an ensemble of video sequences. For learning the parameters of the model, they derived an expectation-maximization algorithm. After that, they proposed the layered dynamic texture (LDT) to represent a video as a collection of stochastic layers of different appearance and dynamics, and then proposed a variation approximation for the layered dynamic texture that provides and enables efficient learning of the model [11].

In this paper, the LBP-WLD based method is used for dynamic texture segmentation, which comes under discriminative model. The general idea is using texture as the key involves identifying areas of similar texture using the histogram and keeping them as a single set. The technique which is used in this paper follows above method with a slightly different approach. The proposed new technique involves use of the texture descriptors described and used in [4], [5] and also using the optical flow proposed by Lucas, B. D., & Kanade, T. [12] which is further improved as in [13]. All these three features are combined together for better visual identification of the dynamic texture. Segmentation of dynamic texture is done by automatic and discriminative way. During segmentation, we do not require any initialization or priori information about dynamic textures in the test set. This paper is extension of the model which is described by Chen J. and Zhao G. [7].

3   Proposed Method

The framework of implementation is illustrated in figure 1. The video is given as input and first this video is pre-processed and then split into suitable equal number of sections. Three different features are calculated for each section. This features include $(LBP/WLD)_{TOP}$ and Lucas-Kanade based optical flow. Using these features, we perform the segmentation by merging and pixel wise classification steps.

3.1   Feature Computation

Two texture descriptors are computed to segment dynamic texture from an input video. These texture descriptors are used as spatial–texture descriptors when utilized in XY plane of a video. Also when these descriptors are used in XT and YT plane, they are called temporal-texture descriptors. Hence these texture descriptors are called spatiotemporal descriptors as these texture descriptors are used in both spatial and temporal domain. XT plane indicates the change/deviation in pixels row-wise over temporal domain. YT plane indicates the change/deviation in pixels column-wise over temporal domain. Local Binary Pattern (LBP) and Weber Local Descriptor (WLD) are computed in all three planes and hence it is called as $(LBP)_{TOP}$ and $(WLD)_{TOP}$ [7] respectively.

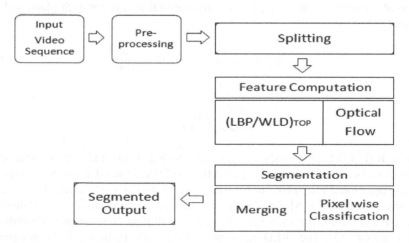

Fig. 1. Flowchart of Proposed Framework

Local Binary Pattern

The computation of LBP is very similar for all the three planes considered (XY, XT, YT) and computation is as follows:

$$LBP_p = \sum_{p=0}^{P} s(g_p - g_c) 2^P \qquad ..(1)$$

$$S(A) = \left\{ \begin{array}{l} 1, \ if \ A \geq 1 \\ 0, \ otherwise \end{array} \right\} \qquad .(2)$$

Where, g_p correspond to the gray values of P number of neighboring pixels on a rectangle of length (2Lx+1) and (2Ly+1) (Lx>0 and Ly>0). All three histograms from 3 different planes are concatenated to obtain the histogram for LBPTOP.

The below figure 2 is similar as given in [6]. The equation 1 is taken from [4] and the number of bins used to map the histogram depends on further experiments. This is based on the concept of uniform pattern same as [4]. Uniform pattern can be defined as follows: A local binary pattern is called uniform sequence if the measure of uniformity is at most 2. For example the patterns 00000000 (0 transitions), 01110000 (2 transitions) and 11001111 (2 transitions) are uniform whereas the patterns 11001001 (4 transitions) and 01010011 (6 transitions) are not Non-uniform patterns. Ojala et al. observed that in their experiments with texture images, uniform patterns account/measure for a bit less than 90% of all patterns when using the (8,1) neighborhood where 8 is number of surrounding pixels and 1 is radius.

Weber Local Descriptor

The actual WLD of [5] comprises of two different components – differential excitation and orientation. But only the differential excitation component is used here. WLD is also computed in three planes in the very same way for each plane and it is computed as follows:

$$WLD \ (Ic) = sigmoid(x) \frac{1-e^{-x}}{1+e^{-x}} \qquad .. \ (3),$$

where

$$x = \sum_{i=0}^{p-1} \frac{(I_i - I_c)}{(I_i + C_0)} \qquad ..(4)$$

Where Ii (I = 0,1,...p-1) indicates the neighboring pixels and p is the number of neighbors. Co is a small constant to avoid the condition of the denominator becoming zero. The equation 2 is as given in [6] and [7] and is based on the method used in [5]. WLD which is given in [5] deals with two different components i.e. orientation and differential excitation. But only the differential excitation component is described in equation number (3). The WLD mapping is done using 16 bins and the mapping is same as given in [5].

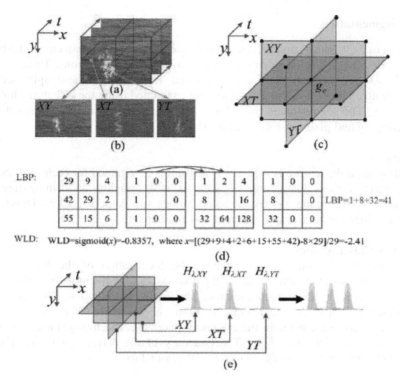

Fig. 2. Computation of $(LBP/WLD)_{TOP}$ for a DT. (a) Sequence of a DT. (b) Three orthogonal planes of the given DT. (c) Vertex coordinates of the three orthogonal planes. (d) Computation of LBP and WLD of a pixel. (e) Computation of sub-histograms for $(LBP/WLD)_{TOP}$, where λ = LBP, WLD

Lucas-Kanade Optical Flow

The optical flow computation from Lucas and Kanade [12] [13] is based on the image brightness constancy assumption which states that for a motion (u, v) of a point in an image(I) the brightness of the point does not change:

$$I(x, y, t) = I(x + u, y + v, t + 1) \qquad .. \qquad (5)$$

Using first order Taylor expansion leads to the gradient constraint equation as:

$$I_x u + I_y v + I_t = 0 \qquad ..(6)$$

This underdetermined system is solved by a least squares estimate [12] over an image patch given by

$$(x, y)^T = [\Sigma(I_x \, I_y)^T (I_x \, I_y)]^{-1} \, \Sigma \, I_t (I_x \, I_y)^T \qquad ..(7)$$

3.2   Segmentation

As shown in the figure 1, preprocessed input video frame is split into equal number of sections. Here we are splitting entire frame into 17×17 equal sections. Three features LBP, WLD and Optical flow are calculated for each of these split sections independently. The Equivalent histograms are calculated using these features for each split section. The segmentation procedure is mainly classified into two distinct steps called merging and pixel-wise classification.

Merging

In splitting stage the texture is split and histogram is determined for each split section. Out of these various histograms of sections with roughly uniform texture, the histograms having same or nearly same values are kept together. Those with completely different set of values are taken as another set.

Pixel Wise Classification

We perform this simple step to improve the localization of the boundaries of segmented regions in result frame. This pixel wise classification is done on the boundary pixel set of the dynamic texture. The Local binary pattern LBP_{TOP}, Weber local descriptor WLD_{TOP} and Optical Flow is computed over an 17 x 17 neighborhood for each pixel in the boundary pixel-set, The histogram is calculated and compared with the threshold. The boundary of the pixels that is classified as dynamic texture is further analyzed to refine the boundary.

4   Results

The implementation of proposed method is divided into four phases- Pre-processing, Splitting, Feature computation and Segmentation. We are performing segmentation on mainly well known dynamic texture dataset called Dyntex [14], UCSD Pedestrian dataset [10] and general CCTV highway traffic [10] videos. We are limiting test video by considering video sequences which are shot by stationary camera only. The experiment is performed on large number of sequences but results of some of them are shown here. Segmentation is performed on Sequence *648ea10*, 645c610 which is taken from the DynTex database [14] and sequence *vidf1_33_000.y* from [10].

The quality of the frame is very essential before performing any video processing operation. In pre-processing phase improvement of the quality of frame is taken into the consideration and accordingly the video is tested for three types of noises. The pre-processed sequence is then passed to feature computation phase and segmentation is done with the help of these features. Figure 3 shows results of each segmentation step on Dyntex dynamic texture sequence. Figure 4 shows experimental results of segmentation of various frames on pedestrian sequence.

Original Frame   Pre-Processed Frame   LBP/WLD   LBP/WLD/LC

Fig. 3. Segmentation Results on Dyntex sequence 648ca10 and sequence 645c610 [23]

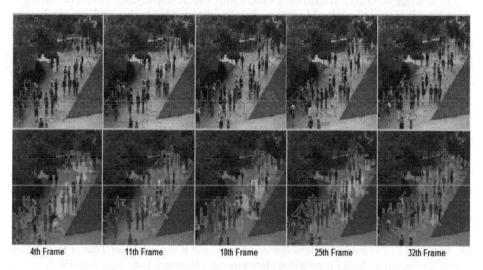

4th Frame   11th Frame   18th Frame   25th Frame   32th Frame

Fig. 4. Segmentation results of different frames of Pedestrian sequence [15] where first and second row shows results of (LBP/WLD) and (LBP/WLD/LC) respectively

5   Conclusion

We propose a new framework for Dynamic texture segmentation based on spatio-temporal features. Three different techniques are combined together for accurate segmentation of dynamic texture. For spatial and temporal domain of dynamic texture we compute LBP, WLD and Lucas-Kanade Optical Flow features.

Most of existing methods require initial contour information for segmentation of dynamic texture. For proposed method, no initial contour information about dynamic texture is needed. Experimental results show that our method gives more accurate segmentation of dynamic texture. In future, by using this method we will measure highway traffic density at various timings in a day.

References

1. Doretto, G., Chiuso, A., Wu, Y.N., Soatto, S.: Dynamic textures. International Journal Computer Vision 51(2), 91–109 (2003)
2. Doretto, G., Cremers, D., Favaro, P., Soatto, S.: Dynamic texture segmentation. In: Proc. IEEE International Conference on Computer Vision, pp. 1236–1242 (2003)
3. Chetverikov, D., Péteri, R.: A brief survey of dynamic texture description and recognition. In: Proc. 4th Int. Conf. Comput. Recognit. Syst., pp. 17–26 (2005)
4. Zhao, G., Pietikäinen, M.: Dynamic texture recognition using local binary patterns with an application to facial expressions. IEEE Trans. Pattern Anal. Mach. Intell. 29(6), 915–928 (2007)
5. Chen, J., Shan, S., He, C., Zhao, G., Pietikainen, M., Chen, X., Gao, W.: WLD: A robust local image descriptor. IEEE Trans. Pattern Anal. Mach. Intell. 32(9), 1705–1720 (2010)
6. Chen, J., Zhao, G., Salo, M., Rahtu, E., Pietikäinen, M.: Automatic Dynamic Texture Segmentation using Local Descriptors and Optical Flow. IEEE Trans. on Image Processing 22(1), 326–339 (2013)
7. Chen, J., Zhao, G., Pietikäinen, M.: An improved local descriptor and threshold learning for unsupervised dynamic texture segmentation. In: Proc. 12th IEEE Int. Conf. Comput. Vis. Workshop, pp. 460–467 (October 2009)
8. Vidal, R., Ravichandran, A.: Optical flow estimation & segmentation of multiple moving dynamic textures. In: Proc. IEEE Int. Conf. Comp. Vis. Pattern Reco., pp. 516–521 (2005)
9. Cooper, L., Liu, J., Huang, K.: Spatial segmentation of temporal texture using mixture linear models. In: Proc. Int. Conf. Dynamical Vis., pp. 142–150 (2005)
10. Chan, A.B., Vasconcelos, N.: Modeling, clustering, and segmenting video with mixtures of dynamic textures. IEEE Trans. Pat. Anal. Mach. Intell. 30(5), 909–926 (2008)
11. Chan, A.B., Vasconcelos, N.: Variational layered dynamic textures. In: Proc. IEEE Int. Conf. Comput. Vis. Pattern Recognit., pp. 1062–1069 (2009)
12. Lucas, B.D., Kanade, T.: An iterative image registration technique with an application to stereo vision. In: Inter. Joint Conf. on Artificial Intelligence, vol. 3, pp. 674–679 (1981)
13. Bouguet, J.: Pyramidal Implementation of the Lucas-Kanade Feature Tracker Description of the algorithm. Intel Corporation, Microprocessor Research Labs 1(2), 1–9 (1999)
14. Pteri, R., Fazekas, S., Huiskes, M.J.: DynTex: A Comprehensive Database of Dynamic Textures. Pattern Recognition Letters 31(12), 1627–1632 (2010), http://rpeteri.free.fr/index.html

Design and Implementation of Brain Computer Interface Based Robot Motion Control

Devashree Tripathy[1,2] and Jagdish Lal Raheja[1]

[1] Advanced Electronics Systems Group,
CSIR - Central Electronics Engineering Research Institute, Pilani, India
[2] Academy of Science and Innovative Research,
Council of Scientific and Industrial Research (CSIR), India
{devashree.tripathy,jagdish.raheja.ceeri}@gmail.com

Abstract. In this paper, a Brain Computer Interactive (BCI) robot motion control system for patients' assistance is designed and implemented. The proposed system acquires data from the patient's brain through a group of sensors using Emotiv Epoc neuroheadset. The acquired signal is processed. From the processed data the BCI system determines the patient's requirements and accordingly issues commands (output signals). The processed data is translated into action using the robot as per the patient's requirement. A Graphics user interface (GUI) is developed by us for the purpose of controlling the motion of the Robot. Our proposed system is quite helpful for persons with severe disabilities and is designed to help persons suffering from spinal cord injuries/ paralytic attacks. It is also helpful to all those who can't move physically and find difficulties in expressing their needs verbally.

1 Introduction

The field of Brain Computer Interaction (BCI) has made possible new avenues to send signals from brain to the external world through non-neurological channels. The research in the BCI field has received a boost in last few years after realizing its necessity to serve people with disabilities as an assistive technology. The BCI devices enable the people, who are paralyzed, to lead their life independently without the help from any care taker. Today, BCI technology is broadly dependent on following principles: P300 [1], SCP (slow cortical potential) [1], SMR (sensorimotor rhythm), etc [1]. They translate the activities in brain in real-time into commands that operate a computer display, wheel chair, prosthetic arm or other device. Hence, both the user and the BCI system need to adapt to each other iteratively using the appropriate feedback mechanism till the optimum performance is obtained. The performance measure of the BCI systems depends on their speed and accuracy.

There are many BCI neuroheadset devices currently available for acquisition of the brain signals. One among those devices is the Epoc neuroheadset built by Emotiv, USA [12]. Though the device was meant to be used primarily for computer games, it could also be used as health equipment. It can assist patients with serious disabilities

© Springer International Publishing Switzerland 2015

S.C. Satapathy et al. (eds.), *Proc. of the 3rd Int. Conf. on Front. of Intell. Comput. (FICTA) 2014*
– *Vol. 2*, Advances in Intelligent Systems and Computing 328, DOI: 10.1007/978-3-319-12012-6_32

to communicate their expression and thoughts, move around independently without any support or care taker to do their daily activities through BCI technology.

2   Related Works

The EEG classification was used for control of application software on computers as far back as 1991[1]. They successfully created a system that allowed a user to control a cursor using EEG signals. Later, Wolpaw explored the area and published his work in a series of papers describing the state of the art throughout the early 2000s [11,17]. These works were further carried forward in [8], where a large, high resolution EEG was used to control a robotic arm. These studies highlighted the importance of EEG, but unfortunately did not address the issues of cost and difficulty of use for the end-user.

The need for an easy-to-use and cost effective neuroheadset has been met out by Emotiv EPOC neuroheadset. The Emotiv EPOC has been used in a variety of applications, due to its accessibility for consumers and researchers. The device was validated [1] where it was found to be useful for the control of computers by patients with no voluntary muscle control. These patients were able to access computer based assistive technology and navigate a virtual space using two discrete thought patterns, all with a greater than random chance of success. The Emotiv EPOC was found to be effective methods for computer control, having the subjects of their experiments perform a 3D rotation task in virtual space [9]. The Emotiv EPOC has also been used in the area of controlling robotic systems through non-invasive BCI. In [10] Emotiv EPOC was used to teleoperate a remote humanoid robot. Their system allowed for the remote control of an Aldebaran Robotics Nao humanoid robot , which the user was able to direct with EEG commands to move forward, turn left, and turn right. It was found that their system was effective to navigate a small course. In [11] the Emotiv EPOC was used to control a Lego NXT robot. They used the Emotiv Emokey software to send key commands to a java program. The Emokey software is provided by Emotiv, and allows a user to select different keystrokes to virtually actuate based on EEG data, with up to four outputs being trained [12]. The authors in [13] showed the use of the Emotiv EPOC for the control of robotic arms through EEG.

3   System Design

3.1   Objective

According to the American Spinal Injury Association Impairment Scale (AIS)-A, the patients suffer from most impairments in which their voluntary muscles don't work. So for those patients, moving around freely and independently without any aid to accomplish their daily needs becomes a big challenge. They need their involuntary muscles to translate their needs or thoughts to a mobile device like electric wheel chair or robotic arm.

The main objective of designing the BCI based Robot motion control system is to help the persons with severe disabilities. This serves as an assistive technology for patients. The robot needs to move in all possible directions and execute all types of motion a patient would require. The same are shown in the written and pictorial form in a GUI containing nine elements in subsequent sections of this paper. When the person needs something he/she will focus on that element. Due to his/her facial expressions and thoughts the specific element being focused on is pressed. Thus, the previously programmed wheel chair or any other mobile device (here, FIRE BIRD V robot) moves as per the person's wish.

3.2   Signal Acquisition

Our system design uses the EPOC neuroheadset for extracting data from the patient's brain. The working principle of the EPOC neuroheadset is described below.

Taking the facial expression and thoughts of the user as input, the device accesses a computer and presses switches on scanning software. The neuroheadset has 16 sensors (out of which there are 2 reference sensors). It is worn on the head in such a way that these make contact with the scalp. It works by detecting electrical signals that are generated naturally by the human brain.

The neuroheadset SDK has three main features like:

1. "Expressiv" suite which analyses the signals captured from the 16 electrodes of the EMOTIV EPOC neuroheadset to recognize 12 separate facial expressions such as "blink", "smile" or "look left' 'etc.

2. "Cognitiv" suite interprets changes in this signal in order to detect 13 specific thoughts (such as "push" and "pull").

3. "Affectiv"suite also analyses the EEG signal, but instead detects the user's level of engagement. [1]

3.3   Design Steps

The main steps of the design include: Signal acquisition, Feature extraction, and Feature classification, Feedback, Testing and Feature Recognition. In other words, the working principle of our BCI system includes: acquiring and processing EEG data, translating that data into device output, and giving a feedback to the user (Figure 1). Implementation of this above said EEG-based BCI system includes the feature extraction and classification of EEG data in real-time.

Step1: This step is concerned with Signal Acquisition from the human brain through a set of sensors. In the current design the EPOC neuroheadset is used for signal acquisition from the patient's brain.

Step 2: The Feature extraction deals with separating relevant EEG data from noise and other disturbances and simplifying that data by reducing the redundancy so that it can be further used for the classification process. Several other feature extraction techniques (i.e. PCA, ICA, CSP etc) [7] can also be used in the system.

Step 3: In the classification part, we decide what output classes the extracted feature corresponds to. Classification is done using the standard techniques such as:

LDA, SVM, ANN algorithms and various other types of pattern recognizers are employed to try to match the input data to known classes of EEG functional model. Recently unsupervised learning algorithms are used predominantly to find natural clusters of EEG data that are indicate certain kind of mental activities with varying degree of successes.

Step 4: Feedback is essential in BCI systems as it allows users to know the types of brainwaves they just produced and to learn behavior that can be effectively classified and controlled. The optimal form which the feedback should take is yet to be found out. [2][3]

Step 5: In the testing part we run some applications. Out of different classifications of EEG based BCIs; Synchronous systems (driven by the computer) are easier to implement as compared to asynchronous (are driven by the user) have so far been the major way of operating the BCI systems. [4] Here, the user is given a clue to perform a certain mental action and then the user's EEG patterns are recorded in a fixed time-window. The accuracy or success rate of using the application depends on the precision of training, the quality of contact of the neuroheadset with the person's scalp(in case of non-invasive EEG), the climatic condition, day and time on which the action is being performed and many more factors.

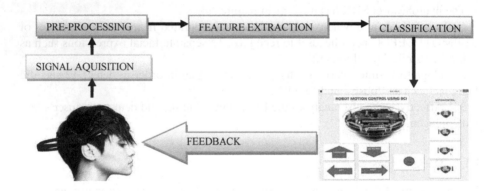

Fig. 1. BCI System

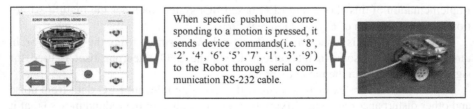

When specific pushbutton corresponding to a motion is pressed, it sends device commands(i.e. '8', '2', '4', '6', '5' ,'7', '1', '3', '9') to the Robot through serial communication RS-232 cable.

Fig. 2. Communication between GUI and Robot

4   Implementation

The implementation of the designed BCI system was done using the Emotiv Epoc neuroheadset, FIRE BIRD V Robot (developed by NEX ROBOTICS) (Refer Fig. 3a, 3b). The DB9 pin male female straight through RS232 Serial cable, computer system with Intel Xeon processor @3.4GHz, 16 GB RAM, and 64-bit Windows 8 operating system was used.

Fig. 3. Fire Bird V Robot

The various soft wares used in this work were: the AVR studio 8 from ATMEL [5], WIN AVR open source C compiler, Khazama AVR programmer, Terminal software for serial communication, Emokey software for mapping different actions with the keypad/mouse buttons. After writing and compiling the program in AVR Studio 8 ".hex" file was generated. This ".hex" file was then loaded to the FLASH buffer and then written on the ATMEGA 2560 microcontroller chip contained in the FIRE BIRD V robot and verified using Khazama AVR In System Programmer. This software does not contain digital signature and hence was blocked by Windows 8 OS. So, the digital signature needs to be deactivated before reading the chip signature. For serial communication between Computer and FIRE BIRD V Robot we used "terminal software". In terminal software, we selected baud rate as 9600 and communication port as COM1 for communication. Then, the program verification was done   by pressing the appropriate keys from the keypad and checking whether the Robot was functioning appropriately or not. For example, as per our program when key '8' is pressed from the keypad the robot should move in the forward direction. The user was trained so as to use the neuroheadset. The quality of contact of each sensor was verified according to the color displayed (black: no contact, red: very poor contact, orange: poor contact, yellow: fair contact, green: good contact) in the Control Panel software supplied by Emotiv. Then the user was trained with each of the 12 facial expressions and 13 thought actions to determine which action could trigger strong signals without being disturbed by the unwanted actions. The sensitivity of the action was changed to get the best performance as per the necessity. It was observed that 'Blink' action has the greatest accuracy of detection as compared to other actions. Here after, the training is done for the user using universal signature. Then the 'left click' button of the mouse was mapped with the 'blink' action using the Emokey software.

The user then accessed a MATLAB GUI –"Robot motion control GUI" which was created using MATLAB R2012b. The cursor control feature is associated with the Emotiv Control panel under 'mouse emulator' option. Thus we can navigate through the computer screen by rolling our eyes. We selected the appropriate button in the GUI and the robot moved accordingly.

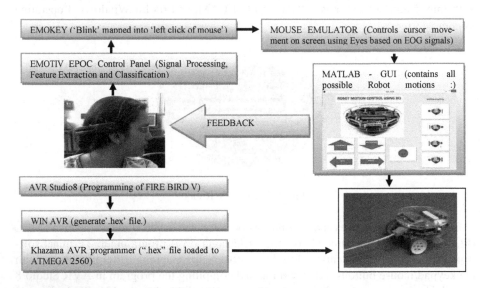

Fig. 4. Flow Diagram of implementation

This GUI has 9 control buttons (forward, backward, left, right, stop, soft left 1, soft right 1, soft left 2, and soft right 2). When the push button is pressed the necessary command is passed on to the device FIRE BIRD V through serial communication (Fig. 5).

Here, the robot motions were controlled using BCIs. All possible robot motions in Fire Bird V include:

  i.   Forward (both the motors move forward),
  ii.   Backward (both the motors move backward),
  iii.   Left (left motor moves backward and Right motor moves forward),
  iv.   Right (left motor moves backward and right motor moves forward),
  v.   Stop (both the motors stop),
  vi.   Soft left 1 (left motor stops and right motor moves forward),
  vii.   Soft right 1 (left motor moves forward and right motor stops),
  viii. Soft left 2 (left motor moves backward and right motor stops),
  ix.   Soft right 2 (left motor stops and right motor moves backward).

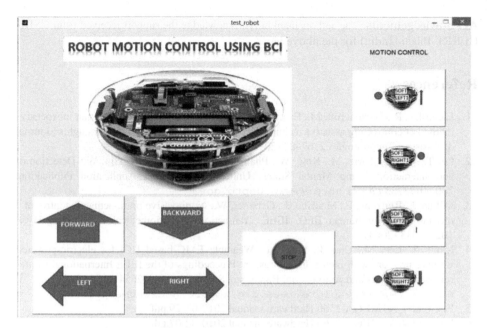

Fig. 5. Robot motion control GUI

The above events were generated as per the pin configurations and hex codes in hardware manual [5]. The pins PA0, PA1, PA2 and PA3 of ATMEGA 2560 micro-controller chip were interfaced with pin7, pin2, pin 15 and pin10 of L293D respec-tively[6]. Then the robot motion Forward, Backward, Left, Right, Stop, Soft Left1, Soft Right1, Soft Left2, SoftRight2 were carried out when the device commands '8', '2', '4', '6', '5' ,'7', '1', '3', '9' were send to the robot through serial communication.

5   Conclusions

This paper presented the design and implantation details of a BCI based robot motion control system for assisting the patients. The system used the Fire Bird robot, AVR studio 8 from ATMEL, WIN AVR open source C compiler, Emotiv EPOC API, Kha-zama AVR programmer, Terminal software for serial communication, Emokey soft wares for the purpose. A GUI was developed. The performance and accuracy of the system after testing was found to be good. Thus, our designed system comes as bliss for all those patients with disabilities who can easily fulfil their immediate needs/wishes without moving any body part. The accuracy of the EMOTIV EPOC for the robot motion control using nine actions (blink, smile, neutral, raise brow, look left, look right, push, pull, smirk) was found out to be less than 40%. But when we reduce the number of actions from nine to two, the accuracy increased to 99%. More-over, this inexpensive system is easy-to-use, fast, and accurate and furthermore, no adjustments are required to make it work.

Acknowledgements. The authors thankfully acknowledge the support received from CSIR-CEERI, Pilani (India) for the above work.

References

1. Lievesley, R., Wozencroft, M., Ewins, D.: The Emotiv EPOC neuroheadset:an inexpensive method of controlling assistive technologies using facial expressions and thoughts? Journal of Assistive Technologies 2, 67–82 (2011)
2. Le, T., Do, N., Torre, M., King, W., Pham, H., Delic, E., Thie, J., Siu, W.: Detection of and Interaction Using Mental States. United States Patent Application Publication, 2007/0173733 (2007), http://www.uspto.gov
3. Millan, J., Renkens, F., Mourino, J., Gerstner, W.: Noninvasive brain-actuated control of a mobile robot by human EEG. IEEE Transactions on Biomedical Engineering 51(6), 1026–1033 (2004)
4. Moon, I., Lee, M., Chu, J., Mun, M.: Wearable EMG-based HCI for electric-powered wheelchair users with motor disabilities. In: Proceedings of the IEEE International Conference on Robotics and Automation, Barcelono, Spain, pp. 2649–2654 (2005), http://www.ieeexplore.ieee.org/xplore/guesthome.jsp
5. Fire bird V ATMEGA-2560 Hardware manual 2010-01-29.pdf
6. Fire bird V ATMEGA-2560 Hardware manual 2010-02-07.pdf
7. Klonovs, J., Petersen, C.K.: Development of a Mobile EEG-Based Feature Extraction and Classification System for Biometric Authentication. Master's thesis, Aalborg University Copenhagen (2012)
8. Iáñez, E., Furió, M.C., Azorín, J.M., Huizzi, J.A., Fernández, E.: Brain-robot interface for controlling a remote robot arm. In: Mira, J., Ferrández, J.M., Álvarez, J.R., de la Paz, F., Toledo, F.J. (eds.) IWINAC 2009, Part II. LNCS, vol. 5602, pp. 353–361. Springer, Heidelberg (2009)
9. Poor, G.M., Leventhal, L.M., Kelley, S., Ringenberg, J., Jaffee, S.D.: Thought cubes: exploring the use of an inexpensive brain-computer interface on a mental rotation task. In: The Proceedings of the 13th International ACM SIGACCESS Conference on Computers and Accessibility (ASSETS 2011), pp. 291–292 (2010)
10. Thobbi, A., Kadam, R., Sheng, W.: Achieving Remote Presence using a Humanoid Robot Controlled by a Non-Invasive BCI Device. International Journal on Artificial Intelligence and Machine Learning 10, 41–45
11. Vourvopoulos, A., Liarokapis, F.: Brain-controlled NXT Robot: Tele-operating a robot through brain electrical activity. In: 2011 Third International Conference on Games and Virtual Worlds for Serious Applications (VS-GAMES), pp. 140–143 (May 2011)
12. Emotiv, Emotiv Software Development Kit User Manual Release 1.0.0.3
13. Ranky, G.N., Adamovich, S.: Analysis of a commercial EEG device for the control of a robot arm. In: Proceedings of the 2010 IEEE 36th Annual Northeast Bioengineering Conference, pp. 1–2 (March 2010)

Advanced Adaptive Algorithms for Double Talk Detection in Echo Cancellers: A Technical Review

Vineeta Das[1], Asutosh Kar[1], and Mahesh Chandra[2]

[1] Dept. of Electronics and Telecommunication Engineering, IIIT, Bhubaneswar, India
[2] Dept. of Electronics and Communication Engineering, BIT, Mesra, India
vineetadas1234@gmail.com, asutosh@iiit-bh.ac.in,
shrotriya@bitmesra.ac.in

Abstract. An acoustic echo cancellation system is one of the most important breakthrough in the field of adaptive systems. Today acoustic echo cancellers (AEC) are an integral part of full duplex hands-free voice communication. Conventional echo cancellers use a linear model to represent the echo path. However many consumer devices include loud-speakers and power amplifiers that generate non-linear distortions. Non-linearity occurs due to the use of low cost electronic loud speakers, microphones and poorly designed enclosures in an AEC system. Non-linearity causes vibration and harmonic distortion and also degrades the speech quality. Double talk detector (DTD) is a key component of an AEC. A DTD is used to sense when the far end signal is corrupted by the near end speech. The DTD freezes the adaptation of model filter to prevent the divergence of the adaptive filter. Various authors have proposed different algorithms for double talk detection. Some of the most popular algorithms are Geigel algorithm, cross-correlation based DTD, normalized cross correlation based DTD, variable impulse response DTD etc. In this paper several double talk detection algorithms in a non-linear platform of an AEC has been discussed.

Keywords: acoustic echo canceller (AEC), double talk detector (DTD), adaptive filter, adaptive algorithm, cross-correlation, non-linear acoustic echo canceller.

1 Introduction

Most tele-conferencing conversations take place in presence of acoustic echo if the delay between speech and its echo is more than few tens of a millisecond, the echo is distinctively noticeable. So AECs are used to nullify the echo produced in the loud speaker microphone environment. In mobile phones and other hands-free communication devices, the loud speaker and the microphone are present in close proximity. This leads to over hearing of the voice of the speaker at the speaker side. Over hearing makes the conversation unpleasant and difficult [1-2].

AECs uses an adaptive filter and an adaptive algorithm to remove the overhearing by cancelling the echo produced in the transmission path. The adaptive filter usually

© Springer International Publishing Switzerland 2015
S.C. Satapathy et al. (eds.), *Proc. of the 3rd Int. Conf. on Front. of Intell. Comput. (FICTA) 2014*
– *Vol. 2*, Advances in Intelligent Systems and Computing 328, DOI: 10.1007/978-3-319-12012-6_33

deploys a traversal Finite Impulse Response (FIR) structure due to its guaranteed stability and ease of implementation. The adaptation of the FIR filter coefficients is controlled by an adaptive algorithm [3].

The adaptive algorithm is the heart of an AEC, which decides the convergence behavior of the AEC. The tracking behavior indicates how fast the adaptive filter can follow dislocations whereas the convergence behavior is studied as an initial adjustment of the adaptive filter to the impulse response of the system. The adaptive algorithms popularly used are Least Mean Square (LMS), Recursive Least Square (RLS), Normalized Least Mean Square (NLMS) etc.[3-5].

The basic model of an AEC is given in fig 1.

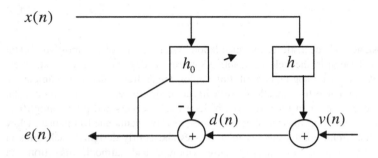

Fig. 1. A Basic AEC model

In the above figure h_0 is the transfer function of modeled echo path, h is the actual echo path, $x(n)$ is the far end signal, $v(n)$ is the near end signal, $d(n)$ is the desired signal, $e(n)$ is the reduced echo (error) signal.

Linear adaptive filters are often used in AEC but sometimes it fails to perform well in hands-free terminals and inexpensive telephony devices. Low quality speakers and poorly designed enclosures that produce vibrations often generate harmonic distortion and this degrades the performance of a linear AEC. In this competitive market and advancement in technology there is a trend towards miniaturization which leads to placing of different components very close to each other. Components placed close to each other catch the hum of the nearby device leading to harmonic distortion and non-linearity [6-9]. Other common sources of non-linearity are the use of cheap electronic components like loud speakers and microphones which degrade the performance of a linear AEC considerably. Various methods have emerged for cancellation of non-linear echo. Non-linear structures like cascaded and parallel loud speaker linearization approach, non-linear orthogonalized power filter approach, multi memory decomposition structure, affine projection based adaptive Volterra filter etc. have proved to be popular. Therefore a non-linear platform has been chosen in this discussion of the DTD algorithms.

Echo canceller works efficiently in presence of only one participant. Problem arises in the case when both the far end and near end speakers indulge in simultaneous conversation with each other. This situation is called double talk. In double talk scenario the near end signal acts as a disturbance in updating the filter coefficients. In situations of double talk, the adaptive filter diverges instead of converging and the error is not reduced.

A DTD is used to tackle the problem of double talk. When double talk is detected by the detector, it prevents the adaptive algorithm from modifying the filter coefficients for a certain time duration [1]. The generic working of a DTD is given as follows-

- A double talk detection variable ξ is calculated using the near end, far end and the received signal.
- It is compared with a preset threshold variable T.
- If double talk is present, the filter coefficients are freezed for a hold time T_{hold}. After the T_{hold} period normal operation is resumed.
- If double talk is not present, DTD does not interrupt in the working of the AEC.

The block diagram of an AEC with a DTD is given in fig 2.

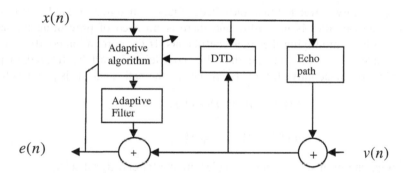

Fig. 2. Block Diagram of AEC with DTD

2   Discussion of DTD Algorithms

2.1   Geigel Algorithm

It is a very primitive algorithm used for double talk detection. In this method the amplitudes of the received signal i.e. the far end echoed signal coupled with the near end signal is compared to obtain the decision threshold [10]. The mathematical expression for the threshold is given as

$$\eta = \frac{|d(n)|}{\max\{|x(n)|,....,|x(n-n_0+1)|\}} \tag{1}$$

The determination of double talk is done by comparing the decision threshold with a preset threshold T. If the decision parameter $\eta > T$, then double talk is present otherwise there is no double talk.

The choice of T should be made very carefully as it affects the performance of the DTD. For line echo cancellers the T is set at 0.5 because the hybrid attenuation is assumed to be 6 dB. However for acoustic echo cancellation environment, not only the back ground noise level but also the echo path characteristics are time varying. Therefore it is not easy to decide a proper value of threshold T.

Geigel algorithm is a rudimentary method of double talk detection which though detects double talk but lacks efficiency and accuracy. As the echo path is completely random in nature, determination of the threshold value becomes difficult.

2.2   Cross Correlation of Far End and Error Signal for Double Talk Detection

In this method the cross correlation between the far end signal $x(n)$ and the error signal $e(n)$ is calculated to obtain the decision threshold. If the far end and the error signal are highly correlated then the effect of near end signal on the healthy working of the adaptive filter is negligible and there is no need of preventing the adaptive algorithm from modifying the filter coefficients. If the correlation coefficient is low, then double talk is present and the filter coefficients must be frozen to prevent divergence of the adaptive filter [11-18]. The mathematic analysis is given as follows.

$$d(n) = h^T x(n) + v(n) \tag{2}$$

$$e(n) = d(n) - h_0^T x(n) \tag{3}$$

Multiplying both sides of equation (3) with the far end signal $x(n)$,

$$e(n)x^T(n) = d(n)x^T(n) - h_0^T x(n)x^T(n) \tag{4}$$

Expanding $d(n)$ from equation (2) in equation (4) we obtain

$$e(n)x^T(n) = (h^T x(n) + v(n))x^T(n) - h_0^T x(n)x^T(n) \tag{5}$$

$$= h^T x(n)x^T(n) + v(n)x^T(n) - h_0^T x(n)x^T(n) \tag{6}$$

Taking expectation $E[.]$ on both sides of equation (6)

$$E[e(n)x^T(n)] = h^T E[x(n)x^T(n)] + E[v(n)x^T(n)] - h_0^T E[x(n)x^T(n)] \tag{7}$$

As the near end signal and the far end signal are uncorrelated, so

$$E[v(n)x^T(n)] = 0 \tag{8}$$

Equation (7) now becomes,

$$E[e(n)x^T(n)] = h^T E[x(n)x^T(n)] - h_0^T E[x(n)x^T(n)] \tag{9}$$

Equation (9) can be rewritten as follows,

$$R_{xe} = (h^T - h_0^T)R_{xx} \tag{10}$$

Where R_{xe} is the cross correlation between the far end and the echo signal and R_{xx} is the auto correlation of the far end signal. The decision parameter is given as-

If $R_{xe} \cong 0$, there is no double talk.

If $R_{xe} \geq 0$, double talk is present.

This method is superior to the Geigel algorithm but it focuses more in determining the variation in echo path than in determination of double talk. So another method based on cross correlation between the far end and the received signal is devised for the double talk detection.

2.3   Cross Correlation of Far End and Received Signal for Double Talk Detection

The cross-correlation between the far end and error signal does not determine the double talk condition explicitly. Instead it determines whether the adaptive filter has converged or not as can be seen from equation (10). Moreover the cross-correlation vector defined in equation (10) is not well normalized. As a result the appropriate value of threshold can vary from one experiment to another [18-26].

In the method of cross correlation between the far end signal and the received signal, the cross-correlation is normalized. This results in the threshold value to be one. The decision parameter is given as below. From equation (2)

$$d(n) = h^T x(n) + v(n)$$

Finding the cross correlation of the received signal and the far end signal, we obtain

$$\rho = \frac{E[d(n)x(n)]}{\sqrt{E[d^2(n)x^2(n)]}} \tag{11}$$

If the value of ρ is very high, it indicates that the received signal and the far end signal are highly correlated and near end signal power is negligible. If the value of ρ is low or zero, then the near end signal is present.

The decision parameter is given as

$$\varphi = norm(\rho) \tag{12}$$

The l_∞ norm is calculated in the above equation.

If $\varphi > T$, the preset threshold then double talk is present.

If $\varphi < T$, then double talk is not present.

2.4   Double Talk Detection Using Variable Impulse Response

In this the variance of the room impulse response h_0 is measured. If the variance is more it is assumed that the room impulse response is varied by some other source that is the near end signal. So if the variance is greater than a particular threshold value T then near end is present and the adaptive algorithm is prevented from modifying the filter coefficients. If variance of the room impulse response is less than T there is no near end and adaptation continues [20].

3   Tabular Review

Table 1. Review of major dtd algorithms

Algorithms	Working mechanism	Remarks
Geigel Algorithm	It is an energy based DTD algorithm. The magnitude of the near end received signal is compared with some recent values of the far end signal.	It is a very simple algorithm and requires less memory. It is not very accurate. Determination of a suitable threshold is very difficult.
Cross-correlation between the far end and error signal	It is a cross-correlation based DTD algorithm. It works on the orthogonality principle between the far end and the error signal. During single talk condition the orthogonality holds. In double talk situation the residual echo gets larger abruptly and the orthogonality principle does not hold.	Coefficient adaptation is done only when the correlation is nonzero as the correlation will be zero after convergence. As the echo path is more or less changing, cross-correlation is likely to take a non-zero value and adaptation is performed even when the near end signal is present. This method does not detect double talk but instead decides whether the adaptive filter has converged or not.
Cross-correlation between the far end and received signal	It is a normalized cross-correlation based DTD.	The threshold is fixed i.e. one. It is a robust algorithm and has high accuracy.
Variable Impulse Response Algorithm (VIRE)	It uses the variance of the room impulse response to determine double talk condition.	It is complex to implement and requires large memory space.

Table 2. Analysis of significant research work on dtd algorithms

Sl. no	Paper Title	Proposed Method	Results	Remarks
1	A New Class of Doubletalk Detection based on Cross-correlation, 2000	A new normalized cross-correlation vector between the far end and the received signal. A normalized cross-correlation vector and matrix is proposed for normalization.	The authors have tried to minimize the response time of the DTD.	The normalized cross-correlation method is superior to the conventional cross-correlation technique. For the simplified form of normalized cross-correlation DTD, it is assumed that the AEC has already converged. So degradation may be seen in performance in a dynamic situation.
2	A multichannel Acoustic Echo Canceller double talk detector based on normalized cross-correlation matrix, 2002	The proposed algorithm consists of using one test statistics that compares all the information of the microphone signals. This approach leads to a nice generalization of the normalized cross-correlation vector to the normalized cross-correlation matrix.	The adaptive identification of echo path and the near end speech detection are combined.	The advantage of this method is that multiple microphone signals are taken into account under one decision variable.
3	Decision of Double Talk and Time-variant Echo Path for Acoustic Echo Cancelation, 2003	In this paper an iterative maximal-length correlation (IMLC) DTD method is proposed. It performs a simplified likelihood ratio test by setting a decision threshold and plot the detection performance by a receiver operating characteristics (ROC).	The ROC of the proposed algorithm along with other DTD algorithms are plotted and compared.	The ROC curve shows a trade-off of double talk detection probability and false alarm probability.
4	Coherence based Double Talk Detector with Adaptive Threshold, 2004	A coherence based DTD is proposed in this paper. When the loudspeaker signal is presented adaptively, the maximal values of the coherence function is tracked and the decision variable is scaled between zero and the maximal value.	The results of the proposed algorithm and the baseline algorithm was compared on the basis of classification error in tracking double talk and ROC was plotted.	This algorithm is more robust to noise and reverberation as compared to the primitive coherence algorithms. The proposed approach improves the precision of the DTD by 5.5% relative to the original version of the algorithm.
5	The fast normalized cross-correlation double talk detector, 2005	In this paper the normalized cross-correlation (NCC) DTD is reviewed and a fast normalized cross-correlation (FNCC) is derived. The recursive least squares (RLS) filtering and double talk detection are combined in this paper.	The adaptive algorithm for the AEC was derived for a multi-channel case.	This algorithm has much higher accuracy, lower false alarm rate and low added complexity.
6	A Robust Double Talk Detector for Acoustic Echo Cancellation, 2010	The DTD proposed in this paper includes a near end voice detector (NEVD), a double to single detector (DSD) and two auxiliary filters. The NEVD and DSD are used to control the adaptive filter either from working or halting. The two auxiliary filters are to save good estimates of echo path and prevent the filter from diverging.	The paper presents experimental data of the proposed algorithm and also compares its performance with Park's and Benesty's method.	It is a robust DTD and performs well in transition of double and single talk situation and is capable of differentiating echo path changes from double talk situation.
7	Frequency-Domain Double-Talk Detection Based on the Gaussian Mixture Model, 2010	In this paper first the frequency domain acoustic echo suppressor (AES) is analyzed. A feature vector is proposed that discriminates between double talk and single talk using Gaussian mixture model.	The performance of the proposed DTD algorithm is measured in terms of DTD error probability which is a sum of the false alarm and miss probabilities.	The proposed approach is efficiently implemented in frequency domain. Experimental results show that this method yields better results than conventional cross-correlation based method.

4   Discussion

Table 1 shows a comprehensive review of the major double talk detection algorithms. The Geigel algorithm though primitive and less accurate but is a very simple algorithm that can be used in line echo cancellers. It may not be a suitable algorithm for acoustic echo cancellers. The cross-correlation and the normalized cross-correlation based DTD are the most popular double talk detection schemes. The cross-correlation based DTD uses orthogonality between the far end and received signal as the detection variable. The drawback of this algorithm is time varying decision threshold. The normalized cross-correlation DTD algorithm is well normalized to have a decision variable as one. The VIRE DTD uses variance of the room impulse response as the double talk detection parameter. It is a complex algorithm so it is not much popular.

Table 2 lists some of the important works in the field of double talk detection. The work on double talk detection started when Ye and Wu [13] proposed the use of cross-correlation coefficient vector between the far end and the error signal. A similar idea using cross-correlation coefficient vector between the far end and the received signal [11] has proven to be more robust and reliable. However the problem is cross-correlation coefficient vectors are not well normalized. Based on cross-correlation Benesty [12] proposed the normalized cross-correlation DTD which eradicates the problem stated above. Many variations on the normalized cross-correlation DTD was proposed by authors in their subsequent works. A new class of algorithm based on coherence function between the loudspeaker and microphone signal was developed [19]. The coherence function is easy to compute and is well normalized. Values close to one means that there is no local speech. Under presence of local speech the values of the coherence function decrease and approach zero, which makes it a good statistical parameter for DTD. The coherence function value decreases under the presence of noise or strong reverberation, which makes this method less suitable for cases when the microphone is away from the loudspeaker or the environment is very noisy. Realization of an efficient double talk detection algorithm in frequency domain has also been worked upon.

5   Conclusion

In this paper some typical double talk detection schemes have been reviewed. The double talk detection schemes can be broadly classified into two types- energy based (Geigel algorithm) and cross-correlation based. The energy based algorithm has an advantage over the cross-correlation based algorithms as it is computationally simple and need very less memory. It does not provide reliable performance in an acoustic echo path environment. Determination of an optimum threshold value is also difficult. On the other hand the cross-correlation based algorithms are extremely reliable in time varying environments. The improved detection performance of the cross-correlation based algorithms are achieved at the cost of increased complexity and high memory storage requirements.

References

[1] Fu, J.: New approaches for nonlinear acoustic echo cancellation. ProQuest Dissertations and Thesis (2008)
[2] Lu, X.J.: Acoustic echo cancellation over nonlinear channels. Ph.D dissertation, McGill University, Canada (2004)
[3] Haykin, S.: Adaptive Filter Theory. Prentice Hall, Englewood Cliffs (1991)
[4] Deb, A., Kar, A., Chandra, M.: A Technical Review on Adaptive Algorithms for Acoustic Echo Cancellation. In: IEEE Sponsored International Conference on Communication and Signal Processing, April 3-5, pp. 638–642 (2014)

[5] Chandra, M., Kar, A., Goel, P.: Performance Evaluation of Adaptive Algorithms for Monophonic Acoustic Echo Cancellation: A Technical Review. International Journal of Applied Engineering Research 9(17), 3781–3806 (2014)

[6] Panda, B., Kar, A., Chandra, M.: Non-linear Adaptive Echo Supression Algorithms: A Technical Survey. In: IEEE Sponsored International Conference on Communication and Signal Processing, April 3-5, pp. 658–662 (2014)

[7] Mossi, M.I., Yemdji, C., Evand, N., Beaugeant, C.: Non-linear acoustic echo cancellation using online loudspeaker linearization. In: IEEE Workshop on Applications of Signal Processing to Audio and Acoustics, New Paltz, NY, October16-19 (2011)

[8] Fermo, A., Carini, A., Sicuranza, G.L.: Analysis of different lowcomplexity non-linear filters for acoustic echo cancellation. Journal of Computing and Information Technology-CIT 8 4, 333–339 (2000)

[9] Seo, J.B., Kim, K.J., Nam, S.W.: Nonlinear acoustic echo cancellation using volterra filtering with a variable step-size GS-PAP algorithm. World Academy of Science, Engineering and Technology 3, 9–20 (2009)

[10] Ding, H., Lau, F.: Double talk detection schemes for echo cancellation. Acoustics Week in Canada-Canadian Acoustical Association (October 2004)

[11] Cho, J., Morgan, D.: An objective technique for evaluating double talk detectors in acoustic echo cancellers. IEEE Transactions on Speech and Audio Processing 7(6) (November 1999)

[12] Benesty, J., Morgan, D.R., Cho, J.H.: A new class of Double Talk Detection based on Cross-Correlation. IEEE Transactions on Speech and Audio Processing 8, 168–172 (2000)

[13] Ye, H., Wu, B.-X.: A new double-talk detection algorithm based on the orthogonality theorem. IEEE Trans. Commun. 39, 1542–1545 (1991)

[14] Benesty, J., Gänsler, T.: The fast cross-correlation double-talk detector. Signal Processing 86(6), 1124–1139 (2006)

[15] Jenq, J.C., Hsieh, S.F.: Decision of double talk and time variant echo path for coustic echo cancellation. IEEE Signal Processing Letters 10(11), 317 (2003)

[16] Benesty, J., Gansler, T.: A Multichannel Acoustic Echo Canceler Double-Talk Detector Based on a Normalized Cross-Correlation Matrix 13(2) (March-April 2002)

[17] Gansler, T.G., Benety, J.: A frequency domain double talk detector based on a normalized cross-correlation vector. Signal Processing 81, 1783–1787 (2001)

[18] Cho, J.H., Morgan, D.R., Benesty, J.: An objective technique for evaluating double talk detectors in acoustic echo cancellers. IEEE Transactions, Speech Audio Processing 7, 718–724 (1999)

[19] Gansler, T.G., Hansson, M., Ivarsson, C.J., Salomonsson, G.: A double talk detector based on coherence. IEEE Trans. Commun. 44, 1421–1427 (1996)

[20] Sonika, Dhull, S.: Double Talk Detection in Acoustic Echo Cancellation based on Variance Impulse Response. International Journal of Electronics and Communication Engineering 4(5), 537–542 (2011)

A Comparative Study of Iterative Solvers for Image De-noising

Subit K. Jain, Rajendra K. Ray, and Arnav Bhavsar

Indian Institute of Technology, Mandi, India
jain.subit@gmail.com, {rajendra,arnav}@iitmandi.ac.in

Abstract. In this paper we propose and compare the use of two iterative solvers using the Crank-Nicolson finite difference method, for image denoising via Partial differential equations (PDE) models such as Bilateral-filter-based model. The solvers considered here are: Successive-over-Relaxation (SOR) and an advanced solver known as Hybrid Bi-Conjugate Gradient Stabilized (Hybrid BiCGStab) method. We demonstrate that proposed hybrid BiCGStab solver for denoising yields better performance in terms of MSSIM and PSNR, and is more efficient than existing SOR solver and a state-of-the-art approach.

Keywords: Bazan model, Crank-Nicolson scheme, hybrid BiCGStab, denoising, anisotropic diffusion.

1 Introduction

One of the most fundamental tasks in image processing applications is that of, image de-noising, which is a significant preprocessing step. Typically, it is assumed that the degradation process for image denoising can be expressed using a mathematical model. Partial differential equations (PDEs) has been a effective tool for denoising. Out of various PDE models, the linear diffusion process is simplest and well considered PDE-based method [1]. But the shortcoming with this is that it fails to preserve edges/textures [2]. To address this issue, various nonlinear PDE models have been proposed for image denoising [3].

In [2], Perona and Malik proposed a nonlinear PDE which consists of an inhomogeneous diffusion coefficient. This model is useful for smoothing, denoising and detection of edges in (digital) image. However, Perona-Malik (P-M) model is ill-posed [4]. Hence to attain a well-posed model Catt et al. [5] proposed a regularization of the P-M model. Recently, a combination of anisotropic diffusion and bilateral filter attracted many researchers, to improve the quality of denoised or filtered image while preserving the edge information [6–8]. In this work we have adopted the Bazan model [6].

In addition to the model, the development of selected numerical technique to discretize the PDE model is a vital factor of PDE-based approaches [9]. Till date only few finite difference schemes are available to solve PDE models. Usually the numerical schemes which are used to discretize the corresponding nonlinear diffusion models are the explicit schemes [9], which can be unstable and require

© Springer International Publishing Switzerland 2015
S.C. Satapathy et al. (eds.), *Proc. of the 3rd Int. Conf. on Front. of Intell. Comput. (FICTA) 2014
– Vol. 2*, Advances in Intelligent Systems and Computing 328, DOI: 10.1007/978-3-319-12012-6_34

more iterations [9,10]. An important alternative is to use the implicit numerical scheme e.g. Crank-Nicolson (C-N) or higher order accurate numerical methods [11]. The C-N method has advantages such as unconditional stability, second order accuracy in time and mathematical/computational competency to find the optimal time step size [11,12].

It is cumbersome to solve such systems using direct methods, because of the requirement of memory space and processing times. Hence, to achieve better accuracy in reasonable time, iterative solvers can be used [13]. The accuracy of a numerical method, in many ways, depends on the choice of iterative solver. In this respect, the Krylov subspace methods are more popular among the iterative solvers to solve non-symmetrical linear systems [14]. In numerical linear algebra, hybrid methods are blend of standard Krylov subspace methods. The Hybrid Bi-conjugate Gradient Stabilized method, often abbreviated as Hybrid BiCGStab is one such combination of Bi Conjugate Gradient method and low degree GM-RES method. To solve the non-symmetric linear systems numerically, Hybrid BiCGStab is developed by H. A. van der Vorst [15]. It has faster and smoother convergence in comparison to the original Bi-CG and its other variants.

This work proposes the application of the Hybrid BiCGStab for the task of denoising using the finite difference discretization of Bazan model and C-N scheme. We compare the Hybrid BiCGStab with another existing iterative solver viz. Successive-over-Relaxation (SOR) method, and with an implicit scheme to solve the Bazan model. In addition to better denoising performance (in terms of quantitative metrics), we also demonstrate a much higher efficiency of the proposed denoising method Hybrid BiCGStab. To best of our knowledge, this type of study in the field of image denoising has not been done yet. The paper is organized as follows. In section 2 we have given PDE model. The proposed numerical scheme for denoising is given in section 3 followed by iterative solvers in section 4.In section 5 we evaluate the experimental results and conclude the work with in section 6.

2   Related PDE-Model

To combine domain and range filtering with nonlinear diffusion, in 2007, Bazan et. al. [6] proposed combination of nonlinear diffusion and bilateral filtering, in which they replaced the Gaussian smoothed image by bilateral filtered image for diffusion coefficient. The model can be expressed as

$$
\left\{
\begin{array}{lll}
I_t & = \nabla(c(|\nabla I_{BF}|^2)\nabla I) & in \ \ \Omega \times (0, +\infty) \\
\frac{\partial I}{\partial n} & = 0 & in \ \ \partial\Omega \times (0, +\infty) \\
I(x, y, 0) & = I_0(x, y) & in \ \ \Omega
\end{array}
\right\}
\tag{1}
$$

$$
where \ \ \ I_{BF} = BF(I_p) = \frac{\sum_{q \in S} G_{\sigma s}(||p - q||)G_{\sigma r}(|I_p - I_q|)I_q}{\sum_{q \in S} G_{\sigma s}(||p - q||)G_{\sigma r}(|I_p - I_q|)}
$$

S is a spatial neighborhood of pixel p. The parameter σ_s and σ_r defines the amount of filtering to filter a pixel of an image I. Here Ω is the picture domain,

I_0 is the observed image, I is the original image to be recovered, n is the unit normal to the boundary of Ω and $c(s^2)$ is diffusion coefficient which diffuses the image, while the boundaries of image are preserved. Here we adopt the diffusion coefficient as [2]

$$c(s^2) = \frac{1}{1 + \frac{s^2}{\lambda^2}} \tag{2}$$

here λ is a contrast parameter and a positive constant.

3   Numerical Scheme

This section presents many aspects related to the numerical implementation of equation (1). To reduce the noise from a noisy image, we have applied Crank-Nicolson Finite Difference method to solve the Bazan model. The C-N method is combination of the forward Euler method at n^{th} step and the backward Euler method at $(n+1)^{th}$ step. The main idea of this method is based on the trapezoidal rule. Hence, our difference scheme is as follows.

Let h represents the spatial step size and τ is the time step. Denote $I_{i,j}{}^n = I(x_i, y_i, t_n)$ where $x_i = ih, y_j = jh$ and $t_n = n\tau$. Since the diffusion term is approximated by central differences, we use the following notations,

$$\frac{\partial I}{\partial x} = \frac{I_{i+1,j}^n - I_{i-1,j}^n}{2h}, \frac{\partial I}{\partial y} = \frac{I_{i,j+1}^n - I_{i,j-1}^n}{2h},$$
$$\frac{\partial^2 I}{\partial^2 x} = \frac{I_{i+1,j}^n - 2I_{i,j}^n + I_{i-1,j}^n}{h^2}, \frac{\partial^2 I}{\partial^2 y} = \frac{I_{i,j+1}^n - 2I_{i,j}^n + I_{i,j-1}^n}{h^2}$$

Then,

$$\nabla(c(|\nabla I_{BF}|^2)\nabla I)|_{i,j}{}^n = (C_N.\nabla_N I + C_S.\nabla_S I + C_W.\nabla_W I + C_E.\nabla_E I)|_{i,j}{}^n$$

$$\nabla(c(|\nabla I_{BF}|^2)\nabla I)|_{i,j}{}^{n+1} = (C_N.\nabla_N I + C_S.\nabla_S I + C_W.\nabla_W I + C_E.\nabla_E I)|_{i,j}{}^{n+1}$$

Our scheme is simply

$$I_{i,j}{}^{n+1} = I_{i,j}{}^n + \tau(\frac{\nabla(c(|\nabla I_{BF}|^2)\nabla I)|_{i,j}{}^n + \nabla(c(|\nabla I_{BF}|^2)\nabla I)|_{i,j}{}^{n+1}}{2})$$

For notations, see [2]. There are large number of unknowns in each system and the sparsity of the coefficient matrix suggests the use of an iterative solver.

4   Iterative Solvers

In this section, we focus on the iterative methods. Nowadays in the iterative methods, basic iterative solvers e.g. Jacobi, Gauss-Seidel methods [13] are frequently used for image denoising. These methods are easy to understand but the convergence is slow, moreover these methods are not able to solve all the linear systems. Whereas the Krylov subspace methods are quite complex to understand but it has the faster convergence [13]. The basic idea of the Krylov subspace method is to search for a good approximate solution of linear system from the subspace $span(b, Ab, A^2b, , A^{k-1}b)$, at iteration k. Hence, as discussed above, to solve the algebraic system of equations arising from the Crank-Nicolson finite difference discretization, we have chosen SOR and Hybrid BiCGStab method.

4.1   SOR Method

The Successive-over-Relaxation (SOR) method is a kind of relaxation methods. This method is generalized or improved form of the Gauss-Seidel (GS) method and formed by adding a relaxation parameter. When the solution at k^{th} iteration is known then we can calculate the solution at $(k + 1)^{th}$ iteration by applying SOR method $x^{k+1} = \omega \widehat{x}^{k+1} + (1 - \omega)x^k$. Where, $\widehat{x}^{k+1}$ is the GS solution at $(k + 1)^{th}$ iteration and ω is the relaxation parameter, which is chosen in such a manner that it will accelerate the convergence of the method towards the solution.

4.2   Hybrid BiCGStab Method

Hybrid Bi-Conjugate Gradient stabilized method [15] is the combination of Krylov subspace methods in which two steps (even and odd) for Bi-CG and one step for low degree GMRES is performed to solve the system of equations. The algorithm of Hybrid BiCGStab to solve the system

$$Ax = b$$

where $A \in \mathbf{R}^{N X N}$ and $x, b \in \mathbf{R}^N$. This method has the combine effect of Bi-CG and GMRES (2). The main idea is that after applying 2 successive Bi-CG steps, it is relatively easy to minimize the residual over that particular Krylov subspace. As a result, the Hybrid BiCGStab method leads only to significant residuals in the even-numbered steps and the odd-numbered steps do not lead necessarily to useful approximations. This method takes less computational memory due to requirement of less number of vector updates, inner-products and vector multiplication for full cycle. Also in the case of break-down, Hybrid BiCGStab method works better than BICGSTAB or its other variants [15].

5   Experimental Results

In this section we evaluate the proposed scheme and compare its results with those of the results obtained by Bazan model, with the stopping criteria proposed by Mrazek and Navara (MN method) [16]. In MN method, the stopping time for diffusion process is chosen in such a way that the correlation between the restored image $I(t)$ and noise is minimized and can be expressed as $T = arg \min_{t} corr(I(0) - I(t), I(t))$. Where $I(0)$ is the initial image. For experiments, we used the set of standard test images of size 256×256.

Fig. 1. Lena noise free image (Top Left) , noisy image with sigma= 0.01 (Top Right), restored image using Bazan model (Down left), SOR solver (middle) and Hybrid BiCGStab solver (Right)

Table 1. PSNR and MSSIM of the test images for Bazan Model with existing and proposed strategy

Images	Measure	Sigma=0.01			Sigma=0.015			Sigma=0.02			Sigma=0.025		
		Bazan Model	Proposed with SOR	Proposed with Hybrid BiCGStab	Bazan Model	Proposed with SOR	Proposed with Hybrid BiCGStab	Bazan Model	Proposed with SOR	Proposed with Hybrid BiCGStab	Bazan Model	Proposed with SOR	Proposed with Hybrid BiCGStab
Lena	MSSIM	0.8272	0.8275	**0.8277**	0.7972	0.7966	**0.7972**	0.7702	0.7711	**0.7712**	0.7569	0.7549	**0.7570**
	PSNR	28.30	28.32	**28.32**	27.28	27.27	**27.28**	26.43	26.46	**26.46**	25.89	25.86	**25.91**
Barb	MSSIM	0.7486	0.7480	**0.7487**	0.7154	0.7162	**0.7162**	0.6835	0.6846	**0.6846**	0.6651	0.6648	**0.6656**
	PSNR	25.96	25.96	**25.97**	25.16	25.19	**25.19**	24.48	24.51	**24.51**	23.92	23.94	**23.95**
Cameraman	MSSIM	0.8036	0.8023	**0.8037**	**0.7756**	0.7751	0.7754	0.7579	0.7589	**0.7589**	0.7322	0.7331	**0.7331**
	PSNR	**27.24**	27.18	27.23	**26.17**	26.14	26.16	25.57	25.59	**25.60**	24.69	24.66	**24.70**
Einstein	MSSIM	0.7026	0.7021	**0.7034**	0.6740	0.6741	**0.6752**	0.6463	0.6455	**0.6469**	0.6355	0.6356	**0.6384**
	PSNR	28.48	28.48	**28.49**	27.54	27.56	**27.56**	26.68	26.68	**26.69**	26.20	26.22	**26.23**

Table 2. Processing Time and Number of Iterations for C-Model

Images	Measure	Sigma=0.01		
		Bazan Model	Proposed with SOR	Proposed with Hybrid BiCGSath
Lena	Time(s)	128.9474	136.1605	30.0455
	Iterations	121	75	43
Barb	Time(s)	89.5359	95.8069	17.9623
	Iterations	112	69	40
Cameraman	Time(s)	143.2059	151.2620	42.6637
	Iterations	132	83	58
Einstein	Time(s)	155.3026	168.2245	38.1891
	Iterations	149	91	56

Fig. 2. Noise level v/s no. of iteration for Einstein image

Fig. 3. Lena noise free image (Top Left) , noisy image with sigma= 0.01 (Top Right), restored image using Bazan model (Down left), SOR solver (middle) and Hybrid BiCGStab solver (Right) for 40 iterations.

Fig. 4. Einstein noise free image (Top Left) , noisy image with sigma= 0.01 (Top Right), restored image using Bazan model (Down left), SOR solver (middle) and Hybrid BiCGStab solver (Right) for 40 iterations

Table 3. MSSIM and PSNR for 40 iterations

Images	Measure	Sigma=0.01		
		Bazan Model	Proposed with SOR	Proposed with Hybrid BiCGSatb
Lena	SSIM	0.6210	0.7726	0.8279
	PSNR	25.21	27.38	28.32
Barb	SSIM	0.6776	0.7427	0.7475
	PSNR	24.52	25.70	25.96
Cameraman	SSIM	0.6297	0.7685	0.8022
	PSNR	25.34	26.95	27.32
Einstein	SSIM	0.5519	0.6848	0.7191
	PSNR	25.44	27.77	28.69

Images are degraded with white Gaussian noise of zero mean and different standard deviations (e.g. $\sigma \in (0.01, 0.025)$). We make comparisons with the Bazan model. In that paper, the authors have used the similar σ values for noise levels [6]. For measuring the denoising performance quantitatively, we compute MSSIM [17] and PSNR [18].

In Fig. 1 , we show the results for $\sigma = 0.01$ for the different methods. For these results all the approaches were run till the stopping criteria is met. We can observe that the Hybrid BiCGStab method shows visually comparable outputs to the existing methods in restoring and edge-preservation. The proposed approach is indeed better in terms of PSNR and MSSIM, as shown in Table 1 for different noise levels. It is also observed that for higher noise standard deviation, the Hybrid BiCGStab method shows a better improvement in the PSNR and MSSIM values.

In addition to improved denoising, from Table 2 and Fig. 2 we can observe that the proposed algorithm satisfies the stopping criteria in the much lesser iterations as well as less computational time. The proposed Hybrid BiCGStab method requires 14 vector updates, 4 matrix-vector multiplications and 9 inner products for each iteration [15]. Hence, from above we may conclude that the proposed strategy takes much more efficient than the existing approaches for denoising for the Bazan model.

To reinforce the argument about denoising quality and efficiency, we show in Fig. 3-4 the denoising for a fixed number of iterations (40) across all the methods. The proposed method greatly outperforms the others, in this respect. This is also verfied quantitatively in Table 3. Thus, the hybrid BiCGStab method for denoising is clearly a favourable option so as to balance the trade-off between the denoising efficacy and computational efficiency.

6   Conclusion

In this paper, we proposed the use of the Hybrid BiCGStab solver for denoising using the Crank-Nicolson numerical scheme and Bazan model, and compared it with the SOR solver. We demonstrated good denoising performance and a high efficiency using the proposed denoising strategy based on Hybrid BiCGStab for different noise levels. These studies establish that advanced iterative solver can be used as an effective tool for image denosing in terms of producing better results with good efficiency.

References

[1] Witkin, A.P.: Scale-space filtering: A new approach to multi-scale description. In: IEEE International Conference on Acoustics, Speech, and Signal Processing, ICASSP 1984, vol. 9, pp. 150–153. IEEE (1984)

[2] Perona, P., Malik, J.: Scale-space and edge detection using anisotropic diffusion. IEEE Transactions on Pattern Analysis and Machine Intelligence 12(7), 629–639 (1990)

[3] Weickert, J.: Anisotropic diffusion in image processing, vol. 1. Teubner Stuttgart (1998)

[4] Kichenassamy, S.: The perona–malik paradox. SIAM Journal on Applied Mathematics 57(5), 1328–1342 (1997)

[5] Catté, F., Lions, P.L., Morel, J.M., Coll, T.: Image selective smoothing and edge detection by nonlinear diffusion. SIAM Journal on Numerical analysis 29(1), 182–193 (1992)

[6] Bazan, C., Blomgren, P.: Image smoothing and edge detection by nonlinear diffusion and bilateral filter. In: CSRCR 2007, vol. 21, pp. 2–15 (2007)

[7] Bazán, C., Miller, M., Blomgren, P.: Structure enhancement diffusion and contour extraction for electron tomography of mitochondria. Journal of Structural Biology 166(2), 144–155 (2009)

[8] Gh, M., Bazan, C., Frey, G.: olume se spatia esholdin xtraction (2010)

[9] Weickert, J., Romeny, B.T.H., Viergever, M.A.: Efficient and reliable schemes for nonlinear diffusion filtering. IEEE Transactions on Image Processing 7(3), 398–410 (1998)

[10] Thomas, J.W.: Numerical partial differential equations: finite difference methods, vol. 22. Springer (1995)

[11] Wang, W., Shi, Y.: An image denoising algorithm in the matrix form. In: 2011 International Conference on Multimedia Technology (ICMT), pp. 176–179. IEEE (2011)

[12] Crank, J., Nicolson, P.: A practical method for numerical evaluation of solutions of partial differential equations of the heat-conduction type. Advances in Computational Mathematics 6(3), 207–226 (1996)

[13] Saad, Y.: Iterative methods for sparse linear systems, 2nd edn. SIAM, Philadelphia (2003)

[14] Sickel, S., Yeung, M.C., Held, M.J.: A comparison of some iterative methods in scientific computing. Summer Reserch Apprentice Program (2005)

[15] Sleijpen, G.L., Van der Vorst, H.A.: Hybrid bi-conjugate gradient methods for cfd problems. Computational Fluid Dynamics Review 1995, 457–476 (1995)

[16] Mrázek, P., Navara, M.: Selection of optimal stopping time for nonlinear diffusion filtering. International Journal of Computer Vision 52(2-3), 189–203 (2003)

[17] Wang, Z., Bovik, A.C., Sheikh, H.R., Simoncelli, E.P.: Image quality assessment: from error visibility to structural similarity. IEEE Transactions on Image Processing 13(4), 600–612 (2004)

[18] Wang, Z., Bovik, A.C.: Mean squared error: love it or leave it? a new look at signal fidelity measures. IEEE Signal Processing Magazine 26(1), 98–117 (2009)

Assessment of Urbanization of an Area with Hyperspectral Image Data

Somdatta Chakravortty[1], Devadatta Sinha[2], and Anil Bhondekar[1]

[1] Govt. College of Engineering & Ceramic Technology, Kolkata, India
csomdatta@rediffmail.com
[2] Calcutta University, Kolkata, India

Abstract. This study attempts to apply time series hyperspectral data to detect change in landcover and assess urbanization of a small town in West Bengal, India. The objective is to utilize the potential of hyperspectral data to extract spectral signatures of the urban components of the study area using automated end member extraction algorithm, classify the area using Linear Spectral Unmixing (LSU) and assess the rate of urbanization that has taken place in the region over a period of 2 years. The automated target generation algorithm has successfully identified the pure spectra of 9 urban features after which their individual abundances in the hyperspectral imageries have been estimated. Post classification, the classes have been compared on a pixel by pixel basis and the increase/decrease in pixels noted. The change thus detected indicates a significant depletion in green cover and water bodies in the study area with increase in concrete cover over the years indicating rapid urbanization.

Keywords: Linear spectral unmixing, Urbanization, Change Detection.

1 Introduction

Urban landscapes are composed of a diverse assemblage of materials(concrete, metal, plastic, glass, water, grass, trees and soil) arranged by humans in complex ways to build housing, transportation system, utilities, commercial and industrial facilities and recreational landscapes. In many cases, urbanization is taking place at a rapid rate, often without planned development. Remote Sensing plays an important role in sensing these urban phenomena by putting into use the temporal, spatial and spectral resolution characteristics of the urban attributes. Urban applications are usually not as time sensitive as those dealing with highly dynamic phenomena such as vegetation where a life cycle might take place during a single season. For analyzing the rate of urbanization, imagery is required to collect data every year or every alternate year.

Hyperspectral remote sensing has been found to have great potential for detailed mapping of urban materials and their condition due to high spectral resolution of its sensors. Hyperspectral data provides images that are capable of resolving 242 spectral bands (from 0.4 to 2.5 μm) with a 10 nm spectral resolution and a spatial resolution of tens of metres. This imagery has helped resolve the spectral confusion that may arise between similar spectral signature types of urban land cover such as specific roof and

© Springer International Publishing Switzerland 2015
S.C. Satapathy et al. (eds.), *Proc. of the 3rd Int. Conf. on Front. of Intell. Comput. (FICTA) 2014*
– *Vol. 2*, Advances in Intelligent Systems and Computing 328, DOI: 10.1007/978-3-319-12012-6_35

road types. However, coarse spatial resolution of hyperspectral sensor may lead to different end members such as concrete, asphalt, vegetation etc. to be present within the Instantaneous Field Of View (IFOV) of a sensor system. This is disentangled using Linear Spectral Unmixing in which a pixel is broken up into its constituent sub pixels (end members) and their individual abundances estimated.

These biophysical materials and man-made features on the earth's surface may be dynamic and changing rapidly. Urbanisation is a result of these changes and need to be inventoried accurately so that the physical and human processes can be fully understood. Significant efforts have been made onto the development of change detection methods using remotely sensed data. Due to the necessity of monitoring change of Earth's surface features, research on change detection techniques has been an active topic in the past three decades and a large number of techniques have been developed, as summarized in the literature review papers [1-5].

The main objective of this study is therefore, to utilize the potential of hyperspectral data to extract spectral signatures of the urban components of the study area using automated end member extraction algorithm, NFINDR, classify the area using Linear Spectral Unmixing (LSU) and assess the rate of urbanization that has taken place in the region over a period of 2 years (2004-2006).

2   Study Area

A small town named Chandpur, Murshidabad (WB), India has been selected for study area(Figure 1) which is located in West Bengal in Eastern India, around 280 kilometers North from state capital Kolkata. It lies between the North Latitude 24°38'03", 24°37'24" and East Longitude 87°54'33", 87°55'15".

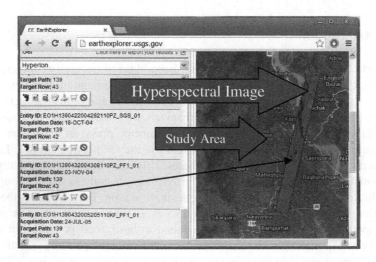

Fig. 1. Image of Study Area

3   Theoretical Background of Algorithms

3.1   N-FINDR Algorithm

The idea of the original N-FINDR algorithm is to find the pixels that can construct a simplex with the maximum volume, and these pixels will be considered end members. [6].
  The algorithm is briefly described as under:

- Consider n to be the number of end members to be generated.
- Let $(e_0^{(0)}, e_1^{(0)}, e_2^{(0)}, ..., e_n^{(0)})$ be a set of initial vectors randomly selected from the data. The volume of the simplex constructed by these vectors can be calculated as

$$V(E^0) = \frac{|det\ E^0|}{n!}$$

  where

$$E^{(0)} = \begin{bmatrix} 1 & 1 & ... & 1 \\ e_0^{(0)} & e_1^{(0)} & ... & e_n^{(0)} \end{bmatrix}$$

- At the i th stage, use the i th pixel r_i to replace each individual end member as a new simplex vertex and compute the resulting volume. If the maximum volume is larger than $V(E^{(i-1)})$ in the previous step and it appears when $e_j^{(i-1)}$ is replaced by r_i then r_i is used as e_j^i ; otherwise, go to the i+1 th stage to test the (i+1)th pixel r_{i+1}.
- The algorithm is stopped when all the pixels are tested.

4   Methodology

4.1   Acquisition and Preprocessing of Hyperspectral Data

An EO -1 Hyperion (hyperspectral) image of the study area has been acquired from the USGS through Data Acquisition Request (DAR). Hyperion data contain 242 spectral bands (from 0.4–2.5 μm) with a 10 nm spectral resolution and a spatial resolution of 30m. A time series hyperspectral data of the study area has been acquired on the dates: November 03, 2004, February 16, 2006 and October 25, 2006. ENVI software based QUAC (QUick Atmospheric Correction) has been applied for atmospheric correction and image to map registration is used for geo-registration of hyperspectral image of study area. In final stage of pre-processing Minimum Noise Fraction (MNF) is used for dimensionality reduction [7] of hyperspectral images.

4.2   Ground Survey

A ground survey of the study area has been made to identify urban features whose image based classification has been carried out. GPS (Global Positioning System) has been used to precisely locate the geographical coordinates of the study trail. Sample plots with 30m x 30 m spatial resolution that are identical to the size of Hyperion image (30 m resolution) in size were established in a pure cover of each end member and plotted on the imagery and spectral profile obtained.

4.3   Identification of Land Cover Classes

This step is to identify land cover classes (end members) of interest to be monitored and eventually place it in the change detection database. The end members of the study area are detected with NFINDR algorithm and different urban features such as concrete, water, soil, vegetation etc are extracted. For this process the set of selected hyperspectral bands obtained after noise removal and band reduction is used for detecting end members in the imagery [8], [9]. The end members identified in an unsupervised manner from the time series hyperspectral data are checked for accuracy by ground survey of the study area.

4.4   Linear Spectral Unmixing

In linear mixing model the spectrum of a mixed pixel is represented as a linear combination of constituent materials (also known as end member) spectra. This model mathematically represented as

$$Z = \sum_{i=1}^{E} a_i * e_i + n$$

Where $Z = \{z_1, z_2, z_3, \dots z_M\}^T$ is the M-dimensional spectra of a pixel, E is the set of end members, $e_i = \{e_1, e_2, e_3, \dots e_M\}^T$ is the set of spectra of end members, a_i denotes a fractional abundance of end member e_i in pixel Z and n is an additive noise [9] [10] .

This is a straightforward and efficient approach to decompose spectral signature of each pixel from the remotely sensed hyperspectral image. This model uses identified end members from the previous step to estimate the fractional abundance of each end member.

4.5   Change Detection

Post classification, the identified classes of time series hyperspectral data will be compared on a pixel by pixel basis and the increase/decrease in pixels noted. A difference between classes of two imageries is obtained after image subtraction and the extent of change in pixel count measured. The change thus detected will indicate a significant increase or decrease in land over the years [11].

5   Result and Analysis

The time series data acquired by the Hyperion sensor have been atmospherically and geometrically corrected[12] and dimensionally reduced using MNF. The locations of extracted end members (urban features) after application of NFINDR algorithm [13] on the dimensionally reduced imageries and their spectral profiles is shown in Figures 2(a), 2(b) respectively. The fractional abundance of end members of the individual time series imageries generated after application of LSU is shown in Fig 3(a-i).

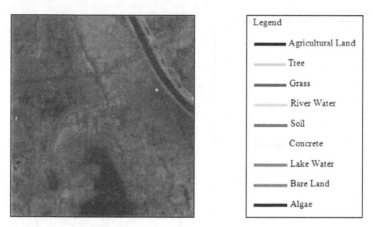

Legend

━━━ Agricultural Land

▬▬▬ Tree

━━━ Grass

━━━ River Water

━━━ Soil

Concrete

▬▬▬ Lake Water

▬▬▬ Bare Land

▬▬▬ Algae

Fig. 2(a). Location of pure end members after NFINDR

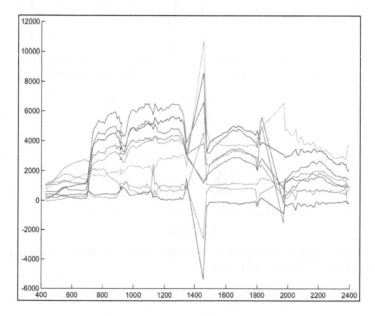

Fig. 2(b). Spectral signature of extracted end member

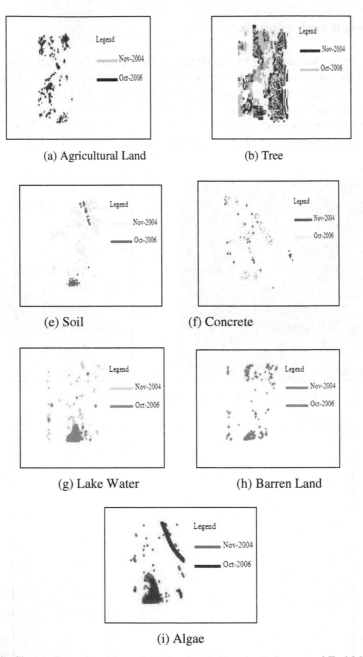

(a) Agricultural Land   (b) Tree

(e) Soil   (f) Concrete

(g) Lake Water   (h) Barren Land

(i) Algae

Fig. 3. (a-i) Change Detection Mapped in Fractional Abundance Images of End Members in Time Series Hyperspectral Data in 2 years

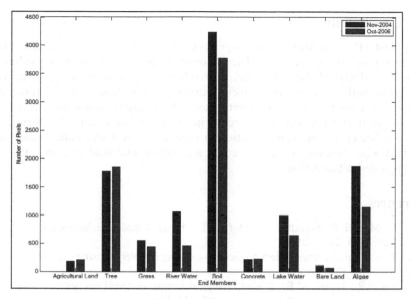

Fig. 4. Increase/Decrease in pixels of end members in 2 years

Table 1. Number of Pixels Comprising each Land Cover in Time Series Data

Contents	B	Increase	Decrease
Agricultural Land	▬▬▬	13.40%	
Tree	▬▬▬	4.62%	
Grass	▬▬▬		20.40%
River water	▬▬▬		56.90%
Soil	▬▬▬		4.33%
Concrete	▬▬▬	8.17%	
Lake Water	▬▬▬		35.49%
Barren Land	▬▬▬		45.87%
Algae	▬▬▬		38.65%

A comparison of change detected in end members from 2004 to 2006 has been summarized in Fig 4 and Table 1. It is observed that there is a subsequent increase in concrete cover in two years indicating urbanization. It is also observed that there is a 4.62% increase in tree cover which signifies afforestation in the study area. End members such as grass, river water, soil, sand, lake water and algae have shown a decreasing trend with increase in agricultural land area over the years.

6   Conclusion

The applied effort on time series hyperspectral imagery has assessed the rate of urbanization of the study area. The automated target generation algorithm has successfully identified the pure spectra of urban features such as concrete, water, green grass, soil etc after which their individual   abundances in the time series hyperspectral imageries have been estimated. Post classification, the classes have been compared and the increase/decrease in pixel number noted. The change thus detected indicates a significant depletion in vegetation cover and water bodies in the study area with increase in concrete cover and agricultural land area over the years indicating rapid urbanization.

References

1. Lu, D., Mausel, P., Brondízio, E., Moran, E.: Change detection techniques. Int. J. Remote Sens. 25, 2365–2407 (2004)
2. Singh, A.: Digital change detection techniques using remotely sensing data. Int. J. Remote Sens. 10, 989–1003 (1989)
3. Coppin, P.R., Bauer, M.E.: Digital change detection in forest ecosystems with remote sensing imagery. Remote Sens. Rev. 13, 207–234 (1996)
4. Coppin, P., Jonckheere, I., Nackaerts, K., Muys, B., Lambin, E.: Digital change detection methods in ecosystem monito ring: a review. Int. J. Remote Sens. 25, 1565–1596 (2004)
5. Kennedy, R.E., Townsend, P.A., Gross, J.E., Cohen, W.B., Bolstad, P., Wang, Y.Q., Adams, P.: Remote sensing change detection tools for natural resource managers: understanding concepts and tradeoffs in the design of landscape monitoring projects. Remote Sens. Environ. 113, 1382–1396 (2009)
6. Du, Q., Younan, N.H., King, R.L.: End-member extraction for hyperspectral image analysis, Doc. ID 93371, Optical Society of America (2008)
7. Maaten, L., Postma, E., Herik, J.: Dimensionality Reduction: A Comparative Review. Tilburg University Technical Report, TiCC-TR-2009-005 (2009)
8. Maselli, F.: Multiclass spectral decomposition of remotely sensed scenes by selective pixel unmixing. IEEE Transaction on Geoscience and Remote Sensing 36(5) (1998)
9. Dobigeon, N., Moussaoui, S., Coulon, M., Tourneret, J., Hero, A.: Joint Bayesian end member extraction and linear unmixing for hyperspectral imagery. IEEE Transaction on Geoscience and Remote Sensing 57(11) (2009)
10. Burazerovic, D., Heylen, R., Greens, B., Sterckx, S., Scheunders, P.: Detecting the adjacency effect in hyperspecpectral imagery with spectral unmixing technique. IEEE Transaction on Geoscience and Remote Sensing 6(3) (2013)
11. Ibrahim, A.: Using remote sensing technique (NDVI) for monitoring vegetation degradation in semi-arid lands and its relation to precipitation: Case Study from Libya. In: 3rd International Conference on Water Resources and Arid Environments and 1st Arab Water Forum (2008)
12. Chakravortty, S., Chakrabarti, S.: Preprocessing of Hyperspectral Data: A case study of Henry and Lothian Islands in Sunderban Region, West Bengal, India. Int. Journ. Geom. Geosc. 2(2), 490–501 (2011)
13. Chakravortty, S.: Analysis of end member detection and subpixel classification algorithms on hyperspectral imagery for tropical mangrove species discrimination in the Sunderbans Delta, India. J. Appl. Remote Sens. 7(1), 073523 (2013), doi:10.1117/1.JRS.7.073523

Range Face Image Registration
Using ERFI from 3D Images

Suranjan Ganguly[*], Debotosh Bhattacharjee, and Mita Nasipuri

Department of Computer Science and Engineering,
Jadavpur University, Kolkata- 700032, India
suranjanganguly@gmail.com, debotoshb@hotmail.com,
mnasipuri@cse.jdvu.ac.in

Abstract. In this paper, we present a novel and robust approach for 3D faces registration based on Energy Range Face Image (ERFI). ERFI is the frontal face model for the individual people from the database. It can be considered as a mean frontal range face image for each person. Thus, the total energy of the frontal range face images has been preserved by ERFI. For registration purpose, an interesting point or a land mark, which is the nose tip (or 'pronasal') from face surface is extracted. Then, this landmark is exploited to correct the oriented faces by applying the 3D geometrical rotation technique with respect to the ERFI model for registration purpose. During the error calculation phase, Manhattan distance metric between the localized 'pronasal' landmark on face image and that of ERFI model is determined on Euclidian space. The accuracy is quantified with selection of cut-points 'T' on measured Manhattan distances along yaw, pitch and roll. The proposed method has been tested on Frav3D database and achieved 82.5% accurate pose registration.

Keywords: 3D face, range image, registration, ERFI, Frav3D, Manhattan distance.

1 Introduction

Face recognition has progressed much from the last decade. But the illumination and pose is still a major drawback. With the advancement of technology, 3D face images can now be captured very easily. Acquisition of 3D face images with different file extension like, '.wrl', '.obj', '.bnt' etc. are captured by different existing 3D scanner such as, MINOLTA VIVID-700, DAVID SLS-1 etc. With the inherent advantages [2] of 3D face images, illumination problem can be eliminated. Now a day's, researchers are also focusing to 3D face images for more accurate and robust recognition purpose for biometric based security system. Recognition of frontal face images is relatively easier than images with pose variations. Orientated face images are required to be aligned as per same co-ordinate space of frontal face image to minimize intra-class variance. This alignment process is termed as registration.

---

[*] Corresponding author.

© Springer International Publishing Switzerland 2015

S.C. Satapathy et al. (eds.), *Proc. of the 3rd Int. Conf. on Front. of Intell. Comput. (FICTA) 2014*
– *Vol. 2,*Advances in Intelligent Systems and Computing 328, DOI: 10.1007/978-3-319-12012-6_36

The rest of the paper has been organized as follows. In section 2, the related work has been described by a set of comparison with proposed working methodology. Section 3 is used for the description of the proposed algorithm. In section 4, experimental results have been illustrated. Conclusion and future scope are summarized in section 5.

2   Related Work

3D face registration is the most significant step towards face recognition. A series of works has been studied and some relevant methods for registration are reported here. In [1], authors have registered the face images considering the properties of symmetry. Authors have considered the maximum distance of the nose tip from the curve ends and registered it using the nose-tip. S. A. Mahmood et.al [3] also proposed only nose tip based landmark detection from Frav3D database with 96% accuracy. In [4] Kakadiaris et al. proposed deformable model framework based expression oriented 3D face recognition on FRGC v.2. Gokberk et al. [5] also proposed non-rigid registration. Registration is performed using AFM (average face model) and different feature descriptor like surface normal, curvature, facial profile curves used to recognize 3D faces from 3D_RMA database. In [6], authors have used nose tip as a reference point, and registration is made by following one to all registration method where pose face images have been registered with the neutral faces corresponding to a particular subject from the FRAV3D database [7]. Thus, it can be concluded that it is an iterative process. The accuracy that has been reported by the authors is 75.84%.

In contrast to the brief study of the literature review, certain key factors of the proposed method have also been highlighted here. At first, it is not an iterative process for accurate facial pose registration. At first the rotated angle is determined then it is registered accordingly. Second, it is free from any type of training-testing phase. Third, It is noticed that the range face images created from original Frav3D database [7] are already having the highest depth value near nose region as because of it is the closest region to 3D scanners. Thus, based on maximum depth value near nose region, 'pronasal' landmark is identified. The measured accurate nose tip detection rate from the database is 97.85%. Fourth, considering ERFI as reference face model for each class, the rotated faces are registered. With the proposed energy level, the depth values of the nose region are highlighted more. Fourth, very small pose changes due to expressions are overlooked by newly generated Energy Range Face Image (ERFI). If a randomly selected frontal face is chosen as reference model for registration of rotated face images then it might so happen that instead of neutral face, frontal face with expression is selected as a reference model.

3   Proposed Method

There are four modules namely; image acquisition, ERFI generation, 'pronasal' landmark localization and face registration in the proposed methodology. The dataflow among these groups is shown in figure 1.

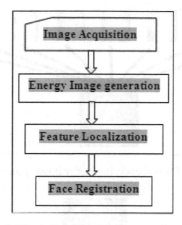

Fig. 1. Block diagram of the proposed method

3.1   3D Image Acquisition

For this research work, Frav3D [7] database is used. The database is having 106 subjects 3D face image in '.wrl' file format acquired by MINOLTA VIVID-700 scanner. Each of the individual dataset contains 16 different face images with varying expression, illumination and also poses. From available 3D face images from Frav3D database, 8 images are in frontal pose, such as neutral, expression as well as illumination and remaining 8 images are rotated 3D face images. From these 3D face dataset, range face images are created. All the 8 frontal posed range face images are used to create Energy Range Face Image (ERFI).

3.2   Energy Range Face Image Generation

The energy range image [12] is created by focusing only 8 frontal images from each person to localize the feature point. It is created from the equation 1. In this equation, 'n' is used to present as number of frontal range face images and $X(x, y)$ is the size of the range face image in terms of number of rows and columns.

$$ERFI(x, y) = \frac{\sum_{i=1}^{n} X(x, y)^i}{n} \tag{1}$$

$$where, x = 1,2,3 \ldots \ldots p \; and \; y = 1,2,3 \ldots \ldots \ldots q;$$

$$and \; p = q = 100 \; and \; n = 8$$

Before applying the equation 1, all the range images have been resized to fixed resolution (100 × 100) using bi-cubic interpolation method [11]. The phenomenon behind the creation of energy image is described in figure 2. The generated ERFI from randomly chosen subject from Frav3D database is illustrated in figure 3. This face model, for each class, is having the total frontal face energy of range face images and it is named as 'Energy Range Face Image' or abbreviated as 'ERFI'.

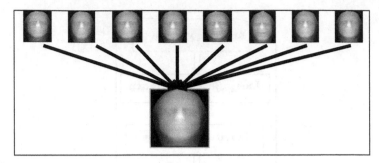

Fig. 2. Phenomenon for energy image generation

Fig. 3. Energy Range Face Image

3.3   Feature Localization

In very recent years, a different technique has been used for localizing the nose tip [8] from face region. Nose region is an obtrude section from face surface and carries a major role for its unique properties as a feature point. For our research work, searching of maximum depth value is done from ERFI paths to localize the nose tip. Feature localization is done using algorithm 1.

Algorithm 1. Nose tip localization

Input: Range Face Image Output: Location of Nose tip Step 1: for i=1 to m Step 2: for j=1 to n Step 3:   find maximum depth value Step 4: end Step 5: end Step 6: return (i, j);

This localization of nose tip is carried out in row major order, which is illustrated in figure 4. With this proposed methodology, the identification rate of 'pronasal' point from face region is 97.85% for Frav3D database.

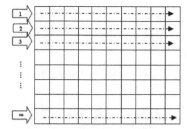

Fig. 4. Pronasal detection phenomenon

3.4   3D Faces Registration

Registering the oriented face images also requires algorithm-1 for feature localization. The localized feature from both the images i.e. ERFI and rotated image are pre-owned by 3D rotation techniques [10]. This phenomenon is established in figure 5.

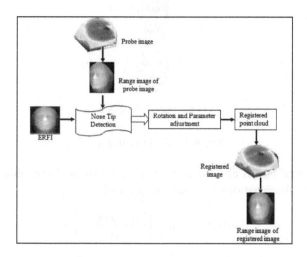

Fig. 5. Energy Image based registration

A face can be rotated along yaw, pitch, and roll. These are the alternative names to present the face orientation in the directions of Y, X and Z axes. In figure 6, the different rotations of 3D face images are shown.

For rotating the non-frontal face image to frontal pose, it is needed to adjust the angle between frontal and rotated face images that will be used by the rotational matrix for face registration purpose. In the proposed methodology, instead of frontal face, ERFI is used as reference face image for pose registration purpose.

Yaw- rotated face along Y axis

Roll- rotated face along Z axis

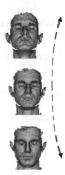

Pitch- rotated along X axis

Fig. 6. Rotated face images

If the non-frontal face is aligned along X-axis, Y-axis or Z-axis, then the angle can be determined by the equations 2, 3 and 4 respectively.

$$\theta = \pm \tan^{-1}\left(\frac{(y_i - y_{ERFI})}{h}\right) \tag{2}$$

$$\alpha = \pm \tan^{-1}\left(\frac{(x_i - x_{ERFI})}{h}\right) \tag{3}$$

$$\beta = \pm \tan^{-1}\left(\frac{(x_i - x_{ERFI})}{(y_i - y_{ERFI})}\right) \tag{4}$$

In all these equations (eq. 2, 3 and 4), (x_i, y_i) and (x_{ERFI}, y_{ERFI}) are the two points pairs, which are obtained from rotated face and ERFI by feature point identification method and 'h' is the depth value. The very next step, after, angle calculation, is to use it for rotation purpose. Rotational matrix to register the non-frontal face along X-axis is

$$R_X = \begin{bmatrix} 1 & 0 & 0 & 0 \\ 0 & \cos\theta & -\sin\theta & 0 \\ 0 & \sin\theta & \cos\theta & 0 \\ 0 & 0 & 0 & 1 \end{bmatrix} \qquad (5)$$

Rotational matrix to register the face orientated along Y-axis is

$$R_Y = \begin{bmatrix} \cos\alpha & 0 & -\sin\alpha & 0 \\ 0 & 1 & 0 & 0 \\ \sin\alpha & 0 & \cos\alpha & 0 \\ 0 & 0 & 0 & 1 \end{bmatrix} \qquad (6)$$

Rotational matrix to register the face for Z-axis is

$$R_Z = \begin{bmatrix} \cos\beta & \sin\beta & 0 & 0 \\ -\sin\beta & \cos\beta & 0 & 0 \\ 0 & 0 & 1 & 0 \\ 0 & 0 & 0 & 1 \end{bmatrix} \qquad (7)$$

4   Experimental Results and Discussion

A hypothesis is followed here for face pose registration. Head is being stiffed with body. For this reason, during different face pose orientation, the location of the pronasal landmark will always shift from a neutral position.

Among available 3D face images from Frav3D database, each subject from 8 frontal face images are used to generate ERFI and remaining face images, which are oriented along yaw, pitch, and roll, are considered for registration purpose (i.e. tested) by our proposed algorithm.

The accuracy measured by quantifying the Minimum Error Distance (MED). MED is based on distance metric. The Manhattan distance is computed between the locations of detected pronasal landmark points from ERFI and registered face range image. The equation that has been formulated is shown in 8.

$$MED = |i_1 - i_2| + |j_1 - j_2| \qquad (8)$$

Where (i_1, j_1) and (i_2, j_2) are the spotted pronasal landmark points from Energy Range Face Image and registered image respectively.

The Manhattan distance metric is dependent on the choice of co-ordinate rotation but independent of translation and rotation of coordinate system. The registered face images along with original unregistered face images are described with MED value in table 1. The overall measured accuracy for face pose registration is 82.5% whereas the average registration rate computed from table 3 is 78.33%. The overall registration accuracy is the ratio between the total number of registered and rotated 3D face images from the database.

Table 1. Face registered from Frav3D database

Unregistered face range image	Registered face range image	Face Orientation	Minimum error distance (MED)
		Along X-axis	1 pixel
			7 pixel
		Along Y-axis	12 pixel
			8 pixel
		Along Z-axis	11 pixel
			15 pixel

The MED value that has been acquired in pixel unit is derived by bringing the range image on Euclidian space and measuring the sum of the differences of the segment of the line between the points from the axes.

The cut-points are selected for accuracy measurement by the investigation made on MED result set. Three different cut-points or threshold values are selected to measure the accuracy of face registration along three axes (X, Y and Z) with respect to energy image. In table 2, threshold values are characterized.

Table 2. Characteristics of threshold value

Face rotation	Selected Threshold Value
Along X-axis (+ve and –ve orientation)	10 pixel
Along Y-axis (+ve and –ve orientation)	20 pixel
Along Z-axis (+ve and –ve orientation)	15 pixel

It is observed that, considering only the neutral face images as a reference face model, during registration accuracy measurement with this selected threshold values reduces the accuracy to 76.25%.

Another set of investigations is carried on to measure the registration accuracy along individual axes (X, Y and Z) i.e. pitch, yaw and roll and it is represented in table 3.

Table 3. Characteristics registration accuracy

Face rotation	Registration Accuracy (%)
Along X-axis	80
(+ve and –ve orientation)	
Along Y-axis	95
(+ve and –ve orientation)	
Along Z-axis	60
(+ve and –ve orientation)	

In figure 7, a summary of the registration accuracy from variation of measurements by the proposed algorithm is depicted.

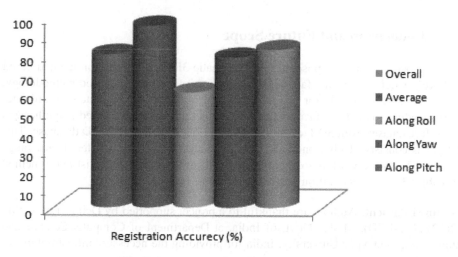

Fig. 7. Summary of the registration accuracy

4.1 Effect of Gaussian Blur on ERFI

The ERFI has also been modified by blurring it to acknowledge another challenging issue towards robust face registration. Especially, Gaussian blur is applied on ERFI (shown in figure 3). Blurring is the technique to remove noise and therefore some image data is also pulled up. For this research work, this challenging issue is also studied. The blurred Energy Range Face Image is displayed in figure 8.

Fig. 8. Blurred ERFI

During this research work, Gaussian kernel with same size of ERFI is convolved with standard deviation 3.0. The equation is shown in equation 9.

$$G(x, y) = \frac{1}{2\pi\sigma^2} e^{-x^2+y^2/2\sigma^2} \tag{9}$$

The 'x' is the distance from origin in the horizontal axis and 'y' is the distance from the origin in the vertical axis. The 'σ' is the standard deviation of the Gaussian blur.

It is observed that, the blur effect does not create any hurdle to our algorithm to locate the control point (i.e. pronasal). There are no differences between the located nose tips from generated ERFI as well as modified ERFI. Therefore, the registration accuracy that has been notated earlier is also preserved during this challenge.

5   Conclusion and Future Scope

In this paper, we have proposed a fully automatic 3D face registration method based on ERFI model. During the face registration process with the proposed methodology, separated ERFI model for each individual subject is generated to register corresponding rotated 3D face images. Thus, the authors have created a synthesized range face dataset from 3D face images that are gathered from Frav3D database. This dataset is consists of all frontal face range images as well as registered face range images. The future work is to explore the application of 3D face registration method for robust 3D face recognition purpose from range images.

Acknowledgment. Authors are thankful to a project supported by DeitY (Letter No.: 12(12)/2012-ESD), MCIT, Govt. of India, at Department of Computer Science and Engineering, Jadavpur University, India for providing the necessary infrastructure for this work.

References

1. Spreeuwers, L.: Fast and Accurate 3D Face Recognition Using registration to an Intrinsic Coordinate System and Fusion of Multiple Region Classifiers. Int. J. Comput. Vis. 93, 389–414 (2011), doi:10.1007/s11263-011-0426-2
2. Ganguly, S., Bhattacharjee, D., Nasipuri, M.: 3D Face Recognition from Range Images Based on Curvature Analysis. ICTACT Journal on Image and Video Processing 04(03), 748–753 (2014), Number (Print): 0976-9099, ISSN Number (Online): 0976-9102

3. Mahmood, S.A., Ghani, R.F., Kerim, A.A.: Nose Tip Detection Using Shape index and Energy Effective for 3d Face Recognition. International Journal of Modern Engineering Research (IJMER) 3(5), 3086–3090 (2013) ISSN: 2249-6645
4. Kakadiaris, I.A., Passalis, G., Toderici, G., Murtuza, M.N., Lu, Y., Karampatziakis, N., Theoharis, T.: Three-dimensional face recognition in the presence of facial expressions: an annotated deformable model approach. IEEE Transactions on Pattern Analysis and Machine Intelligence 29(4), 640–649 (2007), doi:10.1109/TPAMI.2007.1017
5. Gökberk, B., İrfanoğlu, M.O., Akarun, L.: 3D shape-based face representation and feature extraction for face recognition. Image and Vision Computing 24, 857–869 (2006)
6. Bagchi, P., Bhattacharjee, D., Nasipuri, M., Basu, D.K.: A Method for Nose-tip based 3D face registration using Maximum Intensity algorithm. In: Proc of International Conference of Computation and Communication Advancement 2013, JIS College of Engineering, January 11-12 (2013)
7. Frav3D database, http://www.frav.es/databases/FRAV3d/
8. Anuar, L.H., Mashohor, S., Mokhtar, M., Wan Adnan, W.A.: Nose Tip Region Detection in 3D Facial Model across Large Pose Variation and Facial Expression. IJCSI International Journal of Computer Science Issues 7(4(4)) (July 2010) ISSN (Online): 1694-0784, ISSN (Print): 1694-0814
9. Zeptycki, P., Ardabilian, M., Chen, L.: A coarse-to-fine curvature analysis based rotation invariant 3D face landmarking. In: Proceedings of IEEE Conf. Biometrics: Theory, Applications and Systems, Washington (2009)
10. Hearn, D., Baker, M.P.: Computer Graphics, 2nd edn.
11. Gonzalez, R.C., Woods, R.E.: Digital Image Processing, 3rd edn.
12. Ganguly, S., Bhattacharjee, D., Nasipuri, M.: 2.5D Face Images: Acquisition, Processing and Application. In: Computer Networks and Security, International Conference on Communication and Computing (ICC 2014), pp. 36–44. Elsevier Science and Technology (June 2014) ISBN: 9789351072447

Emotion Recognition for Instantaneous Marathi Spoken Words

Vaibhav V. Kamble[1], Ratnadeep R. Deshmukh[2], Anil R. Karwankar[1],
Varsha R. Ratnaparkhe[1], and Suresh A. Annadate[3]

[1] Dept. of Electronics and Tele-communication
Government College of Engineering,
Aurangabad, Maharashtra, India
[2] Dept. of Computer Science & Information Technology
Dr. Babasaheb Ambedkar Marathwada University,
Aurangabad, Maharashtra, India
[3] Dept. of Electronics & Telecommunication
Jawaharlal Nehru Engineering College,
Aurangabad, Maharashtra, India
kamblevv@gmail.com

Abstract. This paper explore on emotion recognition from Marathi speech signals by using feature extraction techniques and classifier to classify Marathi speech utterances according to their emotional contains. A different type of speech feature vectors contains different emotions, due to their corresponding natures. In this we have categorized the emotions as namely Anger, Happy, Sad, Fear, Neutral and Surprise. Mel Frequency Cepstral Coefficient (MFCC) feature parameters extracted from Marathi speech Signals depend on speaker, spoken word as well as emotion. Gaussian mixture Models (GMM) is used to develop Emotion classification model. In this, recently proposed feature extraction technique and classifier is used for Marathi spoken words. In this each subject/Speaker has spoken 7 Marathi words with 6 different emotions that 7 Marathi words are Aathawan, Aayusha, Chamakdar, Iishara, Manav, Namaskar, and Uupay. For experimental work we have created total 924 Marathi speech utterances database and from this we achieved the empirical performance of overall emotion recognition accuracy rate obtained using MFCC and GMM is 84.61% rate of our Emotion Recognition for Marathi Spoken Words (ERFMSW) system. We got average accuracy for male and female is 86.20% and 83.03% respectively.

Keywords: Emotion Recognition, Mel Frequency Cepstral Coefficient, Gaussian mixture models, speaking rate, Marathi Speech Emotional Database.

1   Introduction

Humans express their emotions by speech, body language and facial expression. Speech signal contains not only the linguistic information but also emotions of her or his voice from last decades researchers have work on automatic speech emotion recognition topic in the Human Computer Interaction (HCI) field.

© Springer International Publishing Switzerland 2015   335
S.C. Satapathy et al. (eds.), *Proc. of the 3rd Int. Conf. on Front. of Intell. Comput. (FICTA) 2014*
– *Vol. 2*, Advances in Intelligent Systems and Computing 328, DOI: 10.1007/978-3-319-12012-6_37

As per eighth schedule of Government of India, there are 22 official languages. These are Assamese, Bengali, Gujarati, Hindi, Kannada, Kashmiri, Konkani, Malayalam, Manipuri, Marathi, Nepali, Oriya, Punjabi, Sanskrit, Sindhi, Tamil, Telugu, Urdu, Bodo, Dogri, Maithili and Santhali. Apart from these, there are about 600 dialects spoken in India. Marathi is spoken by about majority population of Maharashtra. All the government work of Maharashtra followed by mainly in Marathi. Devnagari script is used for writing Marathi .Marathi is mostly phonetic in nature i.e. there is one to one correspondence between written symbols and the spoken utterances. Marathi phonemes can be divided into vowels and consonants. There are ten pure oral vowels (a, e, i, o, u). All these vowels have their nasalized forms also. Creaky and whispered vowels are used rarely [1].

Emotion speech recognition from speech is very critical challenge due to speaking rate, speaking style and clear voice quality of speaker. Because of the existence of the unlike sentences, speakers, speaking styles, speaking rates accosting variability was introduced, because of which speech features get directly influence. The same utterance may show different emotions. Therefore it is very complicated to differentiate these portions of utterance. Emotion expression is depending on his or her culture and environment. As the culture and environment gets change the speaking manner also gets change, which is another challenge in front of the speech emotion recognition system [2],[3].

Applications of Speech Emotion Recognition include psychiatric diagnosis, intelligent toys, lie detection, learning environment, educational software, and detection of the emotional state in telephone call center conversations to provide feedback to an operator or a supervisor for monitoring purposes ,the most important application for the automated recognition of emotions from the speech, in car board system where information of the mental state of the driver may provide to the system to start his/her safety [4],[5].

1.1   Speech Emotion Recognition Systems

The proposed human emotion recognition system is of five components: input speech signal, pre-processing, feature extraction and selection, classification and finally emotions recognition. The Architecture of Emotional speech recognition system is as shown in figure 1. The Accuracy of the Emotional speech recognition system is based on the level of naturalness of the database which is used as an input to the speech emotion recognition system. The database as an input to the speech emotion recognition system may contain the real world emotions or the acted ones. It is more practical to use database that is collected from the real life situations [3].

The emotion recognizer system identifies the emotion state of the input speech signal and displays the corresponding emotions of that particular speech. The speech processor, deals with the emotion features selection, speech preprocessing, and extraction algorithm. Determining emotion features is a vital issue in the emotion recognizer design. The emotion recognition result is strongly depending on the emotional features and which provided to various types of classifier that have been used to represent the emotion [6].

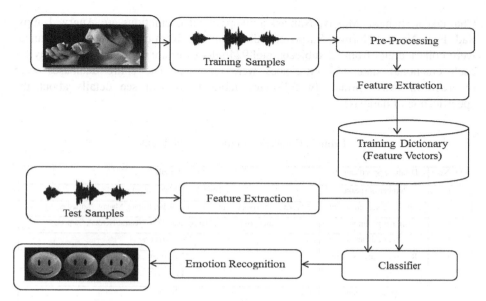

Fig. 1. Architecture of Speech Emotion Recognition System

2   Experimental Work

System Outline: It is observed that most of the studies have been done for English and other languages. There is also a need to study these aspects for Marathi speech. This system having approach to recognize emotions from Marathi spoken words. The project includes stages from the recording of Marathi spoken words emotional database up to the development system and performance evaluation. The block diagram of Emotion Recognition for Marathi Spoken Words is shown in figure 1. This system consists of following stages.

- Data recording and collection
- Pre-processing
- Feature extraction
- Classification
- Recognition.

2.1   Data Collection

The present work aims to investigate the recognition of emotion from Marathi Speech. A headphone-mic, a Laptop computer, the Sony Sound Forge 7 software were used for single channel recording of seven different words of six different emotion from 22 subjects of 11 male and 11 female speakers of Marathi language, in closed-room noise-free environment. For digitization 16000 Hz of sampling frequency and 16-bit quantization were used. Each subject has spoken 7 Marathi words with 6 different emotions i.e. 7 Marathi words are Aathawan, Aayusha,

Chamakdar, Iishara, Manav, Namaskar, Uupay and 6 emotions are Angry, Happy, Sad, Fear, Neutral/Normal, Surprise. This means that from each subject42 utterances were collected and from 22 subjects total 924 utterances are collected. In this way for each emotional state 154 utterances were recorded. The entire databases were recorded in .wav format. In following table 1 we can see details about the specification of database.

Table 1. Database specification and details

Sr. No.	Database Specifications	Details
1	Number of speaker	Total 22;(11Male , 11Female)
2	Number of Marathi Words	7 Words (Aathawan, Aayusha, Chamakdar, Iishara, Manav, Namaskar, Uupay)
3	Number of Emotions	6 Emotions (Angry, Happy, Sad, Fear, Neutral/Normal, Surprise)
4	Size of database	924 utterances
5	Recording environment	Class room
6	Sampling rate	44.1 KHz
7	Speech Coding	16 Bit
8	Recording Device	Microphone / SONY Voice Recorder
9	Channel	Mono

2.2   Feature Extraction Technique

We want to extract Feature vectors from the input speech. We used the (MFCC) algorithm to extract the features. Extraction of MFCC we transform the input waveform into a sequence of acoustic feature vectors, each vector representing the information in a small time window of the signal. Now that we have a digitized, quantized representation of the waveform, now we are prepared to extract MFCC features. The complete pipeline steps for MFCC process in figure 2[7].

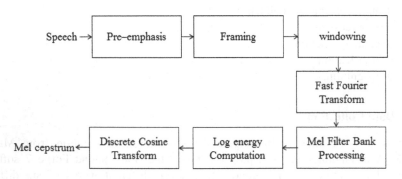

Fig. 2. MFCC Pipeline Steps

Steps 1: Pre–emphasis

In this step signal is passing through a first-order high-pass filter. This process will increase the energy of signal at higher frequency. In the time domain, with input x[n] and $0.9 \leq \alpha \leq 1.0$

$$y[n] = x[n] - \alpha \times x[n-1] \tag{1}$$

Where α is the pre-emphasis parameter.

Steps 2: Framing

Process of segmenting the speech samples obtained from analog to digital conversion (ADC) into a small frame with the length within the range of 25 msec.

The voice signal is divided into frames of N samples. Adjacent frames are being separated by M (M<N). Typical value used are M=100 and N=256.

Step 3: Hamming windowing

In order to reduce the discontinuities of the speech signal at the edges of each frame, a tapered window is applied to each frame. The most common used window is hamming window. Windowing makes waveform more smoother. The Hamming window equation is given as,

If the window is defined as W (n), $0 \leq n \leq N-1$ where
N = number of samples in each frame
Y[n] = Output signal
X (n) = input signal
W (n) = Hamming window, then the result of windowing signal is given below:
Hamming-window

$$Y(n) = X(n) \times W(n) \tag{2}$$

$$w[n] = 0.54 - 0.46 \cos\left(\frac{2\Pi n}{L}\right); \ 0 \leq n \leq L-1 \tag{3}$$

$$= 0; \ otherwise$$

Step 4: Fast Fourier Transform

FFT which convert each frame of N samples from time domain into frequency domain to obtain its magnitude. For each of N discrete frequency bands, is a complex number X[k] representing the magnitude and phase of that frequency component in the original signal.

$$X[k] = \sum_{n=0}^{N-1} x[n] e^{-j2\frac{\Pi}{N}kn} \qquad k = 0,1,2 \dots N-1 \tag{4}$$

FFT is computationally efficient algorithm of Discrete Fourier transform (DFT).

Step 5: Mel Filter Bank Processing

In this step for getting the frequency resolution to a perceptual frequency scale which satisfies the properties of human ears such as a perceptually Mel frequency scale. This issue corresponds to Mel-filter bank stage. Filter bank analysis is consisting of a set of Band pass filter. The filter bank is set of overlapping triangular band pass filter, that according to Mel frequency scale, The center frequency of these filters are linear equally-spaced below 1KHz and logarithmic equally-spaced above 1KHz.

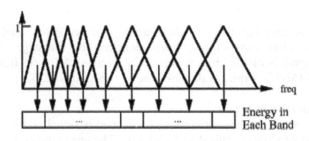

Fig. 3. Mel scale filter bank (young et al, 1997)

Each filter's magnitude frequency response is triangular in shape. Then each filter output is the sum of its filtered spectral components. After that the following equation is used to compute the Mel for given frequency f in HZ. A Mel is unit of pitch.

$$mel(f) = 2595 \times \log 10 \left[\frac{(1 + f)}{700} \right] \quad (5)$$

Step 6: Log energy Computation

Log energy computation is done by computing logarithm of the square magnitude of the output of Mel filter bank. Thus the input to Mel filter bank is the power spectrum of each frame, such that for each frame a log –spectral energy vector is obtained as output of the filter bank analysis.

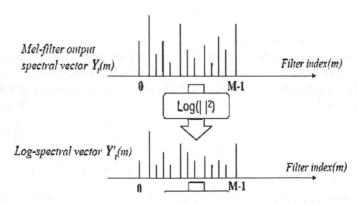

Fig. 4. Log Computation

Step 7: Discrete Cosine Transform

This process converts the log Mel Spectrum into time domain using Discrete Cosine Transform (DCT) or inverse DFT. The result of the conversion is called Mel Frequency Cepstrum Coefficient. The set of coefficient is called acoustic vectors. Therefore, each input utterance is transformed into a sequence of acoustic vector. By taking 1'st order derivative (Delta) and 2'nd order derivative (Double-delta) of acoustic vector of each frame. At last we got total 39 features from each frame. In general we'll just use the first 12 cepstral coefficients.

2.3   Classification

In this experimental work we are used Gaussian mixture Model (GMM) classifiers to identify the emotional state of spoken utterances.

GMM: Gaussian mixture model is a probabilistic model. GMMs are very efficient in modeling multi-modal distributions and their training and testing requirements are not as much of continuous HMM. Therefore we used Gaussian Mixture Model (GMM) classifiers to identify the emotional state of spoken utterances. For identify the emotional state of Marathi spoken utterances. Gaussian mixture density model consists of M component density tuples, each component has tree parameter: Mean vector, covariance matrix, and mixture weight [8].

Here, A, referred to the set of M component density tuples,

$$\lambda = \{p_i, u_i, \Sigma_i\}, \quad = 1, \ldots, M \tag{6}$$

And we define d-dimension basic component densities bi(x), i = 1, ..., M as follows:

$$b_i(x) = \frac{1}{(2\pi)^{d/2}|\Sigma_i|^{1/2}} \exp\left\{-\frac{1}{2}(x-u_i)^T \Sigma_i^{-1}(x-u_i)\right\}, \quad i=1,\ldots,M \tag{7}$$

Given the M component density tuples, we will get one Gaussian mixture model formed by a weighted sum of probability density functions,

$$p(x \mid \lambda) = \sum_{i=1}^{M} p_i b_i(x) \tag{8}$$

GMM with each basic density can imitate some characteristics of phonations and it can smoothly approximate any density of shape. In this, we extract MFCCS feature from emotion speech by plotting the 2D distribution of GMM using 8 mixtures. The main purpose of the emotion model training is to find out a set that collects the weight of the mean vector and mixture from the training speech utterance of each emotion. It can estimate the parameter of GMM by maximum likelihood estimation (ML estimation)[8].

K-means algorithm: Moreover, the k-means clustering is used here to minimize the each square error to each cluster center. Finally, a set of code vectors {C1, C2 ..... Ck} is the representative reference code of the emotional speech. The training process of emotional utterance is to repetition the above all steps as the formation of reference codes of emotional utterances. For emotional identification, the main purpose is to find the representative GMM which identify implicit emotion with the maximum a probability. We estimate the maximum likelihood of GMM parameters in each training emotional utterance, and obtained GMM to represent the different emotions by EM algorithms [8].

2.4   Recognition

Emotion recognition is done by calculating Euclidian distance. Matching test samples feature vector with classified train sample feature vector data base here we are using simple distance classifier i.e. calculating Euclidean distance [9].

Euclidean distance is used as distance parameter to matching with classified template. Let us consider x_r is training feature vector which is stored into database and y_t is the test feature vector of the signature being tested against the system. Euclidean distance between pair of vectors E_d is given by following mathematical equation as:

$$E_d = \sqrt[2]{\sum (x_r - y_t)^2} \tag{9}$$

Minimum Euclidean distance decides the emotional state of the test sample.

3   Experimental Results

The performance of Emotion Recognition for Marathi Spoken Words (ERFMSW) is in terms of emotion accuracy rate and emotion confusion rate. We collected speech emotional Marathi database from 22 subject's i.e. 11 male & 11 female. From each subject we got 42 utterances. In this way we got the total database of 924 utterances. Performance evaluation of this system from out of 924 utterances; 792 utterances for training and 132 utterances for testing are used.

The feature vectors sample speech signal can never be exactly the same as those provided during the enrollment or training phase for same user. This requires effective feature extraction and feature matching algorithm. The performance of Emotion Recognition for Marathi Spoken World system is measured in certain standard terms. These are Emotion Acceptance Rate (EAR), Emotion Confusion (Rejection) Rate (ECR).

3.1   Performance Parameter Evaluation

The performance of system is determined based on the accuracy of classification between the training and testing speech database. Evaluation parameters for Emotion Recognition system are FAR and FCR. Standard definitions of performance evaluation parameters i.e. Emotion Acceptance Rate, Emotion Confusion Rate is as follows:

$$EAR = \frac{Number\ of\ Emotional\ utterances\ are\ accurately\ Recognized}{Total\ number\ of\ utterance\ are\ tested} \tag{10}$$

$$ECR = \frac{Number\ of\ Emotional\ utterances\ are\ confusionly\ Recognized}{Total\ number\ of\ utterance\ are\ tested} \tag{11}$$

Now training all the 792 utterances of training data are trained and shows their training graph for each emotional state. Below we can see Feature Extraction and Training Graph for different Emotions states. Surprise Emotional State: MFCC Feature Extraction graph and training graph for Surprise Emotion is shown in figure 5 and figure 6 respectively.

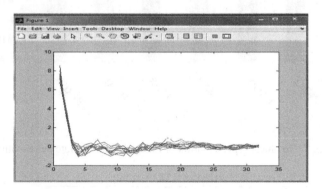

Fig. 5. Feature extraction graph for surprise emotion

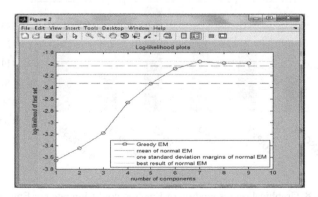

Fig. 6. Training graph for surprise Emotion

In such way we got different feature extraction and training graph for happy, sad, fear, neutral and surprise emotion. We are testing 22 samples for each emotional state. In this way all 132 test samples from test database are tested. Now we are doing performance evaluation of total system. Standard definitions of performance evaluation parameters i.e. Emotion Acceptance Rate, False Confusion Rate is as follows equations 10 and 11. After computing emotion acceptance and confusion rate we achieved 84.61% Emotion Acceptance Rate (EAR) and 15.38% Emotion Confusion Rate (ECR). Overall emotion accuracy rate and confusion rate in percentage for each emotion in statistical and graphical representation in figure 7 and figure 8 respectively. We got average accuracy for male and female is 86.20% and 83.03% respectively it is graphically shown in figure 9.

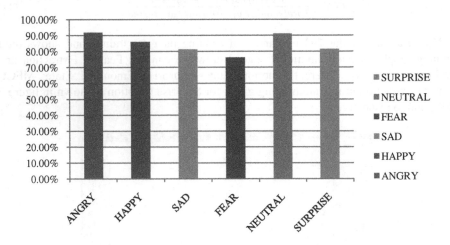

Fig. 7. Graphical representation of Emotion accuracy rate per Emotion

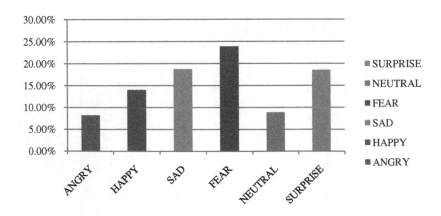

Fig. 8. Graphical representation of Emotion confusion rate per Emotion

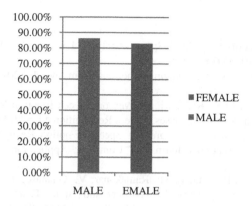

Fig. 9. Graphical representation of Emotion accuracy rate for male and female

This system is used for emotion recognition in Marathi Spoken Words by applied feature extraction techniques as MFCC and classification techniques as GMM. We got 84.61 % average accuracy rate and 15.38% average confusion rate of our system. We got average accuracy for male and female is 86.20% and 83.03% respectively. From the system for each emotion we got different accuracy rate i.e. for Angry, Happy, Sad, Fear, Neutral and Surprise are 91.80%, 85.98%, 81.25%, 76.14%, 91.08%and 81.45% respectively.

4   Conclusion

This proposed work of Emotion Recognition for Marathi Spoken Worlds (ERFMSW) system based on feature extraction and classification techniques. Here we are extracting MFCC feature from Marathi speech signal for different emotional state. For classification or training speech data we use GMM. Resultant Minimum Euclidian distance between training feature vectors and testing feature vectors decides the emotional state of that utterance. Each subject/Speaker has spoken 7 Marathi words with 6 different emotions. This means that from each subject 42 utterances were collected and from 22 subjects total 924 utterances are collected. In this way for each emotional state 154 utterances were recorded. For the evaluation of this system from out of 924 utterances; 792 utterances for training and 132 utterances for testing are used. Evaluation of Emotion Recognition for Marathi Spoken Words (ERFMSW) system is done with calculating EAR (Emotion Accuracy Rate) and ECR (Emotion Confusion Rate). Finally we perceive the high Performance of overall emotion recognition accuracy rate is 84.61% and we got 15.38% emotion confusion rate. We got average accuracy for male and female is 86.20% and 83.03% respectively. From this system for each emotion we got different accuracy rate i.e. for Angry, Happy, Sad, Fear, Neutral and Surprise are 91.80%, 85.98%, 81.25%, 76.14%, 91.08%and 81.45% respectively.

References

1. Agrawal, S.S.: Emotions in Hindi Speech-Analysis, Perception and Recognition (2011) 978-1-4577-0931-9/11© IEEE
2. Pathak, S., Kulkarni, A.: Recognizing emotions from speech (2011) 978-1-4244-8679-3/11© IEEE
3. Ayadi, M.E., Kamel, M.S., Karray, F.: Survey on Speech Emotion Recognition: Features, Classification Schemes, and Databases. Pattern Recognition 44, 572–587 (2011)
4. Chavhan, Y., Dhore, M.L., Yesaware, P.: Speech Emotion Recognition Using Support Vector Machine. International Journal of Computer Applications (0975 - 8887) 1(20) (2010)
5. Chew, L.W., Seng, K.P., Ang, L.-M., Ramakonar, V., Gnanasegaran, A.: Audio-Emotion Recognition System using Parallel Classifiers and Audio Feature Analyzer. In: Third International Conference on Computational Intelligence, Modelling & Simulation (2011)
6. Razak, A.A., Yusof, M.H.M., Komiya, R.: Towards Automatic Recognition of Emotion in Speech. In: Proceedings of the 3rd IEEE International Symposium of Conference on Signal Processing and Information Technology, ISSPIT 2003, pp. 548–551 (2003)
7. Jain, A., Prakash, N., Agrawal, S.S.: Evaluation of MFCC for Emotion Identification in Hindi Speech (2011) 978-1-61284-486-2/111© IEEE
8. LehLuoh, Yu-Zhe Su and Chih-Fan Hsu.: Speech signal processing based emotion recognition. In: International Conference on System Science and Engineering (2010) 978-1-4244-6474-61101, IEEE
9. Attabi, Y., Dumouchel, P.: Anchor Models for Emotion Recognition From Speech (2013), doi:10.1109/T-AFFC, 1949-3045/13© IEEE

Performance Evaluation of Bimodal Hindi Speech Recognition under Adverse Environment

Prashant Upadhyaya[1], Omar Farooq[1], M.R. Abidi[1], and Priyanka Varshney[2]

[1] Department of Electronics, AMU-Aligarh, India
[2] Department of Electronics, GLA University, Mathura, India
{upadhyaya.prashant,abidimr}@rediffmail.com,
{omarfarooq70,priyankavarshney}@gmail.com

Abstract. Designing of a robust Human-Computer Interaction (HCI) system is a challenging task,especially for automatic speech recognition (ASR) when working under unfriendly environment.This paper proposesan ASRsystem which uses bimodal information (i.e. Speech along with the visual input) resulting inimproved robustness. In thisresearch staticand dynamic (Δ) audio features are extracted using the Mel-Frequency Cepstral Coefficients (MFCC).The visual feature isextracted using Two-Dimensional Discrete Wavelet Transform (2D-DWT). Audio-video recognition is performed over different combination of visual feature using HMM (Hidden Markov Model) under clean and noisy environmental conditions.Aligarh Muslim University Audio Visual (AMUAV) Hindi database has been chosen as the baseline data. In addition, noisy speech signal performance is evaluated for different Signal to Noise Ratio (SNR: 30 dB to -20 dB). At last, addition of visual information to ASR is reported to increase the accuracy when working under smart assistive environment, i.e. for applications, which may not have the noise-free background condition.

Keywords: HCI, Bimodal, MFCC, 2D-DWT, ASR, AMUAV.

1 Introduction

Traditionally, the Human-Computer Interaction (HCI) represents the cornerstone of advanced technology making it more natural, free-flowing, and seamlessly integrated into our day-to-day modern life. More frequently used application for HCI system is speech, i.e. the use of speech to recognize the words or to identify the speaker using the audio modality.These application areas take speech as an input tool for increasing their man-machine's rateperformance. Though speech recognition area of research is fairly mature, but there are still problems with performance in the real time environment under highly noisy background conditions [1].

The motivation is to increase the robustness of an ASR system by either using one of the techniques like;speech enhancement, noise reduction orthe addition of visual features[2], [3]. Audio Visual ASR (AVASR) is a very broad and diverse research

© Springer International Publishing Switzerland 2015   347
S.C. Satapathy et al. (eds.), *Proc. of the 3rd Int. Conf. on Front. of Intell. Comput. (FICTA) 2014*
– *Vol. 2*, Advances in Intelligent Systems and Computing 328, DOI: 10.1007/978-3-319-12012-6_38

field, but the versatility of AVASR depends upon the type of feature selection, type of integration process and the classifier used[2], [3].

Benchmark for first actual implementation of an AVASR system was proposed by Petajan [4] in which the authors extracted simple black-and-white images of a speaker's mouth and took the mouth height, width, perimeter and area as his visual feature cues applied to the acoustic waveform to the recognizer.Chen [5] performed experiment using 16^{th}order linear prediction coding (LPC) coefficients as audio features and the sameare used as a visual feature in [4].

Andres et al. [6] reported that due to the high dimensionality of the visual feature, there is inadequate statistical modeling, which results in a small gain in accuracy.An [6] experiment was performed by adding delta (Δ) features to static feature, i.e., Static + Δ features, resulted in a reduction of 12.9% in word error rate (WER) for discrete cosine transform (DCT). Rower et al. [1] reported a robust method for improving the performance of AVASR using the dynamic visual feature technique.

Prashant et al. [2] reported the extended work of [8], [9], [10] to prove that visual feature plays an important role in deciding the robustness of an AVASR system. In [2] it is reported that low dimension visual (LDV) DCT feature outperform static visual DCT feature. In [11] experiment was performed using dynamic stream weighting approach and also compared it with fixed-weighted integration approach in both clean and noisy condition whereas in [12] a compact representation of the visual speech using latent variable is reported using path graph.

In this paper integration of visual features with audio; a bi-modal approach is proposed for incorporating robustness in the Hindi speech recognition system. The experimenthas been conducted in three phases. In the first phase (MFCC+Δ+$\Delta\Delta$) audio features along with twelve static DCT features are concatenated. In the second phase,same audio features along with a combination of 6-static and 6-Δ DCT features are concatenated and in third phases,same audio features along with twelve low dimension static features using singular value decomposition are concatenated.

Recognition for clean signal as well as for noisy signal is evaluated for different type of noise, which is injected in clean audio signal. All speech signals are modeled with left-right HMM [7], with 3 states, except the "sp" silence model which had only one state. The bi-phone and tri-phone model are also used since they account for variations of the observed features caused by co-articulation effects. They were used as speech sub word units, and their state emission probabilities were modeled byGaussian mixture densities.

The Baum-Welch re-estimation algorithm was used for training the HMM models. Iterative mixture splitting and retraining were performed to obtain the final 16 hmm component context dependent models, which were created by simply cloning monophone and then re-estimating using triphone transcription.

A bi-gram language model was created based on the transcriptions of the training data set. Recognition was performed using the Viterbi decoding algorithm, with the bi-gram language model. The same training and testing procedure was used for both audio-only and audio-visual automatic speech recognition experiments.

For recognizing the test data the word insertion penalty (p) and grammar scale factor (s) was kept as 0.0 and 5.0 respectively. To test the algorithm over a wide range of SNRs, noise was added to the audio signals. The experiments for SNRs from 30 dB to -20 dB were performed.

2   Experimental Approach

A major challenge for developing an AVASR is the availability of the database. Aligarh Muslim University Audio Visual (AMUAV) Hindi corpus [2]. AMUAV corpus contains 48 speakers and each speaker,recorded 10 sentences out of which 2 sentences are common for all the speakers. The Video was recorded at 640x380 resolutions with 25 frames per second (fps) in full color. Audio was recorded in 16-bit stereo at 44.1 kHz. Finally, the speech signal was down sampled to 8 kHz as in speech signal most of the energy is concentrated in the low-frequency region.

For the work presented in this paper 10, speakers were selected from AMUAV database, each speaker uttering 10 sentencesresulting into 100 sentences.These sentences, a total 1225 word, were used for performing recognition.

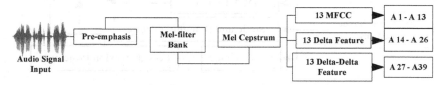

Fig. 1. Audio features extraction using MFCC with d_A=39

2.1   Audio Visual Feature Extraction Methods

The feature extraction was divided into two phases; audio feature and visual feature extraction. During audio feature extraction, pre-emphasis is done to enhance the speech signal, thereby compensate the high-frequency part that was suppressed during the sound production mechanism of humans. Here in this paper 12-dimensional MFCC followed by log energy, Δ and $\Delta\Delta$. They were used as the standard audio features, i.e. d_A=39 as shown in (Fig 1).

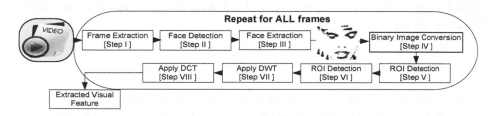

Fig. 2. Complete procedure for visual feature extraction from video sequence

The visual feature extraction method is the challenging task specifying the selection of a region of interest (ROI).Our proposed algorithm is based on the colour conversion method [2] which is used to determine a rough estimate of the face location before going ahead with the actual detection of the lip region. In (Fig 2) step by step procedure for visual feature extraction is reported.Important task was lip region estimation.

In this process extracted face frame is converted into binary image and lip region was determined by opening and closing shapes of the lips; counting lip pixels. After defining the ROI, 2D-DWT [2] is applied using Haar wavelet. Thereafter, wavelet approximated coefficient of the image is calculated and DCT is calculated by taking the difference of the wavelet approximated image.The same procedure is reported for all n numbers of frames.

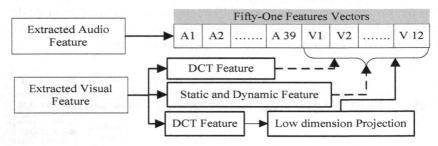

Fig. 3. Concatenating method for audio-visual stream

2.2   Feature Integration

This is the crucial phase of system design in which the audio/visual feature is integrated. At initial; audio and visual information are to be synchronized at the same rate before applying to the classifier. Since audio is extracted at the rate 100 frames/sec and visual features are extracted at rate 25 or 30 frame/sec, therefore, interpolation is needed.

Therefore, for synchronization; interpolation is carried out at visual feature. In this paper, interpolation is done after ROI estimation instead of frame extraction. The key reason is to save the computational time. Finally, the recognition is performed with HTK hidden markov model toolkit [7]. In (Fig 3) three different approaches for visual integration is reported.

3   Result Analysis

In the initial phase, recognition of clean audio signal is performed in which the testing and training both are performed using the same i.e. clean audio signal. Thereafter, the recognition is performed over noisy audio signal using different level of SNRs. Training is performed over clean audio signal, and testing was performed over noisy audio signal. In this research work, two noisy signals, i.e. babble noise and white noiseare selected.

3.1   Audio Only Recognition for Clean Signal

Audio-only speech recognition is performed for a clean audio signal and result obtained is reported in Table 1. Audio only recognition was carried out using 39 (MFCC+Δ+ΔΔ) audio features.

$$\text{Percent Correct}(C) = \frac{N-D-S}{N} \times 100 \qquad (1)$$

$$\text{Percent Accuracy}(A) = \frac{N-D-S-I}{N} \times 100 \qquad (2)$$

Where C = Percentage of the correctly recognized words, A = Percentage of accuracy, H = No of words correctly recognized, the D = Total number of deletions, the S = Total number of substitutions, I = Total number of insertions andN = Total number of evaluated words.

Table 1. Clean Audio-only recognition

% Correct (C)	% Accuracy (A)	H	D	S	I	N
97.88	97.39	1199	11	15	6	1225

A deletion occurs due to removal of words from the recognized word sequence. A substituion occurs when the word in recognizing word sequence is replaced by different word from the dictonary. An insertion occurs when a word is added in the recognized word sequence. From table1, it can be oservedthat out of 1225, word 1199 word wascorrectly recognized, which result in the accuracy of 97.39% and percentage correction of 97.88%. Hence the modeling of HTK with the acoustic feature obtained for Hindi speech performed well.

Table 2. Performance evaluation of audio-only recognition for noisy speech (babble noise and white noise)signal at different level of SNR

SNR (dB)	Babble Noise					White Noise					N
	% C	% A	H	D	S	I	% C	% A	H	D	S	I	
30	97.88	97.39	1199	11	15	6	97.88	97.39	1199	11	15	6	1225
20	97.55	96.9	1195	17	13	8	97.47	97.22	1194	11	20	3	1225
10	92.08	89.88	1128	24	73	27	65.39	64.33	801	154	270	13	1225
5	84.49	79.1	1035	32	158	66	20.65	20.49	253	637	335	2	1225
0	63.92	56.49	783	85	357	91	18.61	18.61	228	824	173	0	1225
-5	31.83	24.98	390	207	628	84	18.61	18.61	228	825	172	0	1225
-10	23.26	22.04	285	479	461	15	18.61	18.61	228	825	172	0	1225
-20	19.02	18.86	223	773	219	2	18.61	18.61	228	825	172	0	1225

3.2   Audio Only Recognition of Noisy Speech Signal

In this section; recognition over noisy audio, speech signal is evaluated. Training is performed over clean audio signal and testing was performed over the noisy audio, speech signal with the different level of SNR's (ranging from 30dB to -20dB). For evaluation, two different noisy signal'sbabble and white noise are selected. Table 2, reports the results obtained for babble and white speech signal.

From Table 2 it can easily be visualized that the percentage of correctly recognized wordsof babble noise (C_{Babble}) tremendously decrease from 10 dB to 0 dB due to the number of substitutions occur in recognizing word sentence. Also,it is to be noticed that below -5 dB; recognition rate is almost half when compared with 10 dB SNR since there is more deletion, and substitution occur in recognizing a word. Reason for the decrease in recognition is HTK get more confused in identifying recognized words due to the presence of noise.

Similarly, the experiment was performed for white noise; speech signal and result are reported in Table 2. Again, the similar case can be visualized from the% of the correctly recognized words(C_{White}) between 10dB to 0 dB; the more substitution and deletion occur in recognizing word sentence. From the Table 2, the SNR of 0 dB downward there is a very sharp decrease in recognition rate and become almost constant at 18.61%.

This is mainly due to the reason that the speech signal is more prominently effect ofthe white noise which has almost the constant spectrumthereby resulting in a worse recognition rating. Therefore, during recognition, the same number of word is recognized, with the same number of substitution and deletion.

3.3   Visual Only Recognition

Visual speech recognition (VSR) is performed using static visual features,a combination of (static+Δ)visual features and low dimension static visual features,i.e. SV features,SDV features and LDV features. Percentage of correction(C) obtained by using SV features is 31.67%, SDV features is 34.69% and using LDV feature is 57.71% [2]. Detail explanation of this is reported by the author in [2]. Therefore, one can clearly note that there is animprovement of 26.04% using LDV featureswhen compared with SV features.

3.4   Audio-Visual Recognitions for Noisy Speech Signal

Here in this section AVASR methods are compared with all possible combinations of visual feature along with audio feature is reported. In Table 3, % of the correctly recognized words is reported with the help of parameter W, X, Y and Z for babble and white noise. Here W stands for the% of the correctly recognized wordsfor audio-only recognition, X stands for the % of the correctly recognized words for audio-only recognition using SV features. Y stands for the % of the correctly recognized words for audio-only recognition using SDV features and Z stands for the % of the correctly recognized words for audio-only recognition using LDV features.

Table 3. Comparsion of % correctly recognized word for babble noise and white noise using the parameter W, X, Y and Z

SNR	Babble Speech Signal				White Speech Signal			
(dB)	W	X	Y	Z	W	X	Y	Z
30	97.88	81.14	88.16	97.71	97.88	80.33	87.84	97.63
20	97.55	76.73	85.96	97.06	97.47	76.16	84	97.47
10	92.08	71.92	79.92	93.22	65.39	44.65	52.16	67.67
5	84.49	64.16	67.43	85.47	20.65	23.27	24.41	23.67
0	63.92	48.24	52.00	70.04	18.61	18.78	20.49	20.65
-5	31.83	32.49	34.04	45.63	18.61	18.78	19.76	20.49
-10	23.26	26.12	26.86	29.06	18.61	18.61	19.67	19.76
-20	19.02	20.08	21.06	21.55	18.61	18.78	18.94	19.51

Table 3 shows the experimental result obtained for audio-visual recognition for babble noise and white noise respectively. For babble speech signal; it can be noticed that there has been improvement in AVASR when LDV features are used. At 0 dB, SNR maximum difference in the recognition rate is reported,resulting in 21.8% increment in the recognition rate when compared to the static audio-visual speech recognition. The improvementis reported using LDV feature which is mainly due to low dimension features space used, hence resulted in improvement in the modeling of the HTK.

Another important issue can be observed from the Table 3 that X, i.e. % of correctly recognized word using static visual feature is reported to give the best performance when used below 0 dB SNR, but no improvement was reported at 0 dB and above.

This is due to the fact that when high dimensional features are integrated, HTK during modeling do not take it into account due to huge dimension,since adding a visual features increase the vector size of the HTK this cause the main reason for the decrease in the AVASR.

Similarly, Y, i.e. % of correctly recognized word using static+Δ feature is reported to give little improvement when compared with X. But no improvement was reported at 0 dB and above. Reason is adding dynamic visual feature have shown little bit improvement.

On the other hand, integration of audio and low dimension visual feature shows tremendous improvement in the recognition rate. This lead to the conclusion that the HTK modeled was better for low dimension visual features, thereby increasing the performance of AVASR. A similar interpretation can be concluded for the white-noise signal.

4   Comparsion with Existing Results

This section deal with the comparative study of audio-visual Hindi speech recognition with those reported in [10]. In [10] same AMUAV database was selected and audio only, audio-video only recognition was performed by injecting different audio noise level by varying signal to noise ratio (SNR) of 10 dB to -10 dB. Audio feature was evaluated using 13 MFCC and 10 visual feature was evaluated using combination of DCT+DWT feature. Finally the performance analysis was carried out using different functions of classifier i.e. Linear, Quadratic and Mahalanobis. Table 4 shows the performance of Hindi AVASR for the combination of 23 features where A= Audio-only recognition and B=Audio-video only recognition.

Table 4. Comparsion of audio-only and audio-video only recognition for different classifier [10] at different level of SNR.

	Clean signal		10 dB		5 dB		0 dB		-5 dB		-10 dB	
Classifier	A	B	A	B	A	B	A	B	A	B	A	B
Linear	34.22	36.49	23.56	25.78	23.56	24.44	20.44	23.56	23.11	28.89	22.22	26.22
Quad	34.22	37.20	27.56	29.78	23.11	26.67	21.33	26.67	24.00	28.44	21.33	29.33
Mahanloabis	35.11	37.07	26.22	35.84	26.22	28.00	25.33	28.44	24.89	31.11	23.56	27.11

From the above table 4 one can easily identify that adding the visual clue have shown the significant improvement. Significant improvement in the recognition rate is increased by 4%, 8% and 3.55% for linear, quadratic and Mahalanobis classifier respectively for the combination of visual information with audio signal at -10dB SNR.

When comparsion with the result from Table 3 for babble speech signal and noisy speech signal of Table 4 is calculated, it is found that at 0 dB maximum improvement reported in this paper is 6.12% whereas for previous paper [10] it is reported to be 5.34% for Quadratic classifier. On the other hand at -5dB maximum improvement is reported in this paper which is 13.8 % where as in [10] maximum improvement of 9.62% in accuracy of audio signal was found at 10 dB SNR by using Mahalanobis classifier.

Thus the result of [10] when compared with the results reported in this paper shows that the improvement in the audio-video only recognition is achieved with respect to that reported in [10]. Finally the proposed scheme in this paper report the maximum recongnition.

5   Conclusion

The main aim of this research was to study the role of visual featuresin designing robustness AVASR systems and their performance in a noisy background.The conclusion that can be drawn is that in audio-visual recognition using static visual feature above 0dB HTK used only the audio features and visual features were discarded, i.e. due to the higher dimension, during modeling, which results in poorer performance.

On the other hand, below 0 dB, highly noisy signal makes HTK more confused and while performing recognition it may have used the visual features which as a result show the improvement over audio-only recognition.

On the other hand, using a combination of (static+Δ) features there was somewhat an improvement in the modeling of the HTK but still the presence of high dimension static visual features havedegraded the performance of the HTK.

At last, it is reported from the result obtained in table 3, that using low dimension audio visual feature, there is a significant improvement in the modeling of the HTK. Finally, improvement in AVASR was obtained using visual features along with the audio features which outperform the ASR.

References

1. Seymour, R., Stewart, D., Ming, J.: Comparison of image transform-based features for visual speech recognition in clean and corrupted videos. EURASIP Journal on Image and Video Processing 2008, 1–9 (2008)
2. Upadhyaya, P., Farooq, O., Varshney, P., Upadhyaya, A.: Enhancement of VSR Using Low Dimension Visual Feature. In: International Conference on Multimedia Signal Processing and Communication Technologies, IMPACT 2013, AMU, Aligarh, India, pp. 71–74. IEEE Press (2013)
3. Potamianos, G., Neti, C., Gravier, G., Garg, A., Senior, A.W.: Recent advances in the automatic recognition of audio-visual speech. Proceedings of the IEEE 91, 1306–1326 (2003)
4. Petajan, E.: Automatic lipreading to enhance speech recognition. In: IEEE Global Telecommunications Conference, Atlanta, GA, USA, pp. 265–272. IEEE Press (1984)
5. Chen, T.: Audiovisual speech processing, Lip Reading and Lip Synchronization. IEEE Signal Processing Magazine, 9–21 (2001)
6. Valles, A., Gurban, M., Thiran, J.: Low Dimensional Motion Features for Audio-Visual Speech Recognition. In: 15th European Signal Processing Conference, EUSIPCO, Poznan, Poland, pp. 297–301 (2007)
7. Young, S.: A review of large vocabulary continuous speech. IEEE Signal Processing Magazine 13(5), 45–57 (1996)
8. Upadhyaya, P., Farooq, O., Varshney, P.: Comparative study of viseme recognition by using DCT feature. In: International Symposium Frontier Research on Speech and Music, FRSM, Gurgaon, Haryana, India, pp. 171–175 (2012)
9. Varshney, P., Farooq, O., Upadhyaya, P.: Hindi viseme recognition using subspace DCT features. International Journal of Applied Pattern Recognition (in press, 2014)
10. Varshney, P., Upadhyaya, P., Farooq, O.: Transform based Visual Features for Bimodal Recognition of Hindi Visemes. International Journal of Electronics and Computer Science Engineering 1(3), 892–897 (2012) ISSN- 2277- 1956
11. Stewart, D., Seymour, R., Pass, A., Ming, J.: Robust Audio Visual Speech Recognition under noisy audio-video conditions. IEEE Transactions on Cybernetics 44(2), 175–184 (2014)
12. Zhou, Z., Hong, X., Zhao, G., Pietikainen, M.: A compact representation of visual speech data using latent variables. IEEE Transactions on Pattern Analysis and Machine Intelligence 36(1), 181–187 (2014)

Extraction of Shape Features Using Multifractal Dimension for Recognition of Stem-Calyx of an Apple

S.H. Mohana and C.J. Prabhakar

Department of P.G. Studies and Research in Computer Science,
Kuvempu University Shankaraghatta-577451, Shimoga, Karnataka, India
mohana.sh43@gmail.com, psajjan@yahoo.com

Abstract. In this paper, we introduce a novel approach to recognize stem-calyx of an apple using multifractal dimension. Our method comprises of steps such as preprocessing using bilateral filter, segmentation of apple using grow-cut method, multi-threshold segmentation is used to detect the candidate objects such as stem-calyx and small defects. The shape features of the detected candidate objects are extracted using Multifractal dimension and finally stem-calyx regions are recognized and differentiated from true defects using SVM classifier. The proposed algorithm is evaluated using experiments conducted on apple image dataset and results exhibit considerable improvement in recognition of stem-calyx region compared to existing techniques.

Keywords: Stem-Calyx Recognition, Multifractal Dimension, Grow-cut, Multi-threshold Segmentation, SVM, k-NN.

1   Introduction

Surface defect detection for inspection of quality of apples has been gaining importance in the area of machine vision. Surface defects are of great concern to farmers and vendors to grade the apples. There are two important stages involved in grading the apples such as surface defect detection and grading the apples based on identified defects. The researchers have adopted image processing and computer vision based techniques to detect the surface defects of apples, and apples are graded based on quantity of surface defects detected using either RGB images [1] [2] or multispectral images [3] [4]. The main drawback of these techniques is that while identifying the defects, stem-calyxes are identified as defects because after applying image processing techniques, stem-calyx region appears to be similar as defect spots. Due to similarity in appearance between stem-calyx and defects, the defect detection algorithms wrongly identify the stem-calyx parts as defects, which reduce the grading accuracy.

Some researchers have proposed techniques to differentiate actual defects and stem-calyx, but these techniques are not able to give the promising results. The X.Cheng et al. [5] have proposed an approach for stem-calyx recognition in multispectral images of apples. Brightness effect is eliminated using normalization

© Springer International Publishing Switzerland 2015   357
S.C. Satapathy et al. (eds.), *Proc. of the 3rd Int. Conf. on Front. of Intell. Comput. (FICTA) 2014*
– *Vol. 2*, Advances in Intelligent Systems and Computing 328, DOI: 10.1007/978-3-319-12012-6_39

and adaptive spherical transform is used to distinguish the low- intensity defective portions from the actual stem-calyx. The global threshold is used to extract the stem-calyx. V. Leemans et al. [4] developed apple grading technique according to their external quality using machine vision. Color grading is done by Fisher's linear discriminant analysis and stem-calyx is detected by using correlation pattern recognition technique. The main drawback of the threshold based segmentation is that misclassification may occur if the threshold of stem-calyx and true defects are similar.

D Unay et al. [6] have proposed an approach to recognize stem-calyx on apples using monochrome images. Fruit area is separated from background using threshold based segmentation. Object segmentation technique is used for stem-calyx detection and it is done by thresholding. Statistical, shape and texture features are extracted. They conclude that SVM classifier gives best performance in recognizing stem-calyx. J Xing et al. [7] proposed an approach for stem-calyx identification for multispectral images. In order to identify the stem/calyx on apples, the technique is developed based on contour features in the PCA image scores. The main drawback of these techniques is that they used MIR camera to capture multispectral images, which are too expensive to use, and it is very difficult to process the huge amount of data.   L Jiang et al. [8] developed a 3D surface reconstruction of apple image and analyzed the depth information of apple surface to identify stem-calyx part. Qingzhong Li et al. [9] proposed an approach to detect the defects in the surface of an apple. Stem-calyx part is identified using fractal features, which are fed into ANN classifier. They considered mono fractal features to classify the stem-calyx part assuming that concave surface. However, this is not true in case where apples are oriented along in any direction and the concave surface may not exist. Monofractals are homogeneous in the sense they have same scaling properties, characterized by a single exponent. But multifractals are inhomogeneous in the sense they have different scaling properties, characterized by many exponents. Multifractals can be decomposed into many possible subsets characterized by different exponents.

The drawback of mono fractal features is that they failed to represent the stem-calyx part exists in the apple image with non-concave surface. In this paper, we proposed an approach to recognize the stem-calyx by extracting the shape features based on multifractal dimension using RGB image of an apple. The motivation to use RGB images is that unlike the MIR camera, digital cameras are inexpensive and easy to afford. Another advantage of digital data is that they can be readily processed using digital computers. The organization of remaining sections of the paper is as follows: In section 2, theoretical description of multifractal dimension is given. The proposed approach for stem-calyx recognition is discussed in the section 3. The experimental results are demonstrated in Section 4. Finally, Section 5 draws the conclusion.

2   Multifractal Dimensions

The fractal dimension is a very popular concept in mathematics due to its wide range of applications. The fractal dimension was developed by Benoit Mandelbrot in 1967 [10]. Fractals are generally self-similar and independent of scale. A fractal is a fragmented geometrical object that can be divided into parts, and each part is a copy of the original in reduced size. The important characteristic of fractal geometry is that ability to represent an irregular or fragmented shape of complex objects, where as traditional Euclidean geometry fails to analyze. A multifractal is a set of intertwined fractals. A multifractal object is more complex in the sense that it is always invariant by translation. The multifractal dimensions were defined based on partition function.

In our approach, we used Box Counting method to compute the multifractal dimension. Consider the mass dimension α at a point x, the box $B_\varepsilon(x)$ is a box of radius ε centered at x. The probability measure of mass in the box is defined as $\mu\left(B_\varepsilon(x)\right)$. It can be shown that $\mu\left(B_\varepsilon(x)\right) \propto \varepsilon^\alpha$. The mass dimension specifies how fast the mass in the box $B_\varepsilon(x)$ decreases as the radius approaches to zero. In the first step, we compute the total number of boxes required to cover the input image. The total number of boxes can be calculated using below equation

$$N_T = \frac{(width)^2}{(\varepsilon)^2}. \tag{1}$$

where, width size of the input image and ε is size of the box. It is necessary to eliminate the boxes which are having zero elements. Therefore, we select the boxes with non-zero elements and are defined as

$$N_S = \sum_{i=1}^{N_T} N_i = not\ null. \tag{2}$$

where N_S represents total number of boxes with non-zero elements. The total sum or mass of pixels in all boxes can be obtained using the following equation

$$M_\varepsilon = \sum_{i=1}^{N_S} m_{[i,\varepsilon]}. \tag{3}$$

where $m_{[i,\varepsilon]}$ represents the mass of ith box with size ε. The probability P of the mass at i^{th} box, is relative to the total mass M_ε and is defined as

$$P_{[i,\varepsilon]} = \frac{m_{[i,\varepsilon]}}{M_\varepsilon}. \tag{4}$$

the qth order normalized probability measures of a variable, $\mu(q,\varepsilon)$ vary with the box size ε and it is defined as

$$\mu_i(q,\varepsilon) = \frac{[p_i(\varepsilon)]^q}{\sum_i [p_i(\varepsilon)]^q}. \tag{5}$$

where $p_i(\varepsilon)$ is the probability of a measure in the ith box of size ε. It can be shown that $P_{[i,\varepsilon]} \propto \varepsilon^\alpha$, the exponent α is called the Holder exponent at the point x, which is defined as

$$\alpha(q,\varepsilon) = \sum_i \mu_i(q,\varepsilon) \log p_i(\varepsilon).\ for\ i=1, 2, 3, \ldots, N_S. \tag{6}$$

For each q value, multifractal dimension is calculated as

$$f(q,\varepsilon) = \sum_i \mu_i(q,\varepsilon) \log \mu_i(q,\varepsilon).\ for\ i=1, 2, 3, \ldots, N_S. \tag{7}$$

3   Our Approach

We detect candidate objects such as stem part, calyx part and defected part and shape features are extracted from the detected candidate objects. As we described in the previous section, we extract and store the description of shape features using Multifractal dimension. The feature extraction step for each candidate object yields a feature vector. The extracted feature vectors from stem part, calyx part and defected part of various training samples captured in different viewpoint are stored independently in the database and these stored feature vectors are treated as reference in order to recognize the stem-calyx part. In the testing phase, we recognize the stem-calyx by extracting the feature vectors from detected candidate objects of testing sample and compared with stored feature vectors using classifier.

3.1   Image Pre-processing

The captured apple images are suffered from illumination variations due to specular reflection and noise. The specular reflection and noise leads to poor performance during candidate objects detection process because the specular reflected part is treated as the candidate object due to its white color compared to its surroundings. We employ bilateral filtering [11] to normalize the uneven distribution of light and to suppress noise. This technique allowed the edges to be preserved while filtering out the peak noise.

Fig. 1. First row: apple images with specular reflection, second row: corresponding preprocessed apple images

3.2   Background Subtraction

It is necessary to extract the ROI i.e. fruit area from the apple image. We used grow-cut approach developed by Vladimir Vezhnevets et al. [12] for background subtraction. It removes background and gives fruit part separately. After extracting the apple fruit part from the image, we segment the candidate objects from fruit area using multi-threshold segmentation. In multi-threshold segmentation, the images were segmented several times at different threshold levels. We experimentally observed that the candidate objects of various types of apples can be segmented accurately using threshold values 30, 50 and 65. The resulting binary images were added to form a so called multi layer image. This in turn was then subjected to threshold segmentation. This segmentation aimed at identifying and separates the healthy skin of an apple from the stem-calyx and defected areas in the original image. The resulting binary image was referred to as a marker image. The final step consists of constructing a binary image based on the marker image and the multi-layer image. With the position of the candidate objects, a simple thresholding routine i.e. a gradient segmentation is employed to determine the area of these candidate objects. The results obtained from the background subtraction and multi-threshold segmentation methods are overlapped to each other. In the resultant image, we observe that the candidate objects are appeared on the surface of the apples.

Fig. 2. Candidate object detection process: First row: The pre-processed Jonagold apple images with stem part in view, Second row: fruit area extracted from images shown in first row, Third row: Detected stem part using multi-threshold segmentation.

3.3   Shape Features Extraction from Candidate Objects

In order to represent stem-calyx part, we extract shape features because of its discriminative characteristics to distinguish stem-calyx part from defects. We extract the shape features from detected stem-calyx part and defect parts where the stem-calyx part and defect parts are captured in different views with variation in scale and views. The extracted shape features represent stem-calyx part and defect parts and invariant to variation in views and scale. We used Multifractal dimension to extract and store the shape features independently for each detected candidate object. We extract multifractal dimension features of detected candidate objects using box counting approach with the box size 2 x 2 (4 pixels). Here we varied the q value from

-1 to +1 (21 points). We calculate the mass distribution of all selected boxes and the probability of each box. Finally, Multifractal dimension is calculated using normalized probability. In the Figure 4, from the plotted multifractal dimension graph for stem-calyx part and defected part, it is observed that the multifractal dimension exhibit discrimination power for stem, calyx and defected part.

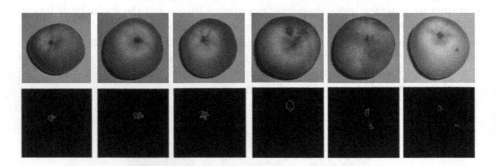

Fig. 3. First row: the apple images with stem/calyx and defected part in view. Second row: detected stem-calyx part with defects.

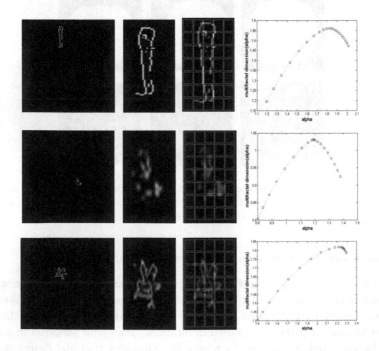

Fig. 4. Results of multifractal dimension: First column- Detected candidate objects, Second column- Cropped image of candidate object, Third column-Schematic diagram of box counting method, Last column- Corresponding graph of multifractal dimension.

3.4   Stem-Calyx Recognition

In the training phase, we extracted the shape features of detected stem/calyx and defected parts using multifractal dimension. In order to differentiate the stem-calyx from true defects, classifier is employed. We considered popular supervised classification algorithms such as Support Vector Machine (SVM), AdaBoost, Artificial Neural Network (ANN), Linear Discriminant Classifier (LDC) and k-Nearest Neighbor classifier (k-NN).

4   Experimental Results

We have conducted the experiments to evaluate the proposed method for recognizing stem-calyx part and differentiated from true defects of an apple. Since, there is no benchmark database of apples is available, we have created a Jonagold apple database for experimentation purpose. Image acquisition device is composed of a resolution 2.0 mega pixel (1240 x 1600). The camera is capable of acquiring only one-view of an apple. The healthy and small defected Jonagold apples were selected for experimentation purpose based on visual inspection. The images of the selected 200 Jonagold apples were captured in indoor environment with variation in lighting condition. Each apple was captured in three views such as stem part in view, calyx part in view and defected part in view. The database contains 600 sample images of each image size 256 x 256 and 200 apple image samples for each view. Among the above-said data samples, we selected randomly 75% (from each view 150 samples) of the apple samples for training purpose, and the remaining 25% (from each view 50) samples are used for testing purpose. In order to eliminate the intersection (duplication) of training and testing sets, drop-one-out approach is used in this work. That is, an apple used for training is excluded from the testing set.

Table 1. Stem-calyx recognition rate (RR %) using multifractal dimension shape features.

Classifiers	RR (%)
SVM (Polynomial degree 3)	94.00
LDC	91.80
AdaBoost	90.50
ANN	92.65
k-NN (k=4)	91.69

Table 2. TPR and FPR for stem-calyx classification using SVM classifier

Different views of an apple	TPR	FPR
Stems (n=50)	49 (98%)	01
Calyxes (n=50)	46 (92%)	04
Defects (n=50)	47 (94%)	03
Stem with small defects (n=50)	48 (96%)	02
Calyx with small defects (n=50)	45 (90%)	05

Table 2 shows the evaluation of experimental results using True Positive Rate (TPR) and False Positive Rate (FPR) obtained using SVM classifier with polynomial degree 3. We used 50 samples for each view such as stem, calyx, defects, stem with small defects and calyx with small defects. From the results shown in the Table 2 indicates that proposed shape descriptors have shown high discrimination between stem-calyx and defects. The TPR for stem is approximately near to 100% classification accuracy and only 2% of classification error. This is not true in case of calyx recognition, only 46 calyx regions were correctly classified and remaining 04 regions were misclassified. This is due to the fact that shape features of small defects and calyx appear to be similar.

5   Conclusion

In this paper, we proposed an approach to recognize the stem-calyx of Jonagold apples and distinguished from the true defects. The preprocessing is done by bilateral filter, which improves the image quality i.e., noise and specular reflections are reduced. The grow-cut approach is used for apple segmentation and multi-threshold segmentation is used to detect the candidate objects. We extracted the shape features from detected candidate objects using Multifractal dimension. The extracted shape features are used to discriminate stem-calyx from true defects. The SVM classifier with polynomial kernel function yields highest recognition rate among the all classifiers. The proposed approach can be incorporated in apple grading technique as one of the steps in order to increase the apple grading accuracy.

Acknowledgments. We would like to thank the anonymous reviewers for their critical comments and suggestions, which helped to improve the previous version of this paper.

References

1. Bennedsen, B.S., Peterson, D.L.: Amy Tabb: Identifying defects in images of rotating apples. Computers and Electronics in Agriculture 48, 92–102 (2005)
2. Prabhakar, C.J., Mohana, S.H., Praveen Kumar, P.U.: Surface Defect Detection and Grading of Apples. In: International Conference on MPCIT, pp. 57–64 (2013)
3. Throop, J.A., Aneshansley, D.J., Anger, W.C., Peterson, D.L.: Quality evaluation of apples based on surface defects - an inspection station design. International Journal Postharvest Biology and Technology 36, 281–290 (2005)
4. Leemans, V., Magein, H., Destain, M.F.: On-line Fruit Grading According to their external quality using machine vision. Biosystems Engineering 83(4), 397–404 (2002)
5. Cheng, X., Tao, Y., Chen, Y.R., Luo, Y.: NIR/MIR Dual sensor machine vision system for online apple Stem-end/ calyx recognition. Transactions of the ASAE 46(2), 551–558 (2003)
6. Unay, D., Gosselin, B.: Stem and Calyx recognition on Jonagold apples by pattern recognition. Journal of Food Engineering 78, 597–605 (2007)

7. Xing, J., Jancso, P., De Baerdemaeker, J.: Stem-end/Calyx Identification on Apples using Contour Analysis in Multispectral Images. Biosystems Engineering 96, 231–237 (2007)
8. Jiang, L., Zhu, B., Cheng, X., Luo, Y., Tao, Y.: 3D Surface reconstruction and analysis in automated apple stem-end/calyx identification. Transactions of ASABE 52(5), 1775–1784 (2009)
9. Li, Q., Wang, M., Gu, W.: Computer vision based system for apple surface defect detection. Computers and Electronics in Agriculture 36, 215–223 (2002)
10. Mandelbrot, B.: How Long Is the Coast of Britain? Statistical Self-Similarity and Fractional Dimension. American Association for the Advancement of Science 156(3775), 636–638 (1967)
11. Tomasi, C., Manduchi, R.: Bilateral filtering for gray and color images. In: Sixth International Conference on Computer Vision, pp. 839–846. IEEE (1998)
12. Vezhnevets, V., Konouchine, V.: GrowCut - Interactive Multi-Label N-D Image Segmentation by Cellular Automata. In: GraphiCon (2005)

An Approach to Design an Intelligent Parametric Synthesizer for Emotional Speech

Soumya Smruti, Jagyanseni Sahoo, Monalisa Dash, and Mihir N. Mohanty[*]

ITER, Siksha 'O' Anusandhan University, Bhubaneswar, Odisha
{soumya9111991,monalisa21dash}@gmail.com,
jagyanseni_sahoo@yahoo.in,
mihirmohanty@soauniversity.ac.in

Abstract. Speech synthesizer is an artificial system to produce speech. But the generation of emotional speech is a difficult task. Though many researchers have been working on this area since a long period, still it is a challenging problem in terms of accuracy. The objective of our work is to design an intelligent model for emotional speech synthesis. An attempt is taken to compute such system using rule based fuzzy model. Initially the required parameters have been considered for the model and are extracted as features. The features are analyzed for each speech segment. At the synthesis level the model has been trained with these parameters properly. Next to it, it has been tested. The tested results show its performance.

Keywords: Speech synthesizer, Feature Extraction, Emotional speech synthesis, Fuzzy model.

1 Introduction

Synthetic or artificial speech has been developed steadily during the last decades. It has a greater impact in modern society. With the development of speech synthesis techniques, the intelligibility has reached an adequate level for most applications such as multimedia and telecommunications and there is still scope of development. In order to enhance the quality of synthetic speech, the implementation of emotional effects is now being progressively considered.

Synthesized speech can be created by concatenating pieces of recorded speech that are stored in a database. The quality of a synthesized speech is defined in terms of naturalness and intelligibility. Naturalness describes how closely the output sounds like human speech, while intelligibility is the ease with which the output is understood. For most applications, the intelligibility and comprehensibility of synthetic speech have reached the acceptable level. In order to increase the acceptability and improve the performance of the synthesized speech in this area of application, the synthesis of attitude and emotion has to be considered as critical. In our work we have considered the emotional speech for synthesis using fuzzy technique as the intelligent

---

[*] Corresponding author.

© Springer International Publishing Switzerland 2015   367
S.C. Satapathy et al. (eds.), *Proc. of the 3rd Int. Conf. on Front. of Intell. Comput. (FICTA) 2014*
– *Vol. 2,* Advances in Intelligent Systems and Computing 328, DOI: 10.1007/978-3-319-12012-6_40

model. We have extracted some standard features from the speech samples taken under different emotional conditions. These features are used for training and testing to design a synthesis model in various ways.

Section 1 introduces the work and its necessity. Section 2 describes the related literature on speech synthesis and emotional speech synthesis. Section 3 contains the work carried out for feature extraction and chosen those as parameters for the model. Section 4 proposes the method of synthesis. Section 5 discusses the results of the synthesis model for different emotions. Section 6 concludes this piece of work.

2   Related Literature

Work on speech synthesis has been carried out since a long period and some of them are discussed in this section and cited as well. In past researches, authors have focused on the development and assessment of an algorithm to interpolate the intended vocal effort in existing databases in order to create new databases with intermediate levels of vocal effort.

For speech analysis and re-synthesis LPC is generally used. LPC has been used by many authors as it is one of the most useful methods for encoding good quality speech at a low bit rate and provides extremely accurate estimates of speech parameters. A procedure for efficient encoding of the speech wave has been described by representing it in terms of time-varying parameters [1-2]. HMM is used for synthesis of speech though it is one of the recognizing models. Many authors have used HMM as a synthesizer. A detailed explanation has been given in [3-4]. In [5], a method named as unit selection synthesis was used for large databases of recorded speech. The studies of flexible approach based on analyzed and modeled the emotional prosody features. A complete review of four majorly researched methods of speech synthesis viz. Articulatory, Concatenative, Formant and Quasi-articulatory Synthesis have been described in [6]. The authors in [7] have analyzed the application of the harmonic plus noise model (HNM) for concatenative text-to-speech (TTS) synthesis. Some examples have been described in [8] related to production of varied style and emotion using existing unit selection synthesis techniques and also highlighted the limitations of generating truly flexible synthetic voices.

The study in [9] investigated how speech pause length influenced listeners ascribe emotional stated to the speaker. Similarly, implementation of TD-PSOLA tools to improve the quality of the Arabic Text-to speech (TTS) system has been described in [10]. In [11] the LPCs were analyzed for each speech segment and the pitch period was detected. At synthesis the speech samples equal to the samples in one pitch period were reconstructed using LPC inverse synthesis. A mathematical tool to build a fuzzy model of a system, where fuzzy implications and reasoning were used, was presented by researchers also [12-13]. The method of identification of a system using its input-output data along with two applications of the method to industrial processes were also discussed [12]. Fuzzy Inference System (FIS) was used for enhancing resultant accuracy in adverse real world noisy environmental conditions [13].

3   Synthesis Parameters

Three emotions have been considered in this paper namely "sad", "happiness", "anger" along with normal speech. The extracted feature of each one is compared with the normal speech. The following parameters have been considered and extracted as features for the synthesis purpose.

3.1   Short Time Energy

Short-Time energy is a simple short-time speech measurement. In this case the short time energy of speech under different emotional conditions have been derived. The relation for finding the short term energy can be derived from the total energy relation defined in signal processing .The total energy of an energy signal is represented by

$$E_T \; = \; \sum_{N=-\infty}^{\infty} s^2\,(n) \qquad (1)$$

This helps to know the type of emotion information through speech.

3.2   Zero Crossing Rate

Zero Crossing Rate gives information about the number of zero-crossings present in a given signal. ZCR gives indirect information about the frequency content of the signal. It has been found under different emotional conditions. The relation for non-stationary signals like speech is termed as short term ZCR. It is defined as

$$Z(n) = \frac{1}{2N} \sum_{m=0}^{N-1} s(m).\,w(n-m) \qquad (2)$$

3.3   Correlation Coefficient

The correlation coefficient is a measure of how well trends in the predicted values follow trends in past actual values. The quantity r, called the linear correlation coefficient, measures the strength and the direction of a linear relationship between two variables. The expression to compute r is:

$$r = \frac{n\sum xy - (\sum x)(\sum y)}{\sqrt{n(\sum x^2)-(\sum x)^2}\,\sqrt{n(\sum y^2)-(\sum y)^2}} \qquad (3)$$

where n is the number of pairs of data.

This has the major role for LPC extraction.

3.4   Linear Prediction

Linear predictor error filtering assumes that the signal to be whitened is a stationary stochastic process that is well modeled by an autoregressive model of order N. The nth sample of a discrete time sequence, $x[n]$, is well modeled by a linear combination of the previous N samples. Given this assumption, we define the predicted sequence in terms of M undetermined coefficients b[m] by the expression

$$\tilde{x}[n]=\sum_{m=1}^{N} b[m]x[n-m] \tag{4}$$

The method used in this experimental study is the autocorrelation method. The autocorrelation method assumes that the signal is identically zero outside the analysis interval ($0<=m<=N$-1) and here N=4455. Then it tries to minimize the prediction error wherever it is nonzero, that is in the interval $0<=m<=N$-1+p, where p is the order of the model. The advantage of this method is that stability of the resulting model is ensured. Based on these parameters, the model is designed and explained in following section.

4   Proposed Model for Synthesis

A Fuzzy based intelligent model has been developed for synthesis. The idea of fuzzy logic is to approximate human decision making using natural language terms instead of quantitative terms. It is a convenient way to map an input space to an output space. A fuzzy set is an extension of a classical set. If X is the universe of discourse and its elements are denoted by x, then a fuzzy set A in X is defined as a set of ordered pairs:

$$A = \{x, \mu A(x) \mid x \in X\} \tag{5}$$

$\mu A(x)$ is called the membership function (or MF) of x in A. A membership function provides a measure of degree of similarity of an element to a fuzzy set. Here it maps each element of X to a membership value between 0 and 1.

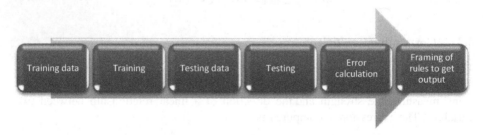

Fig. 1. Stages of synthesis model

The block diagram shown in Figure 1 provides the different stages of synthesis model. The system then undergoes training and testing for all variety of samples (i.e. variety of emotions) and follows the error calculation is done. Sugeno model has been selected for synthesis model in this experimental study, as it is computationally efficient and works well with adaptive and optimization techniques. The trapezoidal membership function as described is suitable than other membership functions for such model. It is defined by a lower limit a, an upper limit d, a lower support limit b, and an upper support limit c, where a < b < c < d and is represented as

$$\mu_{A(X)} = \begin{cases} 0 & (x < a) \, or \, (x > d) \\ \frac{x-a}{b-a} & a \leq x \leq b \\ 1 & b \leq x \leq c \\ \frac{d-x}{d-c} & c \leq x \leq d \end{cases} \quad (6)$$

The importance of the features can be studied from the extracted features in section 3 and are used as the synthesis parameters for the rule based Fuzzy model. The well known sugeno model with trapezoidal membership function has been implemented for this purpose. Using these parameters, the model has been trained accurately. Finally it has been tested. The results have been shown from Figure 2 to Figure 6 in section 5. The If-Then rules are framed to get the output. Here no defuzzification is required in this work as it is meant for synthesis. Output is observed for different rules. Observation and output results are given in following section.

5   Results and Discussion

The features are considered as the parameters for the intelligent parametric model. For different emotions, different parametric values have been given in table 1. It shows the different parameters for model design for three different emotions (i.e. "sad", "happiness" and "anger") and also under normal condition.

Table 1. Features for synthesis model for different emotion

FEATURES	ANGRY EMOTION	HAPPY EMOTION	SAD EMOTION	NEUTRAL
LINEAR PREDICTION COEFFICIENTS	0.0132	0.0165	0.0219	0.0094
SHORT TIME ENERGY	13.2870	8.9610	7.2005	1.9959
AUTOCORRELATION COEFFICIENT	0.0381	0.0011	-0.0038	0.0706
ZERO CROSSING RATE	0.07	0.0569	0.0510	0.0344

6   Training and Testing

The features obtained in the section 3 were passed to the given model. The system then undergoes training where the data is loaded and then an initial FIS model is loaded using generate FIS option. The data is trained for the given number of training epochs by using train now option.

Then the data is tested against the training data as well as testing data by selecting the test now button as shown in Figure 2 and 3. This function plots the test data against the FIS output in the plot region.

The error found when the data was tested against training data (i.e. same emotion) was lesser than the error found when the data was tested against testing data(i.e. different emotion).

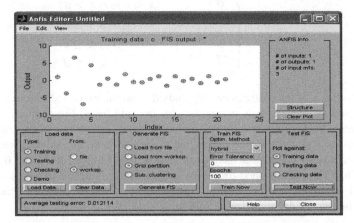

Fig. 2. Test of sentence *"I am sad"* in sorrow against training data

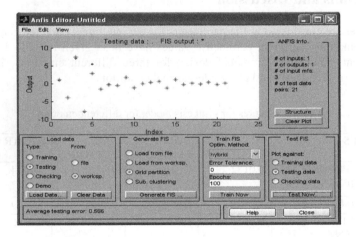

Fig. 3. Test for the sentence *"I am sad"* in sorrow with testing data of angry

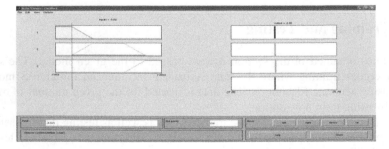

Fig. 4. Synthesis Output for sentence *"Have you gone insane"* for ANGRY

Three rules were set to compute the emotion as an output. The default values of linear prediction coefficients were used to view the output of rules and the crisp value of emotion was observed. In Figure 4, it shows for anger. Similarly Figure 5 shows the output for happiness. Finally, Figure 6 is for sad. From the output it is observed that, the LPC synthesis parameter at the input matches with the output. Similarly other acoustic parameters along with LPC performed well in Fuzzy synthesis system.

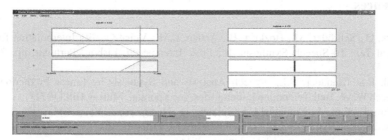

Fig. 5. Synthesis Output for sentence *"I am so happy"* for HAPPY

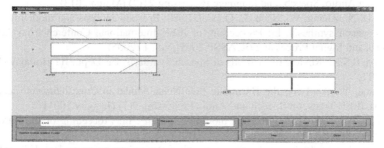

Fig. 6. Synthesis Output for sentence *"I am sad"* for SAD

7   Conclusion

Our database consists of human speech pronounced under different emotional conditions such as anger, sadness, happiness and one neutral emotion were also included: fast, loud and low soft. For this particular work recordings were done and the recordings were saved in wave file format. The database was recorded after the class time by 12 subjects (6 male and 6 female) aged between 20-50 years old. The data base contains 120 utterances all total. The speech samples taken in this experimental study are "I am so happy" which describes happiness, "I am sad" which describes sorrow and "Have you gone insane" which describes anger. The features have been extracted in this work is not new, but their application for such model is very important.

This paper studies a flexible emotional synthesis approach using the intelligent synthesizer. The analysis results show that just manipulating features can obtain meaningful results at least for some emotions. From the observation we concluded that when one emotion was tested against the same emotion the error was less but when it was tested against different emotion, the error was found to be more. Efficient results have been drawn and can be used for practical purpose. Also, the machine learning model can be most suitable in real time environmental application. There is scope to investigate new methods for analyzing Model parameters and synthesizing the emotional speech specifying different Expressions or Emotions.

References

1. Turk, O., Schröder, M., Bozkurt, B., Arslan, L.M.: Voice Quality Interpolation for Emotional Text-To-Speech Synthesis. In: Proc. Interspeech, Lisbon, Portugal, pp. 797–800 (2005)
2. Atal, B.S., Hanauer, S.L.: Speech Analysis and Synthesis by Linear Prediction of the Speech Wave. Bell Telephone Laboratories, Incorporaiai, Murray Hill (1971)
3. Raitio, T., Suni, A., Yamagishi, J., Pulakka, H., Nurminen, J., Vainio, M., Alku, P.: HMM-Based Speech Synthesis Utilizing Glottal Inverse Filtering. IEEE Trans. on Audio, Speech, and Language Processing 19(1) (January 2011)
4. Shannon, M.: Autoregressive Models for Statistical Parametric Speech Synthesis. IEEE Trans. on Audio, Speech, and language Processing 21(3) (March 2013)
5. Jiang, D.-N., Zhang, W., Shen, L.-Q., Cai, L.-H.: Prosody analysis and modeling for emotional speech synthesis. In: Proceeding of IEEE International Conference on Acoustics, Speech, and Signal Processing (ICASSP 2005), vol. 1 (2005)
6. Newton, P.S.R.: Review of methods of Speech Synthesis. EE Dept., IIT Bombay (November 2011)
7. Stylianou, Y.: Applying the Harmonic Plus Noise Model in Concatenative Speech Synthpesis. IEEE Trans. on Speech and Audio Processing 9(1) (January 2001)
8. Black, A.W.: Unit Selection and Emotional Speech". In: Proceeding of the Eurospeech, Geneve (2003)
9. Tisljár-Szabó, E., Pléh, C.: Ascribing emotions depending on pause length in native and foreign language speech. Speech Communication 56, 35–48 (2014)
10. Chabchoub, A., Cherif, A.: High Quality Arabic Concatenative Speech Synthesis. Signal & Image Processing: An International Journal (SIPIJ) 2(4) (December 2011)
11. Bhatlawande, S.N., Apte, S.D.: Emotion Generation using LPC Synthesis. International Journal on Recent and Innovation Trends in Computing and Communication 2(1), 128–134 (2014)
12. Takagi, T., Sugeno, M.: Fuzzy Identification of Systems and Its Applications to Modeling and Control. IEEE Trans. on Systems, Man, and Cybernetics smc-15(1) (January/February 1985)
13. Shrawankar, U., Thakare, V.: Parameters Optimization for Improving ASR Performance in Adverse Real World Noisy Environmental Conditions. International Journal of Human Computer Interaction (IJHCI) 3(3) (2012)

Removal of Defective Products Using Robots

Birender Singh, Mahesh Chandra, and Nikitha Kandru

Department of Electronics and Communication Engineering, BIT, Mesra, India
{birender1332.11,shrotriya}@bitmesra.ac.in,
kandru.nikitha@gmail.com

Abstract. This paper addresses the utility of intelligent autonomous robotic arm for automatic removal of defective products in an industry. The task can be performed in two steps, finding the defective product with digital image processing and removal of defective part from the products. The image is regularly obtained and compared with the standard image. The defective product is sorted out based on threshold value between the real image and standard image. After detection of defective product, it is sorted out with the help of robotic arm and placed in the defective lot.

Keywords: Robotic Arm, Digital Image Processing, Serial Communication.

1   Introduction

Today Robotic arm have become an indispensable tool in modern industries and being used to undertake routine works, often repetitive tasks which are expensive and difficult through highly paid labor force [1-4]. It will be interesting to detect and separate out the defective products using digital image processing and robotic arm. Many researchers have worked either for finding the defect in an item or on working of robotic arm. In this work these both applications will be integrated for industrial applications. The International Organization for Standardization defines a robot in ISO 8373 as: an automatically controlled, reprogrammable, multipurpose, manipulator programmable in three or more axes, which may be either fixed in place or mobile for industrial automation applications.

The use of robotics and automated machinery has been relatively more in industrial and medical purposes, although as technology moves forward more innovations are coming up. Several research articles have recently appeared in literature; like advancement in electronics has empowered engineers to build robots that are capable of operating in unstructured environments [5]. Camera-in-hand robotic systems are becoming increasingly popular wherein a camera mounted on the robot, which provides images of the objects located in the robot's work space [6]. Huge number of robots are being used to separate, grade and package different products of an industry. In [7], the system detects the fruits and classify them into different category while in [8] with the use of machine vision system the irregularities in seed trays are detected. The work detail in this paper is an effort to design and develop a robotic arm for industries where the manufactured products are visually identical and defects can be detected using our eyes. The system was required to have capabilities of sending the images on a regular basis using a camera. The specification for the arm included the ability to move it anywhere in x, y, z plane and is capable of separating out the defective products.

The rest of the paper is organized as follows. Section-2 gives the overview and architecture. The hardware and the software implementation details are given in section-3 and section-4 respectively. The modular testing and systems integration procedures are presented in Section-5. Finally the paper ends with a discussion and enumeration of future work in section-6.

2   Proposed Algorithm for Defect Detection

2.1   Development of Standard Image

Images of 100-1000 identical perfectly manufactured products are taken and stored in a matrix. All the matrices are added and averaged out for creating a reference image. In this way new matrix is produced for reference and the rest of the product's images have been compared with this image. While taking the image of the final products one must keep in mind the intensity of light must not change during whole procedure and placement of object must be same for all objects.

2.2   Error Detection

Two cameras are installed facing each other and in opposite sides across the conveyer belt in order to obtain the complete image of the object from forward as well as backward direction. The product is compared with standard image for defects and processed accordingly. In case of defects, serial command is sent to the robotic arm for separating out the defective product. Two techniques can be used to detect the defective product. First is based on cross correlation [9, 10] between the product and reference product and second is based on matrix subtraction method. In cross correlation method, new image is correlated with the reference image and decision is based on threshold value. If the cross correlation value is more than threshold, the product is taken as defective. In matrix method, new image matrix is subtracted from the reference image matrix and decision is based on threshold value of the difference. If the difference value is more than threshold, the product is taken as defective.

Fig. 1. Overview of Defective Products with Standard One

3   Technical Aspects

In a bottle production factory, when the bottle is traversing the path given by conveyer belt it may or may not be defective. This is decided by the use of digital image processing. After detection of defective product, instructions are sent to microcontroller through serial communication and robotic arm will react accordingly to separate the defective product [11]. Robotic arm comprises of servo motors for different movements as per specified commands.

3.1   Defective Object

If it receives an image of a defective item the defect is detected by the digital image processing and command is sent to the microcontroller.

Step1. Resting Position
The robotic arm must be initially in a resting position. The position is selected in such a way that it must not interfere the movement of objects on conveyer belt. It must be as far as possible from the conveyer belt.

Step2. Gripping of Object
Robotic arm will go to the desired location as coded in the program after gripping the object to avoid collision with rest of the objects (bottles). Finally the product will be dropped in defective compartment.

Step 3: Coming back to Normal Position
After removing the defective object from the conveyer belt it will again come to its rest position and will wait for further signal by the Matlab. After detecting the defective piece, this will be repeated for detection of defective product.

3.2   Robotic Arm Description

The robotic arm consists of 4 servo motors and 3 degree of freedom as shown in the figure 2. Position of all three servo motors is controlled by PWM signal output of the microcontroller and is fed to the servo motor.

Servo s1 is fitted in order to successfully perform gripping and ungripping action. s2, s3 servos are fitted in order to provide the motion in x and y direction in the plane containing s1.

Here are the three sets of angle to conduct the operation of removal of defective object from the conveyer belt.

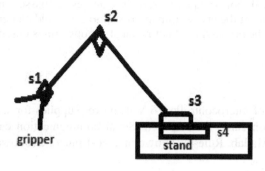

Fig. 2. Robotic Arm Description

Step 1: Rest Position

Assuming theta1, theta2, and theta3 are the initial angles of the robotic arm for the resting position and the robotic arm must be at this position in case of no defect. In absence of defective object, angles are selected so that the robotic arm will not disrupt the motion of the conveyer belt and must avoid collision.

Step 2: Gripping Position

The s4 servo motor in figure 2 is responsible for rotation of complete robotic arm in clockwise or anticlockwise direction and s2 and s3 servo motors are responsible for the motion along the x-y direction in the plane containing s3 and s2. And s1 servo motor is responsible for gripping and un-gripping. In case of defective bottle, servos are adjusted such that the bottle is between the flanges of gripper and gripping action takes place.

Step 3: Placing Position

When the gripper grips the bottle it has to be placed in defective bottles group. Robotic arm places the bottle in defective bottles group with the help of three servo motors. S1 servo grips, s2 lifts up the bottle and s3 places in defective lot.

3.3   Complete Process

With the use of digital image processing first it will detected whether there is defect in the bottle or not. First image is converted in grayscale and then stored in a matrix. The average of different bottles images are taken to make standard image more perfect. After every finite amount of time our camera captures the image and sends it to Matlab for processing. The image is compared with the standard image and subtracts the current image from the standard image. And if there is some error then the determinant of resultant matrix will not be equal to zero. The angles are passed in the Matlab code for specific positions. At the end, command is sent to grip, hold and ungrip the bottle to the required place. The processing of actuation of robotic arm is described in next part

4   Rules

Total 4 PWM pins of microcontroller, 6 V dc power supplies for servo motors and a serial communication cable are required. Serial communication cable connects the microcontroller to Matlab. Rules as shown in table-I must be followed to conduct the process safely.

Rule 1: Gripping of Object
First s4 will take turn so that bottle, s3, s2 and s1 comes in one plane then s3 will rotate to a particular angle so that the y component of the bottle traversed. Then s2 will rotate in order to cover the s direction and bottle comes between the arms of the gripper. Then gripper servo s1 will grip the bottle. The coding for gripping algorithm is done based on these facts.

Rule 2: Separation of Products
Considering the torque requirement and safety motor s3 will rotate first so that the bottle is lifted. Meanwhile the motor s2 will rotate inwards to minimize the torque distance and once the bottle is lifted from the conveyer the motor s4 will rotate. After this motor s3 will rotate to come down at the level of the cartoon and then s2 will adjust it such that the bottle is placed safely and needs just to ungrip.

Rule 3: Resuming of Rest Position
After the ungripping of bottle the robotic arm needs to come again to its resting position and this can be done by feeding the angles of resting position. No matter which motor first rotate they must be set in such a way that they do not collide with any setup.

Table 1. Rules for Robotic Arm Movements

Robotic Arm Position	Theta 1	Theta2	Theta3	Theta 4
Resting Position	0	60	90	90
Gripping Position	180	90	45	0
Ungripping Position	0	60	90	90

5   Simulation Results

In industries the products are manufactured in bulk. Suppose in coca cola industry the bottles are packed in a cartoon and these cartoons are required to be shifted to truck or final position manually. In case of non-defective product, the product will be directed to carton but in case of defective product, the product will be directed to defective lot with the help of Robotic arm. The complete process of picking, carrying and placing the defective product during manufacturing is shown in Figure3, Figure4 and Figure5. For this experiment, I have taken 9 non defective pens and one defective pen (without cap) as shown in Figure6. It has detected the defective pen and picked it up using robotic arm and placed it to another place. In this way, a robotic arm can reduce manual work in industries with the simple usage of image processing and serial communication.

Fig. 3. Gripping Position

Fig. 5. Placing Position

Fig. 4. Resting Position

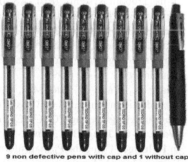

9 non defective pens with cap and 1 without cap

Fig. 6. Sample of a Defective Item with Non-defective Items

6   Conclusion

In the experimental set-ups explained in above paragraphs, the picking up and dropping of defective pen was successful. This work can be extended for industrial applications to save manpower. The robotic arm will be much effective in hazardous places from safety point of view.

References

[1] Chang, C.-Y., Lin, S.-Y., Jeng, M.: Using a Two-layer Competitive Hopfield Neural Network for Semiconductor Wafer Defect Detection. In: Proceedings of the IEEE International Conference on Automation Science and Engineering, Edmonton, Canada, August 1-2, pp. 301–306 (2005)

[2] Robinson, A.P., Lewin, P.L., Swingler, S.G.: Detection of Manufacturing Defects in Polymeric Cable Joint Insulation using X-rays. In: Conference Record of the IEEE International Symposium on Electrical Insulation, pp. 34–37 (2006)

[3] Priya, S., Kumar, T.A., Paul, V.: A Novel Approach to Fabric Defect Detection Using Digital Image Processing. In: Proceedings of International Conference on Signal Processing, Communication, Computing and Networking Technologies (ICSCCN 2011), pp. 228–232 (2011)

[4] Clement, W.I.: An Instructional Robotics and Machine Vision Laboratory. IEEE Transactions on Education 37(I), 87–90 (1994)

[5] Garcia, G.J., Pomares, J., Torres, F.: Automatic robotic tasks in unstructured environments using an image path tracker. Control Engineering Practice 17, 597–608 (2009)

[6] Kelly, R., Carelli, R., Nasisi, O., Kuchen, B., Reyes, F.: Stable visual servoing of camera-in-hand robotic systems. IEEE-ASME Transactions on Mechatronics 5, 39–48 (2000)

[7] Reyes, J.F., Chiang, L.: Location and Classification of Moving Fruits In Real Time with A Single Color Camera. Chilean Journal of Agricultural Research 69, 179–187 (2009)

[8] Wang, H., Cao, Q., Masateru, N., Bao, J.: Image processing and robotic techniques in plug seedling production. Transactions of the Chinese Society of Agricultural Machinery 30, 57–62 (1999)

[9] Zhang, Y., Wang, Y., Zuo, M.J., Wang, X.: Ultrasonic Time Of Flight Diffraction Crack Size Identification Based on Cross Correlation. In: Canadian Conference on Electrical and Computer Engineering CCECE/CCGEI IEEE, pp. 1737–1780 (2008)

[10] Wang, C.-C., Jiang, B.C., Lin, J.-Y., Chu, C.-C.: Machine Vision-Based Defect Detection in IC Images Using the Partial Information Correlation Coefficient. IEEE Transactions on Semiconductor Manufacturing 26(3), 378–384 (2013)

[11] Laddha, N.R., Thakare, A.P.: A Review on Serial Communication by UART. International Journal of Advanced Research in Computer Science and Software Engineering 3(1), 366–369 (2013)

[12] Seelye, M.: Camera-in-hand robotic system for remote monitoring of plant growth in a laboratory. In: 2010 IEEE Instrumentation & Measurement Technology Conference Proceedings (2010)

Contour Extraction and Segmentation of Cerebral Hemorrhage from MRI of Brain by Gamma Transformation Approach

Sudipta Roy[1,3], Piue Ghosh[2,4], and Samir Kumar Bandyopadhyay[1,5]

[1] Department of Computer Science and Engineering
Academy of Technology, Adisaptagram, West Bengal, India
[2] Department of Electronics & Communication Engineering
Academy of Technology, Adisaptagram, West Bengal, India
[3] Academy of Technology, Adisaptagram, West Bengal, India
[4] Sir J. C. Bose School of Engineering, Mankundu, West Bengal- 712139
[5] University of Calcutta, 92 A.P.C. Road, Kolkata-700009, India
sudiptaroy01@yahoo.com, piu.rpe@gmail.com, skb1@vsnl.com

Abstract. Computer-aided diagnosis (CAD) systems have been the focus of several research endeavors and it based on the idea of processing and analyzing images of different hemorrhage of the brain for a quick and accurate diagnosis. We use a gamma transformation approach with a preprocessing step to segment and detect whether a brain hemorrhage exists or not in a MRI scans of the brain with the type and position of the hemorrhage. The implemented system consists of several stages that include artefact and skull elimination as an image preprocessing, image segmentation, and location identification. We compare the results of the conducted experiments with reference image which are very promising visually as well as mathematically.

Keywords: Brain Hemorrhage, Brain MRI Scans, CAD Systems, Image Processing, Image Segmentation, Position of Hemorrhage.

1 Introduction

Nowadays, cerebrovascular diseases are the third cause of death in the world after cancer and heart diseases [1]. One of the cerebrovascular diseases is brain hemorrhage which is caused by a brain blood vessel busting causing bleeding. There are many types of brain hemorrhage such as: epidural, subdural, subarachnoid, cerebral, and intraparenchymal hemorrhage. These types differ in many aspects such as the size of the hemorrhage region, its shape, and its location. MRI and CT are equally helpful for determining when hemorrhage is present [2]. MRI can be more sensitive than CT for identifying other types of neurological disorders that mimic the symptoms of stroke. MRI has a much greater range of available soft tissue contrast, depicts anatomy in greater detail, and is more sensitive and specific for abnormalities within the brain itself. In medical image segmentation, pixels are labeled by tissue type. Image segmentation is required in many medical applications ranging from the education and assessment of medical students to image-guided surgery and surgical simulation.

© Springer International Publishing Switzerland 2015

S.C. Satapathy et al. (eds.), *Proc. of the 3rd Int. Conf. on Front. of Intell. Comput. (FICTA) 2014*
– *Vol. 2*, Advances in Intelligent Systems and Computing 328, DOI: 10.1007/978-3-319-12012-6_42

Appropriate segmentation methods are highly dependent on image acquisition modality and the tissue of interest because it provides superior contrast of soft tissue structures MRI is the method of choice for imaging the brain and most research on brain segmentation focuses on MRI. The detection of hemorrhage is essential in the diagnosis and management of a variety of intracranial diseases, including hypertensive hemorrhage, hemorrhagic infarction, brain tumor, cerebral aneurysm, vascular malformation, trauma, hemorrhagic change following radio- or chemotherapy, and hemorrhagic pial metastasis. A recent report has indicated that the detection of hemorrhage on MR images is useful for the grading of gliomas. The T2-weighted gradient-recalled echo sequence has been reported to be more sensitive than the T2-weighted spin-echo and fast SE sequences to the magnetic susceptibility induced by static field inhomogeneities arising from paramagnetic blood breakdown products. From their name, CAD systems use computers to help physicians reach a fast and accurate diagnosis. CAD systems are usually domain-specific as they are optimized for certain types of diseases, parts of the body, diagnosis methods, etc. They analyze different kinds of input such as symptoms, laboratory tests results, medical images, etc. depending on their domain. One of the most common types of diagnosis is the one that depends on medical images. Such systems are very useful since they can be integrated with the software of the medical imaging machine to provide quick and accurate diagnosis. On the other hand, they can be challenging since they combine the elements of artificial intelligence and digital images processing. This work presents a CAD system to help detect hemorrhages in MRI scans of human brains and identify their types if they exist. Some of the older works [3, 4] addressed the problem of segmenting the region of intracerebral hemorrhages. In the former work, they used a spatially weighted k-means histogram-based clustering algorithm, whereas in the latter work, they applied a multiresolution simulated annealing method. Cheng and Cheng [5] proposed a Fuzzy C-Means (FCM) method based on multiresolution pyramid for brain hemorrhage analysis. They also compared FCM, competitive Hopfield neural network [6] and fuzzy Hopfield neural network [7] in the global thresholding stage. In another work, Liu et al. [8] proposed an Alternative Fuzzy C-Means (AFCM) method for the segmentation phase. A more recent work based on FCM is the work of Li et al. [9]. The authors used a thresholding technique based on FCM clustering to remove all non-brain regions. The results of this clustering were brain regions that were segmented into slices using median filtering to eliminate noise in the image, and then the maximum entropy threshold was computed for each slice chart to determine the potential hemorrhage regions. In the last phase, the hemorrhage regions were determined according to their locations and gray level statistics. The results of this work are extremely encouraging, however, as the authors acknowledge, it is still a premature work. The proposed system can also be helpful for teaching and research purposes. It can be used to train senior medical students as well as resident doctors. Moreover, it can be very helpful in detecting hemorrhage through large sets of MRI of brain scans.

This paper is organized as follows: In this section, related works are briefly discussed. The proposed methodology is discussed in section 2. In section 3, the experiments conducted to test the quality of the results generated by the proposed method is presented. In section 4, accuracy measurement with ground truth image has been discussed. The conclusion part has been discussed in section 5.

2   Proposed Methodology

The Here data set of "Whole Brain Atlas" image data base [10] which consists of T1 weighted, T2 weighted, proton density (PD) MRI image and part of the dataset slice wise has been used. The RGB image has been converted it into grayscale image by forming a weighted sum of the R, G, and B components multiplied by constant and the transformation function is given below

$$g(x, y) = T[f(x, y)]$$

Where f(x, y) is the input image, g(x, y) is the processed image, and T is an operator on f, defined over some neighborhood of f(x, y). In this case, g depends only on the value of f at (x, y), and T becomes a gray-level (also called an intensity or mapping) transformation function of the form

$$s = T(r)$$

Where, for simplicity in notation, r and s are variables denoting, respectively, the gray level of f(x, y) and g(x, y) at any point (x, y). After that we use a binarization as a preprocessing step and to calculate the standard deviation mean and variance derivation is written below, mean is defined by one divided by the number of samples multiplied by the sum of all data points

$$\mu = \frac{1}{MN} \sum_{x=0}^{M-1} \sum_{y=0}^{N-1} f(x, y)$$

Variance, symbolized by v, equals 1 divided by the number of samples minus one, multiplied by the sum of each data point subtracted by the mean then squared.

$$v = \frac{1}{MN} \sum_{x=0}^{M-1} \sum_{y=0}^{N-1} (f(x, y) - \mu)^2$$

Standard deviation, symbolized by σ, equals the square root of the variance s-squared is written below

$$\sigma = \sqrt{v}$$

Depending upon this standard deviation intensity as threshold intensity we binaries the MRI of brain scan image. This standard deviation based binarization is very much helpful for extracting main brain portion to the non brain portion. MRI of brain has the large intensity difference between background and the foreground and the use of standard deviation based binarization has been successfully executed for brain stroke detection purpose.

$$f1(x,y) = \begin{cases} 1 & \text{if } f(x,y) > \sigma \\ 0 & \text{if } f(x,y) \leq \sigma \end{cases}$$

The negative of an image with gray levels in the range [0, L-1] is obtained by using the negative transformation is given by the expression

$$s = L - 1 - r$$

Reversing the intensity levels of an image in this manner produces the equivalent of a photographic negative and binary image complement we use f2(x,y)=1-f1(x,y) which helps us for the next step wavelet decomposition. We begin by defining the wavelet series expansion of function f2(x) $\in$ L2(R) relative to wavelet $\psi(x)$ and scaling function $\varphi(x)$. f2(x) can be represented by a scaling function expansion and some number of wavelet function expansions in subspaces W_{j_0} , W_{j_0+1}, W_{j_0+2} ..... Thus

$$f2(x) = \sum_k c_{j_0}(k)\varphi_{j_0}(x) + \sum_{j=j_0}^{\infty}\sum_k d_j(k)\psi_{j,k}(x)$$

Where j_0 is an arbitrary starting scale and the $c_{j_0}(k)$ and $d_j(k)$ are relabeled. The $c_{j_0}(k)$ normally called approximation or/and scaling coefficients; the $d_j(k)$ are referred to as detail or/and wavelet coefficients. Thus in the above equation first sum uses scaling function to provides an approximation of f2(x) at scale j_0. For each higher scale $j \geq j_0$ in the second sum, a finer resolution function a sum of wavelet is added to the approximation to provide increasing details. If the expansion function forms an orthogonal basis or tight frame, which is often the case, the expansion coefficients are calculated and shown in below

$$c_{j_0}(k) = <f2(x), \varphi_{j_0}(x)> = \int f2(x)\,\varphi_{j_0}(x)\,dx$$

and

$$d_j(k) = <f2(x), \psi_{j,k}(x)> = \int f2(x)\,\psi_{j,k}(x)\,dx$$

Above two coefficients expansion are defined as inner products of function being expanded and the expansion functions being used where φ_{j_0} and $\psi_{j,k}$ are the expansion functions; c_{j_0} and d_j are the expansion coefficients. Two dimensional (2-D) scaling function, $\varphi(x,y)$, which is a product of two 1-D functions and   three two dimensional wavelets, $\psi^H(x,y)$, $\psi^V(x,y)$, and $\psi^D(x,y)$ are required. Excluding products that produce 1-D results, like $\varphi(x)\,\psi(x)$, the four remaining products produce the separable scaling function and separable directionally sensitive wavelets

$$\varphi(x,y) = \varphi(x)\varphi(y)$$
$$\psi^H(x,y) = \psi(x)\psi(y)$$
$$\psi^V(x,y) = \psi(x)\psi(y)$$
$$\psi^D(x,y) = \psi(x)\psi(y)$$

The wavelets measure functional variations intensity variations for images along different directions: ψ^H measures variations along columns, ψ^V measures variations along rows, and ψ^D measures variations along diagonals. The directional sensitivity is a natural consequence of reparability in the above equation and it does not increase the computational complexity. We first define the scaled and translated basis functions:

$$\varphi_{j,m,n}(x,y) = 2^{\frac{j}{2}}\varphi(2^j x - m, 2^j y - n)$$
$$\psi^i_{j,m,n}(x,y) = 2^{\frac{j}{2}}\psi^i(2^j x - m, 2^j y - n), \quad i = \{H,V,D\}$$

Where index i identifies the directional wavelets. The discrete wavelet transform of image $f2(x, y)$ of size $M{\times}N$ is then

$$W_\varphi(j_0, m, n) = \frac{1}{\sqrt{MN}}\sum_{x=0}^{M-1}\sum_{y=0}^{N-1} f2(x,y)\varphi_{j_0,m,n}(x,y)$$

$$W_{\psi^i}(j, m, n) = \frac{1}{\sqrt{MN}}\sum_{x=0}^{M-1}\sum_{y=0}^{N-1} f2(x,y)\psi^i_{j,m,n}(x,y), i = \{H,V,D\}$$

As in the 1-D case, j_0 is an arbitrary starting scale and the $W_\varphi(j_0, m, n)$ coefficients define an approximation $f2(x, y)$ at scale j_0. The $W_{\psi^i}(j, m, n)$ coefficients add horizontal, vertical, and diagonal details for scales $j{\geq}j_0$. normally $j_0{=}0$ and $N{=}M{=}2J$ so that $j{=}0,1,2,....,J{-}1$ and $m{=}n{=}0,1,2,....,2j{-}1$. Thus $f2(x, y)$ is obtained via the inverse discrete transform is given by

$$f2(x,y) = \frac{1}{\sqrt{MN}}\sum_{m}\sum_{n} W_\varphi(j_0, m, n)\varphi_{j_0,m,n}(x,y)$$
$$+ \frac{1}{\sqrt{MN}}\sum_{i=H,V,D}\sum_{j=j_0}^{\infty}\sum_{m}\sum_{n} W_{\psi^i}(j, m, n)\psi^i_{j,m,n}(x,y)$$

After applying wavelet decomposition up to level two non brain portions are totally separated from brain portion in a discrete form which is not useful for our approach, in this case some abnormality may lost and to remove this problem we use a quick hull algorithms [11].The convex hull of a set of points in the plane is the shape taken by a rubber band that is placed "around the points" and allowed to shrink to a state of equilibrium. We can represent a line l as a triple (a;b;c), such that these values are the coefficients a, b, and c of the linear equation ax+by+c = 0 associated with l. Given the Cartesian coordinates $(x_1; y_1)$ of q_1 and $(x_2; y_2)$ of q_2, the equation of the line l through q_1 and q_2 is given by

$$\frac{x - x_1}{x_2 - x_1} = \frac{y - y_1}{y_2 - y_1}$$

From which we derive a = (y_2-y_1); b = $-(x_2-x_1)$; c = $y_1(x_2-x_1) - x_1(y_2-y_1)$. A line segment s1 is typically represented by the pair (p;q) of points in the plane that form s1's end points. The segments between consecutive vertices of P are called the edges of P. A polygon is convex if it is simple and all its internal angles are less than p. Quick-hull [11] is a divide-and-conquer algorithm, similar to quick sort, which divides the problem into two sub-problems and discards some of the points in the given set as interior points, concentrating on remaining points. Quick-hull runs faster than the randomized algorithms because it processes fewer interior points. Now this convexed image is now multiplied with original image without any artefact, any noise and skull which is very much effective for brain abnormality. Then we use power-law transformations which have the basic form

$$f4(x, y) = c * f3(x, y)^\gamma$$

Where c and γ are positive constants, sometimes above equation is written as

$$f4(x, y) = c * (f3(x, y) + \varepsilon)^\gamma$$

To report for a measurable output when the input is zero we use ε. However, offsets typically are an issue of display calibration and as a result they are generally ignored. Transform values of $\gamma>1$ have accurately the opposite effect as those generated with principle values of $\gamma <1$ and to the identity transformation when c= γ =1. Gamma correction is significant if displaying an image exactly on a computer screen is of concern. Thus the final selection is given by

$$T = \frac{1}{MN} \sum_{x=0}^{M-1} \sum_{y=0}^{N-1} f4(x, y) + \sqrt{\frac{1}{MN} \sum_{x=0}^{M-1} \sum_{y=0}^{N-1} (f4(x, y) - \mu)^2}$$

On the basis of the final intensity value we find the hemorrhage portion which is binary output and store it in f5(x, y). We can classify different type of hemorrhage by multiplying original input image or without skull image. A basic definition of the first-order derivative of a one-dimensional function f(x) is the difference and as we deal with binary image to detect the contour thus first derivative can apply for contour detection. Horizontal contour is defined by

$$h_c = \frac{\partial f}{\partial x} = f5(x + 1) - f5(x)$$

Vertical contour is defined by

$$v_c = \frac{\partial f}{\partial y} = f5(y + 1) - f5(y)$$

Horizontal and vertical contour detection does not produce continuous line which is not desire for contour detection thus we combining both horizontal and vertical contour by:

$$c_c = h_c + v_c$$

For abnormality position detection we need to calculate centroid of the brain stroke region by which we can classify the type of hemorrhage, this is done by weighted mean of the pixels and mathematical calculation of centroid is given below:

$$X_{cood} = \sum_{n=1}^{p} x_n I_n / \sum_{n=1}^{p} I_n$$
$$Y_{cood} = \sum_{n=1}^{p} y_n I_n / \sum_{n=1}^{p} I_n$$

We also find the top, bottom, left and right of the abnormal region from the top, bottom, left and right of brain and center of brain to the centroid of the brain abnormality by using simple calculation the distance between two points.

3   Results and Discussion

The researchers conducted to test the superiority of the outcome generated by the proposed method are discussed with proper data set and proper mathematical formulation. The collected dataset [10] consists of 60 MRI images of human brain with different type of hemorrhage and some normal brain image. 20 of these images are for normal brain while the remaining images represent brains that suffer from one of the three types of brain hemorrhage considered in this work. An image of a normal brain shows a distribution of gray matter that appears as clearness in the texture-like fissures, while an abnormal brain has a shape which appears brighter than the normal gray matter. Generally abnormal region of a brain has the different characteristics than the normal brain but brain has very variable characteristics than other organ for T1, T2, PD type of MRI images. To segment brain hemorrhage segmentation we apply several steps and all these steps are shown in Figure 1 below. We have taken

Figure 1(A) as an input MRI of brain and Figure 1(B) is the binarized output which is very much efficient for extracting the skull portion. Figure 1(C) is complemented form of binary image and we use this step to apply the wavelet decomposition. We apply wavelet decomposition up to level 2 shown in Figure 1(D) which is helpful to segment the main brain portion. Figure 1(E) is the complemented form of wavelet decomposition and the image re-composition and Figure 1(F) is the output after applying quick-hull algorithms which is helpful for continuous image. Figure 1(G) is the output image after gamma law transformation which helpful for segment the brain hemorrhages and we segment the brain hemorrhage using expectation maxima. Finally in Figure 1(H) is the segmented hemorrhage portion, Figure 1(I) is the contour detection by horizontal contour detection and Figure 1(J) is the vertical contour detection. Contour line are not continuous for the horizontal and vertical contour that's why we combine the both in to final contour which is continuous as well as appropriate is shown in Figure 1(K). Figure 1(L) is position of the hemorrhage, depending on the position we can classify the different type of brain hemorrhage, it may be appear on the skull or inside the brain.

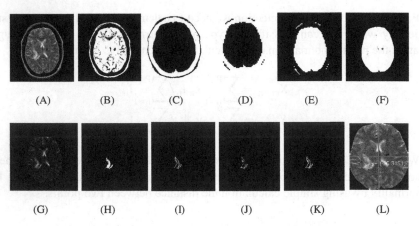

Fig. 1. A) input MRI of brain scan, B) binarized MRI, C) complemented image, D) wavelet decomposition, E) recomposed image, F) convex hull, G) after gamma transformation, H) segmented abnormal portion, I) horizontal contour, J) vertical contour, K) contour of abnormal region, L) position of abnormal region.

Representation of the interior architecture and location of hemorrhagic lesions is important because the appearance of the internal structure helps us to understand the stage of hematomas or to differentiate idiopathic hemorrhage from hemorrhage caused by pathologic conditions, such as intraaxial tumor or vascular malformation. Thus in this case we successfully execute our output by Figure 1(L). The epidural hemorrhage is characterized by its convex shape and its close fitting with the skull. In the above figure hemorrhage has been found on the left lobe of the brain, no relation with the skull portion which represents it is cerebral and intracranial hemorrhage. In our experiment we apply on T2, T1, PD type of MRI image for cerebral hemorrhage detection and our methods successfully detect abnormality but if we apply other kind of hemorrhage we simple avoid the skull elimination steps.

Epidural hematoma locates into potential space between the dura, which is inseparable from cranial periosteum, and the adjacent bone. Subdural hematoma diagnosed on the basis of mass effect, which is depicted as displacement of the blood vessels on angiograms on skull. We use skull removal methodology when we measure cerebral hemorrhage and other type of brain hemorrhage cases does not use the skull removal part. Thus different type of hemorrhage case depends on the quantification and location of the abnormality portion. The location of different kind of abnormality has depicted in Table1 and area of the abnormality shown in Table 2.

Table 1. Position of abnormality for different kind of MRI

Image name	Center to Centroid		Top to Top	Bottom to Bottom	Left to Left	Right to Right
	X-Coordinate	Y-Coordinate				
11_PD	-34.4897	3.7804	72.2011	65.4599	23.2594	83.6780
12_PD	-33.8723	-12.5154	92.7038	60.4401	23.2594	96.1665
13_PD	-30.1475	-5.0902	58.6941	46.0435	39.2938	85.4400
14_PD	-26.6734	-12.0888	67.7791	48.3735	26.4764	78.4474
11_T1	-31.5000	3.8429	78.7909	80.6040	35.9026	98.7320
13_T1	-31.5000	-10.2500	99.3680	75.1864	36.3593	95.5249
11_T2	-35.1268	3.6087	70.4911	63.8905	22.0907	81.4985
12_T2	-43.0661	-15.2562	100.8464	59.1354	16.0000	100.449
13_T2	-33.0759	-27.2188	114.7693	52.4690	38.6005	85.0000
13_T2	-25.3151	-17.7361	80.3617	50.0400	45.8039	80.4114

The objective of this work is to correctly classify brain CT images into one of four classes: normal, epidural hemorrhage (removing skull elimination step), subdural hemorrhage (removing skull elimination step), cerebral hemorrhage (including skull elimination step) and intraparenchymal hemorrhage (including skull elimination step) by finding their location from Table 1 as well as visual inspection. The noteworthy is the system's ability to distinguish between normal and abnormal cases which are also important for automated medical image analysis.

4   Quantification and Accuracy Estimation

Characterizing the performance of image segmentation methods is a challenge mostly in medical image analysis which has a great impact in our community. Thus accuracy is too much important and truthfulness of a segmentation technique refers to the degree to which the segmentation results agree with the true segmentation. The accuracy measures used to evaluate the performance [12] of the proposed methods are the Relative area Error (RE)[12], Kappa Index (Ki), Jacard Index (Ji), correct detection ratio (Cd) and false detection ration (Fd) has been described below. An important difficulty we have to face in developing segmentation methods is the lack of a gold standard for their evaluation. Although physical or digital phantoms can provide a level of known "ground truth", they are still unable to reproduce the full range of imaging characteristics, normal and abnormal anatomical variability observed in clinical data. First let AV and MV denote [12] the area of the automatically and manually segmented objects and |x| represents the cardinality of the set of voxels x. In the following equations $Tp = MV \cap AV$, $Fp = AV - Tp$ and $Fn = MV - Tp$ denote to the "true positive", "false positive" and "false negative" respectively. The Kappa index between two areas is calculated by the following equation:

$$K\_i\,(AV, MV) = ((2|AV \cap MV|))/((|AV| + |MV|)) * 100\%$$

The similarity index is sensitive to both differences in size and location. The Jacard index between two areas is represented as follow:

$$J\_i\,(AV, MV) = (|AV \cap AM|)/(|Tp + Fn + Fp|) * 100\%$$

This metric is more sensitive to differences since both denominator and numerator change with increasing or decreasing overlap. Correct detection ratio or sensitivity is defined by the following equation:

$$C\_d = |AV \cap MV|/MV * 100\%$$

False detection ratio (Fd) is same except 'AV-Tp' in place of 'AV MV'. The Relative Error [8] (RE) for hemorrhage region can be calculated as "AV" hemorrhage area using automated segmentation, "MV" is hemorrhage area using manual segmentation by expert.

$$RE = ((AV - MV))/MV * 100\%$$

Table 2. Different kind performance metric for different type of MRI image

Image name	AV	MV	RE in %	Tp	Fp	Fn	Kappa index (%)	Jacard index (%)	C_d in %	F_d in %
11_PD	667	755	11.65	667	00	88	93.81	88.34	088.34	0.000
12_PD	495	492	00.60	492	03	00	99.70	99.39	100.00	0.609
13_PD	914	930	01.72	912	02	18	98.92	97.85	098.06	0.215
14_PD	537	543	01.10	533	04	10	98.70	97.44	098.15	0.736
11_T1	078	071	09.59	066	12	03	88.59	82.48	092.95	3.846
13_T1	075	079	05.06	073	02	06	94.80	90.12	092.40	2.731
11_T2	611	608	00.49	605	06	03	99.26	98.53	099.50	0.986
12_T2	530	528	00.37	523	07	05	98.86	97.75	099.05	0.946
13_T2	431	428	00.70	422	09	06	98.25	96.56	098.59	1.401
14_T2	415	428	03.03	411	04	17	97.51	95.14	096.02	0.934

From the above table we can find average value of RE (%)=43.95/11=3.99, average of Kappa index (%) =1063.79/11= 96.70, average of Jacard index (%)=1034.8/11=94.07, average of Cd (%)= 1063.06/11=96.64, and average of

Fd (%)= 22.043/11=2.004. The Kappa index or similar is sensitive to both differences in size and location and differences in location are more strongly reflected than differences in size and Ki > 90% indicates a good agreement in our experiment kappa index reaches 96.7%. Jacard index maximum times it gives greater than 90% (average value 94.07%) which indicates our experiment promising. Correct detection ratio indicates the correct detection area normalized by the reference area and is not sensitive to size and it reaches 96.64%. Using this metric with the correct detection ratio can give a good evaluation of the segmentation. Maximum times our methodology gives above 95% correct detection ratio and below 5% false detection ratio with very low relative area error (maximum times below 4%).

5   Conclusion

Accurate measurement of clot thickness, hematoma area, and location on MRI scan have been successfully executed which is important because of need for accurate and rapid diagnosis and treatment, prompt transfer of the patient to a facility capable of MRI scanning and neurological intervention is necessary. Automated systems for analyzing and classifying medical images have gained a great level of attention by our proposed method. The results are encouraging and a higher accuracy for the 3-class classification problem can be attained by obtaining a better dataset with high resolution images taken directly from the MRI of brain. This will allow the future system to reach a level that will allow it to be a significant asset to any medical establishment dealing with brain hemorrhages.

References

1. Li, Y., Hu, Q., Wu, J., Chen, Z.: A hybrid approach to detection of brain hemorrhage candidates from clinical head ct scans. In: Proc. IEEE Sixth International Conference on Fuzzy Systems and Knowledge Discovery (FSKD), vol. 1, pp. 361–365 (2009)
2. (May 2014), http://nihseniorhealth.gov/stroke/preventionanddiagnosis/01.html
3. Loncaric, S., Dhawan, A., Broderick, J., Brott, T.: 3-d image analysis of intra-cerebral brain hemorrhage from digitized ct films. Computer Methods and Programs in Biomedicine 46(3), 207–216 (1995)
4. Loncaric, S., Majcenic, Z.: Multiresolution simulated annealing for brain image analysis. In: Medical Imaging. International Society for Optics and Photonics, pp. 1139–1146 (1999)
5. Cheng, D., Cheng, K.: Multiresolution based fuzzy c-means clustering for brain hemorrhage analysis. In: Proceedings of the 2nd International Conference on Bioelectromagnetism, pp. 35–36. IEEE (1998)
6. Cheng, K., Lin, J., Mao, C.: The application of competitive hopfield neural network to medical image segmentation. IEEE Transactions on Medical Imaging 15(4), 560–567 (1996)
7. Lin, J., Cheng, K., Mao, C.: A fuzzy Hopfield neural network for medical image segmentation. IEEE Transactions on Nuclear Science 43(4), 2389–2398 (1996)

8. Liu, H., Xie, C., Chen, Z., Lei, Y.: Segmentation of ultrasound image based on morphological operation and fuzzy clustering. In: Third IEEE International Workshop on Electronic Design, Test and Applications, DELTA (2006)
9. Li, Y., Hu, Q., Wu, J., Chen, Z.: A hybrid approach to detection of brain hemorrhage candidates from clinical head ct scans. In: Sixth International Conference on Fuzzy Systems and Knowledge Discovery (FSKD), pp. 361–365 (2009)
10. (2014), http://www.med.harvard.edu/AANLiB/cases/
11. Barber, C.B., Dobkin, D.P., Huhdanpaa, H.: The Quickhull Algorithm for Convex Hulls. ACM Transactions on Mathematical Software 22(4), 469–483 (1996)
12. Roy, S., Bandyopadhyay, S.K.: A Review on Volume Calculation of Brain Abnormalities from MRI of Brain using CAD system. International Journal of Information and Communication Technology Research 4(4), 114–120 (2014)
13. Roy, S., Chatterjee, K., Bandyopadhyay, S.K.: Segmentation of Acute Brain Stroke from MRI of Brain Image Using Power Law Transformation with Accuracy Estimation. In: ICACNI 2014, Kolkata. Smart Innovation, Systems and Technologies, vol. 27, pp. 453–461. Springer (2014)
14. Roy, S., Bandyopadhyay, S.K.: Abnormal Regions Detection and Quantification with Accuracy Estimation from MRI of Brain. In: IMSNA 2013, Canada. Proc. IEEE, pp. 611–615 (2013)
15. Roy, S., Nag, S., Maitra, I.K., Bandyopadhyay, S.K.: A Review on Automated Brain Tumor Detection and Segmentation from MRI of Brain. IJARCSSE 3(6), 1706–1746 (2013)

An Innovative Approach to Show the Hidden Surface by Using Image Inpainting Technique

Rajat Sharma and Amit Agarwal

Computer Science and Engineering Department,
ABES Engineering College, Ghaziabad – 201009 (U.P.), India
march10rajat@gmail.com, amitkagarwal@abes.ac.in

Abstract. The research study presented in this paper, focuses on the problems and lacunas of existing image inpainting techniques and shows how proposed approach will prove to be a curing syrup to diseased concepts, available so far for image inpainting. Since, the paper is highlighting image inpainting technique's drawbacks so the former part discusses what actually image inpainting technique means and where this great revolutionary need have its implementations and applications in the real world. Further, the purpose of image inpainting with various existing and latest algorithms/methods which are available so far, to inpaint an image are highlighted as a part of literature survey. The prime focus is to discuss, the innovative approach of the authors to remove disadvantage of existing image inpainting techniques i.e. if an object is small in size and is hidden behind a bigger object then by available inpainting techniques it is next to impossible to generate the image of hidden object as if bigger front object is selected as target region, then whole object along with the hidden object (behind bigger object) will also be removed during the time of object removal phase of inpainting. So, in final phase of paper, various descriptive images and live examples, methodology of whole proposed technology and self-proposed algorithms are discussed to remove this lacuna of the available inpainting techniques. Besides all, the resultant image will have the bigger front object getting transparent and only hidden smaller object as visible on the background image by implementation of proposed concept.

Keywords: Image Inpainting, hidden, damaged portions, target region, transparent, object removal, region filling, Snapshot.

1 Introduction

The term inpainting has been originated from art restoration, which can also be named as re-touching. The purposes for retouching the image could be called as Revert deterioration (e.g. scratches and dust spots on the photographic film), or addition and removal of elements from images .In multimedia signal processing, the technology which is generally applied to the problem of automatic filling-in the missing region in a visually plausible way of an image is defined as image inpainting [1-3]. This technique had seeked its various implementations in electronic imaging applications,

© Springer International Publishing Switzerland 2015   395
S.C. Satapathy et al. (eds.), *Proc. of the 3rd Int. Conf. on Front. of Intell. Comput. (FICTA) 2014*
– *Vol. 2*, Advances in Intelligent Systems and Computing 328, DOI: 10.1007/978-3-319-12012-6_43

such as digital photo editing (e.g., foreground removal), image restoration (e.g., artwork repairing), and multimedia transmission (e.g., lost data recovering) [4-7]. The various conventional schemes to inpaint an image are- texture-oriented [8] and structure-oriented [5]. The texture-oriented scheme (usually known as the texture synthesis scheme) for inpainting an image, generate a target region from the available sample textures amongst its surroundings [8]. Thus the images with large texture areas only found this scheme useful. On the other hand, structure-oriented scheme uses data fusion techniques to obtain the missing regions, such as bilinear interpolation. Nowadays, exemplar-based inpainting scheme is mostly in use as in this, the visible parts of the image serves as a source set of examples to infer the target regions where the missing data is "replicated" rather than "synthesized" from available information [3]. Therefore, the results generated out while comparing above mentioned technique with other image inpainting approaches, it was found that exemplar based approach is very effective in reducing the undesired blurring artefacts and is also applicable to both the small and large image gaps.

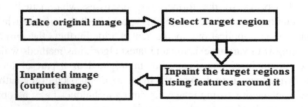

Fig. 1. Procedure to inpaint an image

After discussions of the image inpainting terms, further this paper is divided as in the following sections- Section II throws some light on literature survey of all the image inpainting techniques available so far. Then, Section III shows the inclusion of disadvantages of the existing inpainting techniques as motivation for research study and what is the main proposed methodology of the authors i.e. how they will make a visible object invisible and an invisible object visible from their innovative approach, seek its deep discussions. This section comprises of detailed description of whole procedure along with a self-proposed algorithm showing how the whole proposed methodology will be simulated into a live working module.

Fig. 2. Image showing different phases of object removal and region filling

Also, the methodology is perfectly shown as in live working by some descriptive images portraying the real applications of this concept. At last, later sections include applications of the approach along with references.

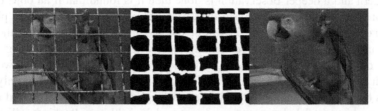

Fig. 3. Image showing the illusion of disappearing cage by using existing image inpainting technique [16]

2 Literature Survey

There have been several numerous approaches proposed for doing image inpainting where its applications could be summarized as- restoration of photographs, films and paintings, removal of occlusions, like text, subtitles, stamps etc. from images. The term digital image inpainting was coined by Bertalmio et al. where inpainting which was based on Partial Differential Equations (PDE) was discussed. The above mentioned idea of suggested method was the smooth completion of isophote lines that arrive at border of the region which is needed to be inpainted, from outside of border to the inner region [11-13]. Two drawbacks of image inpainting methods based on PDE were- it only performs well on small inpainting regions and is not able to fill in the texture. Later on, Criminisi et al. presented a method – called exemplar-based image inpainting - that uses the main idea PDE and is able to fill in texture [9]-[13].

Further by using this approach bigger regions can also be filled. Thus, the result analysis has shown that for image inpainting, the most popular texture synthesis technique is exemplar-based synthesis. Further, after couple of years, Drori et al came with his theory where he iteratively approximates the missing regions [14] from coarse to fine levels using the principle of self-similarity. Also, some fast algorithms were developed and subsequently the visual quality was accordingly degraded. Criminisi et al exploited a patch based algorithm, where filling order is decided by a predefined priority function that could ensure the propagation of linear structures before texture filling so that the connectivity of object boundaries could be preserved. Another algorithm has been proposed for inpainting digital images based on the Poisson equation where the image to be repaired is decomposed into the structure image and the texture image, which are processed respectively according to their image characteristics [15]. So, like this in the literature survey almost all the existing image inpainting techniques so far had been discussed with their innovations, technicalities, advantages as well as disadvantages.

Motivation: The main disadvantage that existing image inpainting technologies included is discussed by the authors as, the available image inpainting techniques remove the object(s) from photographs, mark the target region [9] and then fill the hole (left after removal of target region) with information extracted from the surrounding area by using inpainting techniques (eg. region filling and object removal by exemplar based image inpainting [9]). Now, if the discussed technique is taken into concern and a deeper observation is done then it is found that if an object is kept behind another object (larger in size than previous object) then during the phase of selection of target, the bigger object will be marked and will be removed as the image is two dimensional and the hidden behind object will also be removed alongwith the removal of previous object's image. Thus never ever the hidden object would be visible in the image which is available with us. So, to solve this purpose only, some concepts have been proposed by the authors through which the object placed in front of smaller object will behave as transparent and invisible small object will become visible.

3   The Proposed Methodology

To present the research study, a unique concept by which the visible content of an image will become invisible and invisible content of the image will become visible is proposed. The procedure for accomplishing the task proposed in the research study is as follows-

Firstly, by taking camera, click a Snapshot (i.e. Snapshot 1) where the front view or the image of object coming first in coverage of camera's flash is shown in fig. 4).

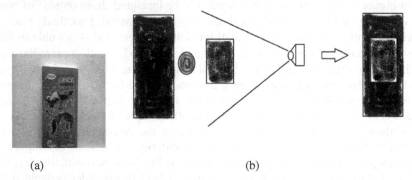

(a)   (b)

Fig. 4. Snapshot 1 showing front view of image (smaller object is hidden in resultant image)

Now, this Snap will show only the screen and larger object, as smaller object will be hidden behind this larger object. So, now taking in concern the objective to show the hidden object or to make the bigger front object transparent, we click second snapshot (i.e. snapshot 2) from the opposite side of camera's snap 1 i.e. hidden object and bigger object are now visible in snapshot 2 (shown in fig.5).

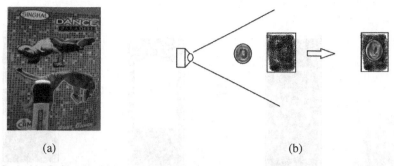

(a)   (b)

Fig. 5. Snapshot 2 showing hidden object and bigger object when snapped from opposite side.

Once both the images are available with us – one showing the bigger object and background, another showing smaller object and bigger object, we will select the portion of smaller object from snapshot 2 by applying the codes designed in matlab for removal of selected smaller object portion so as to get the cropped image of smaller object (as is shown in fig: 6).

Fig. 6. Image showing the cropped portion of snapshot 2

Now, we are having cropped image of smaller object with us, and snapshot 1 is showing bigger object alongwith with background. So, by using existing exemplar based inpainting Technique, we will select the required target region (the bigger (front) object in snapshot 1) and will remove/disappear it (shown in fig.7).

Fig. 7. Image showing targeted region which will be removed from snapshot 1

Subsequently, we will perform region filling Exemplar based image inpainting procedure by using best first algorithm to reconstruct the image by gathering the surrounding information of nearest pixels is shown in the fig.8.

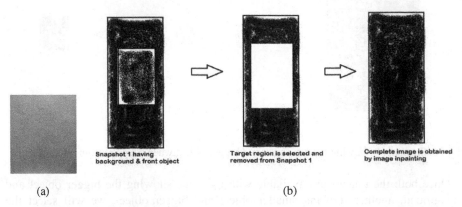

(a)   (b)

Fig. 8. Image showing inpainting process to complete the resulting outcome image

After getting the new image with only background visible (bigger object already removed by inpainting), we will take the cropped hidden object's image from snapshot 2 and perform union operation with inpainted background image obtained in previous step. By doing this, the hidden part will come on background and will not be invisible anymore.

4   Results

Thus, the resultant image will be like (Fig.9) having smaller object visible on the completed background and bigger object will be removed completely from the source image.

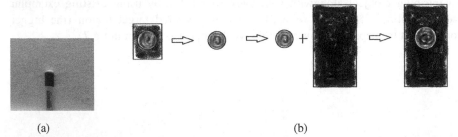

(a)   (b)

Fig. 9. Resultant image showing the whole proposed concept.

So, by the above discussed technique (method) we have successfully made the invisible content visible and visible content invisible.

Applications: This addition in the existing technology may be very useful in the field of medical sciences to detect the internal body parts image like organs/ nerves underneath heart or any other body part can be very easily recognized. Secondly, in satellite image capturing process i.e. now earth's cloud will act as transparent layer and it would be able to view the clear image of earth's surface. On very basic note, it could be applied in exploring out what is hidden behind a particular object by making the obstacle as transparent. Image reconstruction is also possible by using this concept. Future working in this domain will require deployment of image quality assessment (IQA) tools for performance monitoring and benchmarking the restoration quality [17]-[22].

5   Self-proposed Algorithm for Simulation of Whole Concept

1. Start the procedure.
2. Take Snapshot from front side and Determine initial fill front (target region Ω).
3. If target region is Filled, then **goto step 17 else move to step 4**.
4. Initialize Confidence values
5. Find Boundary of target region $\delta\Omega$.
6. Compute patch priorities for points on fill front.
7. Select the valid points on the fill front with maximum priority.
8. Find best match from all the pixels by using best first fill algorithm.
9. Fill destination tile of selected point with best match exemplar.
10. Update Confidence factors for points affected by filling.
11. Continue the process until no pixel is remaining in Ω. *(Repeat from 11 to 2)* until whole target region is completed with nearby high priority pixels.
12. *Output*- Inpainted Complete image.
13. Take snapshot from opposite/different photographing sides clicking image showing small (hidden) and bigger (overlapping) objects.
14. Crop smaller object by using matlab codes.
15. Stick the cropped image on previously inpainted image by union operation in matlab. *(Repeat from 15 to 13)* until all the hidden objects are shown.
16. *Output*- final image with hidden object visible and bigger (overlapping) object invisible from source image.
17. Stop the procedure.

6   Conclusion

Some extracted conclusions from the above research study is discussed as- There are numerous image inpainting techniques available in this era of continuously growing technologies but all are sufferers of some or other limitations as no one is so much advanced and equipped that can take the image of an object which is behind the another object. Thus, authors have discussed the lacunas of existing image inpainting techniques with detailed descriptions and have proposed an innovative method/technique by which the invisible objects(objects behind the bigger objects) can be made visible and visible objects(those objects which are in front of hidden objects) can be made invisible by just using some matlab codes. Also, to show the concept/procedure working at each step, whole process is being explained alongwith self- proposed algorithms that have also played a vital role in understanding the approach more clearly.

References

1. Chen, A.: The Inpainting of Hyperspectral Images: A Survey and Adaptation to Hyperspectral Data. In: SPIE 8537, Image and Signal Processing for Remote Sensing XVIII, pp. 85371K–85371K (2012)
2. Bertalmio, M., Caselles, V., Masnou, S., Sapiro, G.: Inpainting. In: Encyclopedia of Computer Vision. Springer, Berlin (2011),
   http://math.univ-lyon1.fr/~masnou/fichiers/
   publications/survey.pdf
3. Cheng, W.-H., Hsieh, C.-W., Lin, S.-K., Wang, C.-W., Wu, J.-L.: Robust Algorithm for Exemplar-Based Image Inpainting. In: Proc. Int. Conf. Comput. Graphics, Imaging Vis. (CGIV 2005), pp. 64–69 (2005)
4. Gonzalez, R.C., Woods, R.E.: Digital Image Processing. Prentice-Hall, New Jersey (2002)
5. Bornard, R., Lecan, E., Laborelli, L., Chenot, J.-H.: Missing Data Correction in Still Images and Image Sequences. In: ACM Multimedia Conf., pp. 355–361 (2002)
6. Pei, S.-C., Zeng, Y.-C., Chang, C.-H.: Virtual Restoration of Ancient Chinese Paintings Using Color Contrast Enhancement and Lacuna Texture Synthesis. IEEE Trans. on Image Processing 13(3), 416–429 (2004)
7. Park, J., Park, D.-C., Marks, R.J., El-Sharkawi, M.A.: Content-Based Adaptive Spatio-Temporal Methods for MPEG Repair. IEEE Trans. on Image Processing 13(8), 1066–1077 (2004)
8. Efros, A., Leung, T.: Texture Synthesis By NonParametric Sampling. In: Proc. Int. Conf. Computer Vision, Kerkyra, Greece, pp. 1033–1038 (September 1999)
9. Criminisi, A., Perez, P., Toyama, K.: Region Filling and Object Removal by Exemplar-Based Image Inpainting. IEEE Trans. on Image Processing 13(9), 1200–1212 (2004)
10. Goyal, A.P., Diwakar, S.: Fast and Enhanced Algorithm for Exemplar Based Image Inpainting. In: Image and Video Technology, PSIVT, pp. 325–330 (2010)
11. Oliveira, M.M., Bowen, B., McKenna, R., Chang, Y.-S.: Fast Digital Image Inpainting. In: VIIP 2001: Proc. Int. Conf. on Visualization, Imaging, and Image Processing, Marbella, Spain, pp. 261–266 (2001)
12. Bertalmio, M., Sapiro, G., Caselles, V., Ballester, C.: Image inpainting. In: ACM Comput. Graph. (SIGGRAPH 2002), pp. 417–424 (July 2000)
13. Different Methods for image inpainting,
    http://www.caa.tuwien.ac.at/cvl/
    teaching/sommersemester/cvme/template_extended_abstract.pdf
14. Chen, Q., Zhang, Y., Liu, Y.: Image Inpainting with Improved Exemplar-Based Approach. In: Sebe, N., Liu, Y., Zhuang, Y.-t., Huang, T.S. (eds.) MCAM 2007. LNCS, vol. 4577, pp. 242–251. Springer, Heidelberg (2007)
15. Xiaowei, S., Zhengkai, L., Houqiang, L.: An Image Inpainting Approach Based on the Poisson Equation. In: Proc. of the Second International Conference on Document Image Analysis for Libraries, April 27-28, pp. 368–372 (2006)
16. Tschumperlé, D.: Fast Anisotropic Smoothing of Multi-valued Images using Curvature-preserving PDE's. International Journal of Computer Vision 68(1), 65–82 (2006)
17. Gupta, P., Srivastava, P., Bharadwaj, S., Bhateja, V.: A Novel Full-Reference Image Quality Index for Color Images. In: Proc. of the International Conference on Information Systems Design and Intelligent Applications, pp. 245–253 (January 2012)
18. Jain, A., Bhateja, V.: A Full-Reference Image Quality Metric for Objective Evaluation in Spatial Domain. In: Proc. of International Conference on Communication and Industrial Application, vol. (22), pp. 91–95 (December 2011)

19. Trivedi, M., Jaiswal, A., Bhateja, V.: A No-Reference Image Quality Index for Contrast and Sharpness Measurement. In: Proc. of 3rd International Advance Computing Conference, pp. 1234–1239 (February 2013)
20. Trivedi, M., Jaiswal, A., Bhateja, V.: A New Contrast Measurement Index Based on Logarithmic Image Processing Model. In: Satapathy, S.C., Udgata, S.K., Biswal, B.N. (eds.) Proceedings of Int. Conf. on Front. of Intell. Comput. AISC, vol. 199, pp. 715–723. Springer, Heidelberg (2013)
21. Jaiswal, A., Trivedi, M., Bhateja, V.: A No-Reference Contrast Assessment Index based on Foreground and Background. In: Proc. of 2nd Students Conference on Engineering and Systems, pp. 460–464 (April 2013)
22. Bhateja, V., Srivastava, A., Kalsi, A.: Fast SSIM Index for Color Images Employing Reduced-Reference Evaluation. In: Satapathy, S.C., Udgata, S.K., Biswal, B.N. (eds.) FICTA 2013. AISC, vol. 247, pp. 451–458. Springer, Heidelberg (2014)

Fast Mode Decision Algorithm for H.264/SVC

L. Balaji[1,*] and K.K. Thyagharajan[2]

[1] Faculty of Information & Communication, Anna University, Chennai - 600025, India
and Velammal Institute of Technology, Chennai – 601204, India
[2] RMD Engineering College, Chennai - 601206, India
maildhanabal@gmail.com, mailbala81@yahoo.co.in,
kkthyagharajan@yahoo.com
www.tansitresearch@yahoo.com

Abstract. H.264/AVC extension is H.264/SVC which is applicable for environment that demands video streaming. This paper presents an algorithm to reduce computation complexity and maintain coding efficiency by determining the mode quickly. Our algorithm terminates mode search by a probability model for both intra-mode and inter-mode of lower level and higher level layers in a Macro Block (MB). The estimated of Rate Distortion Cost (RDC) for modes among layers is used to determine best mode of each MB. This algorithm achieves about 26.9% of the encoding time when compared with JSVM reference software with minimal degradation in PSNR.

Keywords: Scalable Video Coding, H.264, Fast Mode Decision, Computation Complexity, Macro Block.

1 Introduction

Broadcasting Video Content needs a better video coding strategy with a good compression format; H.264/AVC is one such compression format which is suitable for all forms of transmission, storage and retrieval even with a loss of data. It is a block based form of compression format developed as a standard by ITU-T video coding experts group (VCEG) with the ISO/IEC JTCI moving picture experts group (MPEG) [1]. As an extension to AVC, SVC offers layered approach for scalability, which comprises of one base layer and one or more enhancement layer. Also to provide video services with reduced fidelity, low spatial resolution SVC does single encode and multiple decode. A video is sub divided into more number of subset bitstreams. A subset bitstream is a partial bitstream made by discarding unnecessary packets of video. This partial bitstream is lesser in data size which is more useful in bandwidth constraints. SVC has a unique feature of extracting the video with the help of partial bitstream [1], even from error prone network. The base layer is subset bitstream which decodes low resolution video and enhancement layer is one can decode the bitstreams of base layer along with already encoded enhancement layers.

The bitstream is scalable, if part of it removed and is still decodable with the resultant [1]. It has three forms of scalability – temporal, spatial and quality Scalability. In

---

* Correspsonding author.

© Springer International Publishing Switzerland 2015
S.C. Satapathy et al. (eds.), *Proc. of the 3rd Int. Conf. on Front. of Intell. Comput. (FICTA) 2014*
– *Vol. 2*, Advances in Intelligent Systems and Computing 328, DOI: 10.1007/978-3-319-12012-6_44

Temporal, frame rate is considered, in spatial, resolution is considered and in quality, PSNR (Peak Signal-Noise-Ratio) is considered. In all forms of scalability, either a frame (temporal), size (spatial) or PSNR level in decibel (quality) will be discarded in the subset bitstream. The mode partition of MBs in the current frame is most similar to the reference frame. In Spatial scalability video is coded at different formats such as QCIF, CIF, SDTV, HDTV, etc. With the samples of lower resolution video subset bitstream higher resolution video bitstream can be predicted in spatial scalability. For each MB at the base layer the corresponding upsampled MBs at enhancement layers tend to have the same mode partition. In SNR/Quality/fidelity scalability the multiple levels of quality bitstream for a single spatial resolution bitstream video. High quality video can be obtained by decoding the samples of predicted lower quality subset bitstream.

SVC defines nine prediction modes for INTRA4x4, four prediction modes for INTRA16x16, SKIP mode and seven MB modes for INTER prediction such as INTER16x16, INTER16x8, INTER8x16, INTER8x8, INTER8x4, INTER4x8 and INTER4x4 [2]. Deciding the best mode for a current frame from previous frames introduces complexity in any video compression format; AVC offers single layer approach generates single bitstream, whereas SVC generates multiple bitstream. So the computational complexity is high in SVC because of its unique feature. In order to find a best mode many Fast Mode Decision Algorithm (FMD) are discussed and implemented for AVC but few proposed for SVC. In this paper we focus on deciding the best mode using our proposed model.

2   Background and Related Work

In [3], the computational complexity is reduced based on the correlation between the current frame and reference frame with search range, this however reduces the redundant search range and candidate modes were reduced results in PSNR loss. In [4], best mode for H.264/AVC is achieved based on the RDC with minimal PSNR loss with better coding efficiency with different quantization parameters but not applied for SVC. Enhancement layer MB searching is reduced based on correlation among the base layer and the enhancement layers. The candidate modes of the MB used in Rate Distortion Optimization (RDO) has been reduced significantly at the enhancement layer [5], not in base layer. In [6] a probability model is defined for H.264/AVC which identified the mode by creating a list of modes. The list created based on the highest probability of correlations between the current block and neighboring block. This model effectively identifies the best mode with better coding efficiency and it is summarized as follows:

2.1   Probability Model

A group of modes created based on probability of the mode to be the best with co-located MB is evaluated in [6]. A large correlation found between the current block and a neighboring block in a MB from the model is experimented. Among the correlation, the modes have highest correlation will have more chance to be best mode. Therefore a reference MB set P is described.

$$P = \left\{ \begin{array}{c} MB(j, x - 16, y - 16), MB(j, x - 16, y), \\ MB(j, x, y - 16), \\ MB(j, x + 16, y - 16), MB((j - 1)', x, y) \end{array} \right\} \tag{1}$$

where $MB(j, x, y)$ – denotes the MB located at j^{th} frame with upper left pixel at x, y and $MB((j-1)', x, y)$ – denotes the previous collocated MB with same as current coding MB and neighboring mode set Q is given by,

$$Q = \{M_{MB} | MB \in P\} \tag{2}$$

where M_{MB} – denotes the encoding code of MB. The approximated probability [5] of the mode to be the best possible is given the probability model,

$$P(m = M | M \in Q) \approx P(m \in P) \frac{\sum_{MB \in P, M_{MB}=m} N(m = M_{MB})}{\sum_{P(m \in Q)} \sum_{MB \in P, M_{MB}=m} N(m = M_{MB})}$$

$$= K . \sum_{MB \in P, M_{MB}=M} N(m = M_{MB}) \tag{3}$$

where $N(.)$ – Occurrence time of an event and K – Constant parameter which is same for all modes. Since M is not an element of Q condition may have less probability to be best mode and are considered to be zero.

3   Proposed Algorithm

In H.264/SVC, RDO based mode decision is performed for each MB, which takes most of the encoding computations. To address this problem, many FMD algorithms have been developed to skip unnecessary encoding modes and thus reduce encoding complexities. A FMD algorithm is proposed shown below predicts the best mode with less encoding time. The algorithm makes use of the inter mode information of neighboring blocks to predict an optimal encoding mode list based on a probability model.

Irrespective of the layer levels, the algorithm checks the MB to be encoded is in I slice, then best mode will be the minimum RDC of all the modes. If it is P/B slice a probability model is applied as shown below. A group of modes created based on the highest probability of correlations among the current block mode and neighboring block mode of the MB. The modes in the set are arranged in descending order of their highest probabilities. It checks the first mode in the set, if it is not in P_{8x8} the current mode will be checked, else it checks the modes in the order arranged in set based on minimum RDC.

3.1   Algorithm

```
If(MBencoded== P/Bslice of Lower/Higher layer)
{
     Calculate Probability(P)for all modes using (3) and
     create a group of modes in descending order;

     If(first mode == (P8x8 not in Q) or SKIP/DIRECT or
                         INTER16x16)
          Calculate P(DIRECT8x8,INTER8x8);
          Best = Pmax;
     elseif(firstmode == INTER16x8)
          Calculate P(DIRECT8x8,INTER8x8,INTER8x4);
          Best = Pmax;
     elseif(first mode == INTER8x16)
          Calculate P(DIRECT8x8,INTER 8x8,INTER8x4);
          Best = Pmax;
     elseif(first mode == INTER 8x16)
          Calculate P(DIRECT8x8,INTER8x8,INTER4x8);
          Best = Pmax;
     else
          Calculate P(all sub modes);
          Best = Pmax;
}
else
     Calculate RDC of all INTRA modes;
     Best= RDCmin;
```

4   Experimental Results

The performance of the algorithm is evaluated using the JSVM reference software 9.19.15 [7] with the simulation parameters in Table 1. The system configuration uses Intel Dual Core Processor with 2.4 Ghz clock speed and 320 GB Hard disk with Windows 7 Operating system. We use the following video sequences like Bus, Mobile Ice, Harbour, Foreman, Crew & Football of QCIF for Base layer and CIF for Enhancement layer having a GOP size of 32 with varying quantization parameters from 24 – 36. The frame rate is set to 30 frames per second and 150 frames are encoded for each sequence.

The experimental result for each video sequence in terms of Bit rate, Y-PSNR and Time is compared with JSVM in Table 2. The total encoding time for JSVM is 337.88s whereas the encoding time for the proposed system is 246.99s. The percent reduced average time is given by the relation $((t_{jsvm}-t_{proposed})/t_{jsvm}) \times 100$ and it reduces to 26.90%. So, approximately 27% faster time saving is achieved when compared with the reference software with our algorithm. The Snapshots of crew video sequence for JSVM and proposed algorithm is shown in Fig. 1 and Fig. 2.

Table 1. Simulation Parameters

Codec	JSVM9.19.15
Frame Rate	30Hz
Frame Numbers	150
Base Layer	QCIF
Enhancement Layer	CIF
QP Setting	24~36
GOP Size	32

Table 2. Comparative analysis for Proposed Algorithm vs JSVM

Video Sequence	Average Bit Rate		Average YPSNR		Average Time (s)	
	JSVM	Proposed	JSVM	Proposed	JSVM	Proposed
Bus	511.02	511.67	31.76	31.27	251.27	197.60
Mobile	124.14	125.45	24.29	24.21	217.12	171.23
Ice	262.45	262.54	40.06	40.03	419.20	217.77
Harbour	264.21	264.36	29.01	28.99	333.17	219.44
Foreman	128.91	128.10	32.28	32.17	328.52	185.54
crew	192.26	198.39	33.44	33.41	346.19	278.40
Football	825.22	826.99	36.16	36.11	469.66	458.96
Average	329.74	331.07	32.43	32.31	337.88	246.99

The PSNR level gets bit reduced while applying the proposed algorithm but at utmost a marginal level in which the video does not gets deteriorated when it takes the advantage of reducing the encoding time. The comparative graphs for each video sequence with bit rate Vs PSNR level for proposed algorithm and JSVM is shown in Fig. 3 – 6.

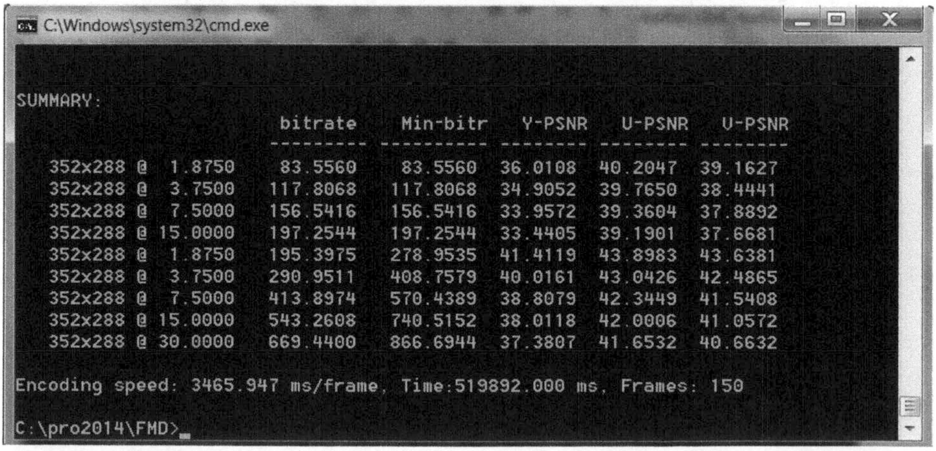

Fig. 1. Snapshot of Crew video Sequence under JSVM

```
C:\Windows\system32\cmd.exe                                                    _  □  X

SUMMARY:
                          bitrate    Min-bitr    Y-PSNR    U-PSNR    U-PSNR
                          ---------  ----------  --------  --------  --------
     352x288 @  1.8750     83.5560    83.5560    36.0108   40.2047   39.1627
     352x288 @  3.7500    117.8068   117.8068    34.8996   39.7593   38.4459
     352x288 @  7.5000    156.5053   156.5053    33.9579   39.3662   37.8934
     352x288 @ 15.0000    197.3888   197.3888    33.4494   39.2018   37.6678
     352x288 @  1.8750     83.9460    83.9460    36.0108   40.2047   39.1627
     352x288 @  3.7500    118.4447   118.4447    34.8996   39.7593   38.4459
     352x288 @  7.5000    157.6105   157.6105    33.9579   39.3662   37.8934
     352x288 @ 15.0000    199.3968   199.3968    33.4494   39.2018   37.6678
     352x288 @ 30.0000    424.7168   424.7168    34.1787   39.4194   38.0296

Encoding speed: 1770.327 ms/frame, Time:265549.000 ms, Frames: 150

C:\pro2014\FMD>
```

Fig. 2. Snapshot of Crew video Sequence under Proposed Algorithm

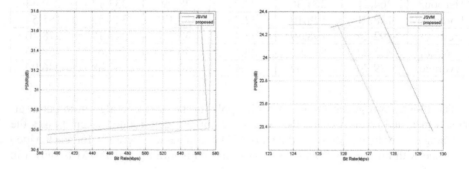

Fig. 3. Comparison in PSNR Vs Bit rate for Bus and Mobile Sequence

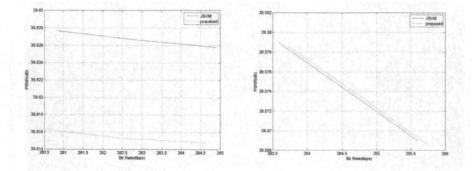

Fig. 4. Comparison in PSNR Vs Bit rate for Ice and Harbour Sequence

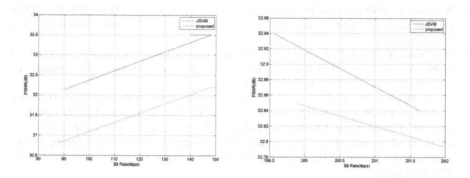

Fig. 5. Comparison in PSNR Vs Bit rate for Foreman and Crew Sequence

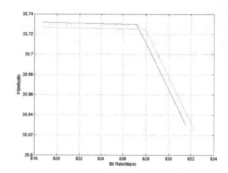

Fig. 6. Comparison in PSNR Vs Bit rate for Football Sequence

5   Conclusion

The algorithm proposed terminates mode search by a statistical analysis based on a probability model for both intra-mode and inter-mode of lower level and higher level layers in a Macro Block (MB). The estimated RDC for modes among layers is used to determine best mode of each MB. Our algorithm achieves 26.9% of the encoding time when compared with JSVM reference software with lesser degradation in PSNR and increase in bit rate. Fast mode search without degradation in PSNR will be considered as a future work.

References

1. Schwarz, H., Marpe, D., Wiegand, T.: Overview of the scalable video coding extension of the H.264/AVC standard. IEEE Trans. on Circuits and Systems for Video Technology 7(9), 1103–1120 (2007)

2. Wiegand, T., Sullivan, G.J., Bjontegard, G., Luthra, A.: Overview of the H.264/AVC Video Coding Standard. IEEE Trans. on Circuits and Systems for Video Technology 13(7), 560–576 (2003)
3. Hamamoto, T., Song, T., Katayama, T., Shimamoto, T.: Complexity reduction algorithm for hierarchical B-picture of H.264/SVC. International Journal of Innovative Computing, Information and Control 7(1), 445–457 (2011)
4. Hu, S., Zhao, T., Wang, H., Kwong, S.: Fast Inter-Mode Decision Based on Rate-Distortion Cost Characteristics. In: Qiu, G., Lam, K.M., Kiya, H., Xue, X.-Y., Kuo, C.-C.J., Lew, M.S. (eds.) PCM 2010, Part II. LNCS, vol. 6298, pp. 145–155. Springer, Heidelberg (2010)
5. Li, H., Li, Z.G., Wen, C.Y.: Fast mode decision algorithm for inter-frame coding in fully scalable video coding. IEEE Trans. on Circuits and Systems for Video Technology 16(7), 889–895 (2006)
6. Zhao, T., Wang, H., Kwong, S., Hu, S.: Probability-based Coding Mode Prediction For H.264/AVC. In: Proc. of IEEE International Conference on Image Processing (ICIP 2010), Hong Kong, pp. 26–29 (2010)
7. JSVM Reference Software,
   http://www.hhi.fraunhofer.de/de/kompetenzfelder/
   image-processing/research-groups/image-video-coding/
   svc-extension-of-h264avc/jsvm-reference-software.html

Recognizing Handwritten Devanagari Words Using Recurrent Neural Network

Sonali G. Oval and Sankirti Shirawale

MMCOE, Pune, Maharashtra
sonalioval@gmail.com, sankirtishrivale@mmcoe.edu.in

Abstract. recognizing lines of handwritten text is a difficult task. Most recent evolution in the field has been made either through better-quality pre processing or through advances in language modeling. Most systems rely on hidden Markov models that have been used for decades in speech and handwriting recognition. So an approach is proposed in this paper which is based on a type of recurrent neural network, in particularly designed for sequence labeling tasks where the data is hard to segment and contains long-range bidirectional interdependencies. Recurrent neural networks (RNN) have been successfully applied for recognition of cursive handwritten documents, in scripts like English and Arabic. A regular recurrent neural network (RNN) is extended to a bidirectional recurrent neural network (BRNN).

Keywords: Devanagari Script, Handwriting recognition, Hidden Markov model, Recurrent neural networks.

1 Introduction

Handwriting is the most natural mode of collecting, storing, and transmitting information which also serves not only for communication among human but also serves for communication of humans and machines [6]. The problem of handwritten word recognition which is one of the challenging problems in pattern recognition has been studied for several decades. The reasons behind this difficulty are large variability in handwriting style, large varieties of pen-type, overlapping wide strokes, and a lack of ordering information of strokes [8]. Handwriting recognition is traditionally divided into online and offline recognition [21]. In online recognition, a time series of coordinates, representing the movement of the pen tip, is captured, while in the offline case [20], only an image of the text is available. Because of the greater ease of extracting relevant features, online recognition generally yields better results [1]. HMMs are able to segment and recognize at the same time, which is one reason for their popularity in unconstrained handwriting recognition. The idea of applying HMMs to handwriting recognition was originally motivated by their success in speech recognition, where a similar conflict exists between recognition and segmentation. Over the years, numerous refinements of the basic HMM approach has been proposed, such as the writer-independent system considered in, which combines point-oriented and stroke-oriented input features [1]. HMMs have several well-known

© Springer International Publishing Switzerland 2015   413
S.C. Satapathy et al. (eds.), *Proc. of the 3rd Int. Conf. on Front. of Intell. Comput. (FICTA) 2014*
– *Vol. 2*, Advances in Intelligent Systems and Computing 328, DOI: 10.1007/978-3-319-12012-6_45

drawbacks. One of these is that they assume that the probability of each observation depends only on the current state, which makes contextual effects difficult to model. Another is that HMMs are generative, while discriminative models generally give better performance in labeling and classification tasks. Recurrent neural networks (RNNs) do not suffer from these limitations and would therefore seem a promising alternative to HMMs. The main reason for this is that traditional neural network objective functions require a separate training signal for every point in the input sequence, which in turn requires presegmented data [15].

The literature survey of different recognition methods for handwritten words is described in section 2. Section 3 describes the Devanagari script. The implementation details of the proposed system, preprocessing and normalization, Feature extraction and Classification are explained in section 4. Section 5 describes the results and data set, and the paper is concluded in section 6.

2   Literature Survey

During recent years, research toward Indian handwritten character recognition is getting increased attention although the first research report on offline handwritten Devanagari characters was published in 1977 [4]. Many approaches have been proposed toward handwritten Devanagari numeral, character, and word recognition in the past decade. Two approaches are mainly used in handwritten character recognition. First is segmentation-based approach and the other is segmentation free approach (holistic approach). In the first approach, the words are initially segmented into characters or pseudocharacters, and then, recognized. As a result, the success of the recognition module depends on the performance of the segmentation technique. The second approach treats the whole word as a single entity and it recognizes without doing explicit segmentation.

Three different classifiers, namely nearest neighbor, k-NN, and SVM were tested independently to recognize handwritten Devanagari numerals in [12]. The performance of SVM in terms of accuracy was better than the other two classifiers. In [17], the feature vector is entered as an input to one of the feedforward backpropagation neural network for the classification of handwritten Devanagari characters. Kumar [9] compared the performances of SVM and MLP classifiers with six different features on handwritten characters and found that the performance of SVM classifier was superior to MLP in all the six cases. But the classification time required for SVM was greater than that of MLP. A modified quadratic classifier is applied by Pal et al. [10] on the features of handwritten characters for recognition. In [11], two classifiers are combined to get higher accuracy of character recognition with the same features. Combined use of SVM and MQDF is applied for the same [14]. The work reported in [16] presents a two-stage classification approach for handwritten Devanagari characters. The first stage is using structural properties like shirorekha and spine in a character. The second stage exploits intersection features of characters, which are then fed to a feedforward neural network (FFNN) for further classification. A segmentation based approach to handwritten Devanagari word recognition is proposed by Shaw et al. [6]. On the basis of the header line, a word image is segmented into pseudocharacters. HMM are proposed to recognize the pseudocharacters. The word level recognition is done on the basis of string edit distance. A continuous density HMM is also proposed by Shaw et al. [8] to recognize a handwritten word images. The states of the HMM are not determined a priori, but are determined automatically based on a database of handwritten word images. An HMM is constructed for each word. To classify an unknown word image, its class conditional probability for each HMM is computed. The class that gives highest such probability is finally selected.

3   Devanagari Script

This script emerged out of Siddham script an immediate descendant of Gupta script ultimately deriving from the Brahmi Script. It follows left to right fashion for writing [8]. The Devanagari alphabet is used for writing Hindi, Sanskrit, Marathi, Nepali and it is closely related to many of the scripts in use today in South Asia, Southeast Asia and Tibet. This script is cursive in nature. Devanagari has 13 independent vowels or "svara", 33 independent consonants or "vyajana" and 12 dependent vowel signs shown in Fig. 1. (a) Most of the consonants can be joined to one or two other consonants so that the inherent vowel is suppressed. The resulting conjunct form is called a ligature or a compound character. Commonly used compound characters appearing in our lexicon of words are shown in Fig. 1.

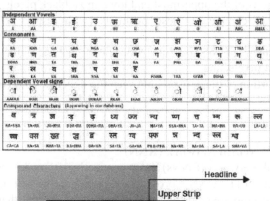

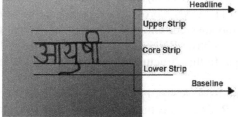

Fig. 1. (a) Devanagari Character Set; (b) Three Strips of a Devanagari word

4   Classification

4.1   Preprocessing and Normalization

First, the image was rotated to account for the overall skew of the document, and the handwritten part was extracted from the form. Then, a histogram of the horizontal black/white transitions was calculated, and the text was split at the local minima to give a series of horizontal lines. Once the line images were extracted, the next stage was to normalize the text with respect to writing skew and slant and character size.

4.2   Feature Extraction

To extract the feature vectors from the normalized images, a sliding window approach is proposed. The width of the window is one pixel, and nine geometrical features are computed at each window position. Each text line image is therefore converted to a sequence of seven-dimensional vectors. The seven features are the following:

- The mean gray value of the pixels.
- The center of gravity of the pixels.
- The second-order vertical moment of the center of gravity.
- The positions of the uppermost and lowermost black pixels.
- The rate of change of these positions (with respect to the neighboring windows).
- The number of black-white transitions between the uppermost and lowermost pixels, and
- The proportion of black pixels between the uppermost and lowermost pixels.

4.3   Classification

4.3.1   Recurrent Neural Network (RNN)

A recurrent neural network (RNNs) is a connectionist model containing a self-connected hidden layer [1]. RNN's provide a very elegant way of dealing with (time) sequential data that embodies correlations between data points that are close in the sequence. Fig. 4 shows a basic RNN architecture with a delay line and unfolded in time for two time steps. In this structure, the input vectors are fed one at a time into the RNN [15].

One of the key benefits of RNNs is their ability to make use of previous context. However, for standard RNN architectures, the range of context that can in practice be accessed is limited. The problem is that the influence of a given input on the hidden layer, and therefore on the network output, either decays or blows up exponentially as it cycles around the recurrent connections. This is often referred to as the vanishing gradient problem [15], [1].

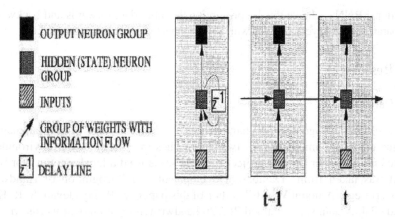

Fig. 2. Structure of RNN

4.3.2   Bidirectional RNN

A bidirectional recurrent neural network (BRNN) is proposed that can be trained using all available input information in the past and future of a specific time frame [18]. Fig. 5 shows a basic BRNN architecture with a delay line and unfolded in time for three time steps. For many tasks, it is useful to have access to future, as well as past, context. In handwriting recognition, for example, the identification of a given letter is helped by knowing the letters both to the right and left of it. Bidirectional RNNs (BRNNs) are able to access context in both directions along the input sequence. BRNNs contain two separate hidden layers, one of which processes the input sequence forward, while the other processes it backward [15]. Both hidden layers are connected to the same output layer, providing it with access to the past and future context of every point in the sequence. Combining BRNNs and LSTM gives BLSTM [2].

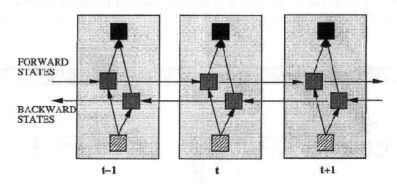

Fig. 3. Structure of BRNN

Long Short-Term Memory (LSTM) [19] is RNN architecture designed to address the vanishing gradient problem. An LSTM layer consists of multiple recurrently connected subnets, known as memory blocks [3]. Each block contains a set of internal units, known as cells, whose activation is controlled by three multiplicative gate units [5], [1].

Bidirectional RNNs achieve this by presenting the input data forwards and backwards to two separate hidden layers, both of which are connected to the same output layer.

5   Results

5.1   Data Set

For the offline recognition of handwritten in Marathi and Hindi languages, a technique that is a combination of two approaches in a single writer environment is presented in this paper. In this project the data set is used a legal amount words. The data set was taken from "Database Development and Recognition of Handwritten Devanagari Legal Amount Words" author of this paper are R. Jayadevan, S. R. Kolhe, Umapada Pal. A data set contained 26,720 handwritten legal amount words written in Hindi and Marathi languages. The database was constructed by taking handwritten data from ninety writers. The sample words are shown in Fig. 4.

Fig. 4. Samples handwritten words

5.2   Results

The results of pre processing steps are as shown in Fig.5 (a) shows the original image. Fig.5 (b) shows the filtered image. For filtration we use median filter. Fig.5(c) shows the output of binarization (we use Otsu's binarization algorithm).Also it shows the skew angle. Fig5 (d) shows the output of corrected skew of the image. Fig5 (e) shows the vertical projection of the image and Fig5 (f) shows the output of horizontal projection.

Fig. 5.

(a) Original Image   (b) Filtered Image

(c) Binarizated Image

(d) Skew correction   (e) Vertical Projection

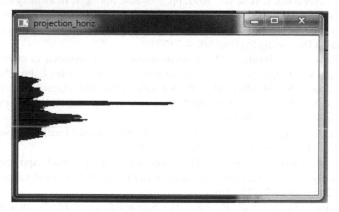

(f) Horizontal Projection

Fig. 5. (*Continued*)

6   Conclusion and Future Work

In this paper we described a proposed approach for recognizing offline handwritten text, using a single recurrent neural network (RNN). The key advance is a recently introduced RNN objective function known as Connectionist Temporal Classification (CTC). CTC uses the network to label the entire input sequence at once. This way the network can be trained with unsegmented input data, and the final label sequence can be read directly from the network output.

In the future work comparison will be done in between HMM and RNN using following parameters.

- ➢ database size and
- ➢ Test set coverage.

It is also planned to overcome the problem of Out-Of-Vocabulary words (OOV). This could be done by using the network probability and the edit distance to the nearest vocabulary word.

References

1. Graves, A., Liwicki, M., Fernandez, S., Bertolami, R., Bunke, H., Schmidhuber, J.: A novel connectionist system for unconstrained handwriting recognition. IEEE Transactions on Pattern Analysis and Machine Intelligence 31, 855–868 (2009)
2. Vinciarelli, A.: Online and offline handwriting recognition: A comprehensive survey. Pattern Recognition 35, 1433–1446 (2002)
3. Graves, A., Fernández, S., Schmidhuber, J.: Bidirectional LSTM Networks for Improved Phoneme Classification and Recognition. In: Duch, W., Kacprzyk, J., Oja, E., Zadrożny, S. (eds.) ICANN 2005. LNCS, vol. 3697, pp. 799–804. Springer, Heidelberg (2005)
4. Graves, A., Fernandez, S., Liwicki, M., Bunke, H., Schmidhuber, J.: Unconstrained Online Handwriting Recognition with Recurrent Neural Networks. Advances in Neural Information Processing Systems 20, 1–7 (2008)
5. Shaw, Parui, S.K., Shridhar, M.: A segmentation based approach to offline handwritten Devanagari word recognition. In: Proc. IEEE Int. Conf. Inf. Technol., pp. 256–257 (2008)
6. Shaw, Parui, S.K., Shridhar, M.: Off-line handwritten Devanagari word recognition: A holistic approach based on directional chain code feature and HMM. In: Proc. Int. Conf. Inf. Technol., pp. 203–208 (2008)
7. Rajput, G.G., Mali, S.M.: Fourier descriptor based isolated Marathi handwritten numeral recognition. Int. J. Comput. Appl. 3(4), 9–13 (2010)
8. Liwicki, M., Graves, A., Bunke, H., Schmidhuber, J.: A Novel Approach to On-Line Handwriting Recognition Based on Bidirectional Long Short-Term Memory Networks. In: ICDAR 2007, pp. 367–371 (2007)
9. Hanmandlu, M., Agrawal, P., Lall, B.: Segmentation of handwritten Hindi text: A structural approach. Int. J. Comput. Process. Lang. 22(1), 1–20 (2009)
10. Schuster, M., Paliwal, K.K.: Bidirectional Recurrent Neural Networks. IEEE Trans. Signal Processing 45, 2673–2681 (1997)
11. Morillot, O., Likforman-Sulem, L., Grosicki, E.: Comparative study of HMM and BLSTM segmentation-free approaches for the recognition of handwritten text-lines, pp. 783–787. IEEE (2013)
12. Agrawal, P., Hanmandlu, M., Lall, B.: Coarse classification of handwritten Hindi characters. Int. J. Advanced Sci. Technol. 10, 43–54 (2009)
13. Jayadevan, R., Kolhe, S.R., Patil, P.M., Pal, U.: Offline Recognition of Devanagari Script: A Survey. IEEE Transactions on Systems, Man, and Cybernetics Part C: Applications and Reviews, 782–796 (November 17, 2011)

14. Arora, S., Bhatcharjee, D., Nasipuri, M., Malik, L.: A two stage classification approach for handwritten Devanagari characters. In: Proc. Int. Conf. Comput. Intell. Multimedia Appl., pp. 399–403 (2007)
15. Kaur, S.: Recognition of handwritten Devanagari script using features based on Zernike moments, zoning and neural network classifier. M.Tech Thesis, Dept. Comput. Sci. Eng., Punjabi University, Patiala, India (2004)
16. Hochreiter, S., Schmidhuber, J.: Long short-term memory. Neural Comput. 9, 1735–1780 (1997)
17. Kumar, S.: Performance comparison of features on Devanagari handprinted dataset. Int. J. Recent Trends 1(2), 33–37 (2009)
18. Steinherz, T., Rivlin, E., Intrator, N.: Offline cursive script word recognition - a survey. International Journal of Document Analysis and Recognition 2(2), 90–110 (1999)
19. Pal, U., Sharma, N., Wakabayashi, T., Kimura, F.: Off-line handwritten character recognition of Devanagari script. In: Proc. 9th Conf. Document Anal. Recognit., pp. 496–500 (2007)
20. Pal, U., Chanda, S., Wakabayashi, T., Kimura, F.: Accuracy improvement of Devanagari character recognition combining SVM and MQDF. In: Proc. 11th Int. Conf. Frontiers Handwrit. Recognit., pp. 367–372 (2008)
21. Frinken, V., Fornés, A., Lladós, J., Ogier, J.-M.: Bidirectional language model for handwriting recognition. In: Gimel'farb, G., Hancock, E., Imiya, A., Kuijper, A., Kudo, M., Omachi, S., Windeatt, T., Yamada, K. (eds.) SSPR & SPR 2012. LNCS, vol. 7626, pp. 611–619. Springer, Heidelberg (2012)

Homomorphic Filtering for Radiographic Image Contrast Enhancement and Artifacts Elimination

Igor Belykh

St. Petersburg State Polytechnical University, St. Petersburg, Russia
igor.belyh@avalon.ru

Abstract. The contrast of radiographic images is provided at physical level by anti-scatter grids and is usually further improved at image processing stage. The known contrast improvement methods are mainly based on image non-linear manipulations that may cause residual artifacts in processed images. In this paper an artifact-free approach is proposed for radiographic image filtering which is still an actual problem. The proposed algorithm is based on homomorphic equalizer design and application for image contrast enhancement, sharpening and artifacts elimination. Experimental results are discussed and concluded with description of advantages over existing approaches.

Keywords: computed radiography, image processing, contrast enhancement, artifacts, homomorphic filtering.

1   Introduction

Image contrast is the key factor in image quality for reliable diagnostics in modern radiography and is provided by hardware and software methods. The factors that affect image quality are image contrast and the absence or minimal amount of different type of noise and artifacts. The subject of this paper is to investigate the status of contrast improvement problem in radiographic imaging and to introduce the artifacts-free approach for contrast enhancement.

Anti-scatter grids [2] are used in radiology stations for image contrast enhancement at acquisition stage but leave visible line artifacts that may cause problems while digital image representation for diagnostics. The second problem is to balance registered image contrast for the diagnostic purposes using image processing methods without or with minimal amount of distortions or artifacts.

The use of anti-scatter grids is based on assumptive decomposition of X-ray radiation propagated through the object of interest into a primary and a secondary component. The primary component is formed by the electron beams with small deviations from initial straight line trajectories while secondary component is formed by scattered electrons which are deflected (or even reflected) by object interior. Scattered component degrades image contrast. Anti-scatter grids are designed to enhance image contrast by means of scattered radiation partial reduction and are located between the object and receiving device, i.e. computed radiography (CR) plate or direct radiography (DR) digital panel.

© Springer International Publishing Switzerland 2015   423

S.C. Satapathy et al. (eds.), *Proc. of the 3rd Int. Conf. on Front. of Intell. Comput. (FICTA) 2014*
– *Vol. 2*, Advances in Intelligent Systems and Computing 328, DOI: 10.1007/978-3-319-12012-6_46

Currently stationary linear grids are the most usable [2, 8] and they are one of the subjects of investigation in this paper. Linear grid structure is similar to jalousie structure with thin stripes made of lead and interspaces usually filled with aluminum. Primary X-ray beams are mostly transmitted through the grid interspace material while scattered beams along with some primary ones are absorbed by the lead stripes. The most important characteristic is the grid stripes' frequency measured in lines per inch (LPI) in a broad range from 85 to 215 LPI. The advantage of the grid use is the higher image contrast but the disadvantages are: the need of higher X-ray dose (up to 3 times) and the visible thin line artifacts that may cause a known Moiré pattern [2] when digital image is resized for display on a diagnostic monitor (Fig.1).

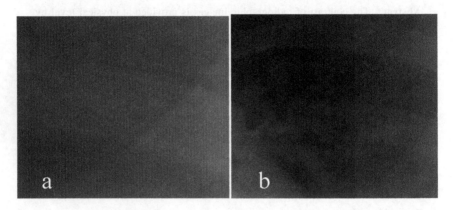

Fig. 1. Grid line artifacts and Moiré pattern on magnified (a) and minified (b) CR image fragments respectively

As shown in [8], there were many efforts made to study the grid artifacts and to develop different methods for their detection and suppression during the past two decades but there is still a need for robust and reliable grid pattern elimination method that will improve but not degrade the diagnostic quality of digital X-ray image. It is proposed that one of the best results for grid artifacts removal can be achieved by bandstop filtering based on Kaiser transfer function [9]. The only problem here is to find an optimal compromise between the filter impulse response length in space domain and the bandstop width in frequency domain to avoid ringing artifacts. The bandwidth can be narrowed using a homomorphic approach [3].

In the last 20 years there were many algorithms developed for clinically acceptable radiographic image contrast enhancement and the most known of them is Multiscale Image Contrast Amplification (MUSICA) [1]. It is based on image multi-resolution analysis and consists of a long chain of image decomposition into low- and high-frequency components at each resolution and reconstruction procedures which include several steps of image lowpass filtering, downsampling, upsampling, filtering, weighting, subtraction and adding, and intensity amplification. Among other known approaches are Enhanced Visualization Processing (EVP) and Radiographic Contrast-enhancement masks (RCM). The latter one [7] is based on physical tissue compensation filters software models development and use of anatomically shaped masks in order to reduce the range of optical densities within images with a large dynamic range to improve visualization of all anatomy within the image. The EVP algorithm [6] is based on a known unsharp masking technique and besides segmentation pre-processing step includes image frequency band decomposition and weighting, image reconstruction and tone scale balance. Either MUSICA and EVP algorithms exploit the idea of different spatial frequency band extraction and weighting, manipulating back and forth with the whole image including image subtraction. The adaptive histogram equalization method [11] relies on a statistical technique and works fine on mid- and small size images with relatively narrow intensity range of diagnostically valuable objects and is appropriate for use in dental X-ray imaging. Another method based on histogram analysis [10] was developed for tone scaling of raw CR images for general diagnostic purpose.

It can be resumed that linear filtering and statistical approaches may not provide enough contrast for a full size and scale images (like chest radiography) and non-linear ones theoretically may not guarantee the artifact absence due to image subtraction operations. Those artifacts may look like "salt and pepper" at some intermediate stage and may be hid by manipulations at later stages in the long processing chain. In [5] the authors made an effort to solve both problems in two steps: (1) they used a class of nonlinear intensity transformations derived from Hurter and Driffield characteristic curves for the contrast enhancement of computed radiography images for different exam types; and (2) they used a scale-dependent frequency filtering to prevent aliasing effect caused by anti-scatter grids while image resizing. In this paper a single stage image processing approach is proposed for contrast enhancement and grid artifacts elimination avoiding possible artifacts generation.

2 Homomorphic Equalizer

As mentioned in previous section a successful approach for the grid artifacts elimination can be achieved by a homomorphic filtering based on Kaiser bandstop transfer function.

The disadvantages of known contrast enhancement algorithms described above led us to develop an alternative method based on a cascade of linear frequency filters/amplifiers applied in homomorphic space.

2.1 Frequency Band Equalizer

A typical radiographic image frequency range of interest is about 0.01-0.1 as a fraction of Nyquist and grid artifact frequency maximum usually [2, 3, 8] is located in a range of 0.3-0.4. This means that the influence of ultra-low frequencies below 0.01 is needed to be reduced to balance the frequencies in the low and mid frequency ranges of interest and to eliminate possible grid artifacts in high-frequency range. This can be done by a highpass filter-amplifier design for the first ultra-low frequency band, i.e. the image dynamic range will be transformed by means of ultra-low frequency amplitudes reduction and mid- and higher frequencies amplification. Then several bandpass filters/amplifiers can be designed for the low- and mid-frequency range, and finally a bandstop filter in high-frequency range is designed. As known [9] the amplification function can be obtained by boosting the central weight of non-recursive filter and several linear filters can be combined into a cascade by consequent convolution operations of their impulse responses or by multiplication of their spectral equivalents. The main problem here is to design a single filter for each sub-band with a transfer function that produces the minimal amount of artifacts, e.g. ringing effects as described in [8].

A variety of approaches in digital filter transfer function design and implementation is known. Among them is digital filter based on Kaiser transfer function [9]. The remarkable feature of Kaiser filter is that besides filter length parameter there is another parameter that controls the shape of smoothing window and what is very important the level of Gibbs events that cause periodic oscillations of intensity around narrow band limited by filter stop and pass frequencies. Let us denote full amplitude of Gibbs oscillations as δ and define their desired level as:

$$A = -20 \lg(\delta) \tag{1}$$

Filter length N was empirically defined by Kaiser as a function of two parameters:

$$N = \frac{(A - 7.95)}{28.72 \, dF} \tag{2}$$

where the width of filter slope can be limited as $dF = \frac{f2-f1}{2}$. The weights of Kaiser window for $|k|>N$ are set to zero and for $|k| \leq N$ are defined as follows:

$$W_k = \frac{I_0(a \sqrt{1 - \left(\frac{k}{N}\right)^2})}{I_0(a)} \tag{3}$$

where I_o – is zero-order Bessel function of the first kind, $a=a(A)$ is a table function of the desired level of Gibbs events empirically defined by Kaiser [9]. Finally, the transfer function is obtained as:

$$H(f) = C_0 + \sum_{k=1}^{N} C_k(f1, f2) \, W_k(I_0, a) \, cos \, (2\pi k f) \tag{4}$$

where C_k are coefficients of a known ideal filter rectangular window and depend on two frequencies $f1$ and $f2$ for each sub-band.

Following the described above technique the weights for each sub-band can be obtained followed by their convolution into a cascade of filters forming the frequency band equalizer (FBE). A typical equalizer may consist of 4-5 filters/amplifiers with a sample transfer function shown in Fig.2 by solid line.

The parameters to control the shown FBE transfer function are cut and pass frequencies and attenuation coefficients for each sub-band.

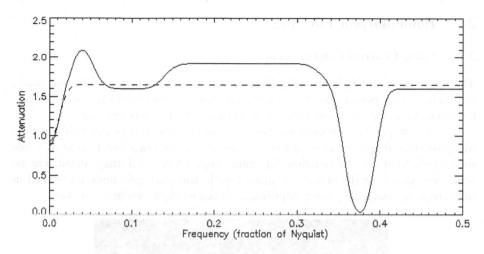

Fig. 2. Typical frequency band equalizer (solid line) and highpass filter/amplifier (dashed line) based on Kaiser transfer function.

2.2   Homomorphic Approach

If a multiplicative image $f(x,y)$ formation model described in fundamental paper [4] is accepted:

$$f(x,y) = i(x,y)\,r(x,y) \tag{5}$$

where $i(x,y)$ – is illumination function and $r(x,y)$ – is reflectivity function of the scene/object. For projection radiography the reflectivity component can be replaced by transmittance component $t(x,y)$. Then a logarithmic function can be used

$$\log[(f(x,y)] = \log[i(x,y)] + \log[t(x,y)] \tag{6}$$

to handle more effectively image dynamic range presented by the first term and image details presented by the second term. In that classical paper the two component image model was processed by a linear filter. It is proposed to use a linear filter cascade, i.e. the FBE described in previous section, in logarithmic space followed by exponentiation to return back into linear space domain. This approach guaranties the absence of such artifacts that can be produced by processed image subtractions and further manipulations used in other non-linear methods as described earlier. As shown in [3] the sub-band with grid artifacts can be narrowed and filtered more effectively using homomorphic approach. Analogically each sub-band can be narrowed and filtered/amplified more accurately using FBE in homomorphic space.

3   Homomorphic Filtering

3.1   Image Contrast Enhancement

The FBE shown in Fig.2 smoothly filters the ultra-low frequencies reducing image luminance component, it also filters and amplifies mid frequencies improving image transmittance component, and suppresses grid artifacts at high frequencies.

The advantages of homomorphic filtering can be estimated in Fig.3. If a simple highpass filter frequency characteristic shown in Fig.2 dashed line is used as a homomorphic filter transfer function then either large objects and image details can be seen more clearly in the processed image Fig.3b. But since grid lines are located in high frequency range they were amplified and became highly visible as artifacts.

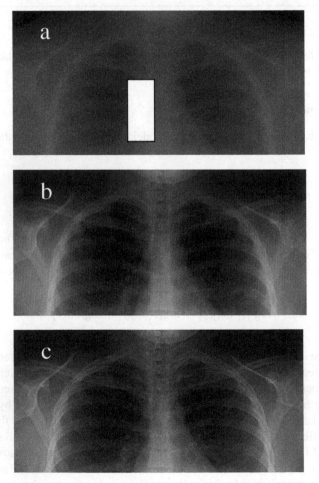

Fig. 3. Minified radiographic images: original (a), filtered by homomorphic function shown in Fig. 2 dashed line (b), filtered by homomorphic FBE with grid artifacts elimination (c).

If the FBE designed in previous section (Fig.2 solid line) is applied in 2-D with the adjusted parameter for ultra-low frequencies in sensitive logarithmic space then the image contrast is enhanced and grid artifacts are eliminated (Fig.3c).

3.2   Grid Artifact Elimination and Image Sharpening

Image sharpening is another important aspect of image contrast and is based on edge enhancement that can be achieved by amplification of mid- and high frequencies. As discussed in [2, 10] grid lines cause artifacts in radiographic images due to their microstructure that may conflict with a pixel size of digitizing hardware. So grid pattern is usually an ultra-high frequency object usually in a range higher than 0.3 while regular edges may have frequencies in a range of 0.1-0.3. Thus there is a need to design a specific FBE to amplify the edge frequencies and to eliminate grid line frequencies. The proposed homomorphic approach can serve to solve this problem by narrowing the sub-bands. The results of the image sharpening with grid artifacts suppression are shown in Fig.4. The advantage of homomorphic approach in grid artifacts elimination and image sharpening can be noticed comparing the quality of image details especially in soft tissue areas in Fig.4c vs. filtering in linear space shown in Fig.4b.

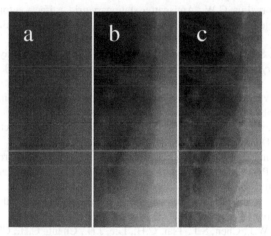

Fig. 4. Fragments of radiographic images denoted by rectangular region in Fig. 3a: original (a), filtered by FBE in linear space (b), filtered by homomorphic FBE (c).

It should be noticed that there are no visible artifacts caused by filtering process neither in linear nor in homomorphic cases.

4   Results and Conclusions

The variety of parameters needed to reach the desired image contrast depend on exam type, body part, patient/body part orientation, modality vendor and radiologist visual perception preferences. They can be combined in equalizer presets, i.e. number of frequency sub-bands, gain factors in each sub-band and image general intensity level. Those presets can be further converted into a radiologist friendly interface as the number of diagnostically valuable objects to be "highlighted" with the selectable strength and general brightness of the image. The process of presets tuning can be similar to audio equalizer presets adjustment for the type of music, e.g. classic, jazz, rock, etc.

The implemented FBE algorithm was tested on a representative set of about 30 CR images acquired for different body parts and using grids with different frequencies. There were no artifacts detected in processed images and enhanced image contrast improved image diagnostic quality as estimated by radiologists based on established clinical metrics. The proposed method is recommended for practical use in CR, DR or PACS systems due to the described advantages over existing approaches on theoretical and practical levels.

References

1. Vuylsteke, P., Schoeters, E.: Multiscale Image Contrast Amplification (MUSICA™). In: Proc. SPIE, vol. 2167, pp. 551–560 (1994)
2. Belykh, I.N., Cornelius, C.W.: Antiscatter stationary grid artifacts automated detection and removal in projection radiography images. In: Proc. SPIE, vol. 4322, pp. 1162–1166 (2001)
3. Kim, D.S., Lee, S.: Grid artifact reduction for direct digital radiography detectors based on rotated stationary grids with homomorphic filtering. Med. Phys. 40(6), 061905-1–061905-14 (2013)
4. Oppenheim, A.V., Schafer, R.W., Stockham, T.G.: Non-linear filtering of multiplied and convolved signals. Proc. IEEE 56(8), 1264–1291 (1968)
5. Bonciu, C., Rezaee, M.R., Edwards, W.: Enhanced visualization methods for computed radiography. J. Dig. Imaging 19, 187–196 (2006)
6. Gallet, J.: The most powerful diagnostic image processing from Carestream Health. DirectView EVP Plus Software. Technical Brief Series, Carestream Health, Inc. CAT No. 120 7091 (2010)
7. Davidson, R.A.: Radiographic Contrast-enhancement masks in digital radiography. Ph.D. thesis, University of Sydney (2007)
8. Lin, C.Y., Lee, W.J., Chen, S.J., Tsai, C.H., Lee, J.H., Chang, C.H., Ching, Y.T.: A Study of Grid Artifacts Formation and Elimination in Computed Radiographic Images. J. Digit. Imaging 19, 351–361 (2006)
9. Hamming, R.W.: Digital filters. Prentice-Hall, Englewood Cliffs (1989)
10. Barski, L.L., Van Metter, R.L., Foos, D.H., Lee, H.-C., Wang, X.: New automatic tone scale method for computed radiography. In: Proc. SPIE, vol. 3335, pp. 164–178 (1998)
11. Ahmad, S.A., Taib, M.N., Khalid, N.E., Ahmad, R., Taib, H.: An Analysis of Image Enhancement Techniques for Dental X-ray Image Interpretation. International J. of Machine Learning and Computing 2(3), 292–297 (2012)

A Framework for Human Recognition
Based on Locomotive Object Extraction

C. Sivasankar and A. Srinivasan

Department of Information Technology,
MNM Jain Engineering College, Chennai, Tamil Nadu, India
{sivayag,asrini30}@gmail.com

Abstract. Moving Object detection based on video, of late has gained momentum in the field of research. Moving object detection has extensive application areas and is used for monitoring intelligence interaction between human and computer, transportation of intelligence, and navigating visual robotics, clarity in steering systems. It is also used in various other fields for diagnosing, compressing images, reconstructing 3D images, retrieving video images and so on. Since surveillance of human movement detection is subjective, the human objects are precisely detected to the framework proposed for human detection based on the Locomotive Object Extraction.The issue of illumination changes and crowded human image is discriminated. The image is detected through the detection feature that identifies head and shoulder and is the loci for the proposed framework. The detection of individual objects has been revamped appreciably over the recent years but even now environmental factors and crowd-scene detection remains significantly difficult for detection of moving object. The proposed framework subtracts the background through Gaussian mixture model and the area of significance is extracted. The area of significance is transformed to white and black picture by picture binarization. Then, Wiener filter is employed to scale the background level for optimizing the results of the object in motion. The object is finally identified. The performance in every stage is measured and is evaluated. The result in each stage is compared and the performance of the proposed framework is that of the existing system proves satisfactory.

Keywords: Background Subtraction, Image Binarization, Adaptive GMM, Sauvola Threshold.

1 Introduction

1.1 Surveillance

Surveillance is the art of monitoring closely or "watching over" from a distance using equipments like CCTV Camera.

© Springer International Publishing Switzerland 2015   431
S.C. Satapathy et al. (eds.), *Proc. of the 3rd Int. Conf. on Front. of Intell. Comput. (FICTA) 2014*
– *Vol. 2*, Advances in Intelligent Systems and Computing 328, DOI: 10.1007/978-3-319-12012-6_47

1.2   Moving Object Detection

Video trailing is that the method of locating a moving object (or multiple objects) over a period of time with a closed circuit camera. Detection of moving object is a herculean task especially in overcrowded and sensitive places like railway stations,airports,traffic junctions etc...Human detection in moving object is vital in observing,interacting,transporting,navigating,diagnosing,compressing,reconstruction,r etrieving image data. Human detection proves crucial in any intelligence video observance system.

2   Proposed Framework

The Proposed framework addresses the issue of illumination changes and crowded individuals. The proposed framework subtracts the background through Gaussian mixture model and the region of interest is extracted. The area of significance is converted to white and black picture by binarization. Then, Wiener filter is employed to scale the background level for optimizing the results of the object in motion. Finally head and shoulder is identified using the detection feature for moving objects.

2.1   Proposed Framework Architecture

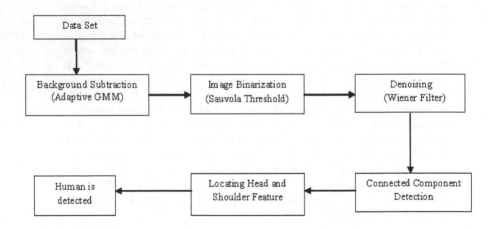

Fig. 1. Proposed Framework Architecture

2.1.1   Background Subtraction Based on Adaptive GMM
The Foreground objects are distinguished from the background by Background subtraction method.

Adaptive GMM
Gaussian mixture model is a strong for separately identifying many varied model distributions. For example, a vehicle passing through lane full of trees. The vehicle

movement becomes the foreground and the trees become the background. GMM separates each and identifies foreground precisely. In GMM, the background is not of fixed values; rather the background model is parametric [6]. Mixture of Gaussian function represents every pixel location and sum total is given by the Function F. The Function F is probably distributed pixel locations.

The pixel value of the next frame is mean denoted by 'μ' of every Gaussian function can be guessed as the pixel value of the next frame. The associated component of the input pixels is compared with the mean 'μ'. if the pixel is nearly equal to the mean 'μi' then it is considered as matched component. The standard deviation of the matched component has to be less than total difference of two pixels. This is scaled by a factor 'D'

$$|i_i - \mu_{i,t-1}| \leq D \cdot \sigma$$

There are five parameters in MOG method and it has to be tweaked accordingly to get accurate optimization. The parameters are the are the weight threshold 'T' of the background element, scaling factor 'D', the studying rate denoted by 'ρ' ,the total sum of all the Gaussian element and the maximum elements 'M' in the background.

The new pixel value is reflected by the updated element variables denoted by (w, μ, σ).The matched element is confirmed if 'w' is high, 'σ' is low, 'μ' becomes closer to the rate of the pixel. When the weights 'w' is lowered exponentially, 'μ' and 'σ' remain in the non matched element. The high 'w' and low 'σ' is rewarded by ordering the elements to a metric 'w/σ'. This enables that 'M' becomes the surest guess. Now 'T' is applied to the element weight 'w'. The threshold 'T' becomes the lowest in the background model. The background model is arranged in a descending order. The element weight 'w' higher than the threshold is arranged first followed by the 'M' elements. This depicts the maximum modes to be expected in the background and is with the help of probability distribution and the function is denoted by 'F',

$$F(i_t = \mu) = \sum_{i=1}^{k} \omega_{i,t} \cdot \eta(\mu, o)$$

Finally the foreground value of the pixels are found. The background is subtracted by marking a foreground pixel and determining, if any pixel is more than 2.5 standard deviations. The Gaussian parts are compared to find out if the pixel is an element of the background. The background is given by[6],

$$B = argmin_b \left(\sum_{k=1}^{b} \omega_k > T \right)$$

The background component's variance is a part of the Gaussian mixture algorithm.

Once the algorithmic program process is complete, it's pivotal that the ensuing binary divided pixel is more enhanced by normal morphological functions. The input image samples will be retained significantly in the lower part of the distributions and the output image has small, noise-generated spots. The false spots have to be erased by a simple area thresholding. The remaining spots are removed through shape based filtering. The hole filling algorithm is used to eliminate the residual false spots.

2.1.2   Image Binarization Using Sauvola Threshold

Binarization of image converts a picture to white picture and black picture. The simplest way to apply picture binarization is to make a decision on a threshold rate, and categorize every one of pixels with values higher than this threshold as white pixel, and every one alternative as black pixel . The next step is to choose the right threshold. In several condions, locating one threshold compatible to the entire picture is extremely tough, and in several conditions not possible. Therefore, adaptation binarization of picture is required wherever the best threshold is chosen for every image space. Image binarization is often treated as merely a thresholding operation on a grayscale image.

The binarization is performed globally or domestically. The calculated value of the threshold is used to divide the pixels as background categories in global way, whereas any tailored values can be used in the native way. In single global thresholds, the main threshold value chosen are not sufficient due to illumination changes.

In native thresholding, the threshold values are decided domestically, example one pixel by and then other pixel, or one region and then other region. The threshold is modified from one region to another region in a single threshold in a given space.

2.1.2.1 Comparison of Sauvola Threshold Method [8]

Evaluation of different method for document binarization has acknowledged some interest in the history. Trier et al.[10] assessed completely diverse domestically adaptation binarization ways for grey level pictures with small distinction, changeable background noise and intensity . The assessed outcome reveals better results in Niblack's technique [9] than alternative native thresholding ways. Badekas et al.[11] assessed diverse techniques to binarize the document. They identified that Sauvola's binarization technique is best over Niblack's technique and works best among the native thresholding techniques; whereas Otsu's technique performs alternate universal binarization techniques. Overall, Sauvola's method of binarization technique works somewhat superior than Otsu's method in lieu of their research.

The Sauvola technique for local binarization is better and the essential plan behind Sauvola is that if there's plenty of native contrast, the brink ought to be chosen near the mean, whereas if there's little or no distinction, the brink ought to be chosen below the mean, by an amount proportional to the normalized native standard deviation.

In Sauvola's method of binarization [7], mean and variance is used for calculating the threshold value and the pixels for windowpane dimension are calculated.

$$T(x,y) = m(x,y)\left[1 + k\left(\frac{\delta(x,y)}{R} - 1\right)\right]$$

Wherever R is the highest value of variance and k is a constant, for the native block.The native mean and variance change accordingly to the threshold rate and contrast within the immediate pixel. The change does not happen only if the area is of high contrast, say the picture $I(x,y) \sim R$ ends up as threshold $\sim$ mean. Similar outcome is accounted in method done by Niblack's [9]. There is a distinction in the region, once the contrast of the image close to each other is very low. In this case, the mean of the threshold 'T' is very low it automatically removes the dark areas in the background. The value of the threshold is controlled by the parameter 'k' provided if the specified upper value k is within the native window. Sauvola and Sezgin[12] has initialized k value as 0.5 where as the technique of Badekas et al.[11] attained the very good outcome by initializing k as 0.34. In common, the technique is not terribly reactive to' k'.

2.1.3   Image De-noising By Wiener Filter

The most vital task in image processing is denoising the image. An excellent de-noising model ensures that maximum background noise is removed. The various types of noise are removed by applying image filtering algorithms. The various noises are part of the image either while capturing the motion of the object or is injected into the image through the transmission. Image noise is the random variation in the brightness information of any electronic equipment like in a scanner or like in a camera.

2.1.3.1 Wiener Filter

After de-noising, the image is recovered through inverse filter. The inverse filter is done through reinstatement techniques for deconvolution.The image is blurred through a notable low pass filter through this technique and the image is recovered. The algorithm for restoring each level of diminished image is developed by reduction of the image one by one and then easily uniting them. The Wiener filter performs optimum exchange between inverse filter and smoothing of noise. Using this filter additive noise is removed and the blur is inverted at the same time. The foremost necessary technique for elimination of blur caused by the linear movement is Wiener filter. The noise corrupted signal is filtered out by the wiener filter. It also minimizes the overall MSE during inverse filter and smoothing of noise. The Wiener filtered image is only a linear inference of the actual image. The filtering is done in different angles using a wiener filter.

2.1.4   Connected Component Detection [14]

The binary foreground image is detected using the connected element detection technique. The technique helps in finding the pixels connected to the same region and same marking value is given to the same region pixels that are connected, even before the object is identified. These interrelated black pixels are perceived as a simple area in a picture that is binary [14]. The score of every pixel is the score of its marked area and the picture of the marked area is determined. The marking counter is set as zero and the pixels are scanned using scanning method based on column. The four pixels

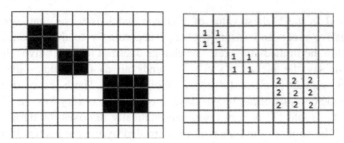

Fig. 2. Image marking [14]

scanned are the lower part in the left side, the left side, the upper part of the left side and the upper part of the every pixel. The algorithm proceeds as follows:

Stage 1: The marking variable adds on one if the values of the four gray area pixels are 255.

Stage 2: If the marking values of all the four pixels are the same but if all pixels are not equivalent to zero then the present pixel is marked.

Stage 3: The marking counter subtracts one if the four pixels have entirely different value of marking and if two kinds of marking excludes marking zero. Judging the size using the two kinds of marking, the lower marking value is assigned for the pixel and big marking value is labeled as small marking value while scanning the image.

Stage 4: Each pixel goes through step 2. When each pixels processing is over, the algorithm ends.

The preliminary detection of the human using combined features such as head feature and shoulder feature is done successfully.

2.1.5   Locating the Head and the Shoulder

The whole image is scanned and the top point of every marked region that is connected is found using vertical projection histogram. The connected marking up region is again scanned through same technique the average changing rate of calvaria is found. After the execution of the last step in the algorithm, important parameters for the detection and determination of the head and shoulder determine if the object is human.

Marked region is scanned to locate the head from the spot in the calvaria to the spot in the neck .Then the majority end of pixel coordinates including the top and bottom of the marked region accumulates to determine the head area 'S' and the coordinates 'S'(X_K, Y_K) of head is calculated. Supposing (x1, y1) denotes the upper side of left part coordinates of the rectangular structure . (X2, y2) indicates the lower side of right part coordinate.The method is

$$X_K = \sum \frac{X_i}{S}, Y_K = \sum \frac{Y_i}{S} \tag{1}$$

$$X_1 = Minimum(X_i), Y_1 = Minimum(Y_i) \qquad (2)$$

$$X_2 = Maximum(X_i), Y_2 = Maximum(Y_i) \qquad (3)$$

If formula (1) is subtracted from formula (3) the barycentric coordinates of the shoulder is got. Then the distance (D) is calculated from the coordinates of head to coordinates of shoulder [14].

$$D_1 = X_2 - X_1$$
$$D_2 = Y_2 - Y_1$$
$$D = \sqrt{(D_1)^2 + (D_2)^2}$$

3   Experimental Results

Table 1.

Bench mark	Existing system[9]	Proposed framework
Test image frame[dataset 1]		
Psnr	32.8962	36.3597
Mse	88.7117	42.3698
Maxerr	245	236
Ltpr	0.9928	0.9986
No.of humans detected	15	26
Detection rate	42.85	74.28
Human detection		

4   Performance Analysis

Table 3.1 shows that proposed framework produces higher results than the prevailing system. The number of human detected by the proposed system is twenty six where as the existing system detected only fifteen individuals .Detection rate of proposed framework is best than the existing system.

5   Conclusion

Our framework focuses on generating higher detection results by concentrating on three areas namely background subtraction, image binarization and noise reduction. Gaussian mixture model is employed for extracting the foreground because it comes up with multimodal distributions in acceptable level. Image binarization is completed by computing mean and variance to come up with accommodative threshold for every component that overcomes illumination drawback. The ensuing image is filtered to cut back the amplitude. The filtered image is employed to find out the region of interest. The performance analysis of proposed framework with the existing system is found to be satisfactory.

6   Future Enhancements

The proposed framework is useful to detect human in relatively crowded places but the limitation arises when changes his pose. Also the deterioration in image increase with the number of frames in the video. In future, the detection of human in various poses can be explored by processing the video through efficient system and employing different techniques to explore increased frames.

References

[1] Lin, S.F., Chen, J.Y., Chao, H.X.: Estimation of number of people in crowded scenes using perspective transformation. IEEE Trans. Syst. Man Cybern. A: Syst. Hum. 33(6), 645–653 (2001)

[2] Li, M., Zhang, Z., Huang, K., Tan, T.: Estimating the number of people in crowded scenes by MID based foreground segmentation and head–shoulder detection. In: 19th International Conference on Pattern Recognition (ICPR), pp. 1–4 (2008)

[3] Viola, P., Jones, M.J., Snow, D.: Detecting pedestrians using patterns of motion and appearance. In: The 9th ICCV, Nice, France, vol. 1, pp. 734–741 (2003)

[4] Mikolajczyk, K., Schmid, C., Zisserman, A.: Human detection based on a probabilistic assembly of robust part detectors. In: Pajdla, T., Matas, J.(G.) (eds.) ECCV 2004. LNCS, vol. 3021, pp. 69–82. Springer, Heidelberg (2004)

[5] Lin, Z., Davis, L.S.: Shape-based human detection and segmentation via hierarchical part-template matching. IEEE Trans. Pattern Anal. Mach. Intel. 32(4), 604–618 (2010)

[6] Mukherjee, S., et al.: An Adaptive GMM Approach To Background Subtraction For Application In Real Time Surveillance. IJRET (January 2013)

[7] Shafait, F., et al.: Efficient Implementation of Local Adaptive Thresholding Techniques Using Integral Images. In: Electronic Imaging 2008, pp. 681510–681510-6 (2008)

[8] Otsu, N.: A threshold selection method from gray-level histograms. IEEE Trans. Systems, Man, and Cybernetics 9(1), 62–66 (1979)

[9] Niblack, W.: An Introduction to Image Processing. Prentice-Hall, Englewood Cliffs (1986)

[10] Trier, O.D., Taxt, T.: Evaluation of binarization methods for document images. IEEE Trans. on Pattern Analysis and Machine Intelligence 17, 312–315 (1995)

[11] Badekas, E., Papamarkos, N.: Automatic evaluation of document binarization results. In: 10th Iberoamerican Congress on Pattern Recognition, Havana, Cuba, pp. 1005–1014 (2005)

[12] Sezgin, Sankur, B.: Survey over image thresholding techniques and quantitative performance evaluation. Journal of Electronic Imaging 13(1), 146–165 (2004)

[13] Kumar, S., et al.: Performance Comparison of Median and Wiener Filter in Image De-noising. International Journal of Computer Applications (2010)

[14] Ye, Q., et al.: Human detection based on motion object extraction and head–shoulder feature. Optik - International Journal for Light and Electron Optics (2013)

Abnormal Event Detection in Crowded Video Scenes

V.K. Gnanavel and A. Srinivasan

Department of Information Technology,
MNM Jain Engineering College, Chennai, Tamil Nadu, India
{vkgnanavel,asrini30}@gmail.com

Abstract. Intelligent Video Investigation is on nice interest in trade applications because of increasing demand to scale back the force of analyzing the large-scale video information. Sleuthing the abnormal events from crowded video scenes offer varied difficulties. Initially, an oversized variety of moving persons will simply distract the native anomaly detector. Secondly it's tough to model the abnormal events in real time. Thirdly, the inaccessibility of ample samples of coaching information for abnormal events ends up in problem in sturdy detection of abnormal events. Our planned system provides a peculiar approach to find anomaly in crowded video scenes. We are initially divide the video frame into patches and apply the Difference-of-Gaussian (DoG) filter to extract edges. Then we work out Multiscale Histogram of Optical Flow (MHOF) and Edge directed bar chart (EOH) for every patch. Then exploitation of Normalized Cuts (NCuts) and Gaussian Expectation-Maximization (GEM) techniques, and to cluster the similar patches into cluster and assign the motion context. Finally exploitation of k-Nearest neighbor (k-NN) search, and establish the abnormal activity at intervals in the crowded scenes. Our spatio-temporal anomaly search system helps to boost the accuracy and computation time for detection of irregular patterns. This technique is helpful for investigation, trade specific and market applications like public transportation, enforcement, etc.

Keywords: Multiscale Histogram of Optical Flow, Gaussian Expectation Maximization, Spatio-temporal anomaly search.

1   Introduction

Intelligent video Surveillance is of great interest in trade applications because of increasing demand to scale back the force of analyzing the large scale video information. Key technologies are developed for intelligent Surveillance like object chase, pedestrian detection, gait analysis, vehicle guide recognition, privacy protection, face and iris recognition, video account and crowd reckoning.

Crowded Scenes involve sizable amount of individuals acquiring irregular directions in an exceedingly vast region. Crowded Scenes offer Brobdingnagian problem in observation and detection of abnormal events. Any deviation from the conventional crowd behavior in an exceedingly vast video is termed as anomaly or

© Springer International Publishing Switzerland 2015   441
S.C. Satapathy et al. (eds.), *Proc. of the 3rd Int. Conf. on Front. of Intell. Comput. (FICTA) 2014*
– *Vol. 2*, Advances in Intelligent Systems and Computing 328, DOI: 10.1007/978-3-319-12012-6_48

abnormal event. Anomaly could be a pattern within the information that does not adjust to the expected behavior. It is conjointly cited as outliers, exceptions, peculiarities, surprise, etc. .

Some of the examples are Circulation of non - pedestrian entry on walkways like bikes, skates small cars, trucks, etc. Further detecting anomaly in crowded scenes is tough.

Recognition and perception of human action in video have gained considerable analysis attention because of the potential application in varied domains, like video Surveillance and observation, human-computer interfaces, content-based video analysis and activity biometry. In video Surveillance, the most important objective is to be able to find events of interest to help security personnel.

More recently, there has been a paradigm shift [8] towards detection of abnormal events. Intuitively, an anomaly could be a pattern that doesn't follow expected traditional behavior in a given context.

In video Surveillance, pattern recognition techniques are typically the specified approach to perform event detection and behavior understanding, by modeling abnormal events from coaching information. Detection is performed by finding patterns in new observations that adjust to the antecedently obtained model. In abnormal event detection, a model of traditional behavior is developed statistically and anomalies are detected by finding patterns that deviate from the model.

2   Proposed System

2.1   Overview

Our planned system provides a peculiar approach to find anomaly in CROWDED video scenes. We have a tendency to initially divide the video frame into patches and apply the Difference-of-Gaussian (DoG) filter to extract edges. Then we have a tendency to work out Multiscale Histogram of Optical Flow (MHOF) and Edge Oriented Histogram (EOH) for every patch. Then using Normalized Cuts (NCuts) and Gaussian Expectation-Maximization (GEM) techniques we have a tendency to cluster the similar patches into cluster and assign the motion context. Finally in exploitation of k-Nearest Neighbor (k-NN) search we have a tendency to establish the abnormal activity at intervals of the crowded scenes. This technique is helpful for investigation, trade specific and market applications like public transportation, enforcement, etc

2.2   Block Diagram

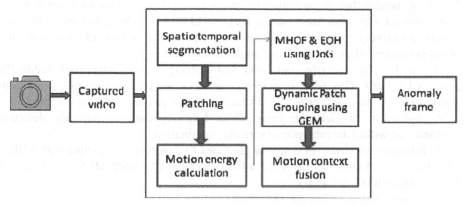

Fig. 1. Block Diagram of Proposed System

2.3   Module Description

The proposed system consists of the following modules:

- Basic Patch Descriptor
  - Partition into 2D patches
  - Evaluating pixel motion energy
  - Appearance Feature – EOH using DoG
  - Motion Feature – MHOF
- Dynamic Patch Grouping using GEM
- Motion Context
- Abnormal Event Detection

2.3.1   Basic Patch Descriptor

2.3.1.1   Partition into 2D Patches
This module remains an equivalent of the existing system. The given check image frame is split into two patches based mostly upon the scale of the input pictures. The quantity of patches depend on the image frame size. The image is split into many patches of 30x30 constituent size.

2.3.1.2   Evaluating Pixel Motion Energy
The motion energy of the constituents of the patches is computed by examining every constituent of the patch. The calculable motion energy is then needed to confirm the foreground patches and discard the background patches. If the motion energy of over 0.5 the pixels is larger than zero then we tend to take that patch as a foreground patch. Whereas if the motion energy is over 0.5 the pixels is lesser than zero then it's taken as background unit and discarded.

2.3.1.3 Appearance Feature – EOH Using DoG

The Edge headed bar graph within the projected system is calculated victimization distinction of Gaussian filter rather than Sobel masks. The sobel mask operator doesn't effectively observe the perimeters of the patches as a result of those anomalies at the perimeters of the image frame might go unobserved.

Hence the effective edge detection and filtering DoG is employed within the projected system. This filter is meant to assist atone for each intensity variations inside a picture domain. Further the differences in image domains were observed.

A distinction of Gaussian image filter helps to enhance anomaly detection performance within the presence of variable illumination.

A distinction of Gaussian image is generated by convolving a picture with a filter. This filter was obtained by detecting a Gaussian filter of breadth θ1 from a Gaussian filter of breadth θ2 (θ2 > θ1).

In existing system, θ1=2 and θ2 =thirty.

The formula for DoG is:

$$D(x, y, \delta_1, \delta_2) = \frac{1}{2\pi\delta_1^2} e^{-\frac{x^2+y^2}{2\delta_1^2}} - \frac{1}{2\pi\delta_2^2} e^{-\frac{x^2+y^2}{2\delta_2^2}}$$

2.3.1.4 Motion Feature – MHOF

For getting the motion feature, Multiscale bar graph of Optical Flow is employed. MHOF has total of sixteen bins. the primary eight bins denote the motion in one direction and also the alternative eight bins denote the motion within the other way. The bar graph worth is obtained supported motion energy worth. If the motion energy is lesser than the brink worth then the bar graph uses the primary eight bins else it makes use of ensuing eight bins. The entropy of the MHOF helps United States of America to any strain the precise foreground patches.

2.3.2 Dynamic Patch Grouping Using GEM

In the projected system the patches area unit sorted on the idea of events by creating use of Normalized Cuts algorithm together with the Gaussian Expectation Maximization rule for bunch. The remainder of the rule employed is same because the existing system except that rather than k-means GEM is employed.

First Construct a similarity graph and Compute unnormalized Laplacian. Then compute 1st k generalized Manfred Eigen vectors as columns. Let U be the matrix containing the Manfred Eigen vectors as columns. Then obtain the row vector for every patch. Finally Cluster victimization GEM.

The EM technique provides some benefits. The power to at the same time optimize an oversized range of variables. The power to seek out smart estimates for any missing info in your knowledge at the same time. The ability to make each the normal "hard" clusters and not-so-traditional "soft" clusters.

The EM rule needs subsequent 2 steps.

1. The Expectation Step: victimization is the best guess for the parameters of the info model. Here we tend to construct associate expression for the log-likelihood for all knowledge, discovered and unobserved, and, later on we interact on the expression with reference to the unobserved knowledge. This expression is shown to rely on each of this best guess for the model parameters and also the model parameters treated as variables.

2. The Maximization Step: Given the expression ensuing from the previous step, for ensuing guess we elect those values for the model parameters that maximize the expectation expression. These represent our greatest new guess for the model parameters.

2.3.3   Fusion of Motion Context

Once the patches are clustered, the motion context is obtained by fusing the spatio – temporal info. The results of MHOF and EOH area unit are combined into a high dimensional feature vector. This as the event descriptor. The results of this module is a picture frame with the foreground units delineated in several colours supported by the direction of their motion. So we tend to get a picture that's dynamically sorted victimization the motion and look info.

2.3.4   Abnormal Event Detection

The sorted image is then compared with the info to seek out however traditional it is. The k-nearest neighbor rule is employed for determinant abnormality. The speed of computation is improved y victimization compact projections.

The rule works as follows; For every x_i, the closest neighbor x_i * is found victimization compact projection that approximates the k-nearest neighbor search. Then the geometer distance $d(x_i, x_i$ * $)$ is calculated. This distance acts because the threshold τx* .

For testing knowledge the closest neighbor x* is calculated. Then work out the geometer distance $d(x$* $, y)$Finally anomaly is detected as follows:

$$y = \begin{cases} normal & d(\mathbf{x}^*, \mathbf{y}) \leq 1.2 \times \tau_{x^*} \\ abnormal & d(\mathbf{x}^*, \mathbf{y}) > 1.2 \times \tau_{x^*} \end{cases}$$

3   Experimantal Results and Analysis

The analysis takes into thought period of each system together with the time taken to work out the Ncuts rule. It additionally includes PSNR values of the systems. The mythical monster curves are drawn and also the space underneath the Curve helps United States of America to estimate the accuracy of the 2 systems. The results section additionally determines the quantity of anomalies detected by every system.

The result and analysis section provides a consolidation value based on performance analysis. This helps to point out the enhancements achieved within the projected system.

3.1   Quantitative Performance Analysis

Table 1. Result Analysis

BENCH MARK	EXISTING SYSTEM	PROPOSED SYSTEM	EXISTING SYSTEM	PROPOSED SYSTEM
TEST IMAGE FRAME				
PSNR	35.7761	42.4610	35.1904	37.7017
MSE	1.0438	0.7674	2.3715	1.1580
RMSE	1.0216	0.8760	1.5399	1.0761
ELAPSED TIME	3.399967	3.161710	3.725921	3.540463
NCUTS TIME	3.0544	2.7865	3.3209	3.1545
ANOMALIES DETECTED	1	2	1	2
AUC	0.0001	0.500	0.0001	0.782
ANOMALY DETECTED IMAGE				

3.2   Performance Analysis

The AUC is equal to the probability that a system will have a higher randomly chosen positive instance than a randomly chosen negative one. Thus greater the AUC greater the accuracy of the system. The value of AUC ranges between 0 to 1. The result analysis shows that the AUC of proposed system is greater than the existing system which indicates that the accuracy of the proposed system is better than the existing system.

The MSE is one of many ways to measure the difference between values implied by an estimator and the true values of the quantity being estimated. The result analysis shows that proposed system is better because the MSE is lesser for the proposed system than existing system.

PSNR is the ratio between the maximum possible power of a signal and the power of corrupting noise that affects the fidelity of its representation. The PSNR value of the proposed system is better than the existing system.

The RMSE is a measure of the differences between values predicted by a model and the values actually observed. The RMSE is reduced for the proposed system indicating better performance.

4 Conclusion and Future Enhancement

4.1 Conclusion

The proposed system recognizes the anomalies in the image frames in a lesser time than the existing system. It also helps to effectively detect the anomalies at the edges of the image which may go undetected in the existing system. The elapsed time and Ncuts computation time of the proposed system is lesser than the existing system. The PSNR is also improved in the proposed system. The ROC curve shows that the accuracy of the proposed system is improved as the AUC is more than the existing system. Also more number of anomalies were detected in the proposed system then the existing system. Thus the proposed system effectively and accurately helps in detecting anomalies in crowded scene video frames.

4.2 Future Enhancement

Though the proposed system effectively recognizes anomaly in the image frames it does not process the complete video together. Also the system performance degrades as patch size increases. As a future enhancement we hope to explore anomaly detection by processing complete video and also improve the efficiency of the system when the patch size increases by employing additional techniques.

References

[1] Chan, A.B., Vasconcelos, N.: Modeling, Clustering, and Segmenting Video with Mixtures of Dynamic Textures. IEEE
[2] Xu, D., Wu, X., Song, D., Li, N., Chen, Y.-L.: Hierarchical Activity Discovery Within Spatio-Temporal Context For Video Anomaly Detection. IEEE (2013)
[3] Jiang, F., Yuan, J., Tsaftaris, S., Katsaggelos, A.: Anomalous videoevent detection using spatiotemporal context. Comput. Vis. Image Understand., 323–333 (2011)
[4] Shrivakshan, G.T., Chandrasekar, C.: A Comparison of various Edge Detection Techniques used in Image Processing. IJCSI International Journal of Computer Science Issues 9(5(1)) (2012)

[5] Min, K., Yang, L., Wright, J., Wu, L., Hua, X., Ma, Y.: Compact Projection: Simple and Efficient Near Neighbor Search with Practical Memory Requirements. In: Proc. CVPR, pp. 3477–3484 (2010)

[6] Meskaldji, K., Boucherkha, S., Chikhi S.: Color Quantization and its Impact on Color Histogram Based Image Retrieval

[7] Cristani, M., Raghavendra, R., Del Bue, A., Murino, V.: Human Behavior Analysis in Video Surveillance: A Social Signal Processing Perspective. Neurocomputing 100, 86–97 (2013)

[8] Thida, M., Yong, Y.L., Climent-Pérez, P., Eng, H.-L., Remagnino, P.: A Literature Review on Video Analytics of Crowded Scenes. In: Intelligent Multimedia Surveillance. Springer, Heidelberg (2013)

[9] Sjarif, N.N.A., Shamsuddin, S.M., Hashim, S.Z.: Detection of Abnormal Behaviors in Crowd Scene: A Review. Int. J. Advance. Soft Comput. Appl. 4(1) (March 2012)

[10] Vaswani, N., Roy-Chowdhury, A.K., Khan, R.C.: Shape activity: A continuous-state hmm for moving/deforming shapes with application to abnormal activity detection. IEEE Trans. Image Process. 14(10), 1603–1616 (2005)

[11] Maini, R., Aggarwal, H.: Study and Comparison of Various Image Edge Detection Techniques. International Journal of Image Processing (IJIP) 3(1)

[12] Mahadevan, V., Li, W., Bhalodia, V., Vasconcelos, N.: Anomaly detection in crowded scenes. In: Proc. CVPR (2010)

[13] Cong, Y., Yuan, J., Liu, J.: Abnormal event detection in crowded scenes using sparse representation. Pattern Recognit. 46, 1851–1864 (2013)

Comparative Analysis and Bandwidth Enhancement with Direct Coupled C Slotted Microstrip Antenna for Dual Wide Band Applications

Rajat Srivastava[1], Vinod Kumar Singh[2], and Shahanaz Ayub[3]

[1,3] Bundelkhand Institute of Engg Technology, Jhansi, U.P. India
[2] S.R. Group of Institutions, Jhansi, U.P., India
{raj.sriv89,singhvinod34}@gmail.com,
shahanaz_ayub@rediffmail.com

Abstract. This paper presents a direct coupled C slot loaded microstrip antenna which can yield wider bandwidth for WLAN/WiMax applications. The different antenna geometries are simulated through IE3D Zeland simulation software for the comparative analysis of bandwidth. The recent development technology of wireless internet access applied to the WLAN (Wireless Local Area Network) frequency bands in the range 2.40-2.50 GHz has forced demand for dual-band antennas, which can be implemented in stationary and mobile devices. The proposed antenna has dual frequency band having fractional bandwidth 3.38% (1.392-1.44 GHz) and 69.5% (1.733-3.58 GHz) which is suitable WLAN/WiMax applications. The gain has been improved up to 5.11dBi, directivity 5.39dBi and efficiency 97.216%. The proposed directly coupled microstrip antenna fed by 50 Ω microstrip feed line.

Keywords: Direct coupling, enhance bandwidth, compact Microstrip (MS) Patch, ground plane, gain, 50Ω feed line.

1 Introduction

Microstrip antennas received considerable attentation starting in the 1970s, although the idea of a microstrip antenna can be traced to 1953 and a patent in 1955[2]. Microstrip patch antennas have drawn the attention of researchers due to its light weight, low profile, low cost and ease of integration with microwave circuit. But the major drawback of rectangular Microstrip antenna is its narrow bandwidth and lower gain. The bandwidth of Microstrip antenna may be increased using several techniques such as use of a thick or foam substrate, cutting slots or notches like U slot, E shaped H shaped patch antenna, introducing the parasitic elements either in coplanar or stack configuration, and modifying the shape of the radiator patch by introducing the slots [3-5].In the present work the bandwidth of Microstrip antenna is increased by direct coupling [6-8] and it is obtained that the bandwidth of direct coupled 'c' slotted rectangular Microstrip antenna is 'ten times' greater than simple rectangular Microstrip antenna.Directly coupled C slotted rectangular Microstrip antenna with

© Springer International Publishing Switzerland 2015

S.C. Satapathy et al. (eds.), *Proc. of the 3rd Int. Conf. on Front. of Intell. Comput. (FICTA) 2014*
– *Vol. 2*, Advances in Intelligent Systems and Computing 328, DOI: 10.1007/978-3-319-12012-6_49

Microstrip line feed is shown in Figure1. The width of the Microstrip line was taken as 2.5 mm and the feed length as 1 mm. The patch is energized electromagnetically using 50 ohm Microstrip feed line [9-13]. There are numerous substrates that can be used for the design of microstrip antennas and their dielectric constants are usually in the range of $2.2 \leq \varepsilon r \leq 12$ [14].The proposed antenna has been designed on glass epoxy substrate ($\varepsilon r = 4.4$). The design frequency of proposed antenna is 2.4 GHz. The frequency band(1.733-3.58 GHz) of proposed antenna is suitable for broad band applications(1.605-3.381GHz) such as military, wireless communication, satellite communication, global positioning system (GPS), RF devices, WLAN/WI -MAX application [14-16]. Broadband devices are mainly used in our daily lives such as mobile phone, radio, laptops and Microstrip patch antennas plays important role in these devices.

2   Optimum Antenna Design Specification and Results

In this paper equations 1, 2, 3, 4, 5 and 6 are used to calculate dimensions of proposed antenna having the design frequency is 2.4 GHz. For designing a rectangular Microstrip patch antenna, the length and width are calculated as below [14-21]. For making a proposed direct coupled antenna one part of the patch is developed as double T shaped design and it is directly coupled with C slotted patch. The proposed antenna structure is compared with four different designs that are shown in figures 5, 7,9,11 having same dimensions and feed at the same position on the patch of the proposed antenna. It is found that rectangular patch antenna having double T and C slot provides highest bandwidth with maximum gain. Figure1 shows the geometry of proposed antenna and Figure 2 shows optimum return loss Vs frequency plot of proposed microstrip antenna giving the maximum bandwidth of 69.5% (1.733-3.58 GHz). Figure 3 shows the gain Vs frequency plot of optimum design having the gain of 5.11dBi. Figure 4 shows 3D Radiation pattern of proposed antenna.

$$W = \frac{c}{2 f \sqrt{(\varepsilon_r + 1)/2}} \tag{1}$$

Where c is the velocity of light, εr (4.4) is the dielectric constant of substrate(glass epoxy), fr(2.4GHz) is the antenna design frequency, W is the patch width, and the effective dielectric constant ε_{reff} is given as [2]

$$\varepsilon_{eff} = \frac{(\varepsilon_r + 1)}{2} + \frac{(\varepsilon_r - 1)}{2} \left[1 + 10 \frac{h}{W} \right]^{-\frac{1}{2}} \tag{2}$$

At h=1.6mm, The extension length ΔL is calculated as [13, 14, 10]

$$\frac{\Delta l}{h} = 0.412 \frac{\left(\varepsilon_{eff} + 0.300\right)\left(\dfrac{W}{h} + 0.262\right)}{\left(\varepsilon_{eff} - 0.258\right)\left(\dfrac{W}{h} + 0.813\right)} \qquad (3)$$

By using the above mentioned equation we can find the value of actual length of the patch as,

$$L = \frac{c}{2f\sqrt{\varepsilon_{eff}}} - 2\Delta l \qquad (4)$$

The length and the width of the ground plane can be calculated as [14-21]

$$L_g = L + 6h \qquad (5)$$

$$W_g = W + 6h$$

Table 1. Antenna design specifications

S. No.	Parameters	Value	S. No.	Parameters	Value
1.	Design frequency (f_r)	2.4	10.	c	5.0
2.	Dielectric constant (ε_r)	4.4	11.	d	10
3.	Substrate height	1.6	12.	e	4.0
4.	Patch width	38.0	13.	f	6.0
5.	Patch length	28.3	14.	g	5.0
6.	Ground plane width	47.6	15.	h	20
7.	Ground plane length	37.9	16.	i	12
8.	a	2.5	17.	j	2.0
9.	b	1.0	18.	1	9.0

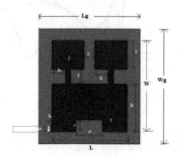

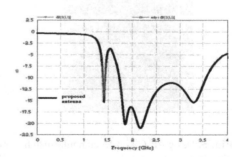

Fig. 1. Geometry of proposed Microstrip antenna

Fig. 2. Return Loss Vs Frequency Plot of proposed Microstrip Antenna

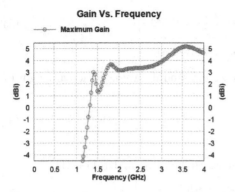

Gain Vs. Frequency

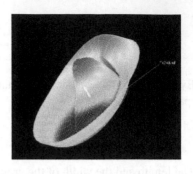

Fig. 3. Gain Vs. frequency plot

Fig. 4. 3D Radiation pattern of proposed antenna

3   Comparative Analysis of Antenna Geometries

For making the proposed direct coupled C slotted antenna, different structures are simulated sequentially through Zeland IE3D simulation software and their bandwidths are compared. By this computation work it is found that the bandwidth of rectangular Microstrip antenna with C slot and double T shaped directly coupled structures is highest among all the designs that are simulated. All five designs and their results such as return loss are compared to get maximum bandwidth, gain and efficiency which are shown in figures 5, 6, 7, 8, 9, 10, 11, 12 respectively and the comparative results have been summarized in table 2.

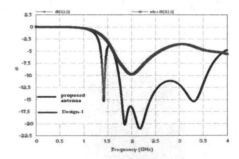

Fig. 5. Design-I: C slot Rectangular patch Antenna

Fig. 6. Comparison of bandwidth of proposed Antenna With the bandwidth of Design-I

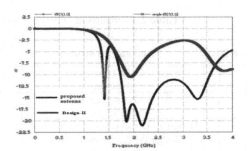

Fig. 7. Design II Rectangular Microstrip patch antenna

Fig. 8. Comparison of bandwidth of proposed Microstrip antenna with the bandwidth of Design II

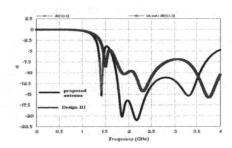

Fig. 9. Design III Direct coupled with T slot

Fig. 10. Comparison of bandwidth of proposed antenna with the bandwidth of Design III

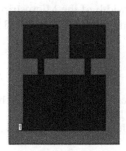

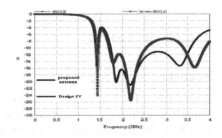

Fig. 11. Design IV Direct coupled with double T slots

Fig. 12. Comparison of bandwidth of proposed antenna with the bandwidth of Design IV

Table 2. Comparison of bandwidth of different antenna Structures

S.No.	Antenna Structures	Frequency band	Band width	Linear Gain	Antenna Efficiency
1.	Design-I: C slot rectangular patch Antenna	1.82-1.92	05.34%	3.11 dBi	89.33%
2.	Design II Rectangular Microstrip patch antenna	2.112-2.563	19.38%	3.24 dBi	95.21%
3.	Design III Direct coupled with T slot.	1.695-2.549	40.24%	3.27 dBi	98.62%
4.	Design IV Direct coupled with double T slot	1.906-3.648	62.72%	4.59 dBi	96.8%
5.	Optimum Design directly coupled with c slot (proposed antenna).	1.733-3.58	69.52%	5.11 dBi	97.21%

4   Simulation Results

In the present work we implemented few novel designs to increase the bandwidth of rectangular Microstrip antenna by direct coupling method [21-23] and the optimal results are obtained by design-V. From the Results given in Table 2 it is clear that rectangular patch antenna having C slot with double direct coupled T structures provides highest bandwidth among all structures that are simulated. The maximum gain of the antenna has been improved up to 5.11dBi, directivity improved up to 5.39dBi, efficiency of the antenna is found to be 97.21%, and the VSWR of the antenna is in between 1 to 2 over the entire frequency band which shows that there is a proper impedance matching.

5   Conclusion

The comparative analysis and characteristics of compact direct coupled C slotted patch antenna is studied. In general, the impedance bandwidth of the traditional Microstrip antenna is only a few percent (2-5%). Therefore, it becomes very important to develop a technique to enhance the bandwidth of the Microstrip antenna. Proposed antenna provides dual frequency band width of 69.52% (1.733-3.58 GHz) and 3.38% (1.392-1.44 GHz).

References

1. Bhattacharya, D., Prasanna, R.: Bandwidth Enrichment for Micro-strip Patch Antenna Using Pendant Techniques. IJER 2(4), 286–289 (2013) ISSN: 2319-6890
2. Balanis, C.A.: Antenna theory, Analysis and Design. John Wiley & Sons, Inc., Hoboken (2005)
3. Srivastava, S., Singh, V.K., Ali, Z., Singh, A.K.: Duo Triangle Shaped Microstrip Patch Antenna Analysis for WiMAX lower band Application. Procedia Technology 10, 554–563 (2013)
4. Singh, P., Chandel, A., Naina, D.: Bandwidth Enhancement of Probe Fed Microstrip Patch Antenna. IJECCT 3(1) (January 2013) ISSN: 2249-7838
5. Jain, S., Singh, V.K., Ayub, S.: Design of Slotted Microstrip Antenna having high efficiency and gain. International Journal of Engineering and Technical Research, 50–52 (April 2014) ISSN: 2321-0869

6. Pavithra, D., Dharani, K.R.: A Design of H-Shape Microstrip Patch Antenna for WLAN Applications. IJESI 2(6), 71–74 (2013) ISSN: 2319-6734

7. Pathak, R.S., Singh, V.K., Ayub, S.: Dual Band Microstrip Antenna for GPS/ WLAN/ WiMAX Applications. IJETED 7(2) (November 2012) ISSN: 2249-6149

8. Singh, V.K., Ali, Z., Singh, A.K., Ayub, S.: Dual Band Triangular Slotted Stacked Microstrip Antenna for Wireless Applications. Central European Journal of Engineering 3(2), 221–225 (2013) ISSN: 1896-1541

9. Jain, S., Singh, V.K., Ayub, S.: Bandwidth and Gain Optimization of a Wide Band Gap Coupled Patch Antenna. IJESRT (March 2013) ISSN: 2277-9655

10. Ali, Z., Singh, V.K., Ayub, S.: A Neural Network Approach to study the Bandwidth of Microstrip Antenna. IJARCSSE 3(1) (January 2013) ISSN: 2277 128X

11. Roy, A., Bhunia, S.: Compact Broad Band Dual Frequency Slot Loaded Microslot Patch antenna with Defecting Ground Plane for WI-MAX and WLAN. IJSCE 1(6) (January 2012) ISSN: 2231-2307

12. Singh, V.K., Ali, Z., Ayub, S., Singh, A.K.: A Compact wide band Microstrip Antenna for GPS/DCS/PCS/WLAN Applications. In Intelligent Computing, Networking, and Informatics, pp. 183–204 (October 2013)

13. Singh, V.K., Jain, S.: Low Profile Slotted Microstrip Patch Antenna for Dual Band Application. International Journal of Advanced Electronics and Communication Systems (IJAECS) 3(2) (April-May 2014)

14. Rop, K.V., Konditi, D.B.O., Ouma, H.A., Musyoki, S.M.: Parameter optimization in design of a rectangular microstrip patch antenna using adaptive neuro-fuzzy inference system technique. IJTPE Journal (September 2012) ISSN 2077-3528

15. Hu, C.-L., Yang, C.-F., Lin, S.-T.: A Compact Inverted-F Antenna to be Embedded in Ultra-thin Laptop Computer for LTE/WWAN/WI-MAX/WLAN Applications. IEEE Trans. AP-S/USRT, 978-1-4244-9561 (2011)

16. Parashar, K.K., Singh, V.K., Sinha, H.P.: Wide Band Circle Slotted Inset Fed Microstrip Patch Antenna for Wireless Application. International Journal of Microwave Engineering & Technologies (STM), 6–10 (March 2014)

17. Singh, V.K., Ali, Z., Ayub, S.: Bandwidth Optimization of Compact Microstrip Antenna for PCS/DCS/Bluetooth Application. Central European Journal of Engineering 4(3), 281–286 (2014) ISSN: 1896-1541

18. Loni, J., Singh, V.K., Ayub, S.: Bandwidth Improvement of Microstrip Patch Antenna for WLAN Application. International Journal of Engineering and Technical Research, 44–46 (April 2014) ISSN: 2321-0869

19. Ali, Z., Singh, V.K., Singhal, A.K.: Meandered Ground Microstrip Patch Antenna for WiMAX/WLAN Application. International Journal of Advanced Electronics and Communication Systems (IJAECS) 3(2) (April-May 2014)

20. Singh, A.K., Kabeer, R.A., Singh, V.K., Ali, Z.: Performance Analysis of First Iteration Koch Curve Fractal Log Periodic Antenna of Varying Angles. Central European Journal of Engineering 3(1), 51–57 (2013) ISSN: 1896-1541

21. Parashar, K.K., Singh, V.K., Pathak, R.S.: Compact Inset Fed Microstrip Patch Antenna for Dual Band Application. International Journal of Engineering and Technical Research, 47–49 (April 2014) ISSN: 2321-0869

22. Lumba, S., Singh, V.K.: Novel Approach for Analysis of Bandwidth of Microstrip Patch Antenna Using Neural Network. IJARCSSE 4(2), 637–642 (2014)

23. Parashar, K.K., Singh, V.K., Sinha, H.P.: Novel Design of Broad Band Microstrip Antenna for WiMax/WLAN/UMTS Applications. International Journal of Microwave and Engineering & Technologies (STM), 1–5 (March 2014)

Quality Assessment of Images Using SSIM Metric and CIEDE2000 Distance Methods in Lab Color Space

T. Chandrakanth and B. Sandhya

Department of Computer Science and Engineering,
M.V.S.R Engineering College, Nadargul, Hyderabad,India
chandra_kanth446@yahoo.co.in, sandhyab16@gmail.com

Abstract. Advances in imaging and computing hardware have led to an explosion in the use of color images in image processing, graphics and computer vision applications across various domains such as medical imaging, satellite imagery, document analysis and biometrics to name a few. However, these images are subjected to wide variety of distortions during its acquisition, subsequent compression, transmission, processing and then reproduction, which degrade their visual quality. Hence objective quality assessment of color images has emerged as one of the essential operation in image processing. During the last two decades, efforts have been put to design such an image quality metric which can be calculated simply but can accurately reflect subjective quality of human perception. In this paper, we evaluated the quality assessment of color images using CIE proposed Lab color space, which is considered to be perceptually uniform space. In addition we have used two different approaches of quality assessment namely, metric based and distance based.

Keywords: Image Quality Assessment, Color Space, SSIM.

1 Introduction

Quality evaluation of digital images is an essential operation with applications across various domains. During each stage of digital image processing like capture, storing, compression, and enhancement, certain distortions are introduced in the images. The amount of distortion can sometimes affect the way an image is perceived by an observer, which is referred to as image quality. Hence it is necessary to measure the overall perceived quality of an image to maintain, control, or to enhance it. One of the essential requirements of an image quality measure is that it should have strong correlation to the human perception. Many efforts have been directed during the last two decades to design such an image quality metric which can be calculated simply but can accurately reflect subjective quality of human perception [1,2,4]. The choice of an adequate metric usually depends on the requirements of the considered application.

Due to the wide application of color images in the recent past, color image quality assessment (IQA) is gaining momentum. Color images can be represented using the three attributes (brightness, hue, and saturation), which form basis of human perception of color. Hence various trichromatic spaces have been proposed for color image processing. At present, most of color image quality assessment methods convert a color

© Springer International Publishing Switzerland 2015   457
S.C. Satapathy et al. (eds.), *Proc. of the 3rd Int. Conf. on Front. of Intell. Comput. (FICTA) 2014*
– *Vol. 2,* Advances in Intelligent Systems and Computing 328, DOI: 10.1007/978-3-319-12012-6_50

image to one of the color spaces like HSV,XYZ etc. and then adopt the gray scale image quality assessment for each channel [13,14,3]. However, this kind of conversion cannot completely describe the nonlinear perception of brightness, hue, and saturation in a color image. It is well known that color image quality assessment results depend on the features of color space and how closely it can characterize visual perception. Thus, it is highly desirable to convert color image to a color space which can reflect the subjective visual characteristics more properly in order to obtain a better quality metric.

In this paper, we evaluated the quality assessment of color images using CIE proposed Lab color space, which is considered to be perceptually uniform space. In addition we have used two different approaches of quality assessment namely, metric based and distance based. Section II describes the previous research done in image quality assessment. Section III describes our approach of using Lab color space in SSIM. Section IV explains the experimental results.

2   Previous Work

Identifying the image quality measures that have highest sensitivity to distortions would help systematic design of coding, communication and imaging systems and of improving or optimizing the picture quality for a desired quality of service at a minimum cost. Image quality measurement basically consists of two approaches:- Subjective measurements and Objective measurements

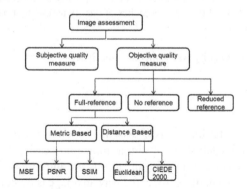

Fig. 1. Classification of image assessment

Figure 1 shows the classification of IQA measures. Subjective measurements are the result of human experts providing their opinion of the image quality and objective measurements are performed with mathematical algorithms. The goal of research in objective image quality assessment is to develop quantitative measures that can automatically predict perceived image quality[15]. An objective image quality metric can play a variety of roles in image processing applications. It can be used to dynamically monitor and adjust image quality. It can be used to optimize algorithms and parameter settings of image processing systems. It can be used to benchmark image processing systems and algorithms [6].

2.1 Metric Based IQA

Most of the objective models belong to full reference (FR) method, among which VIF [10] and SSIM (MS) [20] have much better performance than other algorithms by being evaluated on the image database composed of different types of distortion [11].

SSIM. SSIM(Structural Similarity index) works under the assumption that natural image signals are highly structured, human visual perception is highly adapted for extracting structural information from a scene, so a measurement of structural similarity (or distortion) should provide a good approximation to perceptual image quality [17,16,18].

Suppose x and y are two non-negative image signals,three components are combined to yield an overall similarity measure: luminance $l(x,y)$, contrast $c(x,y)$ and structure comparison $s(x,y)$. The general form of SSIM index between signal x and y is defined as:

$$SSIM = [l(x,y)]^{\alpha}[c(x,y)]^{\beta}[s(x,y)]^{\gamma}$$

usually,$\alpha = \beta = \gamma = 1$.

SSIM indexing algorithm is applied for image quality assessment using a sliding window approach. The window moves pixel-by-pixel across the whole image space. At each step, the SSIM index is calculated within the local window of size 11×11.

Lot of improvements have been made based on SSIM [20,22,19], with its simple formula and high accuracy. SSIM method works with luminance only, while color images are being more widely used as information representing and communicating.

Some of the other metrics in IQA are A Feature Similarity Index (FSSIM) [23] was proposed based on the fact that human visual system (HVS) understands an image mainly according to its low-level features. The phase congruency which is a dimensionless measure of the significance of a local structure is used as the primary feature in FSIM and the image gradient magnitude is employed as the secondary feature. Structure and Hue Similarity (SHSIM) for Color Images [12],extends the concepts of SSIM in Hue color space. SHSIM performs better than SSIM which focus on structural information of images, and better than CD-PSNR which works with color information. Image-Difference Measure (IDM) [5] was proposed based on image-difference framework that comprises image normalization, feature extraction, and feature combination. Using color information to improve the assessment of gamut mapped images. IDM shows higher prediction accuracy on gamut-mapping dataset than all other evaluated measures. Information Content Weighted MultiScale SSIM (IW-SSIM) [21] is proposed to test the hypothesis that when viewing natural images, the optimal perceptual weights for pooling should be proportional to local information content, which can be estimated in units of bit using advanced statistical models of natural images. Color Image-Difference Metric (CID) [8] is proposed to predict gamut-mapping distortions. An algorithm for optimizing gamut mapping employing the CID metric as the objective function is implemented. IQA has also been addressed using difference measure between images wherein color information is effectively used than in metric based methods. .

2.2 Distance Based IQA

Image difference measures are calculated in order to quantify the perceived difference between them. A distance transform, also known as distance map or distance field, is a derived representation of a digital image. In CIELAB color space, the Euclidean distance between two colors is:

$$\Delta E_{ab} = \sqrt{\Delta L^{*2} + \Delta a^{*2} + \Delta b^{*2}}$$

where $\Delta L^*, \Delta a^*, and \Delta b^*$ is the difference in lightness, redness-greenness, and yellow-ness -blueness, respectively. However, it has been tested that in CIELAB color space, the Euclidean distance between two colors poorly relates to visual color difference. Hence there has been considerable research to design a distance measure which can accurately predict the perceived difference of two images.

CIEDE2000. The CIEDE2000 [9] formula provides an improved procedure for the computation of industrial color differences. The color difference formula, such as CIEDE 2000, have been developed to predict color difference under specific illuminat-ing/viewing conditions close to the "reference conditions". The CIEDE2000 formula is considerably more sophisticated and computationally involved than its predecessor color-difference equations for CIELAB ΔE_{ab}^* and the CIE94 color difference ΔE_{94}.

The CIEDE2000 color-difference formula is based on the CIELAB color space. Given a pair of color values in CIELAB space $L_1^*, a_1^*, b_1^* and L_2^*, a_2^*, b_2^*$,denoted by the CIEDE2000 color difference between them as follows:

$$\Delta E_{00}(L_1^* a_1^* b_1^*, L_2^* a_2^* b_2^*) = \Delta E_{00}^{12} = \Delta E_{00}$$

3   IQA Using Lab Color Space

3.1 Lab Based SSIM Value

In SSIM method, the luminance, contrast and structural comparison function are defined as follow:

$$l(x,y) = \frac{2\mu_x \mu_y + C_1}{\mu_x^2 + \mu_y^2 + C_1} \qquad c(x,y) = \frac{2\sigma_x \sigma_y + C_2}{\sigma_x^2 + \sigma_y^2 + C_2} \qquad s(x,y) = \frac{\sigma_{xy} + C_3}{\sigma_x \sigma_y + C_3}$$

For calculating Lab based luminance values, initially images in a dataset are con-verted to Lab color space and divided it into different channels (L-channel, a-channel, b-channel) and SSIM measures are calculated (luminance, contrast and structural mea-sures). For each channel of the image luminance measure, contrast measure and struc-tural measure values are calculated individually. Lab based luminance SSIM values are obtained by calculating the mean of, luminance measure of a-channel and b-channel, contrast and structural measure of L-channel. Block diagram of Lab SSIM is shown in Figure 2.

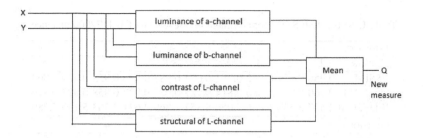

Fig. 2. Block diagram of Lab based luminance measure

3.2   Distance Measure

CIEDE2000 distance values between two images are computed using Lab Color space. We have converted the images into Lab Color space and divided the images into equal windows of size 20x20 and calculated distance between them for all distortions in dataset. Table 2 shows the distance values in 5 levels of an image for each distortion. We have calculated correlation for each distortion of an image from the obtained distance values and calculated the average of correlation value for all distortions for an image. Table 4 shows the average correlation values.

4   Dataset

TID2013 [7] contains 3000 test images (25 reference images, 24 types of distortions for each reference image, 5 different levels of each type of distortion. Mean Opinion Scores (MOS) for this database have been obtained. The obtained MOS can be used for effective testing of different visual quality metrics as well as for the design of new metrics. Set of images are shown in Figure 3.

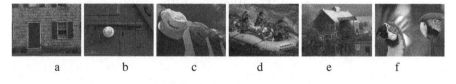

| a | b | c | d | e | f |

Fig. 3. Set of Images

5   Experimental Results

We have calculated SSIM for gray, Lab based SSIM and calculated distance measure CIEDE2000 and obtained the results. Table 1 shows SSIM values. We have conducted the tests on different images for all distortions for all 5 levels and calculated correlation for each image, values are shown in Table 2. Table 3 shows the average correlation values for six images obtained using SSIM with gray and Lab based metric and

Table 1. SSIM, Lab SSIM for an image in 5 levels and CIEDE2000 distance

Distortions	1					2				
Levels	1	2	3	4	5	1	2	3	4	5
SSIM	0.956	0.921	0.862	0.781	0.672	0.987	0.976	0.955	0.920	0.863
Lab SSIM	0.960	0.930	0.880	0.814	0.727	0.988	0.977	0.958	0.927	0.876
CIEDE2000	2.527	3.550	5.046	7.050	9.857	1.869	2.621	3.703	5.186	7.289
Distortions	3					4				
Levels	1	2	3	4	5	1	2	3	4	5
SSIM	0.964	0.934	0.887	0.813	0.714	0.979	0.960	0.924	0.870	0.773
Lab SSIM	0.968	0.942	0.901	0.840	0.761	0.980	0.963	0.931	0.883	0.802
CIEDE2000	2.510	3.555	5.023	7.056	9.829	2.274	3.227	4.569	6.230	8.823

Table 2. Correlation values for each distortion in 5 levels of an image

Distortions	1	2	3	4	5	6	7	8	9	10	11	12
SSIM	0.969	0.970	0.994	0.966	0.982	0.973	0.980	0.995	0.990	0.975	0.985	0.975
Lab SSIM	0.968	0.972	0.995	0.970	0.979	0.972	0.991	0.996	0.989	0.979	0.987	0.973
CIEDE2000	0.955	0.968	0.999	0.981	0.954	0.974	0.985	0.996	0.984	0.972	0.989	0.955
Distortions	13	14	15	16	17	18	19	20	21	22	23	24
SSIM	0.954	0.806	0.974	0.884	0.559	-0.084	0.995	0.837	0.990	0.964	0.980	0.933
Lab SSIM	0.948	0.814	0.978	0.867	0.599	0.806	0.995	0.845	0.991	0.962	0.963	0.939
CIEDE2000	0.961	0.910	0.974	0.848	0.331	0.946	0.998	0.833	0.986	0.978	0.953	0.941

Table 3. Average correlation values of SSIM and Distance Metric for all distortions

Image	SSIM	Lab SSIM	CIEDE2000
a	0.898	0.937	0.932
b	0.906	0.951	0.945
c	0.947	0.952	0.928
d	0.945	0.948	0.930
e	0.849	0.904	0.903
f	0.854	0.927	0.939

CIEDE2000. Figure 4 shows the graph. It can be observed that adding color in terms of a and b channels of Lab color space has certainly improved the average correlation values. If the correlation values are checked distortion wise, it can be observed that CIEDE2000 values are close to gray SSIM for most of the cases and has worked well for some distortions like #14 which is Non eccentricity pattern noise, #18 which is Change of color saturation.

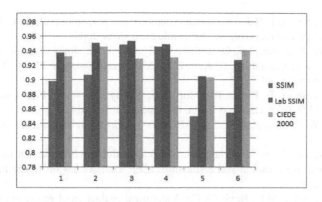

Fig. 4. Graph of average correlation values

6 Conclusion

Image quality assessment has been performed using two major approaches, metric based and distance based. CIE Lab color space which is a perceptually uniform color space has been used for color based IQA and both the approaches have been studied. For metric based we have chosen SSIM measure and added color to the assessment by computing luminance in a and b channels, contrast and structural measurements using L channel which is the achromatic channel of the color space. For distance based approach, we have chosen CIEDE2000 measure, which has been recently proposed and proven to be more closer to perceptual distance than eucledian distance. It is observed that Lab based SSIM has performed better when compared to gray SSIM and CIEDE2000 on average. However for certain distortions of the images, CIEDE2000 performed better than the other two measures. Hence addition of color does not always give better quality assessment but it is dependent on the kind of image, distortion undergone by the image and the suitable measure. Hence in our future work we would like to focus on the automatic selection of an effecient measure (in terms of perception) and a feature to compute the measure, given source and distorted images.

References

1. Bovik, A.C.: Perceptual video processing: seeing the future. Proceedings of the IEEE 98(11), 1799–1803 (2010)
2. Bovik, A.C.: What you see is what you learn. IEEE Signal Processing Magazine 27(5), 117–123 (2010)
3. Le Callet, P., Barba, D.: A robust quality metric for color image quality assessment. In: Proceedings of the International Conference on Image Processing, ICIP 2003 (September 2003)
4. Lee, S.O., Sim, D.G.: Objectification of perceptual image quality for mobile video. Optical Engineering 50(6) (2011)
5. Lissner, I., Preiss, J., Urban, P., Lichtenauer, M.S., Zolliker, P.: Image-difference prediction: From grayscale to color. IEEE Transactions on Image Processing 22(2) (2013)

6. Pappas, T.N., Safranek, R.J.: Perceptual criteria for image quality evaluation. In: Handbook of Image and Video Processing, A. Academic Press (May 2000)
7. Ponomarenko, N., Ieremeiev, O., Lukin, V., Egiazarian, K., Jin, L., Astola, J., Vozel, B., Chehdi, K., Carli, M., Battisti, F., Kuo, C.J.: Color Image Database TID2013: Peculiarities And Preliminary Results
8. Preiss, J., Fernandes, F., Urban, P.: Color-image quality assessment: From prediction to optimization. IEEE Transactions on Image Processing 23(3) (2014)
9. Sharma, G., Wu, W., Dalal, E.N.: The CIEDE2000 color-difference formula: Implementation notes
10. Sheikh, H.R., Bovik, A.C., Veciana, G.: An information fidelity criterion for image quality assessment using natural scene statistics. IEEE Transactions on Image Processing 14, 2117–2128 (2005)
11. Sheikh, H.R., Sabir, M.F., Bovik, A.C.: A statistical evaluation of recent full reference image quality assessment algorithms. IEEE Transactions on Image Processing 15, 3440–3451 (2006)
12. Shi, Y., Ding, Y., Zhang, R., Li, J.: Structure and Hue Similarity for Color Image Quality Assessment (2009)
13. Thakur, N., Devi, S.: A new method for color image quality assessment. International Journal of Computer Applications 15(2), 10–17 (2011)
14. Toet, A., Lucassen, M.P.: A new universal colour image fidelity metric. Displays 24, 197–207 (2003)
15. VQEG. Final report from the video quality experts group on the validation of objective models of video quality assessment (March 2000), http://www.vqeg.org/
16. Wang, Z., Bovik, A.C.: A universal image quality index. IEEE Signal Processing Letters 9, 81–84 (2002)
17. Wang, Z., Bovik, A.C.: Modern Image Quality Assessment. Morgan and Claypool Publisher (2006)
18. Wang, Z., Bovik, A.C., Sheikh, H.R., Simoncelli, E.P.: Image quality assessment: From error visibility to structural similarity. IEEE Transactions on Image Processing 13, 600–612 (2004)
19. Wang, Z., Shang, X.L.: Spatial pooling strategies for perceptual image quality assessment. In: IEEE International Conference on Image Processing (2006)
20. Wang, Z., Simoncelli, E.P., Bovik, A.C.: Multi-scale structural similarity for image quality assessment. In: The 37th IEEE Asilomar Conference on Signals, Systems and Computers, Pacific Grove, CA (2003)
21. Wang, Z., Li, Q.: Information content weighting for perceptual image quality assessment. IEEE Transactions on Image Processing 20(5) (2011)
22. Yang, C.L., Chen, G.H., Xie, S.L.: Gradient information based image quality assessment. Acta Electronica Sinica 35, 1313–1317 (2007) (in Chinese)
23. Zhang, L., Mou, X., Zhang, D.: Fsim: A feature similarity index for image quality assessment. IEEE Transactions on Image Processing 20(8) (2011)

Review Paper on Linear and Nonlinear Acoustic Echo Cancellation

D.K. Gupta[1], V.K. Gupta[2], and Mahesh Chandra[3]

[1] Department of Electronics and Communication Engineering,
Krishna Engineering College, Ghaziabad
[2] Department of Electronics and Communication Engineering,
Inderprastha Engineering College, Ghaziabad
[3] Department of Electronics and Communication Engineering, BIT, Mesra, Ranchi
deepak_gpt@rediffmail.com, guptavk76@gmail.com,
shrotriya@bitmesra.ac.in

Abstract. In this paper a review on acoustic echo cancellation (AEC) systems based on linear AEC and nonlinear AEC are presented. The paper covers advancements in the previous research works related to the acoustic echo cancellation process. Here review is done based on adaptive algorithms used. Since in linear AEC systems, the non-linearity caused by loudspeakers, amplifier and low-quality enclosures are not taken into considerations. The implementation of linear AEC is based on the assumption of linearity. The performance of echo cancellation algorithms are degraded due to non-linearities in the acoustic path. Therefore non-linear AEC systems are evaluated and reviewed along with linear AEC.

Keywords: Acoustic Echo Cancellation (AEC), Nonlinear AEC, LMS, RLS, APA, Volterra Filters.

1   Introduction

The acoustic echo canceller system is used in hands-free systems to reduce the undesired echoes which come due to the loudspeaker and microphone coupling. Echo signals are expressed as delayed and distorted version of a sound which is reflected back to the source of sound. Reflected signal is considered as reverberation if signal arrives after a very short time of direct sound. However, when the reflected signal arrives after a few tens of milliseconds of the direct sound, it is called an echo signal. In hands-free telecommunications, the echo signal makes the conversation impossible.

The Acoustic echo cancellers (AEC) are used to eliminate the echo signals and increase the communication quality. Acoustic echo cancellation system basically employed adaptive filtering techniques for implementation. In the basic structure of AEC, the estimated echo signal is generated by an adaptive filter eliminates the echo signal. The coefficients of the adaptive filter are adjusted iteratively by adaptive algorithms according to the estimated error signals.

© Springer International Publishing Switzerland 2015   465
S.C. Satapathy et al. (eds.), *Proc. of the 3rd Int. Conf. on Front. of Intell. Comput. (FICTA) 2014*
– *Vol. 2*, Advances in Intelligent Systems and Computing 328, DOI: 10.1007/978-3-319-12012-6_51

The problem of non-linear echo cancellation has emerged as an increasingly important problem. There are two main approaches to tackle the problem of nonlinearities. The first approach is based on nonlinear post filtering to suppress the residual non-linear echo. In general the post-filter is preceded by a conventional linear adaptive filter. However, non-linearities have an adverse effect on linear filtering which impacts upon non-linear post processing and thus degrades global performance. The second, more popular approach is based on the use of a Volterra series and non-linear adaptive filtering.

2   Review on Acoustic Echo Cancellation System

In the field of Acoustic echo cancellation, the first novel work for long distance telephony was presented by M M Sondhi in 1967[1]. In this work a replica of echo is synthesized which adapts the transmission characteristics of the echo path and track variations of the path during conversation. Adaptive filtering methods were used to remove the echo. Two types of basic algorithms named least mean square (LMS) and recursive least square (RLS) are generally used for echo cancellation. Efforts are going on by various researchers to yield optimum performance using these basic algorithms and their variants for both linear and nonlinear echo cancellations with and without double talk. Here the algorithms based on both time domain and frequency domain approaches are included.   To provide fast convergence and reduce the effects of the double-talk, V. Reddy *et. al.* [2] in 1983 proposed the application of modified recursive least-square algorithms and least-square lattice algorithms to echo cancellation problems. In their work detection of the double-talk and virtual freezing of the adaptive filter weights were implemented into the algorithm itself. S. Marcos and O. Macchi in 1987[3] studied the tracking capability of a time-varying system by an adaptive filter through the least mean square (LMS) algorithm in the particular case of a white, zero-mean reference input suitable for echo cancellation achieving full-duplex data transmission. Here residual mean square error were calculated for random, white, and zero-mean channel variations, whereas new computations were proposed for the case of deterministic and bounded time variations. Then an optimum step size was derived. Then this study was applied to asynchronous echo canceller to makes the echo channel time varying. In 1990, an exponential step NLMS adaptive algorithm having double convergence speed with same computational complexity as the conventional NLMS was proposed by Makino and Y. Kaneda [4]. Again in 1994, S. Makino and Y. Kaneda [5] proposes exponential step-Recursive Least Square(ES-RLS) algorithm with double convergence speed of the conventional RLS algorithm for acoustic echo canceller. The ES-RLS converged twice as fast as the conventional RLS algorithm.

A paper on adaptive volterra filters for nonlinear acoustic echo cancellation in real time was presented in 1999 by A. Stenger *et. al.* [6]. The loudspeaker nonlinearities in the echo path of hands-free telephone is considered. In this paper an adaptive volterra filter structure with a reduced number of coefficients is proposed. The affine projection algorithm (APA) and its variants are very attractive choices for AEC because it provides- 1) high convergence rates and good tracking capabilities and 2) low misadjustment and robustness against background noise variations and double-talk. In 2008, a variable step-size affine projection algorithm VSS-APA for AEC was proposed by C Paleologu *et. al.*[7]. Chia-Sheng *et. al.* [8] in 2009 gives the concept of optimum step size in a fast converging nonlinear echo cancellation system. Here a tap variant step-size based on the MMSE criterion for second-order volterra filter was implemented to speed up the convergence rate an optimum time. For sparse system identification in network echo cancellation, in 2011, Mohammad Shams *et. al.*[9] proposed several normalized subband adaptive filter based algorithms. The algorithm shows good performance compared to full band adaptive filtering algorithms. A windowing frequency domain adaptive filter and an upsampling block transform preprocessing were presented by Sheng Wu *et. al.*[10] in 2011 to solve the stereo acoustic echo cancellation problems.

In 2012, a nonlinear system identification method for echo cancellation was proposed by Cristian Contan *et. al.*[11]. Adaptation of the system was made using a Normalized Least Mean Square algorithm. They combined the adaptive linear, volterra and power filters. Then the system is applied where several sources of nonlinearities exist, having the overdriven amplifier, the small loudspeaker at high volume and the room with different absorbent walls. Result showed that the overall system performs better than the best single adaptive filter. In the same year, Cristian Contan, *et. al.* [12] implemented nonlinear models for adaptive Volterra kernels using a new updating technique for unknown feedback paths identification a nonlinear echo cancellation system. Here a combination of adaptive linear, Volterra and power filters was used and compared for Echo Return Loss Enhancement in second-order Volterra filter Normalized Least-Mean-Square algorithm for kernel adaptation. The implemented system gives a superior convergence speed.

The concept of stereophonic acoustic echo suppression (SAES) method without preprocessing in a open-loop teleconferencing systems was proposed by Feiran Yang *et. al.*[13] in 2012. They used Wiener Filter based echo suppression in the Short-Time Fourier Transform Domain for estimation of echo spectra impulse responses. The spectral modification technique was used to remove the echo from the microphone signal. In 2012, a paper on State-Space Frequency-Domain Adaptive Filtering for Nonlinear Acoustic Echo Cancellation was presented by Sarmad Malik and Gerald Enzner [14].In this paper they have implemented adaptive acoustic echo cancellation in the presence of an unknown memoryless nonlinearity preceding the echo path. Result shows that the algorithm provides effective nonlinear echo cancellation in the presence of continuous double-talk, varying degree of nonlinear distortion, and changes in the echo path. In 2013, a general framework for designing echo cancellers for full-duplex communication systems with discrete multi-tone (DMT) modulation system in an arbitrary mixed domain were introduced by Neda Ehtiati and Beno Champagne [15]. They derived a new mixed-domain echo cancellation structure, achieved by a generic decomposition of the toeplitz data matrix at the transmitter in terms of arbitrary unitary matrices. This framework proposes a new canceller based on discrete trigonometric transformations. They have shown that this canceller has a faster convergence rate than the existing ones with similar complexity and is more robust. The eigen value spread of the input autocorrelation matrix is reduced using transform domain LMS algorithm to increase the convergence speed of the LMS algorithm. Performance in terms of convergence speed, steady-state MSE and robustness to power-varying inputs is further improved by using a new regularized transform domain normalized LMS (R-TDNLMS) algorithm proposed by S. C. Chan, *et. al.*[16] in 2013. In 2013, a new affine projection algorithm for acoustic echo cancellation and acoustic feedback cancellation applications with intermittent update of the filter coefficients was proposed by Albu, F. *et. al.*[17] where the update interval was determined according to the adaptation state. Result shows improved performance and reduced average computational complexity compared with other similar algorithms. A delay less partitioned block frequency-domain adaptive filter (PBFDAF) algorithm was introduced in 2013 by Feiran Yang *et. al.*[18] to increase the convergence rate as well as computational efficiency. Here a delay compensation method is used to compensate the error path delay to speed up the convergence rate. In 2013, Stanciu, C. and Ciochina, S. [19] introduces a algorithm where RLS is combined with the dichotomous coordinate descent (DCD) algorithm to reduce the matrix inversion arithmetic complexity. The algorithm gave an improve performance in double talk like high disturbance situations. In 2013, Danilo Comminiello [20] introduced a new class of nonlinear adaptive filters based on hammerstein model. Such filters derived from the functional link adaptive filter (FLAF) model, defined by a nonlinear input expansion, which enhanced the representation of the input signal through a projection in a higher dimensional space, and a subsequent adaptive filtering. In order to give robustness against different degrees of nonlinearity, a collaborative FLAF is proposed based on the adaptive combination of filters. Such architecture allows achieving the best performance regardless of the nonlinearity degree in the echo path. In 2013, a Multiple Sub-filter (MSF) parallel structure based on multipath acoustic echo model was proposed by Ravinder Nath [21] using the basis that each sub-filter will compensate the echo contributed by each path of multipath acoustic channel.

In 2014, an algorithm is given for a more accurate estimates of the impulse response between each loudspeaker and microphone by Yellepeddi, A. and Florencio, D. [22]. The algorithm has array structure and it takes the sparsity of the reflections arriving at the array to give improved performance in echo cancellation applications, using both synthetic and real data

3   Tabular Review

Algorithm	Findings	Suitability	Citation
LMS	• Simplicity and convergence guaranty in stationary environments • If step size is small then it may take long time to converge.	For Linear AEC	[23], [24]
TDNLMS	• Exploits the decorrelation properties of transformations such as DFT, DCT and wavelet transform. • Improves the convergence speed of the conventional LMS.	for Linear AEC	[25]
RLS	• Faster convergence rates, but in general require computational resources, often too large for a real-time implementation. • Suitable for Non Stationary environment.	for Linear AEC	[3]

FRLS	• Fast version, i.e., the fast recursive least squares algorithms • It suffers from numerical problems when implemented with finite-word-length arithmetic's.	for Linear AEC	[26]
MSAF	• A multiband-structured subband adaptive filter algorithm • Speed up the convergence time of the NLMS algorithm.	for Linear AEC	[27]
AP	• Include LMS like complexity and memory requirements (low), and RLS like convergence (fast).	Both	[7]
FAP	• Retain the good performance of AP algorithms but with a computational load close to that of LMS. • Offer, in the presence of correlated signals, convergence rates higher than LMS and tracking capabilities better than RLS.	Both	[17], [30],
PBFDAF	• Reduce the computational load remarkably • The low-complexity methods that exploit the fast block convolution techniques in the DFT domain	for Linear AEC	[18]
Adaptive Voletera Filter	• Combat nonlinearities in the acoustic echo path of hands-free telephones, which are caused by low-cost audio components • Echo reduction improvement of 7 dB over conventional linear adaptive filters is achieved	for Non Linear AEC	[6]
Simplified Adaptive Voletera Filter	• Equivalent to the cascade structure for nonlinear AEC. • Has much fewer coefficients than the conventional direct form 2nd order VF does, while enjoys a better convergence performance compared to both the cascade canceller and the conventional 2nd order VF.	for Non Linear AEC	[12], [28],
Dynamic Impluse Response Method	• Focused on the nonlinearity affecting the IR, and propose a dynamic IR model for nonlinear system identification. • Decreases the computational costs of modeling and widely applicable in a real environment.	for Non Linear AEC	[29],
Orthogona-lized Power Filters	• Standard approach in non linear environment • Combine time-domain orthogonalization methods with a DFT-domain implementation of power filters. • Model consists of the cascade of a linear filter, a memoryless nonlinearity, and a second linear filter.	for Non Linear AEC	[30]
Adaptive DFT-domain Volterra Filters	• Method for estimating the optimum number of second-order kernel diagonals of an adaptive Volterra filter. • An efficient version, requiring only minor additional computations as compared to a single Volterra filter.	for Non Linear AEC	[31]
Nonlinear AEC Based on Filter Combinatio n	• Based on the adaptive combination of linear and nonlinear echo cancellers • In this approach two or more adaptive filters adaptively combine their outputs obtaining a combined scheme. • Combination of a linear filter and a Volterra filter	for Non Linear AEC	[32]

Cascaded Nonlinear	• Greater robustness to changes in the acoustic channel than an existing power filter approach. • Model divides the LEM system into two main approaches- • Based on nonlinear post filtering to suppress the residual non-linear echo and Volterra series and non-linear adaptive filtering.	for Non Linear AEC	[34],
Nonlinear AEC Based on Optimum Step Size	• An optimum time- and tap- variant step-size control method for the second-order Volterra filter • Based on an optimum MMSE criterion between coefficients errors of real echo path kernel and adaptive coefficients. • Overcome the slow convergence rate.	for Non Linear AEC	[8]
PWL Approximatin Method	• Based on Amplitude threshold decomposition for memoryless nonlinearity. AEC scheme for the cascade echo path model of memoryless nonlinearity and linear filter.	for Non Linear AEC	[35]
Volterra Filters with EQK	• EQK(Evolutionary Quadratic Kernels) • Exhibits a computational complexity suitable for practical implementations.	for Non Linear AEC	[33]
NRES Based on Partial Channel Modeling	• Nonlinear residual echo suppression (NRES) module is introduced to suppress nonlinear residual echo. • Efficient scheme with low computational complexity compared to adaptive Volterra filter.	for Non Linear AEC	[36]
R-TDNLMS	• Regularized transform domain normalized LMS algorithm • Improve convergence performance, steady-state MSE and robustness to power-varying inputs	for Non Linear AEC	[16]
Spectral Feature Based NRES	• A method for nonlinear residual echo suppression that consists of extracting spectral features from the far-end signal	for Non Linear AEC	[37]
Sliding Window Leaky Kernel AP algorithm	• Outperforms the linear APA, resulting in up to 12 dB of improvement in ERLE at a computational cost that is only 4.6 times higher	for Non Linear AEC	[39]
MSF Based on Multipath AEC Model	• Based on multipath acoustic echo model • MSF can track variations in the channel parameters of the multipath model faster than conventional echo canceller.	for Non Linear AEC	[21]
FLAF Model	• Structure is based on Hammerstein model. • Two FLAF-based architectures- the *split FLAF* and *collaborative FLAF*	for Non Linear AEC	[20]
EMD Based Subband Adaptive Filtering	• Track the change of the room much faster than the normalized adaptive filtering structure. • Best ERLE • Enhance speech under various noise environments.	for Non Linear AEC	[39]

4   Conclusion

The Papers on both linear and non-linear echo-cancellation are studied based on algorithms used. A summary of algorithms used and their outcomes are given in tabulation form. The performance of the algorithms are evaluated in terms of various parameters such as ERLE, Misalignment, Misadjustment, Convergence rate, Robustness (disturbances and numerical), Computational requirements (operations and memory) and stability. It is shown that a further improvement is required in these parameters to improve the effective implementation of echo canceller.

References

1. Sondhi, M.M.: An adaptive echo canceller. Bell Systems Technical Journal 46, 497–510 (1967)
2. Umapathi Reddy, V., Shan, T.-J., Kailath, T.: Application of modified least-square algorithms to adaptive echo cancellation. In: ICASSP, pp. 53–60 (1983)
3. Marcos, S., Macchi, O.: Tracking capabilities of the least mean square algorithm: Application to an asynchronous echo canceller. IEEE transcations on Acoustics, Speech, and Signal Processing 35, 1570–1578 (1987)
4. Makino, S., Kaneda, Y.: Acoustic Echo Canceller Algorithm Based On the Variation Characteristics of a Room Impulse Response. In: International Conference on Acoustics, Speech, and Signal Processing, ICASSP-1990, vol. 2, pp. 1133–1136 (1990)
5. Malkino, S., Kaneda, Y.: A new RLS algorithm based on the variation characteristics of a room impulse response. In: ICASSP-1994, vol. 3, pp. 373–376 (1994)
6. Stenger, A., Rabenstein, R.: Adaptive Volterra Filters For Nonlinear Acoustic Echo Cancellation. In: Proc. NSIP, vol. 2, pp. 679–683 (1999)
7. Paleologu, C., Benesty, J., Ciochin, S.: A Variable Step-Size Affine Projection Algorithm Designed for Acoustic Echo Cancellation. IEEE Transactions on Audio, Speech, and Language Processing 16, 1466–1478 (2008)
8. Shih, C.-S., Hsieh, S.-F.: Fast converging nonlinear echo cancellation based on optimum step size. In: IEEE International Symposium on Signal Processing and Information Technology, pp. 135–139 (2009)
9. Abadi, M.S.E., Kadkhodazadeh, S.: A family of proportionate normalized sub band adaptive filter algorithms. Journal of the Franklin Institute 348, 212–238 (2011)
10. Wu, S., Qiu, X., Wu, M.: Stereo Acoustic Echo Cancellation Employing Frequency-Domain Preprocessing and Adaptive Filter. IEEE Transactions on Audio, Speech, and Language Processing 19(3), 614–623 (2011)
11. Contan, C., Topa, M., Kirei, B., Homana, I.: Nonlinear Acoustic System Identification using a Combination of Volterra and Power Filters. In: IEEE 10th International Symposium onSignals, Circuits and Systems (ISSCS), pp. 1–4 (2011)
12. Contan, C., Zeller, M., Kellermann, W., Topa, M.: Excitation-Dependent Stepsize Control of Adaptive Volterra Filters For Acoustic Echo Cancellation. In: 20th European Signal Processing Conference-EUSIPCO (2012)
13. Yang, F., Wu, M., Yang, J.: Stereophonic Acoustic Echo Suppression Based on Wiener Filter in the Short-Time Fourier Transform Domain. IEEE Signal Processing Letters 19(4), 227–230 (2012)

14. Malik, S., Enzner, G.: State-Space Frequency-Domain Adaptive Filtering for Nonlinear Acoustic Echo Cancellation. IEEE Transactions on Audio, Speech, and Language Processing 20(7), 2065–2079 (2012)
15. Ehtiati, N., Champagne, B.: A General Framework for Mixed-Domain Echo Cancellation in Discrete Multitone Systems. IEEE Transactions on Communications 61(2), 769–780 (2013)
16. Chan, S.C., Chu, Y.J., Zhang, Z.G.: A New Variable Regularized Transform Domain NLMS Adaptive Filtering Algorithm. IEEE Transactions on Audio, Speech, and Language Processing 21(4), 868–878 (2013)
17. Albu, F., Rotaru, M., Arablouei, R., Dogancay, K.: Intermittently-updated affine projection algorithm. In: IEEE International Conference on Acoustics, Speech and Signal Processing (ICASSP), pp. 585–589 (2013)
18. Yang, F., Wu, M., Yang, J.: A Computationally Efficient Delayless Frequency-Domain Adaptive Filter Algorithm. IEEE Transactions on Circuits and Systems II 60, 222–226 (2013)
19. Stancin, C., Ciochine, S.: A robust dual path DCD-RLS algorithm for stereophonic acoustic echo cancellation. In: ISSCS-2013, pp. 1–4 (2013)
20. Comminiello, D., Scarpiniti, M., Azpicueta-Ruiz, L.A., Arenas-García, J., Uncini, A.: Functional Link Adaptive Filters for Nonlinear Acoustic Echo Cancellation. IEEE Transactions on Audio, Speech, And Language Processing 21, 1502–1512 (2013)
21. Nath, R.: Adaptive Echo Cancellation Based on a Multipath Model of Acoustic Channel. Circuits System Signal Process 32, 1673–1698 (2013)
22. Yellepeddi, A., Florencio, D.: Sparse array based room transfer function estimation for echo cancellation. IEEE Signal Processing Letters 21, 230–234 (2014)
23. Haykin, S.: Adaptive Filter Theory, 4th edn. Prentice-Hall, Upper Saddle River (2002)
24. Diniz, P.S.R.: Adaptive Filtering- Algorithms and Practical Implementations, 3rd edn. Springer
25. Beaufays, F.: Transform-domain adaptive filters: An analytical approach. IEEE Trans. Signal Process. 43(2), 422–431 (1995)
26. Gilloire, A., Petillon, T., Theodoridis, S.: Acoustic echo cancellation using Fast RLS adaptive filters with reduced complexity. In: IJCAS 1992, vol. 4, pp. 2065–2068 (1992)
27. Yang, F., Wu, M., Ji, P., Yang, J.: An Improved Multiband-Structure Subband Adaptive Filter Algorithm. IEEE Signal Processing Letters 19, 647–650 (2012)
28. Fermo, A., Carini, A., Sicuranza, G.L.: Simplified Volterra filtersfor acoustic echo cancellation in GSM receivers. In: Proc. Eur. Signal Process. Conf (EUSIPCO 2000), Tampere, Finland (2000)
29. Saito, S., Nakagawa, A., Haneda, Y.: Dynamic Impulse Response Model for Nonlinear Acoustic System and its Application to Acoustic Echo Canceller. In: IEEE Workshop on Applications of Signal Processing to Audio and Acoustics, October 18-21 (2009)
30. Kuech, F., Kellerman, W.: Orthogonalized power filters for nonlinear acoustic echo cancellation. Signal Process 86, 1168–1181 (2006)
31. Zeller, M., Azpicueta-Ruiz, L.A., Arenas-Garcia, J., Kellermann, W.: Efficient Adaptive DFT-Domain Volterra Filters Using An Automatically Controlled Number of Quadratic Kernel Diagonals. In: ICASSP- 2010, pp. 4062–4065 (2010)
32. Azpicueta-Ruiz, L.A., Zeller, M., Figueiras-Vidal, A.R., Arenas-García, J., Kellermann, W.: Adaptive Combination of Volterra Kernels and Its Application to Nonlinear Acoustic Echo Cancellation. IEEE Transactions on Audio, Speech, and Language Processing 19(1), 97–110 (2011)

33. Zeller, M., Azpicueta-Ruiz, L.A., Arenas-García, J., Kellermann, W.: Adaptive Volterra Filters With Evolutionary Quadratic Kernels Using a Combination Scheme for Memory Control. IEEE Transactions on Signal Processing 59(4), 1449–1464 (2011)
34. Mossi, M.I., Yemdji, C., Evans, N., Beaugeant, C., Degry, P.: Robust and Low-Cost Cascaded Non-Linear Acoustic Echo Cancellation. In: IEEE- ICASSP, pp. 89–92 (2011)
35. Shimauchi, S., Haneda, Y.: Nonlinear Acoustic Echo Cancellation Based on Piecewise Linear Approximationwith Amplitude Threshold Decomposition. In: International Workshop on Acoustic Signal Enhancement, September 4-6 (2012)
36. Chao, C., Hui, C., Yi, Z.: Improved Nonlinear Residual Echo Suppression based on Partial Echo Path Modeling DTD. IEEE Region 10 Conference, 1-4 (2013)
37. Schwarz, A., Hofmann, C., Kellermann, W.: Spectral Feature-Based Nonlinear Residual Echo Suppression. In: IEEE Workshop on Applications of Signal Processing to Audio and Acoustics, October 20-23 (2013)
38. Gil-Cacho, J.M., Signoretto, M., van Waterschoot, T., Moonen, M., Jensen, S.: Nonlinear Acoustic Echo Cancellation Based on a Sliding-Window Leaky Kernel Affine Projection Algorithm. IEEE Transactions on Audio, Speech, and Language Processing 21(9), 1867–1876 (2013)
39. He, X., Goubran, R.A., Liu, P.X.: A novel sub-band adaptive filtering for acoustic echo cancellation based on empirical mode decomposition algorithm. Int. J. Speech Technol. (2014)

PCA Based Medical Image Fusion in Ridgelet Domain

Abhinav Krishn, Vikrant Bhateja, Himanshi, and Akanksha Sahu

Department of Electronics and Communication Engineering,
Shri Ramswaroop Memorial Group of Profesional Colleges,
Lucknow-226010 (U. P.), India
{abhinavkrishn01,bhateja.vikrant,
himanshi2593,akankshasahu1708}@gmail.com

Abstract. Medical image fusion facilitates the retrieval of complementary information from medical images and has been employed diversely for computer-aided diagnosis of diseases. This paper presents a combination of Principal Component Analysis (PCA) and ridgelet transform as an improved fusion approach for MRI and CT-scan. The proposed fusion approach involves image decomposition using 2D-Ridgelet transform in order to achieve a compact representation of linear singularities. This is followed by application of PCA as a fusion rule to improve upon the spatial resolution. Fusion Factor (FF) and Structural Similarity Index (SSIM) are used as fusion metrics for performance evaluation of the proposed approach. Simulation results demonstrate an improvement in visual quality of the fused image supported by higher values of fusion metrics.

Keywords: Fusion Factor, MRI, PCA fusion rule, Ridgelet transform.

1   Introduction

'Medical Image Fusion' is the process of combining and merging complementary information into a single image from two or more source images to maximize the information content of images and minimize the distortion and artifacts in the resulting image [1]. The complementary nature of medical imaging sensors of different   modalities, (X-ray, Magnetic Resonance Imaging (MRI), Computed Tomography (CT)) has brought a great need of image fusion for the retrieval of relevant information from medical images. The significance of fusion process is important for multimodal images as images obtained from single modalities provides only specific information; thus it is   not feasible to get all the requisite information from image generated by single modality [2-3]. This highlights the need towards the development of multimodality medical imaging sensors for extracting clinical information to explore the possibility of data reduction along with better visual representation. In the past decades, several fusion algorithms varying from the traditional fusion algorithms like simple averaging and weighted averaging, maximum and minimum selection rule [4] have been proposed. With the advancement of research in this field, algorithms such as Intensity–Hue–Saturation

© Springer International Publishing Switzerland 2015   475
S.C. Satapathy et al. (eds.), *Proc. of the 3rd Int. Conf. on Front. of Intell. Comput. (FICTA) 2014*
– *Vol. 2*, Advances in Intelligent Systems and Computing 328, DOI: 10.1007/978-3-319-12012-6_52

(IHS) [5] and Brovey transform (BT) [6] have been used to fuse medical images. In the recent years multiresolution approaches using Mallat [7] , the à trous [8] transforms, complex wavelet [4], contourlet [9] and non-subsampled contourlet[10] have been proposed for image fusion. Yan Luo et al. [11] used a combination of PCA with à trous wavelet transform which focused on the spatial and spectral resolutions. But, the technique did not laid emphasis on edge or shape detection, which are fundamental structures in natural images and particularly relevant from a visual point of view. V. S. Petrovic et al. [12] proposed a 'fuse then decompose' technique which represented input image in the form of gradient maps at each resolution level. Although, it has been observed that the said approach by authors did not yield satisfactory performance but in turn increased the computational complexity due to the involvement of gradient maps. S. K. Sadhasivam in their work [13] applied PCA along with the selection of maximum pixel intensity to perform fusion. The said combination produced an image which had less structural similarity with the source images along with low contrast and luminescence. On the other hand, work of Z. Xu et al. [14] emphasized only on the contrast fusion rules neglecting other important features such as removing redundancy and representing lines in an image. Further, these fused images are also processed with denoising [15]-[16], contrast [17]-[26] and edge enhancement [27]-[29] techniques to improve upon the visualization of diagnostic information. The proposed work in this paper presents a combination of ridgelet transform and PCA as an improvement to the aforementioned limitations. The obtained results have been evaluated using FF and SSIM as fusion metrics; yielding satisfactory performance. The rest of the paper is organized as follows: The proposed fusion approach is discussed in section 2. Section 3 presents experimental results and finally the conclusions are drawn in section 4.

2   Proposed Fusion Approach

2.1   Motivation

Digital Images are generally described via orthogonal, non-redundant transforms like wavelet or discrete cosine transform for the purpose of multi scale analysis [43]. The Wavelet transform portrays more emphasis on catching zero-dimensional singularities, however the performance of wavelets in mono-dimensional domain is lost when they are applied to images using 2D separable basis. While, ridgelet transform poses to be a powerful instrument in catching and representing mono-dimensional singularities (in bi-dimensional space). Moreover, it achieves very compact representation of linear singularities in images. Instrumental in the implementation of the ridgelet is the Radon transform, which is preferred as a tool to extract lines in edge dominated images [30]. Thus, reliable detection and representation of edges using ridgelets forms the motivation to apply fusion process in this domain.

2.2   Proposed Fusion Algorithm

The first step in the proposed fusion approach involves the pre-processing of the MRI and CT-scan images, i.e. the conversion of image from RGB scale to Gray scale (RGB components of the image are converted into Gray scale components).The next step is to decompose the source images using ridgelet transform. Two dimensional ridgelet transform can be viewed as a wavelet analysis in Radon domain; which is itself a tool of shape detection [31]. The ridgelet coefficients of an image $f(x_1, x_2)$ are represented by Eq. (1).

$$R_f(a,b,\theta) = \int_{-\infty}^{\infty} \int_{-\infty}^{\infty} \psi_{a,b,\theta}(x_1,x_2) f(x_1,x_2) dx_1 dx_2 \tag{1}$$

where, $R_f(a,b,\theta)$ represents ridgelet coefficients and $\Psi_{a,b,\theta}$ denotes the ridgelet basis function given by Eq. (2), for each $a > 0, b \in \mathbf{R}, \theta \in [0\ 2\pi]$.

$$\Psi_{a,b,\theta}(x_1,x_2) = a^{-1/2}\Psi\left(\frac{(x_1\cos\theta + x_2\sin\theta - b)}{a}\right) \tag{2}$$

Once the source images are decomposed using ridgelet transform and ridgelet coefficients are obtained; PCA is applied as a fusion rule to selectively combine the appropriate ridgelet coefficients of input images [32]. PCA serves to transform/project the features from the original domain to a new domain (PCA domain) where they are arranged in order of their variance [44]-[46]. Fusion process is achieved in PCA domain by retaining only those features that contain a significant amount of information. This can be achieved by retaining only those components which have a larger variance. The steps involved in the proposed PCA algorithm are outlined in fig. 1. The Next step involves, the reconstruction of the processed coefficients (after PCA fusion rule) using inverse ridgelet transform to generate the fused image. The reconstruction formula for the inverse ridgelet transform is given by Eq. (3).

$$f(x_1,x_2) = \int_0^{2\pi} \int_{-\infty}^{\infty} \int_0^{\infty} R_f(a,b,\theta)\psi_{a,b,\theta}(x_1,x_2)\frac{da}{a^3} db \frac{d\theta}{4\pi} \tag{3}$$

	BEGIN	
Step 1	:	*Input*: Ridgelet coefficients (of both MRI & CT).
Step 2	:	*Compute*: Column vectors from ridgelet coefficients.
Step 3	:	*Compute*: Covariance matrix using these vectors.
Step 4	:	*Process:* Diagonal elements of the covariance vector.
Step 5	:	*Compute*: Eigen vectors and Eigen values of covariance matrix.
Step 6	:	*Process*: Column vector corresponding to large Eigen value (by dividing each element with the mean of Eigen vector).
Step 7	:	Compute: Multiplication of normalized Eigen vector values by each term of ridgelet coefficient matrix.
Step 8	:	*Process:* Inverse ridgelet transform of two scaled matrices calculated in Step 7.
Step 9	:	Compute: Sum of the images calculated in Step 8.
Step10	:	*Output*: Fused image.
	END	

Fig. 1. PCA Algorithm in Ridgelet Domain

The final step of the proposed approach is its objective evaluation; which requires usage of HVS based image quality assessment (IQA) approaches [33]-[38]. In the present work, which performance evaluation is carried out with the metrics such as Fusion Factor (FF) [4] and Structural Similarity Index (SSIM) [39]-[40]. Higher values of fusion metrics indicate better fusion performance.

3   Experimental Results and Discussions

Simulations in the present work have been performed on images of two different modalities (CT and MRI). This section deals with the qualitative and quantitative analysis of the fused image obtained from the proposed approach. The results of fused images obtained using the proposed approach are given in Fig. 3; while, the quantitative analysis of the same is shown in table 1.

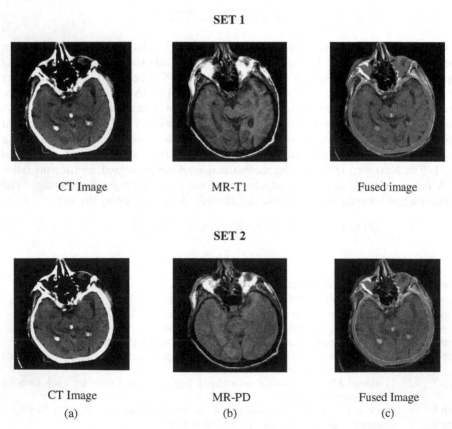

SET 1

CT Image   MR-T1   Fused image

SET 2

CT Image   MR-PD   Fused Image
(a)   (b)   (c)

Fig. 2. Image Fusion Results for Different Sets of MRI and CT-scan Images with the Proposed Approach. (a) Input CT image, (b) Input MRI image, (c) Fused image

Table 1. Quantitative Analysis of Proposed Fusion Approach

SET No.	FF	SSIM
1.	3.8654	0.8129
2.	3.8259	0.8450
3.	3.5491	0.8307

Fusion results in Fig. 2 for Set 1 & Set 2 show that the fused image has a better visual characteristic from the diagnostic point of view. CT-scan images give information about the shape of the tumor which is helpful in determining the extent of disease; whereas MRI image gives soft tissue details. It can be clearly seen that fused image contains complementary information of both the images; i.e. soft tissue details as well as the shape of the tumor. This is further supported by high values of fusion metrics (FF & SSIM). Values of SSIM above 0.8 demonstrate a good structural, contrast, as well as luminance restoration in the fused image. In addition to it, high values of fusion factor shows, the effectiveness of the proposed fusion approach.

3.1   Comparison with Other Fusion Approaches

The present approach has been compared with the DWT based fusion approach proposed by Y. Zheng et al. [41] and sparse representation approaches like Simultaneous Orthogonal Matching Pursuit (SOMP) and Orthogonal Matching Pursuit (OMP) [42] for medical image fusion. The obtained result shows the effectiveness of the proposed approach in visual representation as compared to DWT, SOMP and OMP approaches. The fused image obtained from the proposed approach represents edges more effectively, as the image has better sharpness as compared to images obtained from other approaches. Moreover, the higher values of the fusion metric shown in table 2 validate, that the proposed fusion approach has better diagnostic utility than the other approaches.

Table 2. Performance Comparison of Fused Images

Fusion Approach	SSIM
DWT[41]	0.58
OMP[42]	0.67
SOMP[42]	0.68
Proposed Approach	0.81

4   Conclusion

An image fusion approach employing PCA in ridgelet domain is presented in the paper. This approach is more appropriate for image fusion as it combines the feature enhancement property of PCA and edge detection property of ridgelet. The proposed fusion approach yielded a fused image that represented shapes and soft tissue details of tumor. Thus, it is providing the details of two different modalities in one single image, justifying the purpose of medical image fusion. Significant results relevant from a visual point of view, as well as high values of the fusion metrics (FF/SSIM), have been obtained from the proposed fusion approach. Comparison results indicate improvement in restoration of structural, contrast and luminance features in the obtained fused image; as depicted by high value of SSIM, in comparison to other fused images. Hence, the proposed fusion approach is more precise and can be used more effectively for medical diagnosis than the other methods of fusion.

References

1. Dasarathy, B.V.: Information Fusion in the Realm of Medical Applications – A Bibliographic Glimpse at its Growing Appeal. Information Fusion 13(1), 1–9 (2012)
2. Schoder, H., Yeung, H.W., Gonen, M., Kraus, D., Larson, S.M.: Head and Neck Cancer: Clinical Usefulness and Accuracy of PET/CT Image Fusion. Radiology, 65–72 (2004)
3. Nakamoto, Y., Tarnai, K., Saga, T., Higashi, T., Hara, T., Suga, T., Koyama, T., Togashi, K.: Clinical Value of Image Fusion from MR and PET in Patients with Head and Neck Cancer. Molecular Imaging and Biology, 46–53 (2009)
4. Singh, R., Khare, A.: Fusion of Multimodal Images using Daubechies Complex Wavelet Transform- A Multiresolution Approach. Info. Fusion. 19, 49–60 (2014)
5. Tu, T.M., Su, S.C., Shyu, H.C., Huang, P.S.: A New Look at IHS-Like Image Fusion Methods. Information Fusion 2(3), 177–186 (2001)
6. Gillespie, A.R., Kahle, A.B., Walker, R.E.: Color enhancement of highly correlated images—II. Channel Ratio and 'Chromaticity' Transformation Techniques. Remote Sens. Environ. 22, 343–365 (1987)
7. Mallat, S.: A Theory for Multi-Resolution Signal: The wavelet representation. IEEE Trans. Pattern Anal. Mach. Intell. 11(7), 674–693 (1989)
8. Shensa, M.J.: The discrete wavelet transform: Wedding the à Trous and Mallat algorithms. IEEE Trans. Signal Process. 40(10), 2464–2482 (1992)
9. ALEjaily, A.M.: Fusion of Remote Sensing Images Using Contourlet Transform. Innovations and Advanced Techniques in Systems, Computing Sciences and Software Engineering, 213–218 (2008)
10. Cunha, L.D., Zhou, J.P.: The Non-Subsampled Contourlet Transform: Theory, Design, and Applications. IEEE Tran. on Image Processing 15(10), 3089–3101 (2006)
11. Luo, Y., Liu, R., Zhu, Y.Z.: Fusion of Remote Sensing Image Based on the PCA & Atrous Wavelet Transform. The International Archives of the Photogrammetry, Remote Sensing and Spatial Info. Sciences. XXXVII. Part B7, 1155–1158 (2008)
12. Petrovic, V.S., Costas, S.X.: Gradient-Based Multiresolution Image Fusion. IEEE Transactions on Image Processing 13(2), 228–237 (2004)
13. Sadhasivam, S.K., Keerthivasan, M.K., Muttan, S.: Implementation of Max Principle with PCA in Image Fusion for Surveillance and Navigation Application. Electronic Letters on Computer Vision and Image Analysis 10(1), 1–10 (2011)
14. Xu, Z.: Medical Image Fusion Using Multi-Level Local Extrema. Elsevier-Information Fusion 19, 38–48 (2014)
15. Jain, A., Singh, S., Bhateja, V.: A Robust Approach for Denoising and Enhancement of Mammographic Breast Masses. International Journal on Convergence Computing, Inderscience Publishers 1(1), 38–49 (2013)

16. Srivastava, A., Alankrita, Raj, A., Bhateja, V.: Combination of Wavelet Transform and Morphological Filtering for Enhancement of Magnetic Resonance Images. In: Snasel, V., Platos, J., El-Qawasmeh, E. (eds.) ICDIPC 2011, Part I. Communications in Computer and Information Science, vol. 188, pp. 460–474. Springer, Heidelberg (2011)

17. Bhateja, V., Urooj, S., Pandey, A., Misra, M., Lay-Ekuakille, A.: A Polynomial Filtering Model for Enhancement of Mammogram Lesions. In: Proc. of IEEE Int. Symposium on Medical Measurements and Applications, pp. 97–100 (2013)

18. Siddhartha, Gupta, R., Bhateja, V.: A Log-Ratio based Unsharp Masking (UM) Approach for Enhancement of Digital Mammograms. In: Proc. CUBE Int. Information Tech. Conference & Exhibition, pp. 26–31 (2012)

19. Bhateja, V., Devi, S.: An Improved Non-Linear Transformation Function for Enhancement of Mammographic Breast Masses. In: Proc. 3rd International Conference on Electronics & Computer Technology, vol. 5, pp. 341–346 (2011)

20. Alankrita., Raj, A., Shrivastava, A., Bhateja, V.: Contrast Improvement of Cerebral MRI Features using Combination of Non-Linear Enhancement Operator and Morphological Filter. In: Proc. of (IEEE) International Conference on Network and Computational Intelligence (ICNCI 2011), Zhengzhou, China, vol. 4, pp. 182–187 (2011)

21. Siddhartha., Gupta, R., Bhateja, V.: An Improved Unsharp Masking Algorithm for Enhancement of Mammographic Masses. In: Proc. of IEEE Students Conference on Engineering and Systems (SCES-2012), Allahabad, India, pp. 234–237 (March 2012)

22. Siddhartha., Gupta, R., Bhateja, V.: A New UnSharp Masking Algorithm for Mammography using Non-Linear Enhancement Function. In: Proc. of the (Springer) International Conference on Information Systems Design and Intelligent Applications (INDIA 2012), Vishakhapatnam, India, pp. 779–786 (January 2012)

23. Pandey, A., Yadav, A., Bhateja, V.: Design of New Volterra Filter for Mammogram Enhancement. In: Satapathy, S.C., Udgata, S.K., Biswal, B.N. (eds.) Proceedings of Int. Conf. on Front. of Intell. Comput. AISC, vol. 199, pp. 143–151. Springer, Heidelberg (2013)

24. Bhateja, V., Urooj, S., Pandey, A., Misra, M., Lay-Ekuakille, A.: Improvement of Masses Detection in Digital Mammograms employing Non-Linear Filtering. In: Proc. of (IEEE) International Multi-Conference on Automation, Computing, Control, Communication and Compressed Sensing (iMac4s-2013), Palai-Kottayam, Kerala (India), vol. 119, pp. 406–408 (March 2013)

25. Pandey, A., Yadav, A., Bhateja, V.: Contrast Improvement of Mammographic Masses Using Adaptive Volterra Filter. In: Proc. of (Springer) 4th International Conference on Signal and Image Processing (ICSIP 2012), Coimbatore, India, vol. 2, pp. 583–593 (December 2012)

26. Bhateja, V., Misra, M., Urooj, S., Lay-Ekuakille, A.: A Robust Polynomial Filtering Framework for Mammographic Image Enhancement from Biomedical Sensors. IEEE Sensors Journal 13(11), 4147–4156 (2013)

27. Pandey, A., Yadav, A., Bhateja, V.: Volterra Filter Design for Edge Enhancement of Mammogram Lesions. In: Proc. of (IEEE) 3rd International Advance Computing Conference (IACC 2013), Ghaziabad (U.P.), India, pp. 1219–1222 (February 2013)

28. Bhateja, V., Devi, S.: A Novel Framework for Edge Detection of Microcalcifications using a Non-Linear Enhancement Operator and Morphological Filter. In: Proc. of (IEEE) 3rd International Conference on Electronics & Computer Technology (ICECT-2011), Kanyakumari (India), vol. 5, pp. 419–424 (April 2011)

29. Alankrita., Raj, A., Shrivastava, A., Bhateja, V.: Computer Aided Detection of Brain Tumor in MR Images. International Journal on Engineering and Technology (IACSIT-IJET) 3, 523–532 (2011)

30. Granai, L., Moschetti, F., Vandergheynst, P.: Ridgelet Transform Applied to Motion Compensated Images. In: IEEE International Conference on Acoustics, Speech, & Signal Processing, April 6-10, pp. 561–564 (2003)
31. Ali, F.E., El-Dokany, I.M., Saad, A.A., Abd El-Samie, F.E.: Curvelet Fusion of MR and CT Images. Progress in Electromagnetics Research 3, 215–224 (2008)
32. Naidu, V.P.S., Rao, J.R.: Pixel-level Image Fusion using Wavelets and Principal Component Analysis. Defence Science Journal 58(3), 338–352 (2008)
33. Gupta, P., Tripathi, N., Bhateja, V.: Multiple Distortion Pooling Image Quality Assessment. International Journal on Convergence Computing, Inderscience Publishers 1(1), 60–72 (2013)
34. Gupta, P., Srivastava, P., Bharadwaj, S., Bhateja, V.: A HVS based Perceptual Quality Estimation Measure for Color Images. ACEEE International Journal on Signal & Image Processing (IJSIP) 3(1), 63–68 (2012)
35. Bhateja, V., Devi, S.: A Reconstruction Based Measure for Assessment of Mammogram Edge-Maps. In: Satapathy, S.C., Udgata, S.K., Biswal, B.N. (eds.) Proceedings of Int. Conf. on Front. of Intell. Comput. AISC, vol. 199, pp. 741–746. Springer, Heidelberg (2013)
36. Trivedi, M., Jaiswal, A., Bhateja, V.: A New Contrast Measurement Index Based on Logarithmic Image Processing Model. In: Satapathy, S.C., Udgata, S.K., Biswal, B.N. (eds.) Proceedings of Int. Conf. on Front. of Intell. Comput. AISC, vol. 199, pp. 715–723. Springer, Heidelberg (2013)
37. Jaiswal, A., Trivedi, M., Bhateja, V.: A No-Reference Contrast Assessment Index based on Foreground and Background. In: Proc. 2nd Students Conference on Engineering and Systems, pp. 460–464 (2013)
38. Bhateja, V., Srivastava, A., Kalsi, A.: Reduced Reference IQA based on Structural Dissimilarity. In: Proc. Int. Conf. on Signal Proc. and Integ, pp. 63–68 (2014)
39. Piella, G., Heijmans, H.: A New Quality Metric for Image Fusion. In: 2003 International Conference on Image Processing, Barcelona, Spain, (September 14, 2003)
40. Bhateja, V., Srivastava, A., Kalsi, A.: Fast SSIM Index for Color Images Employing Reduced-Reference Evaluation. In: Satapathy, S.C., Udgata, S.K., Biswal, B.N. (eds.) FICTA 2013. AISC, vol. 247, pp. 451–458. Springer, Heidelberg (2014)
41. Zheng, Y., Essock, E.A., Hansen, B.C.: An Advanced Image Fusion Algorithm Based on Wavelet Transform –Incorporation with PCA and Morphological Processing. In: Proc. SPIE, vol. 5298, pp. 177–187 (2004)
42. Yang, B., Li, S.: Pixel-Level Image fusion with Simultaneous Orthogonal Matching Pursuit. Information Fusion 13, 10–19 (2012)
43. Virmani, J., Kumar, V., Kalra, N., Khandelwal, N.: SVM-Based characterization of liver ultrasound images using wavelet packet texture descriptors. Journal of Digital Imaging 26(3), 530–543 (2013)
44. Virmani, J., Kumar, V., Kalra, N., Khandelwal, N.: A comparative study of computer-aided classification systems for focal hepatic lesions from B-mode ultrasound. Journal of Medical Engineering and Technology 37(4), 292–306 (2013)
45. Virmani, J., Kumar, V., Kalra, N., Khandelwal, N.: PCA-SVM based CAD System for focal liver lesions from B-Mode ultrasound. Defence Science Journal 63(5), doi:10.1007/s10278-014-9685-0
46. Virmani, J., Kumar, V., Kalra, N., Khandelwal, N.: Neural network ensemble based CAD system for focal liver lesions using B-mode ultrasound. Journal of Digital Imaging, doi:10.1007/s10278-014-9685-0

Swarm Optimization Based Dual Transform Algorithm for Secure Transaction of Medical Images

Anusudha Krishnamurthi[1], N. Venkateswaran[2], and J. Valarmathi[3]

[1] Department of Electronics Engineering, Pondicherry University, Pondicherry - 605014, India
[2] Department of Electronics and Communication Engineering,
SSN college of Engineering, Chennai - 603110, India
[3] School of Electronics Engineering, VIT University, Vellore-602014, India
anusudhak@yahoo.co.in, nvenkateswaran@ssn.edu.in,
valarmathij@vit.ac.in

Abstract. Modern healthcare systems are based on managing diagnostic information of patients through e-health. e-health refers to the internet enabled healthcare applications involving transacting personal health records and other internet based services including e-pharmacy etc. This paper introduces a hybrid algorithm which efficiently combines DWT-DCT-PSO for copyright protection and authentication of medical images. Particle swarm optimization is applied on the host image to find the intensities for embedding the watermark bits by gbest solution from objective function and the fitness function. The embedding strategy is adopted with intensity level, so the technique falls under robust blind watermarking technique. The simulation results shows that the proposed scheme yields good results when tested for different images and subjected to various attacks.

Keywords: Discrete Wavelet Transform (DWT), Discrete Cosine Transform (DCT), Particle Swarm Optimization (PSO), gbest solution.

1 Introduction

Telemedicine combines Medical Information System (MIS) with information technology that includes use of computers to receive, store and distribute medical information over long distances. Currently, telemedicine applications in teleconsulting, telediagnosis, telesurgery and remote medical education play a vital role in the evolution of the healthcare domain [1,2]. The transmission, storage and sharing of electronic medical data *via* the networks have many purposes such as diagnosis, finding new drugs and for scientific research. Hence, healthcare industry demands secure, robust and more information hiding techniques promising strict secured authentication and communication through internet or mobile phones.

In general information hiding includes digital watermarking and steganography [3]. A watermarking scheme imperceptibly alters a cover object to embed a message about the cover object (e.g owner's identifier) [4]. Watermarking is used for copyright protection, broadcast monitoring and transaction tracking and thus robustness of

© Springer International Publishing Switzerland 2015 483
S.C. Satapathy et al. (eds.), *Proc. of the 3rd Int. Conf. on Front. of Intell. Comput. (FICTA) 2014*
– *Vol. 2*, Advances in Intelligent Systems and Computing 328, DOI: 10.1007/978-3-319-12012-6_53

digital watermarking schemes becomes critical. In general, digital watermarking algorithms can be divided into two classes depending on the domain of watermark embedding. The first group belongs to the algorithms which uses spatial domain for data hiding [5,6] ,while algorithms of the second group take advantage of transform domain like Discrete Cosine Transform (DCT) [7], Discrete Fourier Transform(DFT) [8] and Discrete Wavelet Transform (DWT) [9] for watermarking purpose. Previous works reveals that transform domain schemes are typically more robust to noise, common image processing tasks and compression when compared with spatial transform schemes [10].

Based on good time frequency features and discrimination that match well with the Human Visual System (HVS) motivates the use of DWT in image watermarking among several techniques [11,12,13].In medical applications, it is very important to maintain the quality of images because of their diagnostic value. The performance of the scheme can be improved by combination of various transforms.

Another issue with digital watermarking refers to making the process involved in the embedding and detection of a watermark more adaptable to variations across different images.

In this paper, a new watermarking scheme based on PSO is introduced. The watermark bits are embedded on intensity of each embedding block within low frequency sub-band in the hybrid DWT-DCT domain. One parameter is determined by exploiting the global solution parameter which is optimized through PSO algorithm. The proposed scheme shows a good visual quality irrespective of the nature of the images chosen and robust to image processing attacks.

2   Proposed Scheme

The proposed system consists of : Embedding mode and Extraction mode

2.1   Embedding Process

The process of embedding is shown in the Fig.1. Both host and the watermark image are compressed by DWT. This compression helps to separate the image into different levels of band. In this paper, watermarking is done on specific intensities selected by the PSO.DCT is applied to the LL band of the host and the watermark image.PSO is applied to the DCT block of the host image to select the intensities for embedding the watermark The group of intensities identified is called a swarm consisting of different particles. Each particle represents a possible solution to the optimization problems in the multi-dimensional problem space. These particles start at a random initial position and search for the minimum or maximum of a given objective function by moving through the search space. After each iteration process in the search space, each particle records its own personal best solution as well as the discovered global best position. The movement of each particle depends only on its velocity and the location where good solutions have already been found by the particle itself or in neighboring particles. When a particle's neighborhood is defined as the whole swarm, the PSO is called the global version, otherwise it is called the local version. The output of PSO is the gbest solution which is obtained from the gbest for the every iteration in the image dimension search space. The gbest solutions are the intensity values, on the condition that watermarking is done in between the ranges of two consequent intensity values. Next stage is framing them intelligently by applying the IDCT and IDWT. Finally, the watermarked image is obtained. This watermarked image is passed through various levels of noise attacks. This process is done to test the robustness in the watermarking process.

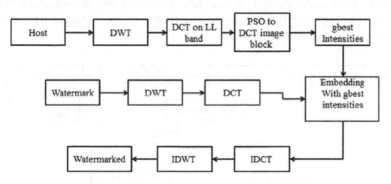

Fig. 1. Block diagram illustration of the Embedding process

2.2   Particle Swarm Optimization

Let a swarm include different particles X i (i = 1,2,..., m) and the ith particle in a d-dimensional space be represented as X i = (xi 1 , xi 2 ,..., xid) .The best previous position of the ith particle pbesti is denoted by Pi = (pi 1 , pi 2 ,..., pid) .The best global location is represented pbestg among all the particles. The velocity for the ith particle is represented as Vi = (vi 1 , vi 2 ,..., vid) .During each iteration process, each particle in the swarm is updated the velocity and location towards its pbesti and pbestg locations each iteration according to following two equations respectively:

$$V_I = w_i v_i + c_1 \xi (pbest_i - X_i) + c_2 \eta (pbest_g - X_i) \tag{1}$$

$$X_i = X_i + V_i \tag{2}$$

where ξ and η are random variables drawn from a uniform distribution in the range [0,1] so as to provide a stochastic weighting of the different components participating in the particle velocity definition. c1 and c2 are two acceleration constants and called ascognitive acceleration and social acceleration respectively.

2.3   Extraction Process

Extracting is the process of retrieving the watermark bits from the watermarked image. The watermarked image is applied with DWT to get the low level band. As in the process of embedding, DCT is applied to the LL block of the host image. The solution in the PSO identifies the intensity which are embedded in the LL block.

When PSO is applied on the DCT block according to equation (1) and equation (2) particles start moving randomly in the image. With the maximum and the minimum velocities assigned to the particles, starts moving in the fixed position points. The solution obtained near the position is taken into consideration as the best solution. This process is continued until the final iteration. For each iteration, PSO leaves out the best solution which can be treated as the *pbest* (previous best) solution. All *pbest* solutions combine to form *gbest* (global best) solution. The gbest value is the intensity value where the watermarking is done. Now watermark bits are separated from the host image. On applying IDCT followed by IDWT to bits obtained from the PSO, final watermark image is obtained.

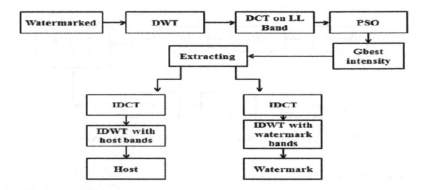

Fig. 2. Block diagram illustration of the Extraction process

3   Embedding and Extraction Algorithm

3.1   Embedding Algorithm

Step1: Perform Discrete Wavelet Transform (DWT) decomposition to the host image and the watermark image.

Step 2: Discrete Cosine Transform is performed on DWT LL sub-band of host and the watermark image.

Step3: PSO is applied to the DCT block to find the gbest intensity to embed the watermark bits.

Step4: The watermark is embedded on the selected intensities of the host image with the following condition:

$$W_t(i,j) = \begin{cases} int(i,j) + W(i,j) \; if \; int > gbest \\ \qquad\qquad and \; if \; int < gbest \\ host \; (i,j) \qquad\qquad others \end{cases} \tag{3}$$

Where $W_t(i,j)$ is the watermarked image and $int(i,j)$ is the intensity at corresponding (i,j) location is the global best solution *gbest* obtained from the PSO.

Step 5: Apply inverse DCT on the embedded LL band of the host image.

Step 6: Perform inverse DWT to obtain the watermarked image.

3.2   Extraction Algorithm

Step 1: Perform Discrete Wavelet Transform to the watermarked image.
Step 2: Discrete Cosine transform is performed on the LL band of the watermarked image.
Step 3: PSO is applied to the DCT block to obtain the gbest intensities.
Step 4: The watermark is extracted from the gbest intensities based on the following equation:

$$W_{t1}(i,j) = \begin{cases} int(i,j) - W(i,j) \ if \ int \ > gbest \\ \qquad and \ if \ int \ < gbest \\ host \ (i,j) \qquad\qquad others \end{cases} \tag{4}$$

Where $W_{t1}(i,j)$ is the watermark image and $int(i,j)$ is the intensity at corresponding (i,j) location is the global best solution obtained from the PSO.
Step 5: Apply inverse DCT on the extracted image.
Step 6: The watermark image is obtained by the inverse DWT.

3.3   Algorithm for PSO

Step1: Initialize the population in the n dimensional search space.
Step2: For each particle i = 1,2 ..., S do:
  ➢ Initialize the particle's position with a uniformly distributed random vector: xi ~ U(blo, bup), where blo and bup are the lower and upper boundaries of the search-space.
  ➢ Initialize the particle's best known position to its initial position:
      pi ← xi
  ➢ If (f(pi) < f(g)) update the swarm's best known position: g ← pi
  ➢ Initialize the particle's velocity: vi ~ U(-|bup-blo|, |bup-blo|)
Step 3: Until a termination criterion is met (e.g. number of iterations performed, or a solution with adequate objective function value is found), repeat:
    For each particle i = 1, ..., S do:
  ➢ Pick random numbers: rp, rg ~ U(0,1)
  ➢ For each dimension d = 1, ..., n do:
      ✓ Update the particle's velocity: vi,d ← ω vi,d + φp rp (pi,d-xi,d) + φg rg (gd-xi,d)
  ➢ Update the particle's position: xi ← xi + vi
  ➢ If (f(xi) < f(pi)) do:
      ✓ Update the particle's best known position: pi ← xi
      ✓ If (f(pi) < f(g)) update the swarm's best known position:
          g ← pi
Step 4: Now g holds the best found solution.
The parameters ω, ϕp, and ϕg are selected by the practitioner to control the behavior of the swarm.

4   Simulation Results and Analysis

The images taken for the analysis of the proposed schemes are a host image and watermark image of dimension 256x256 pixels {Images are taken from DICOM link}. The PSO parameters used for the simulation are:

Table 1. PSO parameters used for the simulation

Parameters	Value
Population	150
Social particle weight c1,c2	0.8,0.8
Internal factor	1.2
Maximum velocity	5
Minimum velocity	-5
Number of iteration	100
Dimension of the search space	256x256

4.1   Performance Analysis

The performance evolution of the watermarking approach is analyzed against various attacks. The Peak-Signal-To-Noise is defined as:

$$PSNR = \frac{10\log_{10}(255)^2}{MSE} \tag{5}$$

Where 'MSE' is the mean squared error between the original and distorted image and is defined as follows:

$$MSE = \frac{1}{mn} \sum_{i,j=0}^{m-1,n-1} [W(i,j) - W\bullet(i,j)] \tag{6}$$

Where m, n gives the size of the image and $W(i,j), W'(i,j)$ are the pixel values at location (i,j) of the original and distorted image respectively.

The robustness of the proposed technique is checked by exposing the watermarked image to various types of removal and geometric attacks. Removal attacks aim at removing the watermark signal from the watermarked image, without attempting to break the security of the watermarking algorithm and geometric attacks intends to distort it. The noise attacks considered are Gaussian noise, Speckle noise, Salt & Pepper noise, Poison noise. Gaussian noise attack is tested with the variance 0 to0.1. salt and pepper noise with the density value ranging from 0 to 0.1. Similarly, for speckle noise with the variance 0 to 0.1.

4.2   Simulation Results

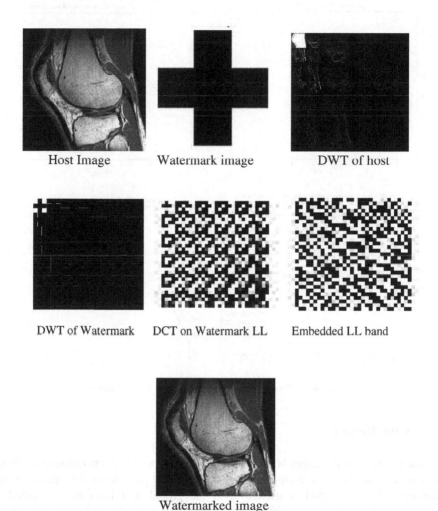

Fig. 3. Simulation results of the proposed scheme

4.3   Analysis in Presence of Noise and Varying PSO Levels

Fig.4 shows the results obtained from subjecting the watermarked image to various attacks. gbest solution helps to easily find the correlation between the original and the watermarked images. From the above analysis it is clear that the variations in the PSO level increases the PSNR value of the image even when subjected to various attacks.

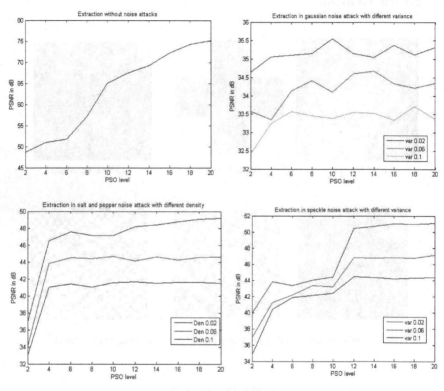

Fig. 4. Analysis in presence of various noise attacks

5   Conclusion

The contribution of the paper is to introduce particle swarm optimization algorithm in watermarking techniques. Watermarking scheme using particle swarm optimization is done on DWT-DCT.PSO is used to find the global best intensity to embed the watermark. The results are verified with various noise attacks and the corresponding PSNR are plotted with respect to the level of PSO. All the parameters are simulated using MATLAB 2012b. The implementation of each algorithm is described, and the performance of them is compared. The proposed model is robust to various noise attacks with an improvement of transparency to watermark. The technique can be extended to other types of PSO methods.

References

1. Youngberry, K.: Telemedicine Research. Journal of Telemedicine and Telecare 10(2), 121–123 (2004)
2. Tachakra, S., Wang, X.H., Istepanian, R.S., Song, Y.H.: Mobile e-health: The Unwired Evaluation of Telemedicine. Telemedicine Journal of e-health 9(3), 247–257 (2003)
3. Swanson, M.D., Koayashi, M., Tew_k, A.H.: Multimedia Data Embedding and Watermarking Technologies. IEEE Transaction on Information Technology in Biomedicine 86, 1064–1087 (1998)
4. Chang, C.C., Tai, W.L., Lin, C.C.: A reversible data hiding scheme based on side match vector quantization. IEEE Transactions on Circuits Systems Video Technology 16(10), 1301–1308 (2000)
5. Berghel, H., O'Gorman, L.: Protecting ownership rights through digital watermarking. IEEE Computational Magazine 29(7), 101–103 (1996)
6. Wang, R.Z., Lin, C.F., Lin, J.C.: Image hiding by optimal LSB substitution and genetic algorithm. Pattern Recognition 34(3), 671–683 (2001)
7. Zkou, X.Q., Huang, H.K., Lou, S.L.: Authenticity and integrity of digital mammography images. IEEE Transactions on Medical Imaging 20(8), 784–791 (2001)
8. Chao, H.M., Hsu, C.M., Miaou, S.G.: A data-hiding technique with authentication, integration, and confidentiality for electronic patients records. IEEE Transactions Information Technology in Biomedicine 6, 46–53 (2002)
9. Giakoumaki, D., Rogers, J., Mazumdar, R., Coutts, D.: Abbott.: An Overview of Wavelets for Image Processing for Wireless Applications. In: Proceedings of SPIE: Smart Structures, Devices and Systems, vol. 4935, pp. 427–435. University of Melbourne, Australia (2003)
10. Shieh, C., Huang, H., Wang, F., Pan, J.: Genetic watermarking based on transform domain techniques. Pattern Recogn. 37, 555–565 (2004)
11. Piva, Bouridane, A., Ibrahim, M., Boussakta, S.: Digital image watermarking using balanced multiwavelets. IEEE Trans. Signal Process 54(4), 1519–1536 (2006)
12. Paul, B., Xiaohu, M.: Image Adaptive Watermarking Using Wavelet Domain Singular Value Decomposition. IEEE Transactions on Circuits and Systems for Video Technology 15, 96–102 (2005)
13. Farina, M., Deb, K., Amota, P.: Dynamic multi-objective optimization problems test cases, approximations, and applications. IEEE Trans. Evol.Comp. 8, 425–442 (2004)

Convolutional Neural Networks for the Recognition of Malayalam Characters

R. Anil, K. Manjusha, S. Sachin Kumar, and K. P. Soman

Centre for Excellence in Computational Engineering and Networking,
Amrita Vishwa Vidyapeetham,
Coimbatore - 641112, India
{anil.soubhagia,manjushagecpkd sachinnme}@gmail.com,
kp_soman@amrita.edu

Abstract. Optical Character Recognition (OCR) has an important role in information retrieval which converts scanned documents into machine editable and searchable text formats. This work is focussing on the recognition part of OCR. LeNet-5, a Convolutional Neural Network (CNN) trained with gradient based learning and backpropagation algorithm is used for classification of Malayalam character images. Result obtained for multi-class classifier shows that CNN performance is dropping down when the number of classes exceeds range of 40. Accuracy is improved by grouping misclassified characters together. Without grouping, CNN is giving an average accuracy of 75% and after grouping the performance is improved upto 92%. Inner level classification is done using multi-class SVM which is giving an average accuracy in the range of 99-100%.

Keywords: Optical Character Recognition (OCR), Deep learning, Multi-stage classification, Convolutional Neural Network (CNN), Gradient based learning, Backpropagation, Multiclass SVM.

1   Introduction

In pattern recognition, Optical Character Recognition is an active and successful area for last few decades. There are many algorithms for pattern recognition. Intelligent Word Recognition (IWR), Optical Character Recognition (OCR), Optical Mark Recognition (OMR), Intelligent Character Recognition (ICR) is some common pattern recognition technologies, but OCR is most prelavent technology [1] [2]. The basic idea of Optical Character Regognition is the convertion of scanned bitmaps of printed or hand written text into data files which can be edited by machine. Using an OCR, a book or article can directly give to a computer to transform it to an editable text format. The ability to store text efficiently and productivity improvement by reducing human involvement are two major advantages of OCR system. The areas where this system can be used are banks, health care, postal departments, education, publication industry, finance, government agencies etc [2] [3].

S.C. Satapathy et al. (eds.), *Proc. of the 3rd Int. Conf. on Front. of Intell. Comput. (FICTA) 2014*
– Vol. 2, Advances in Intelligent Systems and Computing 328, DOI: 10.1007/978-3-319-12012-6_54

Any OCR system contains three main steps: 1. Preprocessing, 2. Feature Extraction and 3. Classification and Recognition. Preprocessing phase enhance and clean up the image by image binarization, noise removal, skew detection and correction, text segmentation etc. Feature extraction is a method used to measure some information from the data which is most relevant to the given classification task. The basic idea of classification is to map the segmented text portion in document image to equivalent text representation [1].

There are different softwares available for Optical Character Recognition. One of the open source OCR engine available is "Tesseract OCR Engine". Wide variety of image formats can read by this software and transform them to text in more than 60 languages. "OCRopus", is a Google sponsored project under IUPR research group. This OCR system use plugins for modular design. Another OCR system used for printed Malayalam documents is "Nayana", which is a product of Centre for Development of Advanced Computing(C-DAC) [4].

Good quality OCR systems are available for different languages like Latin, Arabic, and Chinese etc. Complete and efficient OCR systems for Indian language scripts are not widely available. Because they are more complex with large character set, irregular positioning of characters in a running text, structural similarity among character classes and existence of both old and new version of language scripts [3] [4]. In that Malayalam is one of the famous Dravidian languages. There are total 578 characters in Malayalam script and it contains 52 letters including 16 vowels and 36 consonants. The classification of Malayalam characters is a challenging task because this script is unicase, contains lots of similar structured characters and does not have any inherent symmetry [5].

Over the last few years, neural networks are the most researched area in pattern classification. Multi-layer neural networks trained with gradient decent have the ability to learn complex, high-dimensional, non-linear mappings from very large number of data which makes this a good solution for pattern recognition tasks. Multi layer neural networks with full connection are used as classifier in convolutional neural networks. CNN shows excellent recognition rates for digit recognition which is proposed by Lecun *et. al.*[6] and for Malayalam characters, compared to other classification methods CNN shows better results [2] [7].

In this paper Malayalam character recognition using multi stage classification architecture is presented. I.e primarily the recognition is done using CNN, called outer classification. Then in second stage the misclassified characters from outer classification is further classified using multiclass SVM, called inner classification. Second section discuss about CNN and learning method used in the network. Section 3 depicts a brief description about Multi-class SVM. Classification results are shown in Section 4.

2   Convolutional Neural Networks

In the area of machine learning, deep learning is a set of algorithms that is used for modelling high-level data abstractions by using architectures created with several non-linear transformations. Neural Networks got importance because of their learning ability. In pattern recognition, designed neural network under goes two phases. The first phase is training phase, in which the network will first initialize the weights or trainable parameters with random values. Then the network will update the weights according to the error arouse with each sample provided by the training set. This type of learning method is called stochastic method of learning. In Batch learning weights gets updated only after considering all the samples present in the training set. Gradient descent backpropagation is one of the commonly used weight updation algorithm [7].

2.1   Gradient Based Learning

Gradient-based learning is one of the popularized methods used in automatic machine learning. Here learning machine computes a function

$$Y^p = F(Z^p, W).$$

(1)

Where Z^p is p th input pattern, W is collection of adjustable parameters in the system and Y^p is the output. Training set is created using this input vector Z^p and desired class label vector. By finding error function or loss function the amount of parameter adjustment, W can be identified. There are two error values e_{test} and e_{train} for test set and train set respectively. So the final aim is to minimize the value of $e_{test} - e_{train}$. This value indicates how the classifier performs on input patterns. Parameters $W = W(t)$, where t is the current training step, related to the gradient of the scalar-valued error function, can be modified using the equation

$$W(t+1) = W(t) - \varepsilon \frac{\partial E(W(t))}{\partial W(t)}.$$

(2)

Parameter can be modified after giving all input patters or after the presentation of one pattern [7] [8] [9].

2.2   Multilayer Perceptron

In the area of machine learning, perceptron is an algorithm used for supervised classification which will classify the input into several classes. Single perceptron is a computational model of the biological neuron. In single perceptron connections from other neurons are represented by the input vector l. Then all input vectors are multiplied with corresponding weights W and added. This scalar product is yielded with a threshold or bias, θ:

$$u = \sum_{k=1}^{n} l_k w_k - \theta = \langle l, w \rangle - \theta. \tag{3}$$

Then the perceptron's output is calculated by giving this scalar output to the transfer function or squashing function, i.e $y = f(u)$. For a multilayer perceptron Fermi function or inverse tangent is used as transfer function. Multilayer are capable of classsifying data which is not linear seperable. A multilayer perceptron is obtained by combining two or more layers together, where one layer is fully interconnected with the previous layer. It is also called feed-forward network because data is propagated forward through the network [7] [8] [9].

2.3   Backpropagation

Backpropagation algorithm is used for learning in neural networks. Gradient calculation is the main idea of backpropagation which is done by propagation from output to the input. The error function can be taken as

$$E(W) = \sum_{p} E^{p}(W) = \sum_{p} \left\| T^{p} - y^{p} \right\|_{2}^{2}. \tag{4}$$

Where y^{p} is the output vector from the last layer. To reduce the error rate of a pattern, gradient of E^{p} can be used to adjust the weights accordingly. Then the learning rule for neurons in the output layer can be calculated as

$$w_{ij}(t+1) = w_{ij}(t) + \varepsilon y_i \delta_j. \tag{5}$$

Where $\delta_j = (T_j - y_j)f'(u_j)y_i$, y_j - output of neuron j in the ouput layer, y_i - previous layer output of neuron i. Likewise hidden layers can also be learned. The weight adjustments carried out in hidden layer are responsible for all errors in the hidden layer. Here the error is given back to the hidden layer, so the name backpropagation [8] [9] [10].

2.4   Convolutional Neural Networks

A CNN has one or more convolutional layers which is having full connection. CNN also uses tied weights and pooling layers. LeNet-5 is a typical convolutional network for character recognition and which is used here. Le-Net 5 architecture contains 7 layers, without counting the input. Each unit in a particular layer accepts input from a set of units in the small neighborhood in the previous layer. In this architecture, convolutional layers and sub-sampling layers are organized alternatively. Starting with a convolutional layer, there are three convolutional layers, two sub-sampling layers and an output layer. The Output layer is composed of Euclidean Radial Basis

Function units (RBF), one for each class. That is each output RBF unit computes the Euclidean distance between its input vector and its parameter vector [7] [8] [9] [10].

3   Multiclass SVM

Support Vector Machine (SVM) is usually used for binary classification. One approach for multiclass SVM is building and combining several binary classifiers. Another way is considering all data in one optimization problem. The complexity to solve multiclass SVM problem varies with number of classes [11]. The earliest method for classification using multiclass SVM is one-against-all method. This will create k SVM models, where k is the number of classes. Another popular method is one-against-one method. Here $k(k-1)/2$ classifiers are created where each of them is trained on data from two classes. Directed Acyclic Graph Support Vector Machines (DAGSVM) is another approach for this problem. Training phase is same as one-against-one method. A rooted binary directed acyclic graph with $k(k-1)/2$ internal nodes and k leaves is used in testing phase [12] [13]. These methods can be implemented by using simple and efficient open source software called Libsvm. One-class-SVM, nu-SVM classification, C-SVM classification, nu-SVM regression and epsilon-SVM regression problems can be efficiently solved by this software.

4   Experimental Results and Discussion

This section describes about our attempt to apply the CNN based character classification approach to Malayalam language script. The experiments are conducted in MATLAB programming environment. LeNet-5 network is used for implementing CNN. 119 different character classes (including consonants, vowels, vowel modifiers and compound characters) of Malayalam Language script is used for our experiments. The dataset is created by scanning different Malayalam story, novel books and then segmenting each character images using levelsets [14].

At first for evaluating the performance of CNN on Malayalam language script, we have used an incremental training approach. We started with 5 classes for training first and calculated the accuracy. Then number of classes is slowly increased. When the number of classes reached range of 40 the performance of CNN started to decrease. CNN's performance was decreasing when we increased number of character classes beyond that. The misclassification rate of different set of trials (each set contains different 30 character classes) is shown in Table 1.

After this the network is learned using the 119 characters of Malayalam character dataset. The network shows an average training misclassification as 24% and for testing the misclassification rate is 25%. Therefore for whole Malayalam character dataset average classification accuracy is 76%. The Sample graph showing misclassification rate of set contain 119 characters is shown in Figure 1. It shows root mean square error in the above portion and misclassification rate for train and test set in the below portion. Root mean square error is the measure of difference between predicted values and the actual observed values. Here the algorithm is trained for 5 iterations. Based on the confusion matrix obtained from this result, the characters which are misclassified mostly are grouped into classes. So the total number of classes is reduced to 85 from 119. The network is again learned and tested for these 85 classes. The result shows a performance improvement in the network. The misclassification rate is come down to 10.5% for testing and 11.9% for testing. A graph showing root mean square error and misclassification rate for a sample training and test set is shown in figure 2. Here the algorithm is trained for 10 iterations, but from iteration 5 itself the network shows a good accuracy.

Accuracy is improved when the classes are grouped together and trained. The grouped classes are further classified using multiclass SVM. The inner classification using multiclass SVM is giving accuracy in the range of 99-100%.

Table 1. Misclassification rate of different set of classes

Classes(each set 30 classes)	Error rate (train)	Error rate (test)
Set9	4.16667%	3.33333%
Set13	0.133333%	0%
Set14	0.6%	0.166667%
Set15	5.56667%	4.33333%
Set16	0.266667%	0%
Set17	9.83333%	6.55556%
Set18	10.9333%	6.22222%
Set19	8.4%	6.88889%

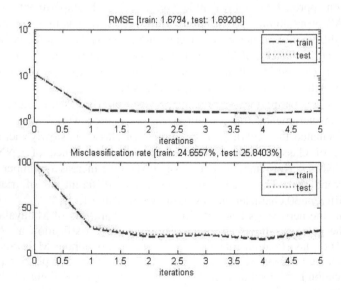

Fig. 1. Misclassification rate for set contain the 119 characters

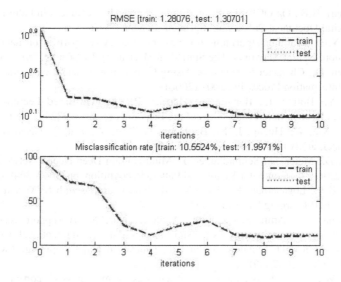

Fig. 2. Misclassification rate for set contain the grouped characters

5   Conclusion

Optical Character Recognition for Malayalam characters is an active research area which always needs an improvement in accuracy. This work is based on recognition of Malayalam characters using convolutional neural networks (CNN) and multiclass SVM. Compare to other deep learning architectures, CNN has better performance in both image and speech applications. To improve the performance of CNN, most misclassified characters are grouped and learned the network. The outer classification performance of proposed method is 92%. Multiclass SVM is used for inner classification and the accuracy is in the range of 99-100%.

References

1. Neeba, N.V., Namboodiri, A., Jawahar, C.V., Narayanan, P.J.: Recognition of Malayalam Documents. In: Advances in Pattern Recognition, Guide to OCR for Indic Scripts. Springer, London (2009)
2. Neeba, N.V., Jawahar, C.V.: Empirical evaluation of character classification schemes. In: Seventh International Conference on ICAPR 2009. Advances in Pattern Recognition, pp. 310–313. IEEE (2009)
3. Anil, R., Pradeep, A., Midhun, E.M., Manjusha, K.: Malayalam Character Recognition using Singular Value Decomposition. International Journal of Computer Applications (0975 – 8887) 92(12) (April 2014)
4. Divakaran, S.: Spectral Analysis of Projection Histogram for En- hancing Close matching character Recognition in Malayalam. International Journal of Computer Science and Information Technology (IJCSIT) 4(2) (April 2012)

5. Chaudhuri, B.B.: On OCR of a Printed Indian Script. In: Advances in Pattern Recognition (ed) Digital Document Processing. Springer, London (2007)
6. Lecun, Y.E.: Learning algorithms for classification: A comparison on handwritten digit recognition. Neural Networks: The Statistical Mechanics Perspective, 261–276 (1995)
7. Bouchain, D.: Character Recognition Using Convolutional Neural Networks. Institute for Neural Information Processing 2007 (2006)
8. LeCun, Y., Bottou, L., Bengio, Y., Haffner, P.: Gradient-Based Learning Applied to Document Recognition. Proceedings of the IEEE 86(11), 2278–2324 (1998)
9. Wei, Y., Xia, W., Huang, J., Ni, B., Dong, J., Zhao, Y., Yan, S.: CNN: Single-label to Multi-label, arXiv preprint arXiv: 1406.5726 (2014)
10. Ciresan, D., Meier, U., Schmidhuber, J.: Multi-column Deep Neural Networks for Image Classification. In: Computer Vision and Pattern Recognition, pp. 3642–3649 (2012)
11. Soman, K.P., Loganathan, R., Ajay, V.: Machine Learning with SVM and other Kernel methods. PHI Learning Pvt. Ltd (2009)
12. Ramanathan, R., Arun, S., Nair, V., Vidhya Sagar, N.: A support vector machines approach for efficient facial expression recognition. In: International Conference on Advances in Recent Technologies in Communication and Computing, ARTCom 2009, pp. 850–854. IEEE (2009)
13. Hsu, C.-W., Lin, C.-J.: A comparison of methods for multiclass support vector machines. IEEE Transactions on Neural Networks 13(2), 415–425 (2002)
14. Cherian, M., Radhika, G., Shajeesh, K.U., Soman, K.P., Sabarimalai Manikandan, M.: A Levelset Based Binarization and Segmentation for Scanned Malayalam Document Image Analysis. In: IEEE International Conference on computational Intelligence and Computing Research (2011)

Modeling of Thorax for Volumetric Computation Using Rotachora Shapes

Shabana Urooj[1], Vikrant Bhateja[2], Pratiksha Saxena[3],
Aime lay Ekuakille[4], and Patrizia Vergalo[4]

[1] School of Engineering, Gautam Buddha University,
Gr. Noida (U.P.), India
[2] Department of Electronics and Communication Engineering,
SRMGPC, Lucknow (U. P.), India
[3] School of Vocational Studies & Applied Sciences, Gautam Buddha University,
Gr. Noida (U.P.), India
[4] Department of Innovation Engineering, University of Salento,
Lecce, Italy
{shabanaurooj,bhateja.vikrant}@ieee.org

Abstract. This paper presents the scope of mathematical modeling by using uncommon geometric shapes for the computation of thoracic volume. The modeling has been done for Rotachora shapes for estimation and computation of fluid volume present in the thoracic area. Proposed extended model based approach demonstrates the scopes of its sensitivity in terms of volumetric variations with the act of breathe. The act of breathe involved inspiration and expiration states. New models have been constructed to compute the thoracic volumes and their variations are shown with respect to the thoracic impedances. Four dimensional Rotachora shapes are taken into consideration. Human thorax is considered as cubinder in the first stage and as duo cylinder in the second phase under Rotachora shape category. It is observed that the volumes are rhythmically varying with the act of breath for the considered thoracic area along with the varying thoracic impedances. The obtained results validates that the chosen models are closely following the act of breath significantly and hence the obtained result could be utilized for clinical purposes.

Keywords: thoracic impedance, thoracic volume computation, rotachora, cubinder, duo cylinder, pulmonary.

1 Introduction

There has been increasing interest using model based approach to understand and monitor the pulmonary disease viz. pulmonary edema. Modeling refers to the measurement or estimation of the amount of fluid present in the lungs related to the physiology and the work of breathe with respect to lungs. S. Urooj et al. developed a cylindrical model of thorax for estimating fluid volume by using anthropometric dimensions [1]. Several numerical simulation based methods for estimation of

© Springer International Publishing Switzerland 2015   501
S.C. Satapathy et al. (eds.), *Proc. of the 3rd Int. Conf. on Front. of Intell. Comput. (FICTA) 2014*
– *Vol. 2*, Advances in Intelligent Systems and Computing 328, DOI: 10.1007/978-3-319-12012-6_55

thoracic fluids [2] and total body water [3] are encountered in literature. These methods are found tedious and cumbersome, that is why another aspect of volumetric estimation i.e. modeling based approach is looked up with non infrequent geometric shapes in this work. Kinnen constructed a cylindrical thorax model to investigate upon the origin of the impedance signal [4]-[6]. Penny et al. summarized a number of studies and estimated the contributions to the impedance signal [7]. The impedance of the thorax can be considered to be divided into two parts; the impedance of both tissue and fluids. If the patient does not breathe, all components forming the impedance of the thorax are constant, except the amount and distribution of blood. The amount of blood in the thorax changes as a function of the heart cycle. During systole, the right ventricle ejects an amount of blood into the lungs which is equal to the stroke volume, at the same time blood flows from the lungs to the left atrium. The effect of these changes in the distribution of blood in the thorax as a function of the heart cycle and distribution of electric current in inhomogeneous volume conductors [8] can be determined by measuring the impedance changes of the thorax. A LabVIEW based approach has been reported to quantify intra thoracic fluid volume [9]. Several other computer based predictions based on transthoracic electrical impedance and anthropometric dimensions are discussed in [10], [11] and [12]. A microprocessor based approach has been discussed for the diagnosis of pulmonary edema [13]. A simulation tool is developed in DOT.NET, interfaced with an optimization procedure based on classical models of assignment system [14]. A model is introduced and a simulation tool (in JAVA) is developed, based on analytical approach to guide optimization strategy [15]. More research in this area is now desirable with various models in carefully phenotype patients as well as in normal subjects. The main aim of this extension to [1] for volumetric estimation for diagnosis of pulmonary diseases is to educate and prove that the volumes are greatly responsive to the act of breathes, i.e. for inspiration and expiration states. Therefore the proposed model can be adopted for chronic disease diagnosis and other clinical purposes. Researchers proven that the variations in thoracic volumes are greatly responsive to the act of breathing [14]. This model based approach shows the extent of its sensitivity in terms of volumetric variations even in the state of inspiration and expiration itself. The practical way to investigate the function of living organisms is to construct a simple model that follows their operation with reasonable accuracy. Thus, to investigate the volumes with respect to different geometric shapes, it is necessary to construct models that accurately describe the thoracic volumes which can be mathematically analyzed. At rural, backward and poor economic places more than 70% of deaths occur due to unavailability of health care systems and monitoring. The model of Rotachora shapes presented in this paper could be used to determine the amount of fluid present in the thorax. This procedure of volumetric estimation is self-governing and do not require costly and advance equipments. The rest of the paper is organized as follows: Material and methods is discussed in section 2 where as section 3 present the results and discussions.

2   Materials and Methods

The same set of data reported in [1] for simple cylindrical model is considered for two more shapes in the present work. The chosen shapes can be assumed as the thorax with different geometrical angles and dimensions. Rotachora shapes are taken into consideration to validate the previous study of the author(s). Thorax is approximated as cubinder and duo cylinder while considering under Rotachora shapes. These two shapes are chosen because of possible similarities with thorax in four dimension consideration. It can also be defined as the Cartesian product of a circle and a square. Its intersection provides different behavior which is based on the way of insertion into realm space. When it passes through realm space flat-side, the intersection is a cylinder shown in Fig. 1. It projects like the cylinder passing through plane space flat-side first, whose intersection is a circle. When a cubinder is passed through realm space round-side first, the intersection will be a square.

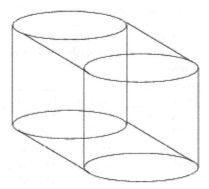

Fig. 1. Cubinder projecting as cylinder

Duo cylinder is a four dimensional shape whose intersection with realm space is a cylinder. As the duo cylinder passes through realm space, the height of the cylindrical intersection changes. Firstly it grows like a circle which converts to a cylinder, then shrinks reverse as a circle again. The dimensions of the chest and thorax are referred as the variables of the cylinder in accordance to the geometrical shapes. The periphery of the chest is referred as the circumference of the cylinder and the length of the thorax or the distance between the outer electrodes are referred as the height of Rotachora. The chest periphery is measured at two positions i.e. at full inspiration and at expiration. The replacement of air inside the thorax and hence lungs is referred as the volume of edema as per the physiology of the lungs. The inspiration impedance is considered when the maximum value of air is filled in, and expiration impedance is recorded when the entire air is breathed out. The lungs are assumed to be empty in this state of expiration.

3   Results and Discussion

The anthropometric dimensions and measured impedances are shown in Table 1. For each of the six subjects three sets of reading were taken and mean of the three readings are used to compute thoracic volume [1]. Four dimensional Rotachora models are considered as cubinder in the phase of study and duo cylinder in the second stage. The standard cubinder dimensions are given as:

$$x^2 + y^2 = 1, \ |z| = 1, \ |w| = 1$$

Standard volume of the cubinder,

$$V = 2\pi r h(h + 2r)$$

Table 1. Anthropometric Parameters & Corresponding Thoracic Impedances

Circumference of chest at resting and breathe hold position		Distance between inner electrodes	Radius of chest corresponds to C1 and C2		Thoracic Impedance in ohms	
C_1(mm)	C_2(mm)	H (mm)	R_1 (mm)	R_2(mm)	Z_1	Z_2
851	889	343	135.441	141.489	55.16	57.66
863	889	355	137.351	141.489	54.66	57.16
978	1003	355	155.654	159.632	55.16	59.33
901	927	381	143.399	147.537	52.00	56.66
938	965	368	149.287	153.585	53.66	57.66
952	978	381	151.516	155.654	49.83	53.16

Four dimensional Rotachora model is considered as cubinder in the first phase of study and as duo cylinder in the second stage. The standard duo cylinder is bounded by two equivalent mutually perpendicular circular tore and is defined as:

$$x^2 + y^2 \le r^2, \ z^2 + w^2 \le r^2$$

Volume of the duo cylinder:

$$V = 4\pi^2 r^3$$

Using the standard cubinder and duo cylinder formulae the thoracic volumes are calculated. Output performance measurements for cubinder and d models are given in Table 2.

Table 2. Output Performance Measurements for Cubinder & Duocylinder Models

Volume of thorax corresponding to (for cubinder		Volume of thorax corresponding and C_2 for duo cylinder	
V_{c1}	V_{c2}	V_{d1}	V_{d2}
179.097208	190.781171	97.987289	111.708783
192.821251	201.240958	102.191498	111.708783
231.218889	239.959487	148.730537	160.427624
229.126782	238.660091	116.294093	126.654940
229.973515	239.645568	131.215536	142.878109
247.981644	257.836416	137.181257	148.730537

The graph of Fig. 2 and Fig. 3 shows the relationship with thoracic volume and thoracic impedance. It is observed from these graphs that the volumes are varying in rhythm with thoracic impedance for both the cases. This proves that the chosen models are greatly responsive to the act of breath as an observable variation is found against thoracic impedances and thoracic volumes by using the new 4-dimensional models. This significantly validates the previous results of [1]. The calculated results may employ for quantifiable investigations and research purposes for the work of breathe and related analysis and control.

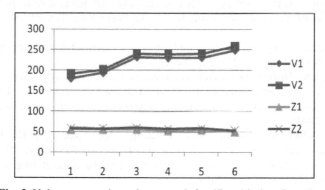

Fig. 2. Volume versus impedance graph for 4D cubinder –Rotachora

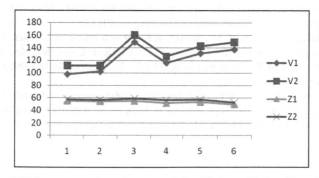

Fig. 3. Volume versus impedance graph for 4D duo cylinder –Rotachora

References

1. Urooj, S., Ekuakille, A.L., Ansari, A.Q., Khan, M., Vergallo, P., Trotta, A.: Volumetric Estimation of Thorax with Cylindrical Model and Anthropometric Measurements. In: IEEE International Symposium on Medical Measurements and Application, Bari, May 30-31, pp. 213-315 (2011) ISBN 978-4244-9337-1/11
2. Iskander, M.F., Maini, R., Durney, C.H., Bragg, D.: A microwave method for measuring changes in lung water content: numerical simulation. IEEE Trans. Biomed. Eng. 28(12), 797–804 (1981)
3. Kushner, R.F., Shoeller, D.: Estimation of total body water by bioelectrical impedance analysis. Am. J. Clin. Nutr. 44, 417–424 (1986)
4. Kinnen, E., Kubicek, W.G., Hill, D.W., Turton, G.: Thoracic cage impedance measurements: Impedance plethysmographic determination of cardiac output (A comparative study). U.S. Air Force School of Aerospace Medicine, Brooks Air Force Base, Texas SAM-TDR-64 8(15) (1964a)
5. Kinnen, E., Kubicek, W.G., Hill, D.W., Turton, G.: Thoracic cage impedance measurements: impedance plethysmographic determination of cardiac output (An interpretative study). U.S. Air Force School of Aerospace Medicine, Brooks Air Force Base, Texas SAM-TDR-64 12(23) (1964b)
6. Kinnen, E., Kubicek, W.G., Hill, D.W., Turton, G.: Thoracic cage impedance measurements, tissue resistivity in vivo and transthoracic impedance at 100 kc. U.S. Air Force School of Aerospace Medicine, Brooks Air Force Base, Texas SAM-TDR-64 14(5) (1964c)
7. Penney, B.C.: Theory and cardiac applications of electrical impedance measurements. CRC Crit. Rev. Bioeng. 13, 227–281 (1986)
8. Malmivuo, J.A.: Distribution of electric current in inhomogeneous volume conductors. In: Lahtinen, T. (ed.) Proceedings of the 8th Internat. Conference on Electrical Bio-Impedance, University of Kuopio, Center for Training and Development, Kuopio, Finland, pp. 18–20 (1992)
9. Urooj, S., Khan, M., Ansari, A.Q., Ekuakille, A.L., Salhan, A.K.: Prediction of Quantitative Intrathoracic Fluid Volume to Diagnose Pulmonary Oedema Using LabVIEW. Computer Methods in Biomechanics and Biomedical Engineering (2011), doi:10.1080/10255842.2011.565054
10. Urooj, S., Khan, M.: A Computer Based Prediction for Diagnosis of Pulmonary Edema. IEEE Xplore Digital Library, 28–31 (2010), doi:10.1109/MEMEA.2010.5480229, ISBN: 978-1-4244-6288-9
11. Urooj, S., Khan, M., Ansari, A.Q.: A Computational Study using Electrical Impedance and Anthropometric Parameters. In: National Conference PICON Februaary (2011)
12. Urooj, S., Ansari, A.Q., Khan, M., Salhan, A.K.: Measurement of Thoracic Impedance and Approximations: A Diagnosis Technique for Clinical Utilization. Indian Journal of Industrial & Applied Mathematics 3(2), 85–93 (2012) Print ISSN: 0973-4317, Online ISSN: 1945-919X
13. Khan, M., Urooj, S., Hara, R.O., Pohman, R.: A Microprocessor Based System for Non Invasive Measurement of High Altitude Pulmonary Edema, Aviation, Space and Environmental Medicine. Official Journal of the Aerospace Medical Association 78(3), 242 (2007)

14. Urooj, S., Khan, M., Ansari, A.Q.: Thorax: Physiological Monitoring and Modeling For Diagnosis of Pulmonary Edema. International Journal of Measurement Technologies& Instrumentation Engineering 1(2), 54–60 (2011)
15. Saxena, P., Sharma, L.: Simulation Tool for Queuing Models: QSIM. International Journal of Computers and Technology 5(2), 74–79 (2013)
16. Saxena, P., Anjali.: Simulation Tool for assignment Models: SIMASI. International Journal of Computers and Technology 13(8), 4723–4728 (2014)

A Review of ROI Image Retrieval Techniques

Nishant Shrivastava and Vipin Tyagi

Department of Computer Science and Engineering,
Jaypee University of Engineering and Technology,
Raghogarh, Guna - (MP), 473226 India
dr.vipin.tyagi@gmail.com

Abstract. Content based image retrieval involves extraction of global and region features of images for improving their retrieval performance in large image databases. Region based feature have shown to be more effective than global features as they are capable of reflecting users specific interest with greater accuracy. However success of region based methods largely depends on the segmentation technique used to automatically specify the region of interest (ROI) in the query. Apart from this user can also specify ROI's in an image. The ROI image retrieval involves the task of formulation of region based query, feature extraction, indexing and retrieval of images containing similar region as specified in the query. In this paper state-of-the-art techniques for ROI image retrieval are discussed. Comparative study of each of these techniques together with pros and cons of each technique are listed. The paper is concluded with our views on challenges faced by researchers and further scope of research in the area. The major goal of the paper is to provide a comprehensive reference source for the researchers involved in image retrieval based on ROI.

Keywords: Regions of interest, Content based image retrieval, similarity measure.

1   Introduction

With the advancement in internet and multimedia technologies, a huge amount of multimedia data in the form of audio, video and images has been used in many fields like medical treatment, satellite data, video and still images repositories, digital forensics and surveillance system. This has created an ongoing demand of systems that can store and retrieve multimedia data in an effective way. Many multimedia information storage and retrieval systems have been developed till now for catering these demands.

The most common retrieval systems are Text Based Image Retrieval (TBIR) systems, where the search is based on automatic or manual annotation of images. A conventional TBIR searches the database for the similar text surrounding the image as given in the query string. The commonly used TBIR system is Google Images. The text based systems are fast as the string matching is computationally less time consuming process. However, it is sometimes difficult to express the whole visual content of images in words and TBIR may end up in producing irrelevant results. In addition annotation of images is not always correct and consumes a lot of time. For finding the alternative way of searching and overcoming the limitations imposed by TBIR systems more intuitive and user friendly content based image retrieval systems (CBIR) were developed. A CBIR system uses visual contents of the images described in the form of low level features like color, texture, shape and spatial locations to represent the images in the databases. The system retrieves similar images when an example image or sketch is presented as input to the system. Querying in this way eliminates the need of describing the visual content of images in words and is close to human perception of visual data. Some of the representative CBIR systems are QBIC [24], Photobook [25], Virage [26], VisualSeek [27], Netra [28] and SIMILIcity[18] etc.

One of the category of CBIR system is Region Based Image Retrieval Systems (RBIR). These system utilizes features extracted from region or part of the image to represent images in the database. RBIR systems can further be classified in to System Designated ROI (SDR) approaches and User Designated ROI(UDR) approaches. In SDR approaches [4, 6, 10, and 17] system automatically specify ROIs by dividing the image in to significant regions and designate each of these regions as ROI's for querying to the database. Whereas UDR methods [1-3,5] facilitate the user to manually select ROI's in the image for query formulation. Success of SDR methods relies heavily on the accuracy of segmentation technique used to divide images in to regions. However segmentation of images is not always reliable as it can introduce unexpected noise in the output and result in reduction of retrieval accuracy. In addition existing segmentation techniques can accurately identify regions but fails to extract objects of interest from the images. If the system automatically designates ROIs, they may not correspond to the regions that the user wishes to retrieve. Due to these reasons SDR approach is limited in reflecting the user intent in the process of retrieval.

If the user selects ROI manually as in UDR approaches it is impossible to tell in advance which part of the image will be selected. To deal with this problem existing studies divide the image in to small number of blocks to extract their feature values and match them with ROIs for retrieval. Here, how to match ROIs and the block is a problem. Since UDR can be of arbitrary size and may encompass more than one block therefore selection of appropriate blocks overlapping with ROI is also important to reflect the user intent accurately. For effective ROI image retrieval reflecting the location of ROIs is also important. In that case blocks having the same location as of ROI are only compared which result in fixed location matching. This approach fails to retrieve similar images when regions similar to ROIs lies in different parts of the database images. For example if the user query for horse in the left corner of image then the system fails to retrieve similar images containing horses in the right corner or other areas of image. This problem can be solved by opting for all blocks matching strategy. However the time complexity of this approach increases with the increase in dimension of layout [3].

Considering relative locations of multiple ROIs is also an important issue for improving the retrieval accuracy. Lee and Nang [7] have devised an algorithm for comparing relative locations of multiple ROIs in which locations of other ROI is determined using predefined location of basis ROI. Finding relative locations is a difficult task and require a complex algorithm which result in increasing the reponse time of retrieval system.

Selection and extraction of features to represent different regions plays an important role in improving the efficacy of SDR and UDR systems. Traditionally color and texture features like MPEG-7 dominant color, color saliency map [14] and Gray level co- occurrence matrix [10], Local binary patterns [19,40] are used in RBIR. An effective region descriptor should be of fewer dimensions and have high discriminating power.

Further in this survey, Traditional framework of RBIR is described in section 2. Section 3 describe the existing techniques together with their pros and cons. Conclusions are summed up in section 4.

2   Framework for ROI Image Retrieval

Fig.1 shows the block diagram of ROI oriented image indexing and retrieval process. In the offline mode images from the dataset are selected one by one and depending upon the type of approach used (SDR or UDR) ROIs are extracted. For SDR system regions or objects recognized by segmentation algorithm acts as potential ROIs. Whereas in UDR systems features of all blocks in the layout images are used to index images in the database. Generally, 3×3 and 5×5 layout is used in UDR systems.

In SDR approaches, an online mode uses same set of steps as were used in offline mode for feature extraction. While UDR approach extract features of only those blocks in the layout which are having full or partial overlap with the UDR(ROBs).The decision of selecting ROI overlapping blocks   depends on many parameters like area of overlap, percentage of overlap etc. Internally these ROBs represent UDR in this approach.

3   System Designated ROI (SDR) Aproaches

In SDR systems, ROIs are extracted automatically using segmentation technique which may result in too many ROIs to be used for effective retrieval. In case of more ROIs global matching is preferred over region based matching. For limiting the number of ROIs Zhang et.al[10], firstly divided image to 32×32 segments with size 4×4, for obtaining accurate image segmentation; and then compute average gray value of every segment; since only three ROIs are considered, so the average gray values are segmented into three groups using K-means clustering algorithm. After segmentation, there are just three values in the segmentation image. Color features based on hue histograms and texture feature based on four directions gray level co-occurrence matrices are extracted for every ROI, and are used for indexing and similarity comparison.

Chan et al. [6] have proposed a ROI image retrieval method based on Color variances among adjacent objects (CVAAO) feature. The color histogram formed using differences of color between two adjacent objects in an image. The feature describe the main color and the texture distribution together with objects of inconsistent contour in the image. However computation of CVAAO feature is time consuming and the method fails when object and background of image have the same color.

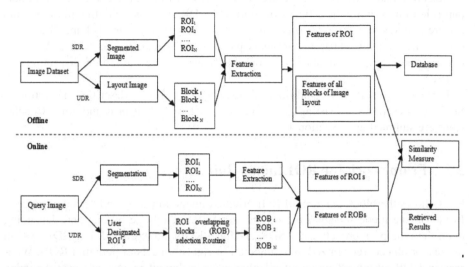

Fig. 1. ROI image retrieval framework for SDR and UDR in both offline and online modes

Reference [4] uses the dominant colors of images to automatically extract regions. This method employs color, shape and location as feature values. 25 colors selected here are mapped to be compared by differences in color histogram value; to locate individual regions, images are divided into 3×3 blocks as illustrated in Figure 2, and the number of block with the largest area of region is designated. For instance, the location of region in Figure 2 would be indexed as "5" In other words; the regions are compared only from fixed locations, as the locations are compared by index number. This method has a problem that is not directly selected by the user, for the regions are automatically classified from the images by color. Also, the method does support multiple regions but merely compares the absolute location of blocks as the locations are compared by index number.

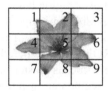

Fig. 2. Image showing location index of different blocks

For reducing the segmentation noise and computation time required for similarity matching between database image regions and ROIs Byoung Chul Ko[8] have proposed adaptive circular filters based on Bayes' theorem and image texture distribution for semantically meaningful image segmentation. Also, optimal feature vectors are extracted from segmented regions and are used for similarity matching using stepwise Boolean AND matching scheme for reducing the computational complexity.

Huang et al [14] uses visual saliency in HSV color space for ROI extraction from color images. Color saliency is calculated by a two-dimensional sigmoid function using the saturation component and brightness component, to identify regions with vivid color. Discrete Moment Transform (DMT)-based saliency can determine large areas of interest. A visual saliency map is obtained by combining color saliency and DMT-based saliency, which is denoted the S image. A criterion for the local homogeneity called the E image is calculated in the image. Based on S image and E image, the high visual saliency object seed points set and low visual saliency object seed points set are determined. The seeded regions growing and merging are used to extract regions of interest. In addition to these there are many region based image retrieval systems, reported in literature; some most widely used systems are Blob world [17] and Simplicity [18].

4   User Designated ROI (UDR) Approaches

UDR approaches enable the user to manually select ROIs in the query image. To implement this images are divided in to fix number of small blocks [3-5]. Reference [3] enables the user to select ROIs on his/her own. It is inefficient to extract the feature values of ROIs randomly selected by the user on a real-time basis. Therefore [3] divides images into blocks of certain size (e.g. 2×2, 3×3, 4×4, 5×5) and defines ROIs as blocks that overlap with user-selected ROIs, in an effort to calculate ROI similarity based on the feature values extracted per block in advance. In this case, the user selected ROIs and the blocks may not be perfectly identical. To address this problem, [3] reflects the proportion of overlap between ROIs and blocks (see Fig. 3). In other words, for blocks overlapping with ROIs in part, their feature values are reflected on similarity measurement by the proportion of overlap.

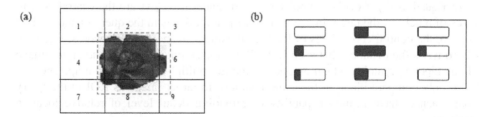

Fig. 3. (a) Division of image in blocks of size 3 X 3 [3]. (b) Reflection of the proportion of ROI overlapping blocks [3].

$$D_j(Q, I^j) = \sum_n \sum_i \lambda W_{n,i} S^j(n,i) \quad j = 1......M \qquad (1)$$

Equation (1) divides the query image Q and the jth image of DB, I j into n blocks and extracts i feature values from each block to calculate the similarity. M is the number of entire images, and S j (n, i) is a function that measures the distance be-tween Q and the i th feature of the nth block of I j . W n, i , is the weight of the ith feature of the nth block; λ ,which is reflected together with W n, i, is the proportion of overlap between the ROIs and the blocks. This method, however, merely measures the distance of blocks in the target image that are in the same location as in the query image, without considering blocks in different locations. In this case, blocks in differ-ent locations that are similar to the ROIs are not retrieved.

Vu et al. [9] proposes a SamMatch based ROI image retrieval technique. The scheme quantifies each image in to 256 colors and   resize them to 256×256. The average color of each 16×16 block of pixels is considered to be the color of the block. The similarity between two arbitrary shaped sub-images Q and S, each represented by n sampled blocks, was given by:

$$Sim\ (Q,S) = \sum_{i=1}^{n} \frac{W_i}{1 + D(C_i^S, C_i^Q)} \tag{2}$$

Where $D(C_i^S, C_i^Q)$ is the distance of C_i^S and C_i^Q. C_i^S is the color of block i of S and C_i^Q is that of Q. The parameter W_i is a weight factor. Since SamMatch com-pares the corresponding sampled blocks of sub images, it involves implicitly the shape, size, and texture features of the image objects. SamMatch based technique share the benefits of region based retrieval and also consider color, texture and shape features of image.

Reference [5] enables the user to select multiple ROIs and retrieves blocks in dif-ferent locations from the ROIs'. The multiple ROIs in query and blocks in target im-ages are also checked for similarity in their spatial location. Here the similarity of spatial layouts for ROIs is compared using Equation (3).

$$S(Q,T) = \sum\sum f(x_i^t - x_j^t) sign(x_i^q - x_j^q) + f(y_i^t - y_j^t) sign(y_i^q - y_j^q) \tag{3}$$

In Equation (3), f (x) is a bipolar sigmoid function. xt , yt are the central coordi-nates of blocks in the target image that correspond to ROIs in the query image, and xq , yq are the central coordinates of ROIs in the query image. In other words, Equation (2) converts the distances between ROIs in the query image (Q) and the target image (T) as bipolar sigmoid and sin function values to multiply and sum them up. The me-thod, only compare block in the same locations in target image as ROI in the query image hence, have limited capability of providing detail level of relative location similarity.

In order to incorporate relative locations of multiple ROIs in the ROI retrieval Lee and Nang [7] have proposed a new method where similarity is computed using rela-tive distribution of ROI in the image layout. The method uses MPEG-7 dominant color to represent each block feature in the layout. The appropriate ROI overlapping blocks are selected by finding the blocks having higher overlapping area than some predefined threshold. The relative location was computed in four directions (i.e. up, down, left and right) from the basis ROI. The similarity between query and target image is calculated as follows

$$MD(R,I^j) = w_1 \sum_{k=1}^{r} SD(R_b^k, I^j) + w_2 \sum_{k=1}^{r-1} LD(R_b^k, I^j) \tag{4}$$

In Equation (4), $MD(R,I^j)$ calculates the degree of similarity between the query image's ROI combination (R) and the jth image of the database (I^j). Here the degree of similarity is calculated as the weighted sum of the distance between feature values and the distance at which the relative location is measured. The similarity measure is higher when the distance is nearer. r refers to the number of ROIs; R_b^k is the list of blocks in the query image that correspond to the kth ROI. In Equation (4) $SD(R_b^k, I^j)$ is calculated by summing up each ROI's similarity measure. A function that measures the relative location of ROIs $LD(R_b^k, I^j)$ is calculated using Equation (5).

$$LD(R_b^k, I^j) = \sum_{s=k+1}^{r} rpos(I_b^k, I_s^j) \tag{5}$$

The value of $rpos(I_b^k, I_s^j)$ is 0 when the relative location is same and is 1 when relative location is different. However considering relative location may results in increase of computation time and complexity. The retrieval performance of each method is compared through experimentation by using 50 MPEG-7 CCQ images. Table 1 shows the results of retrieval performance comparison. It is obvious that method [7] which compares relative location of multiple ROIs perform better than other methods.

Shrivastava et al. [19] have proposed more effective approach considering relative locations of multiple ROI using binary region codes. Initially the image is divided in to 3 × 3, 5 × 5 blocks and region codes are assigned to each block depending on its location with respect to central region. The region codes consists of 4 bits with each bit specify left, right, bottom and top region of the image respectively, starting from the left most least significant bit. Fig.4 shows the region codes assignment for 3×3 layout.

The central region is assigned code 1111 as an exception since its direction cannot be determined. Further to avoid the effect of noise in query formulation ROI overlapping block selection scheme is also proposed. To ensure this blocks having same dominant color as of block with largest ROI overlapping area are only retained for final query formulation.

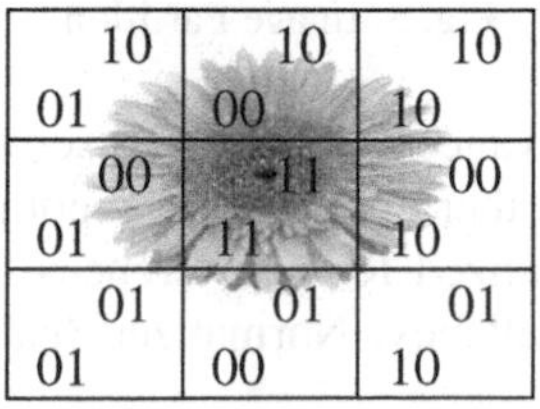

Fig. 4. Region code assignment for image layout 3 × 3

These blocks are compared only with blocks of database images having region code similarity. The similarity between region codes is determined using result of logical AND operation of two or more region codes. If the result contains 1 at any bit position then the region codes are similar. This ensures fewer comparisons than all block matching scheme [5, 7]. The region code scheme allows comparison only in the regions which are related to the location of user designated ROI. The technique was further enhanced for multiple ROI based retrieval by assigning priority for higher bit position similarity. The proposed scheme outperforms others in accuracy and compu-tation time. Table 1 gives the comparison of retrieval accuracy of different scheme on MPEG -7 CCD dataset. The dataset consist of 5000 images coming from consecutive frame of TV shows, newscast and sports show. Averaged Normalized Modified Retrieval Rank (ANMRR) is used as performance measure for comparing the retrieval performance. A set of 50 common color queries (CCQ) together with their specified groundtruth as provided in Mirror [1] image retrieval system are used for performing experiments.

Table 1. Retrieval Performance of ROI overlapping block selection techniques

	Fixed [4]	Moghaddam et al. [5]	Lee and Nang [7]	Shrivastava et al.[19]
ANMRR	0.563	0.528	0.415	0.347

Hsiao et al. [13] approach partitions images into a number of regions with fixed ab-solute locations as shown in Fig. 5 each region is represented by its low-frequency DCT coefficients in the YUV color space. Two policies are provided in the matching procedure: local match and global match. In the local match, the user formulates a query by selecting the interested region in the image. Candidate images are then ana-lyzed, by inspecting each region in turn, to find the best matching region with the query region.

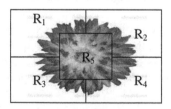

Fig. 5. Image Partition

Zhou et al. [12] utilized relevance feedback to discover subjective ROI perceptions of a particular user and it is further employed to recompute the features representing ROIs with the updated personalized ROI preferences. ROI are represented by color saliency and wavelet feature saliency. Normalized features are selected to represent shape features.

A novel approach to image indexing by incorporating a neural network model, Ko-honen's Self Organising Map (SOM), for content based image retrieval is presented in

[11]. When a user defines the ROI in the image the similarity is calculated by matching similar region in query and target images. The SOM algorithm is used to determine the homogeneous regions in the image adaptively. An unsupervised clustering and classification is used to group pixel level features without prior knowledge of data.

To reduce the effect of noisy descriptors generated in ROI query Wang et al. [15] have proposed a new approach using a general bag of words and auxiliary Gaussian weighting scheme(AGW) for ROI based image retrieval system. Weight of each descriptor is assigned according to its distance from ROI query centre computed with the help of 2-D Gaussian window. The Simiarity score of database images is computed using AGW scheme.

Existing object retrieval methods may not perform satisfactorily in cases where ROI specification is not accurate or its size is too small to identify it using discriminative features. In order to improve the object retrieval performance also in these difficult cases, Yang et al. [16] have suggested to use visual context of the query object to compensate for the inaccurate representation of query ROI. The context of the query object is determined using visual content surrounding it. Here ROI is determined as an uncertain observation of the latent search intent and the saliency map detected for the query image as a prior.All UDR approaches are sensitive to large shifts, cropping, and scaling of regions.

5   Bridging Semantic GAP

RBIR systems can more accurately represent the semantics of an image but still research is going on to improve the existing systems for reducing the semantic gap. Some of the state-of-the-art techniques used for reducing semantic gap are discussed here. Some techniques employ machine learning tools to obtain high level semantic concepts from low level features of segmented image regions [20, 23, 30-34]. For example, in [20], Liu et al applied a decision tree based learning algorithm (DT-ST) to make use of the semantic templates to discretize continuous-valued region features. Fei-Fei et al. [33] proposed an incremental Bayesian algorithm to learn generative models of object categories and tested it on images of 101 widely diverse categories. Methods based on relevance feedback [21, 35-39] are also used to reflect the user intention using feedback loop in the retrieval process. For example in reference [21] query point movement algorithm is proposed for positive example scenario. To speed up the feedback process incremental clustering algorithm is used. Zhang et al. [22] proposed image translation into textual documents which are then indexed and retrieved the same way as the conventional text based search.

6   Conclusions

In this study, we have discussed and presented some state of the art techniques for ROI image retrieval reported in the literature. Further we have generalized the overall framework for ROI image retrieval and discussed various problem faced and there possible solutions as suggested by various researchers. From the study it has been depicted that there are many problems related with ROI image retrieval which have not been answered satisfactorily till now. Some of them are (i) Accurately reflecting the user intent in query formulation (ii) Effective technique for selection of ROI overlapping blocks (iii) Technique for considering relative locations of multiple ROIs and (iv) Reducing the overall computation time for region matching without affecting the accuracy of system. These challenges have to be researched to further improve the existing systems in this area.

References

1. Wong, K.-M., Cheung, K.-W., Po, L.-M.: MIRROR: An Interactive Content Based Image Retrieval System. In: Proc. of IEEE Int. Symposium on Circuits and Systems (ISCAS 2005), vol. 2, pp. 1541–1544 (2005), http://dx.doi.org/10.1109/ISCAS.2005.1464894
2. Broek, E.L., Kisters, P.M.F., Vuurpijl, L.G.: The utilization of human color categorization for content-based image retrieval. In: Proc. of the SPIE, vol. 5292, pp. 351–362 (2004)
3. Tian, Q., Wu, Y., Huang, T.S.: Combine User Defined Region-of-Interest and Spatial Layout for Image Retrieval. In: Proc. of IEEE Int. Conf. on Image Processing (ICIP 2000), vol. 3, pp. 746–749 (2000), http://dx.doi.org/10.1109/ICIP.2000.899562
4. Prasad, B.G., Biswas, K.K., Gupta, S.K.: Region-Based Image Retrieval using Integrated Color, Shape and Location Index. In: Computer Vision and Image Understanding, vol. 94, pp. 193–233 (2004), http://dx.doi.org/10.1016/j.cviu.2003.10.016
5. Moghaddam, B., Biermann, H., Margaritis, D.: Regions-of-Interest and Spatial Layout for Content-Based Image Retrieval. Multimedia Tools and Applications 14(2), 201–210 (2001), http://dx.doi.org/10.1023/A:1011355417880
6. Chan, Y.-K., Ho, Y.-A., Liu, Y.-T., Chen, R.-C.: A ROI image retrieval method based on CVAAO. Image and Vision Computing 26, 1540–1549 (2008)
7. Lee, J., Nang, J.: Content-Based Image Retrieval Method using the Relative Location of Multiple ROIs. Advances in Electrical and Computer Engineering 11(3), 85–90 (2011)
8. Ko, B.C., Byun, H.: FRIP: A Region-Based Image Retrieval Tool Using Automatic Image Segmentation and Stepwise Boolean AND Matching. IEEE Transactions on Multimedia 7(1) (February 2005)
9. Vu, K., Hua, K.A., Tavanapong, W.: Image retrieval based on regions of interest. IEEE Transactions on Knowledge and Data Engineering 15(4), 1045–1049 (2003)
10. Zhang, J., Yoo, C.-W., Ha, S.-W.: ROI Based Natural Image Retrieval using Color and Texture Feature. Fuzzy Systems and Knowledge Discovery (2007)
11. Chen, T., Chen, L.-H., Ma, K.-K.: Colour Image Indexing Using SOM for Region-of-Interest Retrieval. Pattern Analysis & Applications 2, 164–171 (1999)
12. Zhou, Q., Ma, L., Celenk, M., Chelberg, D.: Content-Based Image Retrieval Based on ROI Detection and Relevace Feedback. Multimedia Tools and Application 27, 251–281 (2005)
13. Hsiao, M.-J., Huang, Y.-P., Tsai, T., Chiang, T.-W.: An Efficient and Flexible Matching Strategy for Content-based Image Retrieval. Life Science Journal 7(1) (2010)
14. Huang, C., Liu, Q., Yu, S.: Regions of interest extraction from color image based on visual saliency. Journal of Supercomp., doi:10.1007/s11227-010-0532-x.

15. Wang, Z., Liu, G., Yang, Y.: A New ROI Based Image Retrieval System using an auxiliary Gaussian Weighting Scheme. Multimedia Tools Application (2012), doi:10.1007/s11042-012-1059-3.
16. Yang, L., Geng, B., Cai, Y., Hua, X.-S.: Object Retrieval Using Visual Query Context. IEEE Transactions on Multimedia 13(6) (December 2011)
17. Carson, C., Thomas, M., Belongie, S., Hellerstein, J.M., Malik, J.: Blobworld:Image Segmentation Using Expectation-Maximization and Its Application to Image Querying. IEEE Transactions on Pattern Analysis and Machine Intelligence 24(8), 1026–1038 (2002)
18. Wang, J.Z., Li, J., Wiederhold, G.: SIMPLIcity: Semantics-Sensitive Integrated Matching for Picture Libraries. IEEE Transactions On Pattern Analysis and Machine Intelligence 23(9) (September 2001)
19. Shrivastava, N., Tyagi, V.: Content based image retrieval based on relative locations of multiple regions of interest using selective regions matching. Inform. Sci. 259, 212–224 (2013), http://dx.doi.org/10.1016/j.ins.2013.08.043
20. Liu, Y., Zhang, D., Lu, G.: Region-based image retrieval with high-level semantics using decision tree learning. Pattern Recognition 41, 2554–2570 (2008)
21. Jing, F., Li, M.: Relevance Feedback in Region-Based Image Retrieval. IEEE Transactions on Circuits and Systems for Video Technology 14(5) (May 2004)
22. Zhang, D., Islam, M.M., Lu, G., Hou, J.: Semantic Image Retrieval Using Region Based Inverted File. In: Proceedings of Digital Image Computing: Techniques and Applications, pp. 242–249 (2009)
23. Li, W.-J., Yeung, D.-Y.: Localized Content-Based Image Retrieval Through Evidence Region Identification. In: IEEE Conference on Computer Vision and Pattern Recognition, CVPR 2009, pp. 1666–1673 (2009)
24. Faloutsos, C., Barber, R., Flickner, M., Hafner, J., Niblack, W., Petkovic, D., Equitz, W.: Efficient and effective querying by image content, J. Intell. Inf. Syst. 3(3-4), 231–262 (1994)
25. Pentland, A., Picard, R.W., Scaroff, S.: Photobook: content-based manipulation for image databases. Int. J. Comput. Vision 18(3), 233–254 (1996)
26. Gupta, A., Jain, R.: Visual information retrieval, Commun. ACM 40(5), 70–79 (1997)
27. Smith, J.R., Chang, S.F.: Visualseek: a fully automatic content-based query system. In: Proceedings of ACM International Conference on Multimedia, pp. 87–98 (1996)
28. Ma, W.Y., Manjunath, B.: Netra: a toolbox for navigating large image databases. In: Proceedings of International Conference on Image Processing, pp. 568–571 (1997)
29. Wang, J.Z., Li, J., Wiederhold, G.: Simplicity: semantics-sensitive integrated matching for picture libraries. IEEE Trans. Pattern Mach. Intell. 23(9), 947–963 (2001)
30. Jing, F., Li, M., Zhang, L., Zhang, H., Zhang, B.: Learning in region-based image retrieval. In: Proceedings of International Conference on Image and Video Retrieval (CIVR 2003), pp. 206–215 (2003)
31. Town, C.P., Sinclair, D.: Content-based image retrieval using semantic visual categories, Society for Manufacturing Engineers, Technical Report MV01 211 (2001)
32. Cao, L., Fei-Fei, L.: Spatially coherent latent topic model for concurrent object segmentation and classification. In: Proceedings of IEEE International Conference in Computer Vision, ICCV (2007)
33. Fei-Fei, L., Fergus, R., Perona, P.: Learning generative visual models from few training examples: an incremental Bayesian approach tested on 101 object categories. In: Proceedings of Computer Vision and Pattern Recognition, Workshop on Generative-Model Based Vision, pp. 178–185 (2004)

34. Chang, E., Tong, S.: SVM active—support vector machine active learning for image retrieval. In: Proceedings of ACM International Multimedia Conference, pp. 107–118 (October 2001)
35. Nguyen, G.P., Worring, M.: Relevance feedback based saliency adaptation in CBIR. ACM Multimedia Syst. 10(6), 499–512 (2005)
36. Mezaris, V., Kompatsiaris, I., Strintzis, M.G.: An ontology approach to object-based image retrieval. In: Proceedings of International Conference on Image Processing, pp. 511–514 (2003)
37. Tao, D., Tang, X., Li, X., Wu, X.: Asymmetric bagging and random subspace for support vector machines-based relevance feedback in image retrieval. IEEE Trans. Pattern Anal. and Mach. Intel. (TPAMI) 28(7), 1088–1099 (2006)
38. Tao, D., Tang, X., Li, X., Rui, Y.: Kernel direct biased discriminant analysis: a new content-based image retrieval relevance feedback algorithm. IEEE Trans. Multimedia (TMM) 8(4), 716–727 (2006)
39. Tao, D., Li, X., Maybank, S.J.: Negative samples analysis in relevance feedback, IEEE Trans. Knowl. IEEE Trans. Knowl. Data Eng. 19(4), 568–580 (2007)
40. Shrivastava, N., Tyagi, V.: An effective scheme for image texture classification based on binary local structure pattern. Visual Computer (2013), http://dx.doi.org/10.1007/s00371-013-0887-0

A Novel Algorithm for Suppression of Salt and Pepper Impulse Noise in Fingerprint Images Using B-Spline Interpolation

P. Syamala Jaya Sree[1], Prasanth Kumar Pattnaik[1], and S.P. Ghrera[2]

[1] School of Computer Engg, KIIT University, Bhubaneshwar, India
[2] Jaypee University of IT, Waknaghat, Himachal Pradesh, India
{jayasree.syamala,patnaikprasant,spghrera1}@gmail.com

Abstract. The quality of Finger Print Images in image forensics plays a vital role in the the accuracy of biometric based identification and authentication system. To suppress the salt and pepper noise in fingerprint images, B-Splines have been used for interpolation. In this paper, a two stage novel and efficient algorithm for suppression of salt and pepper impulse noise for noise levels ranging from 15 % to 95 % using B-splines interpolation is being proposed. The algorithm removes salt and pepper impulse noise from the image in the first stage and in second stage, an edge preserving algorithm has been proposed which regularizes the edges that have been deformed during noise removal process.

Keywords: Image Forensics, Fingerprint Images, B-Spline Interpolation.

1 Introduction

The success in Fingerprint based recognition depends on the fact that how accurate is the biometric image system free of noise. In security, automatic identification and authentication systems needs to be noise free so that the efficiency of finger print images used in Biometrics is increased [1]. A biometric system is basically a pattern recognition system that works on extracting biometric data such as fingerprints, iris, facial features etc. from a person, making feature set from the extracted data and comparing this feature set with template set stored in database [2]-[4]. Fingerprint images may get corrupted by noise which degrades the quality of the image by replacing original pixel values with new ones and hence results in the loss of original information. Impulse noise like the Salt and Pepper corrupts the finger print images mainly during acquisition process or transmission process. i.e. salt and pepper noise also known as fixed value noise because the noisy pixel candidate can take value either 0 (Pepper) or 255 (Salt) [6]. To maintain the efficiency and accuracy of biometric systems by extracting correct details from the images, it is necessary to remove the salt and pepper impulse noise from the fingerprint images.

The salt and pepper impulse noise removal in normal images has been dealt very efficiently using Second generation wavelets and B-Splines [5, 7]. An expeditious and

© Springer International Publishing Switzerland 2015   521
S.C. Satapathy et al. (eds.), *Proc. of the 3rd Int. Conf. on Front. of Intell. Comput. (FICTA) 2014*
– *Vol. 2*, Advances in Intelligent Systems and Computing 328, DOI: 10.1007/978-3-319-12012-6_57

novel algorithm of removing salt and pepper noise along with edge preservation is discussed in [6]. Many researchers have tried to solve the problem of removing salt and pepper impulse noise in normal images like Lena, Peppers, Barbara etc, but very few could really propose efficient algorithms to remove the same in Finger Print Images related to Image Forensics. Hence this was the main motivation for this work and in this paper, fingerprint image corrupted with salt and pepper impulse noise is considered and processed using B-Spline Interpolation algorithm for effective noise removal.

The organization of the paper is as follows: In section 2, the proposed novel algorithm for suppression of Salt and pepper Impulse Noise in fingerprint images is discussed. In section 3, performance measures like Peak Signal to Noise Ratio (PSNR) and MSE and SSIM are discussed. Section 4 comprises of simulation results and discussion and Section 5 concludes the work.

2   Proposed B-Spline Interpolation Algorithm

The Cubic B splines [8] are used for interpolation in the proposed algorithm. B-Spline are used by researchers like Unser in signal and image processing [9]. B-Splines are used for interpolation because they have compact support passes through control point and have local propagation property. The idea of interpolation has been taken from Kireeti Bodduna and Rajesh Siddavatam [10] where they have applied the below methodology to solve the problem of random valued Impulse Noise removal using Cardinal Spline Interpolation [10].

Stage 1:
The B-splines [8] are used for the interpolation and removal of salt and pepper impulse noise from the fingerprint image. For every individual pixel in the image, initially the noise free pixels are extracted from the image by checking the pixel value against the threshold value. The noisy pixel candidates have been identified and four control points are placed in a matrix (X) which will be used later for edge restoration. Four pixel values are required for the noisy pixel interpolation and every time the window size used is of 3x3.

Let original fingerprint image be OF(i,j) and fingerprint image corrupted by salt and peppernoise corrupted be N(i,j). We initialize with window W(x,y) ∈ N(i,j), for every pixel of corrupted fingerprint image N(i,j) to get the desired noise free image NF(i,j).

The window of size 3x3 for every element of X is applied. The absolute difference of all the elements of a window with center element is calculated. The Control point selection is done as follows:

Step 1: $\begin{pmatrix} 4 & 8 & 9 \\ 5 & 7 & 2 \\ 3 & 10 & 6 \end{pmatrix}$ (Sample window 3x3)

Step 2: Linear form arrangement will result in 4 8 9 5 7 2 3 10 6

Step 3: Find the absolute difference of each element from center element

to get, 3 1 2 2 5 4 3 1

Sorting above elements ascending order will give 1 1 2 2 3 3 4 5

Step 4: The corresponding elements in original window are , 8 6 9 5 4 10 3 2

Step 5: Absolute difference of corresponding elements with adjacent element.

will give , 2 3 4 1 6 7 1

Step 6: Taking 3 elements together : 234, 341, 416, 167, 671

Step 7: Adding elements of group. We get, 9 8 11 14 14

Now, minimum value of sum is 8, corresponding elements are 6, 9, 5, 4. These are now taken as control points represented by X(1), X(2), X(3), X(4) for B spline interpolation. Also, taking difference between adjacent elements 3,4,1 and marking the maximum difference value as d_i. In this case the maximum difference is 4. So, in case of interior pixel the threshold value of t_i, d_i is used and in the case of edge pixels, threshold value $t_e >= 30$ is used.

Stage 2:

Here, the noise free but edge deformed image of stage 1 is taken as input. The noisy pixels detected in stage 1 are only taken into consideration. These noisy pixels are again interpolated using cubic B spline and final noise free and edge regularized image is obtained.

2.1   Algorithm: Stage I : Suppression of Salt and Pepper Impulse Noise

Input: Original fingerprint image OF(i,j)

Image with noise N(i,i), for all OF(i,j)

The window W_{xy} is of size=3x3, where $W_{xy} \in R$.

for $t_i > d_i$ and $t_e > 30$

As defined in stage I, for four pixels in X, and k=0 to 1

we calculate

{

u0 ← $(1 - k)^3 / 6$;

u1 ← $(3k^3 - 6k^2 + 4) / 6$;

u2 ← $(-3k^3 + 3k^2 + 3k + 1) / 6$;

u3 ← $(k^3) / 6$;

$IP=u0*X(1)+u1*X(2)+u2*X(3)+u3*X(4)$

$F(i,j) \longleftarrow IP$(Replacing old value with Interpolated Pixel IP)
}
Replace $NF(i,j) \longleftarrow F(i,j)$

Output: (NF) Noise free restored image

2.2   Algorithm : Stage II : Edge Preservation

For all NF(i, j) and k=0 to 1
for i = 1 : L4 (Noisy detected pixels)

$X(1)$ = upper element of NF(i ,j)
$X(2)$ = below element of NF(i ,j)
$X(3)$ = element left of NF(i ,j)
$X(4)$ = element right of NF(i, j)

$u0 \longleftarrow (1-k)^3 / 6$;
$u1 \longleftarrow (3k^3 - 6k^2 + 4)/6$;
$u2 \longleftarrow (-3k^3 +3k^2 + 3k + 1)/6$;
$u3 \longleftarrow (k^3)/6$;

$IP=u0*X(1)+u1*X(2)+u2*X(3)+u3*X(4)$
$ER(i,j) \longleftarrow IP$(Replacing noisy pixel with Interpolated Pixel IP)
end

Output: ER(Noise free and edge restored Image)

3   Performance Measures

In this section, the performance such as PSNR, MSE and SSIM, used for performance evaluation of the proposed algorithm are discussed.

Peak-to-Signal Noise Ratio (PSNR) and Mean Square Error (MSE)

It is one of the most commonly used parameter for evaluation and comparison of results of different algorithms. The proposed algorithm has been tested for noise levels starting from 15% and to a maximum of 95%. All the results have been analyzed using MSE and PSNR given below.

$$\text{MSE} = \frac{1}{mn}\sum_{x=0}^{m=1}\sum_{y=0}^{n=1}|A(x,y) - R(x,y)| \quad (1)$$

Where, A represents the original image and R is the restored image of resolution m*n.

$$PSNR = 10 \log_{10} \left(\frac{max^2}{MSE} \right) \qquad (2)$$

Where, max is the maximum possible value of element (pixel) i.e. 255 in case of gray scale images.

4   Experimental Results

A fingerprint grayscale sample image of size 512x512 is used for simulation. The simulation is done on MATLAB 7.14 using image processing toolbox and results obtained are shown in the Table below.

Table 1. Noise density ranging from 15% to 95 % and PSNR(dB)

% Noise Density	PSNR (dB)
15	27.4916
25	25.3113
35	23.5023
45	21.8338
55	20.3208
65	18.5280
75	17.1808
85	14.8466
95	12.6097

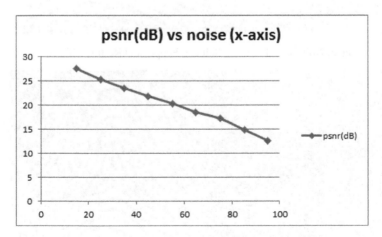

Fig. 1. PSNR Vs Noise levels from 15 % to 95 %.

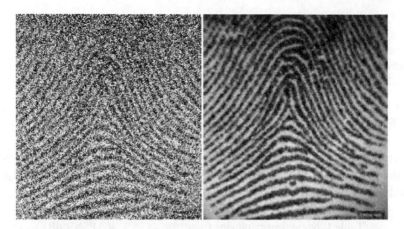

Fig. 2. Noise Level of 55 % Salt and Pepper and Restored Image

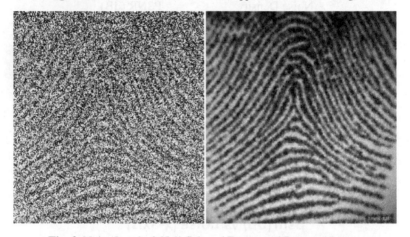

Fig. 3. Noise Level of 65 % Salt and Pepper and Restored Image

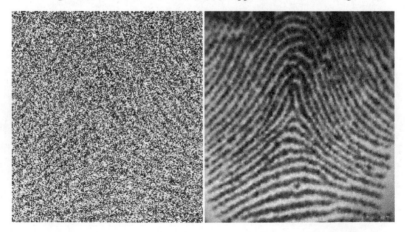

Fig. 4. Noise Level of 75 % Salt and Pepper and Restored Image

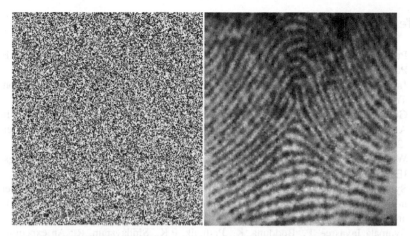

Fig. 5. Noise Level of 85 % Salt and Pepper and Restored Image

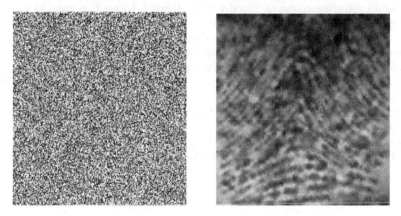

Fig. 6. Noise Level of 85 % Salt and Pepper and Restored Image

5   Conclusion

In this paper, a novel and efficient algorithm for suppression of salt and pepper impulse noise in fingerprint images using B spline interpolation has been proposed. The algorithm not only removes noise from the finger print image but also does efficient edge detail preservation. The quantitative performance of the proposed algorithm is evaluated using measures like PSNR for % Noise density ranging from 15% to 95% as shown in Table 1. The figures 2-6 shows the efficiency of proposed algorithm. The algorithm is pretty fast having time complexity of 45 seconds using MATLAB 7.14 and Core 2 Duo 2.2 GHz processor system.

References

1. Pankanti, S., Prabhakar, S., Jain, A.K.: On the individuality of fingerprints. IEEE Trans. Pattern Analysis and Machine Intelligence 24(8), 1010–1025 (2002)
2. Jain, A.K., Ross, A., Prabhakar, S.: An introduction to biometric recognition. IEEE Transactions on Circuits and Systems for Video Technology 14(1), 4–20 (2004)
3. Goh, A., Ngo, D.C.L.: Computation of cryptographic keys from face biometrics. In: Lioy, A., Mazzocchi, D. (eds.) CMS 2003. LNCS, vol. 2828, pp. 1–13. Springer, Heidelberg (2003)
4. Wildes, R.P.: Iris recognition: an emerging biometric technology. Proceedings of the IEEE 85(9), 1348–1363 (1997)
5. Syamala Jayasree, P., Ghrera, S.P.: A Fast Novel Algorithm for Salt and pepper image noise cancellation using cardinal B-splines. Springer International Journal of Signal, Image and Video Processing 7(6), 1145–1157 (2013)
6. Syamala Jayasree, P., Bodduna, K., Pattnaik, P.K., Siddavatam, R.: An expeditious cum efficient algorithm for salt-and-pepper noise removal and edge-detail preservation using cardinal spline interpolation. Elsevier International J. Visual Communication and Image Representation 25(6), 1349–1365 (2014)
7. Syamala Jayasree, P., Kumar, P., Siddavatam, R., Verma, R.: Salt-and-pepper noise removal by adaptive median-based lifting filter using second-generation wavelets. Springer International Signal, Image and Video Processing 7(1), 111–118 (2013)
8. Hearn., D., Baker, P.: Computer Graphics with OpenGL, 3rd edn. Pearson Publishers (2009)
9. Unser, M.: Splines: A Perfect Fit for Signal and Image Processing. IEEE Signal Processing Magazine 16(6), 24–38 (1999)
10. Bodduna, K., Siddavatam, R.: A novel algorithm for detection and removal of random valued impulse noise using cardinal splines. In: Proceedings of IEEE India Conference INDICON, Kochi, Kerala, pp. 1003–1008 (2012)

Spectral-Subtraction Based Features
for Speaker Identification

Mahesh Chandra, Pratibha Nandi, Aparajita Kumari, and Shipra Mishra

Electronic & Communication Engineering Department, BIT Mesra Ranchi-835215, India
shrotriya69@rediffmail.com,
{pratibha.mou,thakur.aparajita,mishra.shipra.bit}@gmail.com

Abstract. Here wavelet based features in combination with Spectral-Subtraction (SS) are proposed for speaker identification in clean and noisy environment. Gaussian Mixture Models (GMMs) are used as a classifier for classification of speakers. The identification performance of Linear Prediction Coefficient (LPC), Wavelet LPC (WLPC), and Spectral Subtraction WLPC (SS-WLPC) features are computed and compared. WLPC features have shown higher performance over the conventional methods in clean and noisy environment. SS-WLPC features have shown further improvements over WLPC features for speaker identification. Database of fifty speakers for ten Hindi digits are used.

Keywords: LPC, Spectral-Subtraction (SS), Wavelet, Hindi digits, GMM.

1 Introduction

Automatic speaker recognition (ASR) is a technique of speech processing for effective man-machine interface. The main credo of ASR is to automatically identify and verify a person based on their models already being prepared by a machine. In speaker identification, using utterances from a speaker, we determine who the speaker is out of a set of known speakers. Where as in speaker verification, the claimed identity of a speaker is accepted or rejected by a machine using the utterance of a speaker. In today's era ASR have a wide range of applications which includes: computer access control, smart home, bank transfer, fraud detection, forensic labs, and many more. This paper highlights text dependent automatic speaker identification (ASI).

Automatic speaker identification (ASI) basically performs two main tasks: first one is feature extraction which includes, removing unwanted information from the speech signal and retain the useful information, the other includes classification of the extracted features using some machine learning algorithm. Various approaches have been proposed so far by the researchers for speaker identification [1] [2] [3]. There are various feature extraction techniques like LPC, LPCC, MFCC, and PLP etc. [3]. The unvoiced speech samples are not well estimated by LPC [1]. Thus, to overcome the drawback of weak recognition of unvoiced part of the speech sample, and to capture both frequency and location information, Discrete Wavelet Transform (DWT) is used

© Springer International Publishing Switzerland 2015   529
S.C. Satapathy et al. (eds.), *Proc. of the 3rd Int. Conf. on Front. of Intell. Comput. (FICTA) 2014*
– *Vol. 2*, Advances in Intelligent Systems and Computing 328, DOI: 10.1007/978-3-319-12012-6_58

[4-6]. Generally the performance degrades in presence of noise which may be unacceptable on many ocassions. So, in order to reduce the influence of noise and enhance the identification rate, robust feature extraction techniques and denoising techniques are needed. For this purpose spectral subtraction is used to achieve better identification rate by removing disturbances from the signal.

A wavelet based features have already proved their superiority [6-9] over the conventional feature extraction techniques due to their ability of representing a signal in both time and frequency domain. In wavelet transform the size of window is varied with respect to high and low frequency, where short (in time) windows are used at high frequencies and long windows are used for low frequencies [4-5]. The wavelet transform endows better frequency resolution at low frequencies domain and better time localization in the time domain. Here the signals can be decomposed into different levels but by increasing the number of wavelet decomposition levels, the computational complexity increases and it contains redundant data which have no information [6-9]. Using a pair of filters known as quadrature mirror filter we obtain approximation and detailed coefficients by passing the signal through low pass filter and high pass filter respectively. Further at each level of decomposition the approximated signal is decomposed using same LPF and HPF to get the coefficients at each stage.

Spectral subtraction is a technique to enhance the acquired speech signal in presence of noise. It basically operates in the frequency domain to estimate the noise spectrum and use the estimation to filter out the noisy signal. Spectral subtraction is a non-parametric approach [10-12].

Gaussian Mixture Model (GMMs) is used as a classifier due to its capability of representing a large data class [3]. GMM is generally viewed as a hybrid between non parametric and parametric model and widely used for describing the feature distributions of individual speakers [13] [14]. Decisions are usually made using maximum log likelihood ratio.

2   Proposed Features

The objective of feature extraction is to convert the raw signal into an effective and compressed form which carries important information useful for speaker modeling.

2.1   Wavelet Linear Prediction Coefficients (WLPC)

The procedure of our work starts with extracting basic feature from raw signal. For extracting the features, pre-processing is the first step applied which includes: pre-emphasis, filtering and normalization. To spectrally flatten the signal, and to make it less susceptible to finite precision effects [15], the digitized speech signal is passed through first-order FIR filter, having filter coefficient 0.98. Then all the samples are divided by the highest amplitude of the sample, to compensate variation in the amplitudes of speech samples, this process is termed as normalization. After normalization, mean subtraction and silence removal was performed. There after discrete wavelet transform (DWT) was applied on speech signal for decomposing the signal into sub bands, by using two level decomposition. Then by varying the LPC orders of sub band, 5, 6 and 5LPC coefficients are chosen as shown in the figure (1) to make 16- WLPC feature vectors. LPC features were extracted from each sub-band using auto-correlation method that uses Levinson-Durbin recursion algorithm [1]. The transfer function with all pole is shown in equation (1)

$$H(z) = \frac{G}{A(z)} = \frac{G}{(1 + a_1 z^{-1} + a_2 z^{-2} + ... + a_p z^{-p})} \tag{1}$$

The values of a_i are called the prediction coefficients while the gain or the amplitude associated with the vocal tract excitation is represented by G. The linear prediction coefficients a_i are chosen in a way to minimize the mean square prediction error as shown in equation (2).

$$e(n) = x(n) - \hat{x}(n) \tag{2}$$

Then LPC coefficients were obtained using autocorrelation method. Unvoiced sound portion of speech can be modeled by wavelet transform in a better way than by using wavelet with LPC.

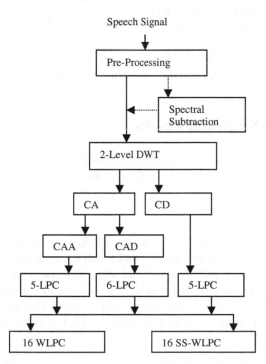

Fig. 1. Feature extraction using WLPC and SS-WLPC

2.2   Spectral Subtraction (SS)

The flowchart shown in figure2 describes the feature extraction using Spectral Subtraction.

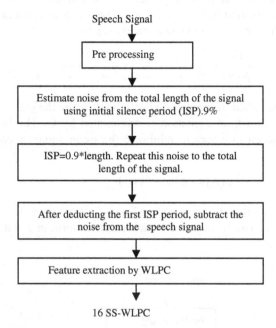

Speech Signal

| Pre processing |

| Estimate noise from the total length of the signal using initial silence period (ISP).9% |

| ISP=0.9*length. Repeat this noise to the total length of the signal. |

| After deducting the first ISP period, subtract the noise from the   speech signal |

| Feature extraction by WLPC |

16 SS-WLPC

Fig. 2. Feature extraction using spectral subtraction

3   Experimental Setup and Results

The experimental setup for speaker recognition is shown in figure 3 which includes database preparation, feature extraction, speaker modelling and decision making.

3.1   Database

A Hindi digit database of fifty speakers has been taken for performing speaker identificaton. Each speaker has spoken ten isolated Hindi digits (shunya, ek, do, teen, chaar, paanch,chheh, saat, aath and nao) and each digit have been repeated with ten different amplitudes.The siganl is sampled at 16kHz and encoded using 16 bits.The voice has been recorded in clean and noisy environment. For preparing noisy database, F16 and speech noise have been artificially added to clean signal at 10dB and 20dB SNR levels [16].

3.2   Feature Extraction and Identification

Speaker identification is the process of identifying a speaker from a set of known speakers. For identifying a person we need to go for feature extraction process then create reference model of each speaker and finally identify the person based on the prepared reference model.

In this paper initially LPC features have been extracted. The feature extraction process includes pre-processing, framing, windowing and finally LPC technique. Here each frame is of 30ms with an overlap of 20ms, then hamming window has been applied on each frame and finally the 16-LPC coefficients have been derived from each frame using Durbin's recursion algorithm. The above process is repeated for extracting the features for both clean and noisy database.

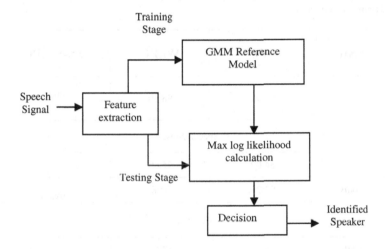

Fig. 3. Experimental set-up for speaker identification

The extracted features of each speaker were stored in fifty different files, one for each speaker and then GMM model was prepared for each speaker using those extracted features, which were further used as a reference model for testing purpose. Clean database was used for training while first twenty frames of fifth sample were used for testing.

WLPC feature extraction process includes pre-processing which is further decomposed into sub-bands using 2 level DWT and feature vectors from the three decomposed sub-bands were concatenated to finally obtain 16-WLPC. The same process is repeated for extracting features in clean and noisy environments. Whereas in case of SS-WLPC after pre-processing, spectral subtraction is performed on the pre-processed signal which is then passed through 2 levels DWT and the features from each sub-bands were concatenated to derive 16 SS-WLPC. In both the above feature extraction technique, training is done using clean database while for testing the fifth sample of each speaker has been used. The process has been repeated for noisy database. Modeling process is same as discussed in above section.

4   Simulation Results and Discussion

The results of speaker identification in clean and noisy environment for text dependent recognition is shown in Table I, figure 4, figure 5 and figure 6 respectively for clean and noisy environment. It is observed from the results that, there is a maximum of 8.34 % improvement in the identification rate with WLPC features over LPC features at 20dB for F16 noise. It is also observed from the results that, there is a maximum of 2.4 % improvement in the identification rate with SS-WLPC features over WLPC features at 20dB for F16 noise. It is clear that spectral-subtraction provides improved recognition rate by minimising noise from noisy speech signal.

Table 1. % Speaker identification using LPC, WLPC and SS-WLC with GMM classifier in clean and noisy environments

SNR	LPC	WLPC	SS-WLPC
∞	93.92	98.80	99.60
Speech Noise			
10dB	84.02	95.80	97.40
20dB	92.28	97.8	99.60
F16 Noise			
10dB	83.09	87.60	91.80
20dB	88.66	97.00	99.40

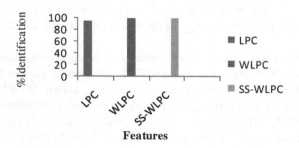

Fig. 4. % Speaker identification in clean environment

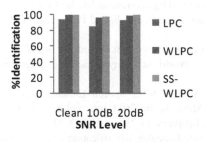 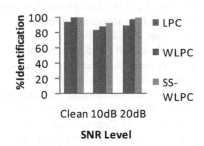

Fig. 5. % Speaker identification in clean and in presence of speech noise

Fig. 6. % Speaker identification in clean and in presence of F16 noise

5   Conclusion

The wavelet based features have shown improved identification performance for both clean and noisy database at all levels. This is due to the fact that wavelet based features study the behavior of signal in small frequency of bands. For noisy speech spectral subtraction has further improved the results. In future, the results can be further improved by using other de-noising techniques at front end.

References

1. Rabiner, L.R., Juang, B.H.: Fundamentals of Speech Recognition, 1st edn. Pearson Education, Delhi (2003)
2. Reynolds, D.A., Rose, R.C.: Robust Text-Independent Speaker Identification using Gaussian Mixture Speaker Models. IEEE Transactions on Speech and Audio Processing 3(1), 74–77 (1995)
3. Makhoul, J.: Linear prediction: A tutorial review. Proc. of IEEE 63(4), 561–580 (1975)
4. Ranjan.: A Discrete Wavelet Transform Based Approach to Hindi Speech Recognition. In: Proceedings of the International Conference on Signal Acquisition and Processing (ICSAP), pp. 345–348 (August 2010)
5. Wang, K., Lee, C.H., Juang, B.H.: Selective feature extraction via signal decomposition. IEEE Signal Processing Letters 4, 8–11 (1997)
6. Tufekci, Z., Gowdy, J.N.: Feature extraction using discrete wavelet transform for speech recognition. In: IEEE International Conference Southeastcon 2000, Nashville, TN, USA, pp. 116–123 (April 2000)
7. Farooq, O., Datta, S.: Mel filter-like admissible wavelet packet structure for speech recognition. IEEE Signal Process. Lett. 8(7), 196–198 (2001)
8. Sharma, A., Shrotriya, M.C., Farooq, O., Abbasi, Z.A.: Hybrid Wavelet based LPC Features for Hindi Speech Recognition. International Journal of Information and Communication Technology 1(3/4), 373–381 (2008)
9. Sharma, R.P., Farooq, O., Khan, I.: Wavelet based sub-band parameters for classification of unaspirated Hindi stop consonants in initial position of CV syllables. Int. J. Speech Technol. 16(3), 323–332 (2012)

10. Berouti, A., et al.: Enhancement of speech corrupted by acoustic noise. In: Proceedings of the IEEE International Conference on Acoustics, Speech, and Signal Processing (ICASSP 1979), pp. 208–211 (1979)

11. Aida–Zade, K.R., Ardil, C., Rustamo, S.S.: Investigation of Combined use of MFCC and LPC Features in Speech Recognition Systems. World Academy of Science, Engineering and Technology 2006, 74–77 (2006)

12. Gupta, V.K., Bhowmick, A., Chandra, M., Sharan, S.N.: Speech Enhancement Using MMSE Estimation and Spectral Subtraction Methods. In: 2011 International Conference on Devices and Communications (ICDeCom), February 24-25, pp. 1–5 (2011)

13. Chowdhury, M.F.R.: Text independent distributed speaker identification and verification using GMM UBM speaker models for mobile communications. In: 10th International Conference on Information Science, Signal Processing and Their Application, pp. 57–60 (2010)

14. Gong, Y.: Noise-robust open-set speaker recognition using noise-dependent Gaussian mixture classifier. In: Proc. ICASSP, pp. 133–136 (2002)

15. Srivastava, S., Nandi, P., Sahoo, G., Chandra, M.: Formant Based Linear Prediction Coefficients for Speaker Identification. In: 2014 International Conference on Signal Processing and Integrated Networks (SPIN), Noida, Delhi-NCR, India, February 20-21, pp. 685–688 (2014)

16. Varga, A., Steeneken, H.J.M., Jones, D.: The noisex-92 study on the effect of additive noise on automatic speech recognition system. Reports of NATO Research Study Group (RSG.10) (June 1992)

Automated System for Detection of Cerebral Aneurysms in Medical CTA Images

M. Vaseemahamed and M. Ravishankar

Department of Information Science and Engineering,
Dayananda Sagar College of Engineering, Bangalore, India
vaseem_cse@yahoo.com, ravishankarmcn@gmail.com

Abstract. In present scenario accurate detection of cerebral aneurysms in medical images plays a crucial role in reducing the incidents of subarachnoid hemorrhage (SAH) which carries a high rate of mortality. Many of the non-traumatic SAH cases are caused by ruptured cerebral aneurysms and accurate detection of these aneurysms can decrease a significant proportion of misdiagnosed cases. A scheme for automated detection of cerebral aneurysms is proposed in this study. The aneurysms are found by applying Normalization and generating the Probability Density Function (PDF) for the input image, local thresholding is used to identify appropriate aneurysm candidate regions. Feature vectors are calculated for the candidate regions based on gray-level, morphological and location based features. Rule based system is used to classify and detect cerebral aneurysms from candidate regions. Accuracy of the system is calculated using the sensitivity parameter.

Keywords: Cerebral Aneurysm detection, Angiography, Probability Density Function.

1   Introduction

Cerebral aneurysm (also known as Intracranial or Intracerebral aneurysm) is a cerebrovascular disorder in which thinning and weakening of arterial blood vessel causes a localized dilation or ballooning of the artery. Aneurysm can press on the nerve or surrounding brain tissue. If it gets ruptured, blood will fill cerebral space to cause subarachnoid hemorrhage. Cerebral aneurysms can occur in any area of the brain, but primarily near the arteries at the base of the brain (Circle of Willis) [1]. The detection and management of unruptured cerebral aneurysms in Computer tomography angiogram (CTA) is a major subject today. Between 3.6 and 6% of population harbor an intracranial aneurysm [2]. The mortality rate after rupture is 50% and 46% who survive have long term impairments. Hence early recognition of these aneurysms is desirable. The aim of this study was to develop an automated scheme for identification of cerebral aneurysms in CTA images in order to assist radiologists' interpretation as a "second opinion".

© Springer International Publishing Switzerland 2015   537
S.C. Satapathy et al. (eds.), *Proc. of the 3rd Int. Conf. on Front. of Intell. Comput. (FICTA) 2014*
– *Vol. 2*, Advances in Intelligent Systems and Computing 328, DOI: 10.1007/978-3-319-12012-6_59

2   Related Work

Clemens M. Hentschke et.al [3] proposed a new feature that can be used to automatically detect cerebral aneurysms in angiographic images. It combines both low-level and high-level features to a feature indicating aneurysms. A sphere-enhancing filter is used to detect VOI. A rule-based system (RBS) excludes FP based on simple rules. The new feature aneurysmness is then derived. The VOI having the highest aneurysmness values of a data set are taken as final aneurysm candidates. H. Zakaria et.al [4] proposed a system that detects cerebral aneurysms using 2D DSA imaging technique, based on the calculation of Time to Peak (TTP) and Time Duration (TD) of flow of contrast agent in the blood vessels. They use two-step processes, Area of interest is selected manually to remove the unwanted area on the image then Multiscale Vessel Enhancement Filtering (MVEF) method is incorporated to improve the vessel segmentation Process. Alexandra Lauric et.al [5] in this study the vessels are segmented and their medial axis is computed. Regions that are small are evaluated along the vessels and are inspected the writhe number is introduced as a new surface descriptor to check how closely a given region approximates a tubular structure. The Aneurysms are then detected as non-tubular regions of the vascular tree. The geometric assumptions underlying the approach are investigated analytically and validated experimentally. H. Prasetya et.al [6] proposed a method that can detect aneurysms based on geometric properties. The method developed was based on curvature analysis of wall vasculatures model in 3D CTA images. It uses Isosurface and voxel grow as the combination of two segmentation techniques: edge based and region growing. Isosurface is a surface that represents the points of constant value, voxel intensity, each value equal to a threshold is projected onto a surface to form a face-vertex triangulation structure. The vasculature reconstructed is segmented based on its curvature to detect aneurysms. Xiaojiang Yang et.al [7] proposed a fully automated computer-aided detection (CAD) scheme for detecting aneurysms. The VOI are identified with the help of segmentation, each suspicious volume of interest (VOI) is assigned a value indicating the likelihood of it representing a true aneurysm. All the VOIs are then arranged in descending order of their value. Finally, clusters of suspicious VOIs are combined to eliminate the overlapping detections. The final clustered suspicious VOIs are used to generate aneurysm suspects in descending order of their values.

3   Methodology

The aim of our proposed system is to automate the detection the cerebral aneurysms in CTA images. Generalization of aneurysms based on shape or grey level distribution is hard.The more information about brain matter or skull is present in data, the more challenging is the detection task. CTA without bone-subtraction is most challenging.

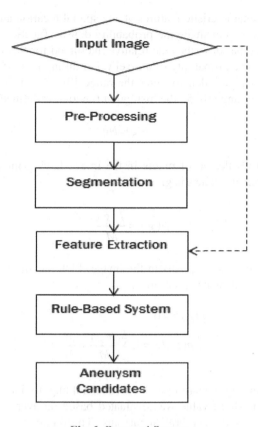

Fig. 1. Proposed System

An overview of our scheme is given in Fig. 1. First the pre-processing is carried out in order to make further process easier. Pre-processing involves conversion of original image to gray scale image. By converting it to gray scale the hidden 3 dimensional and unwanted color components present in the image are converted to 2 dimensional grey components. The pre-processed image is as shown in Fig.2

Segmentation is a three step process, first the grey-scale images are normalized, and probability density function is evaluated for every pixel, and based on the appropriate local threshold value the pixel regions not likely to be aneurysms candidates are eliminated. Normalization, also known as contrast stretching is carried out to alter the range of pixel intensity value and to bring all images to the same intensity level range. The linear normalization of digital image is performed using the equation below.

$$I_n = (I - Min)\frac{newMax - newMin}{Max - Min} + newMin$$

Where the image I having intensity values I:{Min,Max} is changed to new image with intensity values I:{newMin,newMax}. Due to normalization the intensity values are changed but the characteristic features at any spatial location are not altered. Next step in segmentation is to evaluate the probability density function, PDF of a random variable is a function that describes the relative likelihood for this random variable to take an given value. The probability of a pixel value falling in a given range of values is an integral of this pixels density over the range [9]. For an image I(x,y) having intensity values in the range 0-255,the histogram function is defined as

$$h(g)=Ng$$

Where Ng, specifies the no of pixels in the image having intensity values equal to g. The PDF of intensity values is given as

$$p(g) = \frac{h(g)}{M}$$

Where M, is the total no of pixels in the image, PDF has the following properties and it is given by the following equation's

$$p(g) >= 0$$

$$\int_{-\infty}^{\infty} p(g)dg = 1, \sum_{g=0}^{L} \frac{h(g)}{M} = \frac{M}{M} = 1$$

The third step in segmentation is local thresholding to identify the candidate regions [10]. The threshold value was evaluated based on averaging the probability density function values for cerebral aneurysms. The output of segmentation is the segmented image having Region of interests whose PDF values are more likely to be equal to that of the cerebral aneurysms. The threshold value we evaluated is $t=18.75*(4^2)$.

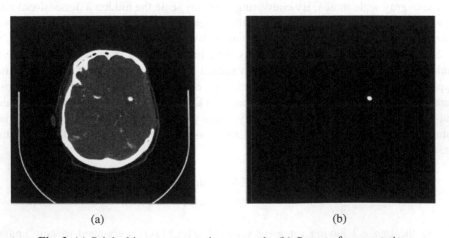

(a)   (b)

Fig. 2. (a) Original image converted to grayscale, (b) Output of segmentation

The output after preprocessing the input image is shown in Fig. 2(a). The result obtained after segmentation process is shown in Fig. 2(b).

In Feature extraction the binary segmented image is mapped on to original image, then for every VOI we calculate 12 specific features [8].

(iavg, imin, imax) we calculate minimum(min), average(avg) and maximum(max) intensity of the VOI this gives information about the grey level value distribution

(bavg, bmin, bmax) minimum(min), average(avg) and maximum(max) circularity as the aneurysm are circular in shape calculating the roundness gives information about its sphericity

(dx, dy) we calculate the distances of the VOI to the image boundary, this can be used to get location information of aneurysm

(p, major, minor) we calculate the perimeter, majoraxis, minoraxis of the VOI, this provides the morphological information

(s) we determine the size of VOI

We noted the different key characteristic features that correspond to cerebral aneurysms [3], (1) a high roundness indicating a high grade of circularity, (2) a large distance to the image boundary as they are located medial than lateral and (3) a minimal diameter of 2 mm.

In Rule-based system we detect the suitable aneurysm candidates using the set of rules that are evaluated on the features obtained [8]. These rules are carefully chosen to obtain the maximum sensitivity. Output of the rule based system is shown in Fig.3.

$0 < iavg < 300$, perimeter > 40

$14 < majoraxis < 39$, $12 < minoraxis < 3$

$s > 5$ mm3, 0 mm$\leq dCoW \leq 10$ mm

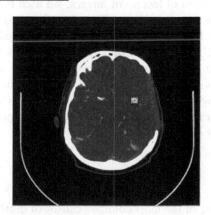

Fig. 3. Aneurysm candidate marked after rule-based system

4   Experimental Results

The automated scheme for detection of cerebral aneurysms was applied to 6 clinical datasets of CTA images containing cerebral aneurysm. (6/6) aneurysms where accurately detected. We achieved a sensitivity of 100%. Sensitivity is percentage of actual cerebral aneurysms detected. The results indicated that the regions of all aneurysms were segmented correctly by use of our method in the pre-processed and normalized image. Because we used probability density function in segmenting the aneurysm candidates, the region corresponding less likely to be aneurysm candidates where removed in segmentation process itself, as the result of using PDF we obtained very negligible amount of false positives (FP) in our evaluation.

Furthermore the gray-level distribution, morphological and location based features are used in our study which produced better results, these features where extracted for all the VOI. To eliminate the FP we used Rule-based system. Few rules where decided based on the characteristic properties of cerebral aneurysm, few where calculated experimentally using trial and error method. The values producing good results were chosen for obtaining the best detection sensitivity.

$$\text{Sensitivity(SN)} = \frac{T_P}{T_p + F_N}$$

We achieved our goal of automating the aneurysm detection system. In our study we are using Rule-based system to detect the aneurysm candidates rather than statistical classification due to less dataset available. Despite our restriction, our quality of results is similar to other publications.

5   Discussion

Because our dataset consisted of less no of images, we used Rule-based system rather than training the data and then classifying. Therefore, we need to evaluate our system with large data set in the further studies. Furthermore, we need to collect many cases acquired from different CTA scanners from different hospitals for clinical application. In addition, we need to conduct an observer performance study for detection of aneurysms without and with the computer output indicating the location of aneurysms in order to investigate whether radiologists 'performance is improved or not. This result will indicate that radiologists are able to use the computer output as a second opinions to improve their diagnostic accuracy

6   Conclusion

Cerebral Aneurysm is one of the most common causes of death in current scenario. It is estimated that people with unruptured aneurysm have an annual 1-2% risk of haemorrhage. Detection of cerebral aneurysms in earlier biological stages before they rupture brings down the number of deaths caused by Brain Haemorrhage throughout the world. Image processing algorithms allow development of automated systems for early detection of cerebral aneurysms. We have presented an efficient method for automatic detection of cerebral aneurysms based on evaluation of probability density function and local thresholding. 12 distinct gray-level, morphological and location based features are calculated leading to the best quality results. Rule based system is used to classify and detect the aneurysms from the candidate regions. This flow is proved robust for the automatic detection of cerebral aneurysms and we intend to check our system with more images.

References

1. Gasparotti, R., Liserre, R.: Intracranial aneurysm. European Radiology 15, 441–447 (2005)
2. Wardlaw, J.M., White, P.M.: The detection and management of unruptured intracranial aneurysms. Brain 123, 205–221 (2000)
3. Clemens, M., Hentschke, K.D., Tönnies, O., Beuing, R.: A New Feature for Automatic Aneurysm Detection. In: International Symposium on Biomedical Imaging 2012, pp. 800–803. IEEE (2012), 978-1-4577-1858-8/12
4. Zakaria, H., Kurniawan, A., Mengko, T.L.R., Santoso, O.S.: Detection of Cerebral Aneurysms by Using Time Based Parametric Color Coded of Cerebral Angiogram. In: International Conference on Electrical Engineering and Informatics. IEEE Press, Indonesia (2011) 978-1-4577-0752-0/11
5. Lauric, A., Miller, E., Frisken, S., Malek, A.M.: Automated detection of intracranial aneurysms based on parent vessel 3D analysis. Medical Image Analysis 14, 149–159 (2009)
6. Prasetya, H., Mengko, T.L.R., Santoso, O.S., Zakaria, H.: Detection Method of Cerebral Aneurysm Based on Curvature Analysis from 3D Medical images. In: International Conference on Instrumentation, Communication, Information Technology and Biomedical Engineering, IEEE Press, Indonesia (2011) 978-1-4577-1166-4/11
7. Yang, X., Blezek, D.J., Cheng, L.T.E., Ryan, W.J., Kallmes, D.F., Erickson, B.J.: Computer-Aided Detection of Intracranial Aneurysms in MR Angiography. Journal of Digital Imaging 24(1), 86–95 (2011)
8. Hentschke, C.M., Beuing, O., Nickl, R., Tonnies, K.D.: Automatic Cerebral Aneurysm Detection in Multimodal Angiographic Images. In: Nuclear Science Symposium Conference Record. IEEE (2011) 978-1-4673-0120-6111.
9. Jiang, H., Tan, H., Yang, B.: A Priori Knowledge and Probability Density Based Segmentation Method for Medical CT Image Sequences. BioMed Research International 2014, article id 769751, 1–11 (2014)
10. Kim, D.Y., Park, J.W.: Connectivity-based local adaptive thresholding for carotid artery segmentation using MRA images. Image and Vision Computing 23, 1277–1287 (2005)

Contrast Enhancement of Mammograms Images Based on Hybrid Processing

Inam Ul Islam Wani[1], M.C. Hanumantharaju[2], and M.T. Gopalkrishna[1]

[1] Department of Information Science and Engg,
Dayananda Sagar College of Engg, Bangalore, India
{inamwani08,gopalmtm}@gmail.com
[2] Department of Electronics and Communication Engg,
BMS Institute of Technology, Bangalore, India
mchanumantharaju@gmail.com

Abstract. This paper introduces a new enhancement algorithm based on combination of different processing techniques. The method uses different methods at different stages of processing. In the beginning input image given to the algorithm is a portable gray map image and then Gaussian low pass filter is used to decompose the input image into low and high frequency components. On low frequency components we apply mathematical morphological operations and on high frequency components we apply edge enhanced algorithm. After this we combine processed low and high frequency components to get an enhanced image. Enhanced image is having better contrast and edge visibility comparing to the original image, but it contains noises. Wavelet transform is used to denoise the noisy image. The denoised image is then processed by using contrast limited adaptive histogram equalization(CLAHE) to have better edge preservation index (EPI) and contrast improvement index (EPI). The resulting image is then smoothed by passing the output image through a guided image filter(GIF). The edge preserve capacity and preservation of the naturalness of the GIF allows us to get better results.

The efficiency of any service or product, especially those related to medical field depends upon its applicability. The applicability for any service or products can be achieved by applying the basic principles of Software Engineering. Applicability of enhanced algorithms depends on parameters like the peak signal to noise ratio(PSNR),edge preservation index(EPI), etc. This paper focuses on introducing a model which highlights on a prototyping approach for highlighting the necessary details that will aid radiologist for the earlier detection of breast cancer.This paper also presents Design of the model, Implementation of the model and finally the results are analyzed by considering the quality metrics values like PSNR, EPI, CII.

Keywords: Gaussian Filter, Mathematical Morphology, Contrast Enhancement, CLAHE, Wavelet, Guided Image Filtering, Mammographic Image, Medical Image Processing.

1   Introduction

Breast cancer is the second most common form of cancer in the world. Women are more likely to develop breast cancer, but it can be developed in men also. Breast cancer is among the top five common causes of death in women around the world [1]. Self-exam can be performed on the patient to detect the initial symptoms of breast cancer. Self-exam can identify the abnormal tissues in the body of a patient that is normally a small lump. The other methods for detecting the abnormal tissues uses mammograms. Mammogram images are used to search the breast for abnormal tissues that may signal the presence of cancer. Mammogram is an X- ray image of breasts and mammography is used as a diagnostic and screening tool to examine the human breasts. Mammography is one of the most effective techniques used in breast cancer detection in it's earlier stages. To diagnose breast cancer we need to find abnormalities like masses and calcifications that indicate breast cancer. Small size of mircocalcifications leads to poor visualization in mammograms. Mammogram contrast needs to be enhanced to improve visibility of the abnormalities to detect breast cancer in mammograms to assist analysts as well as automatic breast cancer detection systems. Removal of noise is essential for enhancement of contrast of an image, specifically for mammograms the microcalcification size is close to noises. Noise should be reduced whereas Microcalcifications in mammograms needs to be enhanced whereas noise should be removed. One more reason for enhancement is that mammograms that show abnormalities, such as masses, microcalcification and their surrounding tissues may be possessing low contrast. Mammogram image enhancement is the process of manipulating mammogram images to increase their contrast and reduce the noise present to facilitate radiologists in the detection of abnormalities.

Outlay of the rest of paper is as follows; In section 2, proposed method is discussed, section 3 presents the experimental results and conclusion is stated in section 4.

2   Proposed Method

Mammogram image enhancement focuses on increasing the contrast of the images and decreasing the noise. Contrast enhanced mammogram aids the radiologists in identifying tumor and other type of abnormalities present in mammographic images. The proposed method enhances the contrast of the image by taking an input mammographic image $i(x, y)$ as input. The image is decomposed into low frequency components $l(x, y)$ and high frequency components $h(x, y)$ using Gaussian low pass filter. Decomposed parts are then enhanced separately. Then both the enhanced components are added to get a contrast enhanced image $c(x, y)$. Noise in the contrast enhanced image is removed by using wavelet transform $w(x, y)$. Contrast limited adaptive histogram equalization(CLAHE) is applied on $c(x, y)$ to get an image which is having strong edges and high edge presevation index $cl(x, y)$. In the end guided image filtering (GIF)

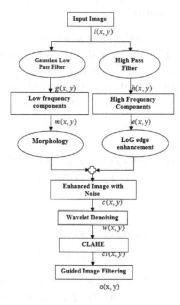

Fig. 1. Flow chart of the proposed method

is applied on the $cl(x, y)$ to get the output image $o(x, y)$ which will preserve the original details of the image. Flow chart of the proposed method is shown in Figure 1.

2.1   Decomposition Using Gaussian Low Pass Filter

Gaussian low pass filters, also known as Gaussian blur or Gaussian smoothing is used to remove high frequency components from the image. The effects of Gaussian filter on image results in removing the details and noise from the image. The use of Gaussian filter is to attenuate high frequency high frequency signals which results in blurring the image. It is similar to mean filter, but it uses a different kernel that represents the shape of Gaussian hump [7]. The equation of Gaussian function in two dimensions has the form given in equation 1.

$$g(x, y) = \frac{1}{2\pi\sigma^2} e^{-\frac{x^2+y^2}{2\sigma^2}} \tag{1}$$

After applying Gaussian low pass filter, we get the high frequency components of the image by subtracting low frequency components $l(x, y)$ from the original input image $i(x, y)$. High frequency components of input image is given by equation 2.

$$h(x, y) = i(x, y) - l(x, y) \tag{2}$$

Low frequency components and high frequency components are then enhanced by using morphological and edge enhanced algorithms respectively.

2.2 Morphological Operation

Mathematical morphology is a methodology for extracting the shape and size information from an image. Morphological processing is applied on the low frequency components of the image. The basic morphological operations are dilation and erosion, from which many operations are derived. For gray scale image and structuring element , dilation (equation 3) and erosion (equation 4) are defined as follows;

Dilation:

$$\delta_B(f)(x,y) = max f(x-s, y-t) + B(s,t)|$$
$$(x-s),(y-t)_f; (s,t)_B \tag{3}$$

Erosion:

$$\delta_B(f)(x,y) = min f(x+s, y+t) + B(s,t)|$$
$$(x+s),(y+t)_f; (s,t)_B \tag{4}$$

Where f is the input image and B is the structuring element, (x, y) and (s, t) are the respective co-ordinate sets and D_f, D_B are the respective domains [2]. The next step in the proposed algorithm is to use morphological operators like opening, closing, top and bottom hat operators. These operators are used to preserve details of the image. Opening and closing of image A by a structuring element B is done by using dilation and erosion as follows:

$$A \circ B = (A \ominus B) \oplus B \tag{5}$$

$$A \bullet B = (A \oplus B) \ominus B \tag{6}$$

We use opening and closing to smooth the contour of an image, but whereas opening breaks narrow isthmuses and eliminates the protrusions, closing fuses narrow breaks, eliminates small holes and fills gaps in the contour. The top and bottom hat filtering extracts the original image from the morphologically closed version of the image. Top hat and bottom hat operations are defined as :

$$T_{hat} = A - (A \circ B) \tag{7}$$

$$B_{hat} = (A \bullet B) - B \tag{8}$$

Appropriate selection of opening and closing filtration, top and bottom hat filtration and proper form of structuring element and its parameters can eradicate the local structures or local geometry of the inspected objects can be customized. In addition we can use each morphological operation multiple times in processing. The elements of morphological processing can be selected on the basis of calculated estimation parameters. Peak signal to noise ratio (PSNR) and contrast improvement index (CII) values are higher for better quality of the mammographic image.

2.3   Edge Enhancement Using Laplacian of Gaussian Edge Detection

Edge enhancement is applied in the high frequency contents of the image to enhance the edges of the image. The edge enhanced high frequency image is then added to the low frequency components that are enhanced by morphological processing. Blob detection refers to mathematical methods that are aimed at detecting regions in a digital image that differ in properties, such as brightness or color, compared to areas surrounding those regions. Informally, a blob is a region of a digital image in which some properties are constant or vary within a prescribed range of values; all the points in a blob can be considered in some sense to be similar to each other [4]. One of the first and also most common blob detectors is based on the Laplacian of the Gaussian (LoG). Given an input image f(x, y), this image is convolved by a Gaussian kernel. Edge detection is done by convolving the image with linear filter that is Laplacian of Gaussian filter LoG (f):

$$LoG = \nabla^2 G(x,y) = \left[\frac{x^2 + y^2 - 2\sigma^2}{\sigma^4} \right] \tag{9}$$

Where the Gaussian is given by

$$G(x,y) = e^{-\frac{x^2+y^2}{2\sigma^2}} \tag{10}$$

2.4   Noise Reduction Based on Wavelet Packet Decomposition

Wavelets are the mathematical functions that decompose the data signals into different frequency components and then analyze each component with a resolution matched to its scale. Wavelets have advantages over traditional Fourier methods, because they can be used in both frequency and time domains. Image is denoised by decomposing the image using wavelet decomposition and then using threshold detail to suppress the noise. After suppressing the noise of the decomposed images wavelet reconstruction is used to reconstruct the original image. The algorithm for noise reduction based on wavelet packet transform contains;Decomposition of the image,Determination of the threshold and thresholding detail coefficients and Restoration of the image. In the proposed method decomposition of the input image is done at two levels by using bi-orthogonal wavelet. Comparing to orthogonal wavelet, Bi-orthogonal wavelet representation has many advantages. The sub band images are invariant under translation and do not have aliasing effect. After decomposition the given image is realized by one approximation coefficient and six detail coefficients. The detailed coefficients obtained by decomposition contains horizontal, vertical and diagonal coefficients. Appropriate threshold value is used to denoise the coefficients [3]. In the proposed method soft thresholding is used. The level dependent threshold is calculated by using the equation given below:

$$T = j/2\sqrt{max(d_i} \tag{11}$$

Where j is the level at which threshold T is calculated.The resulting image after perform wavelet denosing will be a contrast enhanced denoised image which is more detailed and vivid[4].

2.5   Contrast Limited Adaptive Histogram Equalization (CLAHE)

Contrast limited adaptive histogram equalization is a technique that improves contrast of images by generalizing the ordinary histogram equalization and adaptive histogram equalization. CLAHE works on small areas in image rather than operating on whole image like ordinary histogram equalization. Contrast of every tile is enhanced, so that histogram of the output area roughly matches the histogram determined by the 'Distribution' parameter. This parameter can be selected depending on the type of input image. Clip limit is used to limit the maximum slope of all histograms. The clip limit can be related to what is referred to as clip factor, in percent, as follows;

$$\beta = \frac{MN}{L} \left[1 + \frac{\alpha}{100}(s_{max} - 1)\right] \tag{12}$$

Where MxN are number of pixels of each region and L is the number of grayscales.

2.6   Guided Image Filtering Enhancement

Guided image filtering is used to preserve strong edges and also to smooth low gradient regions. Guided image filtering (GIF) is the modification of bilateral filtering (BLF), in which processing is done under the guidance of an image, which can be another image or the input image. If the guided image is similar to then guided image filtering becomes bilateral filtering. Let I_p and G_p be the intensity value at pixel k of the input image and guided image, w_k be the kernel window centered at pixel k , to be consistent with BLF, GIF is then defined by:

$$GIF(I)_p = \frac{1}{\Sigma_{q \in W_k} W_{GIF_{pq}}(G)} \Sigma_{q \in W_k} W_{GIF_{pq}}(G) I_q \tag{13}$$

Where the kernel weights function $W_{GIF}(G)$ can be expressed by:

$$W_{GIF}(G) = \frac{1}{|w|^2} \Sigma_{k:(p,q) \in w_k} + (1 + \frac{(G_p - \mu_k)(G_q - \mu_k)}{\sigma_k^2 + \epsilon}) \tag{14}$$

Where μ_k and σ^2 are the mean and variance of guided image G in local window w_k , $|w|$ is the number of pixels in this window. The term $(1 + \frac{(G_p - \mu_k)(G_q - \mu_k)}{\sigma_k^2 + \epsilon})$ controls the edge preserving property of GIF. The weight assigned to pixel q is large if G_p and G_q are concurrently on the same side of an edge [2]. The filter of GIF can be written as:

$$GIF(I)_p = \Sigma_{q \in w_k} W_{GIF_{pq}}(G) I_q \tag{15}$$

The degree of smoothing of GIF can be adjusted via parameter ϵ. Larger value of ϵ will result in smoother images[5][6].

3   Experimental Results

In this study, the images for experiments are from mini-MIAS (Mammogram Image Analysis Society) database [08]. In the min- MIAS, there are 322 cases, in which we got 113 abnormal cases and 209 normal cases. Effectiveness of the proposed method is verified by selecting images from the mini-MIAS database. All the examples have been produced on Intel i3 2.4 GHz system with 3 GB of RAM. Figure 2 shows the resultsa t different stages of proceesing of the proposed method. For the evaluation of performance analysis of the proposed method, we use contrast improvement index (CII), edge preservation index (EPI) and peak signal to noise ratio (PSNR). CII is defined by:

$$c = \frac{C_{Processed}}{C_{Original}} \tag{16}$$

Where $C_{Processed}$ and $C_{Original}$ are the contrast for the processed and original images, respectively. The contrast C of the original image is defined by:

$$c = \frac{f - b}{f + b} \tag{17}$$

Where, f and b are the mean gray-level value of the foreground and the background, respectively. EPI is defined by:

$$EPI = \frac{\Sigma(|I_p(i_d - I_p(i+1,j)| + |I_p(1,j+1)|)}{\Sigma(|I_o(i_d - I_o(i+1,j)| + |I_o(1,j+1)|)} \tag{18}$$

Where $I_0(i,j)$, is an original image pixel intensity value for the pixel location (x,y), $I_p(i,j)$ is the processed image pixel intensity value for the pixel location (x,y) .The greater value of EPI indicates the better quality of image. Table 1 contains the averaging results from the simulation obtained by applying the method on 40 images selected from mini-MIAS database.

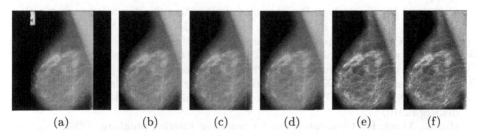

(a)   (b)   (c)   (d)   (e)   (f)

Fig. 2. Result of enhancement on image mdb019.pgm. (a) Original mammographic image, (b) Preprocessed Image(b) Morphologically processed image, (c) Denoised Image, (d) CLAHE Image, (e) GIF Image

Table 1. Simulation results for stages of mammogram enhancement

Stage of Processing	PSNR	CII	EPI
Morphology &LoG	33.0587	0.8396	0.8371
Wavelet Denoising	37.1542	1.1624	1.0793
CLAHE	25.3502	1.5213	1.4529
GIF	28.6304	1.3405	1.2908

4   Conclusion

In this paper, new mammogram enhancement method is presented. Mammography is one of the best known methods in breast cancer detection, but in some cases, radiologists cannot detect tumors despite their experience, mainly because of the noise present in mammogram images and also low contrast present in the images. To aid the radiologists we need to enhance the mammogram images so that they will be able to detect breast cancer and also improve the accuracy of detection. The proposed method is based on mathematical morphology, denoising using wavelet transform and on guided image filtering, which not only enhances the mammogram details but also preserves the naturalness and details in the image. It effectively enhances the contrast of the image and reduces the noise present in the image. The experimental results are encouraging based on the values of CII, EII and PSNR. Future development of the present research concerns on time consumed for enhancing at different levels and also will focus on the construction of enhancement methodologies applied to the classification system that might improve the discrimination between benign and malignant microcalcification clusters.

References

1. Canadian Cancer Society, Facts on Breast Cancer (April 1989)
2. Kimori, Y.: Mathematical Morphology based approach to the Enhancement of Morphological Features in Medical images. Journal of Clinical Bioinformatics 1, 33 (2011)
3. Georgieva, V., Kountchev, R., Draganov, I.: An Adaptive Enhancement of X-Ray Images. In: Kountchev, R., Iantovics, B. (eds.) Adv. in Intell. Anal. of Med. Data & Decis. SCI, vol. 473, pp. 79–88. Springer, Heidelberg (2013)
4. Sakellaropoulos, P., et al.: A Wavelet-based Spatially Adaptive method for Mammographic Contrast Enhancement. Phys. Med. Biol. 48(6), 787–803 (2003)
5. He, K., Sun, J., Tang, X.: Guided image filtering. In: Daniilidis, K., Maragos, P., Paragios, N. (eds.) ECCV 2010, Part I. LNCS, vol. 6311, pp. 1–14. Springer, Heidelberg (2010)
6. Ito, K., Xiong, K.: Gaussian filters for nonlinear filtering problems. IEEE Transactions on Automatic Control 45(5), 910–927 (2000)
7. Zeng, F., Liu, L.: Contrast enhancement of mammographic images using guided image filtering. In: Tan, T., Ruan, Q., Chen, X., Ma, H., Wang, L. (eds.) IGTA 2013. CCIS, vol. 363, pp. 300–306. Springer, Heidelberg (2013), Tomasi, Manduchi, Tang: Bilateral filtering for gray and color images. IGTA 2013 81(1), 24–52 (2009)

An Improved Handwritten Word Recognition Rate of South Indian Kannada Words Using Better Feature Extraction Approach

M.S. Patel[1], Sanjay Linga Reddy[2], and Krupashankari S. Sandyal[1]

[1] Department of Information Science and Engineering,
Dayananda Sagar College of Engineering, Bangalore, India
msr_patel@yahoo.com,
krupasandyal@gmail.com
[2] Department of Computer Science and Engineering,
Alpha College of Engineering, Bangalore, India
sclingareddy@gmail.com

Abstract. Ever since the evolution of communication in human day to day activities, hand writing has gained its own impact and popularity. Therefore, Handwritten Word Recognition (HWR) is quite challenging due to heavy variations of writing style, different size and shape of the character by various writers. Accuracy and efficiency are the major parameters in the field of handwritten character recognition. However, with the progress in technology, human computer interactions have become a mandatory process to carry on the fast and dynamic demanding activities of the everyday cycle. This paper thus throws light on an effective recognition process for the handwritten word recognition. The HWR is carried out in 3 stages. In the first stage, pre-processing removes the unwanted data like noise and the second stage extracts the best features such as the sharp corners, curves and loops and finally the third stage of the process classifies the image under the correct matching class using the Euclidean distance based classifier. This process is implemented and the results indicate an improved accuracy and efficient recognition rate.

Keywords: Handwritten Word Recognition (HWR), Pre-processing, Feature Extraction, Classification.

1   Introduction

Image processing is one of the widely used advent of science and technology due to its varied applications and imperativeness. Handwriting recognition is one of the most challenging research areas in the field of image processing and pattern recognition. This is because, the state of the art in technology not only requires clarity for readers to read the hand written documents but also demands the use of computers for the purpose of reading, processing, for storing and retrieving the documents to perform

© Springer International Publishing Switzerland 2015
S.C. Satapathy et al. (eds.), *Proc. of the 3rd Int. Conf. on Front. of Intell. Comput. (FICTA) 2014*
– *Vol. 2*, Advances in Intelligent Systems and Computing 328, DOI: 10.1007/978-3-319-12012-6_61

varied tasks. Based upon the technique how the input data is obtained, handwritten character recognition can be classified into two categories: **Online** character recognition handwriting is captured using a special pen in conjunction with electronic surface. There is a common consensus among researchers that online handwriting recognition is a much easier task than the offline case because more information is available to the recognizer. **Offline** character recognition Input will be scanned from a surface such as sheet of paper and stored digitally. Offline character recognition include recognition of machine printed, hand printed and handwritten characters.

There exist two popular modes of handwritten word recognition techniques. They are analytical and holistic approach. In analytical approach, the characters from the given input word are segmented and recognized separately and thus obtained results are combined to recognize the given word. Holistic approach considers word as a single entity and extracts the features from the shape of the whole word [1].

The aforementioned stages are performed upon hand written words which are Karnataka district names and further written in the south Indian language specifically in Kannada.

The position of this paper is therefore to perform the HWR in 3 stages. In the first stage, we pre-process the image to remove the noise and unwanted data from the scanned image and then to extract the features from the shape of the word image like the curves, corners and loops. The third stage of this approach is to classify the test image based on the features extracted and try to match it to the trained set of data and recognize it under matching class id.

The organization of the paper is as follows. Section 2 presents the review of literature, Section 3 describes the characteristics of Kannada script, Section 4 gives overview of an effective recognition process for the handwritten word recognition, Section 5 describes the Implementation and Section 6 gives the experimental results, finally in Section 7 concludes the work.

2   Review of Literature

Though many research papers are available in the literature for character and word recognition, still handwritten word recognition has not reached expected level of accuracy. The Literature survey is done for the Kannada HWR since handwritten words are more difficult to recognize compared to the printed words. This is because of the characteristics possessed by the handwritten words like varying styles, strokes and font size.

Rhandley D. Cajote and Rowena Cristina L. Guevara [1] proposed method for offline handwriting recognition using both the local geometric features and the global features. The recognition system takes the isolated handwritten word image as input and the word images were extracted from the IAM (Institute of Computer Science and Applied Mathematics) database. In pre-processing step an optimal thresholding operation based on Otsu's algorithm was used to convert the grayscale image to binary images, Slant correction is performed after the binarization process in order to reduce some of the handwriting variability. The average slant of the word is estimated using the 8-directional chain code method and In order to correct the slant of the image, a shear transformation is applied. In the next step, the local and global features were extracted and these global features were used as input to the Multi Layer Perceptron (MLP) classifier and obtained a recognition rate of 78% for the 20-word vocabulary. The proposed method is ideal for recognition tasks with small vocabularies such as recognition of courtesy amount on bank checks.

Shazia Akram, Dr. Mehraj-Ud-Din Dar and Aasia Quyoum [2] proposed a method on Document Image Processing. The proposed method uses three main components namely Pre-processing, Feature extraction and the Classification. As an initial step, Pre-processing includes Image acquisition, binarization, noise removal and Skew Detection and Correction. Preprocessing is followed by the feature extraction, which reduces the information required to represent an image. Simple feature extraction methods like calculating the difference between the minimum and maximum coordinates of the document image and the shape of the document obtained by comparison of breadth and length of the document image are of prime importance to acquire information regarding the document under process. Finally, classification step identifies each input document image by considering the detected features like spatial arrangements with respect to one another, layout of document, size of the document, color of the paper and texture.

M.M. Kodabagi and Shridevi.B.Kembhavi [3] proposed a method for Recognition of Basic Kannada Characters in Scene Images using Euclidean Distance Classifier. Character recognition from scene images is a very challenging and difficult task due to limitations of digital cameras and variety of applications. The features that were used for extraction were zone based horizontal and vertical profiles followed by knowledge base creation using these feature vectors during training phase. In testing phase, the test image is given for recognition and using Euclidean distance classifier the image is classified under matching class. The method is tested for 460 Kannada character images taken from 2 Mega pixel cameras at various sizes with different styles, shapes and achieved an accuracy of 91%.

Thus, there exist various researches which is carried out in the domain of HWR which motivated this research to focus in the hand written word recognition for the South Indian Kannada language.

3   Characteristics of Kannada Script

The Kannada script (Kannada lipi) is an alphasyllabary of the Brahmic family, used primarily to write the Kannada language, which is one of the Dravidian languages of southern India [12]. Kannada script has forty-nine characters in its alphasyllabary. The characters are classified into three categories namely Swaras (vowels), Vyanjanas (consonants) and Yogavaahakas (part vowel, part consonants).

Swaras (vowels) are the independently existing letters. There are thirteen Swaras and two Yogavaahakas which is as depicted in Figure 1. There are 34 Vyanjanas (consonants) which depend on vowels to take independent form as depicted in Figure 2. They are further classified as Vargeeya and Avargeeya Vyanjanas. Corresponding to each Kannada consonant, there exists a separate and unique glyph, which is specially used to represent the corresponding consonant in a consonant cluster. Figure 3 illustrates the Kannada Consonant Conjuncts. Most of these conjunct consonant glyphs resemble their original consonant forms [12].

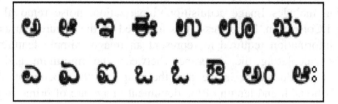

Fig. 1. Kannada Swaras and Yogavaahakas

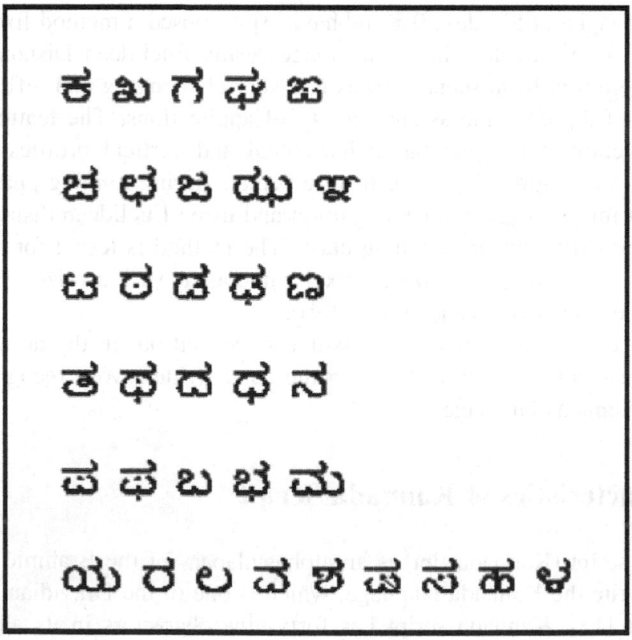

ಕ ಖ ಗ ಘ ಙ

ಚ ಛ ಜ ಝ ಞ

ಟ ಠ ಡ ಢ ಣ

ತ ಥ ದ ಧ ನ

ಪ ಫ ಬ ಭ ಮ

ಯ ರ ಲ ವ ಶ ಷ ಸ ಹ ಳ

Fig. 2. Kannada Consonants

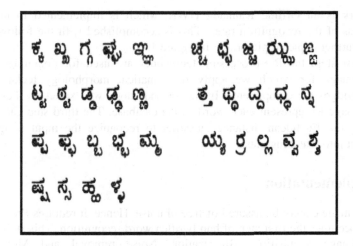

Fig. 3. Kannada Consonant Conjuncts

4   An Effective Recognition Process for the Handwritten Word Recognition

This research work is carried out on the data set which comprises of 1200 bit map images which includes Kannada hand written words. These words are names of various districts of Karnataka. These words are written by writers of different age groups who are residing across the state of Karnataka in India.

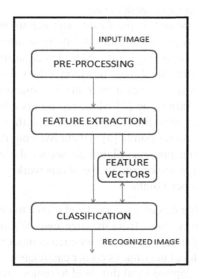

Fig. 4. Proposed method for Kannada Handwritten Word Recognition

The work is an Offline Kannada HWR, which is implemented to uphold the effectiveness of the recognition rate. This is accomplished with the following steps which are further depicted pictorially in Figure 4.

As a first step, the Pre-processing techniques are used for obtaining clear and noiseless image for which we apply mathematical morphology technique. This approach bridges the components. In the second stage, we extract the best features from the image to represent each word in the database. The third step Classifies the word using the Euclidean distance classifier to recognize the input image and to categorize it into matching class.

5   Implementation

Any input image cannot be assured of free of noise. Hence, it requires Pre-processing [2] which becomes the first step of handwritten word recognition. This step generally includes Image acquisition, Binarization, Noise removal and Morphological operations. Image acquisition when performed upon the collected data set converts images in rgb to gray level or binary format. With Binarization, the objective of this research is to automatically choose a threshold that separates the foreground and background information from the images for the data set since setting of a good threshold is often a trial and error process. Due to the existence of noise such as "salt and pepper" noise, which is random occurrences of black and white pixels in the collected images from the data set, it is further required to remove them by applying the median filter. Followed with this operation, is the implementation of Morphological operations where this research is performed using Dilation. It works using a structural element that does a local comparison of shape with the original image that need to be transformed into a new one. Figure 5a shows the sampled preprocessed image in the data set under study.

As a next step of this research is the feature Extraction, which is one of the most vital steps to improve the recognition rate with the least amount of elements. Hence, this work is progressed by applying canny edge detection to extract edges from the image of the data set. Figure 5b shows the canny edges for the sampled image for which the preprocessing was performed from the investigated data set. According to the works of Freeman's Chain Code [4] which represents a boundary by a connected sequence of straight line segments of specified length and direction, we have extracted the features like corner points [5] and curves from the sampled image that is used in training and testing process. Figure 5c shows the corner points which are detected from the sampled preprocessed image of this work.

Algorithm to Detect Corner Points:

Step 1. Apply canny edge detector to gray level image to obtain binary edge-map.
Step 2. Extract the edge contours from the edge-map; fill the gaps in the contours.
Step 3. Calculate curvature at a low scale for each contour to retain all true corners.
Step 4. Find curvature local maxima as corner candidates.
Step 5. Compare with adaptive local threshold to remove round corners.
Step 6. Check corner angle to remove false corners due to boundary noise.
Step 7. Add Endpoints far from the detected corners.

Followed by the completion of feature extraction is the Classification procedure. This process classifies the feature extracted images from the data set to perform the classification based upon the matching classifier technique. In this research, Euclidean distance classifier is used to classify the image into matching class id. Dynamic Time Warping (DTW) [6] is used for matching distance by recovering optimal alignments between sample points in the two time series of the classified image. This alignment is optimal since it minimizes a cumulative distance measure which consists of distances between aligned samples.

5(a)   5(b)   5(c)

Fig. 5. Result of Canny edge detection and corner points

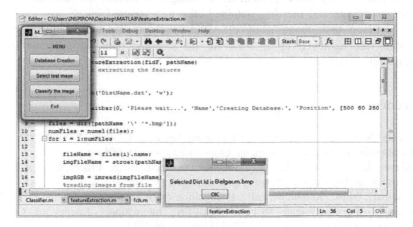

Fig. 6. Result of Classification

6   Experimental Results

This process is completely implemented in the experimental set up comprising of MAT Lab version 12. The HWR process in this investigation is verified using data collected from the 60 different writers who have contributed the words written manually in Kannada language. The database consists of around 1200 bitmap images of scanned Karnataka district names. For carrying out the training and testing phase, a database of 300 bmp images are used. Euclidean distance classifier is applied to classify the test images. The experimental results with the above said HWR process has yielded a recognition rate of 92% for the data set under this research framework The enhancement for this work would be to optimize the recognition rate using minimum number of features in recognition process. The work can be further tested with various words written in other languages.

7   Conclusion

It is indeed a matter of fact that hand written word recognition has become one of the important habitual activities of human computer interactions. Nevertheless, there exists several researches which have advanced in various languages and for various applications, yet there still prevail a lot of challenges. The aim of this research is to improve the recognition rate of Kannada Hand written Word Recognition (HWR). This paper therefore presents an effective mode of conducting HWR upon a data set which is collected from various age grouped Kannada writers. The data set consists of several names of Karnataka districts.

The HWR process follows a holistic approach where a complete word is considered as an input for the recognition process. The HWR process follows three stages for recognizing given test image which includes pre-processing the given input image to remove any noise present in the image, to extract the features from the preprocessed image and the thus obtained feature extracted image are subjected to detect the given test image using classifier. The experimental results indicate a recognition rate achieved up to 92% for the sampled data set collected for this research work.

References

[1] Cajote, R.D., Guevara, R.C.L.: Combining Local and Global Features for Offline Handwriting Recognition. Philippine Engineering Journal (2005)
[2] Akram, S., Dar, M.-U.-D., Quyoum, A.: Document Image Processing - A Review. International Journal of Computer Applications 10 (November 2010)
[3] Kodabagi, M.M., Kembhavi, S.B.: Recognition of Basic Kannada Characters in Scene Images using Euclidean Distance Classifier. International Journal of Computer Engineering and Technology (IJCET) 4 (March-April 2013)
[4] Jusoh, N.A., Zain, J.M.: Application of Freeman Chain Codes: An Alternative Recognition Technique for Malaysian Car Plates. IJCSNS International Journal of Computer Science and Network Security 9(11) (November 2009)
[5] He, X.C., Yung, N.H.C.: Curvature Scale Space Corner Detector with Adaptive Threshold and Dynamic Region of Support. In: Proceedings of the 17th International Conference on Pattern Recognition, vol. 2, pp. 791–794 (August 2004)
[6] Rath, T.M., Manmatha, R.: Word Image Matching Using Dynamic Time Warping. Center for Intelligent Information Retrieval, University of Massachusetts Amherst
[7] Pal, U., Chaudhuri, B.B.: Indian Script Character Recognition: a survey. The Journal of Pattern Recognition Society, 1887–1899 (2004)

[8] Acharyya, A., Rakshit, S., Sarkar, R., Basu, S., Nasipuri, M.: Handwritten Word Recognition Using MLP based Classifier: A Holistic Approach. IJCSI International Journal of Computer Science Issues 10(2(2)) (March 2013)

[9] Park, J.: An Adaptive Approach to Offline Handwritten Word Recognition. IEEE Transactions on Pattern Analysis and Machine Intelligence 24(7) (July 2002)

[10] Gonzalez, R.C., Woods, R.E., Eddins, S.L.: Digital Image Processing Using MATLAB, 2nd edn. Tata McGraw Hill (2010)

[11] Ramappa, M.H., Krishnamurthy, S.: A Comparative Study of Different Feature Extraction and Classification Methods for Recognition of Handwritten Kannada Numerals. International Journal of Database Theory and Application 6(4) (August 2013)

[12] Kannada, K.S.D.: Product Code, KAN0001208. Published Year (2006)

An Efficient Way of Handwritten English Word Recognition

M.S. Patel[1], Sanjay Linga Reddy[2], and Anuja Jana Naik[1]

[1] Department of Information Science and Engineering,
Dayananda Sagar College of Engineering, Bangalore, India
msr_patel@yahoo.com,
anuja2188@gmail.com
[2] Department of Computer Science and Engineering,
Alpha College of Engineering, Bangalore, India
sclingareddy@gmail.com

Abstract. Handwriting recognition has been one of the most fascinating and challenging research areas in the field of image processing and pattern recognition in the recent years, which is motivated by the fact that for severely degraded documents the segmentation based approach will produce very poor recognition rate. The quality of the original documents does not allow one to recognize them with high accuracy. Hence, the aim of this research is to produce system that will allow successful recognition of handwritten words, which is proven to be feasible even in noisy environments. This paper presents a method that performs pre-processing steps on hand written images such as skew and slant correction, baseline estimation, horizontal and vertical scaling. It uses structural features for feature extraction. Further, Euclidean distance method is applied for classification that produces single matching word having minimum difference value. This paper presents a sample of data set which encompasses the names of 30 districts present in the Karnataka state of India. This method is useful for the postal address, script recognition and systems which require handwriting data entry.

Keywords: Holistic word recognition, Skew correction, Slant correction, Feature extraction.

1 Introduction

From the time when the first alphabet was invented handwriting has become a medium for communicating messages and ideas between people, across space and time. Due to ubiquity and convenience of pen and paper in various everyday situations, handwriting has still survived as a useful and versatile communication method. In future, handwriting may only thrive more because of the technological developments under way that intend to establish handwriting as a new mode for

© Springer International Publishing Switzerland 2015
S.C. Satapathy et al. (eds.), *Proc. of the 3rd Int. Conf. on Front. of Intell. Comput. (FICTA) 2014*
– *Vol. 2*, Advances in Intelligent Systems and Computing 328, DOI: 10.1007/978-3-319-12012-6_62

humans to communicate with the computer and thereby enhancing the human-computer interactions.

The automation of handwritten form processing is drawing huge attention towards intensive research due to its wide applications and reduction of the tiresome manual workload. Postal service is an example of one such application of handwritten word recognition task which falls in two broad categories, namely On-line and Off-line. The major difference between Online and Offline Word Recognition is that Online Word Recognition has real time contextual information but offline data does not support real time contextual information [11].

From the literature survey, it is found that there exists popularly two approaches namely the segmentation approach and the holistic or segmentation free approach. In segmentation approach, each word is segmented into characters while holistic approach entails recognition of the whole word. This work mainly focuses on automatic handwritten document processing technology and its application to postal services. In this, we have considered thirty district names in Karnataka state for the purpose data collection which is obtained from [13], and upon which analysis and implementation of the research work is carried out.

The organization of the paper is as follows. Section 2 presents the related work, section 3 indicates the Design of this research. Section 4 presents Implementation and section 5, shows the Experimental results. Section 6 concludes the work.

2   Related Work

Due to the significance of handwritten recognition in the day to day activities, there are several researchers who are contributing in the above said domain.

Ankush Acharya and Sandip Rakshit worked on longest run based holistic feature. Longest run feature is computed in four directions namely row wise (east), column wise (north) and along the directions of two major diagonals (north east and north west).They used neural networks to classify word images. The dataset used to evaluate the technique is CMATER db 1.2.1 and accuracy of the technique is measured using three fold cross validation method. The best case and average case performances of the technique obtained are 89.9% and 83.24% respectively [1].

Mohamad tanvir Parvez,Sabri A.Mahmoud proposed first integrated offline Arabic handwritten text recognition system based on structural features. In implementation of the system, several novel algorithms for Arabic handwriting recognition using structural features are introduced. A novel segmentation algorithm is integrated into recognition phase, which is followed by Arabic character modeling by "fuzzy" polygons. It is further recognized using novel fuzzy polygon matching algorithm [2].

Prashant S. Kolhe, S.G.Shinde proposed a comprehensive machine for Marathi number recognition system. Euclidean distance based K-Nearest Neighbour classifier is used for recognition of numbers and vowels. Feature set is extracted as a moments feature. Accuracy obtained for vowels is 80% and for that of numbers is 81% [3].

3   Design

The design of this part of the research work in order to carry out offline English hand written word recognizer consists of various sequential steps. The data set is collected manually by various people. This work thus comprises of hand written data as obtained by people of various age groups such that the work is more accurate. However, this paper limits to the names of 30 districts of Karnataka state in India for which the entire sequential steps are performed. Figure1 depicts the English handwritten word recognizer.

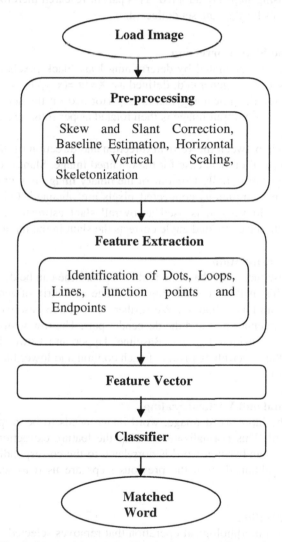

Fig. 1. Sequential steps for Handwritten Word Recognizer

The figure1 depicts the above said research design to enhance the handwritten word recognition, where the entire work is accomplished by following the steps as discussed below.

3.1   Pre-processing

In the process of scanning a word image, there can be aliasing effect on the shapes. Thus, it is necessary to have an effective preprocessing step in order to get high recognition rates. Depending on the application for which the image will be used various preprocessing steps are adopted. This part of research therefore considers pre-processing activities for the data set collected.

3.1.1   Skew and Slant Correction

In this work, skew is calculated by determining least black pixels in every column. Subsequently, a set 'X' is generated, defined as: $X=\{z_i=(x_i,y_i)|$lowest pixel in column $x_i\}$.Further least-squares linear regression is performed on the set 'X' to find skew line of the form y=ax+b. Input image is then rotated as per rotation angle calculated as $\theta=\arctan(a)$.

Slant is the angle between the vertical direction and direction of the strokes that are supposed to be vertical in ideal case for the obtained image. Slant is calculated for the obtained images where initially Contour of the binary image is found. Further, edges of strokes given by continuous connected pixels are obtained. Orientation of those edges with respect to vertical is used as overall slant estimate. Thereafter, Affine transformation with the estimated angle corrects the slant in the input image.

3.1.2   Baseline Estimation

For each word baseline, upper baseline and lower baseline can be detected. To extract the baselines in this method, considering the image as matrix of zeroes and ones we scan the image from top to bottom. We further detect three baselines that divide the input image into ascender part and the descender part. The row with the most number of black pixels is considered as the Baseline. Upper and Lower baselines can be calculated using the upper black pixels of each column and lower black pixels of each column respectively.

3.1.3   Horizontal and Vertical Scaling

To standardize the input word image, word image needs to be cropped sharp at the border of the word. This normalization makes the feature extraction more robust as the location of the handwritten word is correlated to the corresponding characters. In this step, regions obtained from the previous steps are used to scale the image to standard size.

3.1.4   Skeletonization

Skeletonization is a morphological operation that removes selected foreground pixels from the binary images. In our method to remove noise and to correct some of the handwritten mistakes, we first smoothen the image by convolution with a

2-dimensional Gaussian filter. Subsequently, iterative thinning algorithm is applied to reduce the strokes in the word to the width of a pixel.

3.2   Feature Extraction

Feature extraction is transforming the input data into the set of features. In our approach, we employ a fast and simple vertical slicing method to segment the input word into three vertical slices. Further, features from each zone are extracted and saved into feature vector. In this proposed method high level features such as loops, dots, lines, junction points and endpoints are identified and stored in feature vector.

3.2.1   Dots
In alphabets like 'i' and 'j' dots can be interpreted as isolated strokes in the ascender part of the letter. This can be accomplished using the following algorithm.

Step 1. Find connected components in the word image.
Step 2. Assign unique connected components in ascender part of image as strokes.
Step 3. If size of stroke is less than or equal to 15 and size of stroke is greater than or equal to 3 then stroke is considered as dot else continue with next stroke.

3.2.2   Horizontal and Vertical Lines
In this work, Hough Transform is used to detect lines in the word image. It is a global line detection technique with the ability of extracting directional information presenting good tolerance to disconnections, noise and a moderate tolerance to distortions [12].

3.2.3   Loops
Loop is a continuous pair of consecutive strokes that form a hole. Loops are detected from the connected component analysis method. It is done by finding areas of background color not connected to the region surrounding the word [10].

3.2.4   Endpoints
Endpoints are the points with only one neighbor which indicates end of the stroke. It is detected by applying morphological operation 'endpoints'. It returns matrix that contains 1 only if it is an endpoint else returns 0.

3.2.5   Junction points
Junctions are the points where two or more strokes meet. In this implemented method, 3x3 window is moved over the pre-processed image. If white pixel with more than two neighbors with 8-connectivity is found, it is considered as junction point.

3.3   Classification

For the arrangement or division of individual objects into classes or groups based on the characteristics they have in common, a classification method is used. In this method of our research, Euclidean distance based K-Nearest Neighbor classifier is used. It uses feature vectors of trained data set and test input image and produces single matching word based on minimum difference value[3].

4   Experimental Environment

We have implemented offline English handwritten word recognizer using matlab 7.14.0.739(R2012a). The snapshots of each step are as shown in the figures 2, 3, 4.

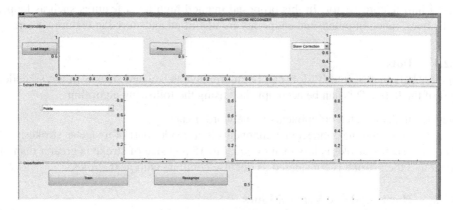

Fig. 2. Main Window

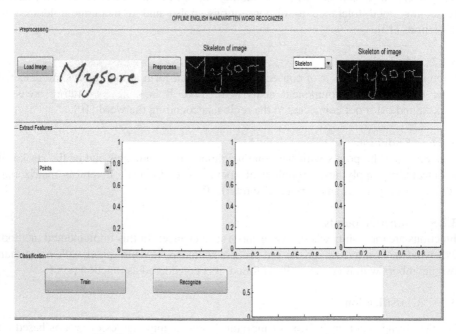

Fig. 3. Loaded image and the final Pre-processed image

Figure 2 indicates the provision for loading the image. Having loaded the image, next step of preprocessing is indicated in figure3.

Figure3. indicates a sample image, Mysore loaded and preprocessed by following all the steps of pre processing.

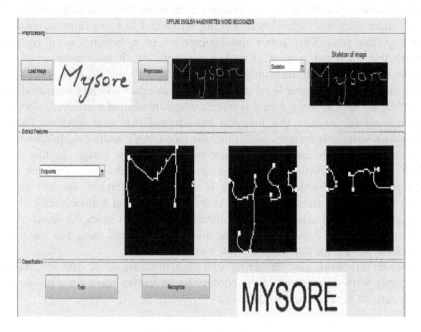

Fig. 4. Recognized image

Figure 4 indicates that the preprocessed image is subjected to zoning where the entire image is divided into three divisions. Further, Figure 4.shows the recognized image which appears in the printed way.

5   Experimental Results

Since this work focused mainly on postal address recognition in Karnataka state in India, the database contains 30 district names of Karnataka. These district names are written by 15 different writers across the state where each writer has written 20 different names, thus forming a database of 300 images. In the training phase, 87% of the total words from the database (262) images are trained and in testing phase 100% images from the database (300) are tested. The test made on the sampled data apparently indicates the classification rate to be 90%. This certainly ensures that the above illustrated approach of hand written recognition technique is effective and more accurate.

6   Conclusion

The significance of hand written word recognition has taken a high popularity in the current days due to the advancement of technology and the call for the day to pay more emphasize upon the human computer interactions. This research therefore leads towards addressing the variability in writing styles and to bring in more accuracy in the recognition rate. This paper thus comprises of a holistic word recognition approach that does not require character segmentation. It uses samples of district names in Karnataka as a bitmap image.

This method yields higher level of recognition accuracy and also increased overall time efficiency due to zone based feature extraction method employed. Among the different types of features, this research has adopted structural features. This work encompasses structural features which include loops, dots, endpoints, junction points, lines considering their number and positions in the word image. In this method, English handwritten word is deemed as input, which goes through several preprocessing steps. Further, the image is divided into three zones and structural features are extracted from each zone. At the end, based on feature vector obtained, the image is classified using Euclidean distance function. It gives the matched image based on minimum distance value. From the tests conducted using the implemented approach has shown the classification rate to be 90%.

However, this work can further by extended to consider as data sets names of taluks, along with other states of the country by collecting them from people spread over the various geographical locations.

References

1. Acharya, A., Rakshit, S.: Handwritten Word Recognition Using MLP based Classifier: A Holistic Approach. International Journal of Computer Science Issues 10, 422–427 (2013)
2. Parvez, M.T., Mahmoud, S.A.: Arabic Handwriting Recognition Using Structural And Syntactic Pattern Attributes. Pattern Recognition 46, 141–154 (2013)
3. Kolhe, P.S., Shinde, S.G.: Devanagari OCR using KNN and Moment. Proceedings of the IJCSE 2, 93–100 (2013)
4. Garg, M., Ahuja, D.: A Novel Approach to Recognize the off-line Handwritten Numerals using MLP and SVM Classifiers. International Journal of Computer Science & Engineering Technology 4 (2013)
5. Sas, J., Markowska-Kaczmar, U.: Similarity-Based Training Set Acquisition for Continuous Handwriting Recognition. Information Sciences 191, 226–244 (2012)
6. Kacem, A., Aouïti, N., Belaïd, A.: Structural Features Extraction for Handwritten Arabic Personal Names Recognition. In: Proceedings of IEEE Transactions on Frontiers in Handwriting Recognition (ICHR), pp. 268–273 (2012)
7. Rodriguez-Serrano, J.A., Perronnin, F.: Synthesizing Queries for Handwritten Word Image Retrieval. Pattern Recognition 45, 3270–3276 (2012)
8. Assabie, Y., Bigun, J.: Offline Handwritten Amharic Word Recognition. Pattern Recognition Letters 32, 1089–1099 (2011)

9. Sagheer, M.W., He, C.L., Nobile, N., Suen, C.Y.: Holistic Urdu Handwritten Word Recognition Using Support Vector Machine. In: Proceedings of IEEE Transactions on Pattern Analysis, pp. 1900–1903 (2010)
10. Steinherz, T., Doermann, D., Rivlin, E., Intrator, N.: Offline Loop Investigation for Handwriting Analysis. Proceedings of IEEE Transactions on Pattern Analysis and Machine Intelligence 31(2) (2009)
11. Negi, Bhagvati, C., Krishna, B.: An OCR System For Telugu. In: The Proceedings of the Sixth International Conference on Document Processing, pp. 1110–1114 (2001)
12. Le, D.X., Thoma, G.R.: Document Skew Angle Detection Algorithm. In: Proceedings of SPIE Symposium on Aerospace and Remote Sensing, vol. 1961, pp. 251–262 (1993)
13. http://www.census2011.co.in/census/state/districtlist/karnataka.html

Text Detection and Recognition
Using Camera Based Images

H.Y. Darshan[1], M.T. Gopalkrishna[1], and M.C. Hanumantharaju[2]

[1] Department of Information Science and Engg,
Dayananda Sagar College of Engg, Bangalore, India
[2] Department of Electronics and Communication Engg,
Sri Bhusanayana Mukundadas Sreenivasaiah Institute of Technology, Bangalore, India
{repsol26.DHY,gopalmtm,mchanumantharaju}@gmail.com

Abstract. The increase in availability of high performance, low-priced, portable digital imaging devices has created an opportunity for supplementing traditional scanning for document image acquisition. Cameras attached to cellular phones, wearable computers, and standalone image or video devices are highly mobile and easy to use; they can capture images making them much more versatile than desktop scanners. Should gain solutions to the analysis of documents captured with such devices become available, there will clearly be a demand in many domains. Images captured from images can suffer from low resolution, perspective distortion, and blur, as well as a complex layout and interaction of the content and background.In this paper, we propose an efficient text detection method based on Maximally Stable Exterme Region (MSER) detector, saying that how to detect regions containing text in an image. It is a common task performed on unstructured scenes, for example when capturing video from a moving vehicle for the purpose of alerting a driver about a road sign . Segmenting out the text from a clutterd scene greatly helps with additional tasks such as optical charater recognition (OCR). The efficiency of any service or product, especially those related to medical field depends upon its applicability. The applicability for any service or products can b achieved by applying thr basic principles of Software Engineering.

Keywords: Text detection, maximally stable extremal re-gions, connected component analysis.

1 Introduction

With the increase in growth of camera-based applications available on smart phones and portable devices, understanding the pictures taken by these devices semantically has gained increasing scope from the computer vision community in these years. Among all the information contained in the image, text, which carries semantic information, could provide valuable clues about the content of the image and thus is very important for human as well as computer to understand the scenes. For character recognition in the scene, these methods directly extract features from the original image and uses various classifiers to the character to recognize. While for scene text recognition, since there are no Binarization and Segmentation stages, most methods that exist take up multi-scale sliding window strategy to get the character detection results. As sliding window strategy does not make use of the special structure information of each character, it will produce many false positives. Thus, these methods mainly depends on the post processing methods such as pictorial structures. As proved by Judd et al., given an image containing text and other objects, viewers tend to fixate on text, suggesting the importance of text to human.

In fact, text recognition is indispensable for a lot of applications such as automatic sign reading, language translation, navigation and so on. Thus, understanding scene text is more important than ever. The following sources of variability still need to be accounted for: (a) font style and thickness; (b) background as well as foreground color and texture; (c) camera position which can introduce geometric distortions; (d) illumination and (e) image resolution. All these factors combine to give the problem a flavor of object recognition. Many problems need to be solved in order to read text in camera based natural images including text localization, character and word segmentation, recognition, integration of language models and context, etc.

2   Review of Literature

Devvrat C. Nigam et al. [2] proposed character extraction and edge detection algorithm for training the neural network to classify and recognize the handwritten characters. In general, handwriting recognition is classified into two types as off-line and on-line handwriting recognition methods. The on-line Methods have been shown to be superior to their off-line counterparts in recognizing handwritten characters due to the temporal information available with the former. There are basically two main phases in our Paper: Pre-processing and Character Recognition. In the first phase, they are preprocessing the given scanned document for separating the Characters from it and normalizing each characters. Initially we specify an input image file, which is opened for reading and preprocessing. The image would be in RGB format (usually), so we convert it into binary format. To do this, it converts the input image to grayscale format (if it is not already an intensity image), and then uses threshold to convert this grayscale image to binary i.e. all the pixels above a certain threshold as 1 and below it as 0.

Mohanad Alata et al. [6] proposed method was based on a combination of an Adaptive Color Reduction (ACR) technique and a Page Layout Analysis (PLA) approach. K. Atul Negi, Nikhil Shanker and Chandra Kanth Chereddi [6] presented a system to locate, extract and recognize Telugu text. The circular nature of Telugu script was exploited for segmenting text regions using the Hough Transform.

Jerod J. Weinman et al. [5] propose a probabilistic graphical model for STR that brings both bottom-up and top-down information as well local and long-distance relationships into a single elegant framework. In addition to individual character appearance, our model integrates appearance similarity, one underused source of information, with local language statistics and a lexicon in a unified probabilistic framework to reduce false matches errors in which the different characters are given the same label by a factor of four and improve overall accuracy by greatly reducing word error. The model adapts to the data present in a small sample of text, as typically encountered when reading signs, while also using higher level knowledge to increase robustness.

Shangxuan Tian et al. [3] propose to recognize the scene text by using an extension of the HOG, namely, co-occurrence HOG (Co- HOG) [13] that captures gradient orientation of neighboring pixel pairs instead of a single image pixel. Co-HOG divides the image into blocks with no overlap which is more efficient than HOG with overlapped blocks [25]. This is essential in the real-time text recognition system. More Importantly, relative location and orientation are considered with each neighboring pixel, respectively, which is more precise to describe the character shape. In addition, Co-HOG Keeps the advantages of HOG, i.e., the robustness to varying illumination and local geometric transformations.

Jain et al. [14] perform a color space reduction followed by color segmentation and spatial regrouping to detect text. Although processing of touching characters is considered by the authors, the segmentation phase presents major problems in the case of low quality documents, especially video sequences.

3   Proposed Method

The proposed text detection and recognition algorithm consists of the following steps: MSER region detection, Edge detection, Filter character candidates, Determine bounding boxes, Perform optical character recognition(OCR)

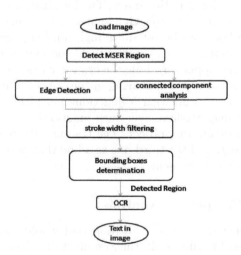

Fig. 1. Flow chart of the proposed method

3.1   Detect MSER Regions

Maximally Stable External Regions(MSER) are used as a method of blob detection in images. This method of extracting a comprehensive number of corressponding image elements contributes to the wid-baseline matching, and it has led to better stereo matching and object recognition algorithm. Becausee the regions are defined exclusively by the intensity functions in the region and the outer border, this leads to many key characteristics of the region which make them useful, and as follows: Invariance to affine transformation of image intensities, Covariance to adjacency preserving (continuous)transformation $T : D \to D$ on the image domain, Stability, Multi-scale detection without any smoothing involved, The set of all external regions can be enumerated in worst-case $O(n)$, where n is the number of pixels in the image. The algorithm -Start from a local intensity extremum point, -Go in every direction until the point of extremum of some function f. The curve connecting the points is the region boundary, -Compute geometric moments of orders up to 2 for this region, -Replace the region with ellipse.

3.2   Edge Detection

The Canny operator was designed to be an optimal edge detector (according to particular criteria — there are other detectors around that also claim to be optimal with respect to slightly different criteria). It takes as input a gray scale image, and produces as output an image showing the positions of tracked intensity discontinuities. The effect of the Canny operator is determined by three parameters — the width of the Gaussian kernel used in the smoothing phase, and the upper and lower thresholds used by the tracker. Increasing the width of the Gaussian kernel reduces the detector's sensitivity to noise, at the expense of losing some of the finer detail in the image. The localization error in the detected edges also increases slightly as the Gaussian width is increased. Usually, the upper tracking threshold can be set quite high, and the lower threshold quite low for good results. Setting the lower threshold too high will cause noisy edges to break up. Setting the upper threshold too low increases the number of spurious and undesirable edge fragments appearing in the output. One problem with the basic Canny operator is to do with Y-junctions i.e. places where three ridges meet in the gradient magnitude image. Such junctions can occur where an edge is partially occluded by another object. The tracker will treat two of the ridges as a single line segment, and the third one as a line that approaches, but doesn't quite connect to, that line segment.

3.3   Connected Component Analysis

First, we were instructed to write a function that would construct a histogram for a grayscale image. In order to do this, we must first be able to read in image data. In MATLAB, the imread() function will return a 3d array that looks like (rows,columns,color channels). Once we have our image read and stored in a 2d

Table 1. Results of Text detection

Algorithm	Precision	Recall
Ashida (ICDAR 2003) [9]	0.55	0.46
Shivakumara (TPAMI 2011) [1]	0.71	0.73
Our method	0.86	0.82

array, we can construct a histogram of relative frequencies. The histogram is an array with a user specified number of bins. The user also specifies the min and max value to use in creating the histogram. For example, if min and max were 100 and 200 respectively, every value lower than 100 would be placed in the smallest bin and every value above 200 would be placed in the largest bin. Finally, each bin is normalized by dividing by the number of pixels in the image. This makes the area under the histogram 1.0, and gives us the probability a pixel has a certain intensity if sampled randomly from the image.Finally we filter out the character candidates using the same.

3.4 Stroke Width Filtering

Another useful discriminator for text in images is the variation in stroke width within each text candidate. Characters in most languages have a similar stroke width or thickness throughout. It is therefore useful to remove regions where the stroke width exhibits too much variation [1]. The stroke width image below is computed using the helperStrokeWidth helper function.

3.5 Determining Bounding Boxes

To compute a bounding box of the text region, we will first merge the individual characters into a single connected component. This can be accomplished using morphological closing followed by opening to clean up any outliers.

4 Experimental Results

In this study, the images for experiments are from ICDAR database [13]. Even we have experimented on NEOCR (Natural Environment Optical Character Recognition), MSER datasets . The precision rate and recall rates are to be as shown in the above table (tab.1). The experimental results for image is as shown below where (a) is the original image taken and followed by the stages to extract text region in image as in (i). Compared with the best algorithm we have listed, the proposed method improved the precision rate by 15 percent and improved the recall rate by 9 percent.

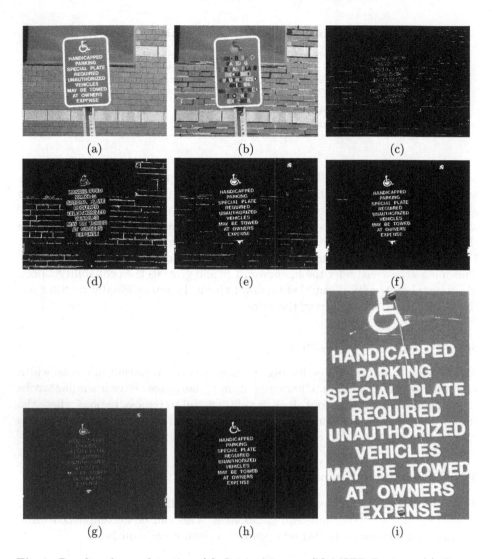

Fig. 2. Results of text detection. (a) Original image, (b) MSER Regions, (c) Canny edge detector, (d) Grown edges, (e) Segmented MSER regions, (f) Region filtered image, (g) Visualization of text candidates using stroke width, (h) Text candidates after stroke width filtering, (i) Text region

5   Conclusion

In this paper, we present a new text detection approach.Our approach uses both MSER and edges information.We have presented a method for locating text within natural images. The algorithm relies on a fundamental feature of text: text is usually surrounded by a contrasting, uniform background. Our proposed method of text segmentation searches for the textâĂŹs background rather than the actual text. This allows for a large variation in the distribution of text features while requiring little computation. We propose a MSER region detector to find the common characteristics of the text in an image. Experimental results on our own database as well as ICDAR 2003 text locating dataset demonstrate that our approach is robust to the orientation, perspective, color, and lighting of the text object, and can detect most text objects successfully and efficiently.

References

1. Wang, K., Babenko, B., Belongie, S.: End-to-end scene text recognition. In: International Conference on Computer Vision, ICCV (2011)
2. Shahab, A., Shafait, F., Dengel: Reading text in scene images. In: International Conference on Document Analysis and Recognition (2011)
3. Felzenszwalb, P., Girshick, R., McAllester, D., Ramanan, D.: Object detection with discriminatively trained partbased models. IEEE Transactions on Pattern Analysis and Machine Intelligence 22(3), 402–413 (2011)
4. de Campos, T., Babu, B., Varma, M.: Character recognition in natural images. In: VISAP (2009)
5. Boureau, Y., Bach, F., LeCun, Y., Ponce, J.: Learning mid-level features for recognition. In: Computer Vision and Pattern Recognition, vol. 2013, Article Id 716948, 8 pages. Hindwani Publication Corporation
6. Weinman, J.J.: Typographical features for scene text recognition. In: IAPR International Conference on Pattern Recognition (August 2010)
7. Sharma, N., Pal, U., Kimura, F.: Recognition of Handwritten Kannada Numerals. In: 9th International Conference on Information Technology (2010)
8. Mishra, A., Alahari, K., Jawahar, C.V.: Top-down and bottom-up cues for scene text recognition. In: Proceedings of IEEE Conference on Computer Vision and Pattern Recognition (CVPR) (2012)
9. Newell, A., Griffin, L.: Multiscale histogram of oriented gradient descriptors for robust character recognition. In: International Conference on Document Analysis and Recognition, ICDAR (2011)
10. Mishra, A., Alahari, K., Jawahar, C.V.: Top-down and bottom-up cues for scene text recognition. In: CVPR (2012)
11. Wang, K., Babenko, B., Belongie, S.: End-to-end scene text recognition. In: International Conference on Computer Vision, ICCV (2011)
12. Watanabe, T., Ito, S., Yokoi, K.: Co-occurrence histograms of oriented gradients for human detection. Information and Media Technologies (2010)
13. Sin, B., Kim, S., Cho, B.: Locating characters in scene images using frequency features. In: Proc. IEEE Int. Conf. Pattern Recognition (2010)
14. Shivakumara, W.H.P., Tan, C.: An efficient edge based technique for text detection in video frames. In: Proc. 8th IAPR Workshop Document Analysis Systems (September 2008)

Retinal Based Image Enhancement Using Contourlet Transform

P. Sharath Chandra, M.C. Hanumantharaju, and M.T. Gopalakrishna

Department of Information Science and Engineering,
Dayananda Sagar College of Engineering, Bangalore, India
{sharath.chandra174,mchanumantharaju,gopalmtm}@gmail.com

Abstract. In medical image processing, retinal image enhancement is the challenging issue to reveal the unseen details of an retinal image, thus, in many applications image enhancement issued to solve the challenges such as, noise reduction, blurring, degradation, etc. To improve the visual grade of retinal images we have many alternative image enhancement techniques that are suitable for specific application. This paper presents an overview of various retinal image enhancement techniques that will process the original Retinal image to obtain enhanced image suitable for a specific application. The method used in this paper has been evaluated with help of PSNR image Quality measure which is applied over several retinal images which is obtained from the datasets such as DRIVE, STARE and few other's provided by local medical experts. The comparative experimental results indicate that our proposed enhanced method has better outcome.

Keywords: Image Processing, Retinal Images, Image analysis, Relative study, Retinal Image Enhancement Techniques, Contourlet Transform, Multi-scale decomposition.

1 Introduction

Image enhancement is used for the refinement of image quality, and to provide detailed information for Retinal image examiner, and also for any automated image processing techniques. During this refinement, one or more attributes of an image is reformed, for a particular task. Depending on the task the attributes are modified without spoiling original image. Spatial Domain method and Frequency Domain method are the two main categories in image enhancement. During the diagnosis and detection of many Retinal diseases, image enhancement techniques are used to recognize Diabetic Retinopathy (DR) and Age Related Muscular degradation (AMD). thus, refinement of an image is mandatory for diagnosis performed either automatically or manually A Retinal image shown in Fig.1, is affected by non-uniform illumination, in the captured image it can be seen that contrast and luminosity is not uniform. Thus the fundamental problem is to improve the quality of an image

© Springer International Publishing Switzerland 2015
S.C. Satapathy et al. (eds.), *Proc. of the 3rd Int. Conf. on Front. of Intell. Comput. (FICTA) 2014*
– *Vol. 2*, Advances in Intelligent Systems and Computing 328, DOI: 10.1007/978-3-319-12012-6_64

during retinal image analysis. There are various filters that are available such as Gaussian filter, median filter, High pass filter and Low pass filter that are used to remove the noise that corrupt the retinal images.

The Basic pre-processing steps for image enhancement are:

1. Input image: - Image can be blur images, medical image etc.
2. Pre-processing of an image: - Before applying various image enhancement techniques on the input images, various pre-processing methods are applied on those images.
3. Applying domain techniques: - By using domain techniques quality of Pre-processed image is enhanced[12]. Such as,
    - Log Transformation Technique
    - Power Law Transformation Technique
    - Alpha rooting Technique

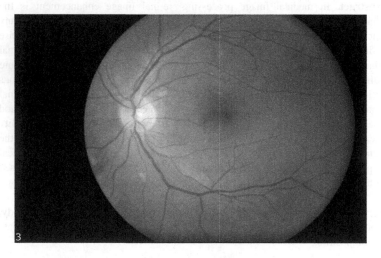

Fig. 1. Retinal Image with Uneven Illumination and Contrast

2   Related Work

M Emre Celebi et al[14]. have proposed the feature preserving contrast enhancement Method for Retinal vascular Images using discrete-shearlet transform (DST) and perceptual uniform color space CIEL*a*b*. The study specify that Diabetic retinopathy and age related macular degeneration are the diseases that cause the retinal impairment and blindness. This method is widely used for both color and gray-scale image enhancement, however this study does not consider the uneven illumination and the noise reduction during reconstruction of the image.

Peng Feng at al[15]. have proposed the concept of Contourlet transform over the retinal images using multi-scale edge enhancement. This study focus on the contrast enhancement which is necessary in pre-processing step for both natural images and ophthalmic images. In this method the image enhancement is performed using Contourlet transform which comprises of the following two steps: Laplacian pyramid (LP) and directional filter bank (DFB).

3   Proposed Method

Today, the Retinal Imaging is widely used in the diagnosis of retinal diseases such as blindness and visual impairments. The images obtained through digital devices contains uneven light illumination and some amount of noise, this results in poor quality of image which is not feasible for medical diagnosis. So far many methods are used world wide to enhance the retinal images, which have marginal drawbacks when compared.

In this proposed method we improve the visibility of image optimally for the detection of various retinal diseases. For experimental purpose, various input retinal images have been obtained from the datasets DRIVE, STARE and few other's provided by local medical experts. As indicated in Fig [2]. Initially the retinal image input obtained can be a gray scale image or color image. In an input image the edges may be brighter than the background area and surrounding edges may be weaker thus Discrete-shearlet Transform(DST) is applied on input retinal image. When the input is color image then Red, Green and Blue components are extracted. Since Green component has the optimal visibility we consider only green component as a primary source for the image enhancement.

Now the green component is converted into gray-scale. Further CLAHE is applied on green component gray-scale image or any suitable HE method is applied as applicable to the type of image. After performing DST, at the time of image reconstruction there will be marginal uneven illumination and noise generated. Hence we use the methods such as Gamma correction which reduces the uneven illumination to obtain optimum image quality and suitable filters to remove different kinds of noise generated. Followed by Contourlet transform is applied.

Finally we verify the image quality by using peak signal to noise ratio (PSNR) quality metric. By following the above proposed architecture, finally we get the optimal enhanced retinal image which has the best PSNR quality measure as experimentally compared.

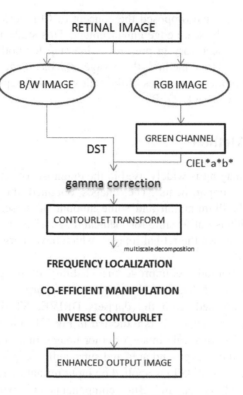

Fig. 2. Proposed architecture

3.1   RGB Color Components

In any given color inputs, the RGB color components are extracted. By extracting these components we can observe the image clarities in all three components, in all the images of the extracted components, the green channel is observed to have significantly better visual quality than others as shown in Fig [3.1] below.

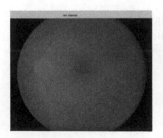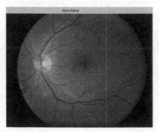

Fig. 3. Color components of RGB Image

The above selected green channel component is converted to gray-scale image, this gray-scale image can be feasible to apply gamma correction in reduction of uneven illumination, and noise reduction is possible by applying suitable filters. Thus we can infer that image is ready for applying contourlet transform. In an obtained input image the RGB color component extraction plays a vital role, further where the green channel component is responsible for image enhancement in getting the optimal result.

3.2   Contourlet Transform

The method Contourlet transform is associated with the decomposition of image into smaller sectors. By Contourlet transform we obtain sector wise smoother image along with smooth contours. The Contourlet transform technique overcome the challenges over traditional methods as wavelet transform and curvelet transform techniques.

In this proposed method we make use of multi-scale decomposition. Multi-scale decomposition technique is best suited to get enhanced image. According to the study, Laplacian decomposition method is widely used for Image enhancement process. There exist a few drawbacks by using Laplacian pyramid in this method. Laplacian pyramid does not focus on oscillations in the images. Since the Laplacian Pyramid give chance of missing a few sub-bands during the reconstruction process image might not have the optimal resolution.

Due to the above mentioned drawbacks, in the proposed method we make use of the multi-scale decomposition where the oscillations in the image is identified and we preserve the edges . By preserving edges the image obtained is smoother. The method of multi-scale decomposition focus on improving the contrast and preserve the contours quality. This makes the retinal image clearly visible and final outcome will have the optimal value. The scaling ratio is maintained uniform and during reconstruction process the quality of image is not lost, where as in traditional methods such as Laplacian pyramid this is a challenge to overcome.

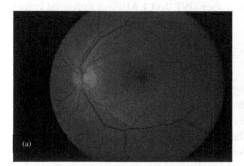 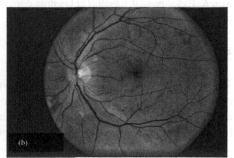

Fig. 4. Contourlet Transform: (a) Input Image (b) Enhanced output Image

4   Experimental Results

For the experimental purpose we consider various retinal image inputs which are obtained from datasets such as DRIVE, STARE and few other's provided by local medical experts. The obtained inputs are either gray scale or color image. The Experiment is carried out for all kind of images, and we could observe optimal results tested under various conditions. Below Fig [4.1] shows pictorial representation of the process where we can see the optimal enhanced image after applying contourlet transform, in which the obtained enhanced image is feasible for medical diagnosis. Compared to traditional methods, this proposed experimental result provides the highest PSNR value.

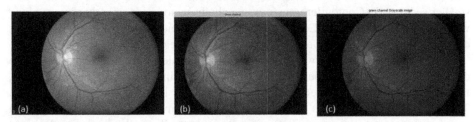

Fig. 5. Experimental Result: (a) Input color image (b) Extracted Green channel of color image (c) gray-scale of green channel image.

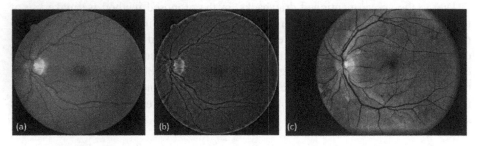

Fig. 6. Experimental Result: (a) Gamma correction after DST (b) CLAHE (c) Enhanced Image obtained using Contourlet Transform

5   Conclusion

The proposed method focuses on improving the visibility of image based on quality metrics like Edge Improvement Index, Contrast Improvement Index and Peak signal to Noise ratio. As evidenced by experiments with the Contourlet transform, there is a better preservation of contours than with other methods. Traditional methods does not consider the uneven illumination and noise reduction, hence the proposed method takes care of both to provide optimal image quality which is very appropriate for medical diagnosis. Finally proposed method tries to maintain a balance between PSNR, EII and CII, which will give us all these values in the accepted range and the comparative experimental results indicate that our proposed enhanced method has better outcome.

References

1. Jamal, I., Akram, M.U., Tariq, A.: Retinal Image Preprocessing: Background and Noise Segmentation. TELKOMNIKA 10(3), 537–544 (2012)
2. Bedi, S.S.: Various Image Enhancement Techniques- A Critical Review. International Journal of Advanced Research in Computer and Communication Engineering 2(3) (2013)
3. Joshi, G.D., Sivaswamy, J.: Colour Retinal Image Enhancement based on Domain Knowledge. In: Sixth Indian Conference on Computer Vision, Graphics and Image Processing (2008)
4. Shaeidi, A.: An Algorithm for Identification of Retinal Microaneurysms. Journal of the Serbian Society for Computational Mechanics 4(1) (2010)
5. Wang, Y., Tian, D., Vetro, A.: A Local Depth Image Enhancement Scheme for View Synthesis. Copyright Mitsubishi Electric Research Laboratories, Inc. (2012)
6. Maini, R., Aggarwal, H.: A Comprehensive Review of Image Enhancement Techniques. Journal of Computing 2(3) (2010) ISSN 2151-9617
7. Yahya, S.R.: Review on Image Enhancement Methods of Old Manuscript with Damaged Background. International Journal on Electrical Engineering and Informatics 2(1) (2010)
8. Vij, K., Singh, Y.: Enhancement of Images Using Histogram Processing Techniques. Int. J. Comp. Tech. Appl. 2(2), 309–313 (2009)
9. Kaur, N.: Image Enhancement Techniques a Selected Review. International Journal of Engineering Research and Technology (IJERT) 2(3) (March 2013)
10. Robinson, P.E., Lau, W.J.: Adaptive Multi-Scale Retinex algorithm for contrast enhancement of real world scenes (2012)
11. Kaur, E.M., Jain, E.K., Lather, E.V.: Study of Image Enhancement Techniques: A Review. International Journal of Advanced Research in Computer Science and Software Engineering 3(4) (2013)
12. Gavet, Y., Fernandes, M., Pinoli, J.-C.: Quantitative evaluation of image registration techniques in the case of retinal images. @JEIGavet 2012 (2012)
13. Mundhada, S.O., Shandilya, V.K.: Spatial and Transformation Domain Techniques for Image Enhancement. International Journal of Engineering Science and Innovative Technology (IJESIT) 1(2) (2012)
14. Abbas, Q., Farooq, A., Khan, M.T.A., Celebi, M.E., Garsia, I.F.: Feature Preserving Contrast Enhancement For Retinal Images. ICIC International 9(9), 3731–3739 (2013) ISSN 1349-1198
15. Feng, P., Pan, Y., Wei, B., Mi, D.: Enhancing: Retinal Image by the Contourlet Transform. Pattern Recognition Letters 28, 516–522 (2007)

Detection and Classification of Microaneurysms Using DTCWT and Log Gabor Features in Retinal Images

Sujay Angadi and M. Ravishankar

Department of Information Science and Engineering,
Dayananda Sagar College of Engineering, Bangalore, India
{sujayangadi90,ravishankarmcn}@gmail.com

Abstract. Diabetic Retinopathy (DR) is one of the major causes of blindness in diabetic patients. Early detection is required to reduce the visual impairment causing damage to eye. Microaneurysms are the first clinical sign of diabetic retinopathy. Robust detection of microaneurysms in retinal fundus images is critical in developing automated system. In this paper we present a new technique for detection and localization of microaneurysms using Dual tree complex wavelet transform and log Gabor features. Retinal blood vessels are eliminated using minor and major axis properties and correlation is performed on images with the Gabor features to detect the microaneurysms. Feature vectors are extracted from candidate regions based on texture properties. Support vector machine classifier classifies the detected regions to determine the findings as microaneurysms or not. Accuracy of the algorithm is evaluated using the sensitivity and specificity parameters.

Keywords: Diabetic Retinopathy, Microaneurysms, Blood vessels, Dual tree complex wavelet transform, Log Gabor.

1   Introduction

Diabetic retinopathy is the name given to retinal disease that is caused due to diabetes. Diabetes is a disorder of metabolism where in the pancreas either produces too little or no insulin or the cells do not react properly to the insulin that is produced. It is estimated that 15 to 25% of the diabetic population have diabetic retinopathy [1]. Diabetic Retinopathy is symptomless in the initial stage, screening is the only way to identify these patients to prevent them from going blind. The various signs like microaneurysms, hemorrhages, and exudates in eye represent the degree of severity of diabetic retinopathy. Microaneurysms are small saccular pouches that are caused by local distension of capillary walls and seem to be small red dots on the surface of the retina. Most of the microaneurysms found near thin blood vessels, but actually not on the blood vessels [2]. The use of image processing techniques on color fundus images is challenging because of distractors such as small vessels, intersection of two thin vessels, choroidal vessels, reflection artifacts and these may lead to misinterpretation [2].

© Springer International Publishing Switzerland 2015   589
S.C. Satapathy et al. (eds.), *Proc. of the 3rd Int. Conf. on Front. of Intell. Comput. (FICTA) 2014*
– *Vol. 2*, Advances in Intelligent Systems and Computing 328, DOI: 10.1007/978-3-319-12012-6_65

2   Related Work

Abhir Bhalero et.al [3] proposed the method for detection of microaneurysms based on complex valued circular symmetric filters and Eigen image. Firstly contrast is normalized on green channel of image using median filter then blob detection is performed using Laplacian of Gaussian (Log). Shape filtering is done using a circular-symmetry operator on the orientation map of the data and thereby combining the outputs with the LoG that produces a candidate set. Small windows are created around candidates are then used as features. Akara Sopharak et.al [4] proposed the automatic method for detection of microaneurysms using morphological operators. Microaneurysms are identified by their diameter and isolated connected red pixels with a constant intensity value, and whose external boundary pixels have a higher value in the green channel of a RGB image. Atsushi Mizutani et.al [5] reported detection of the microaneurysms by applying the double-ring filter on the green channel of the color images and followed by the elimination of lesions in the blood vessels. Next, the shapes of the candidate lesions for an accurate determination of their image features were examined. Finally, the candidate lesions were classified as microaneurysms using an artificial neural network (ANN) and by rule based classifier. Istvan Lazar et.al [6] proposed the method that is capable of constructing a score map to a fundus image, in which the points corresponding to MAs will achieve high scores, while the response to the other regions is minimal. The final microaneurysms are extracted by simple thresholding or by considering all the obtain probability scores obtained. Christy.R.Joseph et.al [7] designed the algorithm that uses Sobel function is used for de-noising the image. This Sobel method is able to detect exact location of optic disk with damaged blood vessels. In this paper, we propose a method that uses color fundus images, achieving a good proportion between the parameters such as sensitivity and specificity. The important contribution of the proposed method is the detection of microaneurysms and blood vessels, where it was used a solution based on DTCWT and log Gabor.

3   Methodology

The idea behind the approach of identifying microaneurysms is shape features that are very different from other regions such as vessels and exudates. This is employed by using matched filter. Microaneurysms occur in circular pattern whose diameter is less than 125um. Fig 1 shows flow of proposed system. First pre-processing is carried out in order to make further process more suitable. Pre-processing involves conversion of RGB image to appropriate grayscale image.

Discrete Wavelet Transform (DWT) has a major drawback i.e lack of shift variance. A small shift of the signal causes major variations in the distribution of energy between wavelet coefficients at different scales. And also lack of directional selectivity complicates processing of geometric image features like ridges and edges. DTCWT overcomes these two drawbacks. It contains trees of which each tree contains purely real filters, but the two trees produce the real and imaginary parts respectively of each complex wavelet coefficient.

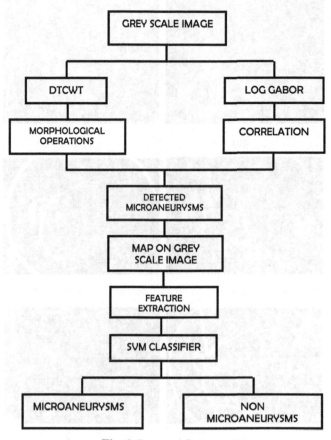

Fig. 1. Proposed System

The DT-CWT has six different wavelets oriented at ±15°, ±45°, and ±75° for $i = 1$, 2, 3 as follows

$$\Psi_i(n,m) = \frac{1}{\sqrt{2}}(\Psi 3,1(n,m) + \Psi 4, i(n,m))$$

$$\Psi_{i+3}(n,m) = \frac{1}{\sqrt{2}}(\Psi 3,1(n,m) + \Psi 4, i(n,m))$$

(1)

Oriented wavelets are produced from 2-D wavelets $\psi(n,m) = \psi(n)\psi(m)$ associated with row-column implementation of wavelet transform, where $\psi(n)$ is the complex wavelet defined by

$$\Psi(n) = \Psi_h(n) + j\Psi_g(n)$$

(2)

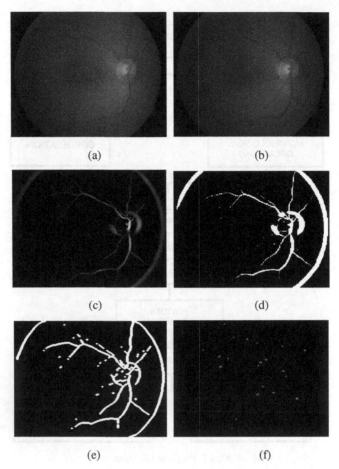

(a)   (b)

(c)   (d)

(e)   (f)

Fig. 2. Detection of microaneurysms. (a) Original Image, (b) Grayscale Image, (c) DTCWT Gradient Image, (d) Binary Image, (e)Morphological operations on Fig. 2(d) and (f) Detected Mircoaneurysms.

DTCWT provides the directional selectivity [8]; therefore it is possible to enhance pixels of vessels and microaneurysms. Effective thresholding is performed by estimating thresholding parameter by experimenting. Microaneurysms and vessels both appear afte thresholding. Morphological processing is carried out such as thinning and dilation to highlight vessels and microaneurysms. Vessels are large in area and have connected components but whereas the microaneurysms are circular in nature and have limited connected components. Taking these factors into consideration we eliminate vessels and other noise using major and minor axis. Another condition also is applied based on number of connected components thereby detecting only microaneurysms. Fig. 2 shows the process of detecting microaneurysms using DTCWT.

Most of the microaneurysms were detected but the microaneurysms that are closer to the retinal blood vessels were not detected because during elimination of vessels microaneurysms were found as a part of vessels. Thus they got eliminated as vessels. In order to overcome this drawback we go for texture Log gabor based identification of microaneurysms. Thereby forming multimodel which uses two methods for detection.

A disadvantage of the Gabor filter is that the even symmetric filter will have a DC component whenever the bandwidth is larger than one octave. However, Log Gabor can cover large frequency space while still maintaining a zero DC component in the even symmetric filter [9], [10]. Therefore the background brightness will not affect the extraction of pure phase information of microaneurysms texture. Filters are constructed in the frequency domain using a polar co-ordinate system. On the linear frequency scale, log Gabor has a transfer function of the form

$$g\,(\,f\,) = e^{\dfrac{-\left(\log\left(\frac{f}{f_0}\right)\right)^2}{2\log\left(\frac{\delta}{f}\right)^2}} \tag{3}$$

Where f_0 is the center frequency of filter and σ is the bandwidth of filter. Convolution is performed on the cropped regions of some microaneurysms using the log gabor filter. Convolution is performed on each row of the image extract the log gabor features using gabor filter. Correlation procedure is applied on test images in order to detect the microaneurysms using obtained range of log gabor features. Once the microaneurysms are detected through both the process, they are mapped on the gray scale image. Fig. 3 shows combined results of both the methods. Mapped regions are again cropped automatically and fed as input for feature extraction and classification stages.

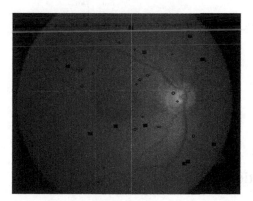

Fig. 3. Combined results

4   Feature Extraction

In statistical based texture analysis, texture features are computed from the statistical distribution of combinations of intensities at specified positions relative to each other in the image. The Gray Level Co-occurrence Matrix (GLCM) method is a way of extracting statistical based texture features.

For an image of size M × N, the gray level co-occurrence matrix (GLCM) is defined as in [11]

$$C_d(i,j) = |\{(P,Q),(P+\Delta X,Q+\Delta Y)\}: \quad (4)$$
$$I(P,Q) = i, I(P,\Delta X,Q+\Delta Y) = j\}|$$

Where (P, Q),(P+ΔX,Q+ ΔY) ∈ M*N, d=(ΔX, ΔY) and |.| denotes the cardinality of a set. From the co-occurrence matrix we calculate the features: Mean, Variance, Contrast, Energy and Dissimilarity.

GLCM mean measures the mean of the probability values from the matrix, this is the mean of all neighbor pixels that have reference pixel with grey level i. Variance is a measure of the dispersion of the values around the mean. It is similar to entropy. Contrast is the measure of local variations present in the image. Energy measures textural uniformity of the image. Dissimilarity is also a measure of variation of grey level pairs but it depends upon the distance from the diagonal weighted by its probability.

$$Mean = \mu_i = \sum_{i,j=0}^{N-1} i(P_{i,j})$$

$$Variance = \sigma_i^2 = \sum_{i=0}^{N-1} \sum_{j=0}^{N-1} P(i,j)(i-\mu_i)^2$$

$$(5)$$

$$Contrast = \sum_i \sum_j (i,j)^2 P_d(i,j)$$

$$Energy = \sqrt{\sum_i \sum_j P_d^2(i,j)}$$

$$Dissimilarity = \sum_{i,j=0}^{N-1} P_i j(i-j)$$

5   Classification

Classification is performed using support vector machine which maps the training data into high-dimensional feature space. We construct a hyperplane maximizing the margin, or distance from the hyperplane to nearest data points [12]. The classification is based on the features extracted from detected microaneurysms. For nonlinear classification of the given data, SVM uses a non-linear kernel function to map the given data into a high dimensional feature space where the given data can be linearly classified. Kernel function K(x,y) represents the inner product in feature space.

We use radial basis function (RBF) as kernel because it can handle the case when the relation between class labels and attributes is nonlinear. The results of the classification procedures are shown in the Fig 4.

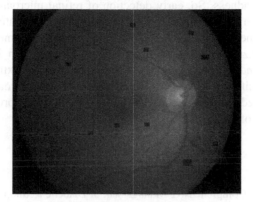

Fig. 4. Classification Results

6   Experimental Results

The features of microaneurysms are obtained from the preprocessed image and are provided as input to the SVM classifier. The implementation of this technique is carried out using Matlab. The sensitivity, specificity and accuracy values are calculated using following formulas.

$$Sensitivity(SN) = \frac{T_P}{T_P + F_N}$$

$$Specificity(SP) = \frac{T_N}{T_N + F_P} \tag{6}$$

$$Accuracy = \frac{T_P + T_N}{T_p + T_N + F_p + F_N}$$

True positive (TP), a number of candidate regions correctly detected as microaneurysms. False positive (FP), a number of candidate regions which are detected wrongly as microaneurysms. False negative (FN), a number of microaneurysms that were not detected. True negative (TN), a number of regions that were correctly identified as non microaneurysms. Sensitivity is the percentage of actual microaneurysms that are detected and specificity is the percentage of non microaneurysms that are classified as non microaneurysms. In our experimentation we performed testing on 20 retinal images. Sensitivity obtained is 91.12%. Specificity is 91.16% and accuracy is 91.42%.

7   Conclusion

Microaneurysm is a common thing in people suffering from diabetic retinopathy. If ignored, leads to blindness. So, early detection of retinal microaneurysm is a crucial task. Image processing algorithms allow development of automated systems for early detection of microaneurysms. We have presented an efficient method for automatic detection and localization of retinal microaneurysm based on Dual tree complex wavelet transform and log gabor features. Texture features are extracted using GLCM method. SVM classifier is used for the classification of microaneurysm and non-microaneurysm based on the extracted features. This flow is proved robust for the automatic detection of retinal microaneurysms.

References

1. Guidelines for the Comprehensive Management of Diabetic Retinopathy in India:A Vision 2020 The Right to Sight INDIA (2008)
2. Antal, B., Hajdu, A.: Improving microaneurysms detection in color fundus images by using an optimal combination of preprocessing methods and candidate extractors. In: 18th European Signal Processing Conference (EUSIPCO 2010), Aalborg, Denmark (2010)
3. Bhalerao, A., Patanaik, A., Anand, S., Saravanan, P.: Robust Detection of Mircoaneurysms for Sight Threatening Retinopathy Screening. In: Proceedings of Sixth Indian Conference of Computer Vision, Graphics & Image Processing, pp. 520–527 (2008)
4. Sopharak, A., Uyyanonvara, B., Barman, S., Williamson, T.: Automatic Microaneurysm Detection from Non-dilated Diabetic Retinopathy Retinal Images. In: Proceedings of the World Congress on Engineering, vol. II, pp. 1583–1586 (2011)
5. Mizutani, A., Muramatsu, C., Hatanaka, Y., Suemori, S., Hara, T., Fujita, H.: Automated Microaneurysm detection method based on double ring filter in retinal fundus images. In: Proceedings of SPIE. Medical Imaging, vol. 7260 (2009)
6. Lazar, I., Hajdu, A.: Microaneurysm Detection in Retinal Images using a Rotating Cross-Section based Model. In: IEEE International Symposium on Biomedical Imaging: From Nano to Macro (2011)
7. Joseph, C.R., Alexander, J.: Microaneurysm Detection and Diabetic Retinopathy Grading Using Candidate Extraction Algorithm. International Journal of Research in Engineering & Advanced Technology 1(1) (2013)
8. Selesnick, I.W., Baraniuk, R.G., Kingsbury, N.G.: The dual-tree complex wavelet transform. IEEE Signal Proc. Mag., 123–151 (2005)
9. Kovesi, P.: Image features from phase congruency. Videre Journal of Computer Vision Research, 1–27 (1999)
10. Kovesi, P.: Phase congruency detects corners and edges. In: DICTA, Sydney (2003)
11. Lee, J., Zee, B., Li, Q.: Segmentation and Texture analysis with Multi-model Inference for the Automatic Detection of Exudates in Early Diabetic Retinopathy. Journal of Biomedical Science and Engineering 6, 298–307 (2013)
12. Wisaeng, K., Hiransakolwong, N., Pothiruk, E.: Automatic Detection of Retinal Exudates using a Support Vector Machine. Applied Medical Informatics 32(1), 33–42 (2013)

Classifying Juxta-Pleural Pulmonary Nodules

K. Sariya and M. Ravishankar

Department of Information Science and Engineering,
Dayananda Sagar College of Engineering, Bangalore, India
{Sariyakazia,ravishankarmcn}@gmail.com

Abstract. Lung cancer is a disease of abnormal cells multiplying and growing into a tumor in the human lung. It is the most dangerous and widespread cancer in the world. According to the stage of discovery of cancer cells in the lung, the process of early detection plays a very important and essential role to avoid serious advanced stages to reduce its percentage of distribution. Our lung cancer detection system basically detects and recognizes Juxta-pleural pulmonary nodules; which are attached to the wall of the lung. It is done in 4 stages such as obtaining ROI (Region Of Interest), Segmentation, Feature extraction and Classification.

CT (Computed Tomography) is considered to be the best modality for the diagnosis of Lung cancer. ROI can be selected either manually or automatically. Automated ROI retrieval is preferred as manual selection is considered to be tedious and time consuming as the operator has to go through the dataset slice by slice and frame by frame. Ray-casting algorithm is used to segment nodule and neural networks are used to classify the nodules appropriately.

Keywords: Classification, Juxta-pleural, Segmentation, Ray-casting.

1   Introduction

Lung cancer is one of the most common causes of deaths among men and women throughout the world. According to a cancer research conducted by Cancer Research Foundation in UK, 28% of deaths occurring due to cancer come under the category of Lung cancer. It can be caused mainly due to excessive smoking or exposure to radon gas and asbestos fibers. Usually Lung cancer can be suspected by the presence of excessive cough, wheezing, chest pain or by the presence of an abscess. It can be treated surgically or by using radiation or chemotherapy. If Pulmonary nodules are identified during earlier stages, survival rate will increase up to 80%.

Many methods are available for the diagnosis of Lung cancer. Some of them are X-Ray, CT (Computed Tomography), MRI (Magnetic Resonance Imaging) and PET (Positron Emission Tomography). CT is considered to be the most efficient modality for Lung cancer detection as it has exceptional contrast resolution which helps in differentiating the densities.

© Springer International Publishing Switzerland 2015   597
S.C. Satapathy et al. (eds.), *Proc. of the 3rd Int. Conf. on Front. of Intell. Comput. (FICTA) 2014*
– *Vol. 2*, Advances in Intelligent Systems and Computing 328, DOI: 10.1007/978-3-319-12012-6_66

The images used for our project are of DICOM (Digital Imaging and Communication in Medicine) format. It is the most common form of images found in the field of Medical image processing.

Different techniques have been applied in order to segment pulmonary nodules. Since the size of a nodule changes before and after treatment, it is possible now to efficiently detect small nodules which are even less than 1cm in diameter. As Juxta-pleural nodules are very difficult to detect, most of the methods fail to identify them.

2   Types of Lung Cancer

Lung nodules are small and approximately spherical abnormal tissues not greater that 3cm in diameter. Pulmonary nodules can be classified based on their microscopic appearance, location or intensity profile.

Two categories of Lung Cancer based on microscopic appearance are SCLC (Small Cell Lung Cancer) and NSCLC (Non Small Cell Lung Cancer). If the criteria are location, they can be classified into four types such as Well circumscribed, Juxta-vascular, Juxta-pleural and Nodule with pleural tail. Depending upon the intensity they can be two types such as Solid and Ground glass.

Various methodologies such as Thresholding, Active contour and Region growing, which have been applied to detect pulmonary nodules, have failed for Juxta-pleural cases.

Ray casting is a unique methodology which has been successful for Juxta-pleural nodule segmentation in an efficient manner. This procedure basically starts by finding the Region of Interest (ROI). It is further enhanced so that contour points can be easily selected. A seed point is selected automatically, from which rays are casted towards the contour points. These rays are then subjected to two tests based on which the final contour points are decided using which nodule is segmented.

In the first test, we identify the path of the ray. If it is entering the background region, then we eliminate those points.

In the second test, we extend the path of the ray along the background to determine its final position. Those points which touch the border are saved, while the rest are eliminated.

3   Literature Survey

Si.Guanglei et.al[10] proposed a method to segment Juxta-pleural nodules by using Ray- casting algorithm. Size of the nodule was evaluated based on which the process was implemented.

Artit.C et.al[7] developed a Surface Fitting method to detect Juxta-Pleural nodules which was based on the growth rate.

Shen shen sun et.al[4] implemented a Region Growing method based on Mean Shift Analysis and Divergence rule which used size of the nodule to determine the class to which it belonged to.

Amal.A et.al[5] used a Variational level set approach based on size and shape of the nodule, to detect Juxta- Pleural nodules in an efficient manner.

Jun Lai et.al[6] developed an Active Contour model to segment Juxta-pleural nodules, considering the shape and size parameters.

Xujiong Ye et.al[8] segmented Juxta-pleural nodules based on Modified Expectation Maximization algorithm which used standard deviation as a measure to determine the efficiency. Mean shift method was also applied for detection, considering the volume as an efficient parameter.

Jinsa Kuruvilla et.al[3] used various morphological operations for segmentation based on morphological features such as area, perimeter etc.

4   Design

The major steps involved in the design are given in the following figure.

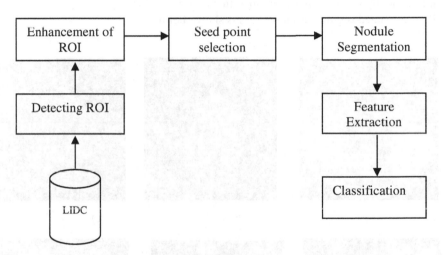

Fig. 1. Steps in detection and recognition of Juxta-pleural nodules

LIDC(Lung Image Database Consortium) is a publicly available reference for medical imaging research community. The database contains 1018 cases of Lung cancer, each providing images from clinical thoracic CT scan verified by a team of experienced radiologists. With the help of Radiologists, we identified the images in which Pulmonary Nodules where attached to the wall of the Lung.

Once we get the chest CT image from the database, the Region Of Interest (ROI) needs to be extracted. In our case it is done automatically by creating a mask and henceforth identifying the intensity values in comparison with other parts of the chest. Before applying the segmentation process, ROI is enhanced by using adaptive histogram equalization.

The segmentation algorithm is then applied on the ROI based on the seed point selection to detect the nodule. Many features of the nodule are evaluated and fed into a feature vector which is passed into the classifier to identify whether the nodule is Juxta-pleural or not.

5   Segmentation

Segmentation is basically the technique of dividing a digital image into multiple parts which makes it easier to represent and analyze. Segmentation algorithms are based on the properties of discontinuity and similarity. Based on the property of discontinuity, we partition an image depending upon sharp changes in intensity, while similarity property segments according to a predefined criterion. Ray casting is the segmentation method used here, to identify the Juxta-pleural pulmonary nodules in the image.

The process starts with the seed point selection, in the nodule area. Rays are casted from this point to the contour points. The number of rays should not be too small or too large. Boundary detection becomes difficult if the number of rays is less and moreover time consumption is huge in case of large number of rays.

The ROI can be selected form the chest CT either manually by drawing a rectangle or automatically based on masking operations.

The automated segmentation process is shown below.

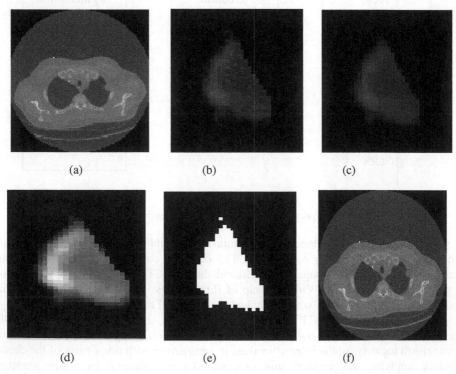

Fig. 2. Steps in Ray-casting segmentation process.(a)Chest CT (b)ROI Selection (c)ROI enhancement (d)Seed point selection (e)Segmented nodule (f) Detected ROI.

Ray Casting is basically a process which begins with the selection of a seed point. After binarizing the ROI, contour points are determined. To each contour point, a ray is drawn from the seed point.

Points which are not needed among these are removed by performing the test of extensibility. Those points which survive the above test will determine the segmentation process to be executed.

6   Feature Extraction

It is the process of transforming the input data to a set of features, so that the dimension of input data is reduced drastically. All the features extracted are stored in the form of a Feature vector.

Different features used for our project are Contrast, Entropy and Correlation. These features are based on texture of the nodule and are calculated from GLCM (Grey Level Co-occurrence Matrix). Besides these morphological features such as Area and Perimeter are also used.

GLCM is a frequency matrix which helps in enhancing the details used in order to interpret an image. It is calculated by determining how often a pixel with grey level intensity value (i) occurs adjacent to a pixel with value(j).

If P(i,j) is used to denote the GLCM matrix, then features based on texture can be determined as given below.

Contrast is calculated as:

$$\sum_{i,j} |i - j|^2 P(i,j) \tag{1}$$

Entropy is evaluated as:

$$\sum_{i,j} P(i,j) \log P(i,j) \tag{2}$$

Correlation is determined by:

$$\sum_{i,j} [(i - \mu_i)(j - \mu_j) P(i,j)]/(\sigma_i \sigma_j) \tag{3}$$

7   Classification

Based on training data set, a classifier is used to identify the class to which the test image belongs to. Depending upon the feature values extracted in the previous step, the nodules are classified as Juxta-Pleural or not.

Nodules are also classified based on the type of cancer as malignant or benign. The size of the nodule, ie its width and height is measured; to determine the type of cancer.

Neural network classifier is used for our method. It has the ability to adapt, generalize, cluster and organize data. It comprises of a group of processing units, which communicate by sending signals to each other, over a large number of weighted connections.

Neuron which acts as a processing unit, receive input from external sources, and computes output which is then propagated to other units. Each connection consists of a weight depending on which the propagation rule is determined.

8   Experimental Results

The accuracy for the proposed method is 83% and the sensitivity is 88%. They can be calculated by using the following formulas:

$$\text{Accuracy} = \frac{TP + TN}{TN + TP + FN + FP} \tag{4}$$

$$\text{Sensitivity} = \frac{TP}{TP + FN} \tag{5}$$

Where,

TP (True Positive) -   Predicts Juxta-pleural as Juxta Pleural.
FN (False Negative) - Predicts Juxta-pleural as Non juxta-pleural.
TN(True Negative -   Predicts Non Juxta-pleural as Non juxta pleural.
FP (False Positive) -   Predicts Non Juxta pleural as Juxta- pleural

9   Conclusion

We successfully proposed and implemented a method to segment Juxta-Pleural pulmonary nodules based on Ray casting. The accuracy of this algorithm is 83% with high processing speed. Hence it will be useful for the radiologists in diagnosing lung cancer. This method lays a foundation in order to diagnose Juxta- pleural nodules and hence provides useful data for the evaluation of treatment effects of Lung cancer.

The work suggested here shows that the above method is a useful feature for nodule detection and classification. In association with other methods, our approach has the advantage of automation. It could be seen from the system proposed that lung cancer detection and classification in CT image is important in medical diagnosis because, it provides information associated to anatomical structures as well as, potential abnormal tissues necessary to treatment planning and patient follow-up.

10   Future Works

Segmenting a nodule is very essential for the measurement of its size based on which a nodule can be categorized as malignant or not. They are also very helpful for the treatment of Lung Cancer.

It will be helpful for the radiologists, if we do further research for the segmentation of other types of nodules. In future we will extend our research by taking more features into consideration for identifying the nodules.

Acknowledgements. We would like to thank Dr. Satish M.R and Dr. Pranitha Rao for providing us their indebt knowledge in diagnosing the pulmonary nodules, which was very valuable for our implementation.

References

[1] Wey, Y., Jia, T., Lin, M.-X.: Autonomous detection of Solitary Pulmonary nodules on CT images for Computer aided diagnosis. In: IEEE Chinese Control and Decision Conference (2011) 978-1-4244-8738-7/11

[2] Wu, S., Wang, J.: Pulmonary nodules 3D detection on serial CT scans. In: IEEE Third Global Congress on Intelligent Systems (2012) 978-0-7695-4860-9/12

[3] Kuruvilla, J., Gunavathi, K.: Detection of Lung cancer using morphological operations. International Journal of Scientific & Engineering Research (2013) ISSN 2229-5518

[4] Sun, S.-S., Li, H., Hou, X.-R., Kang, Y., Zhao, H.: Automatic segmentation of pulmonary nodules in CT images (2007) 1-4424-1120-3/07 IEEE

[5] Amal, A., Hossam, A., James, G., Aly, F., Salwa, E., Sabry, Mohammed, Robert, F., Sahar, H., Rebecca, M.: Variational approach for segmentation of Lung nodules. In: 18th IEEE Conference on Image Processing (2011) 978-1-4577-1033-3/11

[6] Lai, J., Ye, M.: Active contour based Lung field segmentation. In: International Conference on Intelligent Human Machine Systems and Cybernetics IEEE (2009) 978-0-7695-3752-8/09

[7] Artit, C., Yury, D., Anthony, P., David, F., Claudia, H.: Segmentation of Juxta-pleural pulmonary nodules using a Robust surface estimate. International Journal of Biomedical Engineering, 10.1155/2011/632195

[8] Ye, X., Musib, S., Abdel, D., Gareth, B., Greg, S.: Shape based CT lung nodule segmentation using five-dimensional mean shift clustering and mem with shape information. IEEE (2009) 978-1-4244-3932-4/09

[9] Nie, S., Wang, Y., He, C., Ji, F., Liang, J.: A Segmentation Method for Sub-solid pulmonary nodules based on Fuzzy C-means Clustering. In: 5th International Conference on Biomedical Engineering and Informatics, BMEI 2012 (2012) 978-1-4673-1184-7/12

[10] Si, G., Cai, J., Kang, Y.: A three dimensional Ray casting method for Juxta- pleural nodule segmentation in thoracic CT nodules. In: IEEE International Conference on Information and Automation (2012) 978-1-4673-2237-9/12

The Statistical Measurement of an Object-Oriented Programme Using an Object Oriented Metrics

Rasmita Panigrahi[1], Sarada Baboo[2], and Neelamadhab Padhy[3]

[1] Department of Computer Science,
Gandhi Institute of Engineering and Technology, GIET,
Gunupur - 765022, Odisha, India and Sambalpur University, Burla, Odisha, India
rasmi.mcamtech@gmail.com
[2] Department of Computer Science and Application (CSA), Burla
Sambalpur University, Sambalpur, Odisha, India
[3] Department of Computer Science, Gunupur - 765022, Odisha, India
Gandhi Institute of Engineering and Technology
neela.mbamtech@gmail.com

Abstract. Object oriented design is more powerful than function oriented design. Previously the software was developed by using functional or structural approach but due to high quality demand, traditional metrics (i.e. Cyclomatic complexity, lines of code, comment percentage) cannot be applied. Object oriented metric assures to reduce cost and maintenance effort by serving earlier predictors to estimate software faults. The Object Oriented Analysis and Design of software gives the many benefits like reusability, decomposition of problems in to easily understandable objects. This paper presents the different object oriented metrics qualities in different dimensions (i.e. size, complexity, quality, reliability, etc). Object oriented metrics are used to analyze the complexity of any object oriented language (i.e. java, c++, C Sharp).In this paper we have taken the different sets of programs using C++ and Java. It concludes that Java dominants the C++ .The popularity is only due to measuring the software complexity, quality and estimation size of the projects.

Keywords: Object-Oriented Metrics, Object Oriented Programming Languages (C++, Java).

1 Introduction

The quality of the software product increasing popularly day by day and it can be measured in terms of the metrics are used to manage, predict and improve the quality. There are so many different types of metrics are used but in our paper focuses only the Object-Oriented metric(OOM).Now a days the software industry exactly lacking behind the standard metric .This paper presents an object-oriented metrics which can be uses to measure the quality of object oriented design. The metrics for object oriented design focus on measurements that are applied to the class and design characteristics. The main aim of OO metrics is to evaluate macro level assessment of

© Springer International Publishing Switzerland 2015
S.C. Satapathy et al. (eds.), *Proc. of the 3rd Int. Conf. on Front. of Intell. Comput. (FICTA) 2014*
– *Vol. 2*, Advances in Intelligent Systems and Computing 328, DOI: 10.1007/978-3-319-12012-6_67

the systems and finally produce the high-quality results. Now a day an object oriented design is becoming so much popular in the field of software development environment apart from object oriented design metrics is an essential part of software environment. OOAD (Object-Oriented Analysis and Design provides the different benefits like reusability, decomposition the problem into number of parts and each part is termed as an object .That object can easily understood and which support to future modifications .But OOAD life cycle is not much easier then the typical procedural approach. Therefore we must study the object oriented metrics which will helpful to write the reliable code. According to et.al [1] Watts S. Humphery (1996) OOM is an aspect to be considered. During the system design metrics can be used as a set of standards to measure the effectiveness of object oriented analysis technique The main goal of an object oriented metrics is to improve the quality of the software. [2],[3][4]. Object oriented programming is a popular concept in today's software development environment. To evaluate the quality of the object oriented software, we need to assess and analyze its design and implementation using appropriate metrics and evaluation techniques [5].The object-oriented measurements are being used to evaluate and predict the quality of software[6].Sometimes the empirical results supports the theoretical validity of these metrics [7][8][9]. Object Orientation contributes to the solution of many problems associated with the development and quality of software product. This technology promises greater programmer productivity, better quality of software and lesser maintenance cost [10] [11]. The OOM requires an approach which is completely different from the traditional functional decomposition and data flow development approach. Et.al [29] Yeresime Suresh they focused how best we can measure the effectiveness of the software metrics. There are many so many object oriented programming languages are there which supports the object oriented paradigm, these are C++, JAVA, .Net, C# etc., the programmers are considering the java and C#.net is the pure object oriented programming language. Java is highly suited for modeling the real world and solving the real world problems [12]. In this research paper different object oriented programs are studied. We have applied the different object oriented metrics on the same set of 20 programs in C++ and JAVA each. Then, we have calculated the statistical values like mean, median, standard deviation, etc. After that, we have compared the results for both languages.

2   Object Oriented Metrics (Chidamber & Kemerer's Metrics Suite)

Chidamber and Kemerer were developed the object oriented metrics which is popularly known as CK Metrics. [4]. In this section we have presented the metrics along with their structure in a tabular form. There are several factors are there which will help us to develop and design the software system[31] .How object oriented metrics compared with CK metrics that were presented by et.al[32] .In this part of this section we have represented the structured use OOM and CK metrics.

Table 1. For representing the object oriented structure in the CK metrics suites

Types of the Metrics used	Used Metrics	Which Structure Used in Object Oriented Domains
Traditional Metrics or older Metri (STAC)	It is used for understandability of the code i.e (LOC)	Implemented the Functions
	It is used for measuring the complexity of a method which are available in the class i.e CC(Cyclomatic Complexity	Implemented the Functions/Methods
	COM i.e The Comment percentage is used[23]	Implemented the Functions/Methods
New kind of Metrics (Object Oriented Metrics)	It is meant for count the which are implemented within a class (WMC) i.e. Weighted Method Per Class[25]	Used the Class/Function(method
	RFC is used to measure the understandability and testability.The number of functions or procedures that can be potentially be executed in a class i.e. RFC(Response for a class)[4][24]	The class/Message is used in this metrics
	It is the metrics which will indicates the level of cohesion between the methods in the class i.e. LCOM(Lack of Cohesion Methods)[17]	The Class/Cohesion) is used in this new kind of metrics
	It is the metrics which is used to count the total number of classes to which it is coupled i.e CBO(Coupling Between Object)[28],[30]	The coupling factor is used in this new kind of metrics
	Number of children cane be defined as the sum of all the number children's in the class (NOC)[26]	In this new object oriented metrics the Inheritance structure is used.
	The Depth of Inheritance calculated in terms sum of all the ancestors of the class[10],[27]	Inheritance structure is used in this new metrics

3   Motivational Examples for Class Hierarchy Diagram: JAVA

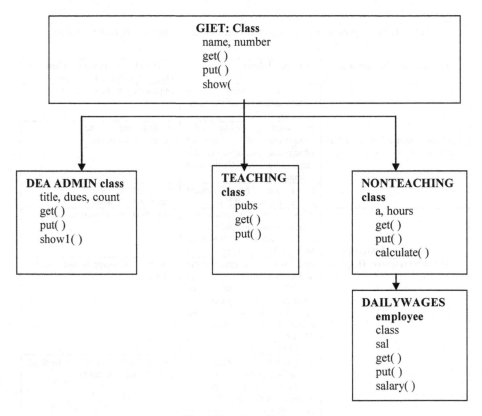

Fig. 1. For class diagram

4   Result Analysis

This portion of the paper presents the four tables are created to study the object-oriented properties of both the languages—C++ and JAVA. We have taken the twenty numbers of the programs and calculate the metrics values of both the languages. Then other two other tables are created for the statistical values such as mean and median for all the programs in C++ and JAVA respectively. By using these properties we study and compare the object-orientation properties of both the languages.

5   Object Oriented Metrics for a JAVA Program

5.1   Weighted Method Per Class

The weighted method per class is measures by counting the number of methods in each class [19] [25]

Table 2. For Weighted Method per Class

Metrics used for this metrics	GIET-Class	DEAN-ADMIN Class	TEACHING Class	NONTEACHING Class	DAILY WAGES
Weighted Method per class (WMC)	3	3	2	3	3

5.2   Response for a Class (RFC)

The response for a class (RFC) can be measures a s the number of functions or procedures that can be potentially be executed in a class. [4][24]. It is simply the number of methods in the set. In mathematically we can say that

$$RFC=M+R \qquad \dots\dots\dots\dots(1)$$

Where M=number of methods in he class and R=Number of remote methods directly called by methods of the class

Table 3. for Response for a class

Metrics used for this metrics	GIET-Class	DEAN-ADMIN Class	TEACHING Class	NONTEACHING Class	DAILY WAGES
Response for a class (RFC)	3	5	4	5	7

5.3   Number of Children (NOC)

The number of children can be measure as the number of immediate subclasses of a class [4].In other ways we can refer that NOC number of immediate child classes which is inherited from the one of the base class .when a high number of children then a large number of child classes, can indicate several things: [14]

Table 4. For Number of Children

Metrics used for this metrics	GIET-Class	DEAN-ADMIN Class	TEACHING Class	NONTEACHING Class	DAILY WAGES
Number of the Children(NOC)	3	0	0	1	0

5.4   Depth of Inheritance Tree (DIT)

DIT can be defined as the relationship between the class that makes the user to use the previous defined objects including the data member and the member functions. Basically it is used for the reusability and efficiency purpose [13] 14].

DIT= maximum inheritance path from the class to the root class

Table 5. for Depth of Inheritance Tree

Metrics used for this metrics	GIET-Class	DEAN-ADMIN Class	TEACHING Class	NONTEACHING Class	DAILY WAGES
Depth of Inheritance	0	1	1	1	2

5.5   Massage Passing Coupling (MPC)

It can measure as the count of total number of function and procedure calls made to external units [14].

Table 6. for Message Passing Coupling

Metrics used for this metrics	GIET-Class	DEAN-ADMIN Class	TEACHING Class	NONTEACHING Class	DAILY WAGES
Message Passing Coupling	0	2	2	2	4

5.6   Number of Subunits (NUS)

It can be defined as the total number of functions and procedures defined for the class [14] [2]21]

Table 7. for Number of Subunits

Metrics used for this metrics	GIET-Class	DEAN-ADMIN Class	TEACHING Class	NONTEACHING Class	DAILY WAGES
Number of subunits	3	3	2	3	3

5.7   Data Abstraction Coupling (DAC)

It can be count of total number of instances of other classes within a given class [14].

Table 8. for Data Abstraction Coupling

Metrics used for this metrics	GIET-Class	DEAN-ADMIN Class	TEACHING Class	NONTEACHING Class	DAILY WAGES
Data Abstraction Coupling	0	1	0	1	0

5.8   Inheritance Dependencies (ID)

The Inheritance metric is calculated using the following equation: [21]
Inheritance tree depth= max (inheritance tree path length)
So referring the above class diagram Inheritance tree depth = 3

5.9   Reuse Ratio (RR)

This metric is calculated using the following equation: [22]
Reuse ratio=Total no. of super classes / Total no of classes
Referring above class diagram Reuse ratio   =2/5
                                                 =0.4

5.10   Specialization Index (SI)

This kind of metrics is calculated as below the equation
Specialization index= Total no. of subclasses / total no. of super classes
From the above class diagram Specialization index =5/2
                                                  = 2.5

5.11   Factoring Effectiveness (FE)

This metric is calculated using the following equation [21]
Factoring effectiveness = No. of unique methods / Total no. of
methods   = 4/14
                 = 0.29

6   Statistical Measurement of the Complexity of C++ and JAVA Program

We have tested simultaneously 20 equal programs of c++ and java then after the statistical measurement have done by using minimum, maximum, mean, median etc.During this we have created the two tables i.e table-1 is meant for c++ and table-2 is meant for java

6.1   Metric Values Calculated for JAVA Program

Types of Metrics used	Number of programs used in this Paper is :20																			
	P1	P2	P3	P4	P5	P6	P7	P8	P9	P10	P11	P12	P13	P14	P15	P16	P17	P18	P19	P20
WMC	2.00	2.25	1.65	2.25	2.00	1.00	2.25	2.00	2.00	1.00	3.33	1.50	2.00	2.00	1.50	1.50	2.00	3.33	2.00	2.00
RFC	3.00	3.00	4.33	1.00	1.00	3.33	2.00	4.00	2.00	1.50	3.00	3.00	2.50	2.50	3.33	1.00	2.00	4.00	3.00	2.00
DIT	2.00	1.00	1.50	.00	1.50	2.00	0.50	1.00	1.00	2.00	0.75	\0.50	0.50	1.00	1.50	2.00	0.33	0.33	0.33	1.00
NOC	1.00	0.75	0.50	0.50	0.50	1.00	1.00	0.50	0.50	0.50	1.00	0.50	1.00	1.00	1.00	1.50	0.67	0.50	\0.50	\0.50
MPC	1.00	0.33	0.20	0.20	0.00	0.00	0.10	0.00	\0.00	0.00	0.20	1.00	0.33	0.33	0.33	0.20	0.00	0.50	0.00	0.00
DAC	0.30	0.00	0.00	0.40	0.50	0.00	0.00	0.33	0.33	0.50	\0.67	0.50	0.20	0.50	0.20	0.33	0.33	0.30	0.20	0.20
NUS	2.00	1.00	1.65	1.65	2.00	2.00	2.50	1.00	1.00	1.33	1.50	1.50	1.00	2.00	2.00	1.67	1.67	2.00	2.00	1.50
ID	2.00	0.50	0.50	1.00	0.50	0.33	2.00	1.00	1.00	1.00	0.20	0.30	0.50	1.00	2.00	2.00	2.00	1.00	0.33	0.33
FE	0.50	0.50	0.30	0.30	0.60	0.67	0.50	1.00	1.50	0.50	0.33	0.50	0.67	0.33	0.33	0.33	0.60	0.50	0.50	0.50
SI	3.00	3.00	1.00	1.00	1.50	3.00	2.00	1.50	1.00	1.00	2.00	3.00	1.00	1.00	\2.00	1.50	2.00	3.00	2.00	\2.00
RR	0.50	0.33	0.25	0.25	0.50	0.50	0.50	0.33	1.00	\0.33	0.25	0.55	0.50	0.50	0.30	0.50	0.25	0.30	0.50	\0.50

6.2   The Values Calculated by Using the Statistical Technique in Java

Types of Metrics used	Minimum values calculated	Max values calculated	Min values calculated	Median values calculated a	Standard Deviation Values calculated
WMC	1.00	3.33	1.98	2.00	0.58
RFC	1.00	4.33	2.57	2.75	1.00
DIT	0.33	2.00	1.08	1.00	0.59
NOC	0.50	1.50	0.75	0.58	0.29
MPC	0.00	1.00	0.24	0.20	0.30
DAC	0.00	0.67	0.29	0.32	0.19
NUS	1.00	2.50	0.65	1.66	0.42
ID	0.20	2.00	0.67	1.00	0.66
FE	0.30	1.50	1.89	0.50	0.28
SI	1.00	3.00	0.43	2.00	0.87
RR	0.25	1.00	0.55	0.50	0.17

6.3   Metrics Value Calculated for C++ Program

Types of Metrics used	Number of programs used in this Paper is :20																			
	P1	P2	P3	P4	P5	P6	P7	P8	P9	P10	P11	P12	P13	P14	P15	P16	P17	P18	P19	P20
WMC	3.00	2.25	1.65	2.25	2.00	1.00	2.25	2.00	2.00	1.00	3.33	115	2.00	2.00	1.00	1.67	2.00	3.33	2.00	2.00
RFC	2.00	3.00	3.33	2.00	3.33	3.33	2..67	3.00	3.00	4.48	3.00	3.33	2.00	2.50	3.33	3.33	2.00	4.11	3.00	2.00
DIT	2.00	1.00	1.00	0..50	1.50	2.00	0.50	0.33	1.00	1.00	0.75	0.50	0.50	1.00	1.00	2.25	0.33	0.33	0.50	0.50
NOC	2.00	0.75	0.50	1.00	0.50	1.50	0.50	0.50	0.50	0.50	0.65	1.00	1.00	1.00	0.75	1.75	0.67	0.50	0.50	0.50
MPC	2.00	0.30	0.20	0.33	0.00	0.00	0.20	0.00	\0.00	0.00	0.50	0.50	0.33	0.33	0.33	0.00	0.00	0.33	0.00	0.00
DAC	0.30	0.00	0.00	0.40	0.67	0.00	0.00	0.00	0.33	0.50	\0.67	0.50	0.50	0.67	0.00	0.30	0.30	0.30	0.30	0.40
NUS	3.00	2.00	1.65	1.65	2.00	2.00	2.00	1.67	1.67	1.33	1.50	1.50	1.50	2.50	2.00	1.67	1.67	2.50	2.00	2.00
ID	2.00	0.10	0.50	0.50	1.00	0.33	2.00	1.00	1.00	1.00	0.50	0.50	1.00	1.00	2.00	2.25	2.00	0.50	0.50	0.33
FE	0.50	0.50	0.30	0.30	0.50	0.67	0.67	0.67	1.25	0.50	0.33	0.67	0.67	0.33	0.33	0.33	0.50	0.67	0.50	0.50
SI	2.00	2.00	1.00	1.00	1.00	3.00	2.00	2.00	1.00	1.00	2.00	3.00	1.00	1.00	1.00	2.00	2.00	3.00	2.00	2.00
RR	0.25	0.33	0.25	0.25	0.50	0.50	0.33	0.33	0.30	0.33	0.25	0.75	0.50	0.50	0.30	0.25	0.25	0.30	0.50	0.30

6.4   The Values Calculated by Using the Statistical Technique in C++

Types of Metrics used	Minimum values calculated	Max values calculated	Min values calculated	Median values calculated a	Standard Deviation Values calculated
WMC	100	3.33	2.04	2.00	0.59
RFC	1.50	4.48	2.86	3.00	0.77
DIT	0.33	2.25	0.89	0.87	0.57
NOC	0.50	2.00	0.83	0.66	0.44
MPC	0.00	2.00	0.26	0.20	0.44
DAC	0.00	0.67	0.31	0.33	0.23
NUS	1.33	3.00	1.89	1.83	0.40
ID	0.33	2.25	1.04	1.00	0.64
FE	0.30	1.25	0.53	0.50	0.21
SI	1.00	3.00	1.75	2.00	0.71
RR	0.25	0.75	0.36	0.31	0.13

6.5   Observation of Figure 1 (WMC)

In this graph Y-axis represents WMC values for C++ and JAVA programs and X-Axis represents 20 programs (C++ & Java). The black square box indicates the java program and gray square box indicates the C++ programs. When the value is higher in WMC there is more chance to occur faults and when the classes are having more number of the methods then there is less chance to reuse it. The mean values of these metrics are greater in Java than C++ as shown in **FIG-1.** This implies that java programs are simpler and less complex. During this of 20 number of programs and observe that an increase the complexity and decrease the quality.

6.6   Observation of Figure 2 (DIT)

In **FIG-2** of the section 6.6, it has been observe that if more numbers of methods are available then more complexity and reusability also get increase due to inheritance. In this graph Y-axis represents DIT values for C++ and JAVA programs and X-Axis represents 20 programs (C++ & Java). The black square box indicates the java program and gray square box indicates the C++ programs.

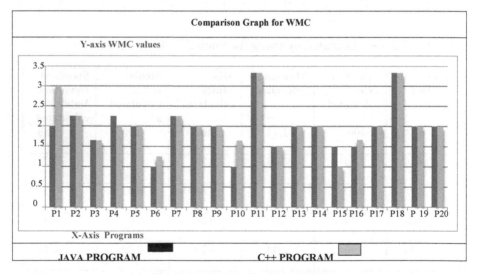

Fig. 2. Comparison Graph for WMC

6.7   Observation of Graph-3 (NOC)

In **FIG-4** of the section 6.7, the more number of children (NOC) indicates high reuse because inheritance is the form of reusability. If the classes are having more number of children then it is required more number of testing as well as more number of children may require more testing. In this graph Y-axis represents the values for C++ and JAVA programs and X-Axis represents 20 programs (C++ & Java).The black square box indicates the java program and gray square box indicates the C++ programs.

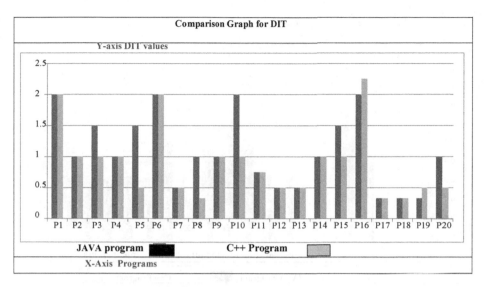

Fig. 3. Comparison Graph for DIT

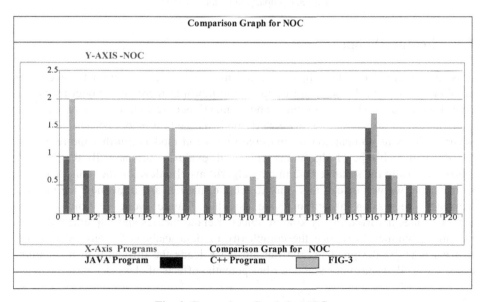

Fig. 4. Comparison Graph for NOC

6.8   Observation of Graph-3 (DAC)

In **FIG-5** of section 6.8 indicates the comparisons graph for DAC(Data abstraction Coupling) which represents the coupling between the classes. The value of this metric is low for JAVA than C++ program. In this graph Y-axis represents the values for C++ and JAVA programs and X-Axis represents 20 programs (C++ & Java).The black square box indicates the java program and gray square box indicates the C++ programs

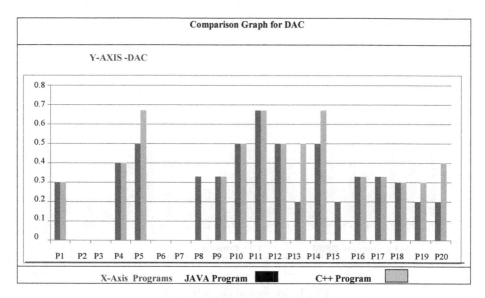

Fig. 5. Comparison Graph for DAC

7   Feature Scope

It is also possible that the object oriented metrics that can be used to predict the number of faults in the software by using the multiple regression model as well as neural network model .From the above mentioned techniques comparison ,the researchers can prove the neural network model is the best model because it predicts accurate faults as comparison to multiple regression models .Further the researcher can develop such type of algorithms which will evaluate automatically the proposed metrics by using the suitable data mining algorithms else develop the tools to achieve it .In feature we will extend this work to reuse to metrics and implement the machine learning   regression algorithm. One possible feature work is point out the data/attributes involved in any kind of object oriented metrics and implement the data mining techniques to reduce the complexity of data analysis for any system. The metrics is to be designed to find the attributes leads the difficulty in maintaining classes as well as attributes that describes potential effects of class changes.

8   Conclusion

In this paper we have implemented by taking 20 programs each in C++ and Java and found that Java program is better then any C++ programs. Our objective of an object oriented metric is to measure the complexity of these programs. C++ and Java is a modern and powerful purely object oriented language. As comparison of other object oriented programming language the java is too good for developing a fault free, reliable and easy to maintain the software In this paper we presented the analysis of existing major object oriented metrics .This paper will helpful for the researchers for better understanding and identifying the software metrics. We have measured the complexity of both C++ and Java in the different dimensions like (function/methods, size, class, cohesion, reusability) by using the object oriented metrics in the field of software engineering. The result which we obtained from the different programs (C++ and java) which leads that the java is the better object oriented programming language then C++.The (OOM) has proven to be an effective technique for improving software quality and productivity.

References

1. Humphery, W.S.: Object oriented metrics concepts (1996)
2. Sastry, B.R., Vijaya Saradi, M.V.: Impact of software metrics on the Object oriented software development life cycle. Int. Journal of Engineering Science and Technology 2(2), 67–76 (2010)
3. Sahik, A., Reddy, C.R.K., Damodaran, A.: Statistical Analysis for Object Oriented Design Software security metrics. Int. Journal of Engineering Science and Tech. 2(5), 1136–1142 (2010)
4. Chidambaer, S.R., Kemerer, C.F.: A metrics suite for object oriented metrics. IEEE Transaction on Software Engineering 20(6) (June 1994)
5. Subharamanyam, R., Krishna, M.S.: Empirical Analysis of CK metrics for object oriented Design Complexity: Implications of software defects. IEEE Transaction on Software Engineering 29(4) (2003)
6. Harrison, R., Counsell, S.J., Nithi, R.V.: IEEE Transaction on software Engineering 24, 491–496 (June 1998)
7. Basili, V.R., Briand, L.C., Melo, W.L.: A validation of Object Oriented Design Metrics as Quality Indicators. IEEE Transactions on Software Engineering 21, 751–761 (1996)
8. Briand, L., Emam, K.E., Morasca, S.: Theoretical and Empirical Validation of Software Metric (1995)
9. Glasberg, D., Emam, K.E., Melo, W., Madhavji, N.: Validating Object-Oriented Design Metrics on a Commercial Java Application. National Research Council 44146 (September 2000)
10. Conte, S.D., Sunsmore, H.E., Shen, V.Y.: Software Engineering Metrics and Models. Benjamin/Cummings Publications, Menlo Park (2003)
11. Wei, L., Salley, H.: Maintenance Metrics for the Object Oriented Paradigm. In: First International Software Metrics Symposium. IEEE Computer Society Press, Los Alamitos (1993)
12. Schildt, H.: Java: The Complete Reference, 6th edn. McGraw Hill Publication (2006) ISBN: 0072263857, 9780072263855
13. Brooks, F.P.: No Silver Bullets: Essence and Accidents of Software Engineering. Computer 20(4), 10–19 (1987)
14. Chidamber, S., Kemerer, C.: Towards a Metrics Suite for Object Oriented Design. In: Object Oriented Programming Systems, Languages and Applications (OOPSLA), vol. 10, pp. 197–211 (1991)
15. Balgurusamy, E.: Object oriented programming language:JAVA
16. Schildt, H., Naughton, P.: Java: The complete Reference. McGraw-Hill Professional, UK (2008)

17. Booch, G.: Object –Oriented Design with applications. ISBN:0-80530091-0. The Benjamin/Cummings Publishing Company, Redwood City (1991) ISBN :0-80530091-0
18. Yourdon, E., Coad, P.: Object – Oriented Design. Youden Press, Englewood Clifs (1991) ISBN-0-13-630070-7
19. Pressman, R.: A Practitioner's Approach to Software Engineering, pp. 658–662. Mc-Graw Hill Publications (2001)
20. Cohn, M.W., Junk, W.S.: Empirical Evaluation of a Proposed Set of Metrics for Determining Class Complexity in Object-Oriented Code. A Thesis, College of Graduate Studies University of Idaho (April 1994)
21. Morris: Metrics for Object-oriented Software Development Environments. Masters Thesis, MIT (1989)
22. Ivar, J.: Object Oriented Software Engineering: A Use Case Driven Approach. Addison-Wesley Publishing Company (1993)
23. Lorenz, Kidd, J.: Object-Oriented Software Metrics. Printice Hall, India (1994)
24. Sharble, R., Cohen, S.: The object –Oriented Brewery: A comparison of two object oriented Development Methods. Software Engineering Notes 18(2), 60–73 (1993)
25. McCabes & Associates, McCabe Object-Oriented Tool Users Instruction (1994)
26. Rosenberg, L.H., Hyatt, L.E.: Software Quality Metrics for Object-Oriented Environments. Crross Talk Journal (April 1997)
27. Shih, T.K., Chung, C., Wang, C.C., Pai, W.C.: Decomposition of Inheritance hierarchy DAGs for Object – Oriented software metrics. In: Workshop on ECBS (1997)
28. Fenton, N., Pfleeger, S.L.: Software Metrics: A rigorous & Practical Approach, 2nd edn. International Thomson Computer Press (1997)
29. Suresh, Y., Pati, J., Rath, S.K.: Effectiveness of software metrics for object-oriented system. In: 2nd International Conference on Communication, Computing & Security (ICCCS 2012), Procedia Technology. Elsevier (2012)
30. Yang, B., Zhao, F.: Study on Measurement of Class Coupling in Object-Oriented Software. In: Yang, Y., Ma, M. (eds.) Proceedings of the 2nd International Conference on Green Communications and Networks 2012 (GCN 2012): Volume 4. LNEE, vol. 226, pp. 71–77. Springer, Heidelberg (2013)
31. Lanza, M., Marinescu, R.: Object Oriented Metrics in Practice, XIV, p. 205. Springer Publication (2006)
32. Kayarvizhy, N., Kanmani, S.: Comparative Analysis of CK Metrics across Object Oriented Languages. In: Das, V.V., Stephen, J., Chaba, Y. (eds.) CNC 2011. CCIS, vol. 142, pp. 397–399. Springer, Heidelberg (2011)

Applicability of Software Defined Networking in Campus Network

Singh Sandeep[1], R.A. Khan[1], and Agrawal Alka[2]

[1] Department of Information Technology,
Babasaheb Bhimrao Ambedkar University, Lucknow, India
[2] Department of Computer Science,
Khwaja Moinuddin Chisti Urdu Arbi Farsi University, Lucknow, India
drsandeep.gbu@gmail.com, khanraees@yahoo.com,
alka_csjmu@yahoo.co.in

Abstract. This research article focuses on application of open flow protocol which is a very useful milestone for researchers to run experimental protocols in their daily used network. Open flow protocol is based on the traditional Ethernet switch, with an internal flow table and a standardized interface for the perspective of adding and removing flow entries. The primary focus of our research is to encourage networking vendors to include open flow applicability into their switch like products for deployment in institute or university level campuses. We also assume that open flow is a pragmatic compromise, in one side it allows researcher to execute their developed experiments and in other side the vendors do not need to disclose the internal working of their product switch. In other words open flow allows researchers to evaluate their ideas in real world traffic setting; hence open flow came into existence with a useful campus component in proposed large scale test beds like Global Environment for Networking Innovations (GENI).

Keywords: Open flow protocol, Software Defined Networking, Ethernet Switch.

1 Introduction

In today's world of high speed networking these devices becomes a crucial part of the critical infrastructure of daily business, home and college life. This huge success has been both a curse and bless for networking researchers. There work is more relevant but the chance of making an impact is very far. The reduction in real life impact of any given network innovation is only because of enormous installed base of equipments, protocols and reluctance to experiment with production traffic which have created an exceedingly high barrier entry for new concepts [1].

Today almost there is no real world practical way to test with new network protocols in sufficiently realistic settings to gain the trust needed for their long term deployment. As the result most of the new ideas and thoughts from the research community remains un trusted and untried. So it is believed that the network

S.C. Satapathy et al. (eds.), *Proc. of the 3rd Int. Conf. on Front. of Intell. Comput. (FICTA) 2014*
– *Vol. 2*, Advances in Intelligent Systems and Computing 328, DOI: 10.1007/978-3-319-12012-6_68

infrastructure has been ossified. After recognizing the issues, the networking researchers are hard at there work on deploying and implementing programmable network. Like Global environment for networking Innovations (GENI) which provides a nationwide research facility for new network architectures and distributed systems. These programmable networks call from programmable switches and routers that can process data packets for experimental networks. For example in GENI it is Envisaged that a researcher will be allocated a slice of resources across the whole network, consisting portion of network links, packet processing elements (e.g. routers) and end-hosts; researchers program their slices to behave as they wish. A slice could extend across the backbone, into access networks, into college campuses, industrial research labs, and include wiring closets, wireless networks, and sensor networks. Virtualized programmable networks could lower the barrier to entry for new ideas, increasing the rate of innovation in the network infrastructure. But the plans for nationwide facilities are ambitious (and costly), and it will take years for them to be deployed.

2   Requirement for Programmable Networks

This research contribution focuses about running experimental protocols in networks. If it can be figured out then we can start giving benefits to the whole research community. To overcome these requirements' we need to answer several questions including; In early days how would college network administrators get comfortable putting experimental equipments (routers, access points and switches etc.) into their network? How will researcher control a specific part of their network in such a way that doesn't disrupt others functionalities depended on it? And finally what specific requirements are needed in network switches to enable experiments. Here our goal is to propose a new switch feature that can help to extend the programmability.

A few open software platforms already exist, but do not have the performance or port-density we need. The simplest example is a PC with several network interfaces and an operating system. All well known operating systems support routing of packets between interfaces, and open source implementations of routing protocols exist (e.g., as part of the Linux distribution, or from XORP [2]); and in most cases it is possible to modify the operating system to process packets in almost any manner e.g., using Click modular [3]. The problem, of course, is performance: A PC can neither support the number of ports needed for a college wiring closet (a fan out of 100+ ports is needed per box), nor the packet-processing performance (wiring closet switches process over 100Gbits/s of data, whereas a typical PC struggles to exceed 1Gbit/s; and the gap between the two is widening). Existing platforms with specialized hardware for line-rate processing are not quite suitable for college wiring closets either. For example, an ATCA-based virtualized programmable router called the Supercharged PlanetLab Platform [4] is under development at Washington University, and can use network processors to process packets from many interfaces simultaneously at line-rate.

This approach is promising in the long-term, but for the time being is targeted at large switching centers and is too expensive for widespread deployment in college wiring closets. At the other extreme is NetFPGA [5] targeted for use in teaching and research labs. NetFPGA is a low-cost PCI card with a user-programmable FPGA for processing packets, and 4 ports of Gigabit Ethernet. NetFPGA is limited to just four network interfaces—insufficient for use in a wiring closet. Thus, the commercial solutions are too closed and inflexible, and the research solutions either have insufficient performance or fan out, or are too expensive. It seems unlikely that the research solutions, with their complete generality, can overcome their performance or cost limitations.

3   The Open Flow Switch Specification

The basic idea is simple: we exploit the fact that most modern Ethernet switches and routers contain flow-tables (typically built from TCAMs) that run at line-rate to implement firewalls, NAT, QoS, and to collect statistics. While each vendor's flowable is different, we've identified an interesting common set of functions that run in many switches and routers. Open Flow exploits this common set of functions. Open Flow provides an open protocol to program the flow table in different switches and routers. A network administrator can partition traffic into production and research flows. Researchers can control their own flows - by choosing the routes their packets follow and the processing they receive. In this way, researchers can try new routing protocols, security models, addressing schemes, and even alternatives to IP. On the same network, the production traffic is isolated and processed in the same way as today. The data path of an Open Flow Switch consists of a Flow Table, and an action associated with each flow entry. The set of actions supported by an Open Flow Switch is extensible, but below we describe a minimum requirement for all switches. For high-performance and low-cost the data path must have a carefully prescribed degree of flexibility. This means forgoing the ability to specify arbitrary handling of each packet and seeking a more limited, but still useful, range of actions. Therefore, later in the paper, define a basic required set of actions for all Open Flow switches. An Open Flow Switch consists of at least three parts: (1) A Flow Table, with an action associated with each flow entry, to tell the switch how to process the flow, (2) A Secure Channel that connects the switch to a remote control process (called the controller), allowing commands and packets to be sent between a controller and the switch using (3) The Open Flow Protocol, which provides an open and standard way for a controller to communicate with a switch. By specifying a standard interface (the Open Flow Protocol) through which entries in the Flow Table can be defined externally, the Open Flow Switch avoids the need for researchers to program the switch.

It is useful to categorize switches into dedicated Open Flow switches that do not support normal Layer 2 and Layer 3 processing, and Open Flow-enabled general purpose commercial Ethernet switches and routers, to which the Open Flow Protocol and interfaces have been added as a new feature. Dedicated Open Flow switches.

A dedicated Open Flow Switch is a dumb data path element that forwards packets between ports, as defined by a remote control process. Figure 1 shows an example of an Open Flow Switch. In this context, flows are broadly defined, and are limited only by the capabilities of the particular implementation of the Flow Table. For example, a flow could be a TCP connection, or all packets from a particular MAC address or IP address, or all packets with the same VLAN tag, or all packets from the same switch port. For experiments involving non-IPv4 packets, a flow could be defined as all packets matching a specific (but non-standard) header. Each flow-entry has a simple action associated with it.

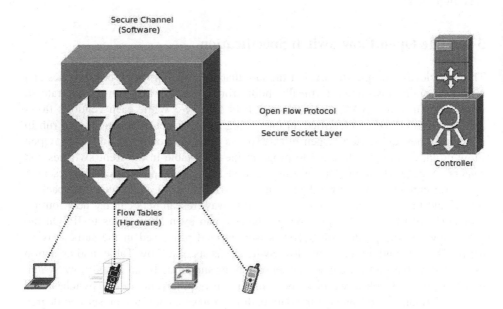

Fig. 1. Open flow Switch Specification

An entry in the Flow-Table has three fields: (1) A packet header that defines the flow, (2) The action, which defines how the packets should be processed, and (3) Statistics, which keep track of the number of packets and bytes for each flow, and the time since the last packet matched the flow (to help with the removal of inactive flows). In the first generation "Type 0" switches, the flow header is a 10- tuple shown in Table 1. A TCP flow could be specified by all ten fields, whereas an IP flow might not include the transport ports in its definition. Each header field can be a wildcard to allow for aggregation of flows, such as flows in which only the VLAN ID is defined would apply to all traffic on a particular VLAN. The detailed requirements of an Open Flow Switch are defined by the Open Flow Switch Specification [6].

Table 1. Open flow switch table; Type 'o' header field matching

In	VLA	Ethernet			IP			TCP	
Po	N	S	D	Typ	S	D	Prot	Sr	D
rt	ID	A	A	E	A	A	o	c	st

3.1 Open Flow Enabled Switch

Some commercial switches, routers and access points will be enhanced with the Open Flow feature by adding the Flow Table, Secure Channel and Open Flow Protocol Typically, the Flow Table will re-use existing hardware, such as a TCAM; the Secure Channel and Protocol will be ported to run on the switch's operating system. Figure 2 shows a network of Open Flow- enabled commercial switches and access points. In this example, all the Flow Tables are managed by the same controller; the Open Flow Protocol allows a switch to be controlled by two or more controllers for increased performance or robustness [7]. Our goal is to enable experiments to take place in an existing production network alongside regular traffic and applications. Therefore, to win the confidence of network administrators, Open Flow-enabled switches must isolate experimental traffic (processed by the Flow Table) from production traffic that is to be processed by the normal Layer 2 and Layer 3 pipeline of the switch. There are two ways to achieve this separation. One is to add a fourth action: Forward this flow's packets through the switch's normal processing pipeline [8].

The other is to define separate sets of VLANs for experimental and production traffic. Both approaches allow normal production traffic that isn't part of an experiment to be processed in the usual way by the switch. All Open Flow enabled switches are required to support one approach or the other; some will support both.

3.2 Additional Features

If a switch supports the header formats and the four basic actions mentioned above (and detailed in the Open Flow Switch Specification), then we call it a "Type 0" switch. We expect that many switches will support additional actions, for example to rewrite portions of the packet header (e.g., for NAT, or to obfuscate addresses on intermediate links), and to map packets to a priority class. Likewise, some Flow Tables will be able to match on arbitrary fields in the packet header, enabling experiments with new non-IP protocols. As a particular set of features emerges, we will define a "Type 1" switch.

3.3 Controllers

A controller adds and removes flow-entries from the Flow Table on behalf of experiments. For example, a static controller might be a simple application running on a PC to statically establish flows to interconnect a set of test computers for the duration of an experiment. In this case the flows resemble VLANs in current networks—providing a simple mechanism to isolate experimental traffic from the production network. Viewed this way, Open Flow is a generalization of VLANs. One can also imagine more sophisticated controllers that dynamically add/remove flows as an experiment progresses. In one usage model, a researcher might control the complete network of Open Flow Switches and be free to decide how all flows are processed [9]. A more sophisticated controller might support multiple researchers, each with different accounts and permissions, enabling them to run multiple independent experiments on different sets of flows. Flows identified as under the control of a particular researcher (e.g., by a policy table running in a controller) could be delivered to a researcher's user-level control program which then decides if a new flow entry should be added to the network of switches.

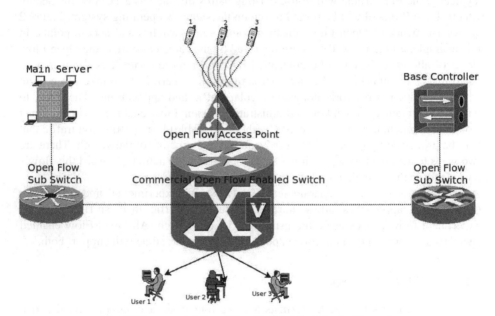

Fig. 2. Commercial Open Flow Enabled Switch; An Example

4   Experiments in a Production Network

Peering between the autonomous systems on the internet today is universally done with Border Gateway Protocol version 4 [12]. Therefore we need a mechanism for Software Defined Network autonomous system to talk with autonomous system of Internet protocol through BGP4. The SDN autonomous system has been controlled by ONOS [10]. Here we have implemented SDN-IP networking as an application on network operating system ONOS. The demonstration is emulated by using Mininet [13].

4.1   Testing of Functionality

Here we created an inter autonomous system flow from SDN S1 to S2 and a transit flow from S3 to S2. Here we successfully obtained the re-convergence when inter operability of SDN and IP fails.

4.2   Testing of Performance

SDN and IP networking applications can be measure up to 10,000 routing information base entries. The Border Gateway protocol of version 4 consumes 480 MB of physical memory in SDN-IP communication application consumes about 360 MB. It can process about 100 RIB per second.

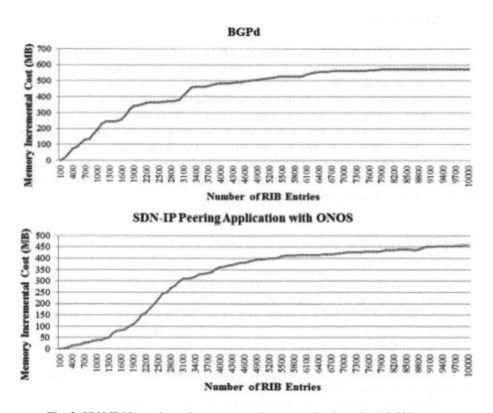

Fig. 3. SDN-IP Network peering system performance after inserting 10,000 routes

5   Deployment

Researchers can control their own traffic and be able to add/remove flow entries. We also expect many different open flow switches to be developed by research community. The open flow website contains "Type o" reference design for several different platforms like Linux, open WRT and NetFPGA.

Route Flow [11] is one of the first implementation of IP routing on open flow switches. Route flow instantiates a VM for each open flow switch with as many virtual network interfaces as there are active ports in corresponding devices, and runs a stake of open source routing protocols on the virtual topology. All control messages are exchanged between virtual machines as if they are running a distributed control plane such as the solution incurs the overhead of distribution without the benefits of scaling. The SDN-IP peering application is a simpler design, better integrated with SDN and easier to implement the advanced features such as policy based routing and Traffic Engineering. More reference designs are created by the community. All reference implementations of Open flow switches posted on the SDN's website will be open source and free for commercial and non commercial use.

6   Conclusion

In this research paper it is observed that open flow is a pragmatic compromise which allows a researcher to test experimental protocols over the heterogeneous switches and routers in a uniform way. The vendors don't need to expose the internal functionality of their products. If this idea became a reality we can gradually explore this into the broad range of other educational centers. It is also observed that a new generation of software controller is emerging very rapidly to reuse the network resources. In collaboration with Google and REANNZ, SDN-IP network peering system to small active SD Network is about to deploy in Willington, New Zealand. After this deployment several issues like improving the scalability and fault tolerance will be open for research perspective.

References

1. Global Environment for Network Innovations, http://geni.net
2. Handley, M., Hodson, O., Kohler, E.: XORP: An Open Platform for Network Research. ACM SIGCOMM Hot Topics in Networking (2002)
3. Kohler, E., Morris, R., Chen, B., Jannotti, J., Kaashoek, M.F.: The Click modular router. ACM Transactions on Computer Systems 18(3), 263–297 (2000)
4. Turner, J., Crowley, P., Dehart, J., Freestone, A., Heller, B., Kuhms, F., Kumar, S., Lockwood, J., Lu, J., Wilson, M., Wiseman, C., Zar, D.: Supercharging PlanetLab - High Performance, Multi-Application, Overlay Network Platform. In: ACM SIGCOMM 2007, Kyoto, Japan (August 2007)
5. NetFPGA: Programmable Networking Hardware, http://netfpga.org
6. The OpenFlow Switch Specification, http://OpenFlowSwitch.org
7. Tennenhouse, D., Wetherall, D.: Towards an active network architecture. In: Proceedings of the DARPA Active Networks Conference and Exposition, pp. 2–15. IEEE (2002)
8. Tootoonchian, A., Ganjali, Y.: Hyperflow: A distributed control plane for openflow. In: Proceedings of the 2010 Internet Network Management Conference on Research on Enterprise Networking, p. 3. USENIX Association (2010)

9. Tootoonchian, A., Gorbunov, S., Ganjali, Y., Casado, M., Sherwood, R.: On controller performance in software-defined networks. In: USENIX Workshop on Hot Topics in Management of Internet, Cloud, and Enterprise Networks and Services, Hot-ICE (2012)
10. ONOS, http://onlab.us/tools.html
11. RoutFlow, http://cpqd.github.io/RouteFlow/
12. BGP [RFC 4271], http://www.ietf.org/rfc/rfc4271.txt
13. Mininet, http://mininet.org

Author-Profile System Development
Based on Software Reuse of Open Source Components

Derrick Nazareth, Kavita Asnani, and Okstynn Rodrigues

Depatment of Information Technology,
Padre Conceicao College of Engineering, Goa University, Goa, India
derricknazareth@yahoo.com, {kavitapcce,mecta2k7}@gmail.com

Abstract. This paper demonstrates the contribution of simple open source tools to the development of a highly efficient author profiling system, which determines the age and gender of the author based on the authored text itself. With the rapid growth of the Web, the number of social websites has increased by twice a fold. Thus it becomes necessary for security agencies and intelligence experts to keep track of any malicious activity by users on the Web (such as pedophiles, security attacks etc.) by monitoring their profiles and flagging them if necessary. Rather than building the system from scratch Software Engineering provides us a Component Based Methodology (CBM) that permits the reuse of various components that will help us in achieving better quality software in a quick span of time, free of cost. Significant differences exist in the way males/females and younger/older people write. We illustrate in detail how the system exploits these differences for its development based on the architecture of the CBM.

Keywords: author profiling, software engineering, age group identification, gender identification, component based methodology.

1   Introduction

In present times, the Internet has abundantly grown by leaps and bounds. Along with it so has the Internet media like emails, blogs, discussion forums, chat rooms, social websites etc. As these media are open to any individual, they have become a norm of life from its original intended purpose of enjoyment. People communicate a lot through these social platforms. They also express their honest and independent views on various entities, ranging from politics, religion, cultures, electronic products, software gadgets, electrical equipment and many such other entities.

Author profiling plays an important role in such a situation in these modern times. Just by analyzing what the author has posted you could tell a lot about the author's age, gender, demographic and cultural background, educational background, his native language etc. Forensic experts, police agencies and business establishments view author profiling as a vital player in their operations of identifying characteristics of the offender in cases when there is very little knowledge about the criminal, tracking pedophiles from the large number of user profiles and identifying user

© Springer International Publishing Switzerland 2015   629
S.C. Satapathy et al. (eds.), *Proc. of the 3rd Int. Conf. on Front. of Intell. Comput. (FICTA) 2014*
– *Vol. 2*, Advances in Intelligent Systems and Computing 328, DOI: 10.1007/978-3-319-12012-6_69

personalities that like the company's product and accordingly modifying the marketing strategy.

The development of an author profiling system from scratch involves considerable amount of time, is very tedious and involves rigorous testing. In our search to develop a profiling system in the minimum amount of time with considerable efficiency we tried to use the Component-Based Methodology [1, 2] that allows us to build our system using components from other applications rather than original development, owing to its principle of software development through software reuse shifting from traditional approach of requirement gathering, analyzing, designing and implementing. The components that are reused can be stand-alone applications, Commercial-Off-The-Shelf (COTS) software etc. The profiling dimensions that our system has considered are age and gender along with content, topic and style features to capture the variations in the expressions of authors.

2   Related Contributions in This Domain

Some of the blog domains being publicly available for processing gives natural language analysts an opportunity to harvest some unknown useful information. This was done by authors in [4] by trying to explore differences between the way males/females and younger/older authors express themselves via blogs. They uncovered that males tend to use more prepositions, numbers and modifiers while women tend to use more pronouns and negation. Older bloggers are primarily male while younger bloggers are largely female.

The Authors in [5] tried to group the content of various publicly available blogs into 20 categories while trying to determine age and gender effects with respect to these categories. They found that male bloggers tend to speak on more general and information based topic categories (like politics) while female bloggers on personal based topic categories (like fun).

Author profiling can be based on various measures of the author (like age, gender, educational background etc.). The authors in [6] have tried to profile the authors based on 4 dimensions (age, gender, native language, personality) using both style and content features. Then have used a machine learning approach that learns from training data and profiles the test data. The classification for native language was native English and non-native English speaking, while for personality, it was the dimension of neuroticism.

Topic features, which give us the topics that a particular blog is talking about, along with content and style features were used by authors of [7] and [8] for profiling. They performed their profiling experiments on the publicly available PAN corpus using 'MALLET' to perform topic modeling along with a 'MaxEnt' classifier ([8]) and a Decision tree classifier of WEKA ([7]) to profile the authors.

Authors of [9] have used only the content and style features to profile the authors, by including a large array of style features in an attempt to improve the accuracy of the Decision tree classifier obtained from WEKA. They have included a new class for words based on their positivity/negativity using the 'RiTaWordNet', while the POS tagging feature is realized using the 'Stanford CoreNLP POS' tagger. Both are publicly available.

3   Component-Based System Architecture

We will explain the basic architecture (Fig. 1) of our system one component at a time.

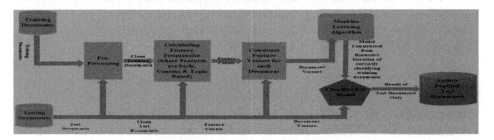

Fig. 1. Building Blocks of our System

3.1   Corpus

We have used publicly available PAN [11] organization's English blog corpus. These blogs are categorized by gender (male/female) and age (10s:13-17, 20s:23-27, 30s:33-47). Our training set is organized as follows: 1000 blog documents of each gender from all age groups, totaling to 6000 blog documents and 500 random blog documents for test set.

3.2   Preprocessing

The corpus consists of blog XML documents with conversations in XML tags. Each blog document is labelled with his/her language, gender and age group. During this preprocessing phase we clean the blog documents, present in .xml format, by removing all the XML tags and storing the content part of the blog into separate text files.

3.3   Features

We have selected three features for our system namely topic, content and style. For our content features we have selected the top-most commonly occurring 1-grams from the male blogs, female blogs, 10s blogs, 20s blogs and 30s blogs. The 1-grams are obtained using a simple N-Gram tokenizer obtained from [14]. The test files are then tested against these content features for the presence of commonly occurring 1-grams that will aid in the classification of these test files based on age and gender.

For the Topic Features we have first performed topic modeling on each of the five categories i.e. males, females, 10s, 20s and 30s separately and we have then generalized the topics by using the 20 categories devised by the authors in [5]. These generalized topics then become the topic features against which the test files are evaluated. We have performed topic modeling using the publicly available java package 'MALLET' [10]. This package contains main subsections. The topic modeling section is made available in the form of a GUI by [3], which we have tweaked based on our use. This GUI is shown in Fig 2.

For the style features we have selected a total of 7 features, namely frequencies of stop words, foreign words, prepositions, punctuations, conjunctions, personal pronouns and determiners. The commonly occurring stop words were obtained from [13]. The list of punctuations was manually prepared using commonly occurring punctuations. For the remaining features, Stanford CoreNLP POS tagger [12], made available by the Stanford University, was used. We are interested in the frequency of POS tags of 'DT', 'IN', 'PRP', 'CC', and 'FW'.

3.4   Feature Vectors

Proceeding from the style, content and topic features, for each of these attributes, every training file will have some values associated with it. Thus for each training file we create an instance of the various attributes and store them all in one file, that is based on the .arff format. The reason for selection of the .arff format will be explained. Similarly attribute value occurrence in all the test files are stored in one file based on the .arff format. A snapshot of the training file is shown in Fig. 3.

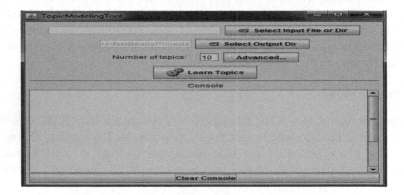

Fig. 2. GUI for the topic modeling

```
@relation Author_Profiling

@attribute Topic_10s_Female numeric
@attribute Topic_10s_Male numeric
@attribute Topic_20s_Female numeric
@attribute Topic_20s_Male numeric
@attribute Topic_30s_Female numeric
@attribute Topic_30s_Male numeric
@attribute no_of_det numeric
@attribute no_of_prep numeric
@attribute no_of_perpro numeric
@attribute no_of_conj numeric
@attribute no_of_fw numeric
@attribute no_of_stopw numeric
@attribute no_of_punct numeric
@attribute @@class@@ {10s-Female,10s-Male}

@data
5,3,4,3,4,4,57,62,34,21,0,0,1,10s-Female
5,3,4,3,4,4,57,62,34,21,0,0,1,10s-Female
5,3,4,3,4,4,57,62,34,21,0,0,1,10s-Female
5,3,4,3,4,4,57,62,34,21,0,0,1,10s-Female
5,3,4,3,4,4,57,62,34,21,0,0,1,10s-Female
5,3,4,3,4,4,57,62,34,21,0,0,1,10s-Female
5,3,4,3,4,4,57,62,34,21,0,0,1,10s-Female
5,3,4,3,4,4,57,62,34,21,0,0,1,10s-Female
5,3,4,3,4,4,57,62,34,21,0,0,1,10s-Female
5,3,4,3,4,4,57,62,34,21,0,0,1,10s-Female
```

Fig. 3. The training file based on the .arff format

3.5   Machine Learning Algorithm

In order to enable our author profiling system to automatically predict the authors' traits (in this case age and gender) we decided to use a Machine Learning Algorithm (MLA) which learns to predict based on whatever it has learnt from the training files. Hence it is very important that the training files that are supplied to the MLA are of good quality. The ensuing fact that comes to mind next, is the accuracy of the machine learning algorithm. Various studies in the past have used different MLAs with the sole intent of achieving better accuracy for the profiling task. Some of the MLAs used were Support Vector Machine (SVM), SMO, Information Gain, Bayesian Regression, Maximum Entropy, Exponential Gradient, LibSVM, etc. We felt that LibSVM would be most appropriate MLA choice for the system; hence we decided to incorporate it in our profiling system.

The LibSVM has a wrapper class that permits its use in WEKA tool. Hence we decided to use WEKA for our classification task. The WEKA tool can be used in Java through its API which is publicly available at its website [15]. Since we are using WEKA for the classification task, all our input training and test files have to be in the .arff format, hence the choice of .arff format is made above. Along with the WEKA API and wrapper class, the libsvm API is also needed that is responsible for the classification task in the author profiling system. Thus a total of 3 APIs in conjunction with the training files, are all needed to build the final classification model [15, 16, 17].

3.6   Author Profiled Test Documents

When the test files are given to the classification model constituted from the previous phase, they basically do not contain any class labels i.e. author age and gender is not known. After the classification model classifies the test files (author profiles them), it basically assigns a class label to each of these files. These class labels basically help us determine approximately the author's gender and age for all the test files.

3.7   System's Relation to Component-Based Methodology

The entire system architecture works on similar lines as the normal functioning of Component Based Methodology, which principally involves the following steps (Fig.4).

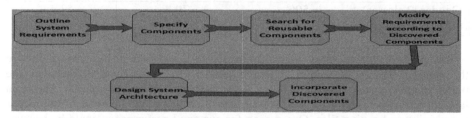

Fig. 4. Basic structure of Component Based Methodology

First we highlight the prerequisites needed to develop the author profiling system. Next a mapping process takes place wherein the requirements are mapped on to individual components. Consequently a check is done to verify whether any of the specified components match any of the existing reusable components. Subsequently if we do find some reusable components from other applications or as a standalone, a quick review is done on how the component fits into our system architecture. Sometimes the components may not interface easily into the system, then either the requirements are modified (and consequently a new system architecture is derived) or the component is modified to meet the system specifications (if possible). Finally the discovered components are incorporated into the system for its final functioning.

4   Experiments and Results

The results of our approach to author profiling are explained below. We have chosen three features (style, content and topic) for the task of profiling, but through our tests and results we have seen that though normally content features play an important role in the profiling task, for this particular corpus that we have chosen, they do not have very much significance. On the contrary, if those features are dropped, the performance of the system improves massively as seen in Fig.5; hence they are dropped from further processing.

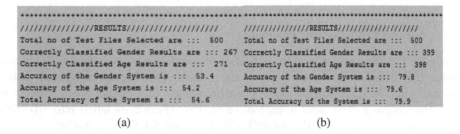

(a)   (b)

Fig. 5. System performance when content features: **(a)** are considered **(b)** not considered

Table 1. Comparison of system performance with the baseline

Category	Baseline	System (No Content Features)	System (Content Features)
Gender (2 classes)	50%	79.8	53.4
Age (3 classes)	33.33%	79.6	54.6

The performance of our system with just the style and topic features is well above the normal baseline for both age and gender. When content is included the performance over the baseline is just a slight increase (shown in Table. 1). Though a large corpus is provided by PAN [11] we have used just a small portion of the corpus for training and testing files. One reason for our good performance could be because we have used different templates for both gender and age classification along with the generalized categories as defined in [5].

5   Conclusion

Through this paper we have shown how Component Based Methodology can be applied to the development of an author profiling system by re-using open source components for its construction. The paper illustrates our approach to developing an author profiling system that automatically classifies the author based on the gender and age. The system has managed to achieve accuracies of 79.8 for gender, and 79.6 for age i.e. given 100 author blogs, our system can correctly predict the gender and age of about 80 authors. The components used in our system can be reused again by other developers by modifying it according to their application requirements.

The performance of the system is quite favorable for it to be used for text analysis by security agencies, forensic experts, industry establishments and natural language analysts, when they have only data content with them (in the form of comments, user reviews etc.) and want to identify the author based on age and gender.

We have profiled the author based on two parameters. Given appropriate training material in the future we could also profile other traits of the author. Since the non-standard language usage in the corpus provided great challenges we could address these issues in the future work to improve the accuracy of the system furthermore. We could also plan to research on the relationship between the demographics such as the gender and age with the emotional profile of the authors and their personality traits, trying to link such tasks in order to build a common framework to allow us to better understand how people use language from a cognitive linguistics viewpoint.

References

1. Bose, D.: Component Based Development. Application in Software Engineering. Indian Statistical Institute (2010)
2. Crnkovic, I.: Component-Based Software Engineering-New Challenge in Software Development. Software Focus 2(4), 127–133 (2001)
3. MALLET GUI, https://code.google.com/p/topic-modeling-tool/
4. Schler, J., Koppel, M., Argamon, S., Pennebaker, J.: Effects of age and gender on blogging. In: AAAI Spring Symposium on Computational Approaches for Analyzing Weblogs, vol. 6, pp. 199–205 (2006)
5. Argamon, S., Koppel, M., Pennebaker, J.W., Schler, J.: Mining the blogosphere: Age, gender, and the varieties of self-expression. First Monday 12(9) (September 2007)
6. Argamon, S., Koppel, M., Pennebaker, J.W., Schler, J.: Automatically profiling the author of an anonymous text. Communications of the ACM 52(2), 119–123 (2009)
7. Santosh, K., Bansal, R., Shekhar, M., Varma, V.: Author Profiling: Predicting Age and Gender from Blogs. Notebook for PAN at CLEF 2013 (2013)
8. Pavan, A., Mogadala, A., Varma, V.: Author Profiling Using LDA and Maximum Entropy. Notebook for PAN at CLEF 2013 (2013)
9. Patra, B.G., Banerjee, S., Das, D., Saikh, T., Bandyopadhyay, S.: Automatic Author Profiling Based on Linguistic and Stylistic Features. Notebook for PAN at CLEF 2013 (2013)
10. McCallum, A.K.: MALLET: A Machine Learning for Language Toolkit (2002), http://mallet.cs.umass.edu

11. PAN Corpus,
    http://www.uni-weimar.de/medien/webis/research/
    events/pan-13/pan13-web/author-profiling.html
12. Toutanova, K., Klein, D., Manning, C., Singer, Y.: Feature-Rich Part-of-Speech Tagging with a Cyclic Dependency Network. In: HLT-NAACL, vol. 1, pp. 173–180 (2003)
13. List of stopwords, http://www.ranks.nl/stopwords
14. Collaborative User Experience group's Java library code for basic natural-language processing capabilities,
    https://github.com/jdf/cue.language#cuelanguage
15. WEKA API,
    http://www.cs.waikato.ac.nz/ml/weka/downloading.html
16. LibSVM API, http://dev.davidsoergel.com/trac/jlibsvm/
17. LibSVM Wrapper Class for WEKA,
    http://www.cs.iastate.edu/~yasser/wlsvm/

Software and Graphical Approach for Understanding Friction on a Body

Molla Ramizur Rahman

Department of Electrical & Electronics Engineering
Dayananda Sagar College of Engineering Bangalore, India
`ramizurscience@yahoo.com`

Abstract. This paper aims to explain the basic physics concept of friction in a better way as it highlights the behavior of the body during motion considering the effect of friction. The paper combines physics and mathematics with equations and graphs. The investigation focuses the characteristics of a body in motion by keeping eye on friction by making graphical approach and also by quantitative study. The research establishes certain facts and points on friction. It establishes certain equations which helps the concept to understand better. A simple demonstration has been made for better understanding the concept. MATLAB software is used to draw the graphs. An algorithm has also been framed for step by step understanding and bringing clarity to the concept.

Keywords: friction, physics, graphical, software.

1 Introduction

Friction is one of the important concepts for studying the motion of any body. It is one of essential topic in the branch of physics. Friction is considered to be a force which opposes motion of a moving body. Various laws have been formulated in friction. There are various types of frictional force namely static friction, kinetic friction, rolling friction.

The friction which occurs when the body is at rest is called static friction. Static friction is responsible in keeping the body at rest even if the force is applied. The friction which occurs when the body is at motion is called kinetic friction. Rolling friction occurs when the body is rolling. Study has been carried out for ages on friction. Friction has advantages as well as disadvantages.

There are various methods to measure friction in laboratory. They may be horizontal table method and inclined plane method. It is very essential to deal with this issue both in the macroscopic and microscopic level or atomic level [1]. The contact force arises due to the contact of a body on another material whose one of the component is friction and the other is normal reaction or normal force. Friction not only occurs between two solids but also occurs between solid and liquid like a stone flowing in river, or between solid and air such as airplane moving in air. It also occurs between two liquids. Friction is sometimes useful and is advantageous. Friction

© Springer International Publishing Switzerland 2015
S.C. Satapathy et al. (eds.), *Proc. of the 3rd Int. Conf. on Front. of Intell. Comput. (FICTA) 2014*
– *Vol. 2*, Advances in Intelligent Systems and Computing 328, DOI: 10.1007/978-3-319-12012-6_70

depends upon the type of surface in contact. Friction experienced in a smooth and polished surface is less as compared to rough surface. Friction produces heat as well as wear and tear.

Physics is the basic and fundamental discipline which deals with the properties, aspects, behavior of natural phenomenon. Physics deals with aspects ranging from the formation of earth to basic level like the motion of a body. Many scientists came up with various hypothesis, theories and laws over years. This paper is closely related to motion as friction and motion are interrelated. Motion basically is rotational motion and translation motion. Motion generally arises due to the application of force. Force may be of different types. They may be frictional force, centripetal force, gravitational force etc. Newton's three laws of motion are notable [2]. The relation between force and acceleration was first proposed by Issac Newton [4] and in this paper an attempt has been made to study the same graphically using software by considering frictional force. Physics is closely related to engineering. Engineering is an application of physics.

Mathematics is used as a tool in various fields to prove various physical phenomenon. The author [3] uses calculus which is a mathematical tool in order to find minimum angle of deviation clearly indicates the essence of mathematics in physics as well as its importance. Friction, a domain of physics is lifeless without mathematics and mathematical approach [5]. The concepts of higher mathematics are far more required in applying the concepts. The paper uses various mathematical principles in fulfilling its aim. This also reveals the close relation of mathematics and friction, on the whole another crystal example of the importance of mathematics in physics.

Demonstration is very much necessary in order to make the result clear and easier. It also helps in confirming facts and convincing. Experimentation is slightly different term which is carried out to proof some results. The author in this paper gives a simple demonstration to establishing the results mentioned.

Graphical approach plays an important role in understanding any concept. Graphs which are again a domain of mathematics is very vital in making the concept easier and again gives a strong evidence of the importance of mathematics. Graphical approach helps in finding solution in a systematic, better and suitable way. The graphs are helpful in understanding the nature of various physical phenomenon and their behavior over different parameters. The graphical approach many a times helps in comparison of results which is again a very important issue in any scientific study. The nature of the graph changes with change of parameters keeping different parameter as constant for different times.

Flowchart is the pictorial method of showing steps of action in a systematic and organized manner. It makes the work more simple and sequential. The algorithm may be suitable for framing a flowchart. The author [6] also dealt of algorithm on the issue of analyzing the problem. Flowchart and algorithm are computer and software approach. Thus, this paper also illustrates that computer and software are essential in every field.

2   Literature Survey

The authors [7] made his effort to find from experimentation the relation between frictional force and displacement. In addition load capacity was found out and the conclusion drawn was that the load capacity was restricted by frictional force. Comparison was made with load capacity and inchworm device.

The author [8] made an attempt to measure the friction existing between the surfaces of graphite and diamond against a sharp silicon nitride tip using frictional force microscope.

The paper [9] verifies experimental results with vibration friction model. The paper greatly deals with vibration friction in ultrasonic motor. The vibration amplitude measured using laser vibrometer is proportional to input voltage. The study further claims frictional coefficient decreases as the input voltage goes on increasing and the experimental results holds good with the theoretical calculations.

The authors [10] find a novel cost effective method for friction co-efficient estimation between vehicle and tire-road by measuring only wheel angular velocity, traction/brake torque and longitudinal acceleration. The method indicates three important steps longitudinal slip ratio estimation, tire longitudinal force estimation and finally the estimation of friction coefficient using recursive least squares method.

The authors [11] by experimentation proposed a dynamic model for friction which highlights various friction behaviors like stribeck effect, hysteresis and various spring like characteristics. The author also presented the stability results by exploring control strategies which includes friction observer.

The authors [12] had made an attempt to directly measure the frictional force between ball-point pen and paper. The accurate measurement of friction force is obtained with the help of this method by measuring the inertial force of mass.

The authors [13] studied the influence of original paper formation and the resin impregnation uniformity of the wet type friction material and establishes the fact that the material formation changes the original paper formation.

The authors [14] formulated robust controller in order to achieve ultimate boundedness in both micro and macro region. The author claims that the method implemented by them gives better performance by making comparison with the results of the existing controllers.

3   Demonstration

Suppose a force $F1$ is applied on a block. It is observed that there is no movement.This is because the value of $F1$ does not overcome the limiting value of static friction. Here the frictional force $f1$= Force applied = $F1$.On increasing the applied force to $F2$, the corresponding frictional force $f2$ = Force applied = $F2$ as there is still no movement on the block. But after that if a slight increase in applied force, the applied force changes to $F3$ and the corresponding frictional force $f3$ decreases slightly and the movement of the block starts. After that on increasing the applied force to any value the frictional force remains constant as $f3$ and increase in applied force cause increase in acceleration.

Table 1. Depicts the applied force and their corresponding frictional force

Applied Force	Frictional Force
$F1$	$f1$
$F2$	$f2$
$F3$	$f3$
$F4$	$f3$

Here, $f2$ is the limiting value of static friction,
    $f3$ is the value of kinetic friction.

Applied Force: $F1 < F2 < F3 < F4$
Frictional Force: $f1 < f2 > f3$

$$F1 = f1$$
$$F2 = f2 = \mu sN$$
$$F3 - f3 = ma1$$
$$F4 - f3 = ma2$$

Where m = mass of the body,
    $a1, a2$ are the accelerations produced by the body when the applied forces are $F3$ and $F4$ respectively.

4   Facts Established

1) The value of static friction increases with the increase in applied force till limiting value of static friction.
    2) The maximum value of static friction= μsN
        Where, μs = co-efficient of static friction and
            N = Normal Reaction
    3) Static friction is always equal to the force applied which causes the tendency of motion. Thus, static friction is self adjusting force. If the static friction was not a self adjusting and always had fixed value then by applying force which is less than maximum value of static friction in one direction, the body would move in the opposite direction. In other words friction at any point of time cannot be greater than the force applied. If it was so, the friction would cause the motion of the body.
    4) Therefore maximum value of force for which there is no movement = Limiting value of static friction
        = μ sN
    5) By declaring the value of static friction, it indicates the maximum value of static friction.
    6) Co-efficient of static friction (μ s) and co-efficient of kinetic friction (μ k) are different.
    7) Kinetic Friction has a constant value and is independent of force applied.

8) *The value of kinetic friction is slightly less than maximum value of static friction.*

9) *Friction increases with the increase in number of contact surfaces.*

10) *Both static friction and kinetic friction cannot exist at the same time.*

11) *The value of friction is zero when no force is applied irrespective of the contact area or the weight of the body.*

5   Algorithm of Behavior of Body under Friction

1) Application of force.

2) The applied force is gradually increased.

3) The body does not move till the applied force is less than maximum value of static friction.

4) As the force overcomes the maximum value of static friction, the body gets into motion, and the friction is no longer static friction and is termed as kinetic friction.

5) The body comes to rest and the process ends up when the applied force is brought to zero.

The sets of equation implies for different friction and acceleration felt by a block under different condition.

$$f = \begin{cases} F, \ if \ F \leq \mu s \ N \\ \\ \mu \ kN, \ otherwise \end{cases} \tag{1}$$

$$a = \begin{cases} 0, \ if \ F \leq \mu s \ N \\ \\ (F - f)/m, \ otherwise \end{cases} \tag{2}$$

Where F is the applied force,
f is the frictional force,
a is the acceleration produced by the body,
m is the mass of the body

The various graphs has been plotted using MATLAB software They are
i)   Force applied vs Friction force.
ii)   Force applied vs acceleration
iii)   Force applied vs velocity

The investigator has control only on the applied force. The applied force which is independent quantity is a common parameter for all the graphs and therefore plotted along the x-axis and various other parameters such as frictional force, acceleration and velocity are observed and plotted along y-axis. The author investigates the effect of various dependent parameters by varying the independent parameter that is the applied force. The graph of two dependent quantity such as acceleration-velocity graph, frictional force-velocity graph, frictional force-acceleration does not make sense as both these parameters are dependent and depends on applied force and the investigator has no control on any one of the parameter.

The various graphs has been drawn by considering a mass of 1 kg being applied with a continuous external force varying from 0 to 100 N in steps of 1N in every 0.5 sec. Clarity is obtained by varying the force gradually thus all the effects are observed. Here, the maximum value of static friction is considered to be as 7 Newton.

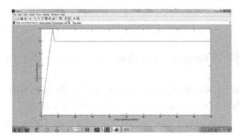

Graph 1. Graph between Force applied and Frictional force

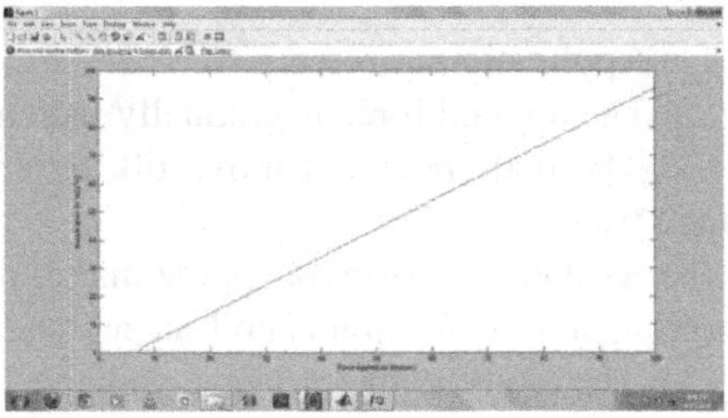

Graph 2. Graph between Force applied and acceleration.

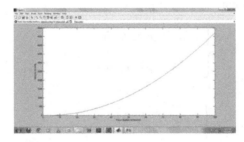

Graph 3. Graph between Force applied and Velocity

In graph 1, friction is plotted against the force applied. The study showed that on increasing the applied force, the friction that is static friction increases till the maximum value of static friction and further on increasing the applied force the transition of static friction into kinetic friction takes place indicated by a sharp fall in friction also indicating kinetic friction is less than static friction which remains constant further throughout. In this particular instant, the frictional force that is static friction increases linearly with the increase in applied force upto 7N which is the maximum value of static friction. When the applied force is 8 N, the frictional force suddenly decreases to 6N which becomes the value of kinetic friction and remains constant throughout.

In graph 2, acceleration is plotted against the force applied and behavior of the body is observed. It is seen that when the force is incapable of making movement on the body, the acceleration was zero. As the movement of the body begins when the force overcomes the maximum value of static friction, acceleration increases with the increase in applied force. Further it was noted that there is a sharp increase in acceleration in the first step as the movement of the body begins and further as the force goes on increasing though the acceleration increases but there is a decrease in increase in acceleration as compared to beginning of motion, which becomes constant in further steps. In other words, the slope of the graph is high as the movement of the body starts and then the slope decreases and becomes constant with the increase in force. In this instant, the value of acceleration remains zero upto applied force is 7N. In the next step as the applied force is increased to 8N, the acceleration becomes 2 m/s^2 so increase in acceleration in the first step of movement is 2 m/s^2 (2 m/s^2 - 0 m/s^2). In the very next step when the applied force is 9N, the acceleration is increased to 3m/s^2. So the increase in acceleration with respect to the previous step is 1m/s^2 (3 m/s^2 -2 m/s^2), which initially during starting of movement of the body was 2m/s^2. Further as the force goes on increasing, the acceleration goes on increasing in each step but the increases in acceleration remains constant as 1m/s^2.

In graph 3, velocity is plotted against force applied. It is observed that the velocity remains zero till there was no movement on the body. Further it was noticed that as the movement of the body begins, the velocity increases with the increase in applied force. It was noted that there is a sharp increase in velocity in the first step of movement and then though the velocity increases, the increase in velocity is reduced for the very next application of force which further increases as the force increase. In other words, the slope of the graph is higher when the movement of the body begins. In the very next stage of application of force, the slope of the graph decreases, and further goes on increasing in the following stages of application of force. In this instant, the velocity was zero upto 7 N. When the applied force becomes 8N, the velocity becomes 8 m/s and the increase in velocity is 8 m/s (8 m/s- 0 m/s). In the next step, when the applied force becomes 9 N, the velocity becomes 13.5 m/s and thus the increase in velocity is 5.5 m/s (13.5 m/s – 8 m/s), which shows that that the increase in velocity is less than that of during the starting. Further it shows the increase in velocity goes on increasing with the increase in applied force.

6   Results

The simulation of the behavior of body under friction is studied. It is noted that there is a sharp increase in acceleration and velocity in the beginning of the motion. This effect cannot be eliminated but can be reduced if the force is applied at smaller interval. The paper also reflects an application of software on friction.

References

1. Verma, H.C.: Concepts of Physics, vol. I, 4th reprint, pp. 85–89. Bharati Bhawan Publishers and Distributaors, ISBN: 81-7709-187-5
2. Sears, F.W., Zemansky, M.W., Young, H.D.: University Physics, 6th edn., p. 59. Narosa Publishing House, ISBN: 81-85015-63-5

3. Rahman, M.R.: Alternate Derivation of Condition for minimum angle of deviation. Undergraduate Academic Reasearch Journal (UARJ) 1(3, 4), 8–10, ISSN: 2278-1129
4. Halliday, Resnick, Walker: Fundamentals of Physics, 8th edn., pp. 88–99. Wiley India, ISBN: 978-81-265-1442-7
5. Singh, D.B.: AIEEE Physics, p. 148. Arihant Prakashan, ISBN: 81-88222-32-1
6. Rahman, M.R.: Enhanced Approach of Computing Least Common Multiple (LCM). Dayananda Sagar International Journal (DSIJ) 1(1), 65–69, ISSN: 2347-1603
7. Kawagoe, K., Furutani, K.: Influence of friction force on seal mechanism with one degree of freedom. In: Proceedings of 2003 International Symposium on Micromechatronics and Human Science, MHS 2003, pp. 317–322 (2003) ISBN: 0-7803-8165-3
8. Ruan, J.-A., Bhushan, B.: Atomic-scale and microscale friction studies of graphite and diamond using friction force microscopy. Journal of Applied Physics 76(9), 5022–5035, ISSN: 0021-8979
9. Liew, J.Y., Chen, Y., Zhou, T.Y.: The measurement on vibration friction coefficient of ultrasonic motor. In: IEEE Ultrasonics Symposium, pp. 154–156 (2008) ISBN: 978-1-4244-2428-3
10. Li, B., Du, H., Li, W.: A novel cost effective method for vehicle tire-road friction coefficient estimation. In: IEEE/ASME International Conference on Advanced Intelligent Mechatronics (AIM), pp. 1528–1533 (2013) ISBN: 978-1-4673-5319-9
11. De Wit, C.C., Olsson, H., Astrom, K.J., Lischinsky, P.: A new model for control of systems with friction. IEEE Transactions on Automatic Control 40(3), 419–425, ISSN: 0018-9286
12. Chigira, Y., Fujii, Y., Valera, J.D.R.: Direct measurement of friction acting between a ballpoint pen and a paper. In: Annual Conference SICE, vol. 2, pp. 1518–1521 (2004) ISBN: 4-907764-22-7
13. Lu, Z., Zhang, D., Huang, X.: Influence of original paper formation and the resin impregnation uniformity on the wearing of the wet type (paper-based) friction material. In: 2011 International Conference on Remote Sensing, Environment and Transportation Engineering (RSETE), pp. 5034–5037, ISBN: 978-1-4244-9172-8
14. Huang, S.-J., Yen, J.-Y., Lu, S.-S.: Dual mode control of a system with friction. IEEE Transactions on Control Systems Technology 7(3), 306–314, ISSN: 1063-6536

An Investigation on Coupling and Cohesion as Contributory Factors for Stable System Design and Hence the Influence on System Maintainability and Reusability

U.S. Poornima[1,2] and V. Suma[1]

[1] Research and Industry Incubation Centre, Dayananda Sagar Institutions, Bangalore, India
[2] Raja Reddy Institute of Technology, Bangalore, India
{uspaims,sumadsce}@gmail.com

Abstract. Complexity is an inherent property. Measuring and keeping it under control is more logical than practical. Since quality is directly proportional to complexity, a quantitative measure is expected. In software industry, software quality is depending on quality of each phases of its development. As the size of the requirement increases, the design phase complexity increases. This has an adverse affect on software stability. The fundamental design-need in Object Oriented Methodology (OOM) is the well-defined modules and their inter-connectivity, namely, cohesion and coupling. The structure of such artefact is expected to be simple since it influences stability and thereon the module reusability and maintainability. This paper encompasses an investigation on coupling and cohesion which are major design decisive factors and their influence on maintainability and reusability through design stability. The paper provides a hypothetical support on the influence of coupling and cohesion on maintainability and reusability. It also focuses on the further research interests in the same field as a part of through literature survey. The work would contribute to design a high quality product by which the industries sustain themselves in the competitive market.

Keywords: Object Oriented Development, Design Quality and Stability, Coupling, Cohesion, Maintainability and Reusability.

1 Introduction

Software engineering for researchers deals with developing analytical models and tools to measure the quality of development process, product and people for quality products. Bottom-up development approach in software engineering is a popular methodology for such a domain with scalable requirements over time. The error percentage in design phase due to the incorporation of new requirements is more when compared to other phases of software development [1]. Since modules are the design artefacts which constitutes the solution domain of the system, the design quality of each module is important [2]. The modules contain well defined classes,

© Springer International Publishing Switzerland 2015
S.C. Satapathy et al. (eds.), *Proc. of the 3rd Int. Conf. on Front. of Intell. Comput. (FICTA) 2014*
– *Vol. 2*, Advances in Intelligent Systems and Computing 328, DOI: 10.1007/978-3-319-12012-6_71

interfaces with interrelationship to provide pre-defined services. The relatedness of elements depicts cohesion and their dependency, coupling. Cohesiveness is the degree of cohesion within a class and coupling is the quantity of dependency within and outside the system [3]. Thus, the design is expected to be flexible as well as less complicated with measurable and controlled cohesion and coupling for future enhancements. We infer such state of design as a stable design through which the abilities of a system such as scalability, reliability, maintainability and reusability is increased. Thus, the quantitative measurement of cohesion and coupling estimate the quality of system design which is served as an input for designers and managers.

Coupling is formed due to dependency. This dependency appears between high level elements and/or the lower level elements. Thus, dependency concept is conceptualised as Package Level Dependency (PLD), Class Level Dependency (CLD) and Interface Level Dependency (IDL). PLD and ILD are static and are established during system high level design. The CLD is code-dependent and is established when the system is implemented. UML depicts overall model of the solution domain as a set of modules and its elements, in which PLD is a subset [4]. We, thus, infer that identifying quantity of PLD dependency at design stage is important to measures the system complexity. Such measurement further supports to find the influence of cohesion and coupling on various abilities of software such as maintainability, reusability, scalability, controllability, simplicity, portability, accessibility, reliability and so on in future.

A package in a solution space has dependencies. Stability of a package is dependent on quantity of coupling. A package is more stable when it is more referred by other packages [5]. Such stable packages are complex in nature than instable packages. Thus, stable packages are more prone to errors than instable packages. High vigil on such packages during design is highly expected since system complexity is depending on package/module complexity. In this paper, we hypothetically relate the contribution of coupling and cohesion in a package, as a stability factor, to its maintenance and reusability respectively.

The paper is organized as below. Section 2 deals with the related work on class coupling and cohesion. In section 3, we relate stability and complexity and set the hypothesis. Section 4 theoretically validates the hypothesis. Section 5 concludes the paper.

2   Related Work

In this section, we summarise the survey of related work on coupling.

Author of [3] defines coupling as quantity of dependency between software part and rest of the system. He proposes import, export coupling of a software part and the principles for defining coupling metrics.

Author of [5] proposes different object oriented design principles. He proposes different pattern for better modelling of the solution domain which encourages system maintainability.

Authors of [6] categorize coupling and their virtue. He proposes a framework in which coupling is measured during runtime of a system. Two variations of coupling such as coupling at class level and object level is measured on different projects and a trade-off is discussed with different cases. The paper also highlights on cohesion and its metrics.

3 Stability v/s Complexity

Software products are expected to be stable and flexible. This is achieved through stable design. Author of [5] states stability as effort for change. *However, the definition of stability is context dependent. We can say that, it is the desired state of a product/module/design with internal or external influences so that optimum result is expected.* When relate stability to the module design, it is the desired state of a module/package design influenced by complexity of coupling and cohesion at either code or objects level of its elements such as a class or interface. When module reaches a stable design state, complexity of addition and deletion of requirements is reduced. Thus, stability and complexity are interwoven as

$$stability \propto 1/complexity \tag{1}$$

Complexity has an inverse effect on stability. As complexity of design increases, stability decreases and vice versa. Since stability factor for a package influences the reusability and maintenance, it becomes decisive factor for quality design.

However, complexity is an inherent property. In design, it is the amount of coupling and cohesion. Thus, complexity is directly proportional to coupling and inversely to cohesion.

So, complexity is expected to be under control when designer has a hold on coupling and cohesion. Thus, two hypotheses can be established to project the interrelationship between stability, complexity, coupling and cohesion.

Hypothesis 1: Maintenance is inversely proportional to stability.
Hypothesis 2: Reusability is directly proportion to stability.

The influence of coupling and cohesion on maintenance and reusability is depicted in the figure 1.

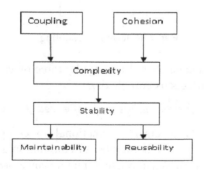

Fig. 1. Hierarchical relatedness of design affecting factors

4 The Validation of Stability v/s Maintainability and Reusability

Two packages are dependent when at least one component of a package such as a class is using the other package component. UML specification for object oriented design depicts such dependency. Package diagram describes system model through different types of interrelationship among them with no implementation of methods. This high level design provides first-hand information on dependency among packages in the system. The complexity measure of such modules is depending on coupling of different types.

$$complexity \propto quantity\ of\ coupling\ types \tag{2}$$

The package coupling is twofold: Refers-to and Referred- by.

1.   Referred-by coupling:
A package could provide high level services by providing services to other modules. The sharing of data or functionality is a decisive factor of the package design complexity. Referred-by coupling is the dependency of other external packages on the one under consideration.

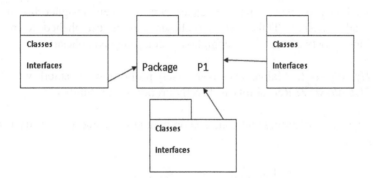

Fig. 2. Referred-by coupling

Figure 2 shows that Package P1 has referred by three other packages. The number of dependencies directly affects its complexity.

$$complexity \propto number\ of\ referred\_by\ coupling \tag{3}$$

In Eq.1, the stability is inversely promotional to complexity. Equating with Equation (3), we can infer that as the number of coupling in a package increases, the stability of package decreases. When relate to maintenance, *we define stability as the minimum effort to be invested for change.* Thus, decrease in package stability increases the maintenance effort. Hence, we relate stability factor with maintenance as

$$maintenance \propto 1/stability \tag{4}$$

Hypothesis 1 is theoretically validated.

From equation (4), it is implied that amount of effort for maintenance for a module is directly proportional to the number of external binding with the module. As complexity increases, stability decreases and requires more effort for system change.

However, the stability factor for reusability is different. When a module is more shared, it is more cohesive. Thus, the stability of reusability can be defined as amount of external access to the module under consideration. As references increases, we can say, module is more stable and cohesive in terms of reusability. The stability factor for reusability is

$$reusability \ \propto \ stability \tag{5}$$

Hypothesis 2 is theoretically validated.

2. Refers-to coupling:
Refers-to relationship depicts a module dependency on other modules. In figure 2, the package M1 has external dependency either at code or object level.

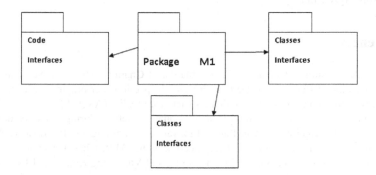

Fig. 3. Refers-to coupling

Since it is using the services offered by other modules, complexity level at its own is less and has different impact on stability for maintenance and reusability.

$$complexity \ \propto \frac{1}{number} of \ referred_{to} coupling) \tag{6}$$

Thus, such modules are more maintainable than reusable. The stability factors of referred-by and refers-to influences amount of reusability and effort for maintainability of that module.

5   Conclusion

Software engineering is a field of developing models, tools and metrics to deliver quality products. Improving the design quality by proper structuring of system modules is a continuous research work in industry and academics. This paper focused on design quality decisive factors such as coupling and cohesion which directly influences design complexity. The design of the solution is expected to be stable to incorporate the new changes over time. In this paper, we infer the stability of a design as state of design with present complexity to incorporate new changes. A stable design then becomes a major factor for measuring abilities such as maintainability and reusability.

Thus, quantity of coupling and degree of cohesion become influential factor for complexity which in turn decides the stability factor for system design. A hypothetical validation on the various design complexity decisive factors, as mentioned, is done with proper hypothesis. The work would be extended to prove the same using data from various projects and tools in future.

Acknowledgements. The authors would like to sincerely acknowledge all the industry personnel for their valuable suggestions, help and guidance in carrying out this part of research. The complete work is undertaken under the framework of Non Disclosure Agreement.

References

1. Nair, T.R.G., Suma, V., Nair, N.G.: Estimation of Characteristics of a Software Team for Implementing Effective Inspection Process Inspection performance Metric. Software Quality Professional Journal, American Society for Quality (ASQ) 13(2), 14–26 (2011)
2. Poornima, U.S., Suma, V.: Significance of quality metrics during software development process. In: International Proceeding of Innovative Computing and Information Processing, ICICIP 2012, Mahendra College of Engineering, India (March 2012), available on dblp
3. Briand, L., Morasca, S., Basili, V.R.: Defining and Validating High-Level Design Metrics. IEEE Transactions (1999)
4. Poornima, U.S., Suma, V.: Visualization of Object Oriented Modeling from the Perspective of Set Theory. In: Presented at International Conference organized by IACSIT(China) in Chennai. Lecture Notes on Software Engineering, vol. 1(3), pp. 214–218 (May 2013), doi:10.7763/LNSE, ISSN:2301-3559
5. Martin, R.C.: Design Principles and Design patterns, visit,
   http://www.objectmentor.com
6. Hitz, M., Montazeri, B.: Measuring coupling and cohesion in object oriented systems. In: Proceedings of the International Symposium on Applied Corporate Computing, pp. 25–27 (1995)

Application of Component-Based Software Engineering in Building a Surveillance Robot

Chaitali More, Louella Colaco, and Razia Sardinha

Depatment of Information Technology,
Padre Conceicao College of Engineering,
Goa University, Goa, India
chaitali.more27@yahoo.com, lmesquita@rediffmail.com,
gal.outlaw@gmail.com

Abstract. In this paper, the application of Component-Based Software Engineering methodology (CBSE) in the development of a robotic system is documented. The robot movements can be controlled remotely with the help of a software application. It is also capable of streaming live video while moving. CBSE methodology emphasizes on developing new system from pre-built components. Therefore, it is suitable for the development of robotic systems where a large number of such components are used and there is also a wider scope for the reuse of these components. This paper gives, in detail, each phase of the robot development and also proves the suitability of CBSE in the development of such systems. The surveillance robot was successfully built using the software development methodology and worked well in accepting instructions from the software application on the direction of movement and capturing the video of the environment.

Keywords: component-based software engineering, software engineering in robotics, surveillance robot.

1   Introduction

Component reuse carries utmost significance in the domain of software engineering called component-based software engineering (CBSE) or component-based development (CBD). It focuses on building systems by combining components, and, deviates from the traditional requirements, analysis, system design and implementation phases.

The robotic system developed in this work is a surveillance robot that is capable of capturing visual information as it moves in the environment in which it is placed. The robot movement is controlled by a system user with the help of a web application that requires Wi-Fi for communication with the robot. The robot can be considered as a system that integrates several components to form an entire system and since CBSE methodology is based on a similar aspect of integrating components, it is best suited in robotics domain.

© Springer International Publishing Switzerland 2015   651
S.C. Satapathy et al. (eds.), *Proc. of the 3rd Int. Conf. on Front. of Intell. Comput. (FICTA) 2014*
– *Vol. 2*, Advances in Intelligent Systems and Computing 328, DOI: 10.1007/978-3-319-12012-6_72

2   Related Work

The component technologies that are extensively used in software development are CORBA, COM/DCOM and Enterprise JavaBeans (EJB). Authors of [1] and [2] have given the architecture of JavaBeans Component Model and CORBA component Model respectively. Component based development process and component architectures of COM and EJB are given by the authors of [3].

Introduction to CBD and component definition are provided by the authors of [4]. The authors of the article [5] have given detailed information on reusable software blocks, component specification and implementation and, also, it gives a working example of CBSE. Some challenges in building robotic software component systems are given by the authors of [6]. More information on CBD is given in [7]. Authors of [8] have given the definition and objectives of CBSE and also advancement in component based approach. So far, robotics software reuse has been possible at the level of libraries. Certain projects maintain repositories of independent components. ORCA is an open-source framework for developing component-based robotic systems [9][10]. Application of ORCA in building software architecture for a human robot team is given by the authors of [11]. Some more frameworks like GenoM, Distributed Control Architecture, Modular Control Architecture and the framework requirements are given by the authors of [12].

CBSE methodology has long been used as a development technique for robotic systems. The authors of [13] have discussed types of software components and how rapid development of robotic applications can be done. Authors of [14] have carried out a survey on component based work in robotics and have provided different architectures available for robotic systems development. A methodology for the development of robot software controller is given by the authors of [15].

3   Component-Based Software Engineering

In CBSE, various components are integrated to form a software system. The components can, however, be produced from scratch or commercial off-the-shelf (COTS) components can be used in the making of software systems. The definition of a component and its properties is given by the authors of [16].

3.1   The CBSE Process

The CBSE process diagram is given by the author of [17]. It begins with the 'requirements specification' phase, in which, the systems requirements and specifications are outlined. 'Component analysis' phase follows next, in which, various components are analyzed. After the components that are to be used are finalized, the requirements are modified as per the availability of the products, their functionality and specifications. This phase is the 'requirements modification' phase of CBSE. The next phase is 'system design with reuse', in which, the preparation of the architectural design is done, keeping in mind the aspect of reuse. 'Development and integration' of the components follows after the design phase. If the components to be used are already developed, then they are integrated to form the entire system. 'System validation' is done to check its conformance to the requirements and interoperability of the components.

4   Application of CBD to the Development of Surveillance Robot

A surveillance robot is a robotic system that is mainly used for the monitoring and surveillance of an area. Such robots usually capture sound, video or images of the area and transmit to a remotely located controller. The development of the surveillance robot follows the steps of the CBSE process given in 3.1.

4.1   Requirements Specification

The requirements of a surveillance robot are: Navigation-The robot needs to move in forward, reverse, left and right directions; Human-robot interaction-A user should be able to command the robot to move in a particular direction with the help of an application on a cell phone; Communication-The communication between cell phone and the robot should use Bluetooth technology; Image capturing-It needs to capture images while on the move and send the images via Bluetooth to a phone or a laptop, and, Obstacle avoidance-While on the move, the robot should be capable of detecting and avoiding any obstacles on its path.

4.2   Component Analysis

Different components were evaluated based on their technical specifications and interoperability. Traditionally, micro-controllers have been used as robot "brains" that give commands to its various other components as to what needs to be done. Initially, ATMega32 was considered as a good choice for the robot. But as the research progressed, a component called Raspberry Pi (RPi) was also considered since it has a CPU mounted on it which offers benefit of faster processing over a micro-controller. Motor drivers are required to control the motors. Out of the different motor drivers available, L298 was considered to be a good option for the robot.

A cell phone based application would be required for controlling the robot movements. Android being the initial choice of technology to develop the application resulted in opting for an Android OS based cell phone. However, after further analysis, a conclusion was drawn as to have a web application instead of a cell phone based application for the benefits of platform independency, memory saving on the device, and ease of access to the web application through any device.

A camera would be installed on the robot to capture images of the environment in which it is placed. Raspberry Pi organization has launched a Raspberry Pi camera that is designed to work with the RPi board. Bluetooth module needs to be installed on the robot that would act as the communication medium between the robot and the cell phone. A Bluetooth module called rs232 Bluetooth adaptor was considered initially. After evaluating Bluetooth on various parameters like range, speed and comparing it with a technology like Wi-Fi, Bluetooth was found to be weaker, and hence, Bluetooth was replaced with Wi-Fi. A nano Wi-Fi adapter could be installed on the robot to make it a Wi-Fi enabled device. An obstacle avoidance module would enable the robot to avoid collision with any obstacles coming in its path. Array of proximity sensors are available, such as ultrasonic range finder, infrared distance sensor, laser range finder and so on. Considering the various options available, infrared (IR) distance sensor was contemplated as the best choice because of its modest cost and easy availability.

4.3   Requirements Modification

The phase of component analysis caused some significant changes in the technologies that were earlier planned to be used and that resulted in changes in the choice of components. The revision of the technology for the robot control application resulted in replacing the wireless technology, Bluetooth, which was initially planned for the robot. It was replaced by Wi-Fi. Another requirement was that of a camera which needs to send images to a laptop/cell phone. The RPi camera, however, comes with several benefits like streaming of live video to laptop/cell phone using Wi-Fi. As a result of which, transferring images proved to be a poor choice for the robot. Live video streaming would not only improve the capability of the robot but also give its users a better experience.

Considering, the above adjustments. The modified requirements are: Human-robot interaction: A web-based application that enables the user to control the robot movements, Communication: The robot needs to be Wi-Fi enabled, and, Image capturing: The camera connected to the robot needs to capture live video and stream it to a laptop or a cell phone.

4.4   System Design with Reuse

A hardware design needs to be created to ensure proper connections between various components of the robot. Referring to fig. 1, the infrared (IR) sensor, which is connected to RPi, takes input from the environment by constantly checking for obstacles and interacts with RPi CPU. If it comes across an obstacle, the CPU mounted on RPi sends commands to the motor driver (L298) to stop the wheel movement and wait for 1sec, and then the wheels reverse for 1sec and stop again. L298 is connected to RPi and camera is mounted on RPi as well.

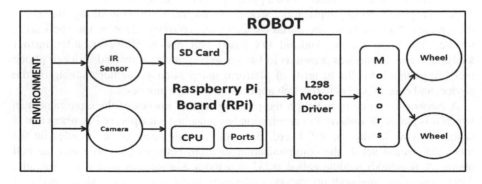

Fig. 1. Hardware Architecture of the Robot

The camera module captures video of the environment and sends it to a laptop or a cell phone through RPi over a Wi-Fi connection. The RPi board is the main component of the robot that is responsible for controlling the other components. The motors are connected to L298 and receive inputs from it to rotate the wheels in either clockwise or anti-clockwise direction or stop rotating. The power source is a rechargeable power bank that is connected to RPi to provide power to the system.

4.5   Development and Integration

Each of the hardware components used can be independently programmed. Also the components can be reused in other systems by modifying certain functionalities if required. The components to be used can be obtained from other systems if available or can be purchased and programmed. The programming of the components needs to be done on the RPi board. The board is connected with a Wi-Fi dongle to connect it to a network. It can then have an IP address assigned to it which will be the IP address of the robot. The files on RPi can be created, accessed and modified through 'putty.exe'. The sensor and motor driver programming was done using Python. The web application, a software component, responsible for controlling of the robot movement was developed using languages like HTML, JavaScript, JQuery and AJAX. The RPi board needs to be installed with Apache2 before the development of application can begin. JQuery and AJAX help to run the python codes in background, when invoked from the web application. The front-end of the web application developed is shown in fig.2 below.

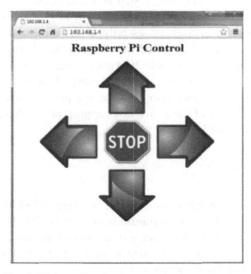

Fig. 2. Web application to control robot movement

The program for IR sensor is started at the time the RPi boots up. After the development stage is complete, all the hardware and software components are integrated to form a complete system based on the hardware design given in fig.1. The L298 motor driver and IR sensor are connected to RPi by connecting their corresponding pins. L298 is then connected to the motors and a battery. The camera is installed on RPi by enabling camera support on it. Fig. 3 shows a picture of the surveillance robot after assembling all the components.

Fig. 3. Assembled Surveillance Robot

4.6   System Validation

The web application can be accessed by entering the robot IP address in the address bar of a web browser. The robot was tested in a real environment to check if it can take commands from the application and move in any of the directions (forward, reverse, left and right). It moved successfully in the directions specified when the corresponding buttons were clicked and halted when 'stop' button was clicked. The robot also captured video which could be viewed over Wi-Fi on a laptop or a cell phone. VLC player was used to get the live streaming of the video. The robot detected obstacles placed on its path and also took the pre-programmed measures to avoid collision. It completed each of the above tasks appropriately, thereby, adhering to the modified requirements stated in 4.3.

5   Results

The video captured was in h264 format with the size 800x400 and 24 frames per second. The protocol used for video transmission was HTTP. The video was viewed on laptop and cell phone but with a delay of about 4-5 seconds. The snapshot of the VLC player window displaying the video is shown in fig.4. The distance at which obstacles are detected by the IR sensor is 1 cm.

Fig. 4. VLC player displaying the video captured by the Robot

6   Conclusion

In this work, the surveillance robot was successfully tested in its capability in receiving the inputs from the web application on the direction of movement. The robot moves with ease in any of the specified directions. The camera module installed on the robot captures live video that can be streamed to a laptop or a cell phone. However, it comes with a slight delay. The IR sensor detects obstacles that are at a minimum distance of 1cm apart. The robot reverses for 1sec and stops, on detecting an obstacle.

The different components of the system work well when integrated. The components can also be re-used in other robotic systems with modifications if needed. CBD approach works well with robotic systems considering the aspect of reuse and use of many off-the-shelf components in robot development.

References

1. Englander, R.: Developing Java Beans, 1st edn. (1997)
2. Wang, N., Schmidt, D.C., O'Ryan, C.: Overview of the CORBA Component Model (2000)
3. Crnkovic, I., Larsson, M.: Component-Based Software Engineering-New Paradigm of Software Development. Malardalen University, Sweden (2001)
4. Schlegel, C., Steck, A., Lotz, A.: Chapter 23 Robotic Software Systems: From Code-Driven to Model-Driven Software Development. University of Applied Sciences, Germany (2012)
5. Brugali, D., Scandurra, P.: Component-based Robotic Engineering (Part I). IEEE Robotics and Automation Magazine (2009)

6. Brugali, D., Shakhimardanov, A.: Component-based Robotic Engineering Part II: Systems and Models. IEEE Robotics and Automation Magazine (2010)
7. Cai, X., Lyu, M.R., Wong, K.F., Ko, R.: Component-Based Software Engineering Technologies, Development Frameworks, and Quality Assurance Schemes (2000)
8. Bose, D.: Component-Based Development: Application in Software Engineering. Indian Statistical Institute (2010)
9. Brooks, A., Kaupp, T., Makarenko, A., Williams, S.: Towards component-based Robotics. In: IEEE/RSJ International Conference on Intelligent Robots and Systems, IROS 2005 (2005)
10. Brooks, A., Kaupp, T., Makarenko, A., Williams, S., Oreback, A.: Orca: A Component Model and Repository. In: Brugali, D. (ed.) Software Eng. for Experimental Robotics. STAR, vol. 30, pp. 231–251. Springer, Heidelberg (2007)
11. Kaupp, T., Brooks, A., Upcroft, B., Makarenko, A.: Building a Software Architecture for a Human-Robot Team using the Orca Framework. In: IEEE (2007)
12. Colon, E.: Robotics Frameworks. Royal Military Academy, Belgium (2014) ISBN: 978-1-463789-442
13. Stewart, D.B., Khosla, P.: Rapid Development of Robotic Applications using Component-based Real-Time Software (1995)
14. Oreback, A.: Components in Intelligent Robotics. Royal Institute of Technology, Sweden (2000)
15. Passama, R., Andreu, D., Dony, C., Libourel, T.: Component based Software Architecture of Robot controllers (2005)
16. Bachman, F., Bass, L., Buhman, C., Comella-Dorda, S., Long, F., Robert, J., Seacord, R., Wallnau, K.: Volume II: Technical Concepts of Component-BasedSoftware Engineering. Software Engineering Institute (2000)
17. Sommerville, I.: Software Engineering, 7th edn. Pearson Education (2004)

Comprehension of Defect Pattern at Code Construction Phase during Software Development Process

Bhagavant Deshpande[1], Jawahar J. Rao[2], and V. Suma[2]

[1] Research Scholar Dept. of Computer Science Engg., JJTU, Rajasthan
deshcapricorn@gmail.com
[2] Dayanandsagar Research and Industry Incubation Centre,
Dayananda Sagar Institutions, Bangalore, India
jawahar_rao@yahoo.com, sumavdsce@gmail.com

Abstract. Ever since the introduction of computers, technological advancement has taken an exponential form. Thus, development of software which attains total customer satisfaction is one of the mandatory needs of any software industry. Delivery of defect free software is one of the primary requisites to achieve the aforementioned objective. In order to comprehend defect facets, it is essential to have knowledge of defect pattern at various phases of software development. This paper therefore provides a comprehensive analysis of occurrence of defect pattern which are obtained through a case study carried out in one of the sampled leading software industry. This empirical investigation is a throw light for the project personnel to formulate effective strategies towards reduction of defect occurrences and thereby improve quality, productivity and sustainability of the software products.

Keywords: Software Engineering, Software Development Life Cycle, Software Quality, Defect Management.

1 Introduction

Introduction of computers has brought tremendous growth in the living style of the human society. Advancement in software has thus made it to be one of the inevitable components in almost all frontiers of mankind. Hence, development of high quality is one of the core demands of the day. Due to this demand, sustainability of any IT organization depends upon the level of quality that is achieved in their software products which ultimately leads to the depth of customer satisfaction.

Quality is not a state and is a continuous process. The term software quality has various definitions based on the viewpoints of the customer, user, and developer. Definition of quality includes fitness for use when the customer's requirements and expectations are considered. Quality definition must also consider the users' perspective. Authors of [1] states that the principal focus of software quality is the user's needs [1]. Authors in [2] define quality as value to some person [2]. However, quality adds value to business too. From the industry standpoint, the term quality has a different definition. Authors of [3] define quality as the existence of characteristics

© Springer International Publishing Switzerland 2015
S.C. Satapathy et al. (eds.), *Proc. of the 3rd Int. Conf. on Front. of Intell. Comput. (FICTA) 2014*
– *Vol. 2*, Advances in Intelligent Systems and Computing 328, DOI: 10.1007/978-3-319-12012-6_73

of a product that can be assigned to meet requirements [3]. Authors in [4] defines quality as an intangible concept that can be discussed, felt, and judged from the customer's perspective [4].He further emphasizes that customer satisfaction can be improved by reducing defects and overall problems.

Defect is any blemish or imperfection that occurs either in the product or in the process of developing a product. Since, defect has the tendency to propagate and magnify, early defect detection prevents defect migration from requirements phase to design and from design phase into implementation phase [5]. It enhances quality by adding value to the most important attributes of software like reliability, maintainability, efficiency and portability [6]. Spiewak and McRitchie suggest that the best practice of identification and fixing of process defects enable one to achieve the product quality. However, all defects are not of same nature and thus do not have same impact on the quality of the product. Hence, in order to understand the various facets of defects that occur at various phases of development, it is required to have awareness of defect pattern and rationale for their occurrence. This knowledge improves the quality and enhances the total productivity of the organization.

This part of the research therefore aims to throw light on existence of defect pattern during various phases of software development. The organization of this paper is as follows. Section 2 provides literature survey carried out in association with defect management and its strategies. Section 3 details the research methodology followed for this investigation. Section 4 is a case study carried out in one of the leading software industries depicting the analysis and its inferences. Section 5 summarizes this part of the research work.

2   Literature Survey

Author of [7] states that development of high quality software product can be realized when the product has few defects [7]. Hence, an awareness of all facets of defects enables the practitioners to achieve effective defect management and develop products with minimal defects.

Author of [8] suggested the curative and preventive approaches to tackle the defect related issues for the development of quality software. He stated that curative approach focuses on testing where both developers and users identify defects and fix them. In preventive approach, he emphasized on three aspects. They are i) anticipation of the defects, ii) use of formal inspections and prototypes to discover defects early in the development cycle iii) use of tools for detection of defects.

Defect Prevention is one of the most significant activities in the software development process. An analysis of defects at the early stage reduces the time, cost, and the resources required for rework [9][10][11][12][13].

Author of [18] states that defect potential and defect removal efficiency are the most critical software quality measurement in industry. Further, it is a fact that DP is one of the reliable components of software quality assurance in any project.

Author of [19] states that defects can be injected at any phase of software development He further suggested the need for reducing residual defects. The objective of their study was to identify the factors that influences defect injection and defect detection to reduce residual defects.

Author of [20] suggest the three approaches to determine the cost of quality. They are prevention, appraisal, failure (PAF) approach, process-costing approach, and activity-based costing approach.

Theoretically, a strong foundation exists on varied benefits of effective defect management. Yet, many leading software industries continue to deploy products with residual defects. However, one of the most important steps toward total customer satisfaction is the generation of nearly zero-defect product [21].

3   Research Methodology

The main objective of this phase of the research work is to have complete awareness on defect pattern from empirical analysis. As such this research comprises of investigations carried out in various leading software industries. Due to the wide spectrum of industries and their production capabilities, this research narrowed down to investigating software industries which are CMMI Level 5 certified and has well adopted defect management strategies in their development process.

Yet another challenge is to sample the type of projects in the deliberate sampled software industries. Hence, this work focused upon similar types of projects developed under a randomly sampled domain. Project data in association with defect profile was collected through various modes such as log history, defect management centres, interviews, face to face communication with various project personnel, emails and through telephonic conversations.

The secondary data thus obtained were analyzed to emerge out with the defect pattern. Analysis infers the existence of defect clusters at various phases of software development due to their associated rationales which will be forthcoming work of this research theme.

4   Case Study

This case study involves a leading software industry which is CMMI Level 5 certified and has well implemented defect management strategies. This research is focused upon software projects developed in telecom domain. All projects that were studied were developed using Windows 7 and Java programming language as their technology support. All projects are either simple projects or medium complex projects where complexity is measured in function points as per the company conventions and standards followed.

Table 1 depicts the randomly sampled projects and their personnel information. Projects P1 to P3 depicts randomly sampled projects of small complexity and Projects P4 to P5 represents medium complexity projects. These projects are enhancement type of projects which are of type change request.

Table 1. Sampled Projects with project type and programmers profile

Projects	P1	P2	P3	P4	P5
Proj Desc	Telecom	Telecom	Telecom	Telecom	Telecom
Technology	Window7	Window7	Window7	Window7	Window7
Prog Lang	Java	Java	Java	Java	Java
Total Proj devp time(*)	800	1200	4000	6200	8400
Project Complexity	Small	Small	Small	Small	Small
Project Type	Change Request	Change Request	Change Request	Change Request	Change Request
Total No. Progm	5	8	16	19	22
No. of Lr.Exp. Progm	1	2	3	3	5
Percentage	20%	25%	18.75%	15.78%	22.72%
No.of Avg.Exp. Progm	2	2	5	7	8
Percentage	40%	25%	31.25%	36.84%	36.36%
No of Fresher	2	4	8	9	9
Percentage	40%	50%	50%	47.36%	40.90%

Proj Desc- Project Description; Prog Lang- Programming Language; Total Proj devp time- Total Project Development Time; (*)- measured in person hours; Total No. Progm – Total Number of Programmers; No. of Lr.Exp. Progm – Number of largely experienced programmers; No. of Avg.Exp. Progm- Number of average experienced programmers

Table 1 infers that as project complexity increases, number of programmers at implementation phase also increases. However, it is worth to note that experience of the programmers has an influential impact on the quality of the code from the product perspective while from the cost and time perspective too, the experience of the programmers is a modulating factor. As per the conventions followed habitually in most of the software industries, a project personnel with an experience of 8 and above is deemed to be largely experienced personnel while an experience ranging between 2 to 8 years are termed to have average experience and fresher is a personnel who has less than 2 years of industry hands on experience.

Table 1 thus infers that up to 25% of experienced programmers are employed in small and medium complexity projects falling under the enhancement type of projects category. Up to 40% of average experienced programmers and up to 50% fresher is the observed pattern allocated in these types of projects.

The inferences thus obtained from Table 1 led towards further investigation on the number and type of defects that were injected by the observed pattern of programmers.

Table 2 illustrates defect profile of the sampled projects. It provides details in terms of total number of defects injected at various phases and number of defects introduced by the programmers. It also depicts information regarding the type of defect that was injected at implementation phase in more specificity.

Table 2. Defect profile for the sampled projects

Projects	P1	P2	P3	P4	P5
Proj Success	92	94	94	96	97
Total no of def count	37	54	87	99	98
Req defects	2	9	13	19	12
Percentage	5.40%	16.67%	14.94%	19.19%	12.24%
Des defects	5	3	6	8	7
Percentage	13.51%	5.55%	6.89%	8.08%	7.07%
Imp def	30	42	68	72	79
Percentage	81.08%	77.77%	78.16%	72.72%	80.61%
Blocker defect	3	2	5	5	6
Percentage	10.00%	4.76%	7.35%	6.94%	7.59%
Critical Defect	12	6	7	9	9
Percentage	40.00%	14.28%	10.29%	12.50%	11.39%
Major Defect	4	9	26	26	25
Percentage	13.33%	21.42%	8.82%	36.11%	31.64%
Minor Defect	8	17	18	12	12
Percentage	26.66%	40.76%	26.47%	16.66%	15.18%
Trivial Defect	3	8	12	20	27
Percentage	10.00%	19.04%	17.64%	27.77%	34.17%

Proj Success -Total Project Success from defect perspective;   Req defects- Total number of requirements defects; Des defects- Total number of design defects; Imp defects- Total number of coding defects.

Table 2. infers that as project complexity increases, total number of defects may not necessarily exponentially increase. Similar inferences may be drawn for the total number of defects injects at requirements or design or at code implementation phases of the software development. The table further provides information of the pattern of defects count that may occur in the projects of the enhancement type category.

Enhancement projects are either change request type of projects which are depicted in this paper or it may be enhancement projects which needs either major or minor functionalities modifications. Thus, requirement will be frozen and hence defects due to rationales such as requirement inconsistency, ambiguity, and amalgamation, misinterpretation, missing requirements or defects due to scope creep hardly occur. Due to above said reasons, it is apparent from the Table 2 that defects at requirements analysis phase is up to 20% and for the same said reasons, design flaws are observed to take a pattern of up to 15%. Table further indicates that percentage of defects that gets usually injected at coding phase is up to 80% in such type of projects.

Table 2 provides an in-depth analysis of pattern of type of defect that occurs at code construction phase of software development which comprises of blocker type , critical types of defects which are having severity of show stopper nature and needs instantaneous addressing. The observation results indicate that up to 10% of coding defects attributes to blocker defects while up to 40% of total defects that appear in coding stage are critical type of defects. Defects of major type occur in 32% while minor type is seen up to 40% out of total number of coding defects. These are medium severity in nature and can be looked into later during developmental activities. The table further explores the pattern of trivial type of defect which has low severity and is of cosmetic type in nature to be comprising of up to 40% of total defect count in this phase. Hence they can be dealt as pending cases.

Above observations focused this research to further leads towards one to explore the root cause for the defect pattern to occur. The forthcoming work of the authors is therefore to discover the root cause analysis for the aforementioned pattern observed.

5   Conclusion

Evolution of software has made software solutions to become one of the inevitable components of every industrial production. Developing high quality software is therefore rudimentary aspects in any organization. Defect is one of the major quality modulating factors. Hence, generation of minimal or zero defects is the core need of the day. The position of this paper is thus to present an empirical investigation carried out in one of the sampled leading software industry in order to understand the defect pattern. This paper provides a case study and the observational inferences indicate the areas that needs to be paid attention in software organizations towards development of desired level of quality software.

Acknowledgement. The authors would like to sincerely acknowledge all the industry people for their immense help in carrying out this work. The authors would like to thank all the referees and the editor for their constructive and helpful comments.

References

1. Humphrey, W.: A Discipline for Software Engineering. Addison-Wesley Professional Publisher (1995) ISBN: 0201546108
2. Weinberg, G.M.: Quality Software Management: Systems Thinking. Dorset House Publisher (1997) ISBN-10: 0932633226, ISBN-13: 978-0932633224
3. Chillarege, R.: The Marriage of Business Dynamics and Software Engineering. IEEE Software 19(6), 43–49 (2002)
4. Kan, S.H.: Metrics and Models in Software Quality Engineering, 2nd edn. Addison-Wesley Publisher, Boston (2003)
5. Tian, J.: Software Quality Engineering: Testing, Quality Assurance and Quantifiable Improvement. Wiley John & Sons Publisher, New Jersey (2005)
6. Walia, G.S., Carver, J.C.: A systematic Literature Review to Identify and Classify Software Requirement Errors, Technical Report MSU-071207
7. Soni, M.: Defect Prevention: Reducing Costs and Enhancing Quality, iSixSigma.com (July 19, 2006),
   http://software.isixsigma.com/library/content/c060719b.asp
8. Iacob, I.M., Constantinescu, R.: Testing- First Step towards software quality. Journal of Applied Quantitative Methods 3(3), 241–253 (2008)
9. Suma, V., Gopalakrishnan Nair, T.R.: Impact of Test Effort in Software Development Life Cycle for Effective Defect Management. International Journal of Productivity and Quality Management (IJPQM) 13(3) (2014)
10. Gopalakrishnan Nair, T.R., Suma, V., Nair, N.G.: Estimation of Characteristics of a Software Team for Implementing Effective Inspection Process Inspection performance Metric. Software Quality Professional Journal, American Society for Quality (ASQ) 13(2), 14–26 (2011)
11. Gopalakrishnan Nair, T.R., Suma, V.: Defect Management Using Pair Metrics, DI and IPM. CrossTalk, The Journal of Defense Software Engineering 24(6), 22–27 (2011)
12. Gopalakrishnan Nair, T.R., Suma, V.: Implementation of Depth of Inspection Metric and Inspection Performance Metric for Quality Management in Software Development Life Cycle. International Journal of Productivity and Quality Management, IJPQM (2011)
13. Gopalakrishnan Nair, T.R., Suma, V., Tiwari, P.: Significance of Depth of Inspection and Inspection Performance Metrics for Consistent Defect Management in Software Industry. IET Software 6(6), 524–535 (2012)
14. Gopalakrishnan Nair, T.R., Suma, V.: Impact Analysis of Inspection Process for Effective Defect Management in Software Development. American Society for Quality (ASQ) Journal 12(2), 2–14 (2010)
15. Suma, V., Gopalakrishnan Nair, T.R.: Better Defect Detection and Prevention Through Improved Inspection and Testing Approach in Small and Medium Scale Software Industry. International Journal of Productivity and Quality Management (IJPQM) 6(1), 71–90 (2010)
16. Gopalakrishnan Nair, T.R., Suma, V.: The Pattern of Software Defects Spanning across Size Complexity. International Journal of Software Engineering (IJSE) 3(2), 53–70 (2010)
17. Gopalakrishnan Nair, T.R., Suma, V.: A Paradigm for Metric Based Inspection Process for Enhancing Defect Management. ACM SIGSOFT Software Engineering Notes 35(3), 1 (2010)
18. Jones, C.: Measuring Defect Potentials and Defect Removal Efficiency. Crosstalk, The Journal of Defense Software Engineering 21(6) (2008)

19. Jacobs, J., Moll, J.V., Kusters, R., Trienekens, J., Brombacher, A.: Identification of Factors that Influence Defect Injection and Detection in Development of Software Intensive Products. Information and Software Technology Journal 49(7), 774–789 (2007)
20. Karg, L.M., Beckhaus, A.: Modeling Software Quality Costs by Adapting Established Methodologies of Mature Industries. In: IEEE International Conference on Industrial Engineering and Engineering Management (IEEM), Singapore, December 2-5, pp. 267–271 (2007)
21. Kan, S.H., Basili, V.R., Shapiro, L.N.: Software quality: An overview from the Perspective of Total Quality Management. IBM Systems Journal 33, 4–19 (1994)

Pattern Analysis of Post Production Defects in Software Industry

Divakar Harekal[1], Jawahar J. Rao[2], and V. Suma[2]

[1] Dept. of Computer Science Engg., JJTU, Rajasthan, India
divakarhv0@gmail.com
[2] Dayanandsagar Research and Industry Incubation Centre,
Dayananda Sagar Institutions, Bangalore, India
jawahar_rao@yahoo.com, sumavdsce@gmail.com

Abstract. Software has laid a strong influence on all occupations. The key challenge of an IT industry is to engineer a software product with minimum post deployment defects. Software Engineering approaches help engineers to develop quality software within the scheduled time, cost, and resources in a systematic manner. In order to incorporate effective defect management strategies using software engineering discipline needs a complete and widespread knowledge of various aspects of defects. The position of this paper is to provide a pattern analysis of post production defects based on empirical observations made on several main frame projects developed in one of the leading software industries. Inferences thus obtained from this investigation indicate the existence of show stopper severity defects and their associated root cause. This awareness enables the developing team to reduce the residual defects and improve the pre production quality. It further aids the attainment of total customer satisfaction.

Keywords: Software Engineering, Software Quality, Defect Management Strategies, Software Testing.

1   Introduction

These Technological transformations have led towards introduction of software in almost all production units. However, factors such as state-of-the-art research topics in software industry increase in hardware complexity, high competitiveness in business world, technological advancement, and frequently changing business requirement, demands for the development of high quality software. Hence, one of the challenging issues in software industry is to deliver high quality software to achieve total customer satisfaction. Hence, development of high quality software is the only solution towards retention of software industries in the high end competitive industrial market.

Since quality has various dimensions, it is imperative to address all the quality concerned issues. One such major concern of quality is engineering and deploying defect free software which upholds existence of effective defect management

© Springer International Publishing Switzerland 2015   667
S.C. Satapathy et al. (eds.), *Proc. of the 3rd Int. Conf. on Front. of Intell. Comput. (FICTA) 2014*
– *Vol. 2*, Advances in Intelligent Systems and Computing 328, DOI: 10.1007/978-3-319-12012-6_74

strategies in the organization. Providentially, advancement in fundamental engineering aspects of software development enables I.T. enterprises to develop a more cost effective and a better quality product through appropriately organized defect detection and prevention strategies quality assurance and testing techniques. Nevertheless, due to the intrinsic nature of software being complex, development of defect free software is always plausibility. Changing trends in defect management enables transition from postproduction detection technique to preproduction detection technique and in site detection during developmental phase [1].

In support of the aforementioned objectives in the current scenarios, authors of [2] recommended to measure quality in terms of defects and further stated that software development process should be capable of fulfilling the needs of the customer with less cost. The main intent of quality cost analysis is not to remove the cost entirely, but to ensure that the investment yields maximum benefit. The knowledge of defect injecting methods and processes enable the Defect Prevention (DP).

Progressive companies therefore claim that over a period, the software quality improves, while the cost of ensuring quality reduces by introducing valuable defect detection and prevention techniques. DP technique reduces the number of defects in the development life cycle. It further enhances business performance. Thus, investing in defect prevention reduces the cost of defect detection and elimination. Defect-free product reduces support cost, programming cost, development time, and competitive advantage. It therefore has a direct and strong impact on the time, cost, and quality of the deliverables [3].

Due to the above stated benefits it is essential to implement effective defect management strategies for which it is required to be completely aware of various types of defects, their occurrence pattern, root cause for their occurrences, impact analysis of these defects. Further, this body of knowledge of defect results in formulation of organizational and technical polices which yields long term return on investment.

Defect is not a state and is a continual process. Knowledge of all facets of defects is imperative for effective defect management and hence this paper narrows down to focus upon post production defects which were captured in a case study conducted on main frame applications in one of the leading software industries. The awareness of defect patterns enables to capture majority of defects close to defect inception point. This knowledge reduces defect injection and helps to develop nearly zero-defect product in software industry.

2   Literature Survey

High quality software attributes to the development of defect-free product, which is competent of producing predictable results and is deliverable within time and cost constraints. It should be manageable with minimum interferences, maintainable, dependable, understandable, and efficient. It should serve the fitness for purpose.

Author in [4] emphasizes on estimating the defect type for the important deliverables to achieve effective defect removal [4]. However, the author [5] has proved that DP is one of the reliable components of software quality assurance in any project. He indicates the expensive nature of defects when identified late in the project [5].

Authors of [6] further described that regardless of advancement in the quality research, the development of quality product is still a challenge. They state that lack of understanding of the source of problems by the developers, inability to learn from mistakes, lack of effective tools and incomplete verification process are the major reasons for the introduction of defects. Hence, they suggest the need for research to provide more insight into the understanding of sources of the faults rather than just the fault themselves [6].

Author in [7] states that developers in United States spend nearly 60% of the development time to fix defects. Therefore integral part of defect prevention life cycle is to develop the product without introduction of defects by performing reviews, causal analysis of the defects and implementing defect prevention activities [7].

While the authors of [8] express that test execution is one of the ways to identify defects in the development process. They state that testing activity detects defects in both functional and non-functional software requirements [8]. Testing is a dynamic activity, which ensures the dynamic behavior of the software performed during the execution of the software. It includes automation testing, regression testing, performance testing, verification, validation testing [9].

Thus, authors of [10] suggested the need for reducing residual defects. They suggested the need for further research focusing on a framework for lowering residuals defects. They recommend the formulation of strategies to reduce residuals defects through creating awareness on defects in organizations, to re-evaluate the verification, validation and other such developmental policies [10].

Authors in [11] [12][13][14][15][16][17] has made an extensive work to analyze defect perspectives and testing effort analysis to reduce pre production defects. This work however emphasizes upon post production defect analysis.

3   Research Methodology

Pleas Since early defect detection and elimination ascertains reduced rework and overheads in industry, this research team focused upon analyzing the defect pattern that occurs at post production period of software development process. In order to accomplish the task, the team visited various software industries which are at CMMI Level 5 maturity level. The main theme behind these maturity level industries to be studied is that latent defect leakage will be less due to the effective defect management techniques such as reviews and testing being followed during the developmental activities.

The next step in this investigation is to explore defect facets in the projects that were developed in the industries under study. However, this paper limits to the post production defects reported in main frame applications. Modes of secondary data collection was through interviews, e mails, questionnaire, over telephone conversations and face to communications in addition to obtaining secondary data from the defect prevention centres and quality assurance departments.

Subsequently, the information is analyzed to capture a pattern for defect occurrence. This awareness enables one to strengthen the pre production defect management strategies.

4   Case Study

This paper presents a case study of a leading software industry which aims at developing applications involving main frames. Table 1 illustrates a sample of 9 projects which are developed in same domain using common technology and programming language. The table further indicates information regarding total project development hours, number of use cases, complexity of the project and number of post production defect count.

Table 1. Sample Projects representing post production defects

Project	P1	P2	P3	P4	P5	P6	P7	P8	P9
Domain	Health Care	Health Care	Health Care	Health Care	Health Care	Health Care	Health Care	Health Care	Health Care
Tech	Main Frame	Main Frame	Main Frame	Main Frame	Main Frame	Main Frame	Main Frame	Main Frame	Main Frame
Prg Lng	COBOL DB2	COBOL DB2	COBOL DB2	COBOL DB2	COBOL DB2	COBOL DB2	COBOL DB2	COBOL DB2	COBOL DB2
Proj Hours	1629	1008	725	2108	597	708	1508	1770	160
# Ucase	15	12	7	15	5	10	18	22	2
cmplx	Low	Low	Medium	Medium	Medium	High	High	High	Low
#PPDeft	2	1	3	3	3	7	7	8	1

Prg Lng – Programming Language; Proj Hours – Total project development hours; #Ucase – Total number of use cases; cmplx – Complexity of the project; #PPDeft – Total number of Post Production Defects.

From the above depicted sampled projects, projects with low complexity are not having lesser effort to develop and vice versa. Hence, Table 1 infers that as project complexity increases, total number of hours required to develop the project need not increase.

However, complexity of the project is determined by use cases or function points. In the above sampled projects, complexity is determined by use cases. It is conventionally accepted that every single use case deems 70 to 80 hours of human effort for completion. From the table, it is clear that lesser the number of use cases, lesser is the total developmental hours required.

Table further indicates that total number of post production defects has an impact on effort, complexity and severity of the defects. Projects whose complexity is low, results in post production defects which are less than 3 in number. However, projects with medium complexity have a tendency to comprise of 3 to 6 post production defects. High complexity projects have a probability to contain greater than 7 post production defects. Additionally, defects with number less than 6 are deemed to be of blocker type while post production defect with count greater than 7 is considered to be of critical type of defects.

The above inferences draw attention towards the existence of post production defect pattern of type blocker and critical to be further explored in order to investigate their associated root cause analysis. Thus, our forthcoming work indicates the root cause analysis for the observed post production defect pattern and their inferences.

5   Conclusion

Production of software which is defect free is all the time the main objective of any software organization. However, accomplishment of the same is only plausibility. Though several defect management strategies exist in all IT industries towards development of nearly defect free software, yet there is a wide possibility of software to contain post production defects. This work therefore comprises of a case study carried out in one of the leading software industry in order to explore the probability of post production defect pattern. This paper puts froths the post production defect pattern as observed in main frame applications. Investigation results indicate the impact of probability of type of software post production defects occurrence with complexity of the project based on use cases and the influence of development hours. Further research aims to explore the root cause for the associated defect pattern and its remedial solutions. This awareness certainly ensures enhanced quality of software projects in addition to reduce overheads and thereby increase productivity of the company.

References

1 Suma, V., Gopalakrishnan Nair, T.R.: Chapter-Defect Management Strategies in Software Development. In: Recent Advances in Technologies. Intec Web Publishers, pp. 379–404 (2009) ISBN 978-953-307-017-9

2 Watts, S.: Humphrey: Defect Prevention. In: Managing the Software Process. Addison-Wesley Publisher, Boston (1989) ISBN 978-953-307-017-9

3 Colligan, T.M.: Nine Steps to Delivering Defect-Free Software (1997), http://www.tenberry.com/errfree/steps.htm

4 Wagner, S.: Defect Classification and Defect Types Revisited. In: International Symposium on Software Testing and Analysis, Seattle, Washington, USA, July 20, pp. 39–40 (2008)

5 Narayan, P.: Software Defect Prevention in a Nutshell, iSixSigma.com (June 11, 2003), http://software.isixsigma.com/library/content/c030611a.asp

6 Walia, G.S., Carver, J.C.: A systematic Literature Review to Identify and Classify Software Requirement Errors. Technical Report MSU-071207

7 Soni, M.: Defect Prevention: Reducing Costs and Enhancing Quality, iSixSigma.com (July 19, 2006),

8 http://software.isixsigma.com/library/content/c060719b.asp

9 Iacob, I.M., Constantinescu, R.: Testing- First Step towards software quality. Journal of Applied Quantitative Methods 3(3), 241–253 (2008)

Secure Efficient Routing against Packet Dropping Attacks in Wireless Mesh Networks

T.M. Navamani[1] and P. Yogesh[2]

[1] Department of Computer Science and Engineering,
Easwari Engineering College, Chennai-89
[2] Department of Information Science and Technology,
Anna University, Chennai-25
navsmi2001@yahoo.com, yogesh @annauniv.edu

Abstract. Wireless Mesh Networks (WMNs) are susceptible to attacks and various other issues like open peer-to-peer network topology, shared wireless medium, stringent resource constraints, and a highly dynamic environment; hence, it becomes critical to detect major attacks against the routing protocols of such networks, and also to provide good network performance. In this paper, we address two severe packet dropping attacks, which cause serious performance degradation in wireless mesh networks. They are misrouting and power control attacks. To mitigate these attacks and to enhance the performance of WMNs, we propose a new secure and efficient routing protocol called Secure Efficient Routing against Packet Dropping Attacks (SERPDA). An extended local monitoring system based on the observing patterns in the behavior of neighboring nodes and checking capacity of the neighboring nodes is implemented to defend against packet dropping attacks. To improve the performance, we use additional metrics besides the usual ones for the selection of routes. With the help of a network simulator, we prove that the proposed protocol efficiently mitigates the attacks and also provides a more optimal path by considering load balancing, link quality, successful transmission rates, and the number of hops.

Keywords: Wireless Mesh Networks (WMN), Routing Metrics, Attacks, Local Monitoring, Packet Dropping.

1 Introduction

Wireless mesh networking is an attractive new way of communication due to its scalable wireless internetworking solutions and its low cost. WMN is being developed actively and deployed widely for a variety of applications, such as public safety, environment monitoring and citywide wireless internet services. In WMN, nodes are comprised of mesh routers and mesh clients. Each node operates not only as a host but also as a router, forwarding packets on behalf of other nodes which may not be within direct wireless transmission range of their destinations. Mesh routers provide network access for both mesh and conventional clients. Fig. 1 shows the architecture of a

© Springer International Publishing Switzerland 2015   673
S.C. Satapathy et al. (eds.), *Proc. of the 3rd Int. Conf. on Front. of Intell. Comput. (FICTA) 2014*
– Vol. 2, Advances in Intelligent Systems and Computing 328, DOI: 10.1007/978-3-319-12012-6_75

WMN. It shows that for every two hops one guard node is maintained. Some of the mesh routers are kept as guard nodes and also as gateway nodes to connect with the Internet. Mesh clients can be either stationary or mobile and can access the network through mesh routers as well as directly meshing with other mesh clients. The advantages of WMN also include simple settings, broad-band capability, self-configuration, self-maintenance and reliable service coverage [1, 2]. To support end to end communication, effective routing protocols are required. Hence secure routing plays an important role in the entire network and the client should have end to end security assurance. However, being different from wired and traditional wireless network, WMN could easily be compromised by various types of attacks. Even the WMN infrastructure like mesh router could be relatively easily reached and modified by attackers. There are various types of packet dropping attacks which include black hole, selective forwarding and delaying, misrouting attack in which the attacker relays packets to the wrong next hop and power control attack in which the attacker controls its transmission power to avoid the next hop. In packet dropping attacks, the attacker achieves the objective of disrupting the packet from reaching the destination by malicious behaviour at an intermediate node. This class of attacks is applicable to packets that are neither acknowledged end to end, nor hop by hop. These attacks could result in significant loss of data or degradation of network functionality, say through disrupting network connectivity by preventing route establishment.

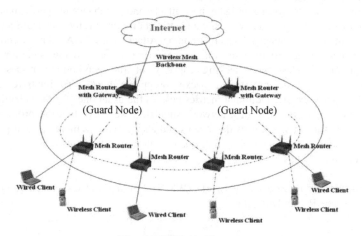

Fig. 1. WMN Architecture

In this paper, we address the severe packet dropping attacks such as misrouting attack and power control attack. To be more specific, assume that the compromised node exists in between source S and destination D at a one hop distance. The compromised node is supposed to forward the incoming packet from S to Destination D. Instead it misdirects the packet to the wrong next hop neighbour. Finally, the packet does not reach the destination. This type of attack is called the misrouting attack. In the power control mode, the compromised node controls its transmission power to relay the packet to a distance less than D. Here also the packet does not reach the destination. To mitigate these attacks, an Extended Local Monitoring technique is used, in which the mesh routers monitor part of the traffic going in and out of their neighbors, by adding some checking responsibility to the neighbor nodes. For monitoring activity, certain nodes are kept as guard nodes, which are normal mesh router nodes, involved in the role of monitoring activity. In wireless mesh networks, the routing metrics have a profound impact over the network performance since the selection of the routing path is dependent on them. The routing path directly affects the factors like delay, packet delivery ratio etc. [3]. Examples of the most popular WMN routing metrics include: Hop count, Expected Transmission Count (ETX), Expected Transmission Time (ETT), and Weighted Cumulative Expected Transmission Time (WCETT). The ETT and WCETT provide more opportunities to select an optimal path even if the ETX value is the same for the multiple paths.

Based on the above review, we present Secure Efficient Routing scheme against Packet Dropping Attacks (SERPDA) which mitigates the attacks by performing extended local monitoring and also provides good network performance using hybrid routing metrics. This paper is organized as follows. In section 2, we discuss the related work. The proposed work is discussed in section 3. The implementations details and performance analysis are discussed in section 4. We conclude this paper in section 5.

2   Related Work

Since security and performance are two extremely important issues in any communication network, researchers have worked on these two areas extensively. However, as compared to MANETs and Wireless Sensor Networks (WSNs), WMN has received very little attention in this regard. This section briefly discusses some of the existing mechanisms for ensuring security and performance in WMNs. A technique proposed to detect malicious behavior involving selective dropping of data, relies on explicit acknowledgement for received data is discussed in [4]. However, the diagnosis process used can only work with a source routing protocol, such as DSR, where the source knows all the intermediate nodes to the destination. In [5], Glass S.M et al. proposed a novel intrusion detection mechanism that identifies man-in-the-middle and wormhole attacks against wireless mesh networks by external adversaries. In [6], Fei Xing investigated the problem of node isolation where the effects of Denial-of-Service (DoS) attacks are considered.

In [7], Devu Manikantan Shila proposed a method to detect a gray hole attack in wireless Mesh network using channel aware detection that adopts two strategies, hop-by-hop loss observation and traffic overhearing, to detect the mesh nodes subject to the attack. Channel aware detection (CAD) detects the attackers effectively even in harsh channel conditions.The main drawback in this method is CAD fails to detect malicious nodes if the downstream and upstream neighbours of malicious nodes set the detection parameters to zero in the PROBE packet. The path diversity techniques increase route robustness by first discovering multipath routes [8] [9] and then using these paths to provide redundancy in the data transmission between the source and the destination. In [11], Issa Khalil, et al. proposed Stealthy Attacks in Wireless Ad hoc Networks: Detection and Countermeasures (SADEC) protocol to detect the stealthy attacks in wireless ad hoc networks. Four types of stealthy attacks are introduced and mitigation solutions are provided. The performance advantages under misrouting create higher false isolation and end to end delay in SADEC. A protocol which defends against stealthy attacks for wireless mesh networks was discussed in our previous work [12], in which, the network performance issues were not considered and detailed throughput analysis was not taken. In [13], Issa Khalil proposed MIMI (Mitigating Misrouting) to detect misrouting attacks in wireless ad hoc networks. Local monitoring has been demonstrated as a powerful technique for mitigating security attacks in multi-hop ad-hoc networks. MIMI protocol designed to detect misrouting and the solution has two forms having nodes maintain additional routing path information, and adding some checking responsibility to each neighbor. But it has higher false isolation and end to end delay for packets.

Single radio, multi radio and hierarchical routing protocols are reviewed, and various routing metrics such as Hop Count, Expected Transmission Count (ETX), Expected Transmission Time (ETT), Weighted Cumulative ETT (WCETT), Metric of Interference and channel switching (MIC), iAware and other routing metrics are discussed in [14][15]. In [16], we proposed an efficient routing protocol for Wireless Mesh Networks with QoS aware reconfiguration. In this work, Self reconfiguration system is integrated with link quality routing metrics to enable QoS support for the WMNs. An extension of this work is presented here. A good routing protocol which provides load balancing within the mesh network is required to provide better performance. The existing routing metric, hop count designed for ad hoc network routing protocols is not applicable for selecting the optimized path in WMN. Hence we introduce hybrid routing metric for WMN to provide optimized path.

Motivated by these factors and also keeping in mind of other possible advantages of using WMNs, in the remainder of this paper, we would like to develop secure efficient routing against packet dropping attacks for wireless mesh networks to provide optimal performance.

3   Secure Efficient Routing against Packet Dropping Attacks

The proposed scheme designs Secure efficient routing protocol to defend against packet dropping attacks and also to provide good network performance. To detect and isolate these attacks, an extended local monitoring technique is implemented. This technique involves two steps. i) By having guard nodes to maintain additional next-hop information gathered during route discovery, and ii) Adding some checking responsibility to each neighbor. The guard nodes are normal mesh router nodes in the network, which perform their basic functionality in addition to monitoring. Monitoring implies verification that the packets are being faithfully forwarded without modification of the immutable parts of the packet, within acceptable delay bounds and to the appropriate next hop[11][12]. To address network performance, multiple routing metrics are integrated during route discovery to find out an optimal path. Hence hybrid routing metrics (WCETT, ETT, ETX and Hop Count) and extended local monitoring technique are implemented in each mesh router node during route discovery, to discover a secure efficient routing path. ETX, ETT and WCETT are computed, based on the information from the MAC layer.

3.1 Performance Based Routing Metrics

Ad hoc networks usually use the hop count as a routing metric. This metric is appropriate for ad hoc networks because new paths must be found rapidly, whereas high-quality routes may not be found in due time. This is important in ad hoc networks because of user mobility. In the WMN, the stationary topology benefits quality-aware routing metrics. To address network performance with link quality parameters, we have used the routing metrics such as ETX, ETT, WCETT and Hop Count. ETT considers the link bandwidth along with packet size and provides load balancing along the path [13]. The metrics ETT and ETX are computed as in equations (1) and (2).

$$ETT = ETX * S/B \qquad (1)$$

Where S represents the size of packet and B represents the bandwidth

$$ETX = 1/d_f * d_r \qquad (2)$$

Where d_f is the forward delivery ratio, i.e., the probability of successful packet transmission in forward direction and d_r is the reverse delivery ratio, i.e., the probability of successful receiving of acknowledgement. To compute ETT at each node, two probe messages are sent back to back one with smaller packet size and other with larger size to each neighbor. Each neighbor calculates the difference between both the received probe packets and reports this to the sender. Consequently, the sender estimates the link capacity by dividing larger probability with smallest delay sample obtained from the neighbors [14]. In our proposed protocol, the routing metrics are updated at each intermediate node and the computed ETT value is cumulatively added in the RREQ packet. To enhance the routing protocol further, we consider intra flow interference along with data rate, packet size, and another routing metric called Weighted Cumulative Expected Transmission Time (WCETT), which is computed for all the paths at the destination node as follows.

$$WCETT (p) = (1-\beta) \sum_{Link\ l\ \in 1\leq i\leq k}^{n} ETT_l + \beta \max_p Xj \qquad (3)$$

Where Xj is the sum of the ETT values of links that are on channel j, k is the number of orthogonal channels and β is the tunable dropping parameter which allows controlling preference over path lengths versus channel diversity. Finally to select the optimal route, the hybrid routing metric, path cost is introduced and it is computed as in equation (4).

$$Path\ cost = Hop\ count * WCETT (p) \qquad (4)$$

Where WCETT (p) denotes that WCETT value which is measured for all the paths through which RREQ arrives.

3.2   Attack Model

In this section, we describe the attack models for the packet dropping attacks. The attacker node achieves the objective of disrupting the packet from reaching the destination by malicious behavior at an intermediate node. In the following sub sections, the two packet dropping attacks which are addressed in this paper, and the sample scenario of these attacks are discussed.

Misrouting Dropping Attack

In the misrouting attack, one compromised node is available in the route between the sender and receiver. In this scenario, we assume that X is source, Y is the Destination and M is the node which is compromised in between X and Y. G1 and G2 are guard nodes.

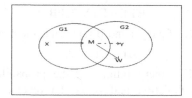

Fig. 2. Misrouting scenario

When X relays the packet, M receives it, since it is within the transmission range of X, but it is one of the compromised nodes. M forwards the packet to wrong next hop W, instead of forwarding to correct hop Y as shown in Fig. 2. This type of attack is called Misrouting attack.

Power Control Dropping Attack

In the Power control attack, one compromised node is in the route between the sender and the receiver. With power control capability, it controls the transmission power and reduces the transmission range to exclude next hop node. The second mode is called the power control attack. In this attack, M controls its transmission power to relay the packet to a distance less than the distance between M and Y. Therefore, the packet does not reach the next hop while the attacker avoids detection by guards.

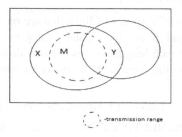

Fig. 3. Power control scenario

Consider the example of power control scenario shown in Fig. 3. Node X relays the packet, which M receives, since it is within the transmission range. This packet has to be forwarded to a correct next hop Y. But M controls its transmission power to reduce the range, and the packet is forwarded to some other nodes within its range, and not to the node Y.

3.3   Assumptions

A network is assumed to consist of mesh clients and wireless mesh backbone. The set of mesh routers, IEEE 802.11-based wireless links, and one control gateway form the backbone. We assume that by performing secure neighbor discovery process, each node in the network has its first hop and second hop information as the approach discussed in [11]. In this work, we are keeping a guard node for every two hops, to monitor the behavior of the neighboring nodes for packet forwarding as shown in Fig. 1.

3.4   Secure Efficient Route Discovery Algorithm

Whenever a mobile client (source) wants to send data to any other mobile client (destination) in the network and source does not have route entry in its routing table for destination, it needs to find route between them. We use the AODV protocol for secure efficient route discovery with slight changes. The first thing is RREQ message modification. To detect and isolate the packet dropping attacks, an extended local monitoring technique is implemented at each of the mesh routers. The functional architecture of Route Discovery and Attack Detection is shown in Fig. 4. Here, the idea is to augment the additional information required for attack detection in the control traffic packets, and ensure that the guard nodes collect the information in the route establishment phase. Among the mesh routers, some nodes are kept as guard nodes to do local monitoring by maintaining additional next hop information.

To collect the next hop identity information in AODV, each forwarding node of the RREQ attaches the previous two hops to the RREQ packet header. As the RREQ packet reaches each intermediate node, it computes the following routing metrics. The ETT value is computed at each node according to equations (1) and (2), and it is cumulatively added at each node. The ETT value is stored in the RREQ packet by using the 11 bits of reserved field in that packet. Initially, this value is set to 0. Each node adds its computed ETT value with its previous value available in the RREQ packet. For illustration purpose, let us assume that there are the sequence of nodes with labels such as S, A, P, Q and D where S, D represent source and destination nodes and other labels represent intermediate nodes. Assume that every node has neighbor nodes which are within its communication range. Here, the previous hop of P is A. When P broadcasts the RREQ received from A, it includes the identity of A and its own identity (P) along with the computed ETT metric in the RREQ header <S, D, RREQ_id, ETT, A, P>. When Q and other neighbors of P get the RREQ from P, they keep in a Verification Table (VT) <S, D, RREQ_id, ETT, A, P, - > (last field blank).

When Q broadcasts the RREQ, the common neighbors of P and Q update their VT to include Q <S, D, RREQ_id, ETT, A, P, Q>. When Q receives RREP from the destination which is to be relayed to P, it includes in that RREP, the identity of the node that P needs to relay the RREP packet to A. Therefore, all the guards of P now know that P not only needs to forward the RREP but also that it forwards it to A. When destination receives RREQ packets through multiple paths, it computes another routing metric Weighted Cumulative Expected Transmission Time (WCETT) for all the paths using the equation (3). Finally, the destination node computes the path cost, using equation (4) given in the above section. And then it selects the path with the minimum path cost. Thus, an optimal path is discovered and it sends RREP (Route Reply) through the optimal path (see Algorithm 1). Each packet sent from the source node S to D is saved in the watch buffer which is available at the guard node. The guard node monitors and verifies that the nodes are sending the packets correctly towards the destination. Each entry in the watch buffer is time stamped with a time thresholds as shown in Fig. 4. Each packet forwarded by some intermediate node Y with X as a previous hop is checked for the corresponding information in the watch buffer. The guard node verifies if the packet is fabricated or duplicated, corrupted, dropped or delayed. A malicious counter is available in each of the guard nodes. The increment of malicious counter depends on the nature of the malicious activity. If the malicious counter value crosses the threshold rate then the neighbors get an alert message. Thus the attacks are detected.

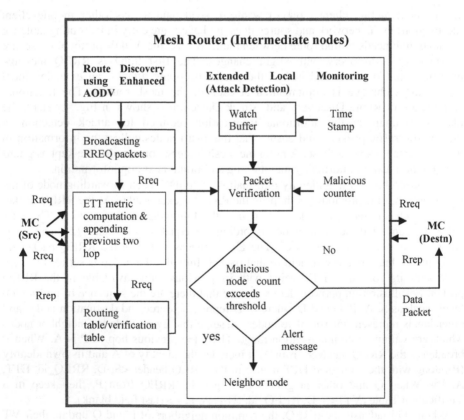

Fig. 4. Functional Architecture

3.5   Security Analysis

By implementing the extended local monitoring technique as explained above in the proposed protocol, the packet dropping attacks are detected and isolated. Here we have analyzed the two attacks such as Misrouting attack and Power control attack.

Detecting Misrouting Attack

As shown in the Fig. 2, the misrouting scenario is taken for implementation. The malicious node M relays the packet to the wrong next hop, which results in a packet drop. A node X relays a packet to the next node M, which is malicious. If the malicious node M selectively misroutes the packet, then the number of packet misroutes occur within particular threshold time. This malicious activity increases the malicious counter by 1. Then route is reestablished.

Algorithm 1: Route Discovery and Attack Detection

Step 1:   The source S broadcasts the RREQ packets

Step 2:   Few nodes among the mesh router nodes are kept as guard nodes to do local monitoring.

Step 3:   Each intermediate node computes ETX and ETT according to the equations (1) and (2) respectively.

Step 4:   Each intermediate node adds previous two hop information in the RREQ packet.

Step 5:   ETT value and Hop information are cumulatively added at each node and stored in the RREQ packet, updated in the route table/verification table.

Step 6:   The Destination node receives RREQ packets through multiple paths.

Step 7:   For each path, WCETT metric is computed according to the Equation (3) at the destination node.

Step 8:   Destination node also computes the path cost for all the paths according to the equation (4).

Step 9:   Finally it selects an Optimal route with minimum path cost.

Step 10: When each intermediate node receives RREP, the guard nodes of intermediate node verify that whether it is correctly forwarding to the next two hops without any modification

Step 11: Thus S obtains Secure Efficient path for data packet Communication with destination D.

To detect these attacks, guard nodes are used. Here guard nodes take the responsibility of checking the packet forwarding in addition to local monitoring. By having some additional information in the guard nodes such as next hop information which is collected during the route establishment process, the misrouting attacks are detected. Let us assume S sends a packet to the destination D through the route <X, M, N, and Y> where M is the malicious node. The malicious node M cannot misroute the packet received from X to a node other than N, because each guard of M over the link X->M has an entry in its Route verification table that, N is the correct next hop. If it forwards to some other node, it is correctly detected by the guard node, by verifying from the verification table. Thus, the misrouting attacks are detected by our proposed protocol.

Detecting Power Control Attack

In power control attack, a malicious node relays the packet by carefully reducing its transmission power. The scenario is shown in Fig. 3. To detect these attacks, each node keeps the count of the number messages sent to each of its neighbors. Each node has to announce the number of packets it has forwarded over the particular time. The nodes which are within the radio range of a node say A are called as Comparators of A. The function of the comparator is to count the number of packets forwarded by the node within particular time interval. Each guard of A over a certain link is called comparator of A. The neighbor node collects the number of forwarded packets by the sender and the comparator node i.e. the guard node compares the result announced by the sender and the neighbor. If the comparator count is not within an acceptable range of announced forward count, the comparator increments its malicious counter value, and identifies it as the attack. Thus the power control attack is detected.

4   Implementation Details and Performance Analysis

We have implemented the SERPDA protocol via ns-2 based simulation. The simulation parameters and values are shown in Table 1. We compared our proposed protocol with the AODV (Ad hoc On demand Distance Vector) protocol and SADEC (Issa Khalil and Saurabh Bagchi, 2011) protocols for performance and security analysis, since the proposed protocol is designed, based on the AODV, and the monitoring algorithm is similar to the SADEC. The nodes used in the simulations are based on IEEE 802.11n, with the normal and malicious nodes. To evaluate the performance and security, we have taken the following quantitative metrics, in both normal and malicious conditions.

Table 1. Simulation Parameters

Parameters	Values
Number of nodes	50
Total Simulation time	500s
Packet size	512 bytes
MAC Protocol	802.11n
Radio transmission range	250, 500m
Area size	500m×500m
Protocols	SERPDA, SADEC, AODV

Packet delivery ratio: The number of packets that are received at the destination to the total number of packets that are sent by the source.

End-to-end delay: The amount of time taken to transmit a packet from the source to the destination.

Here, the SERPDA protocol is compared to the AODV and SADEC for network performance analysis. Fig. 6 shows the PDR performance of SERPDA, SADEC and AODV, keeping the number of malicious nodes from 2 to 12, in the case of a misrouting scenario. The PDR decreases because of the increase in the hacking of packets among the malicious nodes. The SERPDA shows a slight improvement in the PDR, compared to SADEC, since it chooses the more optimal path, and the security mechanisms are the same as those of SADEC. It can be seen that SERPDA and SADEC outperform the AODV, because SERPDA and SADEC choose the more reliable path by avoiding the malicious nodes. When the numbers of malicious nodes increase, the numbers of packets delivered by the AODV are drastically decreased.

Even in adverse environments, the SERPDA almost maintains a better packet delivery ratio, when the number of misrouting attacker nodes is increased from 8 to 10. The end to end delay with respect to the number of misrouting malicious nodes is studied. It is observed that the attackers capture most of the traffic, and increase the delay in reaching the packets to the destination. Fig. 7 shows the variations in the end to end delay for different number of malicious nodes.

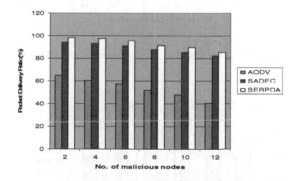

Fig. 5. Packet Delivery Ratio -Misrouting Attack

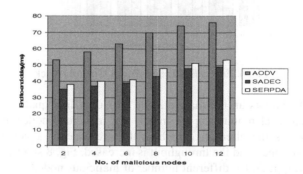

Fig. 6. End to End Delay -Misrouting Attack

It can be observed that SADEC experiences the minimum delay compared to SERPDA and AODV, even when the number of malicious nodes increase. SERPDA shows a slight increase in delay compared to SADEC because, in addition to security mechanisms, routing metrics computation is carried out in SERPDA. The end-to-end delay is the average time taken by a data packet to arrive at the destination. It also includes the delay caused by the route discovery process, and the queue in data packet transmission. When the numbers of malicious nodes increase, the AODV experiences more delay due to lack of quality routing metrics and security. SERPDA and SADEC show the improvement compared to the AODV since the chosen path is the trusted path, by avoiding the malicious nodes.

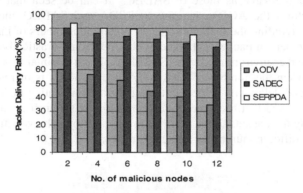

Fig. 7. Packet Delivery Ratio -Power Control Attack

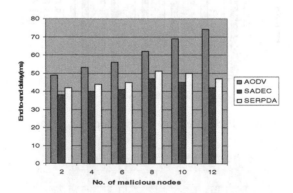

Fig. 8. End to End Delay -Power Control Attack

As SERPDA chooses an optimal path, by providing link quality routing metrics, and an extended local monitoring technique, the malicious nodes are avoided, and every node follows the alternative path in the presence of attackers, most of the packets are not dropped and the throughput is increased. Fig. 8 shows the variations in the packet delivery ratio for different number of malicious nodes in the power control attack scenario. By keeping the guard nodes for every mesh router node, the malicious nodes which trigger the power control attack are easily detected and isolated. In addition to that, during route discovery, the link quality metrics are considered, to select a more optimal path to provide security and better performance.

Hence, SERPDA performs better than SADEC and AODV in the presence of malicious nodes. Here also, when the numbers of malicious nodes increase, the PDR of AODV decreases from 60% to 35%, which is intolerable. Fig. 9 shows the variations in end to end delay for different number of malicious nodes, in the power control attack scenario. As in the misrouting scenario, SADEC shows a minimum delay performance, compared to SERPDA and AODV. Because of the routing metrics computation overhead, SERPDA shows an increase in delay compared to SADEC. As the number of malicious nodes increase, the route reestablishment frequency increases.

5   Conclusion and Future Work

In this paper, we have presented a new Secure Efficient Routing protocol against Packet Dropping Attacks (SERPDA) with improved performance for Wireless Mesh Networks. Our routing scheme addresses two types of packet dropping attacks, viz, the Misrouting attack and the Power Control attack, which interrupt the packet from reaching the destination by malicious interactions. The SERPDA protocol mitigates these attacks by implementing the extended local monitoring mechanism, and at the same time, it provides better network performance by implementing enhanced routing metrics during route discovery. A comparison of SERPDA with well known on-demand routing protocols such as AODV and SADEC protocol, is done using ns-2. The comparison covers the above two attack scenarios for the performance parameters of the Packet Delivery Ratio, and End-to-End Delay. The performance of the protocols is analyzed according to varying malicious nodes. SERPDA provides better performance compared to AODV, by choosing the more optimal path, and it gives a slight increase in delay performance compared to SADEC, because of the routing metrics enhancement. In future work, we consider handling of other packet dropping attacks with better network performance in wireless mesh networks by enhancing the behavior based detection mechanism and routing metrics.

References

1. Akyildiz, X.: Wang, and W. Wang.: Wireless mesh Networks: A survey. Computer Networks 47, 445–487 (2005)
2. Benyamina, D., Hafid, A., Gendreau, M.: Wireless Mesh Networks Design a Survey. IEEE Communications Surveys & Tutorials 14(2), 299–310 (2011)
3. Ajmal, M.M., Mahmood, K., Madani, S.A.: Efficient Routing in Wireless Mesh Networks by Enhanced AODV. In: International Conference on Information And Emerging Technologies (ICIET), pp. 1–7 (June 2010)

4. Shila, D.M., Cheng, Y., Anjali: Mitigating Selective Forwarding Attacks With a Channel-Aware Approach In WMNS. In: IEEE International Conference on Wireless Communications, vol. 9(5), pp. 1661–1675 (May 2010)
5. Glass, S.M., Muthukkumarasamy, V., Portmann, M.: Detecting Man-In-The-Middle And Wormhole Attacks in Wireless Mesh Networks. In: IEEE International Conference on Advanced Information Networking and Applications, pp. 530–538 (May 2009)
6. Xing, F., Wang, W.: On The Survivability of Wireless Adhoc Networks With Node Misbehaviours and Failures. IEEE Transactions on Dependable and Secure Computing 7(3), 284–299 (2010)
7. Shila, D.M., Cheng, Y., Anjali: Channel-Aware Detection of GrayHole Attacks in Wireless Mesh Networks. In: IEEE International Conference on Global Telecommunications, pp. 1–6 (December 2009)
8. Lee, S.J., Gerla, M.: Split Multipath Routing with Maximally Disjoint Paths in Ad Hoc Networks. In: IEEE International Conference on Communication (ICC 2001), pp. 3201–3205 (2001)
9. Awerbuch, B., Curtmola, R., Holmer, D., Nita-Rotaru, C., Rubens, H.: ODSBR: An On-Demand Secure Byzantine Resilient Routing Protocol for Wireless Ad Hoc Networks. ACM Transactions on Information and System Security 10(4) (2008)
10. Hu, Y.C., Perrig, A., Johnson, D.B.: Packet Leashes: A Defense against Wormhole Attacks in Wireless Networks. In: IEEE INFOCOM, pp. 1976–1986 (December 2003)
11. Khalil, I., Bagchi, S.: Stealthy Attacks In Wireless AdHoc networks: Detection and Countermeasure. IEEE Transactions on Mobile Computing 10(8), 1096–1112 (2011)
12. Ramapriya, A., Navamani, T.M.: Defending against Stealthy attacks in Wireless Mesh Networks. In: International Conference on Engineering and Technology (ICET 2K12), April 18-20 (2012)
13. Khalil, I.: MIMI: Mitigating Packet Misrouting in Locally- Monitored Multi-Hop Wireless Ad Hoc Networks. In: IEEE International Conference on Global Telecommunications, pp. 1–5 (2008)
14. Mogaibel, H.A., Othman, M.: Review of Routing Protocols and it's Metrics for Wireless Mesh Routing Protocols, In: IEEE International Association of Computer Science and Information Technology-Spring Conference, IACSITSC 2009 (April 2009)
15. Campista, M.E.M., Esposito, P.M.: Routing Metrics and Protocols for Wireless Mesh Networks. IEEE Network 22(1), 6–12 (2008)
16. Navamani, T.M., Yogesh, P.: Efficient routing in Wireless Mesh Networks with QoS aware Reconfiguration. In: International Conference on Instrumentation, Communication Control and Automation, ICICAC 2013, Kalasalingam University, January 3-5 (2013)

Key Management with Improved Location Aided Cluster Based Routing Protocol in MANETs

Yogita Wankhade, Vidya Dhamdhere, and Pankaj Vidhate

G.H.R.C.E.M, Wagholi, Pune, India
{yogita16.wankhade,vidya.dhamdhere,pankajvidhate}@gmail.com

Abstract. Security is the main challenge in MANETs. The Key management scheme is crucial part of security which is important in MANETs. Authentication with key generation and distribution is a complicated task. In this paper we introduce an ILCRP- improved Location aided Cluster based Routing Protocol with Key Management scheme to make ILCRP secure. This Paper aims to provide better security with ILCRP protocol using Quantum Key Distribution. Quantum key distribution is used to generate a secure communication among the nodes. The ILCRP is a stable clustering protocol and appropriate for large number of nodes where all the nodes are enabled with GPS to achieve higher packet delivery ratio. Simulation result shows the demonstration of ILCRP with ILCRP-IDS in terms of ratio of packet delivery, delay required for an end to end communication and consumption of energy.

1 Introduction

A MANET is comprised set of nodes, mobile in nature those can have interaction with each other with the help of wireless links. It does not rely on predefined infrastructure, nodes freely move.

As ILCRP is cluster Based Routing protocol and many other protocols [1] are projected to increase the packet delivery ratio with the help of GPS technology or clustering schemes. With the help of clusters, MANET is divided into a group of nodes and it will help in selecting the cluster head (also called as CH). Each cluster will have separate GN when clusters are non-overlapping. These gateway nodes will make possible inter Cluster Head communication [2].

Detection of intrusion in MANETs is one of the challenging task because of numbers of causes [3], [4]. The Key management includes generating and distributing as well as updating the keys; it constitutes for secure communication.

2 Related Work

Cooperative IDS proposed by [5] that collect data in bottom up and makes a decision on top down. This IDS system is also referred to as "Dynamic Hierarchy" [26]. There is no static configuration but nodes need to negotiate the hierarchy. It has a drawback

© Springer International Publishing Switzerland 2015   687
S.C. Satapathy et al. (eds.), *Proc. of the 3rd Int. Conf. on Front. of Intell. Comput. (FICTA) 2014*
– *Vol. 2, Advances in Intelligent Systems and Computing* 328, DOI: 10.1007/978-3-319-12012-6_76

as monitoring end to end traffic is difficult. Nodes would loosen their battery power quickly. In MANETs, nodes are distributed and needs cooperation with other nodes. Distributed and cooperative IDS [6] proposed detection of intrusion and response systems which were following both the natures. Local IDS [7] is based on distributed and collaborative architecture. After detecting a local intrusion, [8] LIDS produceds a response. When LIDS receives an alert, it can protect itself from further the intrusion.

A multi-sensor intrusion detection system proposed by [9], and based on mobile agent [10] architecture. It may not check every node which is present in the network because of hop attribute of clusters.

Zone Based IDS [11] introduces division of MANET into non-overlapping zones. In this every node is having an IDS agent which runs on that node itself. It causes the detection and response latency even if there is adequate proof on local node.

Cooperative IDS which is using Cross-Feature Analysis which in tern uses Data-mining techniques to design an anomaly detection model automatically [12], [13]. It has high computational cost. It does not considers cluster-heads capabilities.

Specification-based IDS in MANETs are projected by [14] to detect attacks on AODV.. There is communication overhead under high mobility.Distributed Evidence driven Message exchanging intrusion detection Model [15] (DEMEM) is having distributed nature and which is cooperative too. It states that the distributed and collaborative attacks cannot be detected by the system.

A case-based [17] system which uses reasoning, for monitoring at packet level which is grounded on an architecture of hierarchical IDS. It is noted as the density of network increases the number of dropped packets by cluster heads also goes on increasing and has difficulties in updating case archives in a distributed environment.

In distributed kind of architecture which consists of IDS agents and [18] also proposed a secure database which is stationary (SSD). It has been noted that a stationary node is not following the basic nature of MANET. Clustering is been used in a hierarchy of dynamic intrustion detection system [19] which are extended to large size networks. The clustering method is analogous with the method in [16]; in that system, every node in the network is having responsibility to monitor, examine the activities, and according to that respond or send reports to the cluster heads. Agent based IDS [20] does not maintain description about security of mobile agents. GRID based IDS [21], [16] uses the fundamental beliefs of the Grid computing, However, they cannot properly detect grid intrusions.

We found different IDS systems during the literature survey with some lacunas that are listed below.

1. There is no authentication and key management mechanism is used for better security.
2. Some existing IDS systems uses distributed – collaborative architecture or distributed architecture only.
3. Different types of attacks are not addressed by the existing IDS systems.
4. No cluster based intrusion detection methods is used by some existing IDS system.

The importance of our method is to resolve each of these above cited problems. Our system uses hierarchical based IDS architecture as our protocol need cluster based IDS technique. The Key management includes generating and distributing of keys which leads secure communication.

3   Proposed Approach for ILCRP-IDS

The proposed system introduces Hierarchical IDS architecture for ILCRP protocol with key management in terms of Quantum key distribution. Due to the use of key management it overcomes the problem of active and passive attacks as well as middle attack.

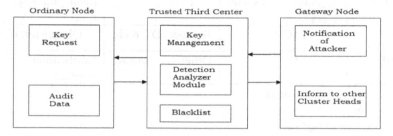

Fig. 1. ILCRP-IDS Architecture

Ordinary node can be considered as any source node, destination node and cluster head. TTC provides a Quantum keys for secure communication between nodes. When a new node broadcast a hello message to join a cluster, Trusted Third Center (TTC) provides a quantum key to both source node and destination node. Trusted Third Center (TTC) introduces a quantum key generation and distribution algorithm.

Source node sends its ID and destination node ID to Trusted Third Center to receive a quantum key and session key for communication. Trusted Third Center will provide a randomly generated key to source node and associate key to destination node. When communication occurs both keys are verified by TTC to identify there is an attacker or not.

When the qubits are mismatching found, TTC will concludes that there is presence of attacker and sends a notification to all the other cluster heads. If the qubits are matched, then connection will be continuing until TTC sends any error notification or source node stops sending packets. When source node's qubits does not matched with destination node's qubits, TTC identifies that key is invalid and concludes that there is presence of attacker. The notification of attacker is inform to all other cluster heads and this notification goes through Gateway nodes to other cluster heads. In the detection action module, the attacker is blocked by Trusted Third Center and attacker's ID is sending it to the blacklist.

Furthermore, secure communication is achieved through Trusted Third Center and notification to cluster heads which prevents further damage.

3.1   Authentication

For every node, certificate key is given to verification its identity. Every node has its certificate key and IP address. To generate a public key, both IP address and Certificate key is encrypted. Based on broadcast request and reply scheme, each authentication occurs between nodes.

node → Cluster_Head : Join _ Request (id),

cluster_Head → node : Join _ Reply (id),

Authentcation → node(id)

Table 1. Detection of Attacker

Message	Sender	destination	content
REQT	Sr_node (Source node)	TTC	(node_id,energy,cluster_id)
REPT	TTC	Sr_node, Ds_node	Qk(quantum key)
Notify	TTC	Other cluster heads	Notification of attacker

1. blacklist ← {malicious node_id}

2. {new_node$_i$}

3. upon receiving <REQT> do

4. msgValid ← Verifiedidentity   \\ new node verification

5. if(new_node != blacklist) then

   new_node is not malicious

   end if

6.if (ds_node != blacklist) then

   \\ ds_node is destination node

   ds_node is not a malicious

   end if

7.{Sr_node, Ds_node} Qk ← TTC   \\ quantum key from trusted third center

8.{audit_data} ← rules

   \\ apply detection analyzer rules explained in proposed system

9. if quantum keys are mismatching

Then

{CH} notify ← TTC

10.ClusterHead ← notify

11. malicious node added to the blacklist.

4   Comparisons between AODV & ILCRP-IDS

4.1   Results Obtained from AODV

The graph shows the performance of packet delivery in increasing number of nodes from 60 nodes to 150 nodes. This performs better with minimum number of nodes while gives lowest percentage with maximum nodes.

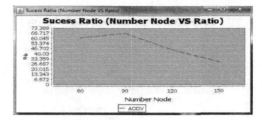

Fig. 2. Success Ratio - Number of Nodes Vs Ratio

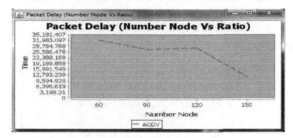

Fig. 3. Packet Delay - Number of Nodes Vs Ratio

The graph shows the packet delays with 60 nodes, 90 nodes, 120 nodes and 150 nodes. The packet delay is more as the quantity of nodes varies.

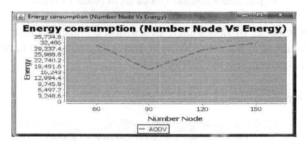

Fig. 4. Energy Consumption - Numbers of Node Vs Ratio

The graph shows the consumption of energy ratio with different quantity of nodes. This results energy required is higher for large number of nodes.

4.2   Results Obtained from ILCRP-IDS

The performance could be measured using the following parameters.

1. Success Ratio (Number of clusters Vs Ratio)

In this graph, number of nodes increases with the constant clusters 6. The graph shows the performance of packet delivery in increasing number of nodes. The success ratio is higher with maximum number of nodes.

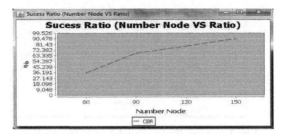

Fig. 5. Success Ratio - Number of Clusters Vs Ratio

2. Success Ratio (Numbers of Node Vs Ratio)

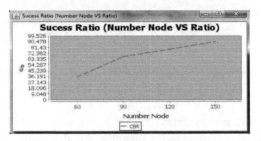

Fig. 6. Success Ratio - Numbers of Node Vs Ratio

In this graph, number of nodes increases with the constant clusters 6. The graph shows the performance of packet delivery in increasing number of nodes. The success ratio is higher with maximum number of nodes.

3. Packet Delay (Number of Cluster Vs Ratio)

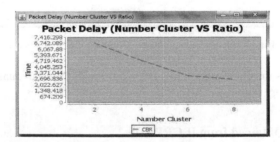

Fig. 7. Packet Delay - Number of Clusters Vs Ratio

This shows packet delay graph when 150 nodes are constant and clusters are increase such as two clusters, four clusters, six clusters and eight clusters.

4. Packet Delay (Numbers of Node Vs Ratio)

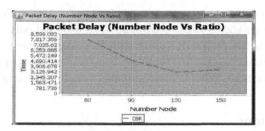

Fig. 8. Packet Delay - Number of Nodes Vs Ratio

This shows packet delay graph when clusters are constant (6 clusters) with increasing number of nodes such as 60, 90, 120 and 150 nodes.

5. Consumption of Energy (Numbers of Cluster Vs Ratio)

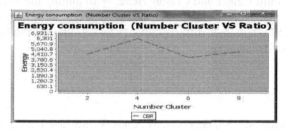

Fig. 9. Energy Consumption - Numbers of Clusters Vs Ratio

It shows the average energy consumed cluster heads with constant number of nodes. When numbers of clusters are six or eight for 150 nodes then it shows the less energy consumption compared to two or four clusters.

6. Energy Consumption (Numbers of Node Vs Ratio)

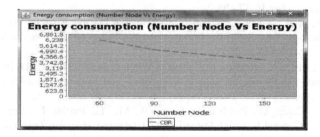

Fig. 10. Energy Consumption Graph - Numbers of Nodes Vs Ratio

It shows the average energy consumed cluster heads with increasing number of nodes with the constant cluster (six). The graph shows less energy consumption with the maximum nodes with clusters six.

4.3   Performance Analysis

Key Management improves the security and prevention of middle attacks. ILCRP-IDS results in higher success ratio compared to AODV and only ILCRP protocol. By combining the benefits of ILCRP along with Key management, this system gives a new direction in scheming ILCRP with key management. This approach gives higher success ratio, reduces packet delay and minimizes energy consumption.

5   Conclusions

Our proposed technique use Key Management with ILCRP protocol to provide the better security. The Implementation of Quantum key distribution allows a secure communication. Key Management improves the security and prevention of middle attacks. The proposed scheme efficiently achieves key management as well as authentication and maintains secret key which is used long term and session key between trusted third Center and Cluster Head. With uniting the advantages of ILCRP along with Key management presents a novel course in designing ILCRP with IDS. Cluster based IDS overcomes different attacks by introducing the Intrusion Detection System with key Management.

References

1. Iwata, A., Chiang, C., Pei, G., Gerla, M., Chen, T.: Scalable routing strategies for ad hoc wireless networks. IEEE J. Select Areas Communication 17(8), 1369–1379 (1999)
2. Mangai, S.V., Tamilarasi, A.: A new approach to geographic routing for location aided cluster based MANETs. EURASIP Journal on Wireless Communications and Networking (2011)
3. Brutch, P., Ko, C.: Challenges in Intrusion Detection for Ad Hoc Networks. In: IEEE Workshop on Security and Assurance in Ad hoc Networks, Orlando, FL, January 28 (2003)
4. Wrona, K.: Distributed Security: Ad Hoc Networks & Beyond. In: PAMPAS Workshop, London, September 16-17 (2002)
5. Sterne, D., et al.: A General Cooperative Intrusion Detection Architecture for MANETs. In: Proceedings of the 3rd IEEE International Workshop on Information Assurance (IWIA 2005), pp. 57–70 (March 2005)
6. Zhang, Y., Lee, W.: Intrusion Detection in Wireless Ad Hoc Networks. In: Proc of the 6th International Conference on Mobile Computing and Network (MobiCom), pp. 275–283 (2000)
7. Albers, P., Camp, O., Percher, J., Jouga, B., Me, L., Puttini, R.: Security in Ad Hoc Networks: a General Intrusion Detection Architecture Enhancing Trust Based Approaches. In: Proceedings of the 1st International Workshop on Wireless Information Systems (WIS 2002), pp. 1–12 (April 2002)
8. Albers, P., Camp, O., Percher, J.M., Jouga, B., Puttini, R.: Security in ad hoc networks: a general intrusion detection architecture enhancing trust based approaches. In: Proceedings of the First International Workshop on Wireless Information Systems, WIS 2002 (2002)

9. Kachirski, O., Guha, R.: Effective Intrusion Detection Using Multiple Sensors in Wireless Ad Hoc Networks. In: Proceedings of the 36th Hawaii International Conference on System Sciences (HICSS 2003). IEEE (2003)

10. Kachirski, O., Guha, R.: Intrusion detection using mobile agents in wireless ad hoc networks. In: Proceedings of the IEEE Workshop on Knowledge Media Networking, pp. 153–158. IEEE Computer Society, Washington, DC (2002)

11. Sun, B., Wu, K., et al.: Zone-Based Intrusion Detection System for Mobile Ad Hoc Networks. International Journal of Ad Hoc and Sensors Wireless Network (2003)

12. Huang, Y., Fan, W., et al.: Cross-Feature Analysis for Detecting Ad-Hoc Routing Anomalies. In: Proc of 23rd IEEE International Conference on Distributed Computing System (ICDCS), pp. 478–487 (2003)

13. Huang, Y., Lee, W.: A Cooperative Intrusion Detection System for Ad Hoc Networks. In: Proceedings of the ACM Workshop on Security in Ad Hoc and Sensor Networks (SASN 2003), pp. 135–147 (October 2003)

14. Tseng, C.-Y., Balasubramayan, P., et al.: A Specification-Based Intrusion Detection System for AODV. In: Proc of the ACM Workshop on Secure in Ad Hoc and Sensor Network, SASN (2003)

15. Tseng, C.H., Wang, S.-H., Ko, C., Levitt, K.N.: DEMEM: Distributed Evidence-Driven Message Exchange Intrusion Detection Model for MANET. In: Zamboni, D., Kruegel, C. (eds.) RAID 2006. LNCS, vol. 4219, pp. 249–271. Springer, Heidelberg (2006)

16. Choon, O.T.: Grid-based intrusion detection system. IEEE (2003)

17. Guha, R., Kachirski, O., et al.: Case-Based Agents for Packet-Level Intrusion Detection in Ad Hoc Networks. In: Proc. of 17th International Symposium on Computer & Info Science, pp. 315–230 (2002)

18. Smith, A.B.: An Examination of Intrusion Detection Architecture for Wireless Ad Hoc Networks. In: Proceeding of 5th National Colloquium for Information Systems Security Education (2001)

19. Sterne, D., Lawler, G.: A dynamic intrusion detection hierarchy for MANETs. In: Proceedings of the 32nd international conference on Sarnoff Symposium, SARNOFF 2009. IEEE Press, Piscataway (2009)

20. Nakkeeran, R., Albert, T.A., Ezumalai, R.: Agent Based Efficient Anomaly Intrusion Detection System in Ad hoc networks. IACSIT International Journal of Engineering and Technology 2 (February 2010)

21. Schulter, A., Reis, J.A., Koch, F., Westphall, C.B.: A Grid-based Intrusion Detection System. In: Proceedings of the International Conference on Networking, International Conference on Systems. IEEE Computer Society (2006)

9. Kamakshi, O., Omisra, K.: Effective Intrusion Detection Using Multiple Sensors in Wireless Ad Hoc Networks. In: Proceedings of the 36th Hawaii International Conference on System Sciences (HICSS 2003), IEEE (2003)

10. Chakeres, I.D., Cuba, J.P.: Intrusion detection using mobile agents. In: Wireless Ad hoc networks. In: Proceedings of the IEEE Workshop on Knowledge Media Networking, pp. 153–158. IEEE Computer Society, Washington, DC (2002)

11. Xu, Y., Wang, J., et al.: Zone-Based Intrusion Detection System for Mobile Ad Hoc networks. International Journal of Ad Hoc and Sensor Wireless Network (2004)

12. Huang, Y., Fan, W., et al.: Cross-feature Analysis for Detecting Ad Hoc Routing Anomalies. In: Proc. of 23rd International Conference on Distributed Computing Systems (ICDCS), pp. 478–487 (2003)

13. Huang, Y., Lee, W.: A Cooperative Intrusion Detection System for Ad Hoc networks. In: Proceedings of the 1st ACM Workshop Security on Ad Hoc and Sensor Networks (SASN 2003), pp. 135–147, October (2003)

14. Yang, H., Shu, J., et al.: Security in Specification-Based Intrusion Detection in Sensor for AODV. In: Proc. of the 12th Workshop on Security Protocols, Hacking Sensor Networks SASN (2004)

15. Tseng, C.H., Vang, Bell, N.L., Co, et al.: A Specification-Based Intrusion Detection System for AODV. In: Setia, S., Swarup, V. (eds.) SASN 2003. pp. 125–134. ACM, New York (2003)

16. Chen, O.P., Dai, and intrusion detection system. IEEE (2003)

17. Deng, H., Agrawal, D.P.: TIDS: threshold and identity-based security scheme for preventing wormhole attacks. Int. J. Ad Hoc network System (2007)

18. Sanzgiri, K., et al.: Authenticated routing for ad hoc networks. IEEE Journal on Selected Areas in Communications (2005)

19. Sun, B., Wu, K., Pooch, U.W.: Zone-based intrusion detection for mobile ad hoc networks. International Journal of Ad Hoc and Sensor Wireless Networks (2003)

20. Sterne, D., Balasubramaniyan, et al.: A general cooperative intrusion detection architecture for MANETs. In: Proceedings of the Third IEEE International Workshop on Information Assurance (IWIA 2005), pp. 57–70. IEEE Computer Society (2005)

21. Nadkarni, K., Mishra, A.: A novel intrusion detection approach for wireless ad hoc networks. In: IEEE Wireless Communications and Networking Conference (WCNC), vol. 2, pp. 831–836 (2004)

22. Marchang, N., Datta, R.: Collaborative techniques for intrusion detection in mobile ad-hoc networks. Elsevier, Ad Hoc Networks (2008)

23. Mishra, A., Nadkarni, K., Patcha, A.: Intrusion detection in wireless ad hoc networks. IEEE Wireless Communications (2004)

24. Albers, P., Camp, O., et al.: Security in ad hoc networks: a general intrusion detection architecture enhancing trust based approaches. In: Proceedings of the First International Workshop on Wireless Information Systems (2002)

Co-operative Shortest Path Relay Selection
for Multihop MANETs

Rama Devi Boddu[1], K. Kishan Rao[2], and M. Asha Rani[3]

[1] Department of E.C.E., Kakatiya Institute of Technology and Science,
Warangal, Telangana, India
[2] Department of E.C.E., Vaagdevi College of Engineering, Warangal, Telangana, India
[3] Department of E.C.E., Jawaharlal Nehru Technological University,
Hyderabad, Telangana, India
ramadevikitsw@gmail.com, prof_kkr@rediffmail.com,
ashajntu1@yahoo.com

Abstract. In this paper, we propose a simple Shortest Distance Path Relay Selection Criteria employing Decode and Forward (DF) Cooperative Protocol. We consider a single source and a single destination network with N candidate relays which are distributed uniformly within the coverage area. Flat Rayleigh fading channel with Log-distance path loss model is considered. In Shortest Distance Path Relay Selection we select the relay which is near to assumed Line of Sight (LOS). In Reactive Best Expectation Relay Selection Criteria relays which minimize total transmission time are selected for cooperation after source transmission. In Proactive Opportunistic Relay Selection Criteria before the source transmission, best relay which maximize mutual information capacity is selected for cooperation. The proposed relay selection criteria was compared with Reactive Best Expectation Relay Selection Criteria and Proactive Opportunistic Relay Selection Criteria. We further analyzed energy consumption, throughput and delay of the proposed system. The simulation results show that the proposed Shortest Distance Path Relay Selection consume less energy and has shortest delay compared to the Reactive Best Expectation and Proactive Opportunistic Relay Selection methods.

Keywords: Decode and Forward, Opportunistic, Proactive, Reactive, Shortest Distance Path.

1 Introduction

Network life time of Mobile Adhoc Networks (MANETs) is limited by the battery power of the mobile nodes. MANETs requires energy efficient, scalable, reliable routing protocols for Multi hop communication.Cooperative techniques provides the spatial diversity via node cooperation, improves the link quality and extensively suitable for emerging wireless networks.

Depending on route calculation routing protocols are classified into Reactive, Proactive and Hybrid routing protocols [1]. In Reactive routing the route is created when source required to send a packet to destination. Example: AODV, DSR. In Proactive routing each node maintains routing tables and updates routing

© Springer International Publishing Switzerland 2015 697
S.C. Satapathy et al. (eds.), *Proc. of the 3rd Int. Conf. on Front. of Intell. Comput. (FICTA) 2014*
– *Vol. 2*, Advances in Intelligent Systems and Computing 328, DOI: 10.1007/978-3-319-12012-6_77

information. Example: DSDV, Distance vector routing. Hybrid routing protocols are mixture of both.

Cooperative relay selection in distributed networks requires space time coding, global CSI which are quite difficult [2]. The simple distributed single Proactive Opportunistic relay selection doesn't require global CSI at each relay and require less cooperation overhead. Proactive Opportunistic relay selection is performed before source transmission [3] and all other relays expect selected relay enters into sleep mode during source transmission and reduces the reception energy. In Reactive Best Expectation Relay Selection method the source has no instantaneous CSI of the channels between relays and destination and adaptively selects the relays to optimize the total transmission time [4].

We propose a simple Shortest Distance Path Relay Selection Criteria in which relay with minimum distance from Line of Sight (LOS) path is selected for transmission. The performance of the proposed technique with Reactive Best Expectation and Proactive Opportunistic relay selection methods were investigated. Various performance metrics like energy consumption, end-to-end delay, and throughput were analyzed.

In this paper, Section 2 describe cooperative transmission, Section 3 describes the reactive and proactive relay selection methods. In Section 4 we present the proposed shortest distance path relay selection criteria, Section 5 describes various performance matrices evaluation and we analyze simulation results in Section 6. Some conclusions were drawn in Section 7.

2   Cooperative Transmission

The conventional relay based cooperative half duplex transmission model is assumed between Source (S), Relay (R_i) and Destination (D). The cooperative transmission employs two phase transmission protocol [5], during Phase-I S broadcasts the message to both R and D. In Phase-II R forwards received signal to D. The relay employs Decode and Forward (DF) protocol. The DF relay decodes received data from S and forwards re-encoded data to D.

The received signal at destination and relay during Phase-I (Broadcasting phase) can be expressed as

$$y_{SD} = \sqrt{P_1} h_{SD} x + n_{SD} \tag{1}$$

$$y_{SR_i} = \sqrt{P_1} h_{SR_i} x + n_{SR_i} \tag{2}$$

The received signal at D from DF relay during Phase-II is

$$y_{R_iD} = \sqrt{P_2} h_{R_iD} \hat{x} + n_{R_iD} \tag{3}$$

where P_1 , P_2 are transmission power of S, R_i respectively. We use equal power allocation $P_1 = P_2 = P$. The channel coefficients $h_{SD}, h_{SR_i}, h_{R_iD}$ between $S-D$, $S-R_i$, R_i-D capture the effect of path loss and Rayleigh fading. h_{ij} has $CN(0, \sigma_{ij}^2)$ under Rayleigh fading channel. The channel gain variance is modeled as $\sigma_{ij}^2 \propto d_{ij}^{-\alpha}$ Log-distance path loss model and d_{ij} is distance between i and j. The noise terms n_{SR_i} , n_{R_iD} and n_{SD} follows the Gaussian distribution n_{ij} has $CN(0, N_O)$.

The destination combines the received signals y_{SD} and y_{R_iD} using Maximal Ratio Combiner (MRC). The Symbol Error Rate (SER) performance under the DF Scheme with M-QAM modulation [6] can be expressed as

$$P_{SER} \approx \frac{1}{b^2 \bar{\gamma}_{SD}} \left(\frac{A^2}{\bar{\gamma}_{SR_i}} + \frac{B}{\bar{\gamma}_{R,D}} \right)$$

(4)

where $\bar{\gamma}_{SD} = P_1\sigma_{SD}^2 / N_o, \bar{\gamma}_{SR_i} = P_1\sigma_{SR_i}^2 / N_o, \bar{\gamma}_{R,D} = P_2\sigma_{R,D}^2 / N_o$ and $b = 3 / 2(M-1)$,

$A = (M-1)/2M + \left(1 - 1/\sqrt{M}\right)^2 / \pi, B = 3(M-1)/8M + \left(1 - 1/\sqrt{M}\right)^2 / \pi$.

Equation (4) depends upon high Signal to Noise Ratio (SNR) regime. $\bar{\gamma}_{ij}$ represents average SNR per symbol between i and j.

3   Relay Selection

The relay selection based on before or after source transmission can be classified into two types: (i) Reactive Relay Selection (ii) Proactive Relay Selection. In reactive type relay selection is done after source transmission and in proactive type relay selection is done before the source transmission.

3.1   Reactive Best Expectation Relay Selection Criteria

Let us consider a network with N relays which are distributed uniformly between S and D. Let R^C is the coverage radius of Source. N candidate relays $R' = \{r_1', r_2', ..., r_N'\}$ are distributed uniformly with in R^C. From R', k relays are selected for cooperation $R_{sel} = \{R_1, R_2, ..., R_k\}$ to minimize total transmission time. $R_{Sel} \subset R'$.

The proposed Best-Expectation algorithm does not require any instantaneous CSI at S of the channel $h_{R,D}$ for relay selection and a set of relays which minimizes total transmission time (T) is selected for relaying as in [4].

$$T = T_L + E\{T_C\} = \frac{N}{\min_{R_i \in R_{Sel}} C_{SR_i} W} + E\left\{ \frac{N}{C_{SR_{Sel}D} W} \right\}$$

(5)

where C_{SR_i}, $C_{SR_{Sel}D}$ are channel capacity between $S - R_i$ and $S - R_{Sel} - D$. W is bandwidth, T_L is worst case decode delay of DF relay, T_C is cooperative transmission time and expectation $E\{.\}$.

$$C_{SR_i} = \log_2\left(1 + \frac{P_1 |h_{SR_i}|^2}{\sigma^2}\right)$$

(6)

$$C_{SR_{Sel}D} = \log_2\left(1 + \frac{P_1}{\sigma^2} H_{SR_{Sel}D}^H H_{SR_{Sel}D}\right)$$

(7)

where $H_{SR_{Sel}D} = \begin{bmatrix} h_{SD} & h_{R_1D} & h_{R_2D} & . & . & h_{R_kD} \end{bmatrix}^T$.

The optimal relay selection can be expressed as

$$R^*_{Sel} = \arg\min_{R_{Sel} \subset R} \left(\frac{1}{\min\limits_{R_i \in R_{Sel}} C_{SR_i}} + E\left\{ \frac{1}{C_{SR_{Sel}D}} \right\} \right) \tag{8}$$

The expectation criteria chooses optimal set of relay R^*_{Sel} based on instantaneous $S - R_{Sel}$ capacity $C_{SR_{Sel}}$. The complexity of relay selection grows exponentially with N. This algorithm is further simplified by minimizing S-Relay Capacity ($C_{SR_{Sel}}$) by selecting set of active co-operative relays

$$C_{S\overline{R}_k} = \min_{R_i \in R_{Sel}} C_{SR_i} \tag{9}$$

$$R^*_{Sel} = \arg\min_{R_{Sel} \subset R'} \left(\frac{1}{C_{S\overline{R}_k (R_{Sel})}} + E\left\{ \frac{1}{C_{SR_{Sel}D}} \right\} \right) \tag{10}$$

The capacity in (10) can be written as

$$C_{SR_{Sel}D} = \log_2 \left(1 + \frac{P_1}{\sigma^2} \left(\sum_{R_i \in R_{Sel}} \left| h_{R_iD} \right|^2 \right) \right) \tag{11}$$

3.2   Proactive Opportunistic Relay Selection Criteria

In this criteria, one best relay among k relays is selected before the source transmission which is equivalent to routing. During source transmission only selected best relay is in active mode and all other relays enter into ideal mode. This reduces the energy consumption and improves the network life time.

The signal received at D is given by

$$y_{R,D} = h_{SR_i} h_{R_iD} x_{R_i} + n \tag{12}$$

where $R_i \in R_{Sel}$.

From (12), the mutual information with $k (\in R_{Sel})$ relays is given by

$$I = \frac{1}{2} \log_2 \left(1 + \left| h_{SR_i} h_{R_iD} \right|^2 \frac{P}{N_o} \right) \tag{13}$$

In (13), we select a single relay with maximum mutual information

$$I_{oop} = \max_{R_i \in R_{Sel}} \frac{1}{2} \log_2 \left\{ 1 + \frac{\left| h_{SR_i} \right|^2 \left| h_{R_iD} \right|^2}{N_o + \left| h_{R_iD} \right|^2 N_o} \frac{P_1}{N_o} \right\} \tag{14}$$

The best opportunistic relay R^*_{Sel} among R_{Sel} is chosen to maximize I_{oop} and can be expressed as

$$R^*_{Sel} = \arg\max_{R_i \in R_{Sel}} W_{R_i} \tag{15}$$

where $W_{R_i} = \dfrac{\left| h_{SR_i} \right|^2 \left| h_{R_iD} \right|^2}{\left(1 + 1/\eta_{SR_i}\right) + \left| h_{R_iD} \right|^2}$ and η_{SR_i} is received SNR.

The individual relay does not require CSI of other relay links. The opportunistic relay selection maximize I_{oop} which minimize outage probability and can be expressed as

$$P_{oop}(outage) = \Pr\{W_{R_{Sel}}^* < T\} = \Pr\left\{\max_{R_i \in R_{Sel}} W_{R_i}\right\}$$

$$= \pi \sum_{i=1}^{k} E_{|h_{R_iD}|^2}\left\{1-\exp\left[-\frac{T\left((1+1/\eta_{SR_i})+|h_{R_iD}|^2\right)}{|h_{R_iD}|^2}\right]\right\}$$

(16)

where $T = (2^{2R}-1)/\eta_{SR_i}$ and R is spectral efficiency.

4   Proposed Shortest Distance Path Relay Selection Criteria

The Mobile Equipment (ME) or Source (S) locks to the strongest received paging signal (set up channel) coming from the Base Station (BS) or Destination (D) as shown in Fig. 1. The direction in which the strongest Received Signal Strength (RSS) indicates the Line of Sight (LOS) between S and BS. The virtual dotted line represents direct LOS assumptions.

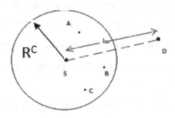

Fig. 1. Radio coverage of Source

Let l_{sd} is the distance between S & D, R^C is the coverage radius of transmitting node which is evaluated based on Target Symbol Error Rate (ρ) for a given transmitting node S & set of relay R_i .The SER guaranteed radio coverage of S is defined with geographic area within which any R_i can meet the SER $\leq \rho$.

$$R^C(S,R_i) = \left\{X \in R^2 | P_{SER}(S,R_i) \leq \rho\right\}$$

(17)

The average received SNR for log-distance path loss model can be expressed as

$$\bar{\gamma}(d_{SR_i}) = \frac{P_1\sigma_{SR_i}^2}{N_o} = \frac{P_1}{N_o d_{SR_i}^\alpha}$$

(18)

The tolerable relay $(z \in R_i)$ is the node which has the minimum distance to the node 'k' i.e,

$$\min\left\{d_{R_i}(R_i,k)\right\} = \left\|x_{R_i} - x_k\right\|$$

(19)

$$R_i \notin BS \;\& \; d_{R_i} \leq R^C$$

The node in the coverage range of the Source and with minimum distance from the assumed LOS is selected and considered for the short distance communication path. The above process is repeated form the selected relay for multihop until destination is reached.

5   Performance Matrices Evaluation

Various performance matrices for the proposed scheme are evaluated as follows:
(i) Energy Calculation: The total number of bits transmitted is 1024, $G_T = G_R = 1$, n_T
      = number of transmitters, number of hops is z, $m = 0,1,...,z-1$. $E_o = 0$ and the
      energy consumption at $m+1$ hop [7,8] is given by

$$E_{m+1} = E_m + \frac{\xi}{\eta'} \frac{n_T \ln|1/\overline{p}_b| N_o}{4\|H\|} \frac{(4\pi)^2 M_l N_f}{G_T G_R \lambda^2} d_{ij}^{\alpha} \qquad (20)$$

where Peak to Average Ratio (PAR) is given by $\xi = 3(\sqrt{M}-1)/(\sqrt{M}+1)$, M_n is Power Spectral Density(PSD) of noise at receiver, drain efficiency of power amplifier $\eta' = \eta'_{max} / \xi$, η'_{max} represents maximum drain efficiency, λ represents carrier wave length, ξ is peak to average ratio, G_T and G_R represents antenna gains, d_{ij} is between transmitting and receiving nodes, M_l represents link margin, noise figure $N_f = M_n / N_o$, α represents path loss coefficient and noise power spectral density $No = -171 dBm / Hz$.

Total energy consumption for 'z' hops can be expressed as

$$E_T = \sum_{m=0}^{z-1} E_m \qquad (21)$$

$$Energy\_per\_bit = E_T / Total\_number\_of\_bits \qquad (22)$$

(ii) Delay: End-to –end delay for the proposed scheme can be expressed as

$$Delay = \frac{dis\tan ce(d)}{propagation\_speed}(2 + Number\_of\_relays)$$

$$(23)$$

(iii) Throughput is the ratio of successful number of packets transmitted to D per unit time required to travel from source to destination. Throughput is measured in bits per second (bps).End to end throughput of a link $i\text{-}j$ is given by

$$\tau_{ij} = P_{ij}^s R_o \qquad (24)$$

where P_{ij}^s is probability of success, R_o is Data transfer rate=2b/s/H, 4b/s/H, 6b/s/H.

$$P_{ij}^s = \exp\left(-\frac{\left(2^{R_o}-1\right)N_o d_{ij}^\alpha}{P_1}\right)R_o \qquad (25)$$

6   Results and Analysis

We considered 802.11g standards with 10MHz channel bandwidth, FFT size 1024, 1024 bits packet size, 0.25 cyclic prefix and other simulation parameters are given in Table 1.The routing path for multi-hop network using shortest distance path, Reactive Best Expectation and Proactive Opportunistic Relay Selection Criteria are shown in Fig. 2 for 50 nodes with node 1 as source and node 15 as destination. Routing path changes as the number of nodes increases and depends on source node and destination node. We simulated for different nodes 30, 50, 80, 100 with different source and destination nodes. Simulation results shows that the energy consumption, delay varies with the distance between S-D, mode density and number relays required for the routing.

Fig. 3 shows that as Signal to Noise Ratio (SNR) increases the energy consumption increases and the proposed scheme consumes less energy compared to other two schemes. In general Proactive routing requires periodic updates of route and consumes more energy. Proposed shortest path considers imaginary LOS line and selects the nodes near to this line and number of intermediate relay nodes selected is less hence consumes less energy compared to other schemes described in this paper.

Fig. 4 compares end to end delay of all schemes which shows proposed scheme is very efficient. In shortest path less number of relays are selected near to LOS distance leads to less delay. Fig. 5 plots the throughput verses SNR and proposed scheme gives better throughput over the other schemes. Shortest path considers LOS imaginary reference line, LOS path has best SNR and gives best throughput.

Table 1. Simulation Parameters

Size of FFT	1024
Number of Pilot Subcarriers	120
Number of user Subcarriers	720
Null subcarriers	184
Maximum Power Amplifier Efficiency	0.35
Link margin	40dB
Path loss coefficient	2.3
Packet size	1024 bits
Physical Layer	802.11g
Radio frequency	2.4GHz
Bandwidth	10 MHz
Modulation	M-QAM
Simulation area	1000x1000 m
Transmitting , Receiving Antenna gain	1

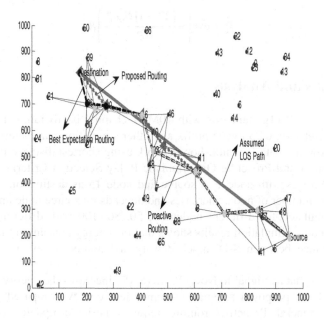

Fig. 2. Routing Path with 50 nodes, Node 1 as Source and Node 15 as destination nodes

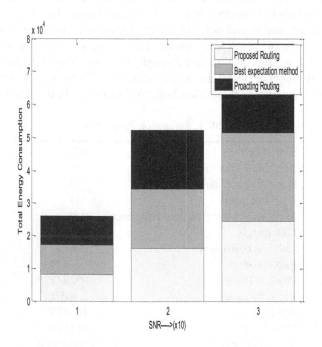

Fig. 3. Energy Consumption Vs SNR for 50 nodes

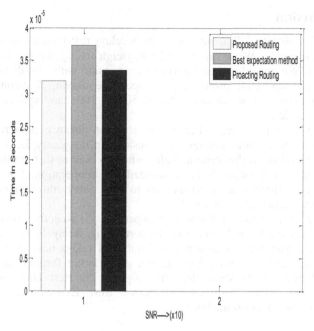

Fig. 4. Delay Vs SNR for 50 nodes

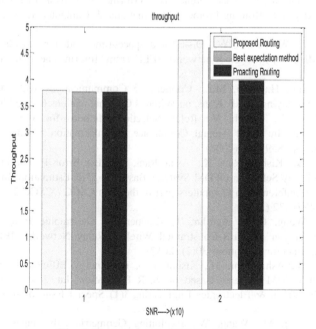

Fig. 5. Throughput Vs SNR for 50 nodes

7   Conclusion

We considered a Mobile Adhoc network with N relays distributed uniformly between Source and Destination. We assumed a Flat Rayleigh fading channel between source and destination. We considered half duplex transmission with Log-distance path loss model and nodes employ DF protocols for cooperative relaying. We consider 802.11g standards with different node densities N=30, 50, 80, 100... and with different source and destination nodes.

In this work, the proposed scheme uses shortest distance path relay selection criteria for multihop communication. We consider an imaginary LOS path between source and destination. In this criteria Relay which is near to this assumed LOS path with SER greater than target SER is selected for cooperation. For the proposed scheme various performance matrices end to end delay, throughput and energy consumption are evaluated in section 5.

Performance of proposed scheme is compared with Reactive Best Expectation Relay Selection Criteria and Proactive Opportunistic Relay Selection Criteria. In shortest distance path relay selection less number of relays near to imaginary LOS path are selected and consumes less energy, gives better throughput and has less delay.   Simulation results show that the proposed Shortest Distance Path Relay Selection Criteria outperforms over Reactive Best Expectation and Proactive Opportunistic relay selection methods.

References

1. Pandey, K., Swaroop, A.: A Comprehensive Performance Analysis of Proactive, Reactive and Hybrid MANETs Routing Protocols. International J. Computer Science 6(3), 432–441 (2011)
2. Laneman, J.N., Wornell, G.W.: Distributed space–time coded protocols for exploiting cooperative diversity in wireless networks. IEEE Trans. Inform. Theory 49(10), 2415–2525 (2003)
3. Bletas, A., Shin, H., Win, M.Z.: Cooperative Communications with Outage Optimal Opportunistic Relaying. IEEE Trans. on Wireless Communications 6(9), 3450–3460 (2007)
4. Nam, S., Vu, M., Tarokh, V.: Relay Selection Methods for Wireless Cooperative Communications. In: 42nd Annual Conference on Information Sciences and Systems, Princeton, NJ, pp. 859–864 (2008)
5. Rama Devi, B., Kishan Rao, K., Asha Rani, M.: Bit Error Probability Analysis of Cooperative Relay Selection OFDM Systems Based on SNR Estimation. In: 16th WSEAS International Conference on Computers(part of the 16th CSCC / CSCC 2012), Kos Island, Greece, pp. 427–432 (2012)
6. Syue, S.-J., Wang, C.-L., Aguilar, T.: Cooperative Geographic Routing with Radio Coverage Extension for SER-Constrained Wireless Relay Networks. IEEE Journal on selected areas in communications 30(2) (2012)
7. Rama Devi, B., Asha Rani, M., Kishan Rao, K.: Energy Efficient Cooperative Node Selection for OFDM Systems Based on SNR Estimation. Int. Journal of Advances in Computer, Electrical & Electronics Engineering 3(1) Special Issue of IC3T 2014, 32–36 (2014)
8. Ahmed, I., Peng, M., Wang, W.: Exploiting Geometric Advantages of Cooperative Communications for Energy Efficient Wireless Sensor Networks. I.J. Communications, Network and System Sciences 1(1), 55–61 (2008)

Intelligent Intrusion Detection System in Wireless Sensor Network

Abdur Rahaman Sardar[1], Rashmi Ranjan Sahoo[2], Moutushi Singh[4],
Souvik Sarkar[5], Jamuna Kanta Singh[3], and Koushik Majumder[6]

[1] Department of CSE, NITMAS, West Bengal, India
[2] Department of ETCE, Jadavpur University, West Bengal, India
[3] Department of CSE, Jadavpur University, West Bengal, India
[4] Department of IT, Institute of Engineering and Managment, Kolkata, India
[5] IBM , Hyderabad, India
[6] Department of IT, West Bengal University of Technology, Kolkata, India
{abdur.sardar,rashmi.cs2005,moutushisingh01}@gmail.com,
souviksarkar@in.ibm.com, jksingh@cse.jdvu.ac.in,
koushikzone@yahoo.com

Abstract. Wireless Sensor Networks (WSN) are formed with small tiny nodes which are sometime densely deployed in open and unprotected environment. In many applications particularly in the military applications WSNs are of interest to adversaries and they are susceptible to different types of attack. Though preventive measures are applied to protect against attacks but some attacks cannot be prevented using known preventive measure. Preventing the intruder from causing damage to the network, the intrusion detection system (IDS) can acquire information related to the attack techniques and helps to develop a preventing system. In this paper we propose an intelligent IDS algorithm and we also simulate our algorithm in castellia simulator. Our simulation results show different scenarios such as the attack period Vs packet dropped and attack period vs packet received. We have seen that under attack our IDS potentially improve the performance of the network in both cases.

Keywords: WSN, intrusion detection system, security threats.

1   Introduction

A sensor network is a computer network Composed of a large number of sensor nodes. [1]. Sensor nodes are usually scattered randomly in the field and will form a sensor network after deployment in an ad hoc manner. There is usually no infrastructure support for sensor networks.   Their low cost provides a means to deploy large sensor arrays in a variety of conditions capable of performing both military and civilian tasks [2].

According to Brutch and Ko, intrusion detection systems in sensor network are partitioned into three types: stand alone, distributed and co operative, and hierarchical.

© Springer International Publishing Switzerland 2015   707
S.C. Satapathy et al. (eds.), *Proc. of the 3rd Int. Conf. on Front. of Intell. Comput. (FICTA) 2014*
– *Vol. 2,* Advances in Intelligent Systems and Computing 328, DOI: 10.1007/978-3-319-12012-6_78

In stand alone architecture all the nodes run an Intrusion Detection System (IDS) independently and without co-operation with each other. Decisions are taken completely on the observation of information collected by the node itself. in the same network. This architecture is not effective due to the fact that no alert information is passed between the nodes when an intrusion is detected.

In distributed and cooperative architecture all the nodes run an IDS agent and intrusion detection is done collaboratively. Every IDS agent is responsible for collecting data and detecting intrusions locally, but they can cooperate with neighbor nodes. This is suitable for flat network infrastructure.

In hierarchical architecture all nodes run an IDS agent and detect intrusions locally but the IDS agent running at cluster head detects intrusion locally as well as globally. This is suitable for multilayered and cluster based infrastructure.

2   Intrusion Detection System in WSN

An intrusion detection system (IDS) is used to detect and to generate some alert when some intrusion activity is attempted into the system or network. An intrusion detection system may produce some alert to take further action when some malicious activity occurs into the network. The main purpose of an intrusion detection system is to detect the malicious nodes into the network as early as possible when some security attack occurs into the network. IDS dynamically monitor the system and user actions in the system to detect intrusion. Generally three basic types of IDS are proposed by different researchers i.e misuse detection, anomaly based detection and specification based detection. In misuse detection techniques some known attack signature are to be compared to check the system vulnerability. Misuse based techniques are not effective to detect new attack because of lack of signature. So it requires to update the signature [3]. Anomaly based IDS create normal profile of the system states or user behavior and compare them with current activities. If significant deviation from the normal behavior occurs, IDS raise an alarm. The technique can detect new types of attack but it is difficult to create exact normal profile. Another technique which combine anomaly and misuse detection techniques is known as specification based detection. This approach is based on manually developed specification. Both anomaly and specification based detection techniques detect an attack by using deviation from a normal profile.

3   Security Threats in WSN

Sensor networks are particularly vulnerable to several key types of attacks both internal and external. Internal attackers are the compromised nodes and they have the authorization on the network. *External* attackers are attackers that are not legally part of the network. They could be part of another network which is linked to the target network using the same infrastructure or same communication technology. The attacker may also be active or passive attacker. Some common attacks are discussed below.

i)   Sinkhole attack: In this type of attack, the attacker goal is to attract all the traffic through a particular compromised node. To launch these types of attack, a compromised node attract all neighboring nodes to forward their packets through the compromised node by showing its routing cost minimum. This can be done either by spoofed or replayed an extremely high quality route to a base station or by using a laptop class attacker which provide a high quality route by transmitting with enough power to reach the base station in a single hop. When all these packets pass through the malicious node it will drop the packets.

ii)   Wormhole attack: In wormhole attack the adversary node crate a virtual tunnel from one end of the network to another end. The two malicious nodes usually claim that they are one hop away from the base station. Wormhole can also be used to convince two distinct nodes that they are the neighbors by relaying packets between two of them. Wormhole attack can also be used in combination of selective forwarding, eavesdropping or Sybil attack[4].

iii)   Selective forwarding: In selective forwarding attack a node may refuse to forward some packets and simply drop it. The simplest form of this type of attack is black hole attack where the malicious node drops all the packets.

iv)   Sybil attack: In Sybil attack a node appears in the network with multiple identities. The Sybil attack also poses a significant threat to geographic routing protocol. Sybil attacks can be prevented by using authentication in link layer protocol [5].

v)   Hello flood attack: In sensor network many protocol broadcast a hello message to inform its existence to its neighbors. The nodes that receive the hello message assume that the source node is within its communication range and add this source node to its neighbor list. The laptop class adversary can spread hello message with sufficient transmission power to convince a group of nodes that they are its neighbor.

vi)   Acknowledgement spoofing: In many sensor networks routing algorithm rely on link layer acknowledgement. An adversary node may spoof link layer acknowledgement of the overheard packet. The main objective of the adversary is to convince the sender that a weak or dead link is a strong link [6]. Any packet sent through that weak or dead link will be lost.

4   Related Work

An Intrusion Detection System generally detects unwanted manipulation to systems. The two most common intrusion detection models used in network security today are misuse detection and anomaly detection.

Anjum et al [7] have used graph theory in order to optimally place the intrusion detection modules around the sensor network. Agah et al [8] proved that game theory techniques can be applied as a defense technique which will outperform intrusion detection techniques based on intuitive metrics i.e. traffic loads and Markov decision processes.

Anjum, Subhadrabandhu and Sarkar [7] have focused on signature based intrusion detection techniques and found that this technique generates better results when coupled with proactive routing algorithms rather than reactive ones. Loo et al [9] have focused on using clustering algorithms and anomaly detection to detect aberrant behavior.

Su et al [10] have researched how to apply intrusion detection techniques in cluster based networks, by making nodes aware of packet forwarding misbehavior of their neighbors and by collectively monitoring the cluster heads.

5   Intelligent Intrusion Detection Algorithm

We have considered that the network is a cluster based network and there is no retransmission mechanism.

Step 1:
During a time interval, each cluster node (CN) within a cluster send node id, number of packet send and number of packet received to the cluster head (CH).
Step 2:
a) Cluster head checks whether the node ids are changing or not. If any node id changes then the cluster head measures the number of packet received and number of packet send within that time interval. The cluster head then find out the total packet dropped ratio i.e.

$$\frac{(total\ packets\ send\ by\ all\ nodes - total\ packets\ received\ by\ all\ nodes)}{total\ packets\ send\ by\ all\ nodes}$$

If this packet dropped ratio is greater than some threshold θ, then it will raise an alarm.
b) Each node also measures the time interval between two successive receptions. If this time interval changes drastically from a fixed time interval t set by the cluster head, it will send a warning message to the cluster head.

    i)   If time interval increases and cluster head receives warning message from most of the nodes (>50%), it will raise an alarm.

    ii)   If time interval decreases and cluster head receives the warning message from very few nodes then it will raise an alarm because of exhaustion attack.

c) For each node, there should be some threshold value set for power consumption rate. If power consumed in a particular time interval is more than the threshold value it will send a warning message to the cluster head. If the cluster head receives the warning message repeatedly, it will raise an alarm.

6   Simulation Result

We have simulated our IDS using Castellia simulator. The Figure 1 shows the attack period vs packet dropped in both cases i.e with our IDS and without IDS. It can be observed from the figure that packet dropped significantly reduced with our proposed IDS. The figure 2 also shows the attack period vs packet received. In this case also packet received in our scheme is significantly increased.

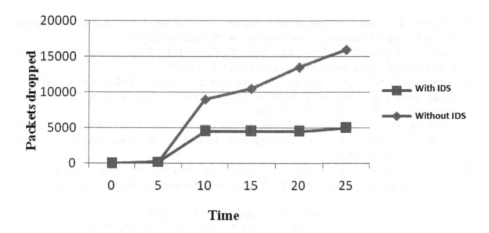

Fig. 1. Attack Period Vs Packet dropped

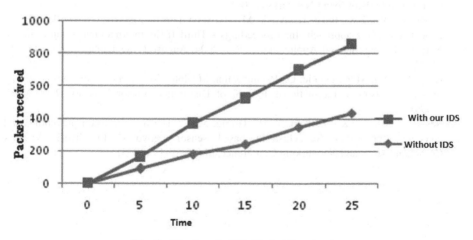

Fig. 2. Attack period Vs Packet Received

7   Conclusion

With the inclusion of intelligent intrusion detection algorithm in network, we found that, our algorithm capable of capturing the intruder nodes. Proposed algorithm also capable to isolate the detected intruder nodes from the network, which enables better performance, in terms of packet receive and packet drop capabilities of entire network. Consequently, this also increases the packet delivery ratio. In future we have planned to extend this work with implementation of both signature and anomaly based IDS as a hybrid model.

References

1. Akyildiz, I.F., Su, W., Sankarasubramaniam, Y., Cayirci, E.: A Survey on Sensor Networks. IEEE Communications Magazine 40(8), 102–114 (2002)
2. Perrig, A., Szewczyk, R., Tygar, J.D., Wen, V., Culler, D.E.: Spins: security protocols for sensor networks. Wireless Networking 8(5), 521–534 (2002)
3. Sardar, A.R., Singh, M., Sarkar, S.K.: Intrusion detection in wireless sensor networks using fuzzy logic. Published in the Proc. of IEMCON 2012 held at Science City Auditoriam, Kolkata, January 17-18 (2012)
4. Singh, M., Das, R.: A Survey of Different Techniques for Detection of Wormhole Attack in Wireless Sensor Network. International Journal of Scientific and Engineering Research 3(10) (October 2012)
5. Sultana, N., Huh, E.-N.: Application driven cluster based group key management with identifier in wireless sensor network. Transaction on Internet and Information System I(1), 5–18 (2007)
6. Karlof, C., Sastry, N., Wagner, D.: TinySec: A Link Layer Security Architecture for Wireless Sensor Networks. In: SenSys 2004, November 3-5, pp. 162–175 (2004)
7. Anjum, F., Subhadrabandhu, D., Sarkar, S., Shetty, R.: On Optimal Placement of Intrusion Detection Modules in Sensor Networks. In: Proceedings of the First International Conference on Broadband Networks (2004)
8. Agah, A., Das, S.K., Basu, K., Asadi, M.: Intrusion detection in sensor networks: A non-cooperative game approach. In: Proceedings - Third IEEE International Symposium on Network Computing and Applications, NCA 2004, August 30-September 1, pp. 343–346 (2004)
9. Loo, C.E., Ng, M.Y., Leckie, C., Palaniswami, M.: Intrusion detection for routing attacks in sensor networks. International Journal of Distributed Sensor Networks 2, 313–332 (2006)
10. Su, C.-C., Chang, K.-M., Kuo, Y.-H., Horng, M.-F.: The new intrusion prevention and detection approaches for clustering-based sensor networks. In: IEEE Wireless Communications and Networking Conference, March 13-17 (2005)

Secure Routing in MANET through Crypt-Biometric Technique

Zafar Sherin and M.K. Soni

Faculty of Engineering, Manav Rachna International University Faridabad, JMI
NCR, India
Sherin_zafar84@yahoo.com, ed.fet@mriu.edu.in

Abstract. A dynamic network self configuring and multi-hop in nature without
having any fixed infrastructure is called a mobile ad-hoc network(MANET).The
main drawback of such types of networks is the occurrence of various attacks
such as unauthorized data modification impersonation etc which affect their
performance. Biometric perception is specified as the most novel method to
protract security in various networks by involving exclusive identification
features. The attainment of biometric perception depends upon image
procurement and biometric perception system. Simulation as well as experimental
results signifies that this proposed method achieves better performance parameters
values for various mobile ad-hoc networks.

Keywords: MANET, security, biometric perception, pattern resemblance,
hamming distance, Hough transform.

1   Introduction

Networks that are autonomous in nature and can be instantaneously developed on
demand to carry out definitive tasks for mission support are referred as Mobile ad-hoc
networks. Communicating data across such a network is carried out through wireless

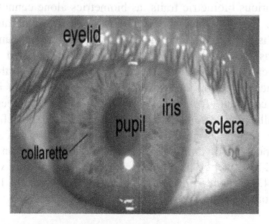

Fig. 1. A front-on view of the human eye depicting its various parts

© Springer International Publishing Switzerland 2015   713
S.C. Satapathy et al. (eds.), *Proc. of the 3rd Int. Conf. on Front. of Intell. Comput. (FICTA) 2014*
– *Vol. 2*, Advances in Intelligent Systems and Computing 328, DOI: 10.1007/978-3-319-12012-6_79

links where nodes develop a temporary infrastructure to route and transmit data. Since mobile ad-hoc networks (MANET) are utilised in various emergency systems like military , weather forecast and disaster management so securing data is the most important feature for such networks. Due to the lack of centralized infrastructure use of various cryptographic algorithms is the most challenging task.

Our novel crypt-biometric technique to provide secure routing in MANET focuses on removing various security breaches in mobile ad-hoc networks.

1.1   Security Demands of MANET

Mobile Ad-hoc networks suffer from various passive and active intrusions such as unauthorized access, modification , deletion or disruption of information flow etc. It is uttermost important and desirable to maintain confidentiality and integrity of data at the application layer itself in-order to develop an eminent secured authentication system. Various traditional methods to provide authentication like cryptography that performs encryption and decryption of messages are used to overcome various attacks on MANET. But for mission critical and highly sensitive applications above mentioned traditional methods cannot be a full proof measure against various intrusions .Therefore the developed method tries to overcome security intrusions of MANET along with providing better performance results from previous traditional security approaches.

1.2   Biometrics

Behavioral lineaments of people for motorized recurrence is referred as biometrics which is considered to be one of the most important security feature added in various networks. Biometrics is immensely strenuous to tantamount as it summates an exclusive qualifier that cannot be acquired, jilted or stolen hence a very comprehensive tool for authentication in MANET.

Crypt-biometrics solutions are analysed in this paper which involves embedding encryption with various biometric traits, as biometrics alone cannot be a wholesome security solution . Fig.1. depicts the headmost perspective of annular component called iris which is permeated adjoining to its mean through an annular circular fissure specified as the pupil. An iris template can be created from 173 out of relatively 266 peculiar inclinations hence iris perception is the most assuring biometric mechanics [2] . This novel study is a cogent exertion in biometrics that employs wavelet commute for the evolution of crypt-iris perception algorithmic design. Additionally, the results validate the accuracy and efficiency of these algorithmic designs serving as a reference in the domain of iris perception.

The upcoming portions of the paper deals with the literature on iris perception in Section 2 followed by novel crypt-biometric perception algorithm which is described in Segment 3 and 4. Segment 5 specifies various empirical results. Lastly conclusions and recommendations for future are considered in Segment 6.

2   Related Works

A strong biometric security system tends to provide a mechanized perception of unique individual traits (iris or fingerprint) of individuals [3] , so the most compelling security task which is dealt in this paper focuses on evolving a system of iris disjuncture residing noises such as varying eyelid, pupil & reflection. The iris disjuncture process undergoes two important steps namely data procurement and iris disjuncture. The first step namely data procurement obtains various iris images through infrared illumination which is employed so as to get better image quality. The iris disjuncture process restrict iris region in the image by suppressing number of noises employing various boundary espial algorithms. Addison construe wavelet(mathematical function) employed to dissect a given function or a continuous-time signal into varying frequency components and consider every component with a resolution that affirm its scale.

A large number of researches are made in the area of iris disjuncture and iris localization[4,5,6,7,8,9,10] in which Daugman and Wilde studies are considered quite fruitful in iris perception algorithms as they achieve a false rate(FAR) of 1 in four million and also false reject rate(FRR) of 0. Here refining undergoes a iris dissolution process which exploits for ascertaining iris as well as pupil contour which is illustrated by the given equation :

$$Max(r, x0, y0)G\sigma(r) * \partial \div \partial x \int r, x0, y0 I(x, y) \div 2 \prod rds \qquad (1)$$

where:

▶ r,x_0,y_0: This is specified as the centre and radius of coarse circle (it is referred for each pupil as well as iris).

▶ G_σ (r): This is the Gaussian function.

▶ ∂_r: This is the radius extent (range) utilised to penetrate I(x, y) of the initial iris image. Iris localized structure is altered through Daugman's dissolution algorithm from Cartesian to polar

3   Crypt-Biometric Perception Algorithm for Secure Routing

Fig.2. outlays novel quartered crypt-biometric perception algorithm which requires hardware as well as software to accomplish iris image acquisition and iris perception. Considering the facts that status of lighting as well as camera focus vary for distinct subjects, an adaptable steel bar backing is devised for the camera and also for the infrared light for the user to set the perfect distance from the camera and light source by the human eye. Every iris image undergoes sampling process once for enlistment and next time for matching. For enlistment as well as matching , the images are taken subsequently acclimatizing the focus only once. A proper chin footing is also interspersed in the device for user adaptation . A digital camera is connected to a frame grabber of the computer by the image acquisition toolbox of MATLAB. All the frames are previewed through a personalized graphical user interface that avow the user for selecting the most pertinent image settings ahead of enlisting it to the iris recognition software.The novel crypt-biometric technique in Fig.2. portray various functions that are used for actuating and authentication of the iris image .

4   Design of Crypt-Biometric Technique

A poor disparity betwixt the iris and pupil leads to out-of obligated values of iris sections causing various difficulties in disjunction process producing poor perception rates. For performing various comparisons, discursive iris image is then subjected to the normalization mechanism that performs transformation of the elicited iris region resulting into a rectangular section having perpetual extensity for overcoming imaging discrepancies, since the iris is elongated from differing levels of radiation. A biometric pattern is developed from the normalized section. Fig.3(a). illustrates the wavelet disintegration applying the wavelet toolbox of MATLAB [11] and Fig.3(b). specifies the application of Hough transform over iris image.

Pattern resemblance of hoarded iris images is the uttermost stage of the iris perception application which is performed by applying the Hamming distance and Correlation specified by the formulae 3 and 4 respectively:

Hamming Distance

$$1 \div c \sum_{i=1}^{c} M_i \oplus N_i \tag{3}$$

where:

M_i and N_i = the i-th bit in the series M and N

c = total number of bits in each sequence.

$\oplus$ =Exclusive or operation.

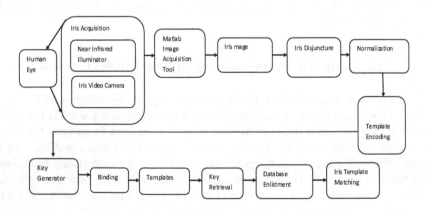

Fig. 2. Novel Crypt-Biometric Technique

Correlation Coefficient

$$r = n\sum xy - (\sum x)(\sum y) \div \sqrt{n\sum x^2 - (\sum x)^2} \wedge 2\sqrt{n\sum y^2 - (\sum y)^2} \wedge 2 \qquad (4)$$

where:

n is the number of pairs of data. The value of r is such that $-1 \le r \le +1$. The + and − signs are used for positive and negative linear correlations respectively.

The propound software was estimated for its pursuance by conducting various biometric evaluation metrics on it so as to actuate whether the purpose of study is fulfilled or not.

Next step of the process is developing crypt-biometrics solutions using elliptic curve cryptography that is chaotic map based which results in reduction of bandwidth overload of MANET since biometrics have some disadvantages dealing with privacy and exposing permanently once lost. Therefore after pattern resemblance a key generation procedure is undertaken providing two levels of security for this model. Biometric template undergoes binding with the generated key for hoarding into the database. Fig.4(a). exemplifies the graphical user interface advanced by applying MATLAB GUI builder, which encompasses all the components for the iris perception application. Template matching with the corresponding hoarded images is performed by calculating the hamming distance and correlation. Correlation parameter is never calculated by any biometric perception software providing uniqueness in this study, also it provides two parameters for pattern resemblance and certification hence improving security .

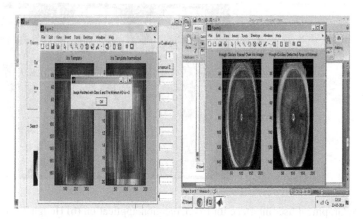

Fig. 3. (a) Bi-orthogonal Wavelet 3.5 Disintegration (b) Hough Transformation

5   Simulation and Experimentation

A database is maintained for storing the iris images of various individuals whose templates are later used for pattern resemblance. The propound technique took total 10 iris classes with 3 images per class and training was done to create iris templates.

Fig.4(b). shows 3 iris images of an individual from three different views. When an image from database was selected it matched with class 6 generating a minimum hamming distance(HD)=0 and Maximum Correlation =1which scrutinized our result. Thus two performance parameters hamming distance and correlation provides a double check on various biometric templates hence enhancing security solutions for mobile ad-hoc networks. Iris disjuncture results were evaluated during the complete trials period as disjuncture results are very important and critical for the completion and success of the iris perception system. The iris image which is encoded and later on crypt-biometric embedded undergoes testing the finest wavelet concomitant for analysis . Several testing methods were considered for evaluating the accomplishment of the propound system. The minimal and maximal Hamming distance and correlation values of various templates accessed in the database were examined.

Since we are dealing with biometric templates sensitivity and specificity classification tests as depicted in Fig.5(a) and 5(b) respectively are performed to analyze the proportion of actual positives; referred as true positive rate(sensitivity) or recall rate. In general, Positive = identified and negative = rejected. Therefore: True positive(TP) = correctly identified, False positive(FP) = incorrectly identified, True negative(TN) = correctly rejected, False negative(FN) = incorrectly rejected,True Positive Rate(TPR) = TP/P, True Negative Rate(TNR) = TN/N, False Positive Rate(FPR) = FP/N, False Negative Rate(FNR) = FN/P, Accuracy = (TP + TN)/(P + N), Precision = (TP)/(TP + FP), Recall = (TP)/(TP + FN)F-measure = 2*Precision*Recall/(Precision + Recall).

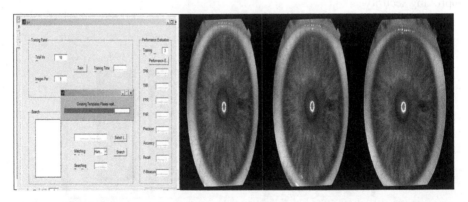

Fig. 4. (a) Graphical user interface for iris perception application (b) Hough Transformation

The example discussed above of iris image matching class 6 having hamming distance =0 and maximum correlation=1 when evaluated using the novel crypt-biometric scheme provided excellent results yielding values:
TPR=1,TNR=0.98765,FPR=0.012346,FNR=0,Precision=0.9444,Accuracy=0.98889, Recall=1,F-Measure=0.96296.

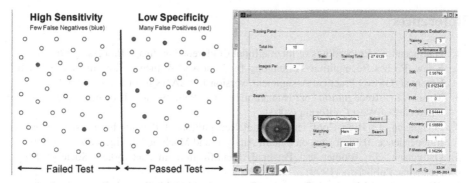

Fig. 5. (a) Sensitivity and Specificity Analysis (b) GUI showing performance evaluation by various parameters

Similarly the generated database with number of iris images was undergone evaluation through our developed technique which led to quite good results through all the above mentioned parameters hence leading to an enhanced security solution for mobile ad-hoc networks.

6   Conclusion and Future Scope

Conclusion specified on the various results shown above specifies notable design and application of a novel crypt-biometric perception algorithm to protract security in mobile ad-hoc networks. Hamming distance and maximum correlation values for pattern resemblance of hoarded iris images results in successful values that enhances the flexibility and usability of the proffered technique. The outcomes of various performance factors such as true positive rate(TPR),true negative rate(TNR),false positive rate (FPR) false negative rate (FNR) ,accuracy , precision recall and f-measure analyzing the performance metrics of biometric also provide good results. Embedding elliptic curve cryptography with biometrics provided two levels of security of the propound system serving as a security protocol or security tool for mobile ad-hoc network. Various simulation results analyzing performance of mobile ad-hoc network comprising the novel crypt-biometric scheme having a specified network length, width, node transmission range ,node speed, node data traffic , speed variation and angle variation resulted in excellent values of average packet drop rate, average packet delivery ratio, average end to end delay and average hop count when compared with other secured approaches for MANET. Hence comparing all the parameters values whether for biometric evaluation or simulating MANET performance ,the developed technique proves to be a novel security solution that helps mobile ad-hoc networks to overcome various security breaches that in turn results in better performance of the system.

References

1. Daugmann, J.: Proceedings of International Conference on Image Processing (2002)
2. Khaw, P.: Iris Recognition Technology for Improved Authentication. In: SANS Security Essential (GSEC) Practical Assignment, version 1.3, SANS Institute, 5–8 (2002)
3. Masek, L.: Recognition of Human Eye Iris Patterns for Biometric Identification. University of West California (2003)
4. Daugmann, J.: High Confidence Visual Recognition of Persons by a test of Statistical Independence. IEEE Trans. on Pattern Analysis and Machine Intelligence 15(11) (1993)
5. Daugmann, J.: Biometrics: Personal Identification in Networked Society. Kluwer Academic Publishers (1999)
6. Sherin Zafar, M.K.: Soni.:Sustaining Security In MANET through Biometric Technique. Engineering Sciences International Research Journal, 41–44 (2014)
7. Daugmann, J.: How Iris Recognition Works. IEEE Transactions on Circuits and Systems for Video Technology (CSVT) 14(1), 21–30 (2004)
8. Zafar, S., Soni, M.K.: Trust Based QOS Protocol (TBQP) using Meta-heuristic Genetic Algorithm for Optimizing and Securing MANET. In: IEEE International Conference on Reliability, Optimization & Information Technology (ICROIT), pp. 173–177 (2015)
9. Wildes, R.P.: Iris Recognition: An emerging biometric technology. Proceedings of the IEEE 85(9), 1348–1363 (1997)
10. Wildes, R.P., Asmuth, J.C., Hsu, C., Kolczynski, R.J., Matey, J.R., McBride, S.E.: Automated,non invasive iris recognition system and method. U.S Patent 5, 572–596 (1996)
11. Misiti, M., Misiti, Y., Oppenheim, G., Poggi, J.: Wavelet Toolbox 4 User's Guide (1997-2009)
12. Xiao, Q.: A Biometric Authentication Approach for High Security Ad hoc Networks. In: Proceedings of IEEE Workshop on Information Assistance, pp. 250–256 (2004)
13. Liu, J., Yu, F.R., Lung, C.-H., Tang, H.: Optimal Biometric-Based Continuous Authentication in Mobile Ad hoc Networks. In: Third IEEE International Conference on Wireless and Mobile Computing, Networking and Communications, pp. 76–81 (2007)

Remote Login Password Authentication Scheme Using Tangent Theorem on Circle

Shipra Kumari and Hari Om

Department of Computer Science and Engineering
Indian School of Mines, Dhanbad, India
shiprakumari18jan@gmail.com, hariom4india@gmail.com

Abstract. - A remote password authentication scheme based on a circle is proposed in this paper. In this scheme, we use some simple tangent theorem like secant tangent theorem to authenticate the user and the server. The security of this scheme depends on the tangent points located in a plane associated with the circle and tangent line. In our scheme, a legal user can freely choose and change his password using his smart card.

Keywords: Authentication, Tangent Theorem, Circle, smart card.

1 Introduction

Now a days, the popularity of computer technologies and computer networks need sharing of resources on a multi-user system. In this network environment, the communication among persons depends on the remote computers. Sharing or communication through remote computers however causes some undesired phenomena such as unauthorized access or unauthorized service. Therefore, the selection of a suitable security scheme to prevent the information to be altered or forged becomes very important. Among the existing security schemes, the password authentication is widely adopted authentication in login procedure because of its easy implementation and user friendliness. Lamport [1] has discussed the first remote password authentication scheme in which the server stores a password table to check the user validity. If a user wishes to enter the system, he needs to login by entering his account ID and password. The system validates the submitted account ID and password according to the information available in the verification table. Other similar schemes have been discussed in [2]-[4]. The very problem with [1] is that the verification table has to be kept inside the system in order to maintain a management system. These schemes however suffer from the problem of server compromise attack or verification table modification attack. Several new methods have been developed that do not require a password table. These methods perform smart card-based authentication. To date, several researchers have developed more secure authentication schemes using different techniques to communicate securely without using any password table. Many use smart card to identify a fake user and to decrease the server overhead, assuming the smart card is maintained securely [5] - [8] . In 2004,Das et al. [9]

© Springer International Publishing Switzerland 2015   721
S.C. Satapathy et al. (eds.), *Proc. of the 3rd Int. Conf. on Front. of Intell. Comput. (FICTA) 2014*
– *Vol. 2*, Advances in Intelligent Systems and Computing 328, DOI: 10.1007/978-3-319-12012-6_80

presented a dynamic ID-based remote user authentication scheme using smart cards. They pointed out that their scheme does not maintain any verifier table and can resist the replay attack, forgery attacks, guessing attacks, and insider attacks. However, in 2009,Wang et al. [10] pointed out that Das et al.s scheme does not achieve mutual authentication and could not resist impersonation attack. In 2012, Wen and Li analyzed Wang et al.s scheme and pointed out that their scheme is vulnerable to impersonation attack; only through intercepting and modifying the messages transmitted in the public networks, the adversary could impersonate the legal user to login the server. Moreover, an insider user who has registered in the remote server can reveal some secret information of the server and the other user and proposed an improved scheme, which can resist impersonation attack, avoiding partial information leakage and providing anonymity for the users[11]. However, recently in 2013, Juan et als pointed that Wen and Lis scheme cannot withstand insider attack and forward secrecy, and, though eavesdropping the users login request message in the public networks, the user can be traced out, and also proposed secure dynamic-ID remote user authentication scheme using ECC[12]. Some schemes are also developed using geometric like, Liaw discusses a password authentication scheme using triangle and straight lines [13]. Liaw and Lei discuss an authentication scheme based on unit circle in which the radius of the circle is one unit, i.e.1 and its center is origin i.e. (0,0) in 2-Dim. [14]. Wang also discusses scheme using geometric triangle by calculating five center points of the triangle [15]. Wang discusses another scheme using an N-dimensional construction based on the circle instead of on the simple Euclidean space in past schemes. The author claimed that this scheme would be more secure than previously proposed schemes[16]. However, Shuhong et al.s show that the scheme of Wang is vulnerable to replaying and off-line password guessing attacks[17]. These schemes also have some limitations, e.g., a user cannot change his password and mutual authentication cannot be performed. Kumar discusses an authentication scheme using unit sphere in which the mutual authentication between server and users can be done [18].

We now discuss our proposed method which fulfill all the security requirements and easy to implement.

2   Proposed Scheme

Our scheme has five phases initialization, registration, login, authentication, password change phase.

2.1   Initialization

Let RS is remote server, that performs the following actions:
Select a Galois field GF(p) for large prime p and assuming g as a its generator.
Choose secret key d and compute
$e = g^d \bmod p$
Server RS stores its parameters: (e,d)

2.2   Registration

Assume that the new user U_i wants to register:

(i) The new user chooses and delivers his own identity ID_i and password $f(PW_i)$ to RS securely.

(ii) RS calculates the following pairs of points on the xy-plane: C=(C_{x_i}, C_{y_i}) and D = (D_{x_i}, D_{y_i}) where
$C_{x_i} = (ID_i)^e$ mod p, $C_{y_i} = (ID_i)^d$ mod p,
$D_{x_i} = e^{ID_i}$ mod p , $D_{y_i} = e^{f(PW_i)}$ mod p

(iii) RS constructs a line Li passing through the points C_i and D_i.

(iv) RS randomly chooses a point $M_i = (M_{x_i}, M_{y_i})$ on line L_i

(v) Compute $HPW_i = f(f(PW_i)\|C_{y_i})$ mod p
(vi) CM stores $\{ID_i, M_i, e, g, p\}$ message in the smart card and delivers it to user U_i .

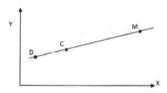

Fig. 1. Registration Phase

2.3   Login

When the user U_i wants to login a set of server RS , he keys his identity ID_i and password PW_i. Then, the smart card performs the following:

(i) Obtain a current login time T from the system.
(ii) Compute the point D_i=(D_{x_i}, D_{y_i}) as
$D_{x_i} = e^{ID_i}$ mod p , $D_{y_i} = e^{f(PW_i)}$ mod p
(iii) Redraw a line L_c passing the points D and M.
(iv) Compute slope (m_i) of line L_c as $m_i = (M_{y_i}$ $D_{y_i})/(M_{x_i}$ $D_{x_i})$
(v) Compute $C_{x_i} = (ID_i)^e$ mod p
(vi) Compute C_{y_i} by substituting C_{x_i} in the equation of line L_c
(vii) Compute $HPW_i^* = f(f(PW_i\|C_{y_i})$ mod p
(viii) Compare HPW_i^* with HPW_i stored in smart card if equal then proceed otherwise terminate the session.
(ix) Compute $V_1 = f(m_i\|C_{y_i}\|D_{y_i}\|T)$
(x) Choose random number r_c
(xi) Compute $V_2 = V_1 \oplus r_c$

(xii) Compute $R_i = e^{V_1} \bmod p$

(xiii) Draw a circle in xy-plane using the point $C_i = (C_{x_i}, C_{y_i})$ as center and R_i as radius.

(xiv) Through the point M draw a tangent on a circle which touches the circle at point N

(xv) Draw a line from the center C to tangent point N. By tangent theorem, the line CN is perpendicular to the tangent line. Thus, $<MNC = 90^0$ (refer Fig.2).

(xvi) By Pythagoras theorem $MN_i^2 = (MC_i^2 - CN_i^2) \bmod p$ where, $CN_i^2 = R_i$

(xvii) Compute $V_3 = f(MN_i^2 \oplus HPW_i^*)$

(xviii) Send the authentication message: $\{ID_i, V_2, V_3, T, HPW_i\}$.

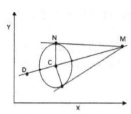

Fig. 2. Login Phase

2.4   Authentication

Upon receiving a login request at time T^*, the server RS performs the following:

(i) Check whether the format of ID is correct. If not, login request is rejected.

(ii) Check whether the transmission time $(T^* - T)$ is within the legal tolerant interval $\triangle T$. If $(T^* - T) > \triangle T$, then the login request is rejected.

(iii) Compute:
$C_{x_i} = (ID_i)^e \bmod p, C_{y_i} = (ID_i)^d \bmod p,$

(iv) Using the points C and M, draw a line L_s.

(v) Compute slope n_i of line L_s
$n_i = (M_{y_i} - C_{y_i})/(M_{x_i} - C_{x_i})$

(vi) Compute $D_{x_i} = e^{ID_i} \bmod p$

(vii) compute D_{y_i} by substituting D_{x_i} in the equation of line L_s

(viii) Compute $V_1^* = f(m_i \| C_{y_i} \| D_{y_i} \| T)$

(ix) Compute $R_i^* = e^{V_1^*} \bmod p$

(x) Draw a circle in xy-plane using the point C_i as center and R_i^* as radius

(xi) Equation of the circle is
$(x - C_{x_i})^2 + (y - C_{y_i})^2 = (R_i^*)^2$ ———(1)

(xii) Equation of the line L_s is
$y - C_{y_i} = n_i(x - C_{x_i})$ ———(2)

(xiii) Let line L_s cut the circle at two points.

(xiv) From (1) and (2), we get two points: A and B

(xv) Draw a tangent on the circle from the point M

(xvi) If N is a tangent point on the circle, then by Secant Tangent theorem $MN_i^2 = MA * MB$, where AB is secant of circle that passes through the center C. Thus, AB is the diameter of circle (refer Fig.3). Now,

$MN_i^2 = MA_i * MB_i = MA_i * (MA_i + MB_i)$
$= (MC_i - CA_i) * ((MC_i - CA_i) + 2 * CA_i)$
$= (MC_i - CA_i) * (MC_i + CA_i)$
$= MC_i^2 - CA_i^2 = MC_i^2 - CN_i^2$
$where, CN_i = CA_i = R_i$

(xvii) Compute $V_3^* = f(MN_i^2 \oplus HPW_i^*)$

(xviii) If $V_3^* = V_3$, then user is authenticated; otherwise the login request is rejected.

(xix) Choose random number r_s

(xx) Compute $V_4 = V_1^* \oplus r_s$

(xxi) Compute $SK = f(ID_i \| C_{y_i} \| D_{y_i} \| r_s \| (V_1^* \oplus V_2))$

(xxii) Compute $V_5 = f(V_1^* \| SK \| T_1)$

(xxiii) Send message $\{ID_i, V_4, V_5, T_1\}$ to the user.

(xxiv) Smart card computes:
$SK = f(ID_i \| C_{y_i} \| D_{y_i} \| (V_4 \oplus V_1) \| r_c)$
$V_5^* = f(V_1 \| SK \| T_1)$ If $V_5^* = V_5$, then Server is authenticated.

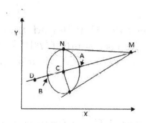

Fig. 3. Authentication Phase

2.5 Password Change Phase

To change the password, the user U_i first enters his smart card and then his Identity ID_i and password PW_i.

(i) Compute the point $D_i = (D_{x_i}, D_{y_i})$ as
$D_{x_i} = e^{ID_i}$ mod p and $D_{y_i} = e^{f(PW_i)}$ mod p

(ii) Redraw a line L_c passing the points D_i and M_i.

(iii) Compute $C_{x_i} = (ID_i)^e$ mod p

(iv) Compute C_{y_i} by substituting C_{x_i} in the equation of line L_c

(v) Compute $HPW_i^* = f(f(PW_i \| C_{y_i})$ mod p

(vi) Compare HPW_i^* with HPW_i stored in smart card if equal then proceed otherwise terminate the session.

(vii) Enter new password PW_i^{new}

(viii) Compute $HPW_i^{new} = f(f(PW_i^{new} \| C_{y_i})$ mod p

(ix) Compute $D_{y_i}^{new}$ as
$$D_{y_i}^{new} = (e)^{f(PW_i^{new})} \bmod p$$

(x) Redraw a line L_c^{new} passing the points D_i^{new} and C.

(xi) Take an arbitrary point $(M)_i^{new}$ on the line L_c^{new}.

(xii) Replace HPW_i by HPW_i^{new}.

(xiii) Replace M_i by M_i^{new}.

(xiv) Password is changed successfully.

2.6   Security Analysis of the Scheme

Replay Attack. The replay attack cannot work on our proposed scheme because of the renewal of V_i and V_2 at different timestamps T for every login session, there is no possibility of replay attacks.

Stolen Verifier. Since the server neither saves any verification table nor stores any entry in its database, no question arises for an attacker to make a way inside the scheme.

Our Scheme Could Withstand Privileged Insider Attack. In the registration phase of our scheme, the user sends $f(PW_i)$ to the Server. So, privileged insider cannot get the users password PW as it is protected by a secure hash function.

Our Scheme Could Withstand User Impersonation Attack. The adversary cannot generate $C_y = ID_i^d \bmod p$ as he does not know the servers secret key d.

Secure in Password Guessing Attack. If the adversary extracts the secured data in smart card though physically monitoring its power consumption. He can also get the authentication message. Let the adversary guess the password PW. He cannot verify the correctness of PW without servers secret key d.

Prompt Detection of the Wrong Password. If the user inputs a wrong password, it will waste unnecessary computation and communication capacity. In the proposed scheme, when a the user inputs IDi, PWi, in login and password phase, the smart card computes $HPW_i = f(f(PW_i)\|C_{y_i})$ and compares with stored HPW in the smart card. If it is wrong that means user entered a wrong password and smart card terminates the session else the smart card perform remaining steps. Hence, entry of wrong password is detected at the beginning of the login phase by the smart card.

Friendly and Efficient Password Change. In the proposed scheme to change the password server is not involved to check the validity of user. However, the smart card first checks the validity of the original password PW_i by and comparing HPW_i^* with stored HPW_i (in smart card). If the password is valid, the user can input a new password and the smart card computes M_i^{new} and HPW_i^{new} and replace old value by new one, to complete the password change. Therefore, the password change phase is efficient and there is also no chance of denial of service because server is not performing in this phase.

Known Session Key Security. A unique secret session key performed in each run of an authentication protocol. If any session key is compromised, it should no impact on other session keys. In the proposed scheme, knowing a session key $SK = f(ID_i \| C_{y_i} \| D_{y_i} \| r_s \| r_c)$ and the random numbers r_c and r_s, it is of no use to compute the other session keys , since it is impossible to compute the session key without knowing correct random values in other session. Therefore, the proposed scheme can provide known session key security.

2.7   Conclusions

In this paper, we have discussed a new remote user authentication scheme based on the tangent theorem of circle. Instead of unit circle, we have used different center of circle on a XY-plane (i.e, center of a circle is not fixed to (0,0) with different size (i.e, radius of circle) for every user in different login session. This scheme provides freedom to choose and change password to a user and achieves mutual authentication between a user and remote server.

References

1. Lamport, L.: Password authentication with in secure communication. Communications of the ACM 24(11), 770–772 (1981)
2. Tan, K., Zhu, H.: Remote password authentication scheme based on cross-product. Computer Communications 22(4), 390–393 (1999)
3. Lemon, R.E., Matyas, S.M., Meyer, C.H.: Cryptographic Authentication of Time-Invariant Quantities. IEEE Trans. on Communications 29(6), 773–777 (1981)
4. Chang, C.C., Wu, T.C.: Remote password authentication with smart cards. IEE Proceedings-E 138(3), 165–168 (1993)
5. Horng, W.B., Lee, C.P., Peng, J.: A secure remote authentication scheme preserving user anonymity with non-tamper resistant smart cards. WSEAS Transactions on Information Science and Applications 7(5), 619–628 (2010)
6. Li, C.T., Lee, C.C., Liu, C.J., Lee, C.W.: Cryptanalysis of Khan et al. dynamic ID based remote user authentication scheme, pp. 1382–1387. IEEE (2010)
7. Awasthi, A.K., Lal, S.: A remote user authentication scheme using smart cards with forward security. IEEE Transactions on Consumer Electronics 49(4), 1246–1248 (2003)
8. Xu, J., Zhu, W.T., Feng, D.G.: An improved smart card based password authentication scheme with provable security. Computer Standards and Interfaces 31(4), 723–728 (2009)

9. Das, M.L., Saxena, A., Gulati, V.P.: A dynamic ID-based remote user authentication scheme. IEEE Transactions on Consumer Electronics 50(2), 629–631 (2004)
10. Wang, Y., Liu, J., Xiao, F., Dan, J.: A more efficient and secure dynamic ID-based remote user authentication scheme. Computer Communications 32(4), 583–585 (2009)
11. Wen, F., Li, X.: An improved dynamic ID-based remote user authentication with key agreement scheme. Computers and Electrical Engineering 38(2), 381–387 (2012)
12. Qu, J., Zou, L.: An Improved Dynamic ID-Based Remote User Authentication with Key Agreement Scheme. Journal of Electrical and Computer Engineering 2013(2013), Article ID 786587 (2013)
13. Liaw, H.T.: Password authentication using Triangle and Straight lines 30(4), 63–71 (1995)
14. Liaw, H.T., Lei, C.L.: An efficient password authentication scheme based on a unit circle. Cryptologia 19(2), 198–208 (1995)
15. Wang, S.J.: Remote table based log-in authentication upon geometric triangle 26(2), 85–92 (2004)
16. Wang, S.J.: Yet Another Log-in Authentication Using N-dimensional Construction Based on Circle Property. IEEE Transactions on Consumer Electronics 49(2), 337–341 (2003)
17. Wang, S., Bao, F., Wang, J.: Comments on Yet Another Log-in Authentication Using N-dimensional Construction. IEEE Transactions on Consumer Electronics 50(2) (2004)
18. Kumar, M., Gupta, M., Kumari, S.: A remote login authentication scheme with smart cards based on unit sphere 1(3), 192–198 (2010)

A Survey of Security Protocols in WSN and Overhead Evaluation

Shiju Sathyadevan, Subi Prabhakaranl, and K. Bipin

Amrita Centre for Cyber Security Systems and Networks, Amrita Vishwa Vidyapeetham,
Amritapuri, Kollam, Kerala, 691572, India
{shiju.s,bipink}@am.amrita.edu,
sooryamsh@gmail.com

Abstract. There has been a widespread growth in the area of wireless sensor networks mainly because of the tremendous possibility of using it in a wide spectrum of applications such as home automation, wildlife monitoring, defense applications, medical applications and so on. However, due to the inherent limitations of sensor networks, commonly used security mechanisms are hard to implement in these networks. For this very reason, security becomes a crucial issue and these networks face a wide variety of attacks right from the physical layer to application layer. This paper present a survey that investigates the overhead due to the implementation of some common security mechanisms viz. SPINS, TinySec and MiniSec and also the computational overhead in the implementation of three popular symmetric encryption algorithms namely RC5 AES and Skipjack.

Keywords: Wireless Sensor Networks, Security Mechanisms, Symmetric Encryption, AES, RC5, Skipjack.

1   Introduction

Wireless Sensor Networks are comprised of large number of tiny sensor nodes, commonly known as motes. These devices have the ability to sense their environment, process the collected data and transmit them over air to nearby devices. These capabilities make them suitable for monitoring the real world environment in which they are deployed. The sensor nodes are essentially cheap devices and hence could be deployed in very large numbers that could cover vast regions. However, the nodes in wireless sensor networks severely lack essential resources such as memory, processing power, energy and hence security mechanisms employed in traditional networks are not suitable for WSN. As a result of which these networks face a variety of attacks such as physical tampering, node fabrication attacks, eavesdropping, hello flood attacks, eavesdropping and so on. This has motivated researchers to come up with various security mechanisms that are suitable for these resource constrained networks. Number of secure communication protocols [8,9,10], secure routing protocols [1,2,3,4] data aggregation protocols [5,6,7] have been proposed over the years by the research community. However, these security mechanisms impose

© Springer International Publishing Switzerland 2015   729
S.C. Satapathy et al. (eds.), *Proc. of the 3rd Int. Conf. on Front. of Intell. Comput. (FICTA) 2014*
– *Vol. 2*, Advances in Intelligent Systems and Computing 328, DOI: 10.1007/978-3-319-12012-6_81

additional overhead on the already resource constrained network. Hence the security mechanisms and cryptographic primitives used to implement these mechanisms have to be chosen with extreme caution. The remainder of this paper is organized as follows. Section2 discusses the resource constraints faced by WSNS. Section 3 discusses major security goals of WSN. This paper studies three very popular security protocols: SPINS [8], TinySec [9] and MiniSec [10]. Important features of these protocols are described in section 4. The result of these studies in terms of overhead is discussed in section 5. Comparison of computational overhead of three popular symmetric algorithms namely RC5, AES and Skipjack is presented in section 6. Finally section 7 concludes the paper.

2   Challenges in WSN

Wireless Sensor Networks faces a number of challenges most of which is unique to them. Stringent resources are one of the primary challenges faced by these networks. Sensor nodes that form the WSN are physically very small in size which is the main reason for the resource constraints and hence a major limitation for these networks. Additionally they also face a host of other challenges discussed in this section.

2.1   Energy Constraints

Energy is one of the most expensive resource as far as wireless sensor networks are concerned and hence the biggest constraint. Most of these devices are battery powered and are deployed in areas such as deep forests, oceans etc where recharging is not practically possible. Some of these devices are also solar powered but still batteries are the main energy sources and hence should be conserved. Energy is spent both for computation and data transmission. Transmitting a single bit of data requires as much as energy in executing 800-1000 instructions[7] and therefore data transmission consumes the largest chunk of available energy.

2.2   Memory Constraints

Sensor devices typically consist of very small amount of storage space. For instance, a commonly used mote TelosB has only 10K RAM, 48K program memory and 1024K flash storage[8] . Almost half of this available memory is consumed by the resident operating system. Memory is also required for storing applications, data sensed by the devices and for storing intermediate results of processing. Hence, not much memory is left for implementing security primitives and therefore heavy weight cryptographic algorithms are not suitable for these platforms.

2.3   Unreliable Communication

Communication in WSN takes place as a result of the nodes transmitting packets to the nodes which are within their communication range, packets hopping from one node to another towards the gateway. This communication follows a connectionless protocol and hence there is always the threat of lost or dropped packets and congested networks. Also the broadcast nature of the communication adds to the unreliability in communication.

2.4 Limited Post Deployment Knowledge

The sensor nodes are deployed in an adhoc manner usually by aerial scattering and hence not much information is available about the topology or structure of the network. This is especially true in the case of networks that are exceptionally large and deployed in large fields. Lack of any a-priori knowledge about the deployment poses a number of challenges.

2.5 Remote Management

Sensor nodes are being largely employed in areas where the operations of these devices cannot be attended physically.[11,12] They remain unattended and have to be handled remotely which make the task of providing security quite challenging. Also it makes these networks highly vulnerable to physical attacks.

2.6 Extensive Scale

In applications such as forest fire monitoring, highway traffic monitoring and management, ocean monitoring, the number of nodes deployed is in the range of thousands or even millions. The sheer size of these networks makes it difficult to manage them efficiently and ensuring security becomes a real challenge.

3   Security Goals in WSN

Even though wireless sensor networks face challenges that are most unique to these kinds of resource starved networks, yet the primary security goals or requirement of a WSN is no different from that of a traditional network. The three primary goals are discussed below.

3.1 Confidentiality

Confidentiality requirement implies that the data that is being sensed and transmitted by these networks should only reach and be understood by the intended recipient and nobody else. No third party should be able to access the information unless they are authorised to do so. This is one of the most critical of the security requirement as these networks are typically employed in applications that deal with highly sensitive data as in medical and traffic monitoring, emergency response management, defence management and battlefield monitoring.

3.2 Integrity

There are certain applications where the information being transmitted may not be very sensitive in nature and therefore confidentiality is not a major concern rather what is more important is to determine and verify that the information has not been modified or altered while in transit. The data as seen by the recipient should be the same as sent by the source. In traditional networks, mechanisms such as Message Authentication Codes (MAC) and hashes are employed to ensure this security requirement.

3.3 Availability

It is necessary to ensure that the services of WSN are available uninterrupted even in wake of attempted attacks to bring down the network. Attackers may try performing denial of service attacks or attacks on the base station so as to make the network unavailable to the users. A robust network should be able to handle such attacks and still provide services to users.

3.4 Authenticity

It is important to verify the authenticity of the source nodes from where the information is believed to have originated. It is possible for an attacker to fabricate false packets and spoof the source address so as make it appear like coming from a legitimate node in the network. To be able to verify the source of information correctly is one of the crucial requirements in any network, be it a traditional network or a WSN.

3.5 Data Freshness

In wireless sensor networks, crucial decisions are made based on the data collected and communicated by the sensor nodes. Decisions are usually based upon the aggregate of all data collected by the participating nodes and hence it is necessary that the data is not stale. Synchronised counters at both ends or nonce are generally used to ensure data freshness.

4   Security Protocols

This section discusses the important features of three popular WSN security protocols namely SPINS, MiniSec and TinySec [8, 9, 10]. All of these security protocols are built over the TinyOS operating system.

4.1 SPINS

It consists of two secure building blocks, one which handles data confidentiality, two party data authentication, data freshness and a second one that ensures authenticated broadcast called SNEP and µTesla respectively [8]. This security protocol for the resource constrained wireless networks uses the same block cipher for implementing its various cryptographic primitives so that the code could be reused and thereby saving precious storage space.

SNEP provides confidentiality by encrypting the data to be transmitted and uses an optimized version of RC5 from OpenSSL to do so [8]. Additionally to ensure semantic security this protocol uses the concept of a counter at both sides that ensure that the same plain text is encrypted to a different cipher text each time. But instead of transmitting these counters, SNEP requires that the sender and receiver maintain the counter at both sides. It does so to save the energy required for transmitting the counters. However, in situations where the counters at both sides fail to remain synchronized in the event of packet losses, an expensive counter resynchronization protocol is required.

µTESLA on the other hand take care of authenticated broadcast. In traditional networks, authenticated broadcast is implemented by means of asymmetric mechanism. However, they are impractical to be used in the severely resource constrained wireless networks. µTESLA manages to achieve the same affect by using a symmetric algorithm. In order to achieve asymmetry, this protocol requires that the keys needed for decryption be provided by the broadcasting device to the nodes after a certain period of predetermined delay [8]. This requires that the base station, which is usually the broadcasting party and the nodes be time synchronized although loosely.

4.2 TinySec

It is the first fully implemented security architecture for wireless sensor networks [9]. TinySec has two modes of operation, one which encrypts the data and additionally authenticates the packet by computing the MAC over the encrypted data and header called TinySec-AE and the second one called TinySec-Auth which only authenticates the packet by computing the MAC and no data encryption takes place.

TinySec uses the cipher block chaining (CBC) mode of operation with SkipJack as its underlying block cipher and an 8 byte Initialization Vector (IV) to introduce randomization and hence ensure semantic security. However, unlike SPINS which does not send the counter with the packet, TinySec-AE transmits the 8 Byte IV along with the packet. But it does so in such a way that it incurs an overhead of 5 bytes for this mode and for TinySec-Auth it incurs an overhead of just 1 byte.

4.3 MiniSec

Similar to the above two protocols, MiniSec also has two modes of operation. One of which secures point to point communication or unicast mode of communication and the second one designed for multicast mode of operation called MiniSec-U and MiniSec-B respectively [10]. The implementation of MiniSec on Telos platform is publicly available.

MiniSec-U makes use of Offset CodeBook (OCB) which is a block cipher mode of operation and uses Skipjack as the underlying block cipher [10]. OCB mode of operation has the added attraction of performing authenticated encryption in a single pass of plain-text. MiniSec-U also uses an incrementing counter as IV to ensure semantic security but it uses an approach that lies between SPINS and TinySec. SPINS that does not transmit the IV at all and TinySec which transmits the entire IV, MiniSec adopts a novel approach of sending few bits of IV and also uses an implicit counter resynchronization protocol that ensures that up to 2^x-1 number of packets lost, no expensive resynchronization protocol is required where x being the number of bits of IV being transmitted.

MiniSec-B uses Bloom Filters and loose time synchronization to achieve authenticated broadcast [10]. The Bloom filter is a space efficient data structure and is well suited for sensor nodes [13]. It also uses a sliding window approach to protect against replay attacks.

5   Comparison of Packet Overhead

In this section we compare the increase in size of packets and the resultant energy overhead of the three protocols discussed below. The comparisons are done with respect to a TinyOS packet with no security implementations. The following figures shows the packet formats of some of the protocols discussed.

2	1	1	1	24	2
Dest	AM	Len	Grp	Data	CRC

a) TinyOS Packet Format

2	1	1	1	24	2
Dest	AM	Len	Grp	Data	CRC

b) TinySec-AE Packet Format

2	1	1	24	4
Dest	AM	Len	Data	MAC

c) TinySec-AUTH Packet Format

2	1	1	2	24	4
Dest	AM	Len (3 bit IV)	Src	Data	MAC

d) MiniSec-U Packet Format

2	1	1	2	24	4
Dest (4 bit ctr)	AM	Len (3 bit ctr)	Src	Data	MAC

e) MiniSec-B Packet Format

Fig. 1. Packet format for the three security protocols and TinyOS which is taken as the reference. The numbers represent the size of the fields in Bytes.

As mentioned in the beginning, every single bit transmitted consumes additional energy and below is a table depicting the overhead in energy consumed due to the additional bytes transmitted for security implementation and the overall percentage overhead in transmission energy for each. The overhead is calculated with respect to the standard TinyOS network stack [10].

Table 1. Comparison of Packet and Transmission Overhead

Protocol	Payload	Packet Overhead	Security Overhead	Total	Energy (mAs)	% Increase
TinyOS	24	7	0	31	0.034	-
SNEP	24	15	8	39	0.0415	22.2
TinySev -AE	24	12	5	36	0.0387	13.9
MiniSe	24	10	3	34	0.368	8.3

According to this table, MiniSec manages to consume the lowest energy in transmitting the packets by keeping the packet overhead to an optimally minimum level.

6   Comparison of Computational Overhead of RC5, AES128, Skipjack

Asymmetric encryption algorithms are highly unsuitable to be used in sensor networks because they are usually designed for powerful processors and requires extensive amount of computation and memory for storing keys and intermediate results. The memory of a typical sensor node is not even capable to hold the keys of commonly used asymmetric algorithms most of which are 1024 bits or higher. Hence the research communities have resorted to symmetric algorithms such as RC5 [16], AES [15,17] and Skipjack [18] for providing security to these networks. SPINS uses an optimized version of RC5 while both TinySec and MiniSec uses Skipjack as its underlying block cipher. Hence in this section, we discuss the computational overhead of these algorithms. Also since Skipjack uses only a 80 bit key and the world is moving towards higher bit keys to ensure security in the long run, this section also analyses AES128 which could be a suitable substitute to Skipjack.

RC5 allows a variable length block size (32, 64, 128 bits) and a variable length key size (up to 2040 bits) and the number of rounds can go up to 255 [14]. AES on the other hand uses a fixed block size of 128 bits and the keys could be either 128, 192 0r 256 bits and accordingly the number of rounds is 10, 12 and 14 respectively. Unlike these two algorithms, Skipjack uses a fixed length block of size 64 bit and an 80 bit key and two rounds named Round A and Round B each of which is executed 16 times in a specific order making a total of 32 cycles. The following table summarizes these parameters as used in the security mechanisms discussed in section 4.

Table 2. Block Cipher Parameters

Algorithm	Block Size(bits)	KeySize(bits)	#Rounds
RC5	64	128	18
SkipJack	64	80	32
AES	128	128	10

The number of CPU cycles per byte for these three block ciphers implemented on ATmega128 processor is shown in the following table.

Table 3. CPU Cycles for Encryption

Algorithm	Block Size(bits)	Cycles/Byte	Cycles/Block
RC5	64	712	5696 [19]
SkipJack	64	186	1488 [20]
AES	128	204	3264 [20]

The energy (E) required by symmetric block ciphers for encryption of N bits of plain text is given by

$$E= (P \times C/f) \times N/u .\tag{1}$$

Where P and f are the power and frequency of the CPU and C is the number of block cycles needed to perform encryption of a block of size u [22]. Table V gives the computational energy cost of encryption operation by the three block ciphers [17].

Table 4. Computational Energy Requirement

Cipher	Energy for Encryption of 128 bit block
RC5	42.5 μJ
SkipJack	31.8 μJ
AES	36.5 μJ

The table shows that the energy efficiency of Skipjack is above the other two ciphers, but from a security point of view AES is considered stronger that Skipjack which has been proved weak against cryptanalysis [21,22].Hence AES seems to be a good choice from a security and energy efficiency point of view.

7   Conclusion

In this paper, we have conducted an extensive research of various types of attack that can be launched against a wireless sensor network. The paper also investigates three very popular security protocols and discusses the overhead as a result of their implementation. A study of three different block ciphers and their computational overhead is also conducted. We conclude from the study that of the three security protocols studied MiniSec seems to fare better than its counterparts and AES is a good choice as the underlying block cipher.

References

1. Deng, Han, R., Mishra, S.: INSENS: Intrusion-tolerant routing in wireless sensor networks., Technical Report CU-CS-939-02, Department of Computer Science, University of Colorado at Boulder (November 2002)
2. Karp, B., Kung, H.T.: GPSR: Greedy perimeter stateless routing for wireless networks. In: Proceedings of the 6th Annual International Conference on Mobile Computing and Networking, pp. 243–254. ACM Press (2000)
3. Papadimitratos, P., Haas, Z.J.: Secure routing for mobile ad hoc networks. In: Proceedings of the SCS Communication Networks and Distributed System Modeling and Simulation Conference, CNDS 2002 (2002)
4. Tanachaiwiwat, S., Dave, P., Bhindwale, R., Helmy, A.: Routing on trust and isolating compromised sensors in location-aware sensor networks. In: Poster paper Proceedings of the 1st International Conference on Embedded Networked Sensor Systems, pp. 324–325. ACM Press (2003)
5. Hu, L., Evans, D.: Secure aggregation for wireless networks. In: Proceedings of the Symposium on Applications and the Internet Workshops, p. 384. IEEE Computer Society (2003)
6. Shrivastava, N., Buragohain, C., Agrawal, D., Suri, S.: Medians and beyond: New aggregation techniques for sensor networks. In: Proceedings of the 2nd International Conference on Embedded Networked Sensor Systems, pp. 239–249. ACM Press (2004)
7. Przydatek, B., Song, D., Perrig, A.: SIA: secure information aggregation in sensor networks. In: Proceedings of the 1st International Conference on Embedded Networked Systems (SenSys 2003), pp. 255–265. ACM Press, New York (2003)
8. Perrig, A., Szewczyk, R., Wen, V., Culler, D.E., Tygar, J.D.: SPINS: Security protocols for sensor networks. Wireless Networks 8(5), 521–534 (2002)
9. Karlof, C., Sastry, N., Wagner, D.: TinySec: a link layer security architecture for wireless sensor networks. In: 2nd ACM Conference on Embedded Networked Sensor Systems (SensSys 2004), Baltimore, MD, pp. 162–175 (November 2004)
10. Luk, M., Mezzour, G., Perrig, A., Gligor, V.: MiniSec: A Secure Sensor Network Communication Architecture. In: IPSN 2007, April 25-27, ACM, Cambridge (2007)
11. Pathan, A.S.K., Lee, H.-W., Hong, C.S.: Security in wireless sensor networks: issues and challenges. Advanced Communication Technology, ICACT (2006)
12. Akyildiz, I.F., Su, W., Sankarasubramaniam, Y., Cayirci, E.: A survey on sensor networks. IEEE Communications Magazine 40(8), 102–114 (2002)
13. Bloom, B.: Space/time trade-offs in hash coding with allowable errors. Communicationsof the ACM (July 1970)
14. Rivest, R.: The RC5 encryption algorithm. In: Preneel, B. (ed.) FSE 1994. LNCS, vol. 1008, pp. 86–96. Springer, Heidelberg (1995)
15. National Institute of Standards and Technology (NIST), Advanced Encryption Standard (AES), Federal Information Processing Standard (FIPS) 197 (November 2001)

16. "Skipjack and KEA algorithm specifications" National Institute of Standards and Technology (May 1998)
17. Law, Y.W., Doumen, J., Hartel, P.: Survey and Benchmark of Block Ciphers for Wireless Sensor Networks. ACM Transactions on Sensor Networks 2(1) (February 2006)
18. Zhang, X., Heys, H.M., Li, C.: Energy Efficiency of Symmetric Key Cryptographic Algorithms in Wireless Sensor Networks
19. Doomun, M.R., Soyjaudah, K.: Analytical Comparison of Cryptographic Techniques for Resource-Constrained Wireless Security. International Journal of Network Security 9(1), 82–94 (2009)
20. Granboulan, L.: Flaws in Differential Cryptanalysis of Skipjack. In: Matsui, M. (ed.) FSE 2001. LNCS, vol. 2355, pp. 328–335. Springer, Heidelberg (2002)
21. Knudsen, L.R., Robshaw, M., Wagner, D.: Truncated differentials and skipjack. In: Wiener, M. (ed.) CRYPTO 1999. LNCS, vol. 1666, p. 165. Springer, Heidelberg (1999)
22. Xing, K., Sundhar, S., Srinivasan, R., Rivera, M., Li, J., Cheng, X.: Attacks and Countermeasures in Sensor Networks: A Survey. Springer (2005)

DFDA: A Distributed Fault Detection Algorithm in Two Tier Wireless Sensor Networks

Kumar Nitesh and Prasanta K. Jana

Department of Computer Science and Engineering,
Indian School of Mines, Dhanbad-826004, India
kumarnitesh.ism@gmail.com, prasantajana@yahoo.com

Abstract. Detection of faulty relay nodes in two tier wireless sensor network (WSN) is an important issue. In this paper, we present a distributed fault detection algorithm for the upper tier of a cluster based WSN. Any faulty relay node is identified by its neighbors on the basis of the neighboring table associated with them. Time redundancy is used to tolerate transient faults and to minimize the false alarms. The algorithm has $O(m)$ message complexity in the worst case for a WSN with m relay nodes. Simulation results are presented and analyzed with various performance metrics, including detection accuracy and false alarm rate.

Keywords: Relay node, Fault Detection, False Alarm Rate, Detection Accuracy.

1 Introduction

Wireless sensor networks (WSNs) are composed of a large number of tiny sensor nodes furnished with limited communication and computational capabilities. WSN has emerged as a popular computing platform for monitoring environments, disaster warning systems, health care, defense, reconnaissance, and surveillance systems [1]. However, the main constraint of the WSNs is the limited power sources of the sensor nodes. Therefore, energy conservation of the WSNs is the most challenging issue for their long run operation. Many researchers have proposed clustering algorithms [2-3] for energy saving of WSNs. In a cluster based WSN, the sensor nodes are grouped into distinct clusters. Each sensor node belongs to one and only one cluster and each cluster is supervised by a special node called cluster head (CH). The sensor nodes senses the local data autonomously in an unattended manner and send it to their CHs. The CHs then aggregate the sensed data and send the aggregated data to a remote base station, called sink through other CHs usually. Thus it forms a two tier network, the sensor nodes within each cluster forms the first tier and CH to CH connectivity forms the second tier.

However, in a two tier WSN, a CH bears some extra work load, that is, receiving sensed data sent by member sensor nodes, data aggregation and data dissemination to the BS. Moreover, in many WSNs, the CHs are usually selected among the normal sensor nodes, which can die quickly as they consume more energy due to such extra work load. In this context, many researchers [4-7] have proposed use of some special nodes called relay nodes which are provisioned with extra energy. These relay nodes

© Springer International Publishing Switzerland 2015
S.C. Satapathy et al. (eds.), *Proc. of the 3rd Int. Conf. on Front. of Intell. Comput. (FICTA) 2014*
– *Vol. 2*, Advances in Intelligent Systems and Computing 328, DOI: 10.1007/978-3-319-12012-6_82

are treated as the cluster heads (CHs) which are responsible for the same functionality of the CHs. Unfortunately, the relay nodes are also battery operated and hence power constrained. Moreover, WSNs are very prone to failure due to several factors such as environmental hazards, energy depletion, malicious attacks and device failure of the nodes. The failure can affect the overall network life time and degrades the overall performance of the network. However, failure of the relay nodes is catastrophic as it can limit the accessibility of the sensor nodes under their supervision and prevents data aggregation and data dissemination. Therefore, in order to keep WSN operational, it is extremely important to consider the fault tolerant aspects of WSNs, especially the failure of the relay nodes.

In this paper, we consider the problem of detection of faulty relay nodes and proposed a distributed algorithm for the same. The algorithm is shown to have constant time complexity and $O(m)$ message complexity, where m is number of relay node employed. We perform rigorous experiments with various scenarios and present experimental results along with the performance evaluation to show the effectiveness of the algorithm.

The organization of the paper is as follows. Section 2 describes some existing methods for fault detection. Section 3 introduces the network model and the fault model. In Section 4, we present the proposed fault detection algorithm. The experimental results and performance analysis are described in section 5 followed conclusion in Section 6.

2   Related Work

Fault tolerance in WSNs is an important topic of research that has been investigated by many researchers [4–13]. A cross validation based technique for online fault detection of sensor nodes have been proposed by Koushanfar et al. [4], where statistical techniques are used to detect the sensor nodes that have the highest probability of fault. The work Proposed in [6] uses a system level comparison diagnosis to locate faulty sensor nodes. An external manager has been presented in [5] as a fault detection scheme for an event-driven wireless sensor network. Irrespective to the theory, the communication overhead due to of the external manager may be problematic. Jaikaeo et al. [8] have addressed the response implosion problem and presented a technique to overcome it. Luo et al. [13] have proposed a fault-tolerant energy-efficient event detection paradigm for wireless sensor networks. For a given detection error bound, minimum neighbors are selected to minimize the communication volume. Ding et al. [10] have proposed a localized fault detection algorithm, where each sensor node compares its own sensed data with the median of neighbor's data to determine its own status. The performance of the localized diagnosis, however, is limited due to the non-uniform nature of node degrees in sensor networks with random deployment. In [11], Chen et al. have proposed a distributed fault detection scheme for sensor networks. They used local comparisons with an amended majority voting. However, this scheme does not cover transient faults in sensor reading and faults in inter-node communications. Considering the problem of transient fault with the nodes, the author in [14] has presented a distributed fault detection scheme for sensor networks which uses local comparisons, where each sensor node makes a decision based on comparisons between its own sensed data and neighbor's data. The scheme is very complex as it increases the computational overhead of each sensor node as well as the communication overhead of the network. Transient faults in sensing and communication have also been investigated in [12] and the authors have proposed a simple distributed algorithm to tolerate transient faults in the fault detection process. All the above proposed work has taken care of fault detection with the sensor node in a single tier only, which cannot be implemented in a two tier architecture. In the proposed algorithm, we consider detection of the faulty relay nodes which is very crucial.

3   Work Model

3.1   Network Model

We consider a WSN in which the sensor nodes together with the relay nodes are deployed in the area of interest, where they form clusters using a standard clustering algorithm such as proposed in [15]. The sensor nodes and the relay nodes deployed in the network can be homogeneous or heterogeneous and a relay node has a communication range larger than that of sensor nodes, say almost four times the communication range of sensor nodes [16]. A sensor node or relay node can communicate with any other node if it is within the communication range. The relay nodes act as cluster heads and form the backbone of the network and each relay node can locate its neighbors within the transmission range during the bootstrapping. These special nodes are capable of generating a very small pulse messages which is negligibly small with respect to the data packet. The characteristics of a pulse message are similar to normal data packet and can encounter noises and can get lost and thus uses Time Division Multiple Access (TDMA) as a channel access method for sharing the networks to minimize the packet collisions. Considering this, we assumed that these packets have different probability to fail in different scenarios and can affect the result and needed to be taken care.

3.2   Fault Model

With the fault detection of a two tier wireless sensor network, we deal with fragile nodes at two different layers, where faults can occur at any time with some probability and to guarantee the performance of the network, any fault occurred is needed to detect as soon as possible. Faults occurring in relay nodes can be either the permanent or the transient faults. A relay node with a permanent defect in transceiver unit will be considered as a faulty node and all the computing units are assumed to be fault-free. We assume that each relay node has equal probability say $p1$ to become faulty. We also assume that, the pulse message used by the relay nodes to communicate with their neighbors can encounter errors with some probability $p2$. As shown in Fig. 1, the complete lifetime of a relay node is divided into small time slices calling it as pulse time in which each neighbor is responsible to communicate with that particular node so that their status can remain as non faulty for next pulse time and can be used as a packet forwarding node.

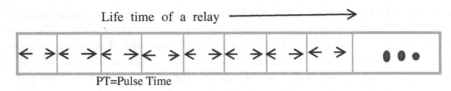

Fig. 1. Pulse time for a relay nodeo

A pulse message can encounter noise or some other physical hindrances and hence a time redundancy is used to detect the erroneous pulse message received from the neighbors.

4   Proposed Method

The basic idea behind the proposed algorithm is as follows. Every fault-free relay node maintains a Neighboring Table (NT) to store the current status of its neighboring nodes. The node forwards the data packets through any of its neighboring nodes only if its status in the neighboring table is identified it to be non-faulty during that pulse time. The neighboring nodes generate a pulse message during each pulse duration to update its status in the Neighboring Tables (NT) as non-faulty, for the next pulse time. In the absence of any pulse message from a neighbor node, the relay node updates its neighboring table entry corresponding to that neighbor as faulty. To cope with such lost pulse messages and transient faults in relay nodes, we employ time redundancy, where the decision is made based on neighbor's status collected for previous "t" consecutive times, where t is a small positive integer.

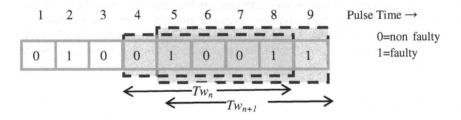

Fig. 2. Time window used to detect faults

A threshold value θ is used to decide about the current status of neighboring node. The value of θ and t can vary with different working scenarios of the sensor network. A sliding window of size t is used to make a decision whether the particular neighbor is faulty or not. This decision is made on the basis of the result obtained after processing the data in the time window and by comparing it with θ. As an illustration, let $Tw= 5$ pulse time, is a time window, storing status of a particular neighbor for past 5 sec then a neighboring node will be faulty if

$$\sum_{i=1}^{t} Tw(n) \geq \theta$$

In this illustration we take the value of θ to be 3. As in Fig. 2, at time window Tw_n, $\sum_{i=1}^{5} Tw(n) = 2$ and that of Tw_{n+1}, $\sum_{i=1}^{t} Tw(n) = 3$, now since the value of θ is set to 3 we can easily recognize that the relay node at Tw_n is non-faulty in spite of the absence of any updating or packet loss in 5th and 8th pulse time while at Tw_{n+1} we detects that it is faulty. The structure of neighboring table for any relay node network is illustrated in Fig. 3.

Neighbor	Pt_1	Pt_2	Pt_3	Pt_4	Pt_5
N2	0	1	0	0	0
N3	0	0	0	0	0
N4	0	0	1	0	0

$Pt_n - n^{th}$ pulse time.

(a)

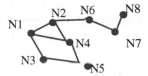

(b)

Fig. 3. (a) Neighboring table for N1 with no fault (b) Upper tier network structure

Our proposed algorithm assumes that the clusters of the sensor nodes have already been formed by using a standard clustering algorithm. The algorithm for the fault detection is presented in Fig. 4.

Algorithm: Distributed Fault Detection Algorithm (DFDA)

Phase 1: Neighboring Table Formation
> *Step* 1: Each relay node broadcast a HELLO message consist of their node id.
> *Step* 2: The relay nodes in the communication range after recceiving the hello packet replies with an acknowledgement.
> *Step* 3: Receiving an acknowledgement each relay node makes an entry in its Neighboring Table.

Phase 2: Fault Detection
> *Step* 4: At each relay node after entering the neighbor's detail its status is up dated to '0' non-faulty.
> *Step* 5: After each pulse time, relay nodes assume that its neighboring node is dead and updates its status to '1' and assume it to be faulty.
> *Step* 6: The pulse message from any specific node updates its status, and announces its existence.
> *Step* 7: Relay nodes check the status of a neighbor node before communicating with it and communicate if
> $$\sum_{i=1}^{t} Tw(n) \le \theta$$
> Else try another path or neighboring node to communicate.

Step 8: Stop

Fig. 4. Fault detection algorithm

Lemma 1: *The DFDA algorithm has worst case message exchange complexity of* $O(m)$ *for the whole network having m relay nodes.*

Proof: Proposed DFDA runs in two phases, first, the neighboring table formation and then the fault detection. During the neighboring table formation, each relay node broadcast a hello message and forms the table on the basis of acknowledgement received and thus has the message complexity of $O(1)$. In the second phase of fault detection, each relay node communicates with each of their neighboring relay nodes via a pulse message. Here, for each relay node the total number of pulse message generated will depend upon the number of neighbors. In the worst case scenario, the total number of neighbors, a relay nodes can have is m and hence the message complexity in the worst case will rise to $O(m)$.

5   Performance Evaluation

The performance evaluation of the suggested fault detection algorithm is done by computer simulation using MATLAB R2013a. It depends on various parameters like the number of relay nodes deployed in a target area, the density of the network, the probability that a relay node to be faulty $p1$, and the probability that a pulse message is faulty p_2. In the simulation, we assume that faults are independent of each other. The following two metrics, detection accuracy (DA) and False Alarm Rate (FAR) are used to evaluate the performance, where DA is defined to as the ratio of the number of faulty sensor nodes detected to the total number of faulty nodes and FAR is the ratio of the number of fault-free sensor nodes diagnosed as faulty to the total number of fault-free nodes. Here nodes with some transient faults and computational unit fault are treated as fault-free nodes.

Computer simulation is carried out in a sensor network, where 1024 sensor nodes are deployed randomly in a rectangular region of size 512 ×512 units. The position and count of minimum number of relay nodes required to cover all the sensor nodes are found by [17]. In the simulation, relay nodes and the pulse message are assumed to be faulty with probabilities of 0.05, 0.10, 0.15, 0.20, 0.25, 0.30 and 0.35.

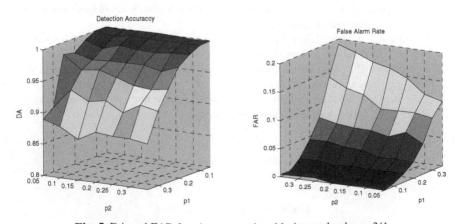

Fig. 5. DA and FAR for given scenario with time redundancy 3/4

For different scenario pulse message probability and faulty node probabilities, we have computed the DA and FAR for 100 times and taken the average as result as shown in   Fig. 5. A complete diagnosis is a diagnosis where all the faulty nodes can be identified and a diagnosis is said to be the correct diagnosis if, no fault-free nodes are identified as faulty [14]. The wireless sensor network is deployed in harsh environment leading the fault probability to be high. Hence a complete and correct diagnosis is very difficult or sometimes might be impossible to make.

The completeness and correctness of a diagnosis are measured by,

$$Completeness = \frac{Total\ number\ of\ faulty\ node\ diagnosed}{Total\ number\ of\ faulty\ nodes} \times 100$$

$$Correctness = \frac{Number\ of\ fault\ free\ node\ diagnosed\ as\ faulty}{Total\ number\ of\ fault\ free\ nodes} \times 100$$

Incomplete diagnosis in sensor networks could be acceptable if the number of faulty sensor is comparatively very small. In addition, it is still safe to use a diagnostic algorithm that might be incorrect, but can identify almost all of the fault-free nodes as long as a negligibly small number of fault-free nodes are excluded from the network. The reason for this is that sensor nodes are generally expected to be cheap and sufficient redundant nodes are typically deployed to achieve fault tolerance and sensing coverage. The scenario has also been studied for its correctness and completeness and the result is found to be satisfactory and the graph plotted for it is shown in Fig. 6. This graph justifies that the completeness and correctness for the scenario for $p1$ and $p2$ less than and equal to 0.2 is more than 98% and is acceptable.

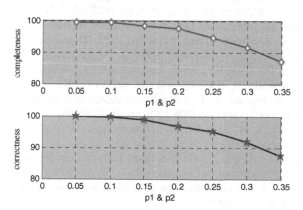

Fig. 6. Completeness and Correctness of the diagnosis

6   Conclusion

We have presented a distributed algorithm for detection of faulty relay nodes in two tier wireless sensor networks. The algorithm has been shown to require $O(m)$ message complexity in the worst situation where m is the number of relay nodes. The algorithm has been rigorously tested with various working scenarios and the experimental results have been presented to demonstrate the strength of the algorithm. However, the algorithm deals only with the detection of the faulty nodes. We like to extend the work in future by proposing a recovery scheme to enable the network operation without any interruption.

References

1. Akyildiz, I.F., Su, W., Sankarasubramaniam, Y., Cyirci, E.: Wireless sensor networks: a survey. Computer Networks 38(4), 393–422 (2002)
2. Abbasi, H., Younis, M.: A Survey on clustering algorithms for wireless sensor networks. Computer Communications 30, 2826–2841 (2007)
3. Akkaya, K., Senel, F., McLaughlan, B.: Clustering of wireless sensor and actor networks based on sensor distribution and connectivity. Journal of Parallel and Distributed Computing 69(6), 573–587 (2009)
4. Koushanfar, F., Potkonjak, M., Sangiovanni-Vincentelli, A.: On-line fault detection of sensor measurements. IEEE Sensors 2, 974–980 (2003)
5. Ruiz, B., Siqueira, I.G., Oliveira, L.B., Wong, H.C., Nogueira, J.M.S., Liureiro, A.A.F.: Fault management in event-driven wireless sensor networks. In: MSWIM 2004 (2004)
6. Chessa, S., Santi, P.: Comparison-based system-level fault diagnosis in ad hoc networks. In: 20th Symp. Reliable Dist. Syst., pp. 257–266 (2001)
7. Elhadef, M., Boukerche, A., Elkadiki, H.: Performance analysis of a distributed comparison-based self-diagnosis protocol for wireless ad hoc networks. In: International Workshop Modeling Analysis and Simulation of Wireless and Mobile Systems, pp. 165–172 (2006)
8. Jaikaeo, C., Srisathapornphat, C., Shen, C.-C.: Diagnosis of sensor networks. In: International Conference on Communications, vol. 5, pp. 1627–1632 (2001)
9. Krishnamachari, B., Iyengar, S.: Distributed Bayesian algorithms for fault tolerant event region detection in wireless sensor networks. IEEE Transactions on Computers 53(3), 241–250 (2004)
10. Ding, M., Chen, D., Xing, K., Cheng, X.: Localized fault-tolerant event boundary detection in sensor networks. In: IEEE Infocom, pp. 902–913 (2005)
11. Chen, J., Kher, S., Somani, A.: Distributed fault detection of wireless sensor networks. In: Proceedings of (2006) Workshop DIWANS, pp. 65–72 (2006)
12. Lee, M.-H., Choi, Y.-H.: Localized detection of faults in wireless sensor networks. In: ICACT, pp. 637–641 (2008)
13. Luo, X., Dong, M., Huang, Y.: On distributed fault-tolerant detection in wireless sensor networks. IEEE Transactions on Computers 55(1), 58–70 (2006)
14. Lee, M.-H., Choi, Y.-H.: Fault detection of wireless sensor networks. Computer Communications 31, 3469–3475 (2008)
15. Kuila, P., Jana, P.K.: Energy efficient clustering and routing algorithms for wireless sensor networks: Particle swarm optimization approach. Engineering Applications of Artificial Intelligence 33, 127–140 (2014)
16. Tang, J., Hao, B., Sen, A.: Relay node placement in large scale wireless sensor networks. Computer Communications 29, 490–501 (2006)
17. Nitesh, K., Jana, P.K.: Relay node placement algorithm in wireless sensor network. In: 2014 IEEE International Advance Computing Conference (IACC), pp. 220–225 (February 2014)

Secured Categorization and Group Encounter Based Dissemination of Post Disaster Situational Data Using Peer-to-Peer Delay Tolerant Network

Souvik Basu and Siuli Roy

Computer Application Centre, Heritage Institute of Technology, Kolkata, India
{souvik.basu,siuli.roy}@heritageit.edu

Abstract. Despite concerted efforts for relaying crucial situational information, disaster relief volunteers experience significant communication challenges owing to failures of critical infrastructure and longstanding power outages in disaster affected areas. Researchers have proposed the use of smart-phones, working in delay tolerant mode, for setting up a peer-to-peer network enabling post disaster communication. In such a network, volunteers, belonging to different rescue groups, relay situational messages containing needs and requirements of different categories to their respective relief camps. Delivery of such messages containing heterogeneous requirements to appropriate relief camps calls for on-the-fly categorization of messages according to their content. But, due to possible presence of malicious and unscrupulous entities in the network, content of sensitive situational messages cannot be made accessible even if that helps in categorization. To address this issue, we, in this paper, propose a secured message categorization technique that enables forwarder nodes to categorize messages without compromising on their confidentiality. Moreover, due to group dynamics and interaction pattern among groups, volunteers of a particular group encounter other volunteers of their own group (or groups offering allied services) more often than volunteers of other groups. Therefore, we also propose a forwarding scheme that routes messages, destined to a particular relief camp, through volunteers of that group or who encounter members of that group most frequently. This expedites the delivery of categorized messages to their appropriate destinations.

Keywords: DTN, Post Disaster Communication, Public Encryption with Keyword Search, Message Categorization, Group-Mobility, PRoPHET.

1 Introduction

In the event of any large-scale disaster, cell phone/ internet connectivity become non-functional due to the failure of the supporting infrastructure [1]. Therefore, the possibility of information exchange is almost ruled out and as a result, relief and rescue operations are often vulnerable to a vacuum of information [2, 3]. This acute need for information exchange demands setting up of a temporary post disaster communication network until the normal communication infrastructure is restored.

© Springer International Publishing Switzerland 2015   747
S.C. Satapathy et al. (eds.), *Proc. of the 3rd Int. Conf. on Front. of Intell. Comput. (FICTA) 2014*
– *Vol. 2*, Advances in Intelligent Systems and Computing 328, DOI: 10.1007/978-3-319-12012-6_83

Recently, the networking research community strongly recommends the use of DTN ([4] -[8]) for setting up post disaster communication networks. Currently, mobile devices like smart phones are having powerful processors and high storage capacity with multi radio interfaces (WiFi, Bluetooth) that can be effectively harnessed during or after a disaster to gather situational information and forward them to the relief camps for necessary actions.

In post-disaster relief operations, volunteers or rescue workers work in groups or units. For example, volunteers of the Health Care Camp provide medical aid and those associated with the Relief Camp provides food stuff, clothes, etc. to the victims. A central control station is set up for coordination of relief work in the different relief camps [18, 19, 20, 21]. These volunteers communicate situational messages (status, needs) of the disaster affected area, to suitable relief camps according to their relevance. Therefore, it becomes imperative for a forwarder to dynamically categorize a received message, based on its content, to identify the relief camp to which the message pertains to. However, in presence of un-trusted and malicious forwarders in the network (who may drop or divert messages with various spiteful interests), content of sensitive situational messages cannot be disclosed for the purpose of such categorization. We, in this paper, propose a public-key encryption with keyword search (PEKS) [9] based message categorization technique where each forwarding node classifies and categorizes messages, at the runtime according to their content, without getting access to the message content. Furthermore, due to interdependence among the groups, a volunteer belonging to a particular group encounters volunteers of its own group or some specific groups more frequently than other groups. Therefore, it is sensible to judiciously exploit this encounter pattern for forwarding categorized situational messages to their respective destinations. PRoPHET [10, 11], one of the benchmark routing protocols for DTN, fits well for such encounter-based forwarding as it uses the history of encounters with other nodes. In this paper, we adapt PRoPHET for proposing a group-encounter based forwarding scheme that routes categorized messages for a particular camp, through volunteers of that group or who encounter members of that group often to accelerate the delivery of such messages. The rest of this paper is organized as follows. Section 2, summarizes related work, Section 3, describes the system architecture. PEKS based message categorization technique is discussed in Section 4 and group encounter based forwarding scheme in Section 5. In Section 6, we provide simulation results and Section 7 concludes the paper.

2   Related Work

In this section, we summarize some related work on PEKS, message prioritization in DTN and DTN routing protocol for disaster scenario. Hsu et al in [12] study six existing security models of PEKS/SCF-PEKS schemes and analyze their efficiency and performance. Shikfa et al in [12] uses PEKS to allow intermediate nodes to discover partial matches between their profile and the destination profile, and uses policy-based encryption to enforce confidentiality of the payload. This scheme suits opportunistic networks well, because it has a low storage and computation overhead and it relies on an offline TTP only. Ramanathan et al in [14] presented a novel mechanism called Prioritized Epidemic (PREP) that expiry time information and topology awareness to decide which bundles to delete or hold back when faced with a resource (buffer, bandwidth) crunch. Joe et al in [15] propose a DTN message priority routing suitable for emergency situations. Mashhadi et al in [16] propose an approach for priority-scheduling in participatory DTNs, whereby messages are being forwarded based on a combination of the likelihood of future encounters. Finally, John Burgess et al in [17] propose MaxProp, a protocol for effective routing of DTN messages, based on prioritizing both the schedule of packets transmitted to other peers and the schedule of packets to be dropped.

3   System Architecture

3.1   Post Disaster Communication Network

We assume three types of nodes in a post disaster communication scenario.

(i)   Shelter-Node: All relevant information about a shelter is stored and transmitted by a shelter-node (SN) (e.g., a laptop in a shelter).

(ii)   Camp-Node: Each relief camp, houses a camp-node (CN) (e.g., a workstation or a server) that receives situational information, pertaining to that camp.

(iii)   Forwarder-Node: Volunteers (carrying smart phones) are forwarder-nodes (FN) that communicate categorized situational messages, depicting status, needs and requirements of the shelters, to suitable relief camps according to their relevance.

3.2   Pin and Keyword Distribution at the Setup Phase

In our scheme, we assume N forwarder-nodes (FN), K shelter-nodes (SN) and n camp-nodes (CN), corresponding to n rescue groups. In the setup phase, each rescue group is allocated a group-id GID_j, $j=1, 2, ..., n$, by the central control station to be stored in each FN. The central control station also creates a pair of pins called control-pin ($CPin$) {private key} and modified control-pin ($MCPin$) {public key} and shares with each SN. SNs and FNs are also provided with lists of prescribed keywords by the central control station. The keywords are related to each specific service group e.g., stranded, evacuation, etc. are keywords related to the Search and Rescue Group; vaccination, first-aid, ambulance, etc. related to the Health Care Group, etc. Thus, the control station provides lists of keywords $\{Ks\}$, $s = 1, ..., M$, classified by their types, as shown in Table 1, to shelter-nodes and forwarder-nodes. Each forwarder-node maintains a priority queue PQ_j corresponding to each group j, for storing messages pertaining to that group.

Table 1. List of keywords classified by service types

Category	Keywords			
Health Care Keywords	K_1	K_2	.....	K_p
Logistics Keywords	K_{p+1}	K_{p+2}	.....	K_r
....		....		
Search and Rescue Keywords	.....	K_{M-2}	K_{M-1}	K_M

4   PEKS Based Message Categorization Technique

In this section, we illustrate the PEKS based message categorization technique. Before describing our proposed scheme, we briefly explain the PEKS method as proposed by Boneh et al [9].

4.1   Public-Key Encryption with Keyword Search (PEKS)

PEKS allows an intermediate node to search for a keyword in an encrypted message that a sender has sent to a receiver. The intermediate node learns nothing except whether the keyword occurs in the message or not. We illustrate the technique with three nodes, say, A, B and C. If B wants to send a message M with keywords W_1, ..., W_k to A through C, B encrypts M using a standard public key encryption technique and then appends to the resulting ciphertext, a PEKS of each keyword. Thus, B computes and sends: $[E_{Apub} (M), PEKS(A_{pub}, W_1), ..., PEKS(A_{pub}, W_k)]$ to A through C, where A_{pub} is the public key of A. For a public key A_{pub} and a keyword W, a searchable encryption of W: $S = PEKS(A_{pub}, W)$ enables searching for specific keywords without revealing any information about the message. C, the intermediate node, cannot infer any information on M. If A wants to enable C to search for the keyword W, A has to provide C with a trapdoor: $T_W = TRAPDOOR(A_{priv}, W)$ where A_{priv} is the private key of A. Given A_{pub}, S and a trapdoor T_W, C proceeds to search for the keyword W' in M using the test function: $TEST(A_{pub}, S, T_W)$ that outputs true if $W = W'$ and false otherwise. If $W = W'$, C only knows that M contains the keyword W and if the result is false it just deduces that M does not contain W.

4.2   Message Categorization Using PEKS

Trapdoor Distribution Phase

In the setup phase, the central control station provides each forwarder-node a trapdoor corresponding to each prescribed keyword K_s, as: $T_{Ks} = TRAPDOOR(CPin, K_s)$, where $CPin$ is the control-pin. Therefore, each forwarder-node carries with it M such trapdoors.

Header Generation & Encryption Phase

Whenever a shelter-node, say SN_t, $t=1, 2, ..., K$, generates a message $MSG_{t,n}$ (n^{th} message from the tth shelter) providing shelter information, it creates a header for the message:

$$H(MSG_{t,n}) = \{K_s\}, s = 1, ..., m$$

containing a set of prescribed keywords extracted from the message content, where $m \subset M$, the total number of keywords prescribed by the central control station. In order to protect privacy, SN_t encrypts $H(MSG_{t,n})$ using a standard public key encryption technique and then appends to the resulting ciphertext, a PEKS of each keyword. Thus, SN_t computes

$$H'(MSG_{t,n}) = E_{MCPin} (H(MSG_{t,n})) \parallel PEKS(MCPin, K_1) \parallel .... \parallel PEKS(MCPin, K_m)$$

where and $MCPin$ is modified control-pin of the control-node.

Message Encryption Phase

In order to provide complete privacy, SN_k encrypts $M_{t,n}$ using a standard public key encryption technique and $MCPin$. Thus, SN_k computes: $MSG'_{t,n} = E_{MCPin} (MSG_{t,n})$

Finally SN_t then concatenates the encrypted header with the encrypted message to form an augmented message:

$$A(MSG_{t,n}) = H'(MSG_{t,n}) \| MSG'_{t,n}$$

and sends it to an appropriate forwarder-node that comes in contact with it.

Keyword Search Phase

When a forwarder-node, say FN, receives the message $A(MSG_{t,n})$, either from a shelter or from another forwarder-node, it proceeds to search for each of the prescribed keywords K_s, in the encrypted message header $H'(MSG_{t,n})$. Given $MCPin$, $PEKS(MCPin, K_s)$ and a trapdoor T_{Ks}, FN can search for the keyword K_s' in $H'(MSG_{t,n})$ using the $TEST$ function: $TEST(MCPin, PEKS(MCPin, K_s))$

that outputs true if $K_s = K_s'$ and false otherwise. If $K_s = K_s'$, FN knows that $MSG_{t,n}$ contains the keyword K_s and deduces that $MSG_{t,n}$ does not contain K_s otherwise.

Message Categorization Phase

The forwarder-node FN executes the $TEST$ function on the message header $H'(MSG_{t,n})$ with each of the prescribed keywords, as stored in table I, and keeps a count of the number of keywords associated with each group in the message. If $FREQ_j(MSG_{t,n})$ denote the frequency of keywords in $MSG_{t,n}$ associated with the group $j, j=1,...,n$, then FN generates a frequency distribution as shown in Table 2.

Table 2. Keywords Frequency Distribution for $MSG_{t,n}$ generated by FN

Group ID	Frequency of Keywords
GID_1	$FREQ_1(MSG_{t,n})$
GID_2	$FREQ_2(MSG_{t,n})$
...	
GID_n	$FREQ_n(MSG_{t,n})$

FN finds the mode of the above frequency distribution, i.e., the group-id for which frequency is highest. If $FREQ_i(MSG_{t,n})$, $i = 1, ..., n$, is highest then it becomes evident that the message contains more information for the group i.

Now, in an opportunistic network like ours, as forwarders have very little contact period with each other, only a few messages can be transferred per contact. Thus, messages that are of high priority and high relevance to a receiver must be forwarded before any other message. To accomplish this, FN inserts $MSG_{t,n}$ in the priority queue PQ_i designated for group i. While inserting, FN puts $MSG_{t,n}$ in its appropriate position so that the prioritized nature of PQ_i is maintained, i.e. if for any other message $MSG_{p,r}$ (r^{th} message from the pth shelter) $FREQ_i(MSG_{p,r}) > FREQ_i(MSG_{t,n})$ then $MSG_{p,r}$ should be placed before $MSG_{t,n}$ so that $MSG_{p,r}$ gets delivered first. This way, high priority messages are transmitted at the first possible opportunity.

If the above frequency distribution turns out to be bimodal i.e. both $FREQ_i(MSG_{t,n})$ and $FREQ_j(MSG_{t,n})$, $(i, j) = 1, ..., n$, are highest, then certainly the message is of relevance to both the groups i and j. Then FN inserts $MSG_{t,n}$ in PQ_i as well as in PQ_j. Situations where there are more than two modes are handled similarly.

Message Decryption Phase

Finally, the camp-node on receiving a message $A(MSG_{t,n})$, decrypts the encrypted message $MSG'_{t,n}$ as: $D_{CPin} (E_{MCPin} (MSG_{t,n}))$ to obtain the original message $MSG_{t,n}$ and allocates relief stock for shelter SN_t according to the requirements specified in $MSG_{t,n}$.

The above technique enables a forwarder-node to categorize received messages and put them in appropriate priority queues without learning anything about the content of the messages.

5   Group Encounter Based Forwarding Scheme

In this section we adapt the PRoPHET routing protocol, as described in [10] and [11], to probabilistically route messages using history of group encounters and transitivity. For group encounter based PRoPHET, we define $P(FN_i , GID_j)$ as the delivery predictability of a forwarder-node FN_i , belonging to group i, with any member of group j. Whenever FN_i and FN_j meet with each other, they exchange DP tables and update their own DP table based on the identity of the group and the received table. For an encountered node FN_j, node FN_i updates its DP using (4)

$$P(FN_i, GID_j)_{new} = P(FN_i, GID_j)_{old} + (1 - P(FN_i, GID_j)_{old}) \times P_{enc} \qquad (4)$$

Similarly, for an encountered node FN_i, node FN_j updates its DP using (5)

$$P(FN_j, GID_i)_{new} = P(FN_j, GID_i)_{old} + (1 - P(FN_j, GID_i)_{old}) \times P_{enc} \qquad (5)$$

Note that, $P(FN_i , GID_j)$ is different from $P(FN_j , GID_i)$. Transitive update of DPs for all other groups k encountered by FN_j is done using (6)

$$P(FN_i, GID_k)_{new} = max(P(FN_i, GID_k)_{old} , P(FN_j, GID_k) \times P(FN_i,GID_j)_{new} \times \beta) \qquad (6)$$

FN_j updates DPs for all other groups k known by FN_i using (7)

$$P(FN_j, GID_k)_{new} = max(P(FN_j, GID_k)_{old} , P(FN_i, GID_k) \times P(FN_j, GID_i)_{new} \times \beta) \qquad (7)$$

In order to eliminate stale information from the network, the DP table is periodically aged for all groups k using (8)

$$P(FN_i, GID_k)_{new} = P(FN_i, GID_k)_{old} \times \gamma^T \qquad (8)$$

Therefore using equations (4) to (8) we can adapt PRoPHET to route messages, destined for a particular relief camp, through volunteers the corresponding group or who encounter members of that group frequently. Moreover, this group encounter based PRoPHET calculates delivery predictabilities corresponding to groups, which reduces the overhead of calculating and storing delivery predictabilities for each and every node in the network.

6   Simulation Results

Our simulation set-up is based on the recent disaster that occurred in the Indian state of Uttarakhand. We have created our simulation environment in the ONE Simulator [23] based on the post-disaster relief operations carried out in the area of Badrinath to Rudraprayag where we have used the data available from Google crisis map [22] and other sources [20, 21]. We have set up 11 shelter nodes, 88 forwarder-nodes belonging to 5 different relief groups and 5 relief camps in the map of the disaster affected area of Uttarakhand. Simulation time limited to 12 hours (43000s). The movement model is Shortest Path Map Based Movement. Number of runs is 20. We evaluate the improvements introduced by our proposed group encounter based PRoPHET (GEPRoPHET), in a disaster scenario, over other well known DTN routing protocols like PRoPHET, PRoPHETv2, MaxProp and Epidemic. We use *Number of Group Messages Delivered*, *Delivery* ratio, *Average Delay* and *Overhead Ratio* as simulation metrics. Fig. 1 shows the percentage of group and non-group messages in each camp for different protocols. GEPRoPHET outperforms all the others. Fig. 2 shows the Delivery ratio, Average Delay and Overhead ratio of different protocols. GEPRoPHET shows the highest delivery ratio but faces largest delay owing to the computational overhead it incurs for message categorization. GEPRoPHET exhibits the least overhead ratio.

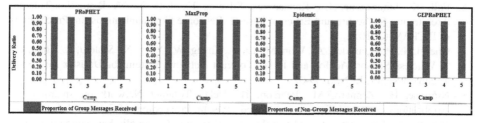

Fig. 1. Percentage of Group and Non-Group Messages in each camp for different protocols

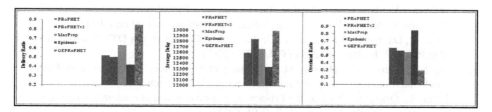

Fig. 2. Delivery Ratio, Average Delay and Overhead Ratio for different protocols

7   Conclusion

In this paper we presented a PEKS based message categorization technique that categorizes messages without compromising on its confidentiality and suggested a group encounter based routing protocol that disseminates categorized messages to appropriate relief camps through the best forwarders so that messages reach their destination at the earliest possible. However, as PEKS has no provision for generating trapdoors using multiple keywords, a forwarder-node in our proposed technique has to tediously execute the *TEST* function on a message header for each of the keywords. Although a few researchers have proposed conjunctive field keyword search schemes for such situations, most of the schemes still have room for improvement [12].

References

1. Luo, H., Kravets, R., Abdelzaher, T.: The-Day-After Networks: A First-Response Edge-Network Architecture for Disaster Relief, NSF NeTS FIND Initiative, 2006-2010
2. World Disasters Report 2013 - Focus on technology and the future of humanitarian action, International Federation of Red Cross and Red Crescent Societies
3. Mehrotra, S., Butts, C., Kalashnikov, D., Venkatasubramanian, Huyck, C.: Project Rescue: Challenges in Responding to the Unexpected. In: Proceedings of EIST 2004 (2004)
4. Fall, K.: A delay-tolerant network architecture for challenged internets. In: Proceedings of SIGCOMM (2003)
5. Cerf, V., Burleigh, S., Hooke, A., Torgerson, L., Durst, R., Scott, K., Fall, K., Weiss, H.: Delay-Tolerant Networking Architecture. RFC 4838 (Informational). Internet Engineering Task Force (April 2007)
6. Fall, K., Iannaccone, G., Kannan, J., Silveira, F., Taft, N.: A Disruption-Tolerant Architecture for Secure and Efficient Disaster Response Communications. In: Proceedings of ISCRAM (2010)
7. Chenji, H., Hassanzadeh, A., Won, M., Li, Y., Zhang, W., Yang, X., Stoleru, R., Zhou, G.: A Wireless Sensor, AdHoc and Delay Tolerant Network System for Disaster Response. Technical Report LENSS-09-02
8. Ntareme, H., Zennaro, M., Pehrson, B.: Delay Tolerant Network on smartphones: Applications or communication challenged areas. In: Proceedings of ExtremeCom (2011)
9. Boneh, D., Crescenzo, G., Ostrovsky, R., Persiano, G.: Public-key encryption with keyword search. In: Eurocrypt (2004)
10. Lindgren, A., Doria, A., Davies, E., Grasic, S.: Probabilistic routing protocol for intermittently connected networks. draft-lindgren-dtnrg-prophet-09.txt (2011)
11. Grasic, S., Davies, E., Lindgren, A., Doria, A.: The Evolution of a DTN Routing Protocol – PRoPHETv2. In: Proceedings of CHANTS (2011)
12. Hsu, S.T., Yang, C.C., Hwang, M.S.: A Study of Public Key Encryption with Keyword Search. International Journal of Network Security 15(2) (March 2013)
13. Shikfa, A., Önen, M., Molva, R.: Privacy in context-based and epidemic forwarding. In: Proceedings of WOWMOM, pp. 1–7. IEEE (2009)
14. Ramanathan, R., Hansen, R., Basu, P., Hain, R.R., Krishnan, R.: Prioritized epidemic routing for opportunistic networks. In: MobiOpp 2007, New York, USA (2007)
15. Joe, I., Kim, S.B.: A Message Priority Routing Protocol for Delay Tolerant Networks in Disaster Areas. In: Future Generation Information Technology, pp. 727–737 (2010)
16. Mashhadi, A., Capra, L.: Priority Scheduling for Participatory DTN. In: Proceedings of WoWMoM (2011)
17. Burgess, J., Gallagher, Jensen, B.D., Levine, B.N.: MaxProp: Routing for Vehicle-Based Disruption- Tolerant Networks. In: proceedings of IEEE Infocom (2006)

18. http://www.ifrc.org/en/what-we-do/
    disastermanagement/responding/disaster-response-system/
    dr-tools-and-systems/eru/types-of-eru/
19. http://www.jica.go.jp/english/our_work/
    types_of_assistance/emergency.html
20. http://www.ndtv.com/article/india/
    uttarakhand-1-350-evacuated-from-badrinath-united-
    nations-says-over-11-000-issing386683
21. http://www.ndtv.com/article/india/uttarakhand-thousands-
    still-stranded-see-over-view-383781
22. http://google.org/crisismap/2013-uttrakhand-floods?gl=in
23. Keranen, A., Ott, J., Karkkainen, T.: The ONE Simulator for DTN Protocol Evaluation. In: SIMUTools 2009 (2009)

Lifetime Maximization in Heterogeneous Wireless Sensor Network Based on Metaheuristic Approach

Manisha Bhende[1], Suvarna Patil[2], and Sanjeev Wagh[3]

University of Pune, Pune,
Maharashtra, India, 411018
{manisha.bhende,suvarnapat,sjwagh1}@gmail.com

Abstract. Increasing the lifetime of heterogeneous wireless sensor network (WSN) and minimizing the run time using improved ACO is an important issue. The computational time is very important in case of such searching algorithms. In this methodology, maximum number of disjoints connected covers are found that fulfill coverage in network and connectivity of the network. A construction graph is designed in which each vertex denotes the a device in a subset. The ants find an optimal path on the construction graph which maximizes the number of connected covers based on pheromone and also heuristic information. The connected covers are built by the pheromone and it is used in the searching. The heuristic information satisfies the desirability of device assignment. The proposed metaheuristic approach is to maximize network lifetime and minimize the computational time for the searching process satisfying both sensing coverage and network connectivity.

Keywords: Heterogeneous Wireless Sensor Networks (HWSNs), Ant colony optimization (ACO), Network Connectivity, Coverage, Network Lifetime.

1   Introduction

As sensor nodes in Wireless Sensor Network (WSN) are powered by batteries having limited energy. To replace or recharge the batteries in many practical scenarios is very difficult. The main issue in the design of wireless sensor network is improving energy efficiency. Energy saving is always related to the total energy consumption, which is required to transfer data to the sink, and also energy balancing is also concerned with the difference in energy consumption between sensor nodes. Number of sensor nodes and energy, storage capacities, and processing constraint requires careful management of resources. Many algorithms have been proposed for the routing in WSNs to improve energy efficiency [10].

In the existing system, use of different sensors with different functionality and capabilities is very difficult. Monitored information is sent to single sink by all the sensors in the network. It leads to overlapping and collision may occur in networks. It is also a difficult to find the maximize number of connected cover and sensing range. Greedy algorithm is used, which can be applied to only in the Homogenous Wireless

© Springer International Publishing Switzerland 2015   757
S.C. Satapathy et al. (eds.), *Proc. of the 3rd Int. Conf. on Front. of Intell. Comput. (FICTA) 2014*
– *Vol. 2*, Advances in Intelligent Systems and Computing 328, DOI: 10.1007/978-3-319-12012-6_84

Sensor Network. Considering all these conditions, finding shortest paths between sensors and sink is difficult.

The proposed ACO-based approach is used for increasing the number of connected covers. The search space of the problem is transformed into a construction graph and device in a subset is represented by vertex in a graph. Heuristic information about each device is associated with each device for calibrating its utility which helps to reduce constraint violations. Pheromone information which is deposited between every two devices is used to record the historical desirability which in turn helps to assign them to the same subset. In each iteration, the number of subsets is determined as one greater than the number of connected covers as in the best-so-far solution. To avoid constructing subsets excessively, the ants find one more connected cover. To further refine the solution. The local search procedure is designed by using reassignment of redundant devices [10] [11]. Using heuristic information, Pheromone and local search procedure, proposed approach find the optimal path which satisfies both network connectivity and sensing coverage. Effectiveness and efficiency of the proposed approach are validated by analyzing proposed approach with greedy algorithm. Ant colony optimization (ACO) is a well-known metaheuristic which is inspired by the foraging behaviour of real ants. In metaheuristic approach, ants are stochastic constructive procedures that build the optimal solution. Thus, with the help of constructive search, ACO becomes suitable for solving combinatorial optimization problems. ACO implements search experiences which are represented by pheromone and domain knowledge. Domain knowledge is expressed in term of heuristic information which accelerates the search process. Many ACO algorithms focus on the routing and energy consumption issue in homogeneous WSNs, this paper has proposed an ACO-based approach for prolonging the network lifetime of heterogeneous WSNs by discovering the maximum number of connected covers which satisfies sensing coverage and network connectivity.

2   Proposed System

This paper proposes ACO based approach for increasing the lifetime of heterogeneous WSNs by discovering the maximum number of connected covers satisfying both sensing coverage and network connectivity. Heterogeneous WSNs deals with sensors having different capabilities, how sensors communicate and how it prolongs the network lifetime by the increasing number of sinks instead of single sink. Ant Colony Optimization (ACO) technique is a metaheuristic optimization technique which is developed to solve any combinatorial optimization problem. Improved ACO is considered as a multi-agent system. In this system the behaviour of each single device called ant, which is inspired by the behaviour of real ants. In ACO, ants utilize stochastic constructive procedures that build optimum solutions with the help of a construction graph. In the area of WSNs, ACO-based algorithms have been used for improving the energy efficiency in unicasting, broadcasting, and also in case of data gathering.

2.1   Proposed System Architecture

The proposed approach increases reliability of sensor network and energy efficiency by reducing energy consumption. In proposed approach more sinks are used to increase the sensor lifetime by choosing the optimal path of nearest sink to send information. This reduces the time of sending monitoring results. By using more than one sink increases the network lifetime of WSNs by finding a nearest optimal path while transmitting the data from sensor to sink, the sensor doesn't have idea to transmit information to which sink. Improved ACO algorithm helps to find optimal path if more than sinks are used to collect information in a given wireless sensor network. The sensors continuously monitor the target and transmit the monitoring results to the nearest sinks, based on improved ACO. The sinks, then relay the monitoring results to the destination, i.e. base station.   To find optimal path, a connected cover in the given heterogeneous WSNs must achieve the following three constraints:

1) The sensors must complete coverage to the target,
2) The sensor must transmit all the monitoring results to the nearest sinks and
3) Once all the information is transmitted to nearest sink using the algorithm. The corresponding sink finally sent all information to the destination which in turn maximize the network lifetime of WSNs.

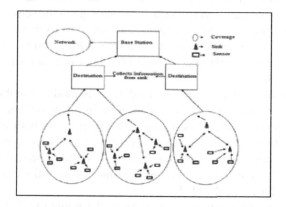

Fig. 1. Proposed System Architecture

Various performance parameters are used for comparing different routing algorithm in WSNs. These are given as follows:

I.   Average Energy consumption: The average energy consumption of all nodes in the given area for transmitting a data packet to the nearest sink.

II.   Network lifetime: The time of the node in the network running out of its energy and how to increase the lifetime.

III. Success Ratio: It calculates the success ratio to send packets from source to destination node.

2.2 Ant Colony Optimization Based Approach for Maximizing Number of Connected Covers

This section describes the improved ACO approach for increasing the lifetime of heterogeneous WSN by increasing the number of connected covers in a heterogeneous WSN. The value of the objective function is calculated first. The ants search behavior is then described.

2.2.1 Objective Function

In order to increase the number of connected cover, the objective function is calculated which has two components. The first component focuses on the constraint violations within subsets. The second component calculates the objective value which are based on the number of connected covers as the main goal of the improved ACO approach is to find a solution which maximizes the number of connected covers. Following are some aspects that are relevant for the design of ACO algorithms,

1) *The coverage constraint*, which requires the sensors in network to fully cover a target area. In other words, for any given point P in target area A, at least one sensor has range less than sensing range of one of the sensors.

2) *The collection constraint*, which needs to satisfy the condition that the sinks to collect all the monitoring results obtained by the sensors in the same subset. Assume that the given sensors cannot transmit data. For each sensor, at least one sink has a less transmission range than giving a transmission range of the sensor

3) *The routing constraint*, which requires the sinks in given network to design a connected network to transmit the collected monitoring results to the destination. Between any two sinks, transmission range of at least one sink should be less given transmission range of sink.

4) *The Random walk algorithm*, In this random walk model, a mobile node (sensor) moves from its current location to a new location by randomly choosing a direction and speed in which to travel. The speed and direction are both chosen from pre-defined ranges which are new, respectively [min-speed, max-speed] and [0, 2*pi] respectively. Each movement in the Random Walk Mobility Model occurs in either a constant traveled d distance or a constant time interval t, at the end of which a new direction and speed are calculated [5].

The values of the above three criteria are in the range of [0, 1]. If the value is larger, less is chance of violation of the given constraints. The average value of the above three constraints, is used to summarize that how well the set satisfies the three constraints[9]. If the average value equals one, then the sensor in given network satisfies all the three constraints and form a connected cover. It is followed by a local search procedure which refines the solution by eliminating redundant devices. Random mobility model enhances energy efficiency and also proved to be used for real time applications. In order to satisfy all three constraints, after generating networks of sensors, values of sensing range, transmission range of sensors and transmission range are set in the given simulation setup as shown in figure 2.

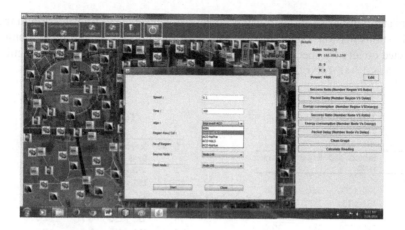

Fig. 2. Simulation setup showing a selection of various algorithms

3   Performance Analysis

The proposed approach is the algorithm for increasing the number of connected covers in the heterogeneous wireless sensor networks and greedy algorithm which is used in homogeneous wireless sensor networks is used for comparison with the improved ACO approach. Heuristic information, Pheromone, and local search are three important components in the proposed improved ACO. In this section, different experiments are performed to evaluate the performance of the improved ACO approach.

3.1   Test Cases

First, improved ACO is compared with greedy algorithm which uses the same heuristic information as ACO approach. The greedy algorithm always assigns a device to a subset where the device implements the greatest improvement in network connectivity or sensing coverage. If two or more subsets have the highest heuristic value, then the devices are randomly assigned to one of them. Heterogeneous WSNs with different scales and redundancy are tested in the experiments. Heuristic information, Pheromone, and local search are three important components in the proposed improved ACO. In this part, we validate their effectiveness by comparing the results of Improved ACO with its variants without these components. These three variants are No_Heu (without heuristic information), No_phe (without pheromone), and No_LS (without local search).

For each case, a maximum number of connected covers have been generated depending upon the values. Based on an optimal solution of the first case, a new case can be generated by removing redundant devices from the connected covers in the solution. In first set, WSNs are generated by randomly deploying sensors and sinks in a 50×50 rectangle. The settings of these networks, including the scale number of sensors and number of sinks, sensing range, r_s of sensors and transmission range r_t of sensors, transmission range R_t of sinks, and the upper bound of the number of connected covers by specifying number of regions. Table I have identical settings in Rs, Rt, Rt as 1400, 1400, 2300 but are differences in the network scale.

Table 1. Test Case 1

Sr. No.	Sensors and Sinks	Regions	Power
1	50 x 50	2x3	1400
2	100 x 100	2x3	1200
3	150 x 150	2x3	1400
4	200x 200	2x3	1200
5	250x 250	2x3	2000
6	300x 300	2x3	1300

Table 2. Test Case 2

Sr. No.	Sensors And Sinks	Regions	Power
1	150 x 150	2x2	1400
2	150 x 150	2x3	1400
3	150 x 150	3x3	1400
4	150 x 150	5x1	1400
5	150 x 150	7x1	1400
6	150 x 150	8x1	1400

Different scenarios are considered while testing success ratio, packet delay and energy consumption with greedy algorithm and Improved ACO. As the greedy algorithm scans the whole network to find the shortest path, it has a success ratio less than the Improved ACO approach, but packet delay and energy consumption parameters have greater value than that of Improved ACO which proves that improved ACO is more efficient and effective in terms of energy consumption. All the scenarios are discussed in Test case 1 and 2. Fig. 3 shows the different energy consumption and Fig. 4 shows success ratio values generated by executing greedy algorithm (WSN), Improved ACO, No-Heu, No-Phe, No_LS.

Start Page	energy_node.txt	
Source	History	
11	60,2679.7,WSN	
12	60,1709.4,Improved ACO	
13	60,2239.6,ACO-NoPhe	
14	60,2145.6,ACO-NoLS	
15	60,1703.7,ACO-NoHue	
16	80,5072.5,WSN	
17	80,1342.4,Improved ACO	
18	80,2146.6,ACO-NoPhe	
19	80,2537.6,ACO-NoLS	
20	80,2542.7,ACO-NoHue	
21	100,5132.4,WSN	
22	100,2560.4,Improved ACO	
23	100,3279.6,ACO-NoPhe	
24	100,2961.6,ACO-NoLS	
25	100,5072.3,ACO-NoHue	
26	120,5205.7,WSN	
27	120,1808.4,Improved ACO	
28	120,3585.6,ACO-NoPhe	
29	120,2739.6,ACO-NoLS	
30	120,2834.7,ACO-NoHue	
31	140,2910.7,WSN	

Fig. 3. Energy Consumption of Greedy, Improved ACO and three variants

Start Page	ratio_node.txt	
Source	History	
11	60,30.03667481662592,WSN	
12	60,60.22349570200573,Improved ACO	
13	60,40.561815336463226,6ACO-NoPhe	
14	60,40.3703007518797,ACO-NoLS	
15	60,30.64841498559078,ACO-NoHue	
16	80,30.188442211055275,WSN	
17	80,60.56649395509499,Improved ACO	
18	80,40.010416666666664,ACO-NoPhe	
19	80,40.371875,ACO-NoLS	
20	80,30.14780600461894,ACO-NoHue	
21	100,30.188442211055275,WSN	
22	100,60.34106728538283,Improved ACO	
23	100,40.177083333333336,ACO-NoPhe	
24	100,40.143487858719645,ACO-NoLS	
25	100,30.195454545454545,ACO-NoHue	
26	120,30.188442211055275,WSN	
27	120,60.01176470588236,Improved ACO	
28	120,40.68405797101449,ACO-NoPhe	
29	120,40.604046242774565,ACO-NoLS	
30	120,30.5622009569378,ACO-NoHue	
31	140,30.0,WSN	

Fig. 4. Success Ratio of Greedy, Improved ACO and three variants

After analysis of greedy algorithm and Improved ACO based on the above figures, it is proved that the greedy algorithm success ratio is always less than improved ACO by 30% and energy consumption is always greater than Improved ACO. Thus, Improved ACO is more efficient and effective in terms of energy consumption and data transmission. After comparison of Improved ACO with three variants No-Phe, No-Heu, No-LS, it is concluded that success ratio for Improved ACO is always greater than 20% in case of No-Phe and No-LS and 30% in case of No-Heu same as that of greedy algorithm. Hence, if heuristic information is not considered in finding optimal path, then the algorithm performs less efficiently. It is shown in figure 5 and 6 for success ratio and energy consumption respectively.

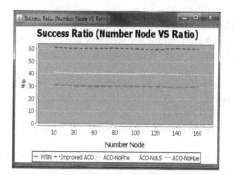
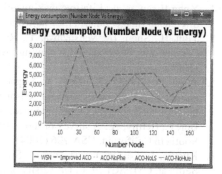

Fig. 5. Analysis in terms of Success Ratio   **Fig. 6** Analysis in terms of Energy Consumption

4   Conclusion

The main objective presented in this paper is to prolong the network lifetime using an improved ACO algorithm to maximize the network lifetime and balance the power consumption of each node by using the concept of multiple sink. The improved ACO incorporates an idea of using the lowest energy path always which may not be best for all types of network. It finds the optimal path quickly most of the time using improved ACO and gets less energy consumption. This approach avoids building excessive subsets by finding connected cover one more than best-so-far solution. To further enhance the search process for prolonging the network lifetime, pheromone and heuristic information are designed. It improves search efficiency by refining given search with the help of local search implementation. Improved ACO is an effective method for maximizing the network lifetime of heterogeneous WSNs. Simulation result proves that the proposed approach provides more effective and efficient way of prolonging the network lifetime of heterogeneous wireless sensor networks which fulfills sensing coverage and network connectivity.

References

1. Zhong, J.-H., Zhang, J.: Ant colony optimization algorithm for lifetime maximization in wireless sensor network with mobile sink. In: Fourteenth International Conference on Genetic and Evolutionary Computation Conference, ACM (2012)
2. Wang, X., Han, S., Wu, Y., Wang, X.: Coverage and Energy Consumption Control in Mobile Heterogeneous Wireless Sensor Networks. IEEE Transactions on Automatic Control 58(4) (April 2013)
3. Singh, A., Behal, S.: "Ant Colony Optimizationfor Improving NetworkLifetime in Wireless Sensor Networks. An International Journal of Engineering Sciences 8 (June 2013) ISSN: 2229-6913
4. Yong-Hwan, et al.: Lifetime maximization considering target coverage and connectivity in directional image/video sensor networks. The Journal of Supercomputing, 1–18 (2013)
5. Tian, H., Shen, H., Matsuzawa, T.: Random walk routing for wireless sensor networks. Parallel and Distributed Computing, Applications and Technologies. In: PDCAT 2005 (2005)
6. Wang, X., Wang, X., Zhao, J.: Impact of mobility and heterogeneity on coverage and energy consumption in wireless sensor networks. In: 2011 31st International Conference on IEEE Distributed Computing Systems, ICDCS (2011)
7. Bai, F., Helmy, A.: A survey of mobility models. Wireless Adhoc Networks. University of Southern California, USA 206 (2004)
8. Kurose, J., Lesser, V., de Sousa e Silva, E., Jayasumana, A., Liu, B.: Sensor Networks Seminar, CMPSCI 791L, University of Massachusetts, Amherst, MA (Fall 2003)
9. Srikanth, Umarani, G., Akilandeswari, M.: Computational Intelligence Routing For LifetimeMaximization in Heterogeneous Wireless Sensor Networks. International Journal of Recent Technology and Engineering (2012) ISSN: 2277-3878
10. Sanjeev, W., Prasad, R.: Heuristic Clustering for Wireless Sensor Networks using Genetic Approach. IJWAMN,International Journal of Wireless and Mobile Networking 1(1) (November 2013s)
11. Wagh, S., Prasad, R.: Energy Optimization in Wireless Sensor Network through Natural Science Computing: A Survey. Journal of Green Engineering 3(4), 383–402 (2013)

Lightweight Trust Model for Clustered WSN

Moutushi Singh[1], Abdur Rahaman Sardar[2], Rashmi Ranjan Sahoo[3],
Koushik Majumder[4], Sudhabindu Ray[3], and Subir Kumar Sarkar[3]

[1] Institute of Engineering & Management, Kolkata, India
[2] Neotia Institute of Technology, Management and Science, WB, India
[3] Department of ETCE, Jadavpur University, Kolkata, India
[4] West Bengal University of Technology, Kolkata, India
{moutushisingh01,abdur.sardar,rashmi.cs2005}@gmail.com,
{koushikzone,sudhabin}@yahoo.com, su_sircir@yahoo.co.in

Abstract. Sensor network's safety measures are built on an unrealistic trusted environment, because proposed trust models for *WSNs* are unsuited for resource supplies and have high computation overhead. This paper proposes a lightweight and realistic trust model for clustered WSNs (LTM). This model has been designed on the dynamic nature of actual trust building mechanism to meet the resource constraint of tiny sensor nodes. A trust metrics priority to emphasize on the important tasks of a sensor node is introduced here. A dynamic trust updating algorithm that ensures brisk drop and sluggish rise of trust is also proposed. Additionally, a self-adaptive weighted method is defined for trust aggregation, to avoid misjudgment of aggregated trust calculation. The proposed trust model also provides better resilience against vulnerabilities. We have tested the feasibility of our trust model with MATLAB.

Keywords: Lightweight trust, WSN, Trust metric priorities, Self adaptive weight.

1   Introduction

Wireless Sensor Networks (*WSNs*) consist of many spatially distributed tiny sovereign devices that cooperatively monitor and react to environment and send the collected data to Base Station (*BS*) using wireless channels. Due to inadequate resources of *WSN*, it is difficult to include basic security functions like authentication, privacy, and key distribution [1-2]. For cluster *WSN*, LEACH [3], EEHC [4] and EC [5] algorithms well improve network scalability and throughput. Here, nodes are grouped into clusters, and within each cluster, a node with strong computing power is designated as Cluster Head (*CH*), which form a higher-level spine network. Trust solves the problem of access control and ensures the reliability of routing paths. With the help of trust factor *CHs* detect malicious nodes. During inter cluster communication, this system also selects trusted routing gateway nodes or other trusted *CHs* via which the sender node communicates with the *BS*. For multi hop clustering a trusted system selects trustworthy routing nodes through which a cluster member (*CM*) sends data to *CH*. [2, 6-12]. In a distributed approach, every node maintains the updated record about the trust values of entire network in a database. Database size and network size

© Springer International Publishing Switzerland 2015   765
S.C. Satapathy et al. (eds.), *Proc. of the 3rd Int. Conf. on Front. of Intell. Comput. (FICTA) 2014*
– Vol. 2, Advances in Intelligent Systems and Computing 328, DOI: 10.1007/978-3-319-12012-6_85

are directly proportional. It is impossible for a single node to store and compute trust value of the entire network. A trust based security of nodes must consider - time, number, amount of interaction and updating of trust value by introducing reward and penalty factor to the trustworthy and faulty nodes.

This paper provides both theoretical essentials and simulation results to validate the design issues. The remnants of this paper are as follows: Section 2: Literature survey, Section 3: Network model besides trust model design, Section 4: Dynamic lightweight trust management scheme for trust decision-making and its assessment, Section 5: Theoretical and simulation-based analyses and evaluation of LTM and Section 6: Conclusion.

2   Literature Survey

Trust evaluation, management and modeling for *WSN* are upcoming research areas. It is based on interaction and reputation of a node. A number of studies like LDTS [1], GTMS [2] and HTMP [8] are proposed for *WSN*s. Yet, *WSN* suffer from many restrictions like incapability to meet the resource constraint requirements. GTMS [2] is based on direct & indirect observation of *CM*s. Based on trust states of all group members; a *CH* detects malicious nodes and forwards a report to the *BS*. It is time based on past interactions or the peer recommendations and requires less memory to store trust records at each node. HTMP [8] considers two aspects of trustworthiness: social trust and QoS. The authors developed a probability model utilizing stochastic Petri net techniques to analyze protocol performance and then validated subjective trust against the objective trust obtained based on ground truth node status. The authors of [11] have defined trust as the ratio of successful transactions to the total transactions made by a node. It is used for a decentralized network. In [12] the authors have given a lightweight trust model for clustering of *WSN*, but they are lacking with the resilience of trust model against various attacks.

3   Network Model

In LTM, we have considered a cluster based *WSN* consisting of multiple clusters each having a *CH* and a number of *CM* (Fig.1).

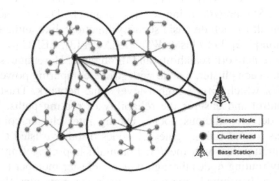

Fig. 1. Scenario of a clustered WSN

Each node monitors its neighbors for a time duration Δt. All monitoring event can be treated as either successful or unsuccessful interaction. If the result of monitoring event matches with the expected result, then it is a successful interaction otherwise it is unsuccessful. At the time of network deployment all nodes are equally trusted. Here direct trust is evaluated by monitoring some events, called trust metric, which is divided into two groups with priority. The trust relationship is classified into two different sheets: Intra cluster and Inter cluster trust formation. In the former, trust relationship is calculated among *CM*s (Direct trust), and between *CH* and *CM*s (Indirect trust). Our trust model effectively recognizes the malicious nodes by the final updated trust values between dealing nodes.

4   Light Weight Trust Management

4.1   Boundary of Trust Value and Trust Metrics Priorities

We have set the domain of trust as unsigned integer between [0, 10] and the initial trust value is 5. This assumption reduces communication and memory overhead.
The most important aspects of trust management are the process of data collection for direct trust calculation. The trust value of a neighboring node can be calculated by using no. of successful and unsuccessful interaction on different trust metrics. Trust metrics are the various QOS characteristics of a node as shown below:

Table 1. Different Trust Metrics and Their Group

Trust Metrics of *Group₁* with high priority (*P=2*)	Data Packet Forwarded	
	Data Packet Message Precession	
	Control Packet Forwarded	
	Control Packet Message Preces.	
	Routing Action	Trust Metrics of *Group₂* with low priority (*P=1*)
	Sensing Communication	
	Range Detection	

Some trust parameters are considered as high priority and some as low. In the above Table no.1 bold faced metrics are grouped into high priority (P_1) with ($P_1=2$) and rest are clubbed into low priority (P_2) with ($P_2=1$). On the basis of successful and unsuccessful interactions of a node, it will be rewarded or penalized. Therefore, we can say priority of group1 is k times of priority of group2.

4.2   Intra Cluster Trust Evaluations

a)   CM to CM Intra cluster direct trust calculation
Considering a reward factor given to a successful interaction and penalty factor given to an unsuccessful interaction the new established trust value between node x & y is:
In eq. (1) the 1st term means the percentage of successful interactions between node x and y at time instant Δt based on the events, belongs to which group of trust metrics (Discussed above). The 2nd term signifies the reward factor. Reward points are given

to every successful interaction between node x and y at time instant Δt based on the events, belonging to a group of trust metrics and its priorities.

$$T_{xy}^{new}(\Delta t) = \left\lceil \begin{pmatrix} 10 * \dfrac{S_{xy}^{G_1}(\Delta t) + S_{xy}^{G_2}(\Delta t)}{S_{xy}^{G_1}(\Delta t) + S_{xy}^{G_2}(\Delta t) + U_{xy}^{G_1}(\Delta t) + U_{xy}^{G_2}(\Delta t)} \end{pmatrix} \times \\ \left(\dfrac{(P_1 * S_{xy}^{G_1}(\Delta t)) + (P_2 * S_{xy}^{G_2}(\Delta t))}{1 + (P_1 * S_{xy}^{G_1}(\Delta t)) + (P_2 * S_{xy}^{G_2}(\Delta t))} \right) \times \\ \left(\dfrac{1}{\sqrt{(P_1 * U_{xy}^{G_1}(\Delta t)) + (P_2 * U_{xy}^{G_2}(\Delta t))}} \right) \right\rceil \quad (1)$$

The reward factor ensures that there is slow rise in trust value with an increase in successful transactions between node x and y. The 3rd term signifies strict penalty factor for unsuccessful interactions between node x and y at time instant Δt based on the events, belonging to that group of trust metrics and its priorities, and ensuring the rapid drop of trust value to zero with an increase in unsuccessful interactions between node x and y. $\lceil . \rceil$ represents the ceiling function, which will give the nearest unsigned integer value of calculated trust.

b)   CH to CM Intra cluster indirect trust calculation
The indirect or recommended trust is computed with trust circulation. In trust propagation, trust transits through third parties. Assume a node x seeks to establish indirect trust on y (When there is no direct trust of node x on y). Node y has to ask for the direct trust of neighbors A_i on node y. Hence the indirect trust is propagated through x - A_i -y. The critical factors for determining the indirect trust are:
   I.   Neighbor nodes which give opinion on the targeted node.
   II.   Calculation of indirect trust value based on recommendations.

$$T = \begin{bmatrix} T_{11} & T_{12} & \cdots & T_{1,n-1} \\ T_{21} & T_{22} & \cdots & T_{2,n-1} \\ \cdots & & \cdots & \cdots \\ T_{n-1} & T_{n-1,2} & \cdots & T_{n-1,n-1} \end{bmatrix}$$

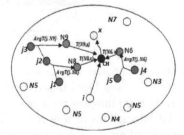

Fig. 2. Trust Matrix   **Fig. 3.** Trust calculation at CM level

Regarding the first factor, our method gathers opinion about the targeted node from all the trusted nodes only.

The *CH* intermittently collects the direct trust value of its *CMs* and maintains these trust values in a matrix T (as shown in Fig. 2).

For the calculation of indirect trust, we are considering, the trusted neighbors and the average trust value of those trusted neighbors, given by their neighbor nodes (Fig. 3). Hence, the recommendation trust value of x given by *CH* is:

$$T_{CH_{,x}}^{indirect} = \frac{\sum_{i \in M} \sqrt{T_{i,x} \times avg_{j \in N}(T_{j,i})}}{|M|} \quad (2)$$

Where M is set of all the trusted neighbors of x. N is the set of neighbors of each trusted node of set M.

4.3 Intra Cluster Trust Evaluations

In cluster *WSNs*, *CHs* forms an inherent vertebra for inter cluster routing where *CHs* can forward the aggregated data to the central *BS* through other *CHs*. For loyal communication, selection of *CHs* is an important step. In this model, the inter cluster trust is evaluated by two information sources: *CH*-to-*CH* direct trust (T_{CH_i,CH_j}^{direct}) and *BS*-to-*CH* indirect recommended trust ($T_{BS,CH_i}^{indirect}$). In agreement with the characteristics of clustered *WSNs*, both *CMs* and *CHs* are resource-restricted nodes, and *BSs* have more computing and storage capability. Hence, energy conservation is a basic requirement for trust calculation at *CHs*. Here also the trust is calculated in the same way as in intra cluster format.

4.4 Self-adaptive Weight Based Trust Aggregation at BS

Equation 3 shows adaptive weight for aggregating the trust at *BS*.

$$T_{CH_i,CH_j}^{Total}(\Delta t) = \left\lceil 10 \times \left((w_1 \times T_{CH_i,CH_j}^{direct}(\Delta t)) + (w_2 * T_{BS,CH_i}^{indirect}(\Delta t)) \right) \right\rceil \quad (3)$$

Where w_1 and w_2 are the weights of the direct trust value of CH_i on CH_j and indirect trust value calculated by *BS* respectively and $w_1 + w_2 = 1$. Moreover, the value of weight factors w_1 and w_2 has been given sensibly with an adaptive weight mechanism. Most of the literature cited here has used expert opinion for assigning the value of weight in trust aggregation manually. Earlier studies have mentioned that trust aggregation is application dependent without providing any detailed insight to assign the proper weight, and lacks adaptability. This weight assignment method may lead to slip-up of aggregated trust calculation. Hence, assignment of proper weight age to both direct and indirect trust is necessary, in order to adopt the dynamic distributed environment. So, without any refinement, we have given a self-adaptability method for weight sharing to both direct and indirect trust at *CH* level. A self-adaptive weight for w_1 and w_2 are

$$w_1 = 1 - \left(S_{BS,CH_i}^{indirect} \Big/ \left(S_{CH_i,CH_j}^{direct} + S_{BS,CH_i}^{indirect} \right) \right) \quad (4)$$

$$w_2 = 1 - \left(S_{CH_i,CH_j}^{direct} \Big/ \left(S_{CH_i,CH_j}^{direct} + S_{BS,CH_i}^{indirect} \right) \right) \quad (5)$$

Where S_{CH_i,CH_j}^{direct} is the total no. of successful interaction between CH_i and CH_j at time ΔT, while calculating the inter cluster direct trust. $S_{BS,CH_j}^{indirect}$ is the total positive recommendation about CH_i collected by (i.e. $s_{BS,CH_i}^{indirect} \geq 5$) BS from other neighboring cluster head of CH_i .

As a requisite property of the self-adaptability weight factors, w_1 gives more weight age to direct trust whenever there are more no. of successful interactions between CH_i and CH_j compared to total no. of positive recommendations of CH_i. Similarly, w_2 gives more weight age to indirect trust whenever there are more no. of positive recommendations about CH_i than no. of successful interactions between CH_i and CH_j.

5   Simulation Results and Overhead Analysis

Definition 1: A node (CMs or CHs) is said to be good, if it interacts successfully for threshold number of time with other nodes and also provides true recommendation value as when necessary. In our protocol a node say CM_y is said to be trusted for another node CM_x if its trust value $T_{xy} \geq 5$ at time Δt .

Definition 2: A node (CMs or CHs) is said to be bad, if it transacts unsuccessfully most of the time with other nodes and deceives other nodes with various attacks. In this protocol a node CM_y is said to be malicious for CM_x , if it has interacted with CM_x at least up to threshold number of transaction Th_{trans} and $U_{xy} > S_{xy}$.

Definition 3: If the updated trusted value between two CM nodes x and y: $T_{x,y}(\Delta t)$ value is greater than or equal to 5 then CM_y is said to be trusted for CM_x . Similarly, a CH_j is said to be trusted for CH_i if $T_{CH_i,CH_j}(\Delta t) \geq 5$.

Proof:
We have analyzed and proved that our proposed trust management system is robust against the malicious node because it can detect malicious behavior and can prevent such nodes from fulfilling their objectives.

Let $U_{xy} \big/ S_{xy} = a > 1$

Given $U_{xy} > S_{xy}$, we can derive $a \geq 1$. Hence given that $U_{xy} + S_{xy} \neq 0$

$$T_{xy}^{new}(\Delta t) = 10 \frac{Sxy}{Sxy + Uxy}\left(1 - \frac{1}{1 + Sxy}\right)\left(\frac{1}{\sqrt{Uxy}}\right)$$

$$= 10 \times \frac{1}{1 + \dfrac{Uxy}{Sxy}} \times \frac{Sxy}{1 + Sxy} \times \frac{1}{\sqrt{a \times Sxy}} (\because \frac{Uxy}{Sxy} = a \therefore Uxy = a \times Sxy)$$

$$= \left(\frac{10}{1 + a}\right)\left(\frac{Sxy}{1 + Sxy}\right)\left(\frac{1}{\sqrt{aSxy}}\right) = \frac{10}{\sqrt{a}(1 + a)\left(\dfrac{1}{\sqrt{Sxy}} + \sqrt{Sxy}\right)}$$

As $a > 1$ and $S_{xy} \geq 1$ so, $(1 + a) \geq 2$ and $\left(\sqrt{Sxy} + \dfrac{1}{\sqrt{Sxy}}\right) \geq 2$ and $\sqrt{a} > 1$

$$\therefore \sqrt{a\,(1+a)\left(\frac{1}{\sqrt{Sxy}}+\sqrt{Sxy}\right)} \geq 4 \quad \therefore T_{xy}(\Delta t) \leq \left\lceil\frac{10}{4}\right\rceil < 5 \quad \text{(Proved)}$$

5.1   Communication Overhead Analysis

If N is the no. of nodes and m is no. of clusters in the network, then the average no. of nodes i.e. average size of the clusters is $n = N/m$.

a)   Intra cluster Communication overhead evaluation
Let node x hunt for the recommended trust about node y within a cluster. x sends one request to its CH and CH sends the recommended trust value about y as a reply. CH will collect the recommendation from $(n-2)$ no. of neighbors of node y. So, the communication overhead is: $[1+ (n-2) +1] = n$ packets. If node x needs the recommendation for all nodes then the communication overhead is $n\,(n-1)$. If all nodes want to communicate with each other then maximum communication overhead is $n \times n(n-1) = n^2(n-1)$.

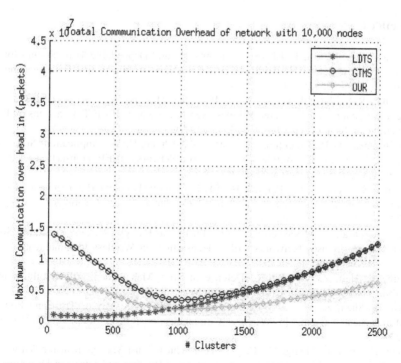

Fig. 4. Maximum communication overheads of entire network with 10,000 nodes

b)   Inter cluster Communication overhead evaluation
Here if a CH_i wants to communicate with another CH_j it will send one request to BS which also sends one reply packet. But BS also has to send the recommendation from

(m-2) CHs. So the communication overhead is: *[1+1+ (m-2)]* =*m*. If CH_i wants to communicate with all other *CHs* then maximum communication overhead is: *m (m-1)*. Thus, maximum communication overhead for entire network is:

$$m\left[n^2(n-1)+m(m-1)\right]$$

6   Conclusion

This paper presents a realistic trust model for a large scale clustered *WSN*, focusing on protecting systems from various attacks. Considering the resource constrained wireless node, we have given a light weight trust model to minimize the computation overhead. Extensive simulation shows that this model has less communication and storage overhead as compared to GTMS and LDTS. Furthermore, our model provides 100% resilience with 30% of malicious node in large scale network. This proves that it is robust against various attacks. Further research can include incorporating the proposed model in a newly designed clustering, routing protocol and study the joint effects of various attacks on this model.

References

1. Li, X., Zhou, F., Du, J.: LDTS: A Lightweight and Dependable Trust-System for Clustered Wireless Sensor Networks. IEEE Transactions on Information Forensics and Security 8(6), 924–935 (2013)
2. Shaikh, R.A., Jameel, H., d'Auriol, B.J., Lee, H., Lee, S.: Group-based Trust Management Scheme for Clustered Wireless Sensor Networks. IEEE Transactions on Parallel Distributed. Systems 20(11), 1698–1712 (2009)
3. Heinzelman, W.B., Chandrakasan, A.P., Balakrishnan, H.: An Application- Specific Protocol Architecture for Wireless Micro sensor Networks. IEEE Transaction on Wireless Communication 1(4), 660–670 (2002)
4. Kumar, D., Aseri, T.C., Patel, R.B.: EEHC: Energy Efficient Heterogeneous Clustered Scheme for wireless Sensor Networks. Computer Communications 32(4), 662–667 (2009)
5. Jin, Y., Vural, S., Moessner, K., Tafazolli, R.: An Energy-Efficient Clustering Solution for Wireless Sensor Networks. IEEE Transaction on Wireless Communication 10(11), 3973–3983 (2011)
6. Sun, Y., Han, Z., Liu, K.J.R.: Defense of Trust Management Vulnerabilities in Distributed Networks. IEEE Communication Magazine 46(2), 112–119 (2009)
7. Yu, H., Shen, Z., Miao, C., Leung, C., Niyato, D.: A Survey of Trust and Reputation Management Systems in Wireless Communications. Proc. IEEE 98(10), 1752–1754 (2010)
8. Bao, F., Chen, I., Chang, M., Cho, J.: Hierarchical Trust Management for Wireless Sensor Networks and its Applications to trust-based Routing and Intrusion Detection. IEEE Transaction on Network Service Management 9(2), 169–183 (2012)
9. Zhan, G., Shi, W., Deng, J.: Design and Implementation of TARF: A Trust-Aware Routing Framework for WSNs. IEEE Transaction on Depend 9(2), 184–197 (2012)

10. Aivaloglou, E., Gritzalis, S.: Hybrid trust and reputation management for sensor networks. Wireless Network 16(5), 1493–1510 (2010)
11. Sahoo, R.R., Singh, M., Sahoo, B.M., Majumder, K., Ray, S.-H., Sarkar, S.K.: A Lightweight Trust Based Secure and Energy Ef-ficient Clustering in Wireless Sensor Network: Honey Bee Mating Intelligence Approach. Procedia Technology, 27–28 (September 2013)
12. Sahoo, R.R., Sardar, A.R., Singh, M., Ray, S., Sarkar, S.K.: Trust Based Secure and Energy Efficient Clustering in Wireless Sensor Network: A Bee Mating Approach. In: Maji, P., Ghosh, A., Murty, M.N., Ghosh, K., Pal, S.K. (eds.) PReMI 2013. LNCS, vol. 8251, pp. 100–107. Springer, Heidelberg (2013)

An Efficient and Secured Routing Protocol for VANET

Indrajit Bhattacharya[1], Subhash Ghosh[2], and Debashis Show[2]

[1] Department of Computer Application, Kalyani Government Engineering College, Kalyani,
Nadia, West Bengal, India
indra51276@gmail.com
[2] Department of Computer Science, Barrackpore Rastraguru Surendranath College,
Barrackpore, West Bengal, India
{subhash.80131,deba.show}@gmail.com

Abstract. Since few years, Vehicular Adhoc Networks(VANET) deserve much
attention. Routing and security are the two most important concerns in this type
of network. A number of routing protocols already exist for vehicular ad hoc
network but none of them are made to handle routing and security issues side
by side. In this paper we propose a new junction-based geographical routing
protocol which is capable of dealing with routing as well as security issues.
This protocol consists of two modules: (i) To implement routing, at first this
protocol selects appropriate junctions dynamically, through which a packet
must transmit in order to reach to its destination. (ii) To deal with security
issues, we have introduced the concept of mix-zones to prevent vehicle tracking
by unauthorised users. At the end, the performance of the proposed work has
been compared with some well known existing routing protocols depending
upon some parameters like packet delivery ratio and normalized routing load in
a simulated environment.

Keywords: Score, Junction selection, GyTAR, mix-zones.

1   Introduction

Inter-vehicle communication is a fast growing research topic in the academic sector
and industry. Through this kind of communication, vehicles are able to communicate
with each other by using wireless technology like WLAN. As a result, they can be
organized in vehicular ad hoc networks (VANETs). VANETs are a special version of
mobile ad hoc networks (MANETs). MANETs have no fixed infrastructure and
instead rely on ordinary nodes to perform routing of messages and network
management functions. However, vehicular ad hoc networks behave in different ways
than conventional MANETs. Driver behaviour, mobility constraints, and high speeds
create unique characteristics of VANETs. These characteristics have important
implications for designing decisions in these networks.

A number of routing protocol already exists for VANET, but many of them uses
the same protocols which was used for MANET earlier and thus they do not perform
well. After a survey it has been noticed that geographical routing protocols performs

© Springer International Publishing Switzerland 2015
S.C. Satapathy et al. (eds.), *Proc. of the 3rd Int. Conf. on Front. of Intell. Comput. (FICTA) 2014*
– *Vol. 2*, Advances in Intelligent Systems and Computing 328, DOI: 10.1007/978-3-319-12012-6_86

very well than proactive and reactive protocols and that is why we concentrate on geographical routing protocols such as GPSR, GSR and GyTAR for our motivation. In our work we have used road junction based concept for establishment of route between source and destination and after that a greedy packet forwarding technique is used to route the data packet towards the destination

Security is also a very important concern for VANET. But after a survey it has been observed that routing and security are implemented separately in the existing protocols. A suitable routing protocol for VANET is still missing which can handle routing as well as security together. In urban vehicular networks where the privacy, especially the location privacy of vehicles should be guaranteed and this is our main area of interest for security implementation in VANET. In this paper we have introduced the concept of mix-zones for location privacy preservation.

The rest of the paper is organized as follows. The next section describes related work. Section 3 describes objective and scope of our work. Section 4 describes the proposed framework for VANET. In section 5 we have described our suggested algorithm and section 6 describes experimental setup and results.

2   Related Work

We have made a survey of various existing routing protocols for VANET. The protocol has been divided into three main categories: (a)   Proactive routing protocols, (b) Reactive routing protocols and (c) Position based routing protocols. Proactive routing protocols like FSR [11, 12, 13] are mostly based on shortest path algorithms. Connected node keep information in form of tables, that is why these protocols are table based. Reactive routing protocols such as AODV [2] and DSR [3] are called on demand routing because it starts route discovery when a node needs to communicate with another node thus it reduces network traffic. Position or geographic routing protocols like GPSR [4, 14], GSR [5] and GyTAR [6] are based on the positional information in the routing process, where the source sends a packet to the destination using its geographic position rather than using the network address. This protocol required each node is able to decide its location and the location of its neighbours through the Geographic Position System (GPS) assistance.

3   Objective and Scope

Our area of interest is the GyTAR protocol because it is one of the effective routing protocols proposed for VANET. It also have some drawbacks like (i) it cannot deal with the local optimal problem, (ii) It cannot handle the sparse network problem. Our objective is to handle those drawbacks belongs to GyTAR routing protocol. We have assumed some input information to be provided by GPS, GLS and IFTIS. We have assumed that every vehicle is aware of its own position using GPS. Furthermore, a sending node needs to know the current geographical position of the destination in order to make the routing decision. This information is assumed to be provided by a location service like GLS [7] (Grid Location Service). Moreover, we consider that each vehicle can determine the position of its neighbouring junctions through pre-loaded digital maps. The presence of such kind of map is a valid assumption, when vehicles are equipped with on-board navigation system. We have also assumed that every vehicle is aware of the vehicular traffic (number of vehicles between two junctions). This information can be provided by IFTIS: a completely decentralized mechanism for the estimation of traffic density in a road traffic network.

4   Proposed Framework

Traditionally, system architecture for VANET can be divided into different forms according to different perspective. From the vehicular communication perspective, it can be categorized into road-vehicle communication (RVC, also called C2I) systems and inter-vehicle communication (IVC, also called C2C) systems. For road-vehicle communication Road side units (RSU) are used. These units are placed at the side of the road and they have higher communication range than the vehicle with higher storage capacity. In our proposed frame work we have placed RSUs at the junctions of the road as well as at the road side. RSUs present at the junctions perform an important role during junction selection mechanism in our proposed work. We have also generated mix-zones only at the junctions of the road, where id of vehicles changes automatically. The reason behind creating mix-zones only at the junctions is that it increases difficulty to a great extent during vehicle tracking by unauthorised users. As soon as a vehicle enters into a mix-zone the id of that vehicle changes. As the mix-zone is located at the junction of the road, hence an unauthorised user cannot determine the direction of the vehicle which he was trying to track. Figure1 depicts our proposed framework for VANET. In the proposed work we like to propose a novel junction selection technique for establishing the path between source and destination. After that, we have used a greedy forwarding technique to route the data packet towards the destination. We have also introduced a concept for dealing with security threats like vehicle tracking by the unauthorized person.

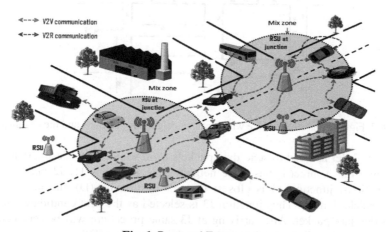

Fig. 1. Proposed Framework

4.1   Junction Selection Technique

In this technique the different junctions that the packet has to traverse in order to reach to the destination are chosen dynamically in a sequential fashion. It considers vehicular traffic density, historical traffic data between junctions and the distance from the source to destination. While selecting the next destination junction, a node (the sending vehicle or an intermediate vehicle in a junction) looks for the position of the neighbouring junctions using the map. A score is given to each junction considering the traffic density, historical traffic data (the average density of the road for a certain duration) of the road between two junction points and the distance from the junction to the destination. The best destination junction (the junction with the highest score) is the geographically closest junction to the destination vehicle having the highest vehicular traffic.

4.2   Technique of Giving Score to a Junction

Let I is the current junction, J is the next candidate junction, Dj and Di are the distance from the candidate junction J to destination and distance from the current junction to destination respectively. Then, $D_p = D_j/D_i$ where D_p determines the closeness of the candidate junction to the destination point. H_p is the probability of historical traffic data between I and J. N_v is the total number of vehicles between junctions I and J.

$$Score(S) = D_p + Nv/H_p \qquad (1)$$

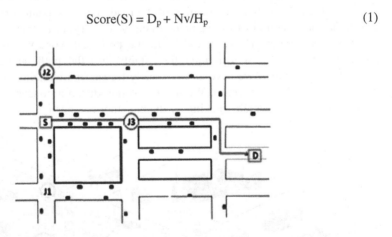

Fig. 2. Figure consisting source with its three neighbour junctions and the destination

In figure 2, let us need to send data packets from source(S) to destination (D). We can observe that source(S) has three neighbouring junctions J1, J2 and J3. Among these junctions, junction J3 is closest towards destination (D) and also it has the highest vehicle density. Thus junction J3 is selected as the next candidate junction to forward the data packet. After arriving at J3 same procedure will be repeated and it will be continued until the packet reaches to the destination (D).

4.3   Packet Forwarding Technique

Once the destination junction is determined, an improved greedy strategy is used to forward packets towards the selected junctions. For that, all data packets are marked by the location of the destination junction. Each vehicle maintains a neighbour table. This table is updated through hello messages exchanged periodically by all vehicles. Thus, when a packet is received, the forwarding vehicle computes the new predicted position of each neighbour using the recorded information in the table, and then selects the next hop neighbour (i.e. the closest to the destination junction) accordingly. Figure 3 shows the flowchart of the proposed scheme.

4.4   Dealing with Local Optimal Problem

It is the problem when a node has no other node to forward the data packet. To deal with situation we are going to include "store and forward" technique where the forwarding vehicle of the packet will carry the packet until the next junction or until another vehicle, closer to the destination junction, enters/reaches its transmission range.

4.5   Dealing with Sparse Network Problem

Sparse network refers to the very low density network where there is not enough number of nodes to satisfy the routing needs. To deal with this situation we include RSUs at the junctions with a high transmission range (500m).

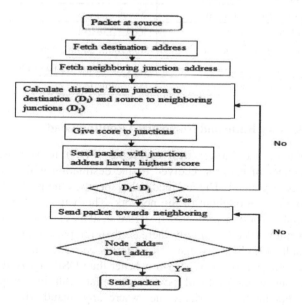

Fig. 3. Flowchart of the packet forwarding procedure

4.6   Location Privacy Preservation

Preserving location privacy can be achieved only when we can prevent unauthorised users from vehicle tracking. Here mix-zones at the junctions prevent unauthorised users from tracking a particular vehicular. As soon as a vehicle enters into a mix-zone the id of that vehicle changes, now as the mix-zone is at the junction of the road, hence an unauthorised user cannot determine the direction of the vehicle which he was trying to track.

5   Experimental Setup and Results

In this paper, various parameters such as Packet Delivery Ratio, Normalized Routing Load, Number of Dropped Packets and Packet Loss are investigated based on a variety of vehicle density and vehicle velocity. The Vanet-Sim simulation framework is used for simulation purposes. Table 1 illustrates the characteristics of the environment in which the simulation is experimented. At first we have investigated the performance of the protocol With respect to packet delivery ratio based on a variety of vehicle speed and vehicles communication distance. Then later we compare our protocol with existing protocols like DSR and AODV.

Table 1. Simulation Parameters

Parameter Type	Value
Network Simulator	Vanet-sim simulator
Routing Protocol tested	AODV, DSR
Maximum simulation time	200sec
Simulation Area	3069 * 3591 m^2
Transmission Range used	50m,100m,200m,300m
Speed of vehicles used	60km/h, 80km/h, 110 km/h
Traffic source/destination	Deterministic

5.1   Packet Delivery Ratio and Normalized Routing Load

In order to calculate the Packet Delivery Ratio (PDR) in velocity and density scenarios, the number of packets received by the destination will be divided by the number of packets originated. The attained value specifies the packet loss rate which confines the maximum throughput of the network. The better PDR implies the more accurate and suitable routing protocol. In figure 4, it is clear that in our proposed protocol, if the vehicle speed is rapidly increase than the packet delivery ratio is also increase as a result in highway scenario it performs better. From figure 5, it can be obtained that, initially our protocol performs better than DSR and AODV. At the end AODV and the suggested protocol performs similar. But the proposed protocol additionally can handle the security issue, where any unauthorized person cannot track the vehicle id. In figure 6 (density diagram), it is observed that our protocol and

DSR is far better than AODV in terms of packet delivery ratio based on vehicle density. It is so because after establishing the path using junction selection mechanism, we have used greedy forwarding technique for sending the packet towards destination and here we maintain a neighbour table for each vehicle. Increasing vehicle density means increasing chance to have more neighbour and thus chance for forming a better stable network. Initially DSR performs same as our protocol. At the end we have seen DSR performs better than the proposed scheme but it does not have any mechanism to implement security issue. Normalized routing load (NRL) is defined as the number of routing packets transmitted per data packet arrived at the destination. Figure 7 depicts the NRL values associated with mentioned routing protocols. It turns out that the AODV routing protocol has the best NRL value in comparison with other routing protocol. But in our proposed work we introduce the concept of mix-zone to improve location privacy of the vehicle which is missing in the case of AODV.

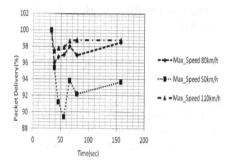

Fig. 4. Packet delivery ratio based on vehicle speed

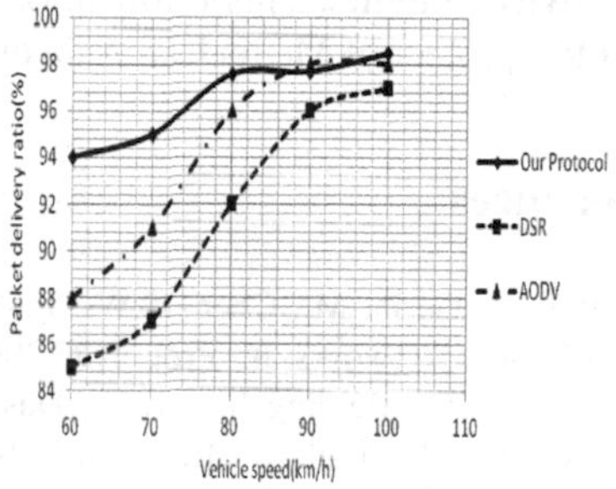

Fig. 5. Comparison with AODV and DSR for packet delivery ratio on vehicle speed

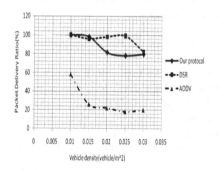

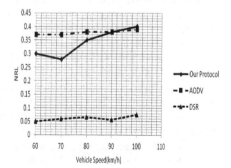

Fig. 6. Comparison with AODV and DSR for packet delivery ratio on vehicle density

Fig. 7. Comparison with AODV and DSR for NRL based on vehicle speed

6   Conclusions

In this paper we have proposed a routing protocol which combines the position-based routing with topological knowledge, as a promising routing strategy for vehicular ad hoc networks in city environments. We have demonstrated by a simulation study that made use of realistic vehicular traffic in a city environment that our protocol outperforms topology-based approaches like DSR and AODV with respect to delivery rate is some cases. We are currently trying to extend the work into the following two directions. First, we are going to perform large scale simulations in order to get results independently from the structure of a given scenario. Second, we are going to work on the more tightly security to prevent different attack like Sybil attack, Reply attack etc.

Acknowledgements. This Publication is an outcome of the R&D work undertaken in the ITRA project of Media Lab Asia entitled "DISARM".

References

1. Pei, G., Gerla, M., Chen, T.W.: Fisheye State Routing: A Routing Scheme for Ad Hoc Wireless Networks. In: Proc. ICC 2000, New Orleans, LA (June 2000)
2. Perkins, C., Belding Royer, E., Das, S.: Ad hoc On-Demand Distance Vector (AODV) Routing. Internet Draft, draft-ietf-manet-aodv-13.txt, Mobile Ad Hoc Networking Working Group (2003)
3. Johnson, D.B., Maltz, D.A.: Dynamic Source Routing in Ad Hoc Wireless Networks. In: Imielinski, T., Korth, H. (eds.) Mobile Computing, vol. ch. 5, pp. 153–181. Kluwer (1996)
4. Karp, B., Kung, H.T.: GPSR: greedy perimeter stateless routing for wireless networks. In: Proc. Mobile Computing and Networking, pp. 243–254 (2000)
5. Lochert, C., Hartenstein, H., Tian, J., Fussler, H., Hermann, D., Mauve, M.: A routing strategy for vehicular ad hoc networks in city environments. In: Proc. Intelligent Vehicles Symposium, June 9-11, pp. 156–161 (2003)
6. Moer, J., Sidi-Mohammed, S., Rabah, M., Yacine, G.: An Improved Vehicular Ad-Hoc Routing Protocol for City Environments. In: Proc. IEEE ICC 2007, pp. 3972–3979 (2007)
7. Li, J., Jannotti, J., De Couto, D., Karger, D., Morris, R.: A scalable location service for geographic ad hoc routing. In: Proc. ACM/IEEE MOBICOM 2000, pp. 120–130 (2000)
8. Network Simulator (NS)-2, http://www.isi.edu/nsnam/ns
9. Fiore, M., Härri, J.: The Networking Shape of Vehicular Mobility. In: Proc. ACM MobiHoc 2008, pp. 261–272 (2008)
10. Moravejosharieh, A., Modares, H., Salleh, R., Mostajeran, E.: Performance Analysis of AODV, AOMDV, DSR, DSDV Routing Protocols in Vehicular Ad Hoc Network. Proc. Research Journal of Recent Sciences 2(7), 66–73 (2013)
11. Lee, K.C., Lee, U., Mario, G.: Survey of Routing Protocols in Vehicular Ad Hoc Networks. Advances in Vehicular Ad-Hoc Networks: Developments and Challenges Reference, 149–170 (2010)

12. Yatendra, M.S., Saurabh, M.: A Contemporary Proportional Exploration of Numerous Routing Protocol in VANET. Proc. International Journal of Computer Applications 50(21), 14–21 (2012)
13. Sandhaya, K., Bandanjot, K., Sabina, B.: A comparative study of Routing Protocols in VANET. Proc. International Journal of Computer Science Issues 8(4) (2011)
14. Al-Doori, M.M.: Directional Routing Techniques in VANET.PhD Thesis, Software Technology Research Laboratory, De Montfort University,Leicester United Kingdom (November 2011)

Authentication of the Message through Hop-by-Hop and Secure the Source Nodes in Wireless Sensor Networks

B. Anil Kumar, N. Bhaskara Rao, and M.S. Sunitha

Computer Network and Engineering Programme,
Department of Computer Science and Engineering,
Dayananda Sagar College of Engineering, Bangalore, India
{abnaidu90,bhaskararaonadahalli,sunithashivu}@gmail.com

Abstract. Message authentication is the most effective way to protect the data from unauthorized access and corrupted messages being forwarded in wireless sensor networks (WSNs). For this reason, many message authentication schemes have been developed, based on either symmetric-key cryptosystems or public-key cryptosystems. Some have the limitations of high computational and communication overhead and lack of scalability to node compromise attacks. To address these issues, a polynomial-based scheme was recently introduced. However, this scheme and its extensions all have the weakness of a built-in threshold determined by the degree of the polynomial when the number of messages transmitted is larger than this threshold, the adversary can fully recover the polynomial. While enabling intermediate nodes authentication, the proposed scheme allows any node to transmit an unlimited number of messages without suffering the threshold problem. In addition VGuard Security framework is used provide source privacy in the network. Both theoretical analysis and simulation results demonstrate that our proposed scheme is more efficient than the polynomial-based approach in terms of computational and communication overhead for various comparable security levels while providing message source privacy.

Keywords: WSN (Wireless Sensor Networks), Source security, Message authentication, Symmetric-key cryptography.

1 Introduction

A sensor network is defined as being composed of a large number of nodes which are deployed densely in close proximity to the phenomenon to be monitored. Each of these nodes collects data and its purpose is to route this information back to a sink.[1] The network must possess self-organizing capabilities since the positions of individual nodes are not predetermined. Cooperation among nodes is the dominant feature of this type of network, where groups of nodes cooperate to disseminate the information gathered in their vicinity to the user.

© Springer International Publishing Switzerland 2015   785
S.C. Satapathy et al. (eds.), *Proc. of the 3rd Int. Conf. on Front. of Intell. Comput. (FICTA) 2014*
– *Vol. 2*, Advances in Intelligent Systems and Computing 328, DOI: 10.1007/978-3-319-12012-6_87

The designer must keep in mind that it is a limited resource that will be exhausted. In this case, the designer's task is more complicated since he Any wireless node there are three major modes of operation: transmitting, receiving and listening. When the node is in listening mode the energy expenditure is minimal. However, if the node spends most of the time listening then this mode is responsible for a large portion of the consumed energy (as is the case in sensor networks). A useful distinction presented in the paper refers to whether energy is treated as a cost function or as a hard constraint. In the former case, the objective of the designer is to minimize the amount of energy per communication task, treating energy as an expensive but inexhaustible resource. However, when energy hassatisfy conflicting objectives: maximizing the longevity of the network vs communication performance (throughput, total data delivered, etc)

The organization of the paper is as follows Section 2 presents the Related work, section 3 presents the Research Work, Section finally in section 4 the Conclusion is presented.

2   Related Work

The symmetric-key based approach requires complex key management, lacks of scalability, and is not resilient to large numbers of node compromise attacks since the message sender and the receiver have [9]to share a secret key. The shared key is used by the sender to generate a message authentication code (MAC) for each transmitted message. However, for this method, the authenticity and integrity of the message can only be verified by the node with the shared secret key, which is generally shared by a group of sensor nodes. [3]An intruder can compromise the key by capturing a single sensor node. In addition, this method does not work in multicast networks.

In these schemes, each symmetric authentication key is shared by a group of sensor nodes. An intruder can compromise the key by capturing a single sensor node. Therefore, these schemes are not resilient to node compromise attacks. Another type of symmetric-key scheme requires synchronization among nodes.

These schemes can also provide message sender authentication. However, this scheme requires initial time synchronization, which is not easy to be implemented in large scale WSNs. In addition, they also introduce delay in message authentication, and the delay increases as the network scales up.

3   Research Work

To facilitate data collection in such a network, sensor nodes on a path from an area of interest to the base station can relay the data to the base station. The unattended nature of the deployed sensor network lends itself to several attacks by the adversary, including physical destruction of sensor nodes, security attacks on the routing and data link protocols, and resource consumption attacks launched to deplete the limited energy resources of the sensor nodes. Unattended sensor node deployment also makes another attack easier: an adversary may compromise several sensor nodes, and then use the compromised nodes to inject false data into the network. This attack falls in the category of insider attacks.[3] Standard authentication mechanisms are not sufficient to prevent such insider attacks, since the adversary knows all the keying material possessed by the compromised nodes. We note that this attack can be launched against many sensor network applications, though we have only given a military scenario.

Here, the new method propose an unconditionally secure and efficient source anonymous message authentication scheme, [1] based on (MES). This MES scheme is secure against no-message attacks and adaptive chosen-message attacks in the random oracle model. Our scheme enables the intermediate nodes to authenticate the message so that all corrupted packets can be dropped at intermediate nodes to conserve sensor power.

While achieving compromise-resiliency, flexible-time authentication and source identity protection, our scheme does not have the threshold problem. Both theoretical analysis and simulation results demonstrate that our proposed scheme ismore efficient than the polynomial-based algorithms under comparable security levels. To the best of our knowledge, this is the first scheme that provides hop-by-hop node authentication without the threshold limitation, while having performance[7][8] better than the symmetric-key based schemes.

Consider a military application of sensor networks for reconnaissance of the opposing forces. Suppose we want to monitor the activities of the opposing forces, e.g., tank movements, ship arrivals or departures, and other relevant events. To achieve this goal, we can deploy a cluster of sensor nodes around each area of interest. [4]We can then deploy a base station in a secure location to control the sensors and collect data reported by the sensors.

ALGORITHM 1: Message Authentication
A SAMA consists of the following two algorithms:

[I] Generate (m,Q1,Q2,..,Qn): Given a message m and the public keys Q1,Q2,..,Qn of the Authentication Set AS = {A1,A2, · · · ,An}, the actual message sender At, $1 \leq t \leq n$, produces an anonymous message[2]S(m).

[II]Verify S(m): Given a message m and an anonymous message S(m), which includes the public keys of all members in the AS, a verifier can determine whether S(m) is generated by a member in the AS.

ALGORITHM 2: Verification[2]

i) Checks that QA = O, otherwise invalid.
ii) Checks that QA lies on the curve.
iii)Checks that n QA = O.

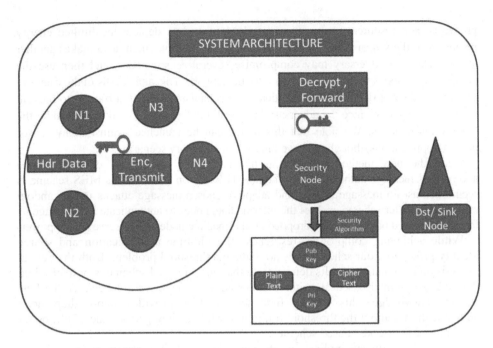

Fig. 1. System Architecture for Proposed Methodology

4   Implementation and Results

The proposed technique has been simulated in NS 2.34. These have been made assuming a network having dimensions 300 x 300 meters. The number of nodes is assumed to be 50. The nodes are generated and placed randomly. The Figure 2 shows node deployment using NS2 network animator.

Table 1. Node configuration

Sl No.	PARAMETER	VALUE
1	Channel Type	Wireless
2	Antenna Type	Omni Antenna
3	MAC Type	MAC 802_11
4	Queue Type	DropTail
5	Propogation Type	Two Way ground
6	MAC Trace	ON
7	Router Trace	ON
8	Traffic Agent	UDP

Table 2. Simulation parameter

Sl No.	PARAMETERS	VALUE
1	Simulator	NS 2.34
2	Number of Nodes	50
3	Simulation time	600 Secs
4	Traffic Source	CBR
5	Delay	5 M/s

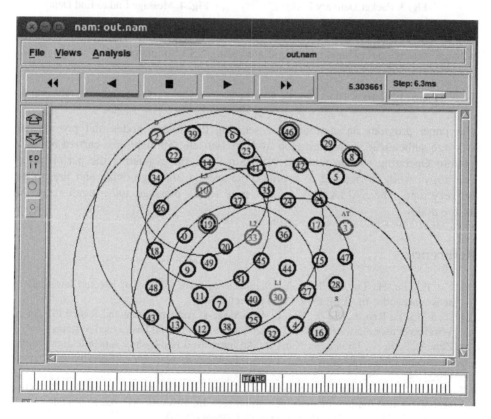

Fig. 2. Simulation Nam Animator in NS 2

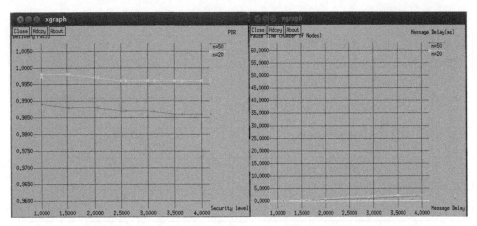

Fig. 3. Packet Delivery Ratio   **Fig. 4.** Message End to End Delay

The above research work Evaluates the packet delivery ratio for 50 nodes and the average message delay have been simulated using NS2 Simulator.

5   Conclusion

The paper provides an approach for securing the source nodes and provide the message authentication through Hop by Hop. Here the Simulation is carried out on a Ubuntu Operating system with the NS2 Simulator for evaluating the performance. The above work carried out to evaluate the average message delay and the packet delivery ratio. The SAMA provides the more secure message authentication for the source nodes.

References

[1] Ye, F., Lou, H., Lu, S., Zhang, L.: Statistical en-route filtering of injected false data in sensor networks. In: IEEE INFOCOM (March 2004)
[2] Li, J., Li, Y., Ren, J., Wu, J.: Hop-by-Hop Message Authentication and Source Privacy in Wireless Sensor Networks. IEEE Transactions on Parallel and Distributed Systems
[3] Zhu, S., Setia, S., Jajodia, S., Ning, P.: An interleaved hop-by-hop authentication scheme for filtering false data in sensor networks. In: IEEE Symposium on Security and Privacy (2004)
[4] Blundo, C., De Santis, A., Herzberg, A., Kutten, S., Vaccaro, U., Yung, M.: Perfectly-secure key distribution for dynamic conferences. In: Brickell, E.F. (ed.) CRYPTO 1992. LNCS, vol. 740, pp. 471–486. Springer, Heidelberg (1993)
[5] Zhang, W., Subramanian, N., Wang, G.: Lightweight and compromiseresilient message authentication in sensor networks. In: IEEE INFOCOM, Phoenix, AZ, April 15-17 (2008)
[6] Albrecht, M., Gentry, C., Halevi, S., Katz, J.: Attacking cryptographicschemes based on "perturbation polynomials". Cryptology ePrintArchive, Report 2009/098 (2009), http://eprint.iacr.org/

[7] Rivest, R., Shamir, A., Adleman, L.: A method for obtaining digitalsignatures and public-key cryptosystems. Communications. of the Assoc.of Comp. Mach. 21(2), 120–126 (1978)

[8] ElGamal, T.A.: A public-key cryptosystem and a signature schemebased on discrete logarithms. IEEE Transactions on Information Theory 31(4), 469–472 (1985)

[9] Wang, H., Sheng, S., Tan, C., Li, Q.: Comparing symmetric-key andpublic-key based security schemes in sensor networks: A case study ofuser access control. In: IEEE ICDCS, Beijing, China, pp. 11–18 (2008)

Rank and Weight Based Protocol for Cluster Head Selection for WSN

S.R. Biradar[1] and Gunjan Jain[2]

[1] ISE Department, SDMCET, Dharwad-580002, Karnataka, India
[2] MITS University, Lakshmangarh-332311, Distt. Sikar, Rajasthan, India
{srbiradar,jgunjan.18}@gmail.com

Abstract. With the evolution of wireless sensor network, the interests in their application have increased considerably. The architecture of the system differs with the application requirement and characteristics. Now days there are number of applications in which hierarchal based networks are highly in demand and key concept of such network is clustering. Some of the most well-known hierarchical routing protocols like LEACH, SEP, TEEN, APTEEN and HEED are discussed in brief. These different conventional protocols have diverse strategies to select their cluster head but still have some limitations. Based on the limitations of these conventional models, a new approach has been proposed on the basis of ranks and weights assignment based protocol known as RWBP. This approach considers not only residual energy but also node's degree and distance of nodes with base station. The node which has higher weight will be chosen as a cluster head. The objective of this approach is to have balance distribution of clusters, enhance lifetime and better efficiency than traditional protocols. The same approach is also applied for multi hop clustering i.e. multi hop RWBP in which the sensing field is divided into more number of areas and the area which lie farther from the base station is sending indirectly via intermediate cluster heads to the base station. The simulations are done in MATLAB with the network size 100x100 meters. The results of the proposed approach are resulting in better lifetime and stability region as compared to LEACH and SEP.

Keywords: Sensor, Wireless, Cluster head, Homogenous, Rank, Energy.

1 Introduction

Wireless Sensor Network (WSN) is entering a new phase of computer networks. Daily a new protocol comes in to market, resulting in improvisation for many applications. This new emerging trend is due to large connectivity leading numerous data exchange. The WSN consists of many autonomous devices called motes used for sensing, communicating and computing services. The sensor nodes collect the data and pass it to gateway or base station from where user can access the event information. The main components of sensor network are assembly of distributed or localized network, an interconnecting network; a central point of information

© Springer International Publishing Switzerland 2015
S.C. Satapathy et al. (eds.), *Proc. of the 3rd Int. Conf. on Front. of Intell. Comput. (FICTA) 2014*
– *Vol. 2*, Advances in Intelligent Systems and Computing 328, DOI: 10.1007/978-3-319-12012-6_88

clustering; and a set of computing resources at the central point (or beyond) to handle data correlation, event trending, status querying, and data mining [1].

There are several applications in this field which require aggregation of data for their better performance. In such cases, the sensors of different regions collaborates their information to provide more accurate reports about their local regions [2]. In order to support data aggregation efficiently nodes are partitioned into small number of groups according to specific criteria describe in the protocol. These small groups are called clusters and each cluster has a cluster head acting as a coordinator, the other nodes of that clusters are called member nodes.

In the WSN field researchers using good cluster head selection strategy in order to prolong network lifetime. To have a global view their basic categorization lies from the following questions like what is the parameter to decide the role of sensor node, which sensor node initializes the cluster head selection, does the load is evenly distributed, does the network require single hop transmission or multi hop transmission. The main aim of different approaches is to reduce the energy consumption and enhancement of network lifetime.

The approach taken is Rank and Weight assignment based Protocol (RWBP) for the WSN. Member nodes are assigned with ranks considering three descriptors i.e. node's degree, distance of nodes from base station and residual energy and the weight is calculated by summing the ranks. The node which is having highest weight is elected as a cluster head. The hierarchal multi hoping techniques is also applied on RWBP. An analysis to the disadvantages of conventional protocols to select the cluster head and comparison of network lifetime has been evaluated.

2   Related Literature

The main aim of wireless sensor network is to gather large amount of data and to enhance the network lifetime with limited battery. In the conventional models like direct data transmission, each sensor nodes transmit its data to the base station directly. In [3] it is observed that if the base station is located far away from sensor nodes then it will drain off the battery quickly and reduces the network lifetime because each node will require large transmit power separately. Similarly, in minimum energy routing protocol where data is transferred to the base station with the help of intermediate nodes. Thus, nodes close to the base station are the ones to die out quickly because these routers have to transmit large amount of data. The hierarchal clustering is now formed the basis of many wireless routing protocol. The advantage of hierarchal routing is that it helps in saving large amount of energy particularly in large networks. The data is passed to the super node until a top level hierarchy is arrived which in turn leads to base station. W. R. Heinzelman, A. P. Chandrakasan and H. Balakrishnan [3] in 2000 proposed a protocol called Low Energy Adaptive Clustering Hierarchy (LEACH) which later turned to be the most popular algorithm in the hierarchal routing for sensor nodes. It is a dense network of sensor nodes grouped in to clusters and utilizes randomized rotation of clusters. These local cluster heads act as a router to send information or knowledge to the base station. Other advantage of LEACH includes it incorporates data fusion into routing protocols; amount of information to the base station is reduced; 4-8 times effective over direct communication in prolonging network lifetime; also the clusters forms grid like area. But the election of cluster head node in LEACH has some deficiency. Like sometimes very big clusters and sometimes very small cluster may exit at the same time. There is unreasonable cluster head selection while nodes have different energy; once the energy of the cluster head node depletes all other nodes fails to function; the algorithm does not take in to account location of nodes, residual energy and other information which may lead cluster head node rapidly fail. Another protocol called HEED [4] works on the basis of cluster head probability which is the function of residual energy and neighbor proximity. The protocol aims to have balanced cluster, works better both for uniform and non-uniform node distribution; required low message overhead but increase in iterations results in complex algorithm. The decrease in residual energy leads to low cluster head probability. HEED outperforms several generic protocols but is deficit in some parameters. Similarly Threshold sensitive energy efficient sensor network [5] protocol is a hierarchical protocol designed to be responsive to sudden changes in the sense attribute. The cluster head broadcast two types of threshold; hard threshold to allow the nodes to transmit only when the sensed attribute is in the range of interest and soft threshold to reduce the number of transmission if there is little or no change in the sensed attribute. APTEEN is an extension of TEEN and captures both periodic data and time critical events [5]. The energy dissipation and network life time is better than LEACH but the main drawback of TEEN and APTEEN are the overhead and complexity of forming cluster head at multiple levels and implementing threshold based function and dealing with attribute based naming of queries [5].

3   Proposed Approach

In the field of wireless sensor network, daily a new approach is putted forward by the researchers to have a good cluster head selection strategy in order to prolong network lifetime. To have a global view their basic categorization lies from the following questions like what is the parameter to decide the role of sensor node, which sensor node initializes the cluster head selection, does the load is evenly distributed, does the network require single hop transmission or multi hop transmission. The main aim of different approaches is to reduce the energy consumption and enhancement of network lifetime. For balanced distribution, the protocol requires that after particular time interval the role of the cluster head is given to different sensor node.

There are different protocols which uses different scheme to choose the cluster head. In the proposed approach the basis of cluster head selection is not the threshold values like LEACH and not only residual energy like SEP [6] and energy aware routing [7]. The proposed approach [8] takes the following parameters in to consideration like-

Node Degree: It is defined as the total number of neighboring node of a specific node. The node degree is one of the important metric to check the connectivity in the wireless and also ad-hoc networks [9]. It can also be defined as to how many nodes the specific node is directly connected or how many nodes are in the range of that particular node. The higher the degree, the better will be the connectivity of the node with other nodes and hence large amount of data can be aggregated at one place.

Distance of node from base station: It has been seen that the energy consumed in transmitting the data is much more than processing it. Larger the distance higher will be the communication cost. This was the reason that for larger distances direct communication does not go up to the mark. In our proposed approach we are choosing the node which have low communication cost. Thus, in this approach there is consideration of location of nodes.

Residual Energy: The lifetime of the network is all about the energy contained in the nodes [10]. The higher the energy remained in the sensor nodes more will be the lifetime. If the cluster heads itself deplete from energy then all other sensor nodes communicating through it also die because of cascading effect. The role of the sensor nodes should be properly defined to have a balanced rate of energy in every grid of nodes.

4   Simulation Scenario

Firstly the nodes in the network are randomly deployed in a 100mt*100mt area then those nodes is differentiated on the bases of grids. For every round of transmission there will be assignment of weights to the nodes. Each node is given a weight on the basis of above parameters like node degree, distance and residual energy.

These nodes are placed at different locations and have different energy level shown in figure 1. The BS in the figure is representing Base station which is positioned in between the sensor nodes. The nodes belonging to one cluster or grid send their information regarding node degree and its distance from base station to the base station (supervisor node). The base station on receiving the data assigns the rank. For example if there are 'n' nodes in the grid then a rank for distance $D(i)= n$ is given to node who is most close to the base station. Similarly $D(i)=1$ if the node is farthest from the base station. Likewise the rank $N(i)=n$ which is having higher number of degree and $N(i)=1$ which is having the least degree. Since for transferring these data some amount of energy is consumed therefore after this process the nodes send their residual energy status. The base station assign the rank $E(i)=n$ to the Node which is having higher energy. In case if the nodes have the same value of parameter, the same rank is given to those nodes. After assignment of ranks the weight is calculated by summing up all the ranks. i.e. $W(i)=D(i)+N(i)+E(i)$.   ....   (1)

For example, There are total 13 nodes in the grid 1.Consider only 3 nodes A, B, C. Considering node C, the distance of node is minimum therefore it is assigned high rank equal to number of nodes and then to node B then Node A. Similarly the degree of node B and energy of node A is high. With the help of equation 1, the weight of these 3 nodes is calculated as: For node A= W (A) =11+12+13=35,
B= W(B) =12+13+12=37   and C = W(C) =11+12+13=36

Thus weight of node B is largest therefore it is selected as cluster head for the round.

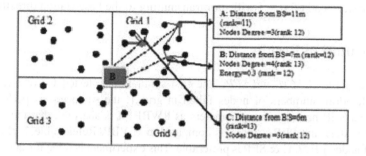

Fig. 1. Example of Proposed approach

Multi hopped RWBP: The RWBP approach has also been tried for multi hoping in which each grid has two areas i.e. area 1 and area 2. Area 1 and 2 will elect cluster head with the same procedure used in RWBP. The cluster head of area 1 which is far away from base station will aggregate all the data of its member nodes and fuse the data with the cluster head of area 2 which reduces redundancy. The cluster head of area 2 will send the combination of area 1 and area 2's data to the base station. Thus the data of area 1 will reach indirectly and data of area 2 will reach directly to the base station. Refer figure 2, the two areas of grid 1 are distinguished on the basis of line, which can be computed as $y - y1 = \dfrac{y2 - y1}{x2 - x1}(x - x1)$ .... (2)

Thus Multi hopped RWBP is using hierarchal clustering and each grid has a two level cluster. The technique helps to reduce the communication cost and as we know that in WSN computational cost is minimum than communication cost this approach will prove to be efficient.

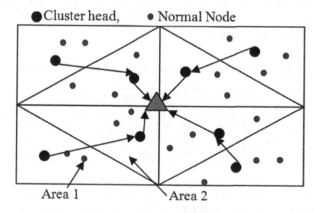

Fig. 2. Multi hoped RWBP

Moreover the fact is considered that there may be a case when all the nodes of area2 will die thus this region will become bottleneck, as the nodes of area 1 also become fail to pass their data since cluster head of area 2 was an intermediate. In such case the cluster head of area 1 will start transmitting to the base station directly.

5   Simulation Results

Comparison with LEACH and SEP: The first simulation is tested for one grid of the network. Random numbers of nodes fallen in grid 1, the same test is performed for LEACH and SEP having same parameters as RWBP. It is analyzed that the total dead nodes in LEACH and SEP is more as compared to RWBP. Refer Table 1. But for less number of nodes LEACH & SEP is preferable. This experiment is done for 2000 rounds.

Table 1. Font Analysis of Total Number of Nodes Dead

Total Nodes	Dead in RWBP	Dead in LEACH	Dead in SEP
6	2	0	0
8	3	0	1
10	3	7	6
16	7	15	14
22	13	21	19
26	13	25	25

The same experiment when performed for all grids shows that total dead node in RWBP is less as compared to LEACH and SEP. Table 2 is the calculation of total nodes die in 2000 rounds and time step of first node die refer to as fdead. The network lifetime can also be evaluated by examining that at which time step (round) the first node of the network will die. Table 2 shows that the first node die more early in LEACH than SEP and the first node dead in SEP is earlier than RWBP. It also evaluates that the results of RWBP is at least 2 times better than LEACH and SEP for total dead nodes. Refer to figure 3.

Table 2. Comparison of Dead Nodes and First Node Dead

Total Nodes	For all grids (Dead nodes in 2000 rounds)		
	RWBP	LEACH	SEP
50	29, fdead=1226	49, fdead=833	49, fdead=1047
75	28, fdead=1171	100, fdead=806	72, fdead=943
100	48, fdead =1244	99, fdead= 803	100, fdead=1077

Stability period is defined as the region up to which all nodes are alive. This period lies between round 1 to round at which the first node dies. Figure 4a is the evaluation of stability period and the duration that how much the network is active. The simulation is done with 50 nodes and other parameters are listed in Table 3.

Multi hop RWBP: The simulation is performed for Multi hop RWBP with the network parameters listed in appendix. In this network 8 different areas are forming, two of them lie in one grid. It is an example of hierarchal clustering. The network environment of Multi hop RWBP is shown in figure 4b. The dead nodes will be turned as red dots.

Table 3. Simulations Parameters

Network Size	100*100 meter
Location of sink	(50,50) meter
Data packet length	4000 bit
Initial energy of nodes	0.5 joule
Transmitter/Receiver	50 nj / bit
Aggregation energy, EDA	5 nj/bit

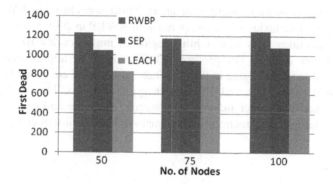

Fig. 3. Comparison of RWBP with LEACH and SEP

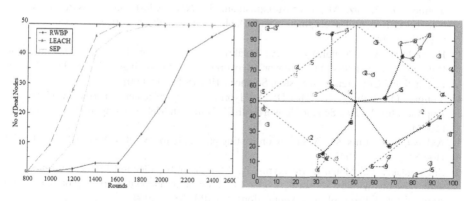

Fig. 4. a) Number of nodes v/s rounds in RWBP, SEP and LEACH   b) Network Environment of Multi hop RWBP

6   Conclusion

Cluster heads aggregate the received data and send them to the sink. Cluster forming is a method that minimizes energy consumption and communication latency. The most well-known hierarchical routing protocols are LEACH, TEEN, APTEEN and HEED. Based on limitations of these conventional models a new approach has been proposed on the basis of weights assignment. The cluster head which will have higher weight will be chosen as cluster head. This approach will consider not only residual energy but also node's degree and distance of nodes with base station. The name given to such protocol is Rank and Weight based Protocol (RWBP). The simulations are done in MATLAB. The analysis concluded that RWBP is at least two times better than SEP and LEACH when the total number of alive nodes is considered for 2000 round. The stability period of the proposed protocol is also more than conventional protocols.

Moreover, RWBP is also implemented with multi hop clustering strategy, dividing the sensing field in to more number of areas. Multi hop clustered RWBP is giving more number of alive nodes when compared with RWBP in 2000 rounds but for large number of nodes taken. Also the stability region of multi hop RWBP is weaker than single hop cluster RWBP. The protocol RWBP and multi hop clustered RWBP is tested for only one network size in this dissertation. It can be enhanced further to view with different network sizes and different energy constraints. The same protocol can also be compared with other number of protocols available and can be designed in enhanced simulators. The functions can be improved so that it will consume less time.

References

1. Kazem, S., Daniel, M., Taieb, Z.: Wireless Sensor Network Technology, Protocol and Application. Wiley Interscience (2007)
2. Younis, O., Krunz, M., Ramasubramanian, S.: Node Clustering in Wireless Sensor Networks: Recent Developments and Deployment Challenges. IEEE Network, 20–25 (May/June 2006)
3. Heinzelman, W., Chandrakasan, A., Balakrishnan, H.: Energy-Efficient Communication Protocol for Wireless Microsensor Networks. In: Proceedings of the 33rd Hawaii International Conference on System Sciences, HICSS 2000 (2000)
4. Younis, O., Fahmy, S.: HEED: A Hybrid Energy-Efficient Distributed Clustering Approach for Ad Hoc Sensor Networks. IEEE Trans. on Mobile Computing 3(4), 366–379 (2004)
5. Akkaya, K., Younis, M.: A Survey on routing protocols for Wireless Sensor Networks. Ad Hoc Networks 3 (2005)
6. Smaragdakis, G., Matta, I., Bestavros, A.: SEP: A Stable Election Protocol for clustered heterogeneous wireless sensor networks. In: Second International Workshop on Sensor and Actor Network Protocols and Applications, SANPA 2004 (2004)
7. Shah, R.C.: Energy Aware Routing for low energy ad-hoc sensor networks. In: Wireless Communications and Networking Conference, WCNC 2002, vol. 1, pp. 350–355 (2002)

8. Jain, G., Biradar, S.R.: Enhanced Approach of Cluster Head Selection Using Rank and Weight Assignment in Wireless Sensor Network. In: International Conference on Recent Trends in Computing and Communication Engineering- RTCCE, April 20-21 (2013)
9. Shorey, R., Ananda, A., Chan, M., Ooi, W.: Coverage and connectivity issues in Wireless Sensor Networks. In: Mobile, Wireless, and Sensor Networks:Technology, Applications, and Future Directions, 1st edn., pp. 221–256. Wiley-IEEE Press (2005)
10. Chan, E., Han, S.: Energy Effiecient Residual Energy Monitoring in Wireless Sensor Networks. International Journal of Distributed Sensor Networks 5, 1–23 (2009)

Author Index

Abidi, M.R.   347
Agarwal, Amit   395
Ahmad, Musheer   75
Alam, Shahzad   75
Alka, Agrawal   619
Angadi, Sujay   589
Anil, R.   493
Anil Kumar, B.   785
Annadate, Suresh A.   335
Artamwar, Rushikesh   57
Asnani, Kavita   629
Ayub, Shahanaz   449

Baboo, Sarada   605
Balaji, L.   405
Bandyopadhyay, Samir Kumar   383
Banerjee, Siddhartha   175
Basak, Saikat   237
Basu, Abhishek   113
Basu, Souvik   747
Belykh, Igor   423
Bhadra, Sajal   65
Bhaskara Rao, N.   785
Bhateja, Vikrant   475
Bhattacharjee, Anup Kumar   255
Bhattacharjee, Debotosh   323
Bhattacharjee, Tamal   207
Bhattacharya, Boudhayan   49
Bhattacharya, Indrajit   775
Bhavsar, Arnav   307
Bhende, Manisha   757
Bhondekar, Anil   315
Bipin, K.   729
Biradar, S.R.   793
Boddu, Rama Devi   697

Chakrabarti, Amlan   13
Chakraborty, Moumita   13
Chakravortty, Somdatta   315
Chandra, Mahesh   297, 375, 465, 529
Chandra, P. Sharath   581
Chandrakanth, T.   457
Chintureena   153
Chopra, Akshay   75
Chowdhury, Arundhuti   237
Colaco, Louella   651

Darshan, H.Y.   573
Das, Jeet   113
Das, Sudhangsu   65
Das, Vineeta   297
Dash, Monalisa   367
Datta, Debdyuti   217
Davalbhakta, Omkar   57
De, Arunava   255
Deshmukh, Ratnadeep R.   335
Deshpande, Bhagavant   659
Dhamdhere, Vidya   687

Ekuakille, Aime lay   501
Eswaraiah, R.   245

Farooq, Omar   347

Gaikwad, Bharatratna P.   225
Ganguly, Suranjan   323
Ghosh, Bibek Ranjan   175
Ghosh, Piue   383
Ghosh, Subhash   775
Ghrera, Satya Prakash   135, 521

Gill, Navneet Kaur   39
Gnanavel, V.K.   441
Goel, Savita   105
Gopalakrishna, M.T.   581
Gopalkrishna, M.T.   545, 573
Guha, Krishnendu   13
Gupta, D.K.   465
Gupta, P.K.   127
Gupta, Punit   127, 135
Gupta, Shilpi   105
Gupta, V.K.   465

Hanumantharaju, M.C.   545, 573, 581
Harekal, Divakar   667
Himanshi   475

Jain, Gunjan   793
Jain, Prakhar   75
Jain, Subit K.   307
Jana, Prasanta K.   739
Jinesh, M.K.   85
Joshy, Lakshmi M.   97

Kalarickal, Boney S.   85
Kamble, Vaibhav V.   335
Kandru, Nikitha   375
Kandul, Akshay   57
Kar, Asutosh   297
Karwankar, Anil R.   335
Kaushik, Nisha   105
Khan, R.A.   619
Kishore, G. Krishna   189
Krishn, Abhinav   475
Krishnamurthi, Anusudha   483
Kulkarni, Dinesh   57
Kumar, B. Naveen   165
Kumar, P.U. Praveen   263
Kumar, S. Sachin   493
kumari, Aparajita   529
Kumari, Shipra   721

Madhulika, Gadang   21
Majumder, Koushik   707, 765
Mal, Indranil   113
Mandal, J.K.   197, 207, 217
Manjusha, K.   493
Manza, Ganesh R.   225
Manza, Ramesh R.   225
Mishra, Shipra   529
Mitra, Ankita   255

Mitra, Sushavan   113
Mohana, S.H.   357
Mohanty, Mihir N.   367
Mona,   273
Mondal, Pritha   197
More, Ashwin   57
More, Chaitali   651

Nagaraja, S.   263
Naik, Anuja Jana   563
Nair, Lekha S.   97
Nandi, Pratibha   529
Nasipuri, Mita   323
Navamani, T.M.   673
Nazareth, Derrick   629
Nitesh, Kumar   739

Om, Hari   721
Oval, Sonali G.   413

Padhy, Neelamadhab   605
Panigrahi, Rasmita   605
Parmar, Nilesh   145
Parwekar, Pritee   29
Patel, M.S.   553, 563
Patil, Suvarna   757
Pattnaik, Prasanth Kumar   521
Pawar, Ambika V.   1
Paygude, Shilpa   281
Poornima, U.S.   165, 645
Prabhakar, C.J.   263, 357
Prabhakaranl, Subi   729
Prasad, B.G.   273

Raheja, Jagdish Lal   289
Rahman, Molla Ramizur   637
Raju, P.V.S.N.   29
Rani, M. Asha   697
Rao, Chinta Seshadri   21
Rao, Jawahar J.   659, 667
Rao, K. Kishan   697
Rathi, Apoorva   145
Ratnaparkhe, Varsha R.   335
Ravishankar, M.   537, 589, 597
Ray, Rajendra K.   307
Ray, Sudhabindu   765
Reddy, E. Sreenivasa   245
Reddy, Sanjay Linga   553, 563
Robert, Phuritshabam   165
Rodrigues, Okstynn   629
Roy, Pratik   175

Roy, Sandipta   113
Roy, Siuli   747
Roy, Sudipta   383
Roychowdhury, Meghdut   121

Saha, Arindam   113
Saha, Banani   49
Saha, Debasri   13
Sahani, Romio Rosan   13
Sahoo, Biswa Mohan   121
Sahoo, Jagyanseni   367
Sahoo, Rashmi Ranjan   707, 765
Sahu, Akanksha   475
Sandeep, Singh   619
Sandhya, B.   457
Sandyal, Krupashankari S.   553
Sanyal, Manas Kumar   65
Sardar, Abdur Rahaman   707, 765
Sardinha, Razia   651
Sariya, K.   597
Sarkar, Arindam   197, 207, 217
Sarkar, Souvik   121, 707
Sarkar, Subir Kumar   113, 765
Sarkar, Suryaday   121
Sathyadevan, Shiju   85, 729
Saxena, Pratiksha   501
Sengupta, Madhumita   207
Shankarwar, Mahesh U.   1
Sharma, Rajat   395
Sherin, Zafar   713
Shirawale, Sankirti   413
Show, Debashis   775
Shrivastava, Nishant   509
Singh, Birender   375
Singh, Jamuna Kanta   707
Singh, Moutushi   707, 765

Singh, Sarbjeet   39
Singh, Vinod Kumar   449
Sinha, Devadatta   315
Sivasankar, C.   431
Smruti, Soumya   367
Soman, K.P.   493
Soni, M.K.   713
Soygaonkar, Pratik   281
Sree, P. Syamala Jaya   521
Srinivasan, A.   431, 441
Srivastava, Rajat   449
Suma, V.   153, 165, 645, 659, 667
Sunitha, M.S.   785

Thakur, Anandita Singh   127
Thyagharajan, K.K.   405
Tripathy, Devashree   289
Tyagi, Vipin   509

Upadhyaya, Prashant   347
Urooj, Shabana   501

Valarmathi, J.   483
Varshney, Priyanka   347
Vaseemahamed, M.   537
Venkateswaran, N.   483
Vergalo, Patrizia   501
Vidhate, Pankaj   687
Vikrant, Bhateja   501
Vyas, Vibha   281

Wagh, Sanjeev   757
Wani, Inam Ul Islam   545
Wankhade, Yogita   687

Yaramasa, Teja   189
Yogesh, P.   673

Printed in the United States
By Bookmasters